Soil and Environmental Science DICTIONARY

Edited by
E.G. Gregorich
L.W. Turchenek
M.R. Carter
D.A. Angers
for the
CANADIAN SOCIETY OF SOIL SCIENCE

CRC Press is an imprint of the
Taylor & Francis Group, an **informa** business

CRC Press
Taylor & Francis Group
6000 Broken Sound Parkway NW, Suite 300
Boca Raton, FL 33487-2742

First issued in paperback 2019

ISBN-13: 978-0-8493-3115-2 (hbk)
ISBN-13: 978-0-367-39724-1 (pbk)

Library of Congress Card Number 2001025292

Library of Congress Cataloging-in-Publication Data

Soil and environmental science dictionary / edited by E. G. Gregorich ... [et al.].
p. cm.
Includes bibliographical references.
ISBN 0-8493-3115-3
1. Soil science—Dictionaries. 2. Soil science—Dictionaries—French. 3. Environmental sciences—Dictionaries. 4. Environmental sciences—Dictionaries—French. 5. English language, Dictionaries—French. 6. French language—Dictionaries—English. I. Gregorich, E. G. II. Canadian Society of Soil Science.
S592 .S59 2001
631.4′03—dc21 2001025292

Visit the Taylor & Francis Web site at
http://www.taylorandfrancis.com

and the CRC Press Web site at
http://www.crcpress.com

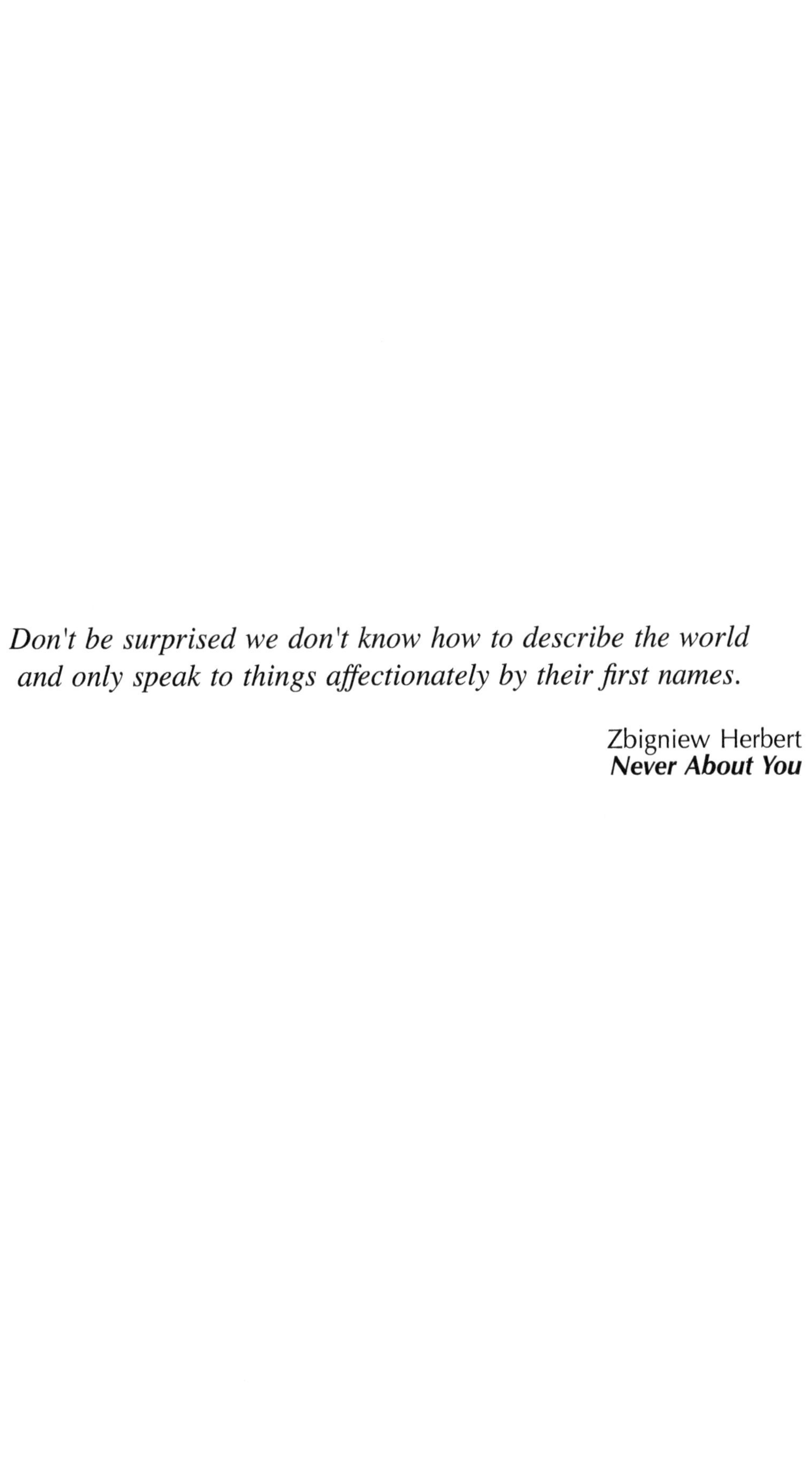

Don't be surprised we don't know how to describe the world
and only speak to things affectionately by their first names.

Zbigniew Herbert
Never About You

Preface

Judicious treatment of environmental questions and challenges requires an integrated, cross-disciplinary approach. Whether working at a local, national, regional, or global scale, the knowledge and expertise resident in a variety of subject areas comes to bear on all questions of sustainable development. Soil scientists now collaborate with colleagues in many fields, and a basic working knowledge of the vocabulary of those fields improves understanding and enhances the flow of information.

This dictionary brings together the conventional vocabulary of soil science with that of many overlapping disciplines such as geology, hydrology, and meteorology. Its purpose is to define and describe technical words for researchers, students of various levels, librarians, policy- and decision-makers, and interested citizens working and studying in a wide variety of disciplines related to soil science.

The terms and definitions for this dictionary were gathered from a wide variety of sources, including several existing glossaries and dictionaries. The editors invited a panel of thirty reviewers, experts in selected subject areas, to help review, select, and update best definitions. The editors acknowledge and thank Professor Alma Mary Anderson, Art Department, Indiana State University, for drafting the illustrations. To reflect Canada's bilingualism and make the dictionary more useful to the international science community, French equivalents are given for English terms.

Canadian Society of Soil Science

The Canadian Society of Soil Science is a non-governmental, non-profit organization for scientists, engineers, technologists, administrators, students, and others interested in soil science. Its three main objectives are:

- To promote the wise use of soil for the benefit of society
- To facilitate the exchange of information and technology among people and organizations involved in soil science
- To promote research and practical application of findings in soil science

The Society quarterly produces the international scientific publication, the *Canadian Journal of Soil Science,* and each year hosts an international soil science conference. Its well-known practical soils methodology book *Soil Sampling and Methods of Analysis* (Lewis Publishers, CRC Press, 1993) is used throughout the world. The Society publishes a newsletter to share information and ideas, and maintains active liaisons and partnerships with other soil science societies. Collaborative projects are currently under way in Sri Lanka, Costa Rica, and Thailand.

For more information about the Canadian Society of Soil Science, please visit the following Web site: *http://www.csss.ca.*

The Editors

E.G. Gregorich, Ph.D. holds degrees in agriculture and soil science from the University of Saskatchewan, and earned a Ph.D. in soil science from the University of Guelph. He has been a research scientist with Agriculture and Agri-Food Canada at the Eastern Cereal and Oilseed Research Centre, Ottawa, Canada, since 1989, focusing on carbon and nitrogen cycling in soil. He is a member of the Canadian Society of Soil Science and currently serves as an associate editor for the *Journal of Environmental Quality.*

L.W. Turchenek, Ph.D. holds degrees in agriculture and soil science from the University of Saskatchewan, and earned a Ph.D. in soil science from the Waite Agricultural Research Institute, University of Adelaide. He held research positions at the University of Alberta and the Alberta Research Council from 1975 to 1994, and is currently a senior soil scientist with AMEC Earth and Environmental Ltd. in Edmonton, Alberta. He is a member of the Canadian Society of Soil Science and has been involved in research and consulting in environmental soil science, reclamation, and soil survey.

M.R. Carter, Ph.D. holds degrees in agriculture and soil science from the University of Alberta, and earned a Ph.D. in soil science from the University of Saskatchewan. He has held agricultural research positions with Agriculture and Agri-Food Canada since 1977 and is currently a research scientist at the Crops and Livestock Research Centre, Charlottetown, Prince Edward Island, working on soil quality evaluation. A fellow of the Canadian Society of Soil Science, he currently serves as editor-in-chief for *Soil and Tillage Research* and *Agriculture Ecosystems and Environment.*

D.A. Angers, Ph.D. holds degrees in agriculture and soil science from Université Laval, Québec and the University of Guelph, and earned a Ph.D. in soil science from McGill University. He has been a research scientist with Agriculture and Agri-Food Canada since 1988 at the Soils and Crops Research and Development Centre, Sainte-Foy, Québec, involved in research in applied soil physics and organic matter dynamics. He is an adjunct professor of soil science at Université Laval and serves on the Canadian Society of Soil Science council, and is an editorial advisor for *Soil and Tillage Research* and *Plant and Soil.*

Subject Reviewers

Manas R. Banerjee

M.H. Beare

G. Bélanger

Claude Bernard

Johanne B. Boisvert

Martin H. Chantigny

Reinder de Jong

J.F. Dormaar

Deena Errampalli

C.A. Fox

L.J. Gregorich

Y.P. Kalra

Claude Lapierre

M.R. Laverdière

David A. Lobb

Douglas G. Maynard

G.R. Mehuys

J.J. Miller

Alan P. Moulin

G.H. Neilsen

Denise Neilsen

Michel C. Nolin

Rock Ouimet

Léon-Étienne Parent

Philippe Rochette

R.R. Simard

Barrie Stanfield

Ulrica F. Stoklas

A.J. VandenBygaart

Karl Volkmar

Table of Contents

Using the Dictionary

Each term is printed in bold type, followed by the definition. In the case of multiple definitions for a single term, definitions are listed in a numbered sequence. Where a definition pertains to a specific discipline, the discipline appears italicized in parentheses at the head of that definition.

The French term equivalent to the defined English term or phrase is shown in bold type following the definition. If the French term differs for multiple English definitions listed in a numbered sequence, the French terms are numbered correspondingly at the end of the list of definitions.

A word in italic font within a definition indicates that it is defined elsewhere in the dictionary, unless it denotes a biological genus and species. Where a term is defined elsewhere in the dictionary under a synonymous term, the reader is directed to this other definition with the instruction, "See...". At the end of a definition, the reader is alerted to synonyms for this term with the instruction, "Also called...".

Two indexes, located at the back of the dictionary, group dictionary terms according to discipline and subject area. These sections enable the English or French reader to determine if a particular term, or group of related terms within a discipline, is defined in the dictionary. Terms are arranged alphabetically within each discipline. The English/French index is sorted alphabetically by English term with the corresponding French equivalent, and the French/English index by French term with the corresponding English equivalent. Several appendices provide additional information to aid in understanding certain terms or the context in which terms are used. The literature used to define some of the terms is cited in References: Sources of Terms, and the sources used for some of the illustrations are given in References: Sources of Illustrations.

α-amino acid A specific organic acid in which an amino group and an alkyl group are attached to the carbon atom residing closest to the carboxyl group. **acide α-aminé**

a axis (*crystallography*) One of the crystallographic axes used as reference in crystal description. It is the axis that is oriented horizontally, from front to rear. See *b axis*, *c axis*. **axe a**

A horizon A mineral soil horizon formed at or near the surface in the zone of removal of materials in solution and suspension, or maximum *in situ* accumulation of organic carbon, or both. The accumulated organic matter is usually expressed morphologically by a darkening of the surface soil (Ah). Conversely, the removal of organic matter is usually expressed by a lightening of the soil color, usually in the upper part of the solum (Ae). The removal of clay from the upper part of the solum (Ae) is expressed by a coarser soil texture as compared to the underlying subsoil layers. The removal of iron is indicated usually by a paler or less red color in the upper part of the solum (Ae) relative to the lower part of the subsoil. The above horizon terms are according to the Canadian system of soil classification. See *Appendix D* for equivalent U.S. Soil Taxonomy and FAO soils terminology. See *B horizon, C horizon, horizon, soil.* **horizon A**

AB horizon A transitional mineral horizon showing properties of both an A and B horizon in which properties of the A predominate. **horizon AB**

abaxial (*botany*) Facing away from the stem of a plant (e.g., the undersurface of a leaf). See *adaxial.* **abaxial**

Abbe refractometer An instrument used for determining the refractive index of liquids, minerals, and gemstones. Its operation is based on the measurement of the *critical angle.* **réfractomètre d'Abbe**

ABC soil A soil that has a complete profile, including an A, a B, and a C horizon. **sol ABC**

abiontic enzyme An enzyme (exclusive of live cells) excreted by live cells during growth and division, attached to cell debris and dead cells, or leaked into soil solution from extant or lyzed cells but whose original functional location was on or within the cell. **enzyme abiontique**

abiotic Non-living, referring to the basic elements and compounds of the environment. **abiotique**

abrasion (*geology*) The mechanical wearing (i.e., scratch, grind, or polish) effect on rocks caused by frictional agents (e.g., sand, pebbles, boulders) transported in various ways: by wind, running water, ocean waves and currents, or glacier ice. **abrasion.**

abscissa The horizontal axis (x axis) in a graph. See *ordinate.* **abscisse**

absolute alcohol Pure alcohol (ethanol). **alcool absolu**

absolute temperature Temperature measured in degrees Celsius from *absolute zero*, –273.16°C. Absolute temperatures are given on a scale of Kelvin (e.g., 150 K). **température absolue**

absolute zero The temperature at which all thermal motion of atoms and molecules theoretically ceases; –273.16°C. **zéro absolu**

absorbance The amount of light absorbed by a solution; the measure is used to determine the concentration of certain ions or molecules in a solution. **absorbance**

absorbed water Water held mechanically in a soil mass and having physical properties similar to ordinary water at the same temperature and pressure. **eau absorbée**

absorptance The ratio of the radiant flux absorbed by a body to that incident upon it. Also called absorption factor. **absorptance, facteur d'absorption**

absorption (*physics*) The process by which the energy of *electromagnetic radiation* is taken up by a molecule and transformed into a different form of energy. (*chemistry*) The process by which one substance is taken up by another substance. **absorption**

absorption band A range of wavelengths over which radiant energy is absorbed by a specific material that may be present on the Earth's surface or in the atmosphere. **bande d'absorption**

absorption of radiation The uptake of *radiation* by a solid body, liquid, or gas. The absorbed energy may be transferred or re-emitted. **absorption du rayonnement**

absorption, active Movement of ions and water into the plant root resulting from the root's metabolic processes, usually against an electrochemical potential gradient. **absorption active**

absorption, passive Movement of ions and water into the plant root as a result of diffusion along an activity gradient. **absorption passive**

absorptive power The total flux of radiant energy absorbed in a unit area of absorbing material; measured in watts per square centimeter. **pouvoir d'absorption, pouvoir absorbant**

absorptivity The ratio of the amount of radiation absorbed by a body to the maximum amount it can absorb. A surface that is a poor reflector is a good absorber. If no radiation is reflected, the surface acts as a black body and has an absorptivity and *emissivity* of 1. **absorptivité**

AC horizon Analogous to an *AB horizon*, except the transition is between an A and a C horizon in a profile lacking a B horizon. **horizon AC**

AC soil A soil that has an incomplete profile, including an A and a C horizon, but no clearly developed B horizon. Commonly, such soils are young, like those developing from alluvium or on steep, rocky slopes. **sol AC**

access tube Small diameter tube (typically about 50 mm) inserted through the soil root zone for passage of a *neutron probe* to determine the water content of soil at various depths. **tube d'accès**

accelerated erosion See *erosion*. **érosion accélérée**

acclimation See *acclimatization*. **acclimatation**

acclimatization Physiological and behavioral adjustments of an organism in response to a change in environment. **acclimatation**

accuracy The degree to which calculation, measurement, or set of measurements agrees with a true value or an accepted reference value. Accuracy includes a combination of random error (*precision*) and systematic error (*bias*) components that are due to sampling and analytical operations. **exactitude**

acetylene-block assay A technique used to estimate *denitrification* by determining release of nitrous oxide from acetylene-treated soil. **test de blocage à l'acétylène**

acetylene-reduction assay An estimation of nitrogenase activity accomplished by measuring the rate of acetylene reduction to ethylene. **test de réduction de l'acétylène**

acid A substance that contains hydrogen and dissociates in water to produce positive hydrogen ions (or H_3O^+) (i.e., Arrhenius theory). A substance that exhibits a tendency to release a proton (i.e.,

Lowry-Brønsted theory). An acid is a compound that can accept a pair of electrons, and a *base* is one that can donate an electron pair (i.e., Lewis theory). **acide**

acid deposition Acidic material introduced to the ground or surface waters including wet deposition from precipitation, dry deposition from particle fallout, and acid fog. Air contaminants, such as sulfur oxides and nitrogen oxides, from both *anthropogenic* and natural sources react with water in the atmosphere to form acids. Often called acid rain. **déposition acide**

acid detergent fiber (ADF) Insoluble residue remaining after extraction of herbage with acid detergent; cell wall constituents minus hemicellulose. **fibre au détergent acide (ADF)**

acid detergent fiber digestibility The digestibility of *acid detergent fiber* (ADF) of a forage, calculated as the percent difference ADF measured before and after *in vitro* or *in vivo* digestion. **digestibilité de la fibre au détergent acide**

acid dissociation constant (K_a) The equilibrium constant for a reaction in which a proton is removed from an acid by H_2O to form the conjugate base and H_3O^+; a measure of the strength of the acid. **constante de dissociation d'un acide**

acid gas The anhydrous gaseous form of an acid (e.g., hydrogen chloride). **gaz acide**

acid mine drainage Water contamination by sulfuric acid produced by seepage through sulfur-bearing spoil and tailings from coal and metal mining. **drainage minier acide**

acid rain See *acid deposition*. **pluie acide**

acid soil A soil having a pH of less than 7.0. See *reaction, soil*. **sol acide**

acid spoil Coal and metal mine tailings that contain sulfur and generate acidity. **déblais acides**

acid-base indicator A substance that marks the end point of an acid-base titration by changing color. **indicateur acido-basique**

acid-forming fertilizer See *fertilizer*. **engrais acidifiant**

acidic Having a low pH value (less than 7); the opposite of alkaline. **acide**

acidic cation A cation that, when added to water, undergoes hydrolysis resulting in an acidic solution. Hydrated acidic cations donate protons to water to form hydronium ions (H_3O^+) and thus in aqueous solution are acids (Bronsted definition). Examples in soils are H^+, Al^{3+}, and Fe^{3+}. **cation acidique**

acidic rock Igneous rock that is high in silica, generally greater than 52%. One of four subdivisions of a commonly used system for classifying igneous rocks based on silica content (e.g., acidic, *intermediate rock*, *basic rock*, and *ultrabasic rock*). **roche acide**

acidic solution A liquid whose hydrogen ion concentration is greater than its hydroxyl ion concentration, or whose pH is less than 7.0. **solution acide**

acidimetry Volumetric analysis in which a standard solution of an acid is added to the unknown (base) solution to determine the amount of base present. **acidimétrie**

acidity constant See *acid dissociation constant*. **constante d'acidité**

acidity, exchangeable The amount of exchangeable hydrogen and aluminum ions in soil, as estimated by replacement from a soil by an unbuffered salt solution such as KCl or NaCl. Also called "salt-replaceable acidity." **acidité d'échange**

acidity, residual Soil acidity that is neutralized by lime or other alkaline materials, but which cannot be replaced by an unbuffered salt solution; calculated by subtraction of salt replaceable acidity from total acidity. **acidité résiduelle**

acidity, salt-replaceable The aluminum and hydrogen that can be replaced from an acid soil by an unbuffered salt solution such as KCl or NaCl. **acidité échangeable par un sel**

A

acidity, total The total acidity in a soil or clay, usually estimated by a buffered salt determination of [*cation exchange capacity – exchangeable bases*] = total acidity. Also approximated by the sum of salt replaceable acidity + residual acidity. Often calculated by subtraction of exchangeable bases from the cation exchange capacity determined by ammonium exchange at pH 7.0. It can be determined directly using pH buffer-salt mixtures (e.g., $BaCl_2$ plus triethanolamine, pH 8.0 or 8.2) and titrating the basicity neutralized after reaction with a soil. **acidité totale**

acidophilic Preferring or thriving in a relatively acid environment. **acidiphile, acidophile**

acidulation The process of treating a fertilizer source with an acid or mixture of acids (e.g., treating phosphate rock with sulfuric, nitric, or phosphoric acid). **acidulation**

actinometer An instrument which measures solar radiation. The corresponding term for a recording instrument is actinograph. **pyranomètre, actinomètre**

actinomycetes Gram-positive bacteria that form branching filaments. They may form true *mycelia* or produce conidiospores. The pleasant odor of freshly plowed ground comes from actinomycetes in the soil. See *figure*. **actinomycète**

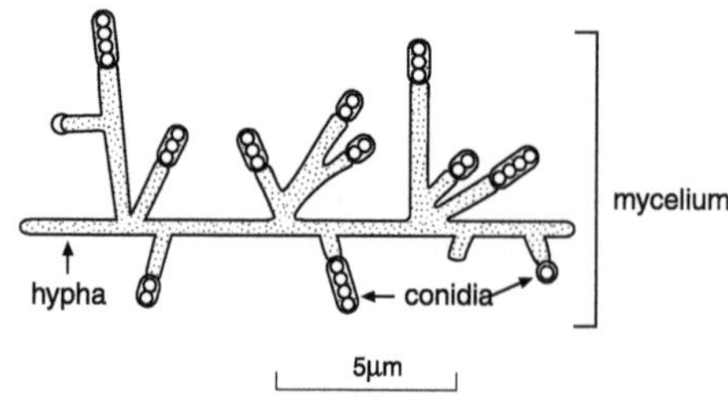

Actinomycete

activated carbon A highly absorbent form of carbon, used to remove odors and toxic substances from gaseous emissions and dissolved organic matter from wastewater. See *carbon filtration*. **charbon activé ou actif**

activation energy The minimum amount of energy required for a chemical reaction to take place. **énergie d'activation**

active organic matter The portion of soil organic matter composed of material that is relatively easy to decompose by soil microorganisms. Also called active fraction of organic matter. **matière organique active**

active ingredient The chemical component(s) in a pesticide product or formulation that causes the desired effect on the specific pest. Usually expressed as a percent and abbreviated as a.i. **matière active**

active layer The top layer of soil in a permafrost zone, subjected to seasonal freezing and thawing which, during the melt season, becomes very mobile. **couche active**

activity *(chemistry)* (1) A dimensionless measure of the deviation of the chemical potential of a substance from its value in some state which, for convenience, is chosen as a standard state. Defined by the equation: $\mu = \mu° + RT \ln a$, where μ is the chemical potential in a state in which the activity is a, $\mu°$ is the chemical potential in the standard state (where $a = 1.0$), R is the molar gas constant, and T is the absolute temperature. (2) The effective concentration of a substance in a solution. **activité**

actual use (range-pasture) The use of forage on any area by livestock and/or game animals without reference to permitted or recommended use; usually expressed in terms of animal unit months or animal units. **utilisation courante (parcours-pâturage)**

adaptation A change in the structure, physiology, or behavior of an organism resulting from natural selection or variation of genetic characteristics by which the organism becomes better fitted to survive in its environment. **adaptation**

adaptive enzyme (enzyme induction) An enzyme produced by an organism in response to the presence of a specific substrate or a related substance. Also called an induced or inducible enzyme. **enzyme induite**

adaptive management Management practice in natural resource exploitation that rigorously combines management, research, monitoring, and means of changing practices so that credible information is gained and management activities are modified by experience. **gestion adaptative**

adaptive zone A unit of environment occupied by a single type of organism, because particular environmental opportunities require similar adaptations for diverse species. Species in different adaptive zones usually differ by major morphologic or physiologic characteristics. **zone adaptative**

adaxial (*botany*) Facing toward the stem of a plant (e.g., the upper surface of a leaf). See *abaxial*. **adaxial**

additive effects (1) The combination of reactions or substances, acting together or independently, to cause a total response equal to or greater than the sum of the separate reactions or substances (e.g., the combined toxic effects of more than one pollutant). (2) Effects on biota of stress imposed by one mechanism, contributed from more than one source (e.g., sediment-related stress on fish imposed by sediment derived from streambank and land surface sources). See also *cumulative effects*. **effets additifs**

adenosine diphosphate (ADP) On hydrolysis, adenosine triphosphate (ATP) loses one phosphate to become adenosine diphosphate (ADP), releasing usable energy. **adénosine diphosphate (ADP)**

adenosine triphosphate (ATP) An energy storage compound common to all biological systems. The high-energy intermediate is formed during photosynthesis or by the breakdown of energy-containing material, such as glucose. Supplies the energy for all cellular reactions and functions. **adénosine triphosphate (ATP)**

adenylate energy charge ratio (EC) A measure of the metabolic state of microorganisms and state of growth of natural microbial communities. The energy charge ratio is calculated using the formula: EC = (ATP + 1/2ADP)/(ATP + ADP + AMP). The denominator represents the total adenylate pool; the numerator, the portion charged with high energy phosphate bonds. **charge énergétique**

adhesion (*chemistry*) A force that acts to hold the molecules of dissimilar substances together. The static attractive force at the contact surface between two bodies of different substances. (*soil mechanics*) The shearing resistance between soil and another material under zero externally applied pressure. **adhésion**

adiabatic process A process that occurs without heat entering or leaving a system. Generally involves a rise or fall in the temperature of the system. **transformation adiabatique**

adobe soil Clayey and silty deposits found in the desert basins of southwestern North America and in Mexico; used extensively for making sun-dried brick. **terre à briques**

adsorbed water Water held in a soil by physicochemical forces and having physical properties substantially different from *absorbed water* or chemically combined water at the same temperature and pressure. **eau adsorbée**

adsorption The process by which atoms, molecules, or ions are taken up and retained on the surfaces of solids by chemical or physical binding (e.g., the adsorption of cations by negatively charged minerals). The two types of adsorption are physisorption, in which the attractive forces are purely *van der Waals*, and chemisorption, where chemical bonds are actually formed between the adsorbent (the material doing the adsorbing) and adsorbate (the material being adsorbed). **adsorption**

adsorption complex The group of substances in the soil capable of adsorbing ions and molecules. Organic and inorganic colloidal substances form the greater part of the adsorption complex. The noncolloidal materials, such as silt and sand, exhibit adsorption to a much lesser extent than the colloidal materials. **complexe d'adsorption**

adsorption isotherm A graph of the quantity of a given chemical species bound to an adsorption complex (e.g., soil) at fixed temperature, as a function of the concentration of the species in a solution in equilibrium with the complex. See *Freundlich isotherm, Langmuir isotherm.* **isotherme d'adsorption**

advection The movement of air, water, and other fluids in a horizontal plane. **advection**

adventitious roots Roots that arise from unusual parts of a plant, usually forming on aerial organs, rhizomes, and older parts of the root body. They can develop under normal environmental conditions or in response to pathogens and wounding. They are found among all vascular plants, and in some cases may be essential to normal growth and development. **racines adventives**

aerate To impregnate with a gas, usually air. **aérer**

aerial photograph (*remote sensing*) A photograph of the Earth's surface taken from airborne equipment; sometimes called aerial photo or air photograph. An oblique aerial photograph is taken with the camera axis directed between the horizontal and vertical. In a high oblique photograph, the apparent horizon is shown, and in a low oblique photograph the apparent horizon is not shown. A vertical aerial photograph is made with the optical axis of the camera approximately perpendicular to the Earth's surface and with the film as nearly horizontal as practical. See *remote sensing*. **photographie aérienne**

aerial reconnaissance The collection of information by visual, electronic, or photographic means from the air. **reconnaissance aérienne**

aerial survey A survey using photographic, electronic, or other data obtained from an airborne platform. **levé aérophotogrammétrique**

aerial triangulation See *phototriangulation.* **aérotriangulation**

aerobe Organism requiring oxygen for growth. **organisme aérobie**

aerobic (1) Having molecular oxygen as a part of the environment. (2) Growing only in the presence of molecular oxygen, such as aerobic organisms. (3) Occurring only in the presence of molecular oxygen, as applied to certain chemical or biochemical processes such as aerobic decomposition. **aérobie**

aerobic decomposition The biodegradation of materials by aerobic microorganisms; the process produces carbon dioxide, water, and other mineral products. Generally a faster breakdown than anaerobic decomposition. **décomposition aérobie**

aerosols Particles of matter, solid or liquid, larger than a molecule but small enough to remain suspended in the *atmosphere*. Particles can come from natural sources (e.g., particles from sea spray or clay particles from the weathering of rocks, both of which are carried upward by the wind) or result from human activities. (Such particles are often considered pollutants.) **aérosols**

afforestation The artificial establishment of forest crops by planting or sowing on land that has not previously, or recently, grown tree crops. See also *reforestation*. **boisement**

aflatoxin Toxins produced by the fungus *Aspergillus flavus* in grains or grainmeals stored under moist conditions; a known carcinogen. **aflatoxine**

after-ripening A curing process sometimes required by seeds, bulbs, and related structures of various plants before germination will take place. **post-maturation**

agar A complex polysaccharide, derived from a particular marine algae, that is a gelling agent in the preparation of nutrient media for growing micro-organisms. Consists of about 70% agarose and 30% agaropectin. Can be melted at temperatures above 100°C; gelling temperature is 40 to 50°C. **agar**

age structure The distribution of individuals in a population into age classes. **structure d'âge**

aggregate A soil structure unit formed by biological and physical agents in which soil primary particles (i.e., sand, silt, clay), along with colloidal and particulate organic and inorganic materials, are grouped together to form larger secondary particles. A group of soil particles cohering in such a way that they behave mechanically as a discrete unit. **agrégat**

aggregate distribution The characterization of soil *aggregates*, usually on the basis of a sieving procedure, based on size range (e.g., 5 to 0.25 mm) or specific order (e.g., *micro-aggregate* and *macro-aggregate*). **distribution des agrégats**

aggregate stability The ability of soil *aggregates* to resist rearrangement and breakdown into primary particles by various disruptive forces, especially the effects of water. The stability of aggregates to disruptive processes is related to soil particle size distribution, type of clay mineral, specific ions associated with the clay, the kind and amount of organic matter present, and nature of the microbial population. **stabilité des agrégats**

aggregated A broad category of *soil structure*, in which primary particles (e.g., sand, silt, clay) unite to form secondary particles or *aggregates*. **structure agrégée ou fragmentaire**

aggregation The process whereby primary soil particles (i.e, sand, silt, clay) are bonded together to form *aggregates*, usually by natural forces and substances derived from root exudates and microbial activity. **agrégation**

agric horizon A diagnostic subsurface illuvial horizon (U.S. system of soil taxonomy) formed under cultivation, containing significant amounts of illuvial silt, clay, and humus. **horizon agrique**

agri-environmental indicator A measure of a key environmental condition, risk, or change resulting from agriculture; a measure of management practices used by producers. **indicateur agro-environnemental**

agrichemical A chemical used in agriculture (e.g., fertilizers, pesticides, and other chemicals used in crop production). **produit agrochimique**

agroclimate A compilation of the average and extreme weather as it affects agricultural cropping in a given area. Agroclimatic classification in Canada is based on limitations of available heat and/or moisture. **agroclimat**

agroecological resource area A natural landscape area more or less uniform in terms of ecoclimate, landform, soils, and general agricultural potential. Agroecological resource areas are subdivisions of *agroecological resource regions*. The extent of an agroecological resource area is in the order of tens to hundreds of square km. Introduced in Canada as convenient planning units upon which to develop databases for use in agricultural research. These units can be used to study agricultural systems, land use, conservation, and the impacts of various management and socio-economic practices. **aire de ressource agroécologique**

agroecological resource region A large area with broadly similar agricultural potential and types of farming. Separated on the basis of general

agro-climatic and physiographic characteristics, landform, soils, and general agricultural potential. The extent of an agroecological resource region is large, in the order of hundreds of square km. See *agroecological resource area.* **région de ressource agroécologique**

agroecological zone A geographic mapping unit developed by the Food and Agriculture Organization of the United Nations, based on climatic conditions and landforms that determine relatively homogeneous crop growing environments. Characterization of agroecological zones permits a quantitative assessment of the biophysical resources upon which agriculture and forestry research depend. The classification system distinguishes between tropical regions, subtropical regions with summer or winter rainfall, and temperate regions. These major regions are further subdivided into rainfed moisture zones, lengths of growing periods, and thermal zones, based on the prevailing temperature regime during the growing season. Classifications include tropical, subtropical, temperate, warm, cool, warm/cool, arid, semiarid, subhumid, and humid. **zone de ressource agroécologique**

agroecosystem A dynamic *landscape* association of crops, pastures, livestock, other flora and fauna, atmosphere, soils, and water. Agroecosystems are contained within larger landscapes that include uncultivated land, natural *ecosystems*, and rural communities. They are open dynamic systems connected to other ecosystems through the transfer of energy and material. **agroécosystème, écosystème agricole**

agroecosystem complexity Classification of agricultural systems on the basis of biodiversity of species (i.e., crops, livestock, pests, trees), and spatial (e.g., field size) and temporal (e.g., type and length of crop rotation) dimensions. **diversité ou complexité d'un agroécosystème**

agroforestry Land use system in which woody perennials are grown for wood production with agricultural crops, with or without livestock production. **agroforesterie, agrosylviculture**

agrohydrology The science dealing with the distribution and movement of rainfall and/or irrigation water and soil solution to and from the root zone in agricultural land. See *hydrology.* **hydrologie agricole**

agrology The study of applied phases of soil science and soil management. A broader term than *agronomy.* **agrologie**

agronomic practices Soil and crop activities used in the production of farm crops (e.g., seed selection, seedbed preparation, fertilizing, liming, manuring, seeding, cultivation, harvesting, curing, crop sequence, crop rotations, cover crops, stripcropping, pasture development). **pratiques agronomiques ou agricoles**

agronomically sustainable yield The maximum yield that can be achieved by a given crop cultivar in a given area, taking account of climatic, soil, and other physical or biological constraints. **rendement agronomique durable**

agronomy The branch of agriculture that deals with the theory and practice of field-crop production and soil management. **agronomie**

agro-silvo-pastoral Land use system in which woody perennials are grown with agricultural crops, forage crops, and livestock production. **agro-sylvo-pastoral**

agrostology Study of grasses; classification, management, and utilization of grasses. **agrostologie**

air dry The state of dryness (water content) of a soil at equilibrium with the moisture contained in the surrounding atmosphere. **séché à l'air**

air entry value The value of water content or potential at which air first enters a porous medium. **point d'entrée d'air**

air frost Air at *Stevenson screen* level (1.2 m) with a temperature at or below 0°C. **gel atmosphérique**

air pollution The presence of contaminants in the air in concentrations that prevent the normal dispersive ability of air and interfere with human health, safety, or comfort. Air contaminants can have a human origin (e.g., smokestacks) or a natural origin (e.g., dust storms). **pollution de l'air**

air porosity The portion of the bulk volume of soil that is filled with air at any given time or under a given condition (e.g., a specified soil water potential). Usually, this portion is made up of large pores (i.e., those drained by a potential of more than about -100 cm of water). **porosité d'air, teneur en air**

air dry (1) The state of dryness of a soil at equilibrium with the moisture content of the surrounding atmosphere. The moisture content depends on the relative humidity and the temperature of the surrounding atmosphere. (2) To allow a soil sample to reach equilibrium in moisture content with the surrounding atmosphere. See *humidity, relative*. **séché à l'air**

air-dry mass Mass of a substance (e.g., soil) after it has been allowed to dry to equilibrium with the atmosphere. **masse sèche à l'air**

air-filled porosity A measure of the relative air content of a soil. As an index of *soil aeration* it is related negatively to *degree of saturation*. See *air porosity*. **porosité d'air, teneur en air**

akinete A nonmotile spore formed singly within a cell and with the spore wall fused with the parent cell wall. **akinète**

alban (*soil micromorphology*) A light-colored *cutan* composed of materials that have been strongly reduced. **albane**

albedo The ratio of reflected to incident light, expressed as a percentage or a fraction of 1. Snow-covered areas have a high albedo (up to about 0.9 or 90%) due to their white color, whereas vegetation has a low albedo (generally about 0.1 or 10%) due to the dark color and the absorption of light for *photosynthesis*. Clouds have an intermediate albedo and are the most important contributor to the Earth's albedo, which is about 0.3. **albédo**

albic horizon A subsurface diagnostic mineral horizon in the U.S. system of soil taxonomy from which clay and free iron has been removed, or in which the oxides of iron have been segregated to the extent that the horizon color is determined primarily by the color of the primary sand and silt particles. An *eluvial* horizon. **horizon albique**

Albolls A suborder in the U.S. system of soil taxonomy. *Mollisols* that have an *albic horizon* immediately below the *mollic epipedon*. These soils have an *argillic* or *natric horizon* and mottles, iron-manganese concretions, or both, within the albic, argillic, or natric horizon. **Albolls**

alcohol An organic compound that contains the –OH group (e.g., simple alcohols are methanol (CH_3OH) and ethanol (C_2H_5OH)). **alcool**

aldehyde An organic compound that contains the –CHO group (the aldehyde group) which consists of a carbonyl group attached to a hydrogen atom (e.g., methanol or formaldehyde (HCHO) and ethanol or acetaldehyde (CH_3CHO)). **aldéhyde**

Alfisols An order in the U. S. system of soil taxonomy. Mineral soils that have *umbric* or *ochric* epipedons, *argillic* horizons, and hold water at <1.5 MPa tension during at least 90 days when the soil is warm enough for plants to grow outdoors. Alfisols have a mean

A

annual soil temperature of <8°C or a base saturation in the lower part of the *argillic horizon* of 35% or more when measured at pH 8.2. **Alfisols**

alga (plural algae) Phototrophic eukaryotic microorganism. Algae contain chlorophyll and are unicellular or multicellular. They form the base of the food chain in aquatic environments; some species may create a nuisance when environmental conditions are suitable for prolific growth. **algue**

algal bloom A proliferation of living *algae* on the surface of lakes, streams, or ponds. Algal blooms are stimulated by phosphate enrichment. **prolifération algale**

algicide A chemical compound used to kill filamentous *algae* and *phytoplankton.* **algicide**

algology The study of *algae.* **algologie**

algorithm A series of well-defined steps used in carrying out a specific process (e.g., a classification algorithm). An algorithm may be in the form of a word description, explanatory note, diagram, labeled flow chart, or computer code. **algorithme**

aliphatic compound An organic compound in which carbon and hydrogen molecules are arranged in straight or branched chains (i.e., no ring structures present); a type of hydrocarbon. An organic compound that is an *alkane, alkene,* or *alkyne*, or a derivative of these. **composé aliphatique**

aliquot A subsample resulting from dividing a liquid sample into a number of equal parts. Generally used to define any representative portion of the sample. **aliquote**

alkali Any substance capable of furnishing to its solution or other substances the hydroxyl ion (OH^-); a substance having marked basic properties in contrast to acid. **alcali**

alkali metals (group 1 elements) A group of soft reactive metals, each representing the start of a new period in the periodic table. The alkali metals are lithium (Li), sodium (Na), potassium (K), rubidium, (RB), cesium (Cs), and francium (Fr). **métaux alcalins**

alkali soil (1) A soil having a high degree of alkalinity (pH of 8.5 or higher), or having a high exchangeable sodium content (15% or more of the exchange capacity), or both. (2) A soil that contains enough alkali (i.e., sodium) to interfere with the growth of most crop plants. **sol à alcalis, sol alcalin**

alkaline Having a high pH value (greater than 7); also basic; the opposite of acidic. **alcalin**

alkaline soil A soil that has a pH greater than 7.0. See *reaction, soil.* **sol alcalin**

alkaline solution A liquid whose hydroxyl ion concentration is greater than its hydrogen ion concentration, or whose pH is greater than 7.0. **solution alcaline**

alkaline-earth metals (group 2 elements) A group of moderately reactive metals, harder and less volatile than the *alkali metals.* The term alkaline earth strictly refers to the oxides, but is often used loosely for the elements themselves. The elements are beryllium (Be), magnesium (Mg), calcium (Ca), strontium (Sr), barium (Ba), and radium (Ra). **métaux alcalino-terreux**

alkalinity (1) The quality or state of being *alkaline*; the concentration of OH^- ions. (2) A measure of the ability of water to neutralize acids. It is measured by determining the amount of acid required to lower the pH of water to 4.5. In natural waters, the alkalinity is effectively the bicarbonate ion concentration plus twice the carbonate ion concentration, expressed as milligrams per liter calcium carbonate. **alcalinité**

alkalinity, soil The degree or intensity of alkalinity of a soil expressed by a soil pH value greater than 7.0. **alcalinité du sol**

alkalinity, total The total measurable bases (OH, HCO_3, CO_3) in a volume of water; a measure of the material's capacity to neutralize acids. **alcalinité totale**

alkalization The accumulation of sodium ions on the exchange sites in a soil. **alcalinisation**

alkaloid One of a class of basic organic compounds with nitrogen in their structure; a secondary product of plant metabolism. **alcaloïde**

alkane A type of *saturated hydrocarbon* with the general formula C_nH_{2n+2}. The straight-chain alkanes form a *homologous series* methane (CH_4), ethane (C_2H_6), propane (C_3H_8), butane (C_4H_{10}), pentane (C_5H_{12}), etc. Low molecular weight alkanes are gases; high molecular weight alkanes are waxy solids. Alkanes are fairly unreactive. **alcane**

alkene A type of unsaturated hydrocarbon with the general formula C_nH_{2n}. Alkenes contain one or more carbon–carbon double bonds and form a *homologous series* beginning with ethene (C_2H_4) and propene (C_3H_6). **alcène**

alkyl group A group obtained by removing a hydrogen atom from an *alkane* or other *aliphatic compound.* **groupement alcoyle ou alkyle**

alkyne A type of unsaturated hydrocarbon with the general formula C_nH_{2n-2} that contains one or more triple carbon-carbon bonds. The simplest member of the *homologous series* is ethyne (acetylene, C_2H_2). **alkine**

allele Any of a group of alternative forms of the same gene. **allèle**

allelopathy Any direct or indirect harmful effect of one plant or microorganism on one or more other organisms through the production and release of chemical compounds into the environment. **allélopathie**

allitization See *desilication.* **allitisation**

allochthonous Non-native or transient; referring to organisms that are not indigenous to a specific habitat but enter or are transported into the habitat by various means (e.g., in precipitation, diseased tissues, manure, or sewage). They may persist for some time but do not contribute in a significant way to ecological transformations or interactions, reproduce, or occupy the habitat permanently. **allochtone**

allochthonous peat Peat formed from the remains of plants brought in, mainly by water, from outside the site of deposition. Constitutes an integral part of *peat* deposits that develop from the filling in of water bodies by lateral or vertical transport or both. See *autochthonous peat, limnetic.* **tourbe allochtone**

allogenic succession Changes in species composition and environmental properties due to changes in extrinsic environmental factors (e.g., fire). **succession allogène**

allopatric speciation Separation of a population into two or more evolutionary units as a result of reproductive isolation caused by geographic separation of two subpopulations. **spéciation allopatrique**

allophane An aluminosilicate with primarily short-range structural order; occurs as very small spherical particles, especially in soils formed from volcanic ash. Also occurs in podzolic soils formed on weathered granite in a cool, moist climate. **allophane**

alloy A substance that contains a mixture of elements and has metallic properties. **alliage**

alluvial Pertaining to *alluvium.* **alluvial**

alluvial terrace See *river terrace.* **terrasse alluviale**

alluvium Material (e.g., clay, silt, sand, and gravel) deposited by running water, including the sediments laid down in riverbeds, *flood plains*, lakes, and *estuaries.* **alluvion**

alpha decay The spontaneous decomposition of the atom nuclei resulting in the emission of *alpha particles.* **désintégration alpha**

alpha diversity Diversity within a specific community. **diversité alpha**

alpha error See *Type I error.* **erreur de première espèce, erreur de type I**

alpha particle A positively charged particle emitted by some radioactive materials. It is the least penetrating of the three common types of radiation (alpha, beta, and gamma) and is usually not dangerous to plants, animals, or humans. A high-energy helium nucleus (two protons, two neutrons) emitted by some heavy radioactive nuclei. **particule alpha**

alpine Of, pertaining to, or like any high mountain; implying high elevation and cold climate; referring to that portion of mountains above tree growth, or the organisms living there. **alpin**

alpine biome Considered as a *biome* or a subtype of the *tundra* biome. A mountain area above the timberline characterized by permanently frozen subsoil and a dominant vegetation of mosses, lichens, herbs, and dwarf shrubs. **biome alpin**

alpine soil A mountain soil occurring above the tree line. **sol alpin**

alpine tundra The grassland area found above the tree line on mountain ranges. **toundra alpine**

altimeter An instrument that indicates height above sea level, based on the average decrease in atmospheric pressure with increasing height. This averages 3.4 kPa for every 300 m, but variations occur owing to differences of air temperature and latitude. Used mainly in aircraft but also by ground surveyors. An altigraph is a self-recording altimeter. **altimètre**

aluminosilicate See *silicate.* **silicate d'aluminium**

aluminum A soft, moderately reactive metal; the second element in group III of the periodic table. There are numerous minerals of aluminum; it is the third most abundant element in the Earth's crust (8.1% by weight). Commercially important minerals are bauxite (hydrated Al_2O_3), corundum (anhydrous Al_2O_3), cryolite (Na_3AlF_6), and clays and mica (aluminosilicates). **aluminium**

amendment, soil The addition of materials (e.g., lime, gypsum, sawdust, compost, animal manures, or synthetic soil conditioners) to soil to enhance plant growth. Fertilizers constitute a special group of soil amendments. **amendement du sol**

amensalism The suppression of one organism by another, often involving toxins. **allélopathie**

amictic lake A lake that does not experience mixing or turnover on a seasonal basis. See *dimictic lake.* **lac amictique**

amide Any of a group of *herbicides* designed to retard root and shoot growth, mostly effective on grassy weeds, causing stunted and malformed seedlings. **amide**

amines Organic compounds derived by replacing one or more of the hydrogen atoms attached to nitrogen by one or more organic groups. The different types of amines are named for the number groups attached to the nitrogen atom: primary amines have one hydrogen atom replaced (e.g., methyline [CH_3NH_2]); secondary amines have two hydrogen atoms replaced (e.g., dimethylamine [$(CH_3)_2NH$]); and tertiary amines have all three hydrogen atoms replaced (e.g., trimethylamine [$(CH_3)_3N$]). Amines are produced during the decomposition of organic nitrogen. **amines**

amino acid An organic acid containing both an amino (NH_2) and a carboxyl (COOH) group. Amino acid molecules combine to form proteins. **acide aminé**

ammate A chemical compound, ammonium sulfamate, used as a relatively short-lived *herbicide*. **ammate**

ammonia The gaseous compound of nitrogen and hydrogen (e.g., NH_3); commonly known as anhydrous

A

ammonia in the fertilizer industry. **ammoniaque**

ammonia volatilization Mass transfer of nitrogen as ammonia gas from soil, plant, or liquid systems to the atmosphere. **volatilisation de l'ammoniaque**

ammoniation The process of introducing various ammonium sources into other fertilizer sources forming ammoniated compounds (e.g., ammonium polyphosphates, ammoniated superphosphate). **ammoniation**

ammonification A biochemical process carried out by microorganisms in which nitrogen-containing organic compounds are degraded or mineralized and ammoniacal nitrogen is formed. **ammonification**

ammonium entrapment See *ammonium fixation*. **fixation ou intégration de l'ammonium en position interfeuillet**

ammonium fixation The process of entrapment of ammonium ions in interlayer spaces of phyllosilicates, in sites similar to K^+ in micas. Smectites, illites, and vermiculites all can fix ammonium, but vermiculite has the greatest capacity. The fixation may occur spontaneously in aqueous suspensions, or as a result of heating to remove interlayer water. Ammonium ions in collapsed interlayer spaces are exchangeable only after expansion of the interlayer. See *potassium fixation*. **fixation d'ammonium**

ammonium nitrate A colorless crystalline compound readily soluble in water (971 g per 100 g of water at 100°C), produced by combining anhydrous ammonia with nitric acid; NH_4NO_3. It is used in the manufacture of explosives and, because of its high nitrogen content, as a fertilizer. **nitrate d'ammonium**

ammonium phosphate A general class of compounds used as phosphorus fertilizers; manufactured by the reaction of anhydrous ammonia with orthophosphoric acid or superphosphoric acid to produce either solid or liquid products; $(NH_4)_3PO_4$. **phosphate d'ammonium**

ammonium sulfate A colorless crystalline solid that is soluble in water; $(NH_4)_2SO_4$. When carefully heated it gives ammonium hydrogen sulfate, which on stronger heating yields nitrogen, ammonia, sulfur dioxide, and water. Manufactured by the reaction of ammonia with sulfuric acid. A disadvantage as a fertilizer is that it tends to leave an acidic residue in the soil. **sulfate d'ammonium**

amoeba (plural amoebae) Protozoa that can alter their cell shape, usually by extrusion of one or more *pseudopodium*. Existing in soil in large numbers; many live as parasites, and some species are pathogenic to humans. See *figure*. **amibe**

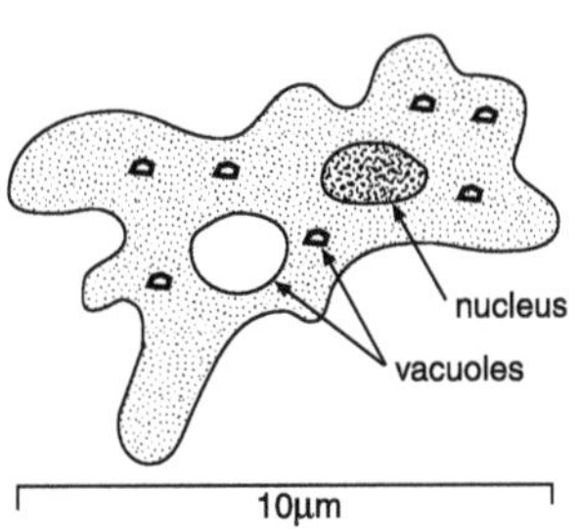

Amoeba

amorphous mineral (1) A mineral that has no definite crystalline structure. (2) A mineral that has a definite crystalline structure but appears amorphous because of the small crystallite size. (3) A noncrystalline constituent that either does not fit the definition of allophane or it is uncertain that the constituent meets allophane criteria. **minéral amorphe**

amorphous peat The structureless portion of an organic deposit in which plant remains are decomposed to sizes too small to be visually recognized. **tourbe amorphe**

amphibole One of the ferromagnesian silicate mineral group, characterized by prismatic, columnar, or fibrous crystals with a structure of cross-linked double chains of tetrahedra (e.g., hornblende). **amphibole**

amphoteric substance A substance that can behave either as an *acid* or as a *base*. **substance amphotère**

amplitude (*physics*) The vertical distance between the crest of a wave and the base of the adjacent trough. (*ecology*) The range of tolerance of a species. **amplitude**

anabolism The metabolic process involving the conversion of simpler substances to more complex substances or the storage of energy. More generally, the synthesis of organic compounds within an organism. Also called *assimilation*, *biosynthesis*, or constructive metabolism. See *catabolism*. **anabolisme**

anaerobe Organism that lives in the absence of air (oxygen). **organisme anaérobie**

anaerobic (1) The absence of molecular oxygen. (2) Growing in the absence of molecular oxygen (e.g., anaerobic bacteria). (3) Occurring in the absence of molecular oxygen (e.g., a biochemical process). **anaérobie**

anaerobic (soil) The absence of molecular oxygen (O_2); a condition that exists in soils when they are flooded or compacted. **anaérobie (sol)**

anaerobic decomposition The degradation of materials by anaerobic microorganisms living in oxygen-depleted soil or water to form reduced compounds such as methane or hydrogen sulfide. Generally slower than aerobic. **décomposition anaérobie**

anaerobic respiration A metabolic process in which electrons are transferred from an organic compound to an inorganic acceptor molecule other than oxygen. The most common acceptors are carbonate, sulfate, and nitrate. **respiration anaérobie**

anaglyph A *stereogram* in which the two views are printed or projected superimposed in complementary colors, usually red and blue. By viewing through filter spectacles of corresponding complementary colors, a stereoscopic image is formed. **anaglyphe**

analog A derivative of a naturally occurring compound that an organism cannot distinguish from the natural product. The uptake of this derivative results in the formation of a biological molecule incapable of carrying out its proper function. **substance analogue**

analysis The process of determining the constituents or components of a material (i.e., sample). The two broad major classes of analysis are, *qualitative analysis* and *quantitative analysis*. **analyse**

analysis of variance, ANOVA (*statistics*) The analysis of the total variability of a set of data (measured by their total sums of squares) into components which can be attributed to different sources of variation. A table that lists the various sources of variation together with the corresponding degrees of freedom, sums of squares, mean squares, and values of F is called an analysis of variance table. See *one-way analysis of variance, two-way analysis of variance*. **analyse de variance**

analyte The substance which is to be measured by chemical analysis. **analyte**

analytical model A model in which all functional relationships can be expressed in closed form and the parameters fixed so that the equations can be solved by the classical methods of analytical mathematics. **modèle analytique**

anastomosing (*geomorphology*) Stream patterns in which the channels bifurcate, branch, and rejoin irregularly to create a net-like formation. **anastomosé**

A

anchor See *tillage, anchor*. **enfouir, incorporer**

Andepts An obsolete term in the U.S. system of soil taxonomy for *Inceptisols* that have formed in volcanic materials. **Andepts**

andesite A fine-grained volcanic rock composed of andesine, similar in mineralogy to a diorite. **andésite**

andic Soil properties related to volcanic origin of materials. The properties include organic carbon content, bulk density, phosphate retention, and iron and aluminum extractable with ammonium oxalate. **andique**

Andisols An order in the U.S. system of soil taxonomy. Mineral soils dominated by andic soil properties in 60% or more of their thickness. **Andisols**

Andosol Term used in some countries for a soil developed on volcanic material. Characterized by a dark, organic A-horizon and a barely altered B-horizon. See Inceptisol. **Andosol**

anemometer Any instrument that measures wind speed. **anémomètre**

angiosperm A plant with true flowers, in which the seeds are enclosed in an ovary, comprising the fruit. The two subclasses are: the Monocotyledoneae (monocots), which are herbaceous and include such economically important plants as grasses (e.g., maize, wheat, barley); and the Dicotyledoneae (dicots), which include both woody and herbaceous members. **angiosperme**

angle of dip (*geology*) The angle that a bed, vein, or stratum makes with the horizontal. See *dip*. **angle de chute, pendage**

angle of emergence The angle formed between a ray of energy (e.g., optic, acoustic, or electromagnetic) and the horizontal. It is the complement of the *angle of incidence*. **angle d'émergence**

angle of incidence The angle that a ray of energy (e.g., optic, acoustic, or electromagnetic) makes with the normal to a boundary surface. It is the complement of the *angle of emergence*. **angle d'incidence**

angle of reflection The angle between a ray of light reflected from a surface and the normal to that surface. **angle de réflexion**

angle of refraction The angle between a ray of light that has been refracted (e.g., through water) and the normal from the surface at which the ray is refracted. **angle de réfraction**

angle of repose (rest) (*geology*) The maximum slope gradient at which a mass of unconsolidated material (e.g., *scree, talus*) will remain stable. The angle varies according to the character of the material. If the slope angle becomes steeper than the angle of repose the slope becomes unstable, leading to a landslide or earth-flow until the slope angle returns to a state of stability. The coarser the material, the higher the angle of repose. **angle de repos**

angular Having sharp angles or borders; especially applied to sedimentary particles showing little or no evidence of abrasion, with all edges and corners sharp. Also, applied to the roundness class containing angular particles. **anguleux**

angularity See *roundness*. **forme anguleuse**

anhedral A form of minerals that are not bounded by their own crystal faces but whose forms are controlled by mineral grains next to them. Also called allotriomorphic. **allotriomorphe**

anhydrite A mineral, anhydrous calcium sulfate, $CaSO_4$. Orthorhombic, commonly massive in evaporite beds. Under natural conditions it slowly hydrates to form gypsum. Used in the manufacture of cements and fertilizers. **anhydrite**

animal unit (A.U.) A measurement of livestock based on the equivalent of a mature cow (about 454 kg live weight); roughly one cow, one horse, one mule, five sheep, five swine, or six goats. **unité animale (U.A.), unité-gros-bétail (U.G.B.)**

A

animal unit month (A.U.M.) A measure of forage or feed requirement to maintain one animal unit for one month. **unité animale-mois (U.A.M.)**

anion A negatively charged ion. **anion**

anion exchange capacity The total amount of exchangeable anions that a soil can adsorb. It is expressed as centimoles, or millimoles, of charge per kilogram of soil or other adsorbing material such as clay. **capacité d'échange anionique**

anion exclusion The exclusion or repulsion of anions from the vicinity of negatively charged soil particle surfaces. **exclusion anionique**

anionic resin An ion-exchange material that can exchange an *anion*, such as Cl^- and OH^-, for anions in the surrounding medium; used for a wide range of analytical and purification purposes. **résine anionique**

anisotropic soils Soils not having the same physical properties when the direction of measurement is changed. **sols anisotropes**

anisotropy The condition of having different properties in different directions. For example, in soil there is vertical anisotropy due to the presence of different horizons; in geology, geologic strata transmit sound waves with different velocities in the vertical and horizontal directions; in crystallography, crystalline materials not belonging to the cubic crystal system are optically anisotropic. **anisotropie**

annelid (*zoology*) Member of the phylum Annelida containing red-blooded worms such as earthworms. **annélide**

annual plant A plant that completes its life cycle (from germination to flowering to seed production and the death of its vegetative parts) within a single growing season. See *biennial plant, perennial plant*. **plante annuelle**

anomaly A deviation from the norm; a deviation of an observed value from a theoretical value due to an abnormality in the observed quantity. **anomalie**

ANOVA See *analysis of variance*. **analyse de variance**

antagonism (*ecology*) Production of a substance by an organism that can destroy or prevent the growth of one or more organisms. (*plant nutrition*) The interference of one element with the absorption or utilization of an essential nutrient by plants. (*chemistry*) Strong competition between pairs of similar ions (e.g., Ca^{2+} and Sr^{2+}, K^+ and Rb^+). (*pollution*) The combined reaction of several pollutants acting simultaneously or independently to cause a total response less than their separate effects. **antagonisme**

antecedent moisture The soil water content prior to the sampling or measurement event. **humidité antérieure**

anther The pollen-bearing portion of a stamen. **anthère**

anthesis The time of flowering in a plant, which may respond to a combination of environmental factors such as day length, temperature, and rainfall. **anthèse**

anthocyanins Group of water soluble flavonoid pigments that occur in vacuoles of leaves during the autumn after chlorophyll is destroyed due to environmental changes. Their color (e.g., red to purple) is related to the solution acidity. **anthocyanines**

anthracite A hard, black, lustrous coal containing a high percentage of volatile matter; often called hard coal. **anthracite**

anthric saturation A variation of soil saturation associated with controlled flooding, which causes reduction in the surface soil layer and oxidation of mobilized iron and manganese in a lower unsaturated subsoil. **saturation anthrique**

anthropic epipedon A surface layer of mineral soil (U.S. system of soil taxonomy) that has the same requirements as the *mollic epipedon* with respect to color, thickness, organic carbon content, consistence, and base saturation, but has >110 mg P kg^{-1} soluble in 0.05 M citric acid, or is dry >300 days (cumulative) during the period when not irrigated. The anthropic epipedon forms under long continued cultivation and fertilization. **épipédon anthropique**

anthropic soil A soil body that has been constructed by human actions. Excluded are natural soils that have been altered by humans in a progressive or accidental manner. **sol anthropique**

anthropogenic Derived from human activities. **anthropogénique**

antibiosis The production of an organic compound by one species that is toxic at low concentrations to another species. **antibiose**

antibiotic A substance produced by a species of microorganism and, in dilute solution, has the capacity to inhibit the growth of or kill certain other organisms. **antibiotique**

antibody A protein produced by the immune system in response to the presence of an antigen which interacts with that specific antigen to remove or inactivate it. **anticorps**

anticline (*geology*) A *fold* of rock beds that is convex upward. See *fold*, *syncline*. **anticlinal**

antigen A substance that can initiate production of a specific antibody and is capable of inducing an immunological response whereby the antibody binds to the antigen. **antigène**

anti-oxidant An organic compound that prevents or retards the damage caused by oxidation to living cells, food, and other material such as rubber and plastic. It acts by scavenging free radicals generated during the oxidation process. **antioxydant**

antitoxin An antibody that is formed in response to bacterial toxins or toxoids and can neutralize or inactivate those toxins. **antitoxine**

apatetic (*zoology*) The coloration of an animal that causes it to resemble physical features of its habitat. **apatétique**

apatite A group of hexagonal minerals consisting of calcium phosphate together with fluorine, chlorine, hydroxyl, or carbonate in varying amounts and having the general formula $Ca_5(PO_4,CO_3)_3(F,OH, Cl)$. Also, any mineral of this group (e.g., fluorapatite, chlorapatite, hydroxylapatite, carbonate-apatite, and francolite); when not specified, the term usually refers to fluorapatite. The apatite minerals occur as accessory minerals in igneous rocks, metamorphic rocks, and ore deposits, most commonly as fine-grained and often impure masses as the chief constituent of phosphate rock and of bones and teeth. Also called calcium phosphate. **apatite**

apedal Condition of a soil that has no structure, i.e., having no peds, but rather is massive or composed of single grains. **apédal**

apical meristem The growing point, composed of meristematic tissue, at the tip of the root or shoot in a vascular plant. See *figure*. **méristème apical**

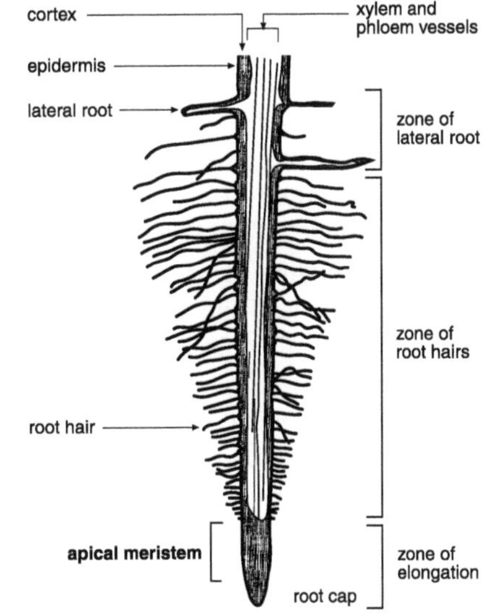

Apical meristem adapted from Dunster and Dunster, 1996).

apogee (*remote sensing*) The point in the orbit of a satellite or in a missile trajectory that is farthest from the center

of the Earth; opposite of perigee. **apogée**

apparent cohesion Cohesion in granular soils due to capillary forces associated with water. **cohésion apparente**

apparent density See *bulk density*. **masse volumique apparente, densité apparente**

apparent specific gravity See *bulk specific gravity*. **densité spécifique apparente**

applanation (*geology*) Reduction of the relief of an area, causing it to become more and more plain-like, including lowering of the high parts by erosion and raising of the low parts by the addition of material. **aplanissement**

approximate original contour (*land reclamation*) The surface configuration achieved by backfilling and grading of mined areas so that the reclaimed area, including any terracing or access roads, closely resembles the general surface configuration of the land prior to strip mining and blends into and complements the drainage pattern of the surrounding terrain. **configuration approximative d'origine**

apron (*geology*) An extensive, blanket-like accumulation of alluvial, glacial, or other unconsolidated material commonly formed at the base of a mountain or in front of a glacier. **plaine d'épandage**

aqua regia A mixture of one part concentrated nitric acid and three parts concentrated hydrochloric acid. It is a powerful oxidizing mixture and will dissolve all metals (except silver), including gold and platinum, hence the name ("royal water"). **eau régale**

Aqualfs A suborder in the U.S. system of soil taxonomy. *Alfisols* that are saturated with water for periods long enough to limit their use for most crops other than pasture or woodland unless they are artificially drained. Aqualfs have mottles, iron-manganese concretions, and gley colors immediately below the A horizon and gray colors in the *argillic horizon*. **Aqualfs**

Aquands A suborder in the U.S. system of soil taxonomy. *Andisols* that are saturated with water for periods long enough to limit their use for most crops other than pasture unless they are artificially drained. Aquands have low chromas in redox depletions or on ped faces. **Aquands**

aquatic Related to environments that contain liquid water. **aquatique**

aquatic plant A plant other than algae, growing in or near water, with true roots, stems, and leaves. **plante aquatique**

Aquents A suborder in the U.S. system of soil taxonomy. Entisols that are saturated with water for periods long enough to limit their use for most crops other than pasture unless they are artificially drained. Aquents have low chromas or distinct mottles within 50 cm of the surface, or are saturated with water at all times. **Aquents**

aqueous (1) Of or pertaining to water. (2) Made from or with water (e.g., aqueous solutions). (3) Produced by the action of water (e.g., aqueous sediments). **aqueux**

aqueous solution A solution in which water is the dissolving medium or solvent. **solution aqueuse**

Aquepts A suborder in the U.S. system of soil taxonomy. *Inceptisols* that are saturated with water for periods long enough to limit their use for most crops other than pasture or woodland unless they are artificially drained. Aquepts have either a *histic* or *umbric epipedon* and gray colors within 50 cm of the surface, or an *ochric epipedon* underlaid by a *cambic horizon* with gray colors, or have sodium saturation of 15% or more. **Aquepts**

Aquerts A suborder in the U.S. system of soil taxonomy. Vertisols that are saturated with water for periods long enough to limit their use for most

A

crops other than pasture and woodland unless they are artificially drained. Aquerts have in one or more horizons between 40 and 50 cm from the surface, aquic conditions for some time in most years and chromas of two or less in 50 percent of the pedon or evidence of active ferrous iron. **Aquerts**

aquic A mostly reducing soil moisture regime nearly free of dissolved oxygen due to saturation by groundwater or its capillary fringe and occurring at periods when the soil temperature at 80 cm below the surface is >5°C. **aquique**

aquiclude (*hydrology*) A low-permeability formation located above and/or below an aquifer. A formation which contains water but cannot transmit it rapidly enough to furnish a significant supply to a well or spring. **aquiclude**

aquifer A geologic formation or stratum which contains sufficient saturated permeable material to yield significant quantities of water to wells and springs. **formation aquifère**

aquifer, confined Aquifer located between two relatively impermeable layers of material and under pressure significantly greater than atmospheric pressure. Also called an *artesian aquifer*. **nappe artésienne**

aquifer, unconfined An aquifer with no low-permeability zones between the zone of saturation and the surface. **aquifère non captive**

aquifuge A geologic formation or structure that contains no interconnected openings or interstices and therefore neither absorbs, holds, nor transmits water. **aquifuge**

aquitard (*hydrology*) A low-permeability underground formation that retards but does not prevent the flow of water to or from an adjacent aquifer. **couche semi-perméable capacitive**

Aquods A suborder in the U.S. system of soil taxonomy. *Spodosols* that are saturated with water for periods long enough to limit their use for most crops other than pasture or woodland unless they are artificially drained. Aquods may have a *histic epipedon*, an *albic horizon* that is mottled, or a duripan, mottling or gray color within or immediately below the spodic horizon. **Aquods**

Aquolls A suborder in the U.S. system of soil taxonomy. *Mollisols* that are saturated with water for periods long enough to limit their use for most crops other than pasture unless they are artificially drained. Aquolls may have a *histic epipedon*, a sodium saturation in the upper part of the *mollic epipedon* of > 15% that decreases with depth, or mottles or gray colors within or immediately below the mollic epipedon. **Aquolls**

Aquox A suborder in the U.S. system of soil taxonomy. *Oxisols* that have continuous *plinthite* near the surface, or that are saturated with water sometime during the year if not artificially drained. Aquox have either a *histic epipedon*, or mottles or colors indicative of poor drainage within the *oxic horizon*, or both. **Aquox**

Aquults A suborder in the U.S. system of soil taxonomy. *Ultisols* that are saturated with water for periods long enough to limit their use for most crops other than pasture or woodland unless they are artificially drained. Aquults have mottles, iron-manganese concretions, and gray colors immediately below the A horizon and gray colors in the *argillic horizon*. **Aquults**

arable Pertaining to land suited for tillage and cultivation of crops. **arable**

arboretum A collection of plants, trees, and shrubs grown for public exhibition, public enjoyment, recreation, education, or research. **arboretum**

arbuscule Special dendritic (highly branched) structure formed within root cortical cells by *endomycorrhizal* fungus. See *vesicular arbuscular*. See *figure*. **arbuscule**

arctic (1) The region within the Arctic Circle (66° 30' N). (2) Lands north of the 10°C July isotherm (or that

A

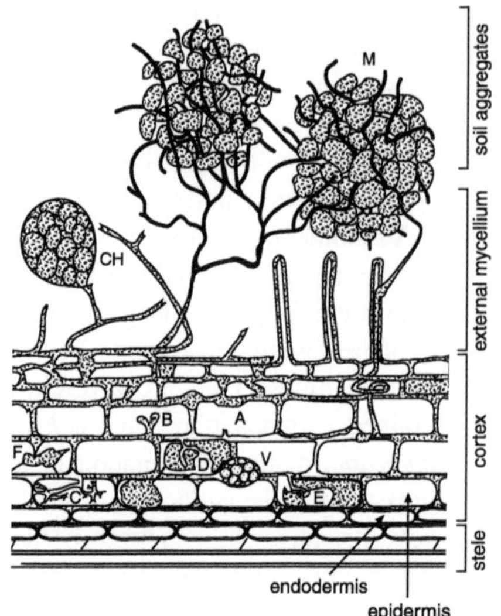

Association of **arbuscular** mycorrhizal fungi and soil aggregate of a plant root. The external mycelium bears large chlamydospores (CH). Infection of the plant can occur through root hairs or between epidermal cells. Arbusculae at progressive stages in development and senescence are shown (A–F), as is a vesicle (V), and macroaggregate (M) (adapted from Paul and Clark, 1996).

of whichever month is warmest), provided the mean temperature for the coldest month is not higher than 0°C. (3) Pertaining to cold, frigid temperature, or to features, climate, vegetation, and animals characteristic of the arctic region. **arctique**

arctic tundra The grassland *biome* characterized by *permafrost* (subsurface soil that remains frozen throughout the year); found in Alaska, Canada, Russia, and other regions near the Arctic Circle. **toundra arctique**

area mining The type of mining used to extract mineral resources close to the surface in relatively flat terrain. Overburden is removed in a series of parallel trenches to allow extraction of the resource, and the overburden removed from one trench is used to fill in the adjacent trench after removal of the resource. Also called area strip mining. **exploitation minière de surface**

area source The geographic source from which air pollution originates. **source diffuse ou étendue**

areal Pertaining to an area. **aréal**

areal map (*geology*) A geologic map showing the horizontal extent and distribution of rock units exposed at the surface. **carte de surface**

Arents A suborder in the U.S. system of soil taxonomy. *Entisols* that contain recognizable fragments of pedogenic horizons which have been mixed by mechanical disturbance. Arents are not saturated with water for periods long enough to limit their use for most crops. **Arents**

Argids A suborder in the U.S. system of soil taxonomy. *Aridisols* that have an *argillic* or a *natric horizon*. **Argids**

argillaceous Pertaining to rocks or substances composed of clay minerals, or having a notable proportion of clay in their composition, especially sedimentary materials such as shale. **argileux**

argillan (*soil micromorphology*) A *cutan* composed dominantly of clay minerals. **argilane**

argillic Pertaining to clay or clay minerals. Argillic alteration is a process in which certain minerals are converted to minerals of the clay group. **argilique**

argillic horizon A diagnostic subsurface illuvial horizon (U.S. system of soil taxonomy) that is characterized by the accumulation of layer-lattice silicate clays. The argillic horizon has a certain minimum thickness depending on the thickness of the solum, a minimum quantity of clay in comparison with an overlying *eluvial horizon* depending on the clay content of the eluvial horizon, and usually has coatings of oriented clay on the surface of pores, peds, or bridging sand grains. **horizon argilique**

argillipedoturbation Disruption and mixing of soil material caused by shrinking and swelling, as indicated by the presence of irregular shaped, randomly oriented intrusions of displaced materials within the solum, and by vertical cracks, often containing sloughed-in surface materials. **argilipédoturbation**

arid Pertaining to a climate in which the limits of precipitation vary considerably according to temperature conditions, with an upper annual limit of about 25 cm for cool regions and 50 cm for tropical regions. Regions with this climate lack sufficient moisture for crop production without irrigation. See *semi-arid, humid, subhumid.* **aride**

aridic A soil moisture regime that has no water available for plants for more than half the cumulative time that the soil temperature at 50 cm below the surface is > 5°C, and has no period as long as 90 consecutive days when there is water for plants while the soil temperature at 50 cm is continuously > 8°C. **aridique**

Aridisols An order in the U.S. system of soil taxonomy. Mineral soils that have an *aridic* moisture regime, an *ochric epipedon*, and other pedogenic horizons but no *oxic horizon.* **Aridisols**

arithmetic mean *(statistics)* The sum of all the values of a set of measurements divided by the number of values in the set, usually denoted by $\bar{x}$; a measure of central tendency; also called average. See *measure of central tendency.* **moyenne arithmétique**

aromatic compound An organic compound that contains a *benzene* ring in its molecules. Aromatic compounds have a planar ring of atoms linked by alternate single and double bonds. See *aliphatic compound.* **composé aromatique**

Arrhenius equation An equation with the form $k = A\exp(-E_a/RT)$ where k is the rate constant of a given reaction, A represents the product of the collision frequency and the steric factor, and E_a/RT is the fraction of collisions with sufficient energy to produce a reaction. **équation d'Arrhénius**

artesian aquifer A water-bearing bed that contains water under hydrostatic pressure. **nappe artésienne**

artesian spring A spring from which water issues under artesian pressure, generally through a fissure in the confining bed that overlies the aquifer. **source artésienne**

artesian water Pertaining to ground water under sufficient hydrostatic pressure to rise above the aquifer containing it, but not necessarily to or above the ground surface. **eau artésienne**

artesian well A water well drilled into a confined aquifer where enough hydraulic pressure exists for the water to flow to the surface, or above the top of the aquifer without pumping. Also called a flowing well. **puits artésien**

arthropod (*zoology*) Any one of a group of invertebrates belonging to the phylum Arthropoda, characterized chiefly by jointed appendages and segmented bodies. It includes, insects, spiders, centipedes, and beetles. **arthropode**

artificial brine Brine produced from an underground deposit of salt or other soluble rock material in the process of solution mining. **saumure artificielle**

artificial manure See *compost.* In European usage, artificial manure may denote commercial fertilizers. **fumier artificiel, compost**

artificial regeneration (*silviculture*) Establishing a new forest by planting seedlings or by direct seeding (as opposed to natural regeneration). **régénération artificielle**

artificial soil body A soil body that has been constructed by human actions (e.g., the construction of artificial sequences of horizons in land reclaimed from strip-mined areas, and similar activities to build human-made soils in lands reclaimed from the sea). Natural soils that have been

A

altered by humans in a progressive or accidental manner are generally excluded. **sol anthropique**

asbestos A white, gray, green-gray, or blue-gray fibrous variety of *amphibole*, usually tremolite or actinolite, or of chrysotile. Blue asbestos is crocidolite. Asbestos has several industrial uses. It is a hazardous air pollutant when inhaled. **amiante**

ascospores The spores produced by *Ascomycetes*. See *figure*. **ascospores**

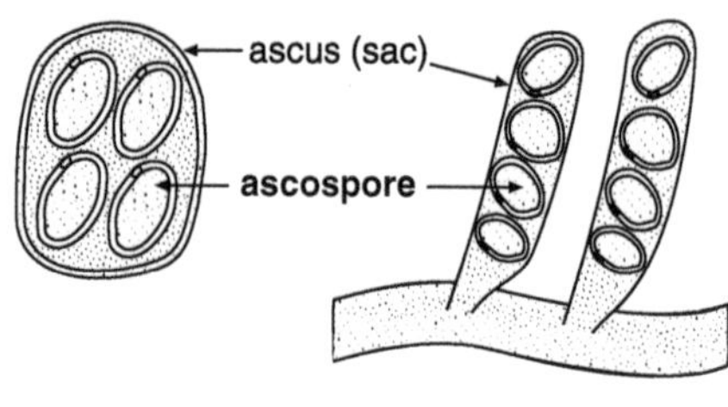

Ascospore

aseptic Procedure that maintains sterility. **aseptique**

ash The residue remaining after complete burning of combustible organic matter; consists mainly of minerals in oxidized form. **cendre**

aspect The compass direction toward which a slope faces. **orientation, exposition**

asphalt A dark brown to black viscous liquid or low-melting solid *bitumen* that consists almost entirely of carbon and hydrogen, melts between 65°C and 95°C, and is soluble in carbon disulfide. Natural asphalt is formed as a residual deposit by evaporation of volatiles from oil, and is also obtained as a residue from the refining of certain petroleum types. **asphalte**

assimilation (*biology*) Conversion or incorporation of absorbed nutrients into protoplasm; the uptake of food material for production of new biomass. (*pollution*) The ability of a body of water to purify itself of organic pollution. (*geology*) The incorporation into a magma of material originally present in the wall rock; the "assimilated" material may be present as crystals, including wall rock elements, or as a true solution in the liquid phase of the magma. The resulting rock is called hybrid. Also called magmatic assimilation. **assimilation**

assimilation efficiency The proportion of energy in ingested food that is assimilated into the bloodstream of an organism. **efficacité d'assimilation**

assimilatory nitrate reduction Conversion of nitrate to reduced forms of nitrogen, generally ammonium, for the synthesis of amino acids and proteins. **réduction assimilatoire des nitrates**

associate, soil A non-taxonomic, cartographic grouping of soils or land segments which combines related soils into units having similar geomorphic position, landform, edaphic, and mechanical properties of soils (e.g., climate, drainage, particle size), and some similarity in the geological nature of the soil materials and taxonomic classes. **sol associé**

association A grouping or combination of entities. (*ecology*) A group of species occurring in the same place because of environmental requirements or tolerances. (*soil science*) A grouping of soils based on similarities in climatic or physiographic factors and soil parent materials. **association**

associative dinitrogen fixation An enhanced rate of dinitrogen fixation resulting from a close interaction between a free-living diazotrophic organism and a higher plant. **fixation associative d'azote**

asymmetrical Without proper proportion of parts; unsymmetrical. (*crystallography*) Having no center, plane, or axis of symmetry. **asymétrique**

atmophile elements (1) The most typical elements of the atmosphere (H, C, N, O, I, He, and inert gases). (2) Elements that occur in the uncombined state or were concentrated in the gaseous primordial atmosphere. **éléments atmosphiles ou atmophiles**

atmosphere The mixture of gases surrounding the Earth. The Earth's atmosphere consists of about 79.1% nitrogen (by volume), 20.9% oxygen, 0.036% carbon dioxide, and trace amounts of other gases. The atmosphere can be divided into a number of layers according to its mixing or chemical characteristics, generally determined by its thermal properties (temperature), as follows. *Troposphere*, the layer nearest the Earth, reaching to an altitude of about 8 km in the polar regions and 17 km above the equator. *Stratosphere*, beyond the troposphere and reaching to an altitude of about 50 km. *Mesosphere*, beyond the stratosphere and extending up to 80–90 km. *Thermosphere*, or *ionosphere*, beyond the mesosphere and gradually diminishing to form a fuzzy border with outer space. Relatively little mixing of gases occurs between layers. **atmosphère**

atmospheric attenuation (*remote sensing*) The reduction of radiation intensity due to absorption and/or scattering of energy by the atmosphere; usually wavelength dependent; may affect both solar radiation traveling to the Earth and reflected/emitted radiation traveling to the sensor from the Earth's surface. **atténuation atmosphérique**

atmospheric pressure The pressure exerted by the weight of the atmosphere on the Earth's surface. The average pressure at sea level is 101.325 kPa. **pression atmosphérique**

atmospheric windows (*remote sensing*) Those wavelength ranges where radiation can pass through the atmosphere with relatively little attenuation; in the optical portion of the spectrum, approximately 0.3 to 2.5, 3.0 to 4.0, 4.2 to 5.0, and 7.0 to 15.0 micrometers. **fenêtres atmosphériques**

atom The smallest part of an element that can exist as a stable entity. Atoms consist of a small dense nucleus, made up of *neutrons* and *protons*, surrounded by *electrons*. The chemical reactions of an element are determined by the number of electrons (which is equal to the number of protons in the nucleus). All atoms of a given element have the same number of protons (the proton number). A given element may have two or more isotopes, which differ in the number of neutrons in the nucleus. **atome**

atom percent The percentage of an atomic species in a substance, calculated with reference to number of atoms rather than weight, number of molecules, or other criteria. **pourcentage atomique**

atomic absorption (AA) spectroscopy A technique that uses the absorption of light to measure the concentration of gas-phase atoms. Samples are usually liquids or solids and the *analyte* atoms or ions must be vaporized in a flame or graphite furnace. The atoms absorb ultraviolet or visible light and make transitions to higher electronic energy levels. The analyte concentration is determined from the amount of absorption. **spectroscopie d'absorption atomique**

atomic bond Attraction exerted between atoms and ions. Four types are; metallic, *ionic* or polar, homopolar or *coordinate*, and residual or *van der Waals*. Bonding may be intermediate between these types. **liaison atomique**

atomic emission spectroscopy (AES) A technique that uses quantitative measurement of the optical emission from excited atoms to determine analyte concentration. Analyte atoms in a solution are aspirated into the excitation region where they are desolvated, vaporized, and atomized by a flame, discharge, or plasma. These high-temperature atomization sources provide sufficient energy to promote the atoms into high energy levels. The atoms decay back to lower levels by emitting light. Since the transitions are between distinct atomic energy levels, the emission lines in the spectra are narrow. The spectra of samples containing many elements can be very congested, and spectral separation of nearby atomic transitions requires a high-resolution spectrometer. Since all atoms in a sample are excited simultaneously, they can be detected simultaneously using a polychromator with multiple detectors. This ability to simultaneously measure multiple elements is a major advantage of AES compared to *atomic absorption (AA) spectroscopy.* **spectroscopie d'émission atomique**

atomic mass unit A unit of mass used for atoms and molecules; equal to 1/12 of the mass of an atom of the isotope carbon-12 and is equal to 1.66033 x 10^{-27} kg. **unité de masse atomique**

atomic number The number of protons in the nucleus of an atom. Each element has a unique atomic number. **nombre atomique**

atomic radius Half the distance between the nuclei in a molecule consisting of identical atoms. **rayon atomique**

atomic waste Radioactive byproducts produced during activities such as the mining and processing of radioactive materials, fabrication of nuclear weapons, and operation of nuclear reactors. **déchet atomique**

atomic weight For each element, the weighted sum of the masses of the protons and neutrons composing the isotopes of that element. Approximately equal to the sum of the number of protons and neutrons found in the most abundant isotope. **masse atomique**

atomizer An instrument used to produce a fine spray or mist from a liquid. **atomiseur**

attapulgite See *palygorskite.* **attapulgite**

attenuation (1) A reduction in the amplitude or energy of a signal, such as might be produced by passage through a filter. (2) A reduction in the amplitude of seismic waves, as produced by divergence, reflection and scattering, and adsorption. (3) That portion of the decrease in seismic or sonar signal strength with distance that is not dependent on geometrical divergence, but on the physical characteristics of the transmitting medium. (4) Lessening or reduction of the virulence of microorganisms. **atténuation**

Atterberg limits See *liquid limit* and *plastic limit.* **limites d'Atterberg**

attitude (1) (*remote sensing*) The position of a body as determined by the inclination of the axes to some frame of reference. If not otherwise specified, this frame of reference is fixed to the Earth. (2) (*geology*) The relation of some directional features to a rock in a horizontal surface. **(1) orientation (2) disposition**

attribute (1) Any property, quality, or characteristic of a sampling unit. The indicators and other measures used to characterize a sampling site or resource unit are representations of the attributes of that unit or site. (2) A characteristic of a map feature (point, line, or polygon) described by numbers or text; for example, attributes of a tree represented by a point might include height and species. **(1) variable qualitative, attribut (2) attribut**

attribute of soil quality Properties that reflect or characterize a soil process or processes that support a specific *soil function.* **attribut de la qualité des sols**

aufwuchs (*liminology*) Slimy aquatic community consisting of diatoms and gelatinous green and blue-green algae attached to substrate. **algue épiphytique**

A

auger, soil A tool for boring into the soil and withdrawing a small sample for observation in the field or laboratory. The different kinds of augers include those having worm-type bits that are unenclosed, those having worm-type bits enclosed in a hollow cylinder, and those having a hollow half-cylinder with a cutting edge on the side that rotates around a stabilizing vane. **tarière**

autecology A subdivision of ecology that deals with the study of the relation of an individual species or population to its environment. See *dynecology, synecology.* **auto-écologie**

authigenic (*geology*) Formed or generated *in situ*. Rock constituents that formed at the spot where they are now found; minerals that came into existence at the same time as, or later than, the rock of which they constitute a part. See *autochthonous*. **authigène**

autocatalysis Catalysis in which one of the products of the reaction is a catalyst for the reaction. **autocatalyse**

autochthonous (1) Native, or indigenous, to a locale; the opposite of allochthonous. (2) Used to describe soil microorganisms that have the ability to grow under low nutrient supply. They are thought to subsist on the more resistant organic matter and are little affected by the addition of fresh organic materials. **autochtone**

autochtonous peat Peat formed *in situ*. See *allochthonous peat*. **tourbe autochtone**

autoclave A chamber in which steam under pressure is used to sterilize objects and solutions. **autoclave**

autocorrelation (*statistics*) The internal correlation (relationship) between elements of a stationary time series; usually expressed as a function of the time lag between observations. See *correlation, statistics*. **autocorrélation**

autolysis Self-destruction of cells by the action of autolytic or intracellular enzymes, resulting in cellular death. **autolyse**

autopoiesis The organizing principle of life by which an entity's boundary structure (e.g., cell membrane), processes of metabolism, and energy exchange are determined by its internal organization and the interchange with its immediate surroundings. **autopoièse**

autotoxic Chemicals that may be detrimental to individuals of the species that make the chemical. **autotoxique**

autotroph An organism that manufactures its own food from inorganic compounds (e.g., carbon dioxide) in the environment, obtaining energy from light (*photoautotroph*) or another inorganic compounds (*chemautotroph*). See *heterotroph*. **autotrophe**

autotrophic nitrification Oxidation of ammonium to nitrate through the combined action of two *chemoautotrophic* organisms, one forming nitrite from ammonium and the other oxidizing nitrite to nitrate. **nitrification autotrophe**

autotrophy Self-production of food and fixation of energy by green plants and some bacteria. See *heterotrophy*. **autotrophie**

auxin A hormone that promotes the longitudinal growth in the cells of higher plants by increasing the rate of cell elongation rather than the rate of cell division; produced at the growing points of stems and roots and involved in the curvature of plant parts towards light or gravity. Synthetic auxins are used as herbicides. **auxine**

available moisture See *available water*. **eau disponible**

available nutrient The amount of a nutrient element or compound in the soil that can be readily absorbed and assimilated by growing plants. **élément assimilable**

available water (capacity) The amount of water released between *in situ field capacity* and the *permanent wilting point* (usually estimated by water content at soil matric potential of −1.5 MPa). **eau disponible**

average See *arithmetic mean*. **moyenne**

Avogadro's law Equal volumes of gases at the same temperature and pressure contain the same number of molecules. **loi d'Avogadro**

Avogadro's constant The number of atoms or molecules in a mole of a substance, equal to 6.022×10^{23}. **constante d'Avogadro**

axenic A system in which all biological populations are defined, such as a pure culture. **axénique**

azimuth The horizontal angle or bearing of a point measured from the true (astronomic) north. Used to refer to a compass on which the movable dial (used to read direction) is numbered in 360°. See *bearing*. **azimut**

azonal soil Soil without distinct genetic horizons. Such soils may have non-Chernozemic Ah horizons or thin Chernozemic Ah horizons, and lack B horizons. In Canada, they are included in the *Regosolic* order. **sol azonal**

B

B

b axis (*crystallography*) One of the crystallographic axes used as reference in crystal description. It is the axis that is oriented horizontally, right to left. See *a axis*, *c axis*. **axe b**

B horizon A subsoil horizon characterized by one of the following: an enrichment in clay, iron, aluminum, or humus (Bt or Bf); a prismatic or columnar structure that exhibits pronounced coatings or stainings associated with significant amounts of exchangeable sodium (Bn or Bnt); an alteration by hydrolysis, reduction, or oxidation to give a change in color or structure from the horizons above or below, or both (Bm). Other types and combinations of types of B horizons include: Bg, Bfg, Btg, (characterized by gleying); Bmk (characterized by presence of carbonate); Bs, Bsa (presence of salts); Bss (presence of slickensides); Bv (characterized by *argillipedoturbation*); By (affected by *cryoturbation*); and Bmz (a frozen Bm). The above horizon terms are according to the Canadian system of soil classifiaction. See *Appendix D* for equivalent U.S. Soil Taxonomy and FAO soils terminology. See *A horizon, C horizon, horizon, soil*. **horizon B**

BP An abbreviation for "before (the) present (day)" (as opposed to BC). It is used in geochronology, especially in *carbon dating* of the later Quaternary period. **BP, avant le présent**

bacillus A rod-shaped bacterium with specific physiologic characteristics. **bacille**

back end system The extraction of recyclable materials from incinerator ash (in a solid waste resource recovery operation). **système de récupération**

back furrow See *tillage, back furrow*. **ados**

backfill (*land reclamation*) (1) The placement of spoils (waste soil and rock) in the notches cut in the hills, and restoration of the original slope. This process reduces soil erosion and allows for the re-establishment of vegetation. (2) The operation of refilling an excavation. Also, the material placed in an excavation in the process of backfilling. **remblayer, remblai**

background (1) The normal slight radioactivity of the environment, due to cosmic rays and the Earth's naturally radioactive substances. (2) The range in values representing the normal concentration of a given element in a material under investigation, such as rock, soil, plants, and water. (3) The amount of pollutants present in the ambient air owing to natural resources. **bruit de fond, bruit**

background samples (*analytical chemistry*) Samples that do not contain the *analyte* of interest (*blank*) and that are subjected to steps of the analytical procedure (e.g., reagents, glassware, preparations, and analytical instrumental) to account for the presence of spurious analytes, interferences, and background concentrations of the analyte of interest. **échantillons de référence, échantillons témoins**

backscatter (*remote sensing*) Radiation reflected back toward the source; the opposite of forward scatter; also called backscattering. **rétrodiffusion**

B

backslope See *slope morphological classification.* **revers**

back-swamp A marshy area occurring outside a river channel's *levee* on a *flood plain.* **dépression latérale humide**

backwashing The movement of clean water in a direction opposite (upward) to the normal flow of raw water through rapid sand filters to clean the filters, in a water treatment facility. **lavage à contre-courant, lavage par retour d'eau, nettoyage inversé**

bacterium (plural bacteria) Single-cell microorganism that lacks *chlorophyll* and an evident nucleus; may be *aerobic*, *anaerobic*, or *facultative*; some bacteria can cause disease. See *figure.* **bactérie**

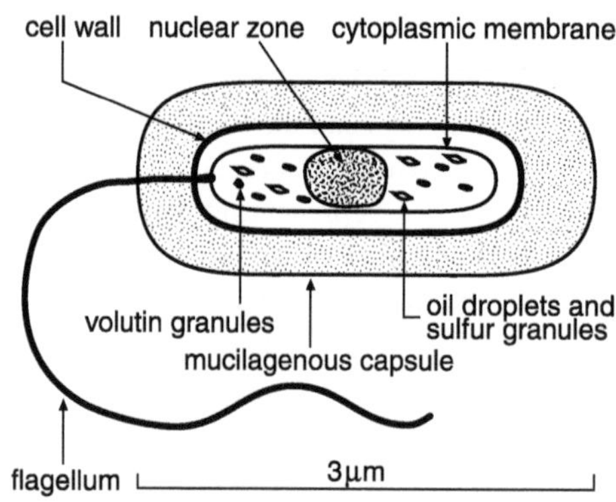

A soil **bacterium** (adapted from Killham, 1994).

bacterial plate count A system of quantifying the number of bacteria in a sample of solid or liquid substance. The sample is mixed and/or diluted with a suitable sterile diluent, and a measured, small portion of the mixture is placed on the surface of a solid substrate suitable for the growth of bacteria (a medium). Following incubation, the number of bacterial colonies that have grown on the medium is counted, and the necessary mathematical estimations are made on the assumption that each colony represents one bacterium that was present in the original sample. **numération bactérienne sur plaque**

bactericidal Able to kill bacteria. **bactéricide**

bacteriophage A virus that infects bacteria and causes lysis of bacterial cells. **bactériophage**

bacteriostatic A substance inhibitory to bacterial growth but not necessarily lethal. If a bacteriostatic material dissipates or lowers in concentration, bacterial growth may resume. **bactériostatique**

bacteroid An irregular form of cells of certain bacteria. Refers particularly to the swollen, vacuolated cells of *Rhizobium* and *Bradyrhizobium* in legume nodules. **bactéroïde**

badland A land type generally devoid of vegetation and broken by an intricate maze of narrow ravines, sharp crests, and pinnacles resulting from severe erosion of soft geologic materials. Badlands occur most commonly in arid or semiarid regions. **badland**

bagasse The fibrous residue remaining after juice is extracted from sugar cane or sugar beets. **bagasse**

ballistic separator A machine that separates inorganic from organic matter in a composting process. **séparateur par projection**

band application See *banding.* **traitement en bandes**

band-elimination filter (*remote sensing*) A wave filter that attenuates one frequency band, neither the critical nor cutoff frequencies being zero or infinite. **filtre à élimination de bande, filtre coupe-bande**

banding (*geology*) (1) A pattern of alternating layers in igneous or metamorphic rock that differ in color or texture and may or may not differ in mineral composition. (2) Thin bedding in sedimentary rocks consisting of different materials in alternating layers, and conspicuous in cross-section. (3) (*agronomy*) A method of applying fertilizers or agrichemicals above, below, or alongside the planted seed row. Refers to either the placement of fertilizers close to the seed at planting or the surface or subsurface application of solids or fluids in strips before or after planting. Also called band application. **(1) litage métamorphique (2) litage (3) application en bande**

band-pass filter (*remote sensing*) A wave filter that has a single transmission band extending from a lower cutoff frequency greater than zero to a finite upper cutoff frequency. **filtre passe-bande**

bandwidth (*remote sensing*) (1) In an antenna, the range of frequencies within which its performance, in respect to some characteristic, conforms to a specified standard. (2) In a wave, the least frequency interval outside of which the power spectrum of a time-varying quantity is everywhere less than some specified fraction of its value at a reference frequency. (3) The number of cycles per second between the limits of a frequency band. **largeur de bande**

bar (1) (*geomorphology*) A deposit of sand or mud in a river channel. Also, an elongate offshore ridge, bank, or mound of sand or gravel in the sea, built by waves and currents. (2) (*meteorology*) A unit of pressure equal to 10^6 dynes cm^{-2}; equivalent to a mercurial barometer reading of 750.076 mm at 0°C and gravity being equal to 980.616 cm sec^{-2}. A bar is equal to the mean *atmospheric pressure* at about 100 meters above mean sea level. **(1) barre, levée (2) bar, megabarye**

barchan (*geomorphology*) A crescent-shaped sand dune with the convex side facing the wind. The gentler slope is on the convex side, and the steeper slope is on the concave side between the horns of the crescent. A dune type characteristic of very dry, inland desert regions. **barkhane**

bareroot seedling (*silviculture*) Tree seedlings to be used for reforestation, whose roots are exposed at the time of planting (as opposed to container or *plug seedling*). Seedlings grown in nursery seedbeds are lifted from the soil and planted in the field. **plant à racines nues**

barometer A device for measuring atmospheric pressure. **baromètre**

B

barometric pressure The force exerted on a surface by the acceleration of gravity acting on a column of air above that surface. **pression barométrique**

barophile A microorganism that resides at high pressures, usually at the bottom of the ocean. **barophile**

barren An area relatively barren of vegetation compared to adjacent areas because of adverse soil or climatic conditions, wind, or other adverse environmental factors (e.g., sand barrens or rock barrens). **dénudé ou stérile**

barren lands An early description of the tundra regions of Northern Canada, characterized by permafrost, sparse vegetation, long winters, and inhospitable climate. Now rarely used. **terrains improductifs**

barrier beach (*geomorphology*) An offshore *bar* consisting of a single elongated ridge rising slightly above the high-tide level and extending generally parallel with the coast, but separated from it by a lagoon. **cordon littoral**

barrier lake (*geomorphology*) Any lake formed by impounding of a river by a natural dam such as a landslide, a morainic dam, deltaic deposits, a vegetation dam, a calcium carbonate deposit in a karstic region, or an avalanche. **lac de barrage**

basal application The application of agrichemical on plant stems or trunks just above the soil line, usually in a field. **traitement à la base des tiges, traitement basal**

basal metabolism The energy used by an organism while at rest. **métabolisme de base**

basal till Till carried at or deposited from the under-surface of a glacier. See *till*. **till de fond**

basalt A dark-colored, fine-grained igneous rock forming lava flows or minor intrusions. The fine-grained equivalent of gabbro, it is composed of

B

plagioclase, augite, and magnetite, with or without olivine. **basalte**

base (1) (*chemistry*) A substance that produces OH^- when dissolved in water (i.e., *Arrhenius theory*). A proton acceptor (i.e., Lowry-Brønsted theory). An electron-pair donor (i.e., Lewis theory). (2) A rock with a high proportion of base-metal oxides. (3) A substance capable of combining with silica in a rock. (4) An initial length of measurement in a survey (*base line*). (5) In statistics, the number of digits used in a positional method of counting (e.g., base ten = zero plus nine digits). **base**

base course A layer of specified or selected material of planned thickness constructed on the subgrade or sub-base for distributing load, providing drainage, minimizing frost action, and other such purposes. **base, assise**

base dissociation constant (k_b) The equilibrium constant for the reaction of a base with water to produce the conjugate acid and hydroxide ion. **constante de dissociation d'une base**

base flow (*hydrology*) The flow in a stream arising from groundwater, and excluding surface runoff. **débit de base**

base level The theoretical limit or lowest level toward which erosion constantly tends to reduce the land but is seldom, if ever, reached. Sea level is the permanent base level, but there may be many temporary base levels provided by lakes, other rivers, or resistant rock strata. **niveau de base d'érosion**

base line A surveyed line established with more than usual care, to which surveys are referred for coordination and correlation. **bande géodésique**

base map A map of any kind showing essential outlines necessary for adequate geographic reference, on which additional or specialized information is overlayed. **carte de base, géobase, fond de carte**

base metal Any of the more common and more chemically active metals (e.g., lead, copper). **métal**

base saturation percentage The extent to which the adsorption complex of a soil is saturated with exchangeable cations other than hydrogen and aluminum; expressed as a percentage of the total cation exchange capacity. **pourcentage ou taux de saturation en bases**

basement (*geology*) Any mass of ancient igneous and metamorphic rocks underlying unconformable strata of unmetamorphosed sedimentary rocks. **socle**

basic Having a tendency to release OH^- ions. Thus, any solution in which the concentration of OH^- ions is greater than that in pure water at the same temperature is described as basic (i.e., the pH is greater than 7). **basique**

basic fertilizer A fertilizer that reduces residual acidity and increases soil pH after application to and reaction with soil. **engrais de base**

basic intake rate The nearly constant infiltration rate of water into the soil after the soil has been thoroughly wetted. See *infiltration*. **taux d'infiltration limite**

basic rock (*geology*) Igneous rock having low silica content, usually between 45 and 52%. One of four subdivisions of a commonly used system for classifying igneous rocks based on silica content (e.g., acidic, intermediate, basic, and ultrabasic). **roche basique**

basic slag A byproduct in the manufacture of steel that contains lime, phosphorus, and small quantities of other plant nutrients (e.g., sulfur, manganese, and iron). **scories**

basidiospore A sexual spore produced by the *Basidiomycetes*. See *figure*. **basidiospore**

basidium The specialized part of the mycelium of *Basidiomycetes* on which *basidiospores* are formed. **baside**

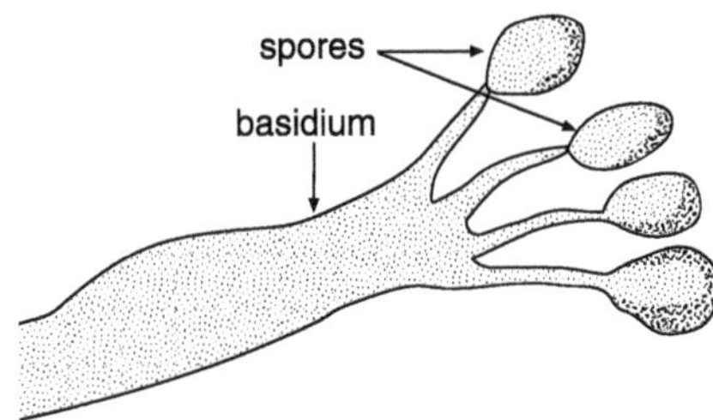

Basidiospore

basin (1) A depressed area with no surface outlet. (2) The drainage area of a stream or river. (3) A low area in the Earth's crust, of tectonic origin, in which sediments have accumulated. These features were drainage basins at the time of sedimentation but are not necessarily so today. **bassin**

batch culture A method for culturing organisms in which the biological component and supporting nutritive medium are added to a closed system. See *chemostat.* **culture en batch, culture discontinue**

batch method The procedure, used in industry or laboratory, that involves adding all raw materials or reagents and allowing the process to be completed before the products are removed. In a batch method for composting, the solid materials (e.g., leaves) are mixed and prepared, placed in a vessel or static pile, allowed to compost for a period of time, and then removed. See *continuous-feed reactor.* **méthode en lots**

bauxite A gray, yellow, or reddish-brown rock composed of a mixture of various aluminum oxides and hydroxides (principally gibbsite, boehmite, and diaspore), along with free silica, silt, iron hydroxides, and clay minerals; a highly aluminous *laterite.* It is a common residual or transported constituent of clay deposits in tropical and subtropical regions, and occurs in concretionary, compact, earthy, pisolitic, or oolitic forms. Bauxite is the principal commercial source of aluminum. **bauxite**

beach The gently sloping shore of a body of water which is washed by waves or tides, especially the parts covered by sand or pebbles. **plage**

beach deposit Sediment that is modified in its degree of sorting, surface relief, or both, by the action of waves in forming beaches. **dépôt de plage**

beach ridge Linear accumulation of beach and dune material formed by the action of waves and currents on the backshore of a beach beyond the present limit of storm waves or ordinary tides. They occur singly or as one of a series of approximately parallel deposits which represent successive positions of an advancing shoreline. **crête de plage**

beaded drainage (*hydrology*) Drainage pattern developed in a permafrost environment owing to differential surface melting, characterized by interconnected pools and short linking streams. **lacs orientés**

bearing The direction of a line with reference to the cardinal points of the compass. True bearing is the horizontal angle between a ground line and a geographic meridian. A bearing may be referred to either the south or north point (N 30°E, or S 30°W). Magnetic bearing is the horizontal angle between a ground line and the magnetic meridian. A magnetic bearing differs from a true bearing by the exact angle of magnetic declination of the locality. **azimut vrai, azimut géographique, relèvement**

bearing capacity The average load per unit area required to rupture a supporting soil mass; measured in kg m^{-2}. **capacité portante, capacité de charge**

Beccari process (*compost management*) A process in which anaerobic fermentation is followed by a final stage in which decomposition proceeds under partially aerobic conditions. **procédé de Beccari**

becquerel The SI unit of radioactivity. **becquerel**

B

bed (1) (*soil science*) A unit layer 1 cm or more thick that is visually or physically distinct from other layers above and below in a stratified sequence. (2) (*geology*) The smallest layer (or stratum) of a stratified sedimentary rock that can be separated from its adjacent layers by a planar surface. It is the smallest of the *lithostratigraphic units*. (3) A layer of pyroclastic or volcanic material, such as an ash-bed or a bed of lava. **lit**

bed load (*hydrology*) The part of a stream's load (e.g., sand, silt, gravel, or soil and rock detritus) carried by a stream on or immediately above its bed. The particles of this material have a density or grain size such as to preclude movement in suspension, or far above, or for a long distance out of contact with the stream bed under natural conditions of flow. **charriage**

bed material (*hydrology*) The material of which a stream bed is composed. **matériau du lit**

bed planting See *tillage, bed planting*. **semis sur buttes ou sur lits surélevés**

bed shaper See *tillage, bed shaper*. **billonneuse, buttoir**

bedded Arranged or deposited in layers or *beds*; refers especially to sedimentary rocks, but also to stratified material of other origin (e.g., volcanic ash). See *cross-bedding*. **lité**

bedding (1) (*agriculture*) The mounding of soil into elevated parallel strips or rows performed with tillage tools such as sweeps, shovels, disks, or moldboard plows. See *tillage, bedding*. (2) (*geology*) The arrangement of sediment and sedimentary rock in layers, strata, or beds more than one cm thick. (3) (*engineering*) The process of installing and securing a drain in a trench. **(1) billonnage, (2) litage, (3) enterrage**

bedding, animal Any material, usually organic, that is placed on the floor of livestock quarters for animal comfort and to absorb animal excreta. **litière**.

bedrock The solid rock that underlies soil and the regolith, or that is exposed at the surface. **assise rocheuse, roc sous-jacent, roc**

bedrock spoil Bedrock material that has been mined and dumped. It may consist of hard fragments of varying size or may be soil-sized particles. See *spoil*. **déblais rocheux**

Beer-Lambert law Law allowing the quantitative determination of the concentration of chemicals in solution by measuring the light absorbed upon passage through the solution. It is expressed as A = ebc, where A is the *absorbance*, e is the molar *absorptivity*, b is the length of the light path through the sample in cm, and c is the chemical concentration in moles per liter. **loi de Beer-Lambert**

beidellite An aluminum-rich member of the smectite group of clay minerals. It is a common constituent of soils, and of certain clay deposits. **beidellite**

belt (1) A distinctive zone of terrain, vegetation, or climate. (2) A narrow strip of seawater. **zone**

bench (*land reclamation*) (1) In mining, the surface of an excavated area at some point between the material being mined and the original surface of the ground on which equipment can be set, moved, or operated; a working road or base below a highwall, as in *contour stripping* for coal. (*gemorphology*) (2) A narrow, flat, or gently sloping ledge or step, bounded above and below by steeper slopes. It may be formed by structural movement (faults), agencies of erosion (e.g., wave-cut bench), or human quarrying. (*geology*) (3) A coal bed or seam separated from adjacent seams by a parting of slate (shale), or one of several layers within a coal seam that may be mined separately from the others. **(1) gradin (2) replat (3) plate-forme**

bench terrace See *terrace*. **terrasse en gradins**

benchmark (*general*) A measurement or standard against which similar units can be compared. (*surveying*) A virtually permanent point of reference used by a surveyor during a topographical survey. A specific point or level on the benchmark has a known elevation related to a datum at a fixed level. **repère, point géodésique**

benchmark site A fixed locale or site at which certain soil properties and processes can be determined at regular intervals. **site de référence, site repère**

benthic Aquatic organisms that grow or inhabit the bottoms of lakes and oceans. **benthique**

benthic region The bottom of a body of water, supporting the *benthos*. **région benthique**

benthos The plant and animal life whose habitat is the bottom of a sea, lake, or river. **benthos**

bentonite A soft, plastic, light-colored clay formed by chemical alteration of volcanic ash composed of montmorillonite and related minerals of the smectite group. The properties of bentonite depend largely on its ion-exchange characteristics. It shows extensive swelling in water and has a high specific surface area. **bentonite**

benzene The simplest aromatic hydrocarbon; formula C_6H_6. The benzene molecule is a closed ring of six carbon atoms connected by bonds that resonate between single and double bonds. Benzene and its derivatives are known as *aromatic compounds*. The liquid is volatile, insoluble in water but miscible in all proportions with organic solvents. Benzene has a long history of use in the chemical industry as a solvent and as a starting compound for the synthesis of a variety of other materials, and is now also used extensively in the rubber, paint, and plastic industries. The liquid is volatile, and emissions are regulated as a toxic air pollutant. **benzène**

benzene hexachloride (BHC) A chemical compound also known as hexachlorocyclohexane (HCH) and present in the insecticide *lindane*; BHC has carcinogenic properties, bioaccumulates in the fatty tissue of humans, and may cause depression of the central nervous system. **hexachlorure de benzène**

benzine A mixture of aliphatic hydrocarbons, such as gasoline or certain cleaning solvents. Not related to benzene. See *aliphatic compound*. **benzine**

benzoic Any of a group of *herbicides* that cause abnormal shoot and root growth by upsetting plant *hormone* (i.e., *auxin*) balance. **herbicide benzoïque**

berm (*landscaping*) An elongated mound in a naturally level land area or one made artificially by a landscaper to gain privacy or interest in a private or public area. (*mining*) A strip of coal left in place temporarily for use in hauling or stripping. A layer of large rock or other relatively heavy, stable material placed at the outside bottom of the spoil pile to help hold the pile in position (i.e., a toe walk). (*erosion control*) A shelf or relatively flat area that breaks the continuity of a slope; used on earth dams and cut slopes to increase stability. **berme**

beta counter A device used to measure ionizing radiation. The counter is configured in such a way that the ionizing events caused by a *beta particle* are selectively measured. **compteur bêta**

beta diversity Diversity across several communities on an environmental gradient. **diversité bêta**

beta error See *Type II error*. **erreur bêta, erreur de type II**

beta particle Particles equivalent in mass to an electron emitted by certain substances undergoing *radioactive decay*. Beta particles can carry a positive or negative charge. Some emissions result when a neutron is converted to

B

a proton in the atomic nucleus. This increase in the number of protons in the nucleus changes the *atomic number* of the substance, and it therefore becomes a different element (e.g., beta emission by a radioactive isotope of phosphorus [atomic number 15] converts the atom to sulfur [atomic number 16]). **particule bêta**

beta radiation The emission of a *beta particle* by an unstable atomic nucleus. **radiation bêta**

bias *(statistics)* An error in data gathering or analysis caused by faulty program design, mistakes on the part of personnel, or limitations imposed by available instrumentation. Bias deprives a statistical result of representativeness by systematically distorting it, as distinct from a random error that may distort on any one occasion but balances out on the average. Bias is indicated by a difference between the conceptual weighted average value of an estimator over all possible samples and the true value of the quantity being estimated. **biais, erreur systématique**

biaxial compression Compression caused by the application of normal stresses in two perpendicular directions. **compression biaxe**

bicarbonate A compound containing the HCO_3^- group (e.g., sodium bicarbonate, $NaHCO_3$), which ionizes in solution to produce HCO_3^-. See *carbonate buffer system.* **bicarbonate**

bicarbonate alkalinity *Alkalinity* caused by the *bicarbonate* ions. **alcalinité bicarbonatée**

bidentate ligand A *ligand* that can form two bonds to a metal ion. **ligand bidenté**

biennial plant A plant that normally requires two growing seasons to complete its life cycle. Only vegetative growth occurs the first year, often resulting in the formation of an over-wintering rosette; flowering and fruiting occur in the second year. See *annual plant, perennial plant.* **plante bisannuelle**

bimodal distribution *(statistics)* A collection of observations with a large number of values found around each of two points (i.e., a frequency distribution with two *modes*). A frequency curve with more than two maxima has a multimodal distribution. See *unimodal distribution.* **distribution bimodale**

binary fission Division of one cell into two cells by the formation of a septum; the most common form of cell division in bacteria. **fission binaire**

binomial nomenclature The system of naming individuals and groups of organisms in a taxonomic classification. Every species is given two names: a generic name (the *genus*), which has the first letter capitalized, and a specific name (the *species*); both names are italicized or underlined. The author of the name is often given last. Most genus and species names have Greek or Latin origins. No two species have the same specific name. **nomenclature binominale**

bioassay The use of living organisms to estimate the amount of biologically active substances present in a sample. **test biologique, épreuve biologique**

biochemical oxygen demand (BOD) The measure or quantity of oxygen used in aquatic environments for the biochemical oxidation of organic matter under specified conditions. An indirect measure of biologically degradable organic material. Also called biological oxygen demand. **demande biochimique en oxygène (DBO)**

biochemistry The study of chemical compounds and reactions occurring in living organisms. **biochimie**

biochemistry, soil The branch of soil science that deals with enzymes and reactions, functions, and products of soil microorganisms. **biochimie du sol**

biochore (1) Any geographic area supporting a distinctive plant or animal life. (2) More precisely, used to describe each of four major vegetation types: *forest, grassland, savanna,* and *desert.* **biochore**

biochronology (1) A geologic time-scale based on fossils. (2) The study of the relationships between organic evolution and geologic time. (3) The dating of geologic events by evidence from *biostratigraphy.* **stratigraphie paléontologique**

biocide A toxic chemical compound capable of killing living organisms, desirable as well as undesirable. See *algicide, fungicide, herbicide, insecticide, pesticide, rodenticide.* **biocide**

biocoenosis (1) A group of organisms that live closely together and form a natural ecologic unit. (2) A set of fossil remains found in the same place as where the organisms lived. **biocénose**

bioconversion Conversion of one form of energy into another by plants or microorganisms (e.g., production of biogas from biomass). **transformation biologique**

biodiversity The diversity of plants, animals, and other living organisms in all their forms and levels of organization, including genes, species, ecosystems, and the evolutionary and functional processes that link them; also called biological diversity. **biodiversité**

biofacies (1) Lateral changes or variations in the biologic aspect of a stratigraphic unit. (2) Assemblages of plants or animals that formed at the same time but under different conditions. **biofaciès**

biogenic Resulting from the activities of living organisms (e.g., biogenic changes in the environment). **biogénique**

biogenic sediment Sediment that is formed by living organisms, either animal or plant (e.g., coral limestone). **sédiment organogène**

B

biogeochemical cycle The cycling or flow of chemical elements (in various chemical forms) through the major environmental reservoirs: atmosphere, hydrosphere, lithosphere, and bodies of living organisms. The three major cycles are gaseous, hydrologic, and sedimentary. **cycle biogéochimique**

biogeochemistry The study of the transformation and movement of chemical materials to and from the atmosphere, hydrosphere, lithosphere, and the bodies of living organisms. A branch of geochemistry that deals with the effects of life processes on the distribution and fixation of chemical elements in the biosphere. **biogéochimie**

biogeoclimatic classification system A hierarchical classification system of ecosystems that integrates regional, local. and chronological factors and combines climatic, vegetation, and site factors. See *ecological classification.* **système de classification biogéoclimatique**

biogeoclimatic unit Part of the *biogeoclimatic classification system.* The recognized units are a synthesis of climate, vegetation, and soil data, and are defined as classes of geographically related ecosystems that are distributed within a vegetationally inferred climatic space. **unité biogéoclimatique**

biogeoclimatic zone A geographic area having similar patterns of energy flow, vegetation, and soils as a result of a broadly homogenous macroclimate. **zone biogéoclimatique**

biogeographic province Geographic area characterized by specific plant formations and associated fauna. **province biogéographique**

biogeography The study of the adaptations of an organism to its environment, systematically considering the origins, migrations, and associations of living things. **biogéographie**

bioindicator A biological attribute, living organism, or group of organisms that denotes a specific environmental condition. **bioindicateur**

biological additives (*compost management*) Organisms or chemical compounds introduced into organic wastes or contaminants (e.g., compost, hydrocarbon-contaminated soil) to promote decomposition. These can be cultures of bacteria, enzymes, nutrients, or combinations of these. **activateurs ou additifs biologiques**

biological availability The readiness of a chemical compound or element to be taken up by living organisms. **biodisponibilité**

biological control A method of controlling pests or undesirable species by introducing or manipulating naturally occurring predatory organisms or their products, or by sterilizing them. In agriculture, biological control is used to reduce or replace mechanical or chemical means of controlling pests. **lutte biologique**

biological denitrification See *denitrification*. **dénitrification biologique**

biological herbicide A naturally occurring substance or organism that kills or controls undesirable vegetation. Preferred over synthetic chemicals because of reduced toxic effect on the environment. **herbicide biologique**

biological interchange The interchange of elements between organic and inorganic states in a soil or other substrate through the action of living organisms. Biological decomposition of organic compounds with the liberation of inorganic materials (mineralization), and the utilization of inorganic materials with synthesis of microbial tissue (immobilization), result in biological interchange. **transformation biologique**

biological methylation The addition of a methyl group ($-CH_3$) to elemental or inorganic mercury (i.e., a mercury atom or ion) by anaerobic bacteria, usually occurring in sediments within a water body. **méthylation biologique**

biological productivity The rate at which growth processes occur either in an ecosystem or an organism, usually expressed as the weight of dry matter/unit area/unit time (e.g., kg ha^{-1} $year^{-1}$). Occasionally it is expressed as grams of carbon/unit area/unit time (e.g., g C m^{-2} day^{-1}). See *net primary productivity*. **productivité biologique**

biological wastewater treatment (*wastewater management*) The use of bacteria to degrade organic materials in wastewater. See *secondary treatment*. **traitement biologique secondaire**

bioluminescence The production of light by living organisms as a result of a chemical reaction; light energy produced by luminescent organisms. **bioluminescence**

biomanipulation Biological, chemical, or physical methods used to control an organism in a specific environment (e.g., the introduction of new animal species to consume excessive algae growth in an aquatic environment). **biomanipulation**

biomass The total mass of living organisms in a given area, usually expressed as weight or volume per unit area. **biomasse**

biomass energy Energy produced by burning renewable biomass materials such as wood. The carbon dioxide emitted during this process will not increase total atmospheric carbon dioxide if this consumption is done sustainably (i.e., if regrowth of biomass takes up as much carbon dioxide as is released from biomass combustion over a given period). **énergie de la biomasse**

biome A major biotic unit consisting of plant and animal communities formed by the interaction of regional climates with regional biota and substrates. A number of biomes and subtypes are recognized, including *alpine biome, chapparral biome (or Mediterranean scrub biome), desert biome, freshwater biome, grassland biome, marine biome, savannah biome, taiga biome, tundra biome, temperate deciduous forest biome, and tropical rainforest biome.* **biome**

biometrics See *biostatistics.* **biométrie**

biophile An element required by or found in the bodies of living organisms. The list of such elements includes C, H, O, N, I, S, Cl, I, Br, Ca, Mg, K, Na V, Fe, Mn, and Cu. All may belong also to the chalcophile or lithophile groups. **élément biophile**

bioremediation The use of biological agents (e.g., microorganisms or plants) to remove or detoxify toxic or unwanted chemicals from an environment. **biorestauration**

biosequence A group of related soils that differ primarily because of differences in kinds and numbers of plants and soil organisms as a soil forming factor. **bioséquence**

biosphere That part of the Earth and atmosphere capable of supporting living organisms. **biosphère**

biostabilizer A machine used to convert solid waste into compost by grinding and aerating. **bioréacteur**

biostasis A state of biological steady-state between soil, vegetation, and climate. **biostasie**

biostatistics The methods for the mathematical analysis of data gathered relative to biological organisms. Also called biometrics. **biostatistique**

biostratigraphy The part of *stratigraphy* in which rock units are separated and differentiated on the basis of their fossil assemblages. See *chronostratigraphy, lithostratigraphy.* **biostratigraphie**

biosynthesis The use of chemical energy by plants or animals to make carbohydrates, fats, or proteins. See *anabolism.* **biosynthèse**

biota All living organisms of an area, taken collectively. **biote**

biota influence The influence of animals and plants on associated plant or animal life (as contrasted with the influences of climate and soil). **influence biotique**

biotic factors Environmental factors that are the result of living organisms and their activities (e.g., competition and predation), and which are distinct from physical and chemical factors. See *ecological factors.* **facteurs biotiques**

biotic potential The upper limit of a species ability to increase in number; the maximum reproductive rate, assuming no limits on food supply or environmental conditions. (A theoretical number that is not observed in nature for extended periods). **potentiel biotique**

biotite A trioctahedral layer silicate of the mica group $K(Mg, Fe^{+4})_3(Al, Fe^{+3})Si_3O_{10}(OH)_2$. It is black in a hand specimen, brown or green in a thin section, and has perfect basal (001) cleavage. **biotite**

biotope An area in which all the faunal and floral elements are uniformly adapted to the environment in which they occur; the ultimate subdivision of the habitat. See *biota.* **biotope**

biotransformation The metabolic conversion of an absorbed chemical substance by an organism, usually resulting in a less toxic and more excretable form. **détoxication**

bioturbation The breakdown, fragmentation, and reworking of soil and organic materials by soil biota (e.g., earthworms, arthropods, microbes). **bioturbation**

biotype A group of individuals occurring in nature, all with essentially the same genetic constitution. A species usually consists of many biotypes. See *habitat, ecotype.* **biotype**

biozone A biostratigraphic unit including all strata deposited during the

B

B

existence of a particular kind of fossil. **biozone**

birefringence (*crystallography*) The ability of crystals other than those of the isometric system to split a beam of ordinary light into two beams of unequal velocities. The difference between the highest and lowest refractive index of a crystal. Also called *double refraction.* **biréfringence**

birefringent (*crystallography*) Said of a crystal that displays *birefringence;* such a crystal has more than one *index of refraction.* **biréfringent**

birnessite A black manganese oxide, $(Na_{0.7}Ca_{0.3})Mn_7O_{14} \cdot 2.8H_2O$, common in iron-manganese nodules of soils. It has a layer structure. **birnessite**

bisect A profile of plants and soil showing the vertical and lateral distribution of roots and tops in their natural position. **coupe, profil racinaire**

bisequa Two sequa in one soil; that is, two sequences of an eluvial horizon and its related illuvial horizon. **biséquums**

bisiallitization The formation in the soil of secondary 2:1 clay minerals such as smectite and vermiculite. **bisiallitisation**

bitumen Various solid and semi-solid hydrocarbons that are fusible and are soluble in carbon bisulfide (e.g., petroleums, asphalts, natural mineral waxes, and asphaltites). Most commonly applied to the heavy viscous hydrocarbons associated with the tar that sand deposits. **bitume**

bituminous (1) Material composed in part of *bitumen,* in the form of tarry hydrocarbons. (2) Certain varieties of *coal* that burn freely with flames, although they contain no bitumen. **bitumineux**

biuret A byproduct formed at high temperature during the manufacture of urea; $H_2NCONHCONH_2$. Detrimental to plants when present at more than 1.2% concentration in urea fertilizers. **biuret**

black box (*modeling*) A part of a living or nonliving system that has a known or described function or role, but the details of its operation or parts are unknown or omitted. **boîte noire**

black box system (*modeling*) A system whose structure is completely unknown except for that deduced from its behavior. Thus, by manipulating the inputs to a system one can discover statistical relationships between *inputs and outputs.* See *gray box system, white box system.* **système de type boîte noire**

Black Chernozem A great group of soils in the *Chernozemic* order (Canadian system of soil classification). The soils occur in the cool to cold subhumid grassland and parkland regions. They have a very dark surface (Ah or Ap) horizon and ordinarily a brownish B (Bm, Btj, or Bt) horizon, which may be absent, over a highly base-saturated, usually calcareous C horizon. **chernozem noir**

black mud Mud formed in lagoons, sounds, or bays in which there is poor circulation or a weak tide. The color is generally black because of black sulfides of iron and organic matter. **boue noire**

blackbody (*remote sensing*) An ideal body, which, if it existed, would be a perfect absorber and a perfect radiator, absorbing all incident radiation, reflecting none, and emitting radiation at all wavelengths. Emittance curves of blackbodies at various temperatures can be used to model naturally occurring phenomena such as solar radiation and terrestrial *emissivity.* **corps noir**

blank An artificial sample designed to monitor the introduction of artifacts into the process (e.g., for aqueous samples, reagent water is used as a blank). The blank is taken through all the appropriate steps of the sample preparation and analysis process. See *background samples, instrument* *blanks, laboratory blank, reagent blank.* **blanc, essai à blanc**

blanket (*geomorphology*) A mantle of unconsolidated materials thick enough (usually at least 100 cm thick) to mask minor irregularities in the underlying unit but which still conforms to the general underlying topography. **couverture**

blanket deposit (*geology*) (1) Sedimentary deposit of great lateral extent and relatively uniform thickness. (2) A flat deposit of ore of which the length and breadth are relatively great as compared with the thickness. Also called flat sheets, bedded veins, beds, or flat masses. **gisement stratoïde**

blind drain (*soil engineering*) A type of drain consisting of an excavated trench refilled with pervious materials (e.g., coarse sand, gravel, crushed stones) through which water percolates and flows toward an outlet; also called French drain. **drain en pierres, pierrée**

blind inlet Inlet to a drain into which water enters by percolation rather than open flow channels. **orifice de percolation**

blinding material (*soil engineering*) Material placed above and around a closed drain to improve the flow of water to the drain and prevent displacement during back filling of the trench. **pierrée**

bloat (*range management*) Excessive accumulation of gases in the rumen of animals because loss through the esophagus is impaired, causing distension of the rumen. **ballonnement, météorisation**

block An angular rock fragment more than 256 mm in diameter showing little evidence of modification by transportation. **bloc**

block-cut method (*reclamation*) A surface coal mining method in which overburden is removed from and placed around the periphery of a box-shaped cut. After mining is completed, the spoil is pushed back into the cut and the surface is blended into the topography. **foudroyage**

blocky See *soil structure types*. See *Appendix B, Table 1* and *Figure 1*. **polyédrique**

bloodworm (*zoology*) A midge fly larvae. Many of the species have hemoglobin in the blood causing a red color and are often associated with soils rich in organic deposits. **ver de vase**

bloom (*limnology*) A readily visible concentrated growth or aggregation of minute organisms, usually algae, in bodies of water. **fleur d'eau**

bloom, early (*botany*) Initial flowering (*anthesis*) in the uppermost portion of the inflorescence. **début floraison**

bloom, full (*botany*) Essentially all florets in the inflorescence in anthesis. **pleine floraison**

blooming Refers to *anthesis* in the grass family or to the period during which florets are open and anthers extended. **floraison**

blowdown Uprooting by the wind; also refers to a tree or trees so uprooted. Also called windthrow. **chablis**

blown-out land An area from which all or almost all the soil material has been removed by wind erosion; usually a barren, shallow depression that has a flat or irregular floor consisting of a resistant layer or accumulation of pebbles, or both, or a wet zone immediately above a water table. The land is usually unfit for crop production. **terrain de déflation**

blowout A small area from which soil material (often sand) has been removed by wind. **cuvette de déflation**

BOD See *biochemical oxygen demand.* **DBO**

BOD_5 The amount of dissolved oxygen consumed in five days by biologic processes breaking down organic matter in an effluent. See *biochemical oxygen demand.* **DBO_5**

boehmite A gray, brown, or reddish orthorhombic mineral, AlO(OH). It is a major constituent of some bauxites. **boehmite**

B

B

bog A *peatland*, generally with the water table at or near the surface. The bog surface, which may be raised or level with the surrounding terrain, is virtually unaffected by the nutrient-rich groundwaters from the surrounding mineral soils, and is thus generally acidic and low in nutrients. The dominant peat materials are weakly to moderately decomposed *Sphagnum* and woody peat, underlaid at times with sedge peat. The soils are mainly Fibrisols, Mesisols, and Organic Cryosols (Fibrists, Hemists, and Histels). Bogs may be treed or treeless and are usually covered with *Sphagnum* spp. and *ericaceous* shrubs. (A *wetland class* in the *Canadian wetland classification system*). **tourbière ombrotrophe**

bog iron Impure iron deposits that develop in bogs or swamps by the chemical or biochemical oxidation of iron carried in solution. **fer des marais**

bog peat Peat consisting mainly of *Sphagnum* spp., usually poorly decomposed and raw; may also contain *Eriophorum* spp., *Carex* spp., and *ericaceous* species. (Sphagnum, or bog peat is an organic genetic material in the Canadian, U.S., and other systems of soil classification). See *sedge peat, sedge-reed peat, woody peat, sedimentary peat.* **tourbe de sphaigne**

boiling point The temperature at which a liquid starts to boil. **point d'ébullition**

bole The trunk of a tree. **tronc**

bomb calorimeter A laboratory device used to measure the *heat of combustion* (e.g., *calorific value* of fuels and food). **bombe calorimétrique**

bond energy The energy required to break a given chemical bond. **énergie de liaison**

bond length The distance between the nuclei of the two atoms connected by a bond; the distance where the total energy of a diatomic molecule is minimal. **longueur de liaison**

bone meal A soil amendment or manure derived from animal bones used as a high phosphorus fertilizer, containing more than of 30% P. **poudre d'os**

boot stage Growth stage when a grass inflorescence is enclosed by the sheath of the uppermost leaf. **gonflement**

Boralfs A suborder in the U.S. system of soil taxonomy. *Alfisols* that have formed in cool places. Boralfs have frigid or cryic but not *pergelic* temperature regimes, and have *udic* moisture regimes. They are not saturated with water for periods long enough to limit their use for most crops. **Boralfs**

border (*wildlife management*) A strip of low-growing herbaceous or woody vegetation, usually more than three meters wide, established along the edges of fields, woodlands, or streams. **bande**

border dike An earthen ridge built to guide or hold irrigation water within prescribed limits in a field; a small levee. **levée de planche**

border ditch A ditch used to border an irrigated strip or plot, with water spread from one or both sides of the ditch along its entire length. **fossé de planche**

boreal Northern; often refers to the *coniferous* forest regions that stretch across Canada, northern Europe, and Asia. **boréal**

boreal forest A plant formation type associated with cold temperature climates (cool summers and long winters); also called *taiga* and *coniferous* evergreen forest. Dominant forest species are spruces, firs, larches, and pines. **forêt boréale**

borehole An exploratory hole drilled into the Earth or ice to gather geophysical data. **puits de forage**

boresight (*remote sensing*) To align the axis of an instrument with another, generally by optical means. (*radar*) The direction of maximum transmitted signal from the antenna. **axe de visée, alignement de capteur**

Borolls A suborder in the U.S. system of soil taxonomy. *Mollisols* with a mean annual soil temperature of <8°C that are never dry for 60 consecutive days or more within the 90 days following the summer solstice. Borolls do not contain material that has a $CaCO_3$ equivalent >400 g kg^{-1} unless they have a calcic horizon; they are not saturated with water for periods long enough to limit their use for most crops. **Borolls**

borrow pit A bank or pit from which earth is taken for use in filling or embanking. **carrière d'emprunt**

botanical forest products Prescribed plants or fungi that occur naturally on forest land. The seven recognized categories are wild edible mushrooms, floral greenery, medicinal products, fruits and berries, herbs and vegetables, landscaping products, and craft products. **produits forestiers botaniques**

botanical pesticide A plant-derived chemical used to control pests (e.g., nicotine, strychnine, or pyrethrum). **pesticide botanique ou d'origine végétale**

bottom load See *bed load*. **charriage**

botulism A disease caused by the ingestion of food containing a *toxin* produced by the *anaerobic* microorganism *Clostridium botulinum*. The toxin attacks the nervous system; the fatality rate is about 60%. Primary risk is from home-canned food, especially improperly prepared beef or pork products, string beans, maize, olives, beans, and spinach. **botulisme**

boulder Rock fragment greater than 60 cm in diameter. See *coarse fragments*. In engineering practice, boulders are greater than 20 cm in diameter. **bloc rocheux, bloc**

boulder pavement (1) An accumulation of glacial boulders once contained in a moraine and remaining nearly in their original positions after removal of finer material by water movement. (2) A surface boulder-rich till abraded to flatness by the passage of a glacier. **pavage de pierres, dallage de pierres**

B

bound water Water molecules that are held tightly to soil or other solids. This water is not easily removed by normal drying and is not available for plant growth. **eau liée**

boundary screen Plant or construction materials on the boundary of a site that provide protection or concealment. **écran de protection**

bovine Of, or relating to, cattle. **bovin**

box model The representation of stocks and flows of material or energy in and through systems, by boxes for the stocks and arrows from box to box, for flow directions and amounts. At equilibrium, the stock in a box divided by the rate of flow in or out is called the residence time. The model can be applied to biological and biophysical systems (e.g., nutrient cycling models). **modèle de la boîte**

Boyle's law The volume of a given mass of gas at a constant temperature is inversely proportional to its pressure at a constant temperature. Expressed as $pV=K$, where p is the pressure, V is the volume, and K is a constant. **loi de Boyle**

brackish Water with a salinity intermediate between that of normal seawater and normal fresh water. **saumâtre**

Bragg's equation (*crystallography*) An equation that describes the x-ray diffractions from a three-dimensional lattice as $nl = 2d\ sin(q)$, in which n is any integer, l is the wavelength of the x-ray, d is the crystal plane separation, also known as *d*-spacing, and q is the angle between the crystal plane and the diffracted beam, also known as the Bragg angle. Also called Bragg's law. **équation de Bragg**

break of slope An abrupt change of slope. **rupture de pente**

B

breakthrough curve The relative solute concentration in the outflow from a column of soil or porous medium after a step change in solute concentration has been applied to the inlet end of the column, plotted against the volume of outflow (often in number of pore volumes). **courbe de fuite**

breccia (*geology*) A rock composed of coarse angular fragments cemented in a fine-grained matrix. **brèche**

brightness range (*remote sensing*) Ratio of the apparent brightness of highlights to the deepest shadow in the actual scene (imagery). **gamme de luminance**

brightness value (*remote sensing*) A number in the range of 0–63, 0–127, or 0–255 (in units of watts cm^{-2}) related to the amount of radiance striking a detector in a *multispectral scanner*. **valeur de luminance**

brittleness A property of solid material that ruptures easily, with little or no plastic flow. **fragilité**

broad-base terrace See *terrace*. **terrasse à large base**

broadcast To scatter abroad over the surface as opposed to being sown in drills or rows. **à la volée**

broadcast application The application of fertilizer or an agrichemical on the soil surface. Usually done prior to planting; normally the fertilizer or agrichemical is incorporated into the soil by tillage, but it may not be in no-till systems. **application à la volée**

broadcast planting See *tillage, broadcast planting*. **semis à la volée**

broadcast tillage See *tillage, broadcast tillage*. **travail sur toute la surface de sol**

broad-leaved deciduous forest A plant formation type associated with cool temperature climates with a short period of frost; also called nemoral forests. Dominant forest species are oak, beech, maple, and ash. **forêt feuillue**

broad-leaved evergreen forest A plant formation type associated with warm temperature climates that have no periods of frost. **forêt feuillue à feuillage persistant**

broken-edge bond The exposed bonds at the edges of mineral fragments. These bonds can have either a negative or positive charge, or some of each, and they account for a portion of the *anion-* and *cation-exchange capacities* of soils. Broken-edge bonds occur in all minerals but are significant only in *clay*-size particles. They are also important in the *card-house structure* formed by certain clay minerals. **liaison attribuable à un bris de lien en périphérie de minéral**

brow The top of a hill or mountain slope. Also, the point where a steep slope eases into a gentle slope. **front de colline, sommet**

Brown Chernozem A great group of soils in the *Chernozemic* order (Canadian system of soil classification). The soils occur in the cool, subarid to semiarid grassland regions and consist of a brown (dry) surface (Ah or Ap) horizon, and ordinarily a lighter-colored brownish B (Bm, Btj, or Bt) horizon, which may be absent, over a highly base-saturated, usually calcareous, C horizon. **chernozem brun**

brown coal See *lignite*. **lignite, houille brune**

brown moss peat Peat composed of various proportions of mosses of Amblystegiaceae (*Scorpidium, Drepanocladus, Calliergon, Campylium*), *Hypnum*, and *Tomenthypnum*, and some of the more nutrient-demanding *Sphagnum* species such as *S. subsecundum*, *S. teres*, and *S. warnstorfii*. **tourbe de mousse brune**

brown rot (1) A common disease of fruit (e.g., apples, pears, plums) caused by fungi of the genus *Sclerotinia*. (2) A type of timber decay in which the wood turns reddish-brown in color and becomes cracked and eventually crumbly in texture. Fungi that cause brown rot are not able to break down lignin in the wood. See *white rot.* **(1) pourriture brune (2) carie brune**

Brownian motion The random, nondirectional motion of small particles suspended in a gaseous or liquid medium, caused by movement of the molecules constituting the medium. First described by the British botanist, Robert Brown (1773–1858). **mouvement Brownien**

browse That part of leaf and twig growth of shrubs, woody vines, and trees available for animal consumption. **brout**

browse line The line on woody plants marking the height to which browsing animals have removed *browse.* **ligne d'abroutissement**

brucellosis A contagious bacterial disease especially of humans and cattle, caused by the bacteria of the genus *Brucella*, that results in abortion; also called Bang's disease. **brucellose**

brunification The formation of brown color in soil. **brunification**

Brunisolic An order of soils in the Canadian system of soil classification in which the horizons are developed sufficiently to exclude the soils from the *Regosolic* order, but lack the degrees or kinds of horizon development specified for soils of the other orders. These soils, which occur under a wide variety of climatic and vegetative conditions, all have Bm or Btj horizons. The great groups *Melanic Brunisol*, *Eutric Brunisol*, *Sombric Brunisol*, and *Dystric Brunisol* belong to this order. **brunisolique**

brush matting (1) A pile of branches placed on sloping land to conserve moisture and reduce erosion while trees or other vegetative covers are established. (2) A pile of mesh wire and brush used to retard streambank erosion. **paillassonnage en branches**

brushing A silvicultural activity by chemical, manual, grazing, or mechanical means to reduce competition of other forest vegetation with crop trees or seedlings for space, light, moisture, and nutrients. **dégagement**

bryophyte A nonvascular, green plant that may have differentiated stems and leaves, but no true roots (e.g., liverworts, hornworts, and mosses). **bryophyte**

bud (1) An embryonic shoot of a plant. (2) A vegetative outgrowth of yeasts and some bacteria as a means of asexual reproduction. **bourgeon**

buffalo wallow A small natural depression of prairie occasionally containing standing water and having vegetation different from that of the surrounding area. **dépression à bisons**

buffer A chemical substance in solution that neutralizes acids or bases. Buffered solutions resist changes in pH. Surface waters and soils with chemical buffers are not as susceptible to acid deposition as those with poor buffering capacity. **tampon**

buffer capacity The ability of a buffered solution to absorb protons or hydroxide ions without a significant change in pH. **capacité tampon**

buffer compounds, soil The clay, organic matter, and materials such as carbonates and phosphates that enable soil to resist appreciable change in pH. **complexe tampon du sol**

buffer power The ability of ions associated with the solid phase to buffer changes in ion concentration in the solution phase. Expressed as dC_S/dC_l, where C_S represents the concentration of ions on the solid phase in equilibrium with C_l, the concentration of ions in the solution phase. **pouvoir tampon**

buffer strip A small area or strip of land (often in undisturbed or permanent vegetation), or an area of land along the side of a watercourse or between cultivated areas or fields, designed to reduce erosion, intercept pollutants, provide a habitat for wildlife, and manage other environmental concerns. Buffer strips include riparian buffers, filter strips, grassed waterways, shelterbelts, windbreaks, living snow fences, contour grass strips, cross-wind trap strips, shallow water areas for wildlife, field borders, alley cropping, herbaceous wind barriers, and vegetative barriers. **bande de protection**

buffer zone See *buffer strip*. **zone tampon**

buffered (soil) A soil that contains enough lime or base saturation to resist change in pH, especially toward acidity. Sandy soils with low *cation exchange capacity* are low in buffering capacity; clay soils with high *cation exchange capacity* are highly buffered. Soils that contain excess lime are highly buffered against acidity. **sol tamponné**

buffered solution A solution containing a *buffer*. **solution tamponnée**

bulb (*botany*) In plants, an underground storage organ surrounded by protective scale leaves. **bulbe**

bulk area The total area, including solid particles and pores, of a cross-section through an arbitrary quantity of soil; the area counterpart of bulk volume. **aire brute**

bulk-blend fertilizer See *fertilizer, bulk-blended*. **engrais de mélange**

bulk density, soil The mass of dry soil per unit *bulk volume*, thus often termed *dry bulk density*. *Bulk volume* is determined before the soil is dried to constant mass at 105°C. Also called *apparent density*. **masse volumique apparente, densité apparente**

bulk fertilizer Commercial fertilizer delivered to the purchaser, either in a solid or liquid state, in a nonpackaged form to which a label cannot be affixed. **engrais en vrac**

bulk specific gravity The ratio of the *bulk density* of a soil to the mass of a unit volume of water. Also called *apparent specific gravity*. **poids spécifique apparent**

bulk volume The volume, including the solids and the pores, of a given soil mass. **volume brut**

bunch grass A grass having the characteristic growth habit of forming a bunch; lacking *stolons* or *rhizomes*. **graminée cespiteuse**

buried soil Soil covered by an alluvial, loessial, or other deposit, usually to a depth greater than the thickness of the solum. See *figure*. **sol fossile, sol enfoui**

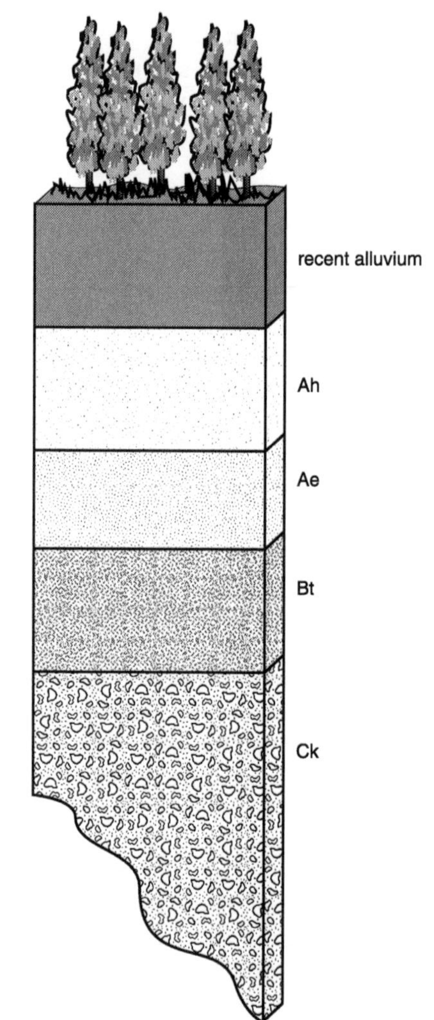

Buried Soil (A.J. Vanden Bygaart, 2001).

buried topography (*geology*) Any terrain that has been buried by sedimentary deposits upon inundation by transgressing seas. **paléorelief enfoui**

burying See *tillage, burying.* **enfouissement**

butte (*geomorphology*) An isolated hill rising abruptly above the surrounding area and having steep sides and a flat top. Most of the top has been removed by erosion but may be protected by *cap-rock;* has a smaller summit area than a *mesa.* **butte**

bypass flow See *preferential flow.* **écoulement préférentiel**

byproduct A substance, other than the intended product, generated by an industrial process. **sous-produit**

C

c axis (*crystallography*) One of the crystallographic axes used for reference in crystal description. It is oriented vertically. See *a axis*, *b axis*. **axe c**

C horizon A mineral soil horizon generally beneath the solum which is relatively unaffected by biological activity and pedogenic processes operating in A and B horizons, except the process of gleying (Cg) and the accumulation of calcium and magnesium salts (Cca) and more soluble salts (Cs, Csa). Marl, diatomaceous earth, and rock with a hardness ≤3 on the Mohs scale are considered to be C horizons. It may or may not be like the material from which the A and B horizons have formed. The above horizon terms are according to the Canadian system of soil classification. See *Appendix D* for equivalent U.S. Soil Taxonomy and FAO soils terminology. See *A horizon, B horizon, horizon, soil*. **horizon C**

C3 plant Plant with a photosynthetic pathway in which carbon dioxide is reduced to phosphoglycerate (PGA), a three-carbon compound, via the enzyme ribulose bisphosphate carboxylase (RuBisCO). Approximately 85% of all plant species possess the C3 pathway of photosynthesis. The $\delta^{13}C$ values for C3 plants range from approximately –32 to –22 per mil with a mean of –27 per mil. See *figure*. **plante ou végétation de type C3**

C4 plant Plant with a photosynthetic pathway in which carbon dioxide is reduced to aspartic or malic acid, both four-carbon compounds, via the enzyme phosphoenolpyruvate (PEP) carboxylase. C4 plants comprise about 5% of all plant species and are most common in tropical and subtropical regions. Plants with C4 have $\delta^{13}C$ values ranging from approximately –17 to –9 per mil with a mean of –13 per mil. See *figure*. **plante ou végétation de type C4**

cadastral map A map showing ownership of land, usually drawn at a large scale so that all individual land holdings may be accurately depicted. **carte cadastrale**

calcan A *cutan* composed of carbonates. **calcane**

calcareous Descriptive of materials containing calcium carbonate. **calcaire**

calcareous algae *Algae* that remove calcium carbonate from the shallow water in which they live and deposit it as a more or less solid calcareous structure. **algue calcaire**

calcareous classes Six classes that represent the amount of carbonates, expressed as percent calcium carbonate ($CaCO_3$) equivalent, present in the soil or parent material. The classes are noncalcareous (<1%), weakly calcareous (1 to 5%), moderately calcareous (6 to 15%), strongly calcareous (16 to 25%), very strongly calcareous (26 to 40%), and extremely calcareous (>40%). At the family level of soil taxonomy, strongly calcareous means 5 to 40% $CaCO_3$ equivalent. **classes calcaires**

calcareous rock A rock containing high proportions of calcium carbonate, $CaCO_3$ (e.g., limestone). **roche calcaire**

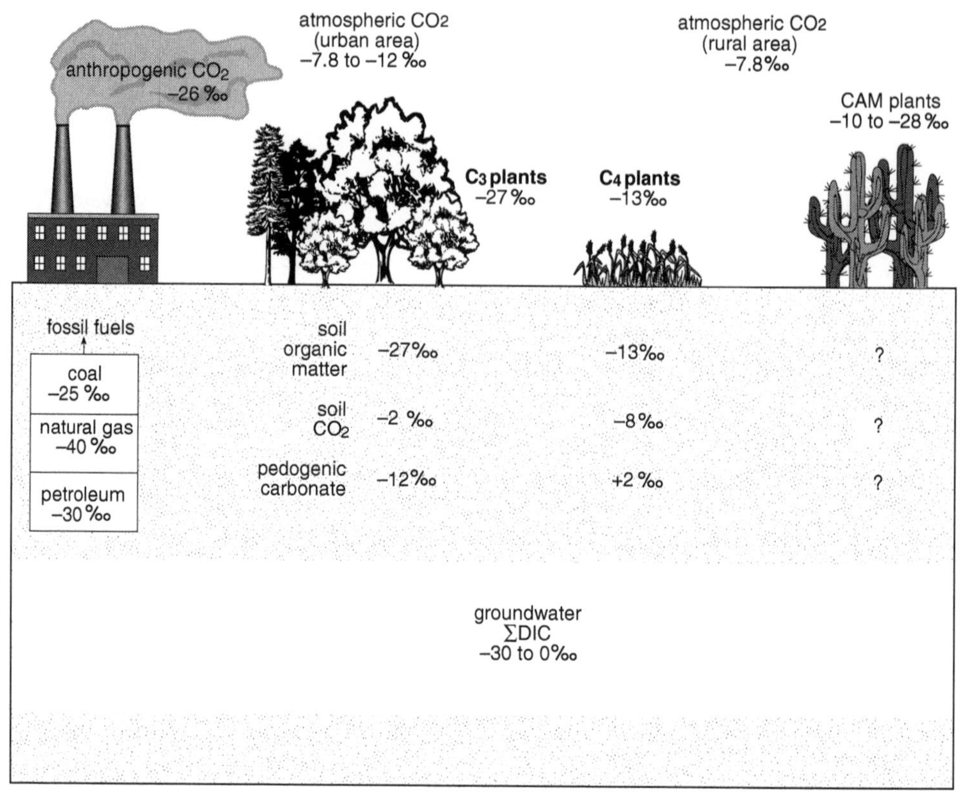

Stable **carbon** isotopes ratios of major components of terrestrial ecosystems (adapted from Boutton, 1991).

calcareous soil Soil containing sufficient calcium carbonate, often with magnesium carbonate, to effervesce visibly when treated with cold 0.1 M hydrochloric acid. **sol calcaire**

calcic horizon A diagnostic subsurface illuvial horizon (U.S. system of soil taxonomy) in which secondary calcium carbonate or other carbonates have accumulated to a significant extent. **horizon calcique**

Calcids A suborder in the U.S. system of soil taxonomy. Aridisols that have a calcic or petrocalcic horizon that has its upper boundary within 100 cm of the soil surface. **calcids**

calcification Process of soil formation involving accumulation of calcium carbonate in the C, and possibly other, soil horizons. **calcification**

calcify To make or become hard or stony by the deposit of calcium salts. **calcifier**

calcine To heat a substance at a temperature below its *melting point* but high enough to cause the loss of water or volatile components, making the material crumbly. The process is often performed in a rotating cylindrical kiln. **calcine**

calciphobe, calcifuge A plant that cannot tolerate lime in the soil and flourishes only on *acid soils* (e.g., azalea). A calcifuge can survive on a lime-rich soil but will not flourish; a calciphobe will rapidly die under similar circumstances. **calcifuge**

calciphyte A plant that requires or tolerates large amounts of calcium or is associated with soils rich in calcium. **calciphyte**

calcite The crystalline form of calcium carbonate ($CaCO_3$) and the principal constituent of limestone and marble. Commonly white or gray, it has perfect rhombohedral cleavage and reacts readily with cold dilute hydrochloric acid. It forms on stream beds and in caves as *stalactites, stalagmites,* and *tufa*. **calcite**

calcitic dolomite Dolomitic rock with significant *calcite* content; generally a dolomite rock containing 10 to 50% *calcite* and 50 to 90% *dolomite*. **dolomie calcaire**

calcitic limestone Limestone containing mostly $CaCO_3$. **calcaire calcitique**

calcium bentonite Bentonite in which Ca^{+2} is the dominant exchangeable ion. Calcium bentonites swell little more than ordinary clays but strongly absorb water. **bentonite calcaire**

calcium carbonate equivalent The carbonate content of a liming material or *calcareous soil* calculated as if all of the carbonate is in the form of $CaCO_3$. For a liming material the effective calcium carbonate equivalent (ECCE), or effective neutralizing value (ENV), depends on both the fineness and its chemical nature. If a lime sample is ground fine enough to have 70% effectiveness and has a calcium carbonate equivalent of 90%, its ECCE is 63% (i.e., 70% of 90%). The best way to determine the relative worth of liming materials for raising soil pH is to compare their ECCE (or ENV) values. **équivalent carbonate de calcium**

calcium sulfate extractant Reagent used to extract nitrate from soils. Since nitrates are very soluble in water, the calcium sulfate allows easy filtration. **sulfate de calcium**

calcrete (1) A conglomerate consisting of surficial sand and gravel cemented into a hard mass by calcium carbonate precipitated from solution by infiltrating waters, or deposited by the escape of carbon dioxide from *vadose water*. (2) A calcareous *duricrust*. See *silcrete*. **calcrète**

caldera (*geology*) A large *crater*, more or less circular, having a diameter many times greater than that of the included vent or vents, irrespective of steepness of the walls or form of the floor. It usually results from collapse of underground lava pools. **caldeira, caldera**

C

calibration (1) (*analytical chemistry*) The act of introducing known quantities to a measuring device to determine the response of the instrument. (2) (*agriculture*) The adjustment of measurement of the output (i.e., rate per area) of application equipment (e.g., planter, fertilizer, manure or pesticide spreader) to match a prescribed standard under typical operating conditions. **(1) étalonnage (2) réglage**

calibration curve A set of responses given by a measuring instrument plotted against a set of known quantities introduced to the instrument. Used to translate unknown quantities to measured quantities. **courbe d'étalonnage**

caliche A layer near the soil surface that is cemented by secondary carbonates of calcium or magnesium precipitated from the soil solution. It may be a soft thin soil horizon, a hard thick bed just beneath the solum, or a surface layer exposed by erosion. It is not a geologic deposit. **caliche**

California bearing ratio The load-supporting capacity of a soil as compared to that of a standard crushed limestone, expressed as a ratio and multiplied by 100. A soil with a ratio of 16 will support 16% of the load that would be supported by the standard crushed limestone per unit area, and with the same degree of distortion. **indice portant californien**

callus A tissue of thin-walled cells developed on plant wound surfaces; most evident, in time, after cutting off a tree limb flush with a main trunk or limb. **cal**

calorie The amount of heat required to raise the temperature of 1g of water one degree Celsius. **calorie**

calorific value For solid fuels and liquid fuels of low volatility, the heat per unit mass produced by complete combustion of a given substance. Also used to measure the energy content of foodstuffs; the heat produced when food is oxidized in the body. **pouvoir calorifique**

C

CAM plant Plant with a Crassulacean Acid Metabolism (CAM). CAM plants fix carbon dioxide at night using carbon-fixation methods similar to C4 plants (i.e., via the enzyme phosphoenolpyruvate (PEP) carboxylase); under particular environmental conditions, some facultative species are able to switch daytime C3 photosynthesis (i.e., fix carbon dioxide via ribulose bisphosphate carboxylase (RuBisCO)). CAM plants comprise about 10% of all plant species. The $\delta^{13}C$ values for CAM plants range from approximately –28 to –10 per mil. See *C3 plant, C4 plant.* **plante ou végétation de type CAM**

cambic horizon A mineral soil horizon with a texture of loamy very fine sand or finer, has soil structure, contains some weatherable minerals, and is characterized by the alteration or removal of mineral material as indicated by mottling or gray colors, stronger chromas or redder hues than in underlying horizons, or the removal of carbonates. The cambic horizon lacks cementation or induration and has little evidence of illuviation to meet the requirements of the *argillic* or *spodic* horizon. A subsurface diagnostic horizon in the U.S. system of soil taxonomy, and analogous to the *Bm horizon* in the Canadian system of soil classification. **horizon cambique**

Cambids A suborder in the U.S. system of soil taxonomy. Aridisols that are not in cryic temperature regimes and do not have the following diagnostic subsurface horizons or features: argillic, salic, duripan, gypsic, petrogypsic, calcic, petrocalcic. **cambids**

cambium Layer of cells that separate the living *xylem* cells from the *phloem.* As the tree grows and develops, the cambium forms new phloem and xylem cells. See *figure.* **cambium**

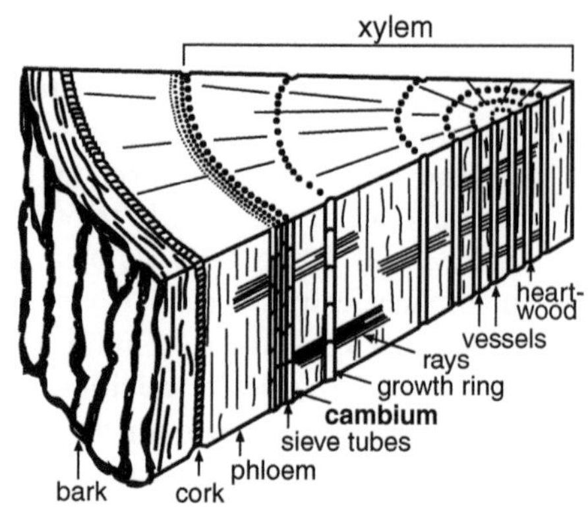

Cambium (adapted from Dunster and Dunster, 1996).

Canadian ecological land classification system See *ecological land classification system, Canadian.* **système canadien de classification écologique du territoire**

Canadian forest fire weather index system See *forest fire weather index system, Canadian.* **système canadien de l'indice forêt-météo**

Canadian system of soil classification See *soil classification, Canadian system.* **système canadien de classification des sols**

Canadian wetland classification system See *wetland classification system, Canadian.* **système canadien de classification des milieux humides**

canker A disease that kills relatively localized lesion areas of bark, and sometimes the underlying *cambium* and wood, on branches or trunks of trees. **chancre**

canopy The aerial portion of plants in their natural growth position; usually expressed as percent of ground so occupied, or as leaf area index. **couvert**

canopy closure The progressive reduction of space between crowns as they spread laterally, increasing canopy cover. **fermeture du couvert**

canopy cover The maximum aerial extent of the foliage of a plant. **surface de la cime**

canyon A steep-walled gorge, ravine, or chasm cut by river action, in which the depth exceeds the width. Canyons are characteristic of arid or semiarid regions where down-cutting by streams greatly exceeds weathering. **canyon**

cap (*waste management*) The final, permanent layer of impermeable material placed on the surface of a *landfill* as part of the closure process. (e.g., compacted clay or a synthetic material designed to resist liquid infiltration). **couche imperméabilisante**

capability The potential of an area of land or other resource such as water to produce food and fiber, supply goods and services, and allow resource uses. Capability is determined by conditions such as climate, landform, slope, soils, geology, and current vegetation. The capability of a water body, for example, is affected by ambient temperature, pollution inputs, buffering capacity, and biochemical oxygen demand. See also *capability class, soil.* There are two forms of capability; intrinsic capability refers to capability of a resource without management considerations, and managed capability is the potential capability expected after human-induced changes have been made. **possibilité**

capability class, soil A rating that indicates the *capability* of soil for some use such as agriculture, forestry, recreation, wildlife, or other use. It most commonly is a grouping of lands that have the same relative degree of limitation or hazard. Also called land capability class. **classe de possibilités des sols**

capability mapping The depiction of *capability* classes of soils, waters, wildlife habitat, or other resource on a map. **cartographie du potentiel**

capillarity The action or condition by which a fluid, such as water, is drawn up in small interstices or tubes (i.e., soil *pores*) as a result of surface tension. Water molecules in an unsaturated soil are subject to both adsorption and capillarity, which combine to produce *matric potential.* The surface tension of water and its contact angle with solid particles result in capillarity. **capillarité**

C

capillary (1) A mineral that forms hairlike or threadlike crystals (e.g., millerite). (2) Tubes or interstices with such small openings that they can retain fluids by *capillarity.* **capillaire**

capillary action The spontaneous rising of a liquid (e.g., water) in a narrow tube (e.g., soil pore) as a result of surface tension. See *capillarity.* **action capillaire**

capillary fringe A zone in the soil just above the plane of zero gauge pressure that remains saturated or almost saturated with water. **frange capillaire**

capillary interstice An opening small enough to hold water by surface tension at an appreciable height above a free water surface. **interstice capillaire**

capillary rise The flow and movement of soil water due to *capillarity*, specifically the rise of groundwater above the water table. **remontée capillaire**

capillary water That amount of water that is capable of movement after the soil has drained. It is held by adhesion and surface tension as films around particles, and in the finer pore spaces. **eau capillaire**

capillary zone See *capillary fringe.* **zone capillaire**

capping (*land reclamation*) The disposal of oilfield drilling waste by burial of the undisturbed sump contents. Sump contents are left in place and a layer of subsoil is carefully placed over the contents, which remain essentially undisturbed. Topsoil is

C

then spread to cover the capped sump area. **recouvrement**

cap-rock (1) A layer of hard, resistant rock that forms the flat summit of a hill and protects underlying, less-resistant strata. (2) A virtually impermeable stratum overlying an oil or natural gas reservoir, aquifer, or salt dome. (3) A mass of barren rock overlying an ore-body. **roche couverture**

capsule A layer of mucoid material surrounding a bacterial cell. **capsule**

carbohydrate Any compound of carbon, hydrogen, and oxygen in the ratio of CH_2O (e.g., sugar, starch, and cellulose). **glucide, hydrate de carbone**

carbohydrates, nonstructural Soluble *carbohydrates* found in the cell contents of plants and other organisms, as contrasted with structural carbohydrates in the cell walls. **glucides non-structuraux**

carbohydrates, structural *Carbohydrates* found in the cell walls of plants and other organisms (e.g., hemicellulose, cellulose). **glucides structuraux**

carbon (C) An element present in all materials of biological origin. The element contains six protons and usually six neutrons in the nucleus. Carbon atoms will bind with each other as well as with a host of other elements. Deposits derived from living sources, such as limestone, coal, oil, and natural gas, contain carbon as a principal element. **carbone**

carbon cycle The sequence of transformations whereby carbon dioxide is converted to organic forms by photosynthesis or chemosynthesis, recycled through the biosphere (with partial incorporation into sediments), usually via microbes and herbivores, and ultimately returned to its original state through respiration or combustion. See *figure*. **cycle du carbone**

carbon dating A method of estimating the age of a material of biological origin; also known as radiocarbon dating. As a result of the interaction between atmospheric nitrogen in the air with cosmic rays, radioactive carbon, ^{14}C, is produced and combines with atmospheric oxygen to form to carbon dioxide ($^{14}CO_2$). During the process of *photosynthesis* these radioactive forms of carbon are incorporated into living plants along with the non-radioactive, stable form of carbon, ^{12}C. While the plant is alive, the ratio of ^{14}C to ^{12}C remains about the same. When the plant dies and photosynthesis stops, the ratio of radioactive to stable carbon changes as the radioactive form of carbon decays with a *half-life* of about 5760 years. **datation au carbone**

carbon dioxide (CO_2) (1) A molecule made up of one carbon and two oxygen atoms. (2) A *greenhouse gas* that serves as the reference for all other greenhouse gases (see *carbon dioxide equivalent*). Carbon dioxide *emissions* derive mainly from *fossil fuel* combustion but also from forest clearing, *biomass* burning, and non-energy production processes such as cement production. Atmospheric concentrations of CO_2 have been increasing at a rate of about 0.5% per year and are now about 30% above pre-industrial levels. **dioxyde de carbone**

carbon dioxide equivalent (CDE) A measure used to compare the *emission* of various *greenhouse gases* based on their *global warming potential (GWP)*. The carbon dioxide equivalent for a gas is derived by multiplying the tonnes of the gas by the associated GWP, commonly expressed as million metric tonnes of carbon dioxide equivalents (MMTCDE) (e.g., the GWP for methane is 24.5, meaning that one million metric tonnes of *methane* emissions are equivalent to 24.5 million metric tonnes of carbon dioxide emissions). See *carbon equivalent*. **équivalent en dioxyde de carbone**

C

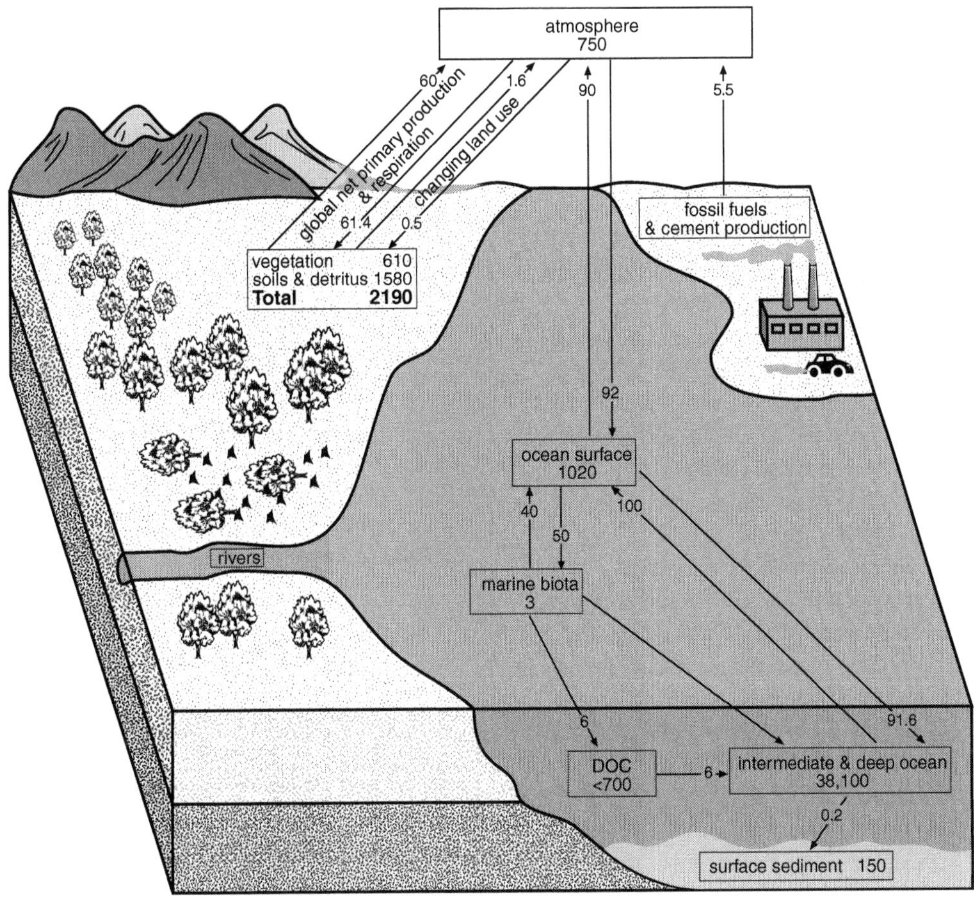

The global **carbon cycle** showing reservoirs (in Gt C; in boxes) and fluxes (in Gt C yr^{-1}) (adapted from Schimel, 1995).

carbon dioxide fertilization Increased plant growth as a result of a higher *carbon dioxide* concentration. Sometimes referred to as fertilization. **effet fertilisant du dioxyde de carbone**

carbon equivalent (CE) A measure used to compare the *emission* of the different *greenhouse gases* based upon their *global warming potential (GWP)*. Global warming potentials are used to convert greenhouse gases to *carbon dioxide equivalents*, which can then be multiplied by the ratio of the molecular weight of carbon to that of carbon dioxide to give the carbon equivalents. Thus, the formula to derive carbon equivalents is: MMTCE (million metric tonnes of carbon equivalent) = (million metric tonnes of a gas) × (GWP of the gas) × (12/44). **équivalent en carbone**

carbon filtration The passage of treated wastewater or domestic water supplies through *activated charcoal* to remove dissolved chemicals present in low concentrations. **filtration au carbone**

carbon pool Reservoir containing stocks of inorganic and organic forms of carbon. **bassin ou pool de carbone**

carbon ratio The ratio of the most common carbon isotope (C^{12}) to either of the less common isotopes (C^{13} and C^{14}), or the reciprocal of one of these ratios. **rapport isotopique du carbone**

C

carbon sequestration The uptake and storage of carbon (e.g., trees and other plants absorb carbon dioxide, release the oxygen, and store the carbon. This carbon eventually reaches the soil and is stored there. *Fossil fuels* were at one time plant biomass, and continue to store carbon until burned). **séquestration, captage ou immobilisation du carbone**

carbon sinks Carbon reservoirs and conditions that take in and store more carbon than they release (e.g., forests and oceans); they can serve to partially offset *greenhouse gas emissions*. See *carbon sequestration*. **puits de carbone**

carbon:nitrogen ratio (C:N ratio) The ratio of the mass of organic carbon to the mass of total nitrogen in a soil or organic material. It is obtained by dividing the percentage of organic carbon (C) by the percentage of total nitrogen (N). **rapport carbone:azote**

carbon-14 (^{14}C) A radioactive *isotope* of carbon that emits *beta particles* when it undergoes *radioactive decay*. The nucleus of a ^{14}C atom contains eight neutrons rather than the six neutrons found in ^{12}C, the most abundant isotope in nature; also known as radiocarbon. **carbone 14 (^{14}C)**

carbonaceous Pertaining to or containing carbon derived from plant and animal residues. **carboné**

carbonate (1) The CO_3^{-2} ion in the *carbonate buffer system*. Combined with one proton, it becomes *bicarbonate*; with two protons, *carbonic acid*. The carbonate ion forms a solid precipitant when combined with dissolved ions of calcium or magnesium. (2) A mineral compound characterized by a fundamental anionic structure of CO_3^{-2} (e.g., calcite and aragonite, $CaCO_3$). (3) A sedimentary rock formed of the carbonates of calcium, magnesium, and/or iron (e.g., limestone and dolomite). **carbonate**

carbonate buffer system An important buffer system in natural surface waters and wastewater treatment, consisting of a carbon dioxide–water–carbonic acid–bicarbonate–carbonate ion equilibrium that resists changes in pH. For example, if acids are added to this buffer system, the equilibrium is shifted and carbonate ions combine with the hydrogen ions to form bicarbonate; bicarbonate then combines with hydrogen ions to form carbonic acid, which can dissociate into carbon dioxide and water; thus, the pH of the system remains unaltered. **système de tampon carbonate**

carbonate hardness Water hardness caused by the presence of carbonate and bicarbonates of calcium and magnesium. **dureté carbonatée**

carbonation (1) A process of chemical weathering involving the transformation of minerals containing calcium, magnesium, potassium, sodium, and iron into carbonates or bicarbonates of these metals by weak carbonic acid (H_2CO_3) formed by dissolved carbon dioxide (CO_2) in water. (2) The introduction of carbon dioxide into a fluid. **carbonatation**

carbonic acid A mild acid, H_2CO_3, formed in solution when carbon dioxide is dissolved in water. The carbonic acid content of natural, unpolluted rainfall lowers its pH to about 5.6. **acide carbonique**

carboxyl group The –COOH group in an organic acid. **carboxyle**

carboxylic acid An organic compound containing the *carboxyl group*; generally a weak acid. Many long-chain carboxylic acids occur naturally as *esters* in fats and oils and are known as *fatty acids*. **acide carboxylique**

cardhouse structure A clay structure resulting from the presence of positive charges at the edges of clay *micelles* (both positive and negative charges occur in edge positions). This charge arrangement permits the edge of one micelle to be attracted to the negatively charged surface of the next. The resulting three-dimensional arrangement has abundant small pore spaces favorable for storing water but is fragile and easy to break down if the soil is worked. **structure en château de cartes**

cardinal points (*surveying*) Any of the four main compass directions: north, east, south, west. (*optics*) Those points of a lens used as reference for determining object and image distances. **points cardinaux**

carex peat Peat composed mostly of the stalks, leaves, rhizomes, and roots of sedges (*Carex* spp.). Also called *sedge peat*. See *reed peat*, *sedge-reed peat*. **tourbe de *Carex***

carotenoids A class of pigments that includes the carotenes (orange pigments) and the xanthophylls (yellow pigments); found in chloroplasts of plants. During the *senescence* of leaves, *chlorophyll* breaks down faster than carotenoids and the carotenoid colors are revealed. **caroténoïdes**

carpet A type of microhabitat in a *peatland*. A flat wet feature in or surrounding pools of open water in the peatland. The water table is at the moss capitulum and is visible around the moss stems. **tapis, radeau flottant**

carr *Transition peatland* with *peat* usually rich in wood, bark, and cone remains of coniferous or deciduous woods. The term is often used to describe shrub-covered wetlands. **fourré marécageux, carr**

carrier A host that harbors infectious microorganisms and can transmit them to others but shows no disease symptoms. **porteur**

carrier gas An inert gas, such as helium, employed as a necessary medium for the transport of low concentrations of some active ingredient. **gaz vecteur**

cartography The art and science of making maps and charts. **cartographie**

caryopsis (*botany*) Small, one-seeded, dry fruit with a thin pericarp surrounding and adhering to the seed; the seed (grain) or fruit of grasses. **caryopse**

cascading system A type of system made up of a chain of subsystems, each having a spatial location and a magnitude, that are dynamically linked by a cascade of mass or energy. See *black box system, general systems theory, gray box system, white box system*. **système en cascade**

casual species A species that occurs rarely or without regularity in an ecosystem. **espèce sporadique**

catabolism Metabolic processes involved in the breakdown of organic compounds, usually leading to the production of energy. **catabolisme**

catalyst A chemical substance that increases the rate of a chemical reaction without itself being consumed or altered by the reaction. *Enzymes* are highly specific biological catalysts, enhancing reactions within living organisms. **catalyseur**

catalytic converter An air pollution abatement device that removes organic contaminants from the air by oxidizing them into carbon dioxide and water through chemical reaction; used to reduce nitrogen oxide emissions from motor vehicles. **convertisseur catalytique**

catch crop An incidental crop that is planted, usually for a short period of time, either between the rows of a main crop or in the fall after the main crop has been harvested, to take up excess nutrients such as nitrogen from the soil. **culture dérobée**

catchment A reservoir or basin developed for flood control or water management related to livestock and/or wildlife. An area having a common outlet for its surface runoff. Same as *drainage basin*. **bassin hydrographique, bassin versant, bassin d'alimentation**

categorical variable (*statistics*) A qualitative variable created by classifying

C

observations into categories. For example, a series of temperature measurements could be classified into the categorical variables low, normal, and high, with low defined as less than 10°C, normal between 10 and 30°C, and high greater than 30°C. See *quantitative variable.* **variable qualitative**

category (*soil taxonomy*) A grouping of related soils defined at approximately the same level of abstraction. In the Canadian system of soil classification the categories are *order, great group, subgroup, family,* and *series*. In the U.S. system of soil taxonomy the categories are *order, suborder, great group, subgroup, family,* and *series.* **catégorie**

catena A repeated sequence of soil profiles geographically related to and associated with relief features. The term is derived from the Latin word for chain because the soil profiles are linked together in the same fashion as a hanging chain when traced laterally down a slope. The profiles change character along the traverse with changing slope angle and hydrological conditions, so that different degrees of soil development are found. See *figure*. Geomorphic history of the land surface also plays a role in developing a catena through its effect on stability of the soil surface and differing ages of soils along the slope. **caténa, chaîne de sols**

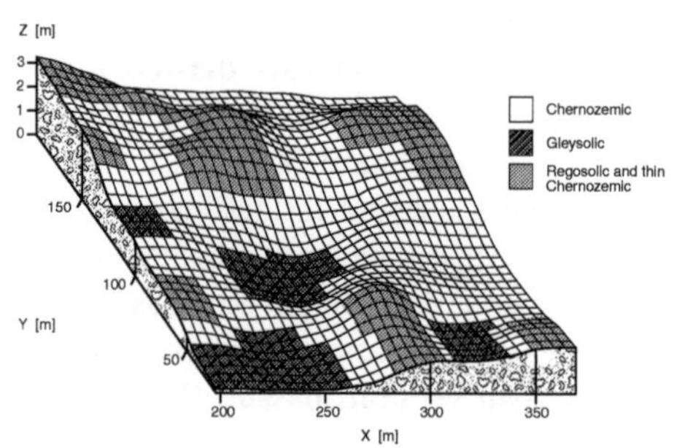

Different degrees of soil development found in a catena (D.J. Pennock, 2001).

cathode The positive pole of an electrolytic cell or a battery. When the battery is connected in a circuit, electrons flow from the anode to the cathode. **cathode**

cathodic protection The protection of an iron or steel from corrosion by using a more reactive metal. Iron or steel corrosion is caused by oxidation, i.e., loss of electron. A common form of cathodic protection is galvanizing, in which the iron surface is coated with a layer of zinc that serves a source of replacement electrons for those lost by the iron. **protecteur cathodique**

cation A positively charged *ion*. **cation**

cation exchange capacity (CEC) The total amount of exchangeable cations that a soil can adsorb; sometimes called "total exchange capacity," "base exchange capacity," or "cation adsorption capacity". It is expressed in centimoles of charge per kilogram of soil or of other adsorbing material (e.g., clay). See *effective cation exchange capacity* and *pH-dependent cation exchange capacity.* **capacité d'échange cationique**

catkin A spikelike, *inflorescence* of unisexual flowers; found only in woody plants. **chaton**

causality The relation between a cause and its effects; relationship between two objects or events or sets of objects and sets of events in which one is explained in terms of the other. **causalité**

caustic soda Sodium hydroxide (NaOH). A strongly alkaline, caustic substance used as the cleaning agent in some detergents. **soude caustique**

cavitation The formation of gas- or water-vapor-filled cavities in a liquid volume when the pressure is reduced (tension is increased) to a critical level. **cavitation**

cecum Intestinal pouch located at the junction of the large and small intestines in non-ruminants with functions somewhat similar to those of a rumen. Usually it is much larger in the herbivorous horse than in the nonherbivorous monogastrics. **caecum**

cell The fundamental unit of biological structure and function that can stand alone and carry out all the life processes. A cell is an independent unit that can obtain nutrients from the environment, derive energy from organic materials, reproduce exact copies of itself, and release waste products into the surrounding environment. No structure derived from a cell, such as the nucleus, mitochondria, or chloroplast, can carry out all of the life functions independently. All cells are bound by a plasma membrane composed of lipids, proteins, and nucleic acids (DNA) necessary for inheritance functions. Bacteria and simple plants and animals consist of only one cell, capable of independent existence. In higher plants and animals, similar cell types form tissues, which in turn form organs. **cellule**

cell division (*botany*) The division of the cytoplasm into two equal parts; brought about in higher plants by formation of a cell plate, which grows outward to the margins of the cell; also referred to as cytokinesis. **division cellulaire**

cell potential (electromotive force) The driving force in a galvanic cell that pulls electrons from the reducing agent in one compartment to the oxidizing agent in the other. **force électromotrice**

cell wall The rigid outermost layer of the cells found in plants, in some protista, and in most prokaryotes. **paroi cellulaire**

cell wall constituents Compounds that constitute the cell wall (e.g., cellulose, hemicellulose, lignin, and minerals). **constituants de la paroi cellulaire**

cellulase An enzyme that digests cellulose to hexose units. **cellulase**

cellulose A polysaccharide that consists of a long, unbranched chain of glucose molecules that are linked by beta–1, 4 bonds; a major constituent of cell walls of all plants and some algae and fungi. It is responsible for providing rigidity to the cell wall and is an important constituent of dietary fiber. **cellulose**

C

cement (1) Siliceous, ferruginous, or calcareous material that has been chemically precipitated in the spaces among the grains of a sedimentary rock, thus binding the grains into a rigid mass. (2) An industrial product obtained by burning a mixture of pulverized materials containing lime, silica, and alumina in varying proportions and finely grinding the resulting fused clinker. When mixed with water, the material makes a plastic mass that will set or harden. **ciment**

cementation The process by which clastic sediments are converted into rock by precipitation of a mineral *cement* among the grains of the sediment. See *diagenesis*. **cimentation**

cemented Having a hard, brittle consistence because the particles are held together by cementing substances (e.g., humus, calcium carbonate, or the oxides of silicon, iron, and aluminum); hardness and brittleness persist even when wet. See *indurated, indurated layer*. **cimenté**

central meridian The line of longitude upon which a *map projection* is constructed. In general, it is the central axis of the projection. See *meridian* **méridien central**

central tendency *(statistics)* The tendency of individual members in a population to cluster around some variate value. The position of the central value is usually determined by one of the measures of location such as *mean, median,* or *mode*. The closeness with which values cluster around the central value is measured by one of the measures of dispersion such as *mean deviation* or *standard deviation*. **tendance centrale**

C

centrifugal pump (*irrigation*) A commonly used device that converts mechanical energy to pressure, or kinetic energy, in a fluid by imparting centrifugal force on the fluid through a rapidly rotating impeller. **pompe centrifuge**

centrifuge A device that employs centrifugal force to separate a mixture into components by relative densities, especially to separate suspended solids from liquids. **centrifuge**

centripetal drainage pattern (*hydrology*) A drainage pattern in which the streams drain radially inwards, either towards a single river trunk or to a lake which may or may not have an outlet. **tracé centripète de drainage**

ceramic A material made from clay and hardened by firing at high temperature. It contains minute silicate crystals suspended in a glassy cement. **céramique**

cesspool A lined and covered excavation in the ground that receives the discharge of domestic sewage or other organic wastes from a drainage system. It retains organic matter and solids but permits liquids to seep through the bottom and sides. **fosse septique, puisard, puits absorbant**

chalk A soft, fine-textured, usually white to light gray limestone composed mainly of the calcareous shells of various marine microorganisms and possibly some chemically precipitated calcium carbonate. **craie**

chambers (*soil micromorphology*) Vesicles or vughs connected by a channel or channels. **vacuoles reliées, chambres**

change of state The change of a chemical substance from one physical state (solid, liquid, or gas) to another as a result of temperature or pressure changes; in the environment, usually caused by a temperature change. The changes of state of water and the associated heat transfers are especially important in environmental heat regulation, including the moderation of ecosystem weather extremes and, on a larger scale, the movement of heat poleward at the surface of the Earth. It is also vital for human body heat regulation and the removal of heat from industrial processes. **changement d'état**

channel A natural stream that conveys water; a ditch or channel excavated for the flow of water. An open conduit either naturally or artificially created which periodically or continuously contains moving water. The terms watercourse, river, creek, run, branch, and tributary are used to describe natural channels. **canal, chenal, lit**

channel bar An elongated deposit of sand and gravel located in the course of a stream. **levée de chenal**

channel density (*hydrology*) Ratio of the length of stream channels in a given basin to the area of the basin. **densité de drainage**

channel flow Part of the surface runoff of water that is confined in a river channel. **écoulement en chenal**

channel stabilization The use of engineering structures, vegetation, and other measures to prevent stream bank *erosion* and stabilize the velocity of flowing water in channels. **stabilisation des berges**

channel storage Water temporarily stored in *channels* while enroute to an outlet. **emmagasinement dans un cours d'eau**

channel terrace See *terrace*. **terrasse en canaux**

channeled (eroded) Landscapes having surfaces crossed by a series of abandoned channels. **cannelé, raviné**

channel-fill deposit Sedimentary deposit formed in a river channel when the transporting capacity of the river is incapable of removing the detrital material as rapidly as it has been delivered. **remplissage de chenaux**

channelization The straightening and deepening of a stream *channel* to permit the water to move faster, to reduce flooding, or to drain marshy areas for farming. **canalisation**

channery A thin and flat limestone, sandstone, or schist fragment up to 15 cm in length. See *coarse fragments*. **en plaquettes**

chapparral (1) A thicket of dwarf evergreen oaks, or more broadly, a dense thicket of shrubs or dwarf trees. (2) A type of ecosystem with hot, dry summers, and precipitation occurring mainly in the winter months. The vegetation consists of shrubs and evergreens; found in central and southern California, along the coast of Chile, in southern Australia, around the Mediterranean Sea, and in southern Africa. **chaparral**

character plant Unique, atypical, or distinctly different plant, either in form or density, created by pruning or other cultural manipulation. **plante spécimen**

charcoal filter See *carbon filtration*. **filtre au charbon**

Charles' law The volume (V) of a fixed mass of dry ideal gas is directly proportional to the absolute temperature (T, measured in K, Kelvin) provided the amount of gas and the pressure remain fixed; $V = kT$. **loi de Charles**

check dam Small dam constructed in a gully or other small watercourse to decrease the stream-flow velocity, minimize channel scour, and promote deposition of sediment. **barrage submersible**

chelate An organic chemical with two or more functional groups that can bind with metals to form a ring structure. The process is known as chelation. This protects the metal ions from being tied up in the soil in forms that plants cannot use. **chélat**

cheluviation A term derived from the combination of chelation and *eluviation,* whereby water containing organic extracts combines with metallic cations in the soil to form a *chelate*. This sesquioxide-rich solution then moves downward through the soil profile (hence eluviation) and moves the aluminum and iron into the lower horizons. See *leaching, podzolic*. **chéluviation**

C

chemical bond The strong force of attraction holding atoms together in a molecule or crystal. The different types are ionic bonds, which are formed by the transfer of electrons, and covalent bonds, formed by the sharing of valence electrons. **liaison chimique**

chemical equilibrium A reversible chemical reaction in which the concentrations of all reactants and products remain constant as a function of time. **équilibre chimique**

chemical fallow See *tillage, chemical fallow*. **jachère chimique**

chemical fertilizer See *fertilizer, chemical*. **engrais minéral**

chemical oxygen demand (COD) A measure of the amount of oxygen required to oxidize organic and oxidizable inorganic compounds in water. Used to determine the degree of pollution in an effluent. **demande chimique en oxygène (DCO)**

chemical potential The rate of change of Gibbs free energy with respect to the mass variable (e.g., the number of moles) in a mixed chemical system, with pressure, temperature, and amounts of other variables constant. **potentiel chimique**

chemical rooting conditions The chemical characteristics of soil that control root growth. **conditions chimiques d'enracinement**

chemical stoichiometry Calculation of the quantities of material consumed and produced in chemical reactions. **stoechiométrie chimique**

chemical weathering The processes which lead to decomposition, transformation, or breakdown of minerals and rocks by means of chemical reactions, such as hydrolysis, hydration, oxidation, carbonation, ion exchange, and solution. These reactions remove natural cementing agents and also cause the formation of secondary minerals that are less resistant to erosion than those of unweathered rocks. **altération chimique, météorisation chimique.**

chemigation The process by which fertilizers and other agrichemicals are applied in irrigation water to fertilize crops and control pests. **fertigation**

chemisorption See *adsorption.* **adsorption chimique, chimisorption**

chemistry, soil The branch of soil science that deals with the chemical constitution, chemical properties, and chemical reactions of soils. **chimie du sol**

chemoautotroph An organism that obtains energy from the oxidation of inorganic compounds and carbon from carbon dioxide. **chimiotrophe**

chemodenitrification The nonbiological process leading to the production of molecular nitrogen or an oxide of nitrogen from oxidized nitrogen. **dénitrification chimique**

chemoheterotroph An organism that derives its energy and carbon from the oxidation of organic compounds. **chimio-hétérotrophe**

chemolithotroph An organism capable of using CO_2 or carbonates as the sole source of carbon for cell biosynthesis, and deriving energy from the oxidation of reduced inorganic or organic compounds. Also called chemolithoautotroph and chemotroph. **chimiolithotrophe**

chemoorganotroph An organism that obtains energy and electrons (reducing power) from organic compounds. Used synonymously with *heterotroph.* **chimio-organotrophe**

chemostat An apparatus used to grow bacteria continuously in a specific growth phase. **chémostat**

chemosterilant A chemical compound that controls pests or other organisms by destroying their ability to reproduce. **stérilisant chimique**

chemotaxis The oriented movement of a motile organism with reference to a chemical agent. May be positive (toward) or negative (away) with respect to the chemical gradient. **chimiotaxie, chimiotactisme**

chemotrophy Nutritional mode of organisms that obtain their energy from inorganic compounds. **chimiotrophie**

Chernozemic An order of soils in the Canadian system of soil classification that have developed under *xerophytic* or *mesophytic* grasses and forbs, or under grassland-forest transition vegetation, in cool to cold, subarid to subhumid climates. The soils have a dark-colored surface (Ah, Ahe, or Ap) horizon and a B or C horizon, or both, of high base saturation. The order consists of the *Brown, Dark Brown, Black,* and *Dark Gray* great groups. **chernozémique**

chernozemic A A *diagnostic horizon* (Canadian system of soil taxonomy) with characteristics including thickness of at least 10 cm, organic carbon content of 1 to 17%, C:N ratio less than 17, base saturation percentage (neutral salt) more than 80%, and occurrence in soils with mean annual soil temperature of 0°C or higher. **A chernozémique**

chert A hard, extremely dense or compact, dull to semilustrous, *cryptocrystalline* sedimentary rock consisting mostly of cryptocrystalline silica, with lesser amounts of micro- or cryptocrystalline quartz and amorphous silica (opal). It is black or dull in color, splinters easily, and fractures along flat planes in contrast to the conchoidal fracture of flint, which is a variety of chert. It sometimes contains calcite, iron oxide, and the remains of siliceous and other organisms. Chert occurs principally as nodules in limestones and dolomites, and less commonly as thick, layered deposits. **chert**

cherty An adjective used in the soil textural class designations of horizons when the soil mass contains between 15 and

90% by volume of chert fragments. See *coarse fragments*. **cherteux**

china clay A commercial term for *kaolin,* which, after processing, is suitable for use in the manufacture of chinaware. **kaolinite**

chirality The property of an object or molecule to exist in left- and right-handed structural forms, i.e., to be nonsuperimposable on its mirror image. See *optical activity*. **chiralité**

chisel See *tillage, chisel*. **scarifier, sous-soler**

chisel cultivator A tillage implement with one or more shanks to which are attached chisel, spike, or narrow-shovel points (i.e., feet) used to shatter the soil without inversion. When used to shatter or loosen hard, compact layers or subsoils and penetrate deeper than 30 cm, the shanks are stronger and spaced farther apart. Also called chisel plow. See *subsoiling*. **cultivateur, chisel, charrue scarificatrice**

chisel planting Seedbed preparation by *chiseling* without inversion of the soil, leaving a protective cover of *crop residue* on the surface for *erosion* control. **travail à l'aide d'un cultivateur ou d'un chisel**

chiseling (1) Breaking or loosening the soil, without inversion, with a *chisel cultivator* or chisel plow. (2) A method of tillage in which hard, compact layers, usually in the subsoil, are shattered or loosened to depths below normal tillage depth. See *ripping*, *subsoiling*. **(1) préparation de sol avec cultivateur ou chisel, scarification (2) sous-solage**

chi-square test *(statistics)* A statistical test that employs the sum of values given by the quotients of the squared difference between observed and expected theoretical frequencies divided by the expected frequency. It enables assessment of the association or commonalty in a population and is used to determine equivalency of observed sample and expected population. **test du khi-carré**

chitin A resistant nitrogen-containing polysaccharide present in the covering layer of insects and in the cell walls of many fungi. **chitine**

chlamydospore A thick-walled multinucleate asexual spore developed from *hyphae* but not on basidia or conidiophores; produced by many parasitic and mycorrhizal fungi. **chlamydospore**

chloramine A compound containing nitrogen, hydrogen, and chlorine, formed by the reaction between hypochlorous acid (HOCl) and ammonia and/or organic amines in water. **chloramine**

chlorides Negative chlorine ions, Cl^-, found naturally in some surface- and groundwaters and in high concentrations in seawater. Higher than normal chloride concentrations in fresh water, due to sodium chloride that is used on foods and present in body wastes, can indicate sewage pollution. The use of highway deicing salts can also introduce chlorides to surface water or groundwater. Elevated groundwater chlorides in drinking water wells near coastlines may indicate saltwater intrusion. **chlorures**

chlorinated (1) An organic compound to which atoms of chlorine have been added. (2) Water or wastewater that has been treated with either chlorine gas or a chlorine-containing compound. **chloré**

chlorinated hydrocarbon Any of a group of generally long-lasting, broad-spectrum insecticides that contain chlorine, hydrogen, and carbon. Chlorinated hydrocarbons may present a potential hazard through accumulation in the food chain and persistence in the environment (e.g., *DDT*). **hydrocarbure chloré**

chlorination (*wastewater management*) A disinfection process in which chlorine is added to water or wastewater to kill or inactivate dangerous microorganisms or viruses. **chloration, javellisation.**

C

chlorine residual The quantity of chlorine added to drinking water in excess of the amount needed to react with organic and inorganic materials suspended or dissolved in the water. **chlore résiduel**

chlorite A layer-structured group of nonexpanding clay minerals of the 2:1 type that have the interlayer filled with a positively charged metal-hydroxide octahedral sheet (in place of exchangeable cations and water). There are both trioctahedral and dioctahedral varieties. Chemically similar to vermiculite, the general formula is $(Mg, Fe^{+2}, Fe^{-3})_6\ AlSi_3O_{10}$ $(OH)_8$. Chlorites are green, platy, and monoclinic or triclinic. They are found in low-grade metamorphic rocks and as alteration products of mafic minerals. **chlorite**

chlorofluorocarbons and related compounds A family of *anthropogenic* compounds, including chlorofluorocarbons (CFCs), bromofluorcarbons (halons), methyl chloroform, carbon tetrachloride, methyl bromide, and hydrochlorofluorcarbons (HCFCs); shown to deplete stratospheric *ozone*, and therefore typically referred to as ozone-depleting substances. **chlorofluorocarbures**

chloroform A simple halogenated *hydrocarbon* ($CHCl_3$) obtained from the chlorination of methane (CH_4). Once used in human anesthesia, it remains an important industrial chemical. Chloroform is one of the more common halomethanes produced during the chlorination of water. Low concentrations in drinking water promote kidney and liver damage in animals. **chloroforme**

chlorophyll The green pigment found in plants, some algae, and some bacteria that absorbs and channels the energy of sunlight into chemical energy through the process of photosynthesis. Chlorophyll absorbs energy mainly in the blue (435 to 438 nm) and red (670 to 680 nm) regions of the spectrum. The light removes an electron from the chlorophyll molecule, which is used to produce either ATP (adenosine triphosphate) or NADP (nicotinamide adenine dinucleotide phosphate) for carbon dioxide fixation. The central ion in chlorophyll is magnesium, and the large organic molecule is a porphyrin. The porphyrin contains four nitrogen atoms that form bonds to magnesium in a square planar arrangement. There are several forms of chlorophyll. See *figure*. **chlorophylle**

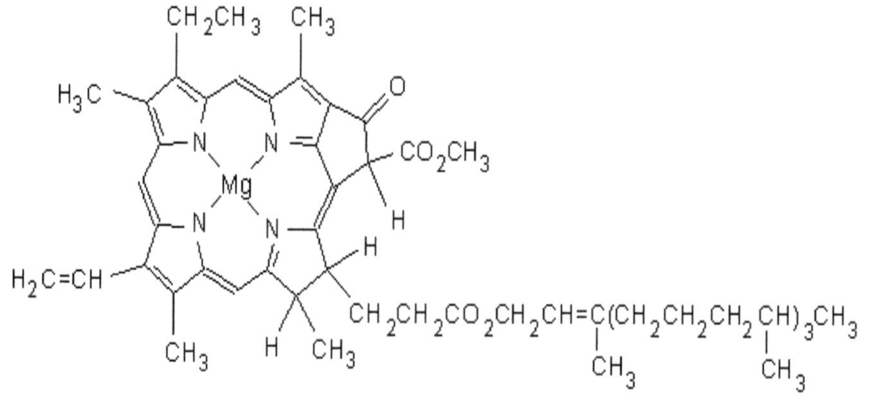

The structure of **chlorophyll** a.

chloroplast An organelle found only in photosynthetic organisms and carrying *chlorophyll.* **chloroplaste**

chlorosis Blanched or yellowish coloring in plant leaves resulting from the failure of chlorophyll to develop. Often caused by a nutrient or light deficiency. **chlorose**

chorochromatic map A type of thematic map in which qualitative spatial distributions (especially those relating to resources) are depicted by color-tinting. See *choropleth map.* **carte chorochromatique.**

chorographic map Any map representing large regions, countries, or continents on a small scale. **carte chorographique**

choromorphographic map A type of map that depicts the terrain units and morphological classes of any land area, based on shape rather than landform genesis. **carte choromorphographique**

choropleth map A type of thematic map in which quantitative spatial distributions are depicted from data that have been computed from mean values per unit area. **carte choroplèthe**

chroma The relative purity, strength, or saturation of a color. Directly related to the dominance of the determining wavelength of light. It is one of the three variables of color. See *Munsell color system*, *hue*, and *value, color.* **chroma, saturation**

chromatid Half a chromosome in prophase and metaphase; becomes a chromosome in anaphase of *mitosis.* **chromatide**

chromatin The complex of nucleic acids (DNA and RNA) and proteins (histones and nonhistones) comprising chromosomes, the component fibril comprised of nucleosomes and spacers (limited to eukaryotes). **chromatine**

chromatography A technique for analyzing or separating mixtures of gases, liquids, or dissolved substances. The components in a sample to be separated are distributed between two phases: a stationary phase, made of an adsorbent material, and a mobile phase that percolates through the stationary phase. The separation depends on competition for molecules of the sample between the mobile and stationary phases. A column or other support holds the stationary phase, and the mobile phase carries the sample through it. Sample components that partition strongly into the stationary phase spend a greater amount of time in the column and are separated from components that stay predominantly in the mobile phase and pass through the column faster. As the components elute from the column, they can be quantified by a detector and/or collected for further analysis. See *gas chromatography, high performance liquid chromatography, size-exclusion chromatography, thin-layer chromatography.* **chromatographie**

chromosome A threadlike structure in the cell nucleus of all plant and animal cells, composed of nucleic acids and proteins. The chromosome contains the nucleic acid deoxyribonucleic acid (DNA), which contains genes that determine the hereditary characteristics of the cell, animal, or plant. The number of chromosomes in the nucleus is characteristic of the species. Humans have 23 pairs of chromosomes, for a total of 46. **chromosome**

chronological Pertaining to matters of time, especially in order of occurrence (e.g., chronological order). **chronologique**

chronosequence A sequence of related soils that differ from one another in certain properties primarily as a result of time as a soil-forming factor. **chronoséquence**

chronostratigraphy The branch of *stratigraphy* that deals with the age of strata

C

and their time relations. **chronostratigraphie**

chute A narrow, sloping *channel* of steep gradient through which a stream descends rapidly to a lower level. **rapide**

ciliate Protozoan that moves by means of *cilia* on the surface of the cell. **cilié**

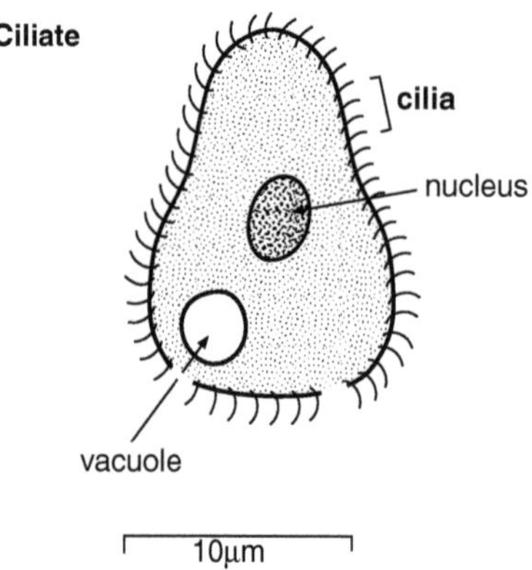

Ciliate, cilia.

cilium (plural cilia) (*microbiology*) Small hairlike projections found on the surface of some microorganisms and used for movement. See *figure*. **cil**

citrate-insoluble phosphorus That portion of phosphorus (P) in fertilizer remaining after water (water-soluble phosphate) and ammonium citrate extractions (see *citrate-soluble phosphorus*). The P content of fertilizers that is considered to be unavailable to plants in the guaranteed analysis of the fertilizer. See *phosphate*. **phosphore insoluble au citrate**

citrate-soluble phosphorus That part of the total phosphorus (P) in fertilizer that is insoluble in water but soluble in neutral 0.33 M ammonium citrate and which, together with water-soluble P, represents the readily available P content of the fertilizer. See *phosphate*. **phosphore soluble au citrate**

clarification (*wastewater management*) The process of removing suspended solids from wastewater; the water is usually allowed to stand, allowing the particles to settle. **clarification, décantation**

clarifier (*wastewater management*) A settling tank used to mechanically remove settled solids from wastes. **clarificateur, décanteur**

class A unit or group in the taxonomic classification of organisms composed of one or more orders; subdivision of a *phylum*. **classe**

class, soil A group of soils having a range in a particular property such as acidity, degree of slope, texture, structure, land-use capability, degree of erosion, or drainage. **classe de sols**

classification The assignment of objects or units to groups within a system of categories distinguished by their properties or characteristics. **classification, classement**

classification, soil See *soil classification*. **classification des sols**

clast (1) An individual constituent, grain, or fragment of a detrital sediment or sedimentary rock, produced by the physical disintegration of a larger rock mass. (2) A piece of fragmented rock from a volcanic explosion. See *pyroclastics*. **claste**

clastation (*geology*) The breaking up of rock masses *in situ* by physical or chemical means. **fragmentation**

clastic Composed of broken fragments of rocks and minerals. **clastique**

clastic rock (*geology*) Sedimentary rock composed mainly of fragments derived from pre-existing rocks and transported to their places of deposition (e.g., a sandstone, conglomerate, or shale, or a limestone consisting of particles derived from a pre-existing limestone). **roche clastique**

clastic sediment Detrital material consisting of rock fragments (i.e., *clasts*) that have been eroded, transported, and redeposited at a different site. They range in particle size from boulders to silt. **sédiment clastique**

clay (1) A soil separate consisting of particles <0.002 mm in equivalent diameter. See *soil separates*. (2) A textural class. See *texture*. (3) A naturally occurring material, composed primarily of fine-grained minerals, which is generally plastic at appropriate water contents and will harden when dried or fired. Although clay usually contains phyllosilicates, it may contain other materials that impart plasticity and harden when dried or fired. Associated phases in clay may include materials that do not impart plasticity and organic matter. **argile**

clay (1:1) Any *clay mineral* consisting of a *tetrahedral* layer of silica (in which four oxygen atoms are arranged around a silicon ion) paired with an *octahedral* layer (in which oxygen and hydroxyl ions are grouped around metal cations such as aluminum); i.e., any clay having one tetrahedral layer for each octahedral layer. *Kaolinite* is an example of a 1:1 clay. **argile (1:1)**

clay (2:1) Any *clay mineral* consisting of *octahedral* layers sandwiched between two *tetrahedral* layers; i.e., any clay having two tetrahedral layers for each octahedral layer. *Illite* is an example of a 2:1 clay. **argile (2:1)**

clay domain A bundle of clay particles that are only visible in crossed polarized light. **domaine argileux, champ des argiles**

clay film (skin) Coating of oriented clays on the surfaces of soil peds and mineral grains, and in soil pores. **pellicule argileuse, enrobement, film argileux**

clay loam Soil material that contains 27 to 40% clay and 20 to 45% sand. See *texture*. **loam argileux**

clay mineral A phyllosilicate mineral or a mineral that imparts plasticity to clay and hardens upon drying or firing. See *phyllosilicate* mineral terminology. **minéral argileux**

clay mineralogy See *phyllosilicate* mineral terminology. **minéralogie des argiles**

clayey Containing large amounts of clay, or having properties similar to those of clay. **argileux**

clay-organo complex A microstructure in soil composed of clay particles and organic matter stabilized by humic materials and inorganic ions (e.g., Ca^{2+}). Often considered to be the basic building blocks of soil *aggregation*. See *organo-mineral complex*. **complexe argilo-organique**

claypan A dense, compact layer in the profile having a much higher clay content than the overlying material, from which it is separated by a sharply defined boundary. **claypan**

claystone An *indurated* clay having the texture and composition of shale but lacking its fine lamination or fissility. Also, a concretionary body of clay in alluvium or of calcareous material in clay. **argilite**

clean tillage See *tillage, clean tillage (clean culture, clean cultivation)*. **labour nettoyant**

clearcut An area of forest land from which all merchantable trees have recently been harvested. **coupe totale, rase ou à blanc**

clearcutting silvicultural system System in which merchantable trees are cleared from an area at one time and an even-aged replacement stand is established. It does not include *clearcutting with reserves*. Clearcutting is designed so that most of the opening has full light exposure and is not dominated by the canopy of adjacent trees (this produces an open area climate). The minimum size of a clearcut opening is generally considered to be 1 ha. **système sylvicole par coupe totale**

clearcutting with reserves A variation of the clearcut silvicultural system in which trees are retained, either uniformly or in small groups, for purposes other than regeneration. **coupe totale avec réserves**

cleavage The splitting, or tendency of a mineral to split, along crystallographic planes. As applied to rocks, it is the property of splitting into thin parallel sheets that may be highly inclined to the bedding planes, as in slate. **clivage**

C

cleavage plane The plane of mechanical fracture in a rock or mineral. See *cleavage.* **plan de clivage**

Clerici solution A solution of thallium malonate and thallium formate in water that is used as a *heavy liquid;* its specific gravity is 4.15. See *heavy liquid, heavy mineral, light mineral.* **solution de Clérici**

climate The average *weather* (usually taken over a 30-year period) for a particular region and time period. The properties that characterize climate are thermal (temperatures of the surface air, water, land, and ice), kinetic (wind and ocean currents, together with associated vertical motions and the motions of air masses, humidity, cloudiness and cloud water content, groundwater, lake winds, and water content of snow on land and sea ice), and static (pressure and density of the atmosphere and ocean, composition of the dry air, salinity of the oceans, and the geometric boundaries and physical constants of the system). These properties are interconnected by various physical processes such as precipitation, evaporation, infrared radiation, convection, advection, and turbulence. **climat**

climate change A significant change from one climatic condition to another; a long-term change in temperature, precipitation, wind, and all other aspects of the Earth's climate, as influenced by external processes (e.g., solar-irradiance variations, variations of the Earth's orbital parameters, lithosphere motions, and volcanic activity) and internal variations of the climate system (e.g., changes in the abundance of greenhouse gases). **changements climatiques**

climate feedback An atmospheric, oceanic, terrestrial, or other process that is activated by the direct *climate change* induced by changes in radiative forcing. Climate feedbacks may increase (positive feedback) or diminish (negative feedback) the magnitude of the direct climate change. **rétroaction climatique**

climate lag The delay that occurs in *climate change* as a result of some factor that changes very slowly (e.g., the effects of releasing more carbon dioxide into the *atmosphere* may not be known for some time, because a large fraction is dissolved in the ocean and released to the atmosphere many years later). **retard climatique**

climate model A quantitative way of representing the interactions of the atmosphere, oceans, land surface, and ice. **modèle climatique**

climate modeling The simulation of climate using computer-based models. See *general circulation model.* **modélisation du climat**

climate system The system of the Earth's *atmosphere*, oceans, *biosphere*, *cryosphere*, and *geosphere*. **système climatique**

climatic region A specific area in which various combinations of climatic elements (e.g., precipitation, humidity, temperature, atmospheric pressure, and wind) can be recognized. **région climatique**

climatology The study of climate. **climatologie**

climax A plant community of the most advanced type capable of development under, and in dynamic equilibrium with, the prevailing environment. **climax**

climax forest A forest community that represents the final stage of natural forest succession for its environment. **forêt climacique**

climosequence A sequence of related soils that differ from one another in certain properties primarily as a result of the effect of climate as a soil-forming factor. **climoséquence**

C

cline A change in population characteristics over a geographic area, usually related to a corresponding environmental change. **gradient géographique**

clinker Solid residue formed in an incinerator from various noncombustible materials such as glass or metal. **scories, mâchefer**

clinometer A simple instrument for measuring vertical angles or slopes. In forestry a clinometer is used to measure distance and tree heights. **clinomètre**

clod A compact, coherent mass of soil produced by digging or tillage, in contrast to *aggregates* or peds that form naturally in soil. Clods usually slake easily with repeated wetting and drying. **motte**

clone An organism descended from a single individual by nonsexual reproduction; such organisms are therefore of the same genetic constitution. **clone**

closed basin Area draining to some depression or lake, from which water escapes only by evaporation. **bassin fermé**

closed drain Subsurface drain, tile, or perforated pipe that receives surface water through surface inlets. **tuyau fermé**

closed handling system A system for transferring chemicals or fertilizer directly from the storage container to the applicator so that humans and the environment are not exposed to the chemicals. **système de manutention fermé**

closed system *(physics)* A system that does not exchange matter or energy with its surroundings. *(ecology)* A system exchanging energy, but not matter, with its surroundings. The planet Earth is a closed system in the ecological sense, absorbing and radiating solar energy, but recycling matter within the biosphere. See *open system*. **système fermé**

cluster analysis *(statistics)* A multivariate approach for determining whether individuals fall into groups or clusters. There are several methods of procedure, all of which depend on setting up a metric to define the "closeness" of individuals. **analyse en classification automatique, analyse de groupement**

coagulation The grouping together of solids suspended in air or water, resulting in their precipitation. Coagulation is encouraged in wastewater treatment plants by the addition of alum, ferrous sulfate, and other materials. See *flocculation*. **coagulation**

coal A solid, brittle, distinctly stratified, combustible carbonaceous rock, formed by compaction and induration of partially decomposed plant remains. Coal is a fossil fuel used primarily to generate steam for the production of electricity. Coal is graded on the basis of heat content and classified as *anthracite*, *bituminous*, *subbituminous*, or *lignite*. **charbon, houille**

coal ash The residual inorganic matter left after coal has been completely burned. **cendre de houille**

coal gas Fuel gas produced from a high-volatile *bituminous* coal. Its average composition by volume is 50% hydrogen, 30% methane, 8% carbon monoxide, 4% other hydrocarbons, and 8% carbon dioxide, nitrogen, and oxygen. **gaz de houille**

coal gasification Conversion of coal to a gas, suitable for use as a fuel. **gazéification du charbon**

coal processing waste Earth materials that are combustible or physically unstable, form acids or toxins, and are wasted or otherwise separated from coal after it is physically or chemically processed, cleaned, or concentrated. **résidus de conditionnement de charbon**

coal seam A layer, vein, or deposit of coal. A stratigraphic layer of the Earth's surface containing coal. **filon de charbon**

C

coalification The biochemical processes of diagenesis and the geochemical processes of metamorphism in the formation of coal. **houillification, carbonification**

coarse fragments Rock or mineral particles (harder than 3 on *Mohs scale* of hardness) larger than 2 mm in diameter. Coarse fragments in soils are *gravel* or *channery* (up to 8 cm in diameter or 15 cm in length), *cobbles* or flags (8 to 25 cm diameter or 15 to 38 cm length), and stones (greater than 25 cm diameter or 38 cm length). **fragments grossiers**

coarse sand Sand with particle diameters between 0.5 mm and 1 mm. **sable grossier**

coarse texture A broad textural grouping of soils or materials dominated by sand, loamy sand, and sandy loam (except very fine sandy loam) textural classes. **texture grossière**

coarse woody debris Dead woody material (sound and rotting logs and stumps) that is not self-supporting, above the soil in various stages of decomposition; provides habitat for plants, animals, and insects, and is a source of nutrients for soil development. **débris ligneux grossiers**

coarse-filter analysis An analysis of aggregates of elements (e.g., cover type plant community). **analyse par filtre brut**

coarse-grained (*geology*) Pertaining to crystalline or sedimentary rock or sediment, or its texture, in which the individual minerals are relatively large and can be easily seen with the unaided eye. **à grain grossier**

coast A strip of land of indefinite width (may be several kilometers) that extends from the seashore inland to the first major change in terrain features. **côte**

coastal erosion The actions of marine waves through the processes of hydraulic action, corrasion, attrition, and solution, in attacking a coastline, thereby causing it to retreat. **érosion côtière**

coastal plain Any gently sloping plain or lowland that has its margin on the landward side of a *coast* and that often continues offshore as a submarine continental shelf. **plaine côtière**

coastal zone Coastal waters and adjacent lands that exert a measurable influence on the uses of the seas and their resources and ecology. **zone côtière**

coated fertilizer See *fertilizer, coated.* **engrais enrobé**

coated seed A seed that has been covered with a layer of material to make it uniform in size and shape and free-flowing, or that has been covered with fertilizer, pesticides, nitrogen-fixing microorganisms, coloring, or other additives. **semence pelliculée**

coating (*soil micromorphology*) Soil material observed microscopically that occurs on the surface of voids, grains, and aggregates and is different from the adjacent soil. Similar meaning to *cutan.* **revêtement, enduit, enrobement**

cobble Rounded or partially rounded rock or mineral fragment 7.5 to 25 cm in diameter. In engineering practice, cobbles are greater than 7.5 cm but less than 20 cm in diameter. **galet, caillou**

cobblestone See *cobble.* **caillou roulé, galet**

cobbly Containing appreciable quantities of rounded or subrounded coarse rock or mineral fragments 8 to 25 cm in diameter, (i.e., *cobbles*) The term angular cobbly is used when the fragments are less rounded. See *coarse fragments.* **caillouteux**

coccus A spherical bacterium. **coque**

CO_2 fertilization See *carbon dioxide fertilization.* **effet fertilisant du dioxyde de carbone**

COD See *chemical oxygen demand.* **DCO**

co-dominant (*soil survey*) Two or more soils (or other features) of roughly equal proportion that together constitute the majority of a soil (or soil landscape) mapping unit. **composantes significatives (d'une unité cartographique)**

co-dominant trees Trees with crowns forming the general level of the forest canopy and receiving full light from above but comparatively little from the sides; usually with medium-sized crowns and crowded on the sides. **arbres codominants**

coefficient of correlation (r) See *correlation, statistical.* **coefficient de corrélation**

coefficient of determination (r^2) See *correlation, statistical.* **coefficient de détermination**

coefficient of friction A constant proportionality factor relating normal stress and the corresponding critical shear stress at which sliding begins between two surfaces. **coefficient de frottement**

coefficient of linear extensibility (COLE) The percent shrinkage in one dimension of a molded soil between two water contents (e.g., between the *plastic limit* and *air dry).* **coefficient d'extensibilité linéaire (indice COLE)**

coefficient of multiple correlation (R) See *correlation, statistical.* **coefficient de corrélation multiple**

coefficient of multiple determination (R^2) See *correlation, statistical.* **coefficient de détermination multiple**

coefficient of variation *(statistics)* A measure of variability within a sample or population calculated as 100 times the standard deviation divided by the mean. **coefficient de variation**

coenocyte A multinucleate cell or protoplast in which nuclear divisions have not been followed by cytoplasmic cleavage. **coenocyte**

coenocytic organism An organism that consists of protoplasm containing many nuclei. Slime molds are coenocytic at one stage in their development. **organisme coenocytique**

coenzyme An organic molecule, or nonprotein organic cofactor, which plays an accessory role in enzyme-catalyzed processes, often by acting as a donor or acceptor of a substance involved in the reaction. **coenzyme**

coevolution The simultaneous evolution of two or more species of organisms that interact in significant ways. **coévolution**

coexistence The occurrence of two or more species in the same habitat; usually applied to potentially competing species and often taken to imply a stable situation. **coexistence**

cohesion The force holding a solid or liquid together, owing to attraction between like molecules. Cohesion and intergranular friction are the forces that combine to give soil *shear strength.* **cohésion**

cohort A group of organisms; in demographic studies, often taken as a group of similar age. **cohorte**

cold desert A polar region where plant life is inhibited by low temperatures and physiological drought; synonymous with *tundra.* Also used occasionally to define enclosed basins of central Asia, cut off by high relief from maritime influences and having at least one month with a mean temperature below 6°C (43°F). See *desert.* **désert froid**

coldwater fish A fish that requires relatively cool water for survival; optimum temperatures for species vary, but most are found in water where temperatures are 20°C or less. **poisson d'eaux froides**

coliform A group of bacteria common to the intestinal tracts of humans and animals, whose presence in water is an indicator of fecal contamination. Includes all aerobic and facultative anaerobic, gram negative, nonsporulating bacilli that produce acid and gas from the fermentation of lactose (e.g., *Escherichia coli, Aerobacter aerogenes).* **coliforme, colibacillle**

coliform index An index of the purity of bacteriological quality water based

C

on a count of its *coliform* bacteria. **niveau de coliformes**

coliphage A virus that infects *Escherichia coli.* **coliphage**

C

collapse scar bog, collapse scar fen A depression produced by the thawing of *permafrost* within a *palsa* or *peat plateau.* It is usually water-saturated, treeless, and *minerotrophic*. The thawing permafrost edge appears as a steep bank with leaning, mostly dead trees. **tourbière effondrée, fen effondré**

collapsible soil A soil that undergoes reorientation of particles and reduction in volume under constant load when its water content is increased. **sol sensible au tassement ou à l'affaissement**

colligative properties Those properties that vary with the number of chemical elements in a solution and not with the composition of the elements. In sea water with increasing salinity, boiling point and osmotic pressure increase, and freezing point and vapor pressure decrease. **propriétés colligatives**

colloid A particle 0.1 to 0.001 μm in diameter. Soil clays and soil organic matter are often called colloids because they fall within these size dimensions. **colloïde**

colluvial Pertaining to *colluvium* (e.g., colluvial deposits). **colluvial**

colluvial slope Sloping land at the foot of steep hills or mountains made up of deposits of unconsolidated material that have been moved over short distances by gravity, water, or both, and which includes *talus* material and local alluvium. **pente colluviale**

colluvium A heterogeneous mixture of material that, as a result of gravitational action, has moved down a slope and settled at its base. See *creep*. **colluvion**

colonization Establishment of a community of microorganisms at a specific site or ecosystem. **colonisation**

colony (*microbiology*) A clump of microorganisms on a solid culture medium visible to the naked eye. **colonie**

colony count A method for counting the number of bacteria in an environmental sample. A portion of a liquid sample or a dilution thereof is spread across the surface of a suitable solid nutrient medium and allowed to incubate. The number of bacterial colonies that develop on the surface are counted, and the necessary mathematical calculations are made to compute the number of bacteria per unit volume of the sample. The technique is based on the assumption that one bacterial cell will grow and divide to produce one colony on the surface of the nutrient medium. **numération de colonies bactériennes**

color See *Munsell color system.* **couleur**

colorimetry Method of chemical analysis in which a change in color and/or color intensity is the indication of the presence, concentration, or both of a particular material. **colorimétrie**

columnar soil structure A soil structural type with a vertical axis much longer than the horizontal axes and a distinctly rounded upper surface. See *structure*. See *Appendix B, Table 1* and *Figure 1.* **structure colomnaire**

combined tillage operations See *tillage, combined tillage operations.* **opérations combinées de travail du sol**

combustible liquid A liquid with a *flash point* above 100°F and below 200°F. **liquide combustible**

combustion A chemical reaction in which a substance reacts rapidly with oxygen and produces heat and light. Often these reactions are free-radical chain reactions which result in the oxidation of carbon to form its oxides and the oxidation of hydrogen to form water. **combustion**

cometabolism Conversion of a substate by a microorganism without deriving energy, carbon, or nutrients from the substrate. The organism is able to convert the substrate into intermediate degradation products but fails to multiply at its expense. **cométabolísme**

commensalism Interaction between two species in which one species derives benefit while the other is neither benefited nor harmed. **commensalisme**

comminution The reduction of a substance (rock, mineral, organic material) by weathering, breakdown, and erosion to form progressively smaller particles. Also a means of preparing (i.e., pulverization) stone, coal, or ore for various uses. **fragmentation**

common ion effect The decrease in solubility of an ion salt (i.e., one that dissociates in solution into its ions) caused by the presence in solution of another solute that contains one of the same ions as salt. The effect is important in the removal of ions from solution in water treatment. **effet d'ion commun**

community All the organisms that occupy a specific habitat and interact with one another in time and space. **communauté**

community-unit hypothesis A hypothesis about how groups of co-occurring species are functionally related which assumes communities are groups of species with interdependent functional relationships. **hypothèse des communautés**

compactibility A property of a soil or sedimentary material that permits it to decrease in volume or thickness under load; it is a function of the size, shape, hardness, and brittleness of the constituent particles. The standard method for determining soil compactibility is the Proctor test. **compactibilité**

compaction (1) Increasing soil bulk density, and concomitantly decreasing soil porosity, by the application of mechanical or other forces to a soil. As soil compaction increases, a state of excessive compaction can be reached that adversely affects plant growth. (2) The mechanical reduction of the volume of solid waste (an increase in density) by the application of pressure. **compaction, compactage, tassement**

companion crop A crop (e.g., a small grain) that is sown with another crop, especially one that will emerge and develop slowly (e.g., a forage crop). Also called nurse crop. **culture associée, culture compagne, plante-abri**

competition The rivalry between two or more individuals or species that occurs when (a) a necessary resource is in limited supply relative to organism demands, (b) resource quality varies and demand is greater for higher-quality resources, or (c) one organism uses a resource at a greater rate than another organism. **compétition**

competitive exclusion A hypothesis stating that when organisms of different species compete for the same resources in the same habitat, one species will commonly be more successful in this competition and will exclude the second from the habitat. **exclusion compétitive**

compiled map A map produced by the transfer of information, usually by reduction of scale, from an existing map rather than by original survey work. **carte dérivée, carte synthèse**

complete block design *(statistics)* An experimental design in which every treatment appears the same number of times in each block. **plan d'experience à blocs complets**

completed test (*wastewater management*) See *confirmed test*. **épreuve de complétion**

completely randomized design *(statistics)* An experimental design in which the treatments are allocated to the experimental units (e.g., plots) entirely at random. **plan d'experience complètement aléatoire**

complex A substance formed when one or more anions or neutral molecules become bonded to a metal atom or ion. **complexe**

complex ion A charged species consisting of a metal ion surrounded by *ligands*. **ion complexe**

complex, soil (*soil survey*) A mapping unit used in detailed and reconnaissance soil surveys where two or more defined soil units are so intimately intermixed geographically that it is impractical, because of the scale used, to separate them. **complexe de sols**

C

composite map A map that portrays information of two or more general types; usually a *compiled map*, bringing together on one map, for purposes of comparison, data that were originally portrayed on separate maps. **carte composée, plurifactorielle ou multicouche**

composite sample A sample comprised of two or more subsamples. **échantillon composite**

compost Soil conditioner and fertilizers produced from organic residues or a mixture of organic residues and soil, that have been piled, moistened, and allowed to decompose; often called *artificial manure* or synthetic manure. **compost**

compound A substance made up of two or more elements which are in a fixed proportion by weight. The various elements can only be separated by chemical reactions and not by physical means. The physical and chemical properties of a compound are a result of the chemical combination of its elements and are not properties of the individual elements. See *mixture*. **composé**

compound unit (*soil survey*) A soil or map unit that is characterized by at least two and usually not more than four major soils or groups of soils. **unité composée**

compressibility The susceptibility of a soil to decrease in volume when subjected to load. **compressibilité**

compressibility index The ratio of pressure to *void ratio* on the linear portion of the curve relating these two variables. **indice de compressibilité**

compression A system of forces or stresses that tend to decrease the volume of, or compact, a substance, or the change of volume produced by such a system of forces. **compression**

compressional wave A moving disturbance in an elastic medium characterized by volume and density changes caused by pressure. Any displacement of particles is in the direction of wave propagation. **onde de compression**

compressive strength The maximum *compression* that can be applied to a material, under given conditions, before failure occurs. **résistance à la compression**

computer simulation model A mathematical model, processed on a computer, used to describe soil and environmental processes and predict the output from these processes if certain variables are altered. **modèle de simulation par ordinateur**

concavity One of the morphological characteristics of a slope, in which the gradient becomes progressively more gentle as it is traced downwards towards the foot of the slope. **concavité**

concentrated flow A relatively large water flow over or through a relatively narrow course. It often causes serious erosion and gullying. **écoulement concentré**

concentration The amount of a particular substance in a given quantity of a mixture, solution, or ore. Concentration is usually stated as a percentage by weight or volume, as weight per unit volume, molarity (a molar solution contains one gram-mole of solute per liter of solution), or normality (a normal solution contains one gram equivalent weight of solute per liter of solution). For gaseous air contaminant concentrations, two expressions used are volume/volume ratio (usually in parts per million) and mass/volume ratio (usually micrograms per cubic meter of air). In water, mass/volume and mass/mass ratios are used; the volume and mass of the aqueous medium are easily interchanged because one liter of water has a mass of one kilogram. Typical units are milligrams of pollutant per liter of water or micrograms of pollutant per liter of water, which equals parts per billion (mass). Soil and food concentrations are mass/mass ratios in milligrams of chemical per kilogram of medium, which is the same as parts per million (mass), or micrograms of chemical per kilogram of medium, equal to parts per billion (mass). **concentration**

concentration gradient The change in the concentration of a material over distance. Chemicals will diffuse from areas of higher concentration toward areas of lower concentration; the diffusion rate increases with an increase in the concentration gradient. This principle is used in removal of pollutants from exhaust gases (e.g., scrubbers) and water effluents (e.g., packed tower aeration). The gaseous or liquid material into which the pollutant is diffusing is replenished rapidly to maintain its low concentration and, thus, a high collection efficiency. **gradient de concentration**

conceptual model A description of a series of working hypotheses. **modèle conceptuel**

concordant (*geomorphology, geology*) (1) The harmonious correspondence between a morphological feature and its underlying geological structure (e.g., a drainage pattern that follows the geologic structure in a region, as opposed to one that doesn't, which is called discordant). See *discordance*. (2) Strata displaying parallelism of bedding or structure. (3) Agreement in radiometric ages determined by more than one method, or by the same method from more than one material. **concordant**

concrete frost Ice in the soil in such quantity as to form virtually a solid block. **masse cryoconsolidée**

concretion A mass or concentration of a chemical compound, such as calcium carbonate or iron oxide, in the form of a grain or nodule of varying size, shape, hardness, and color, found in soil and in rock. The term is sometimes restricted to concentrations having concentric fabric. The composition of some concretions is unlike that of the surrounding material. **concrétion**

condensation The physical process by which a vapor becomes a liquid or solid; the opposite of evaporation. **condensation**

condensation reaction A chemical reaction in which two molecules combine to form a larger molecule with elimination of a smaller molecule (e.g., water). **réaction de condensation**

conditioned fertilizer See *fertilizer, conditioned*. **engrais conditionné**

conductance The ability of a material to transmit electricity; the opposite of resistance. **conductance**

conduction, thermal or heat The process whereby heat is transferred through matter from a point of high temperature to a point of low temperature. **conduction thermique**

cone (*geomorphology*) A conical mass of which the base is a circle and the summit a point; applied mostly in connection with a volcano (e.g., ash cone, cinder cone), or with reference to the surface morphology of a glacier (e.g., dirt cone). Commonly used in the sense of a section of a cone, with straight or concave long profile and slope greater than 15° (26%) (e.g., *talus* cone, alluvial cone). (*botany*) Specialized seed-bearing structures unique to *coniferous* trees (e.g., firs, cedars, pines, cypress, and spruces). The seeds develop within the cones and are attached to a central axis. In the pine tree the developmental period may be as long as three years. Shortly after the seeds mature, the protective scales of the cone open up, and the seeds are released. **cône**

cone index The force per unit basal area required to push a *cone penetrometer* (i.e., penetrometer resistance) through a specified increment of soil. **indice de résistance à la pénétration de la pointe**

C

cone of depression (*hydrology*) The depression, roughly conical in shape, produced in a water table or other *piezometric surface* by the extraction of water from a well at a given rate. The volume of the cone will vary with the rate of withdrawal of water. It is caused by the extraction of water, by pumping or artesian flow, at a rate greater than that at which the *aquifer* is being recharged. Also called the cone of influence. **cône de dépression**

cone penetrometer An instrument in the form of a cylindrical rod with a cone-shaped tip designed to penetrate soil and measure the end-bearing component of penetration resistance. Penetration resistance developed by the cone equals the vertical force applied to the cone divided by its horizontally projected area. **pénétromètre à cône**

confidence coefficient (*statistics*) The probability statement that accompanies a confidence interval which expresses the probability of the truth of a statement that the interval will include the parameter value. A confidence coefficient of 0.10 implies that 90% of the intervals resulting from repeated sampling of a population will include the unknown (true) population parameter. **coefficient de confiance**

confidence interval (*statistics*) An interval defined by two values, called *confidence limits,* calculated from sample data using a procedure which ensures that the unknown true value of the quantity of interest falls between such calculated values in a specified percentage of samples. Commonly, the specified percentage is 95%; the resulting confidence interval is then called a 95% confidence interval. A one-sided confidence interval is defined by a single calculated value called an upper (or lower) confidence limit. **intervalle de confiance**

confidence limits (*statistics*) The endpoints of a confidence interval. See *confidence interval.* **limites de confiance**

confined ground water See *artesian water.* **nappe captive**

confining layer (*hydrology*) A body of impermeable or distinctly less permeable material located above and/or below an *aquifer.* Also called a confining bed. **couche encaissante**

confirmed test (*wastewater management*) The second stage in examining water for the presence of bacteria of fecal origin. Cultures that are positive on the first portion of the testing procedure (the presumptive test) are used to inoculate tubes of brilliant green lactose bile broth and examined for fermentation after incubation at 35°C for 48 hours. If fermentation is present, a third stage (the completed test) is performed. **épreuve de confirmation**

conformable (*geology*) The unbroken relationship between rock strata that lie one above another in parallel sequence. This geological arrangement demonstrates that there has been a lengthy period of deposition, undisturbed or uninterrupted by Earth movements or by denudation. Where such disturbance has ocurred, the sequence will be broken and is therefore said to be unconformable. See *unconformity.* **concordant**

conformity (*geology*) The relationship between adjacent sedimentary strata that have been deposited in parallel sequence with little or no evidence of time lapse. **conforme**

congelifluction The progressive flow of earth in conditions of permanently frozen subsoil. It differs from *gelifluction,* which also includes seasonally frozen ground. See *permafrost.* **gélifluxion**

congelifraction (*geomorphology*) Mechanical weathering by the freezing of interstitial water, leading to expansion, fracturing, and disintegration of rock material. **gélifraction, gélivation, cryoclastie**

conglomerate (*geology*) A coarse-grained *clastic* sedimentary rock composed of rounded to subangular fragments (e.g., granules, pebbles, cobbles, boulders) set in a fine-grained matrix of sand or silt, and commonly cemented by calcium carbonate, iron oxide, silica, or hardened clay. **conglomérat**

conidium (plural conidia) A thin-walled secondary asexual spore borne terminally on a specialized hypha called a conidiophore. See *figure*. **conidie**

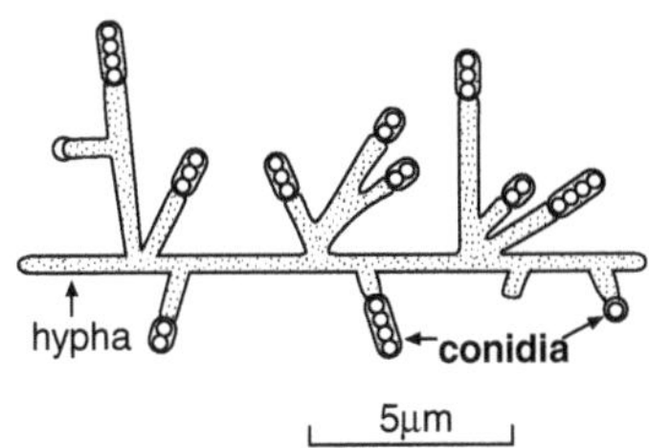

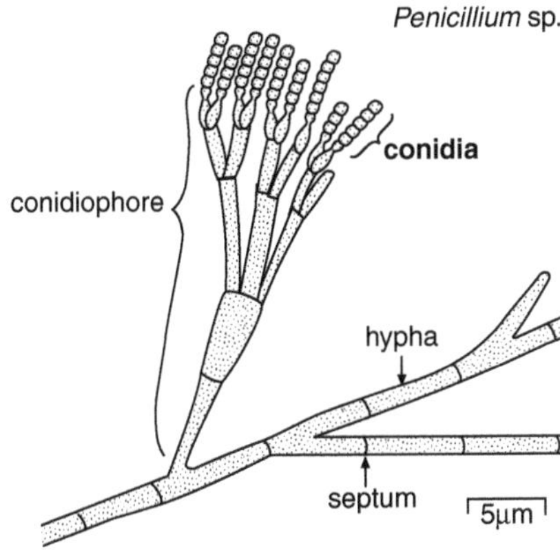

Conidia (adapted from Killham, 1994).

conifer A gymnosperm, member of the class Coniferae, having needlelike or scalelike leaves and naked seeds borne in cones (e.g., pines, firs, and spruces). **conifère**

coniferous Bearing cones, as in *conifer*. **coniférien**

coniferous forest A forest consisting predominantly of cone-bearing trees with needle-shaped leaves; usually evergreen, but some are deciduous, such as the larch forests *(Larix dehurica)* of central Siberia. Their greatest extent is in the wide belt across northern Canada and northern Eurasia. Coniferous forests produce soft wood, which has a large number of industrial applications including paper making. See *boreal forest*. **forêt coniférienne**

conjugate acid The species formed when a proton is added to a base. **acide conjugué**

conjugate acid-base pair Two species related to each other by the donating and accepting of a single proton. **paire acide-base conjuguée**

conjugate base What remains of an acid molecule after a proton is lost. **base conjuguée**

conjugated metabolites Metabolically produced compounds linked together by covalent binding (complex formation). **métabolites conjugués**

conjugation The act of joining together; in bacteria, a process in which genetic information is transferred from one cell to another. **conjugaison**

conjunctive water use The joining together of two sources of irrigation water (e.g., groundwater and surface water) to serve a particular piece of land. **utilisation combinée des eaux souterraines et de surface**

conservation (1) Careful, organized management, use, and protection of some natural resource, organism, or system to prevent its depletion, exploitation, waste, or destruction. (2) The planning, management, and implementation of an activity with the objective of protecting and preserving the essential physical, chemical, and biological characteristics of the environment against degradation. (3) The protection, improvement, and use of natural resources according to principles that will assure their highest economic or social benefits and prevent harm to the environment. **conservation**

conservation biology The discipline that treats the content of biodiversity, the natural processes that produce it, and the techniques used to sustain it in the face of human-caused environmental disturbance. **biologie de conservation**

conservation plan A collection of material, including maps (e.g., aerial photographs), containing land user information for making decisions regarding the conservation of soil, water, and related plant and animal resources for all or part of an operating unit. **plan ou programme de conservation**

conservation practice A technique or measure (e.g., cropping and/or tillage practice) used to meet a specific need in planning and carrying out soil and water conservation programs for which standards and specifications have been developed. **pratique de conservation**

conservation tillage See *tillage, conservation tillage*. **pratique de conservation**

consilience The "jumping together" of knowledge by linking facts and fact-based theory across disciplines to create a common groundwork of explanation. **connexité**

consistence (1) The resistance of a material to deformation or rupture. (2) The degree of cohesion or adhesion of the soil mass. Terms used for describing consistence depend on soil moisture content: wet soil (non-sticky, slightly sticky, sticky, and very sticky; non-plastic, slightly plastic, plastic, and very plastic); moist soil (loose, very friable, friable, firm, very firm, compact, very compact, and extremely compact); dry soil (loose, soft, slightly hard, hard, very hard, and extremely hard); cementation (weakly cemented, strongly cemented, and indurated). In engineering practice, consistency has the same meaning as consistence. **consistance**

consociation *(biogeography)* Any unit of vegetation dominated by a single species. See *association*. **consociation**

consolidate To become firm and hard (e.g., the hardening of loose, soft, or liquid earth materials). **consolider**

consolidation (*soil science*) The gradual reduction in volume of a soil mass as a result of increased compression or load. (*geology*) Any or all of the processes that cause loose or soft earth materials to become coherent or firm. In both uses it involves a decrease in *void ratio* by the squeezing of water from the pores. **consolidation**

consolidation test A test in which the soil specimen is laterally confined in a ring and is compressed between porous plates. **essai de compressibilité, essai oedométrique**

constant A value in an equation which does not vary. Thus, constancy refers to a state in which values do not change. **constante**

constant-charge surface A mineral surface carrying a net electrical charge whose magnitude depends only on the structure and chemical composition of the mineral itself. Constant-charge surfaces usually arise from *isomorphous substitution* in phyllosilicate clay structures. **surface à charge constante**

constant-potential surface A variable charge surface; at constant activity of the potential determining ion (e.g., constant pH) the electrical potential difference between the solid surface and the bulk solution is constant. See *pH-dependent charge*, and *constant-charge surface*. **surface à potentiel variable**

constitutive enzyme An enzyme whose formation does not depend on the presence of a specific substrate. **enzyme de constitution**

consumer (*ecology*) An organism that gains energy by eating another organism. The place for a consumer in the food chain is defined by what it eats. Herbivores eat plants, and are *primary consumers*; a human can be a primary consumer by eating plants or a secondary consumer by eating an animal that feeds on plants. **consommateur**

consumptive use (*irrigation*) The quantity of water used and transpired by vegetation plus the quantity evaporated. **évapotranspiration**

contact herbicide A *herbicide* that kills the foliage of weed plants on contact. See *systemic pesticide*. **herbicide de contact**

contact pesticide A *pesticide* that kills pests on contact with the plant, animal, or human body, rather than through *systemic* means or by ingestion (stomach poison). **pesticide de contact**

container seedling Seedling grown in a small container in a controlled environment. See *plug seedling* and *bare-root seedling*. **semis en récipient**

contaminate To introduce external microorganisms to a sample that is either sterile or under controlled growth conditions. **contaminer**

continent A large landmass rising more or less abruptly above the deep ocean floor, including marginal areas that are shallowly submerged. **continent**

continental crust That part of the Earth's crust composed of *sial,* which constitutes the continents. It averages 33 km in thickness and is over 50 km beneath high mountain regions. **croûte continentale**

continental drift The movement of the Earth's continents as the crustal plates on which they rest move across the surface of the semi-liquid mantle of the Earth. See *plate tectonics*. **dérive des continents**

continental glacier An ice sheet covering a large part of a continent, as in the Antarctic. Also called *icesheet*. **inlandsis, glacier continental**

continuous cropping Growing crops every growing season or maintaining continuous crop growth on the same piece of land. **culture continue, monoculture**

continuous permafrost A zone of permafrost that, for the most part, is uninterrupted by pockets or patches of unfrozen ground. **pergélisol continu**

continuous reaction series A reaction series in which early-formed crystals react with later liquids without abrupt phase changes. For example, the plagioclase feldspars form a continuous reaction series. **suite réactionnelle continue**

C

continuous stocking (*range management*) A method of grazing livestock on a given unit of land where animals have unrestricted and uninterrupted access throughout the time period when grazing is allowed. **pâturage continu, pâturage libre**

continuous variable A variable which, in principle, can take any value within its range, but in practice the values are limited by the accuracy of the measurement method (e.g., temperature, volume, mass of a soil sample). See *discrete variable*. **variable continue**

continuous-feed reactor (composting) An apparatus used to convert solid organic wastes into compost. The waste is moved through the reactor in a way that the conversion is done rapidly, with the solid refuse added to the front end of the reactor and humified materials removed from the back in a continuous fashion. See *batch method*. **digesteur à alimentation en continu**

continuous-flow microbiological system An operation in which liquid wastes or media are added to a decomposition or growth vessel to allow for the growth and metabolism of bacteria, and the products are removed continuously. **système microbiologique à régime continu**

continuous-flow system (*modeling*) A system having an uninterrupted, time-varying input and output of material in one of two ways: (a) the plug-flow model assumes that the effluent from the system is discharged in the same order as the material entering the

C

system, with little mixing; and (b) the completely mixed-flow model assumes that material flowing through the system, is uniformly mixed while in the system, and thus the discharge is physically and chemically the same as the system contents. **système à écoulement continu**

continuum A continuously changing pattern of species abundance. **continuum**

contorted drift (*geology*) The crumpled and folded appearance of glacial till where it has been deformed by an overriding icesheet or by the pressure of an ice front. See *push moraine*. **moraine de poussée**

contour A fixed level or elevation across a slope. **courbe de niveau, élévation**

contour cultivation Cultivation or tillage along or nearly parallel to the contour of the land, rather than up and down the slope, to prevent erosive flow of water along the tillage lines. Used to facilitate both soil and fuel conservation. **culture en courbes de niveau**

contour ditch An irrigation ditch that follows the elevation of the land. **rigole de niveau**

contour flooding A method of irrigating by flooding from contour ditches. **inondation en contour**

contour furrows Furrows plowed approximately on the contour of the land, especially on pasture and rangeland, to prevent water runoff and increase infiltration; often used in irrigation schemes. **canaux de dérivation**

contour interval The vertical distance between contour lines. **intervalle d'élévation, équidistance**

contour line A line connecting points of equal elevation on the land surface. Also called *contour*. **courbe de niveau, isohypse**

contour map A topographic map that portrays relief by means of lines which connect points of equal elevation. **carte topographique**

contour mining A type of surface mining of coal in which coal is exposed and removed from relatively shallow deposits by removing the earth and rock covering the deposit. The mining proceeds around the natural topographical features of a hill or mountain. See *surface mining*. **exploitation suivant les courbes de niveau**

contour strip cropping Layout of crops in comparatively narrow strips in which the farming operations are performed approximately on the contour. See *strip cropping*. **culture en bandes suivant les courbes de niveau**

contour stripping In coal mining, the removal of overburden and mining from a seam that outcrops or approaches the surface at approximately the same elevation, in steep or mountainous areas. **décapage suivant les courbes de niveau**

contour surface mining See *mining*. **mine à ciel ouvert exploitée suivant les courbes de niveau**

contour tillage See *tillage, contour tillage*. **travail du sol suivant les courbes de niveau**

contraction Linear strain associated with a decrease in length. **contraction**

control (1) Something under study, either untreated or given a standard treatment, which is used as a standard for comparison against the results of other treatments. (2) A collective term for a system of marks or objects on the Earth, a map, or a photograph, whose positions or elevations, or both, have been or will be determined. **(1) témoin (2) réseau de points de contrôle**

control group In experimentation for the testing of a new method, process, or factor against an accepted standard, the part of the test which involves that standard for comparison is known as the control, and the set of individuals is the control group. **groupe témoin, groupe contrôle**

control point (*remote sensing*) A reference point precisely located on a photograph and on the ground; used in assembling photographs for map compilation. **point de référence**

control section, soil The vertical section upon which the taxonomic classification of soil is based. The control section usually extends to a depth of 100 cm in mineral materials and to 160 cm in organic materials (Canadian system of soil classification). **coupe témoin d'un sol, profil témoin**

control system A *process-response system* in which key components are controlled by some intelligence, thereby causing the system to function in a manner controlled or determined by that intelligence (e.g., the introduction of a river management scheme in a drainage basin in which runoff and discharge may be controlled). See *systems analysis*. **système de contrôle**

controlled drainage Regulation of the water table to maintain water level at a depth favorable for optimum crop growth. **drainage contrôlé**

controlled traffic See *tillage, controlled traffic*. **circulation limitée**

controlling variable Under a specific set of conditions, the most important factor that determines the response of a process or experiment (e.g., where nitrogen is limiting for crop growth, it becomes the controlling variable). See *limiting factor*. **facteur limitatif**

convection A process by which heat, solutes, or particles are transferred from one part of a fluid to another by movement of the fluid; also called advection. **convection**

conventional tillage See *tillage, conventional tillage*. **travail du sol conventionnel**

convexity A slope in which the gradient becomes progressively steeper as it is traced downwards from the top. **convexité**

cool-season plant A plant that generally grows most during the late fall, winter, and early spring. Cool-season species generally exhibit the C3 photosynthetic pathway. **plante de climat frais**

coordinate Any of the magnitudes that serve to define the position of a point, line, or plane by reference to a fixed figure or system of lines. **coordonnées**

coordinate bond A chemical bond between two atoms, or an atom and a group, which share two electrons that are both supplied by one of the atoms or groups. See *atomic bond*. **liaison de coordination**

coppice (coppicing) The tendency of certain tree and brush species (e.g., red alder and bigleaf maple) to produce a large number of shoots when a single or few stems are mechanically removed but the root system is left intact. **taillis**

coppice mound A small mound of stabilized soil material around desert shrubs. See *microrelief*. **tertre avec taillis**

coprogenic Sediment originating as excrement. **coprogène**

coprogenous earth A material in some organic soils that contains at least 50% by volume of fecal pellets less than 0.5 mm in diameter. **terre coprogène**

coprolite Fecal pellet or casting; often composed mostly of calcium phosphate. **coprolithe, coprolite**

coprophagous (*zoology*) Describing organisms other than bacteria that feed on fecal matter, such as certain insects. **coprophage**

cordillera A group of mountain ranges (together with their associated valleys, basins, plains, plateaus, rivers, and lakes), the component parts having various trends but the mass itself having one general direction. **cordillère**

core (1) The central part of the Earth, probably consisting of a dense nickel-iron alloy with a temperature estimated at about 2700 K. It begins at a depth of

C

about 2900 km and is divisible into an outer core that may be liquid, and an inner core about 1300 km in radius that may be solid. (2) A cylindrical section of rock, soil, or sediment, usually 5 to 10 cm in diameter and up to several meters in length, taken for examination and analysis. (3) A mass of impervious material (e.g., clay) forming the central part of an embankment, dike, or dam. **(1, 3) noyau (2) carotte de sondage**

core drilling The process by which a cylindrical sample of soil or rock is obtained through the use of a hollow drilling bit that cuts and retains a section of the penetated strata. **carottage**

core sample A relatively undisturbed sample of soil, peat, rock, or ice that is withdrawn from the solid material under examination. The samples are obtained by driving a hollow metal tube or cylinder into the material and withdrawing it in the form of a lengthy tubular section or core. **échantillon non dérangé, carotte de sondage**

core trench Excavation for a core wall in the construction of an earth embankment. **tranchée pour mur écran**

Coriolis force The apparent force caused by the Earth's rotation that causes moving objects to turn to the right in the northern hemisphere and to the left in the southern hemisphere. **force de Coriolis**

corm In plants, an underground storage organ formed from a swollen stem base, bearing adventitious roots and scale leaves. The top of the corm has one or more growing points, or eyes; roots grow from a basal plate on the underside of the corm. **cormus**

corrasion (*geology*) A process of mechanical erosion of rock surfaces and soil by the abrasive action (cutting, scraping, scratching, and scouring) of solid materials moved along by wind, waves, running water, glaciers, or gravity movement. The resulting effect on a rock surface is called *abrasion*. **corrasion**

correlation (*statistics*) See *correlation, statistical*. (*surveying*) The removal of discrepancies among survey data so that all parts are interrelated without apparent error. (*soil survey*) See *soil correlation*. (*geology*) Demonstration of the equivalence of two or more geologic phenomena in different areas; it may be lithologic or chronologic. (*seismology*) Identification of a phase of a seismic record as representing the same phase on another record. **corrélation**

correlation, statistical The degree of relationship which exists between two or more variables such that when one changes, the other also changes. **corrélation statistique**

Types of statistical correlations include:

autocorrelation – a relationship between members of a series of samples ordered in space and in time, which is due to dependence between them. ***autocorrélation***

coefficient of correlation (r) – an index of interdependence between two variables. ***coefficient de corrélation***

coefficient of determination (r^2) – the square of the coefficient of correlation, expressing the proportion of variation in the dependent variable explained by the association with the independent variable. ***coefficient de détermination***

coefficient of multiple correlation (R) – an index or measure similar to *r* but giving a precise value to the relationship involving more than one independent variable. ***coefficient de corrélation multiple***

coefficient of multiple determination (R^2) – the square of the coefficient of multiple correlation, expressing the proportion of variation in the dependent variable explained by the joint association with the independent variables. ***coefficient de détermination multiple***

multiple correlation – the correlation between two or more independent variables and one dependent variable. ***corrélation multiple***

negative correlation – the relationship in which one variable increases as the other decreases. ***corrélation négative***

partial correlation – the correlation between two variables when a third, which is related to both, is controlled. ***corrélation partielle***

product-moment correlation – a parametric test for correlation between two variables on a ratio or interval scale. ***corrélation du moment des produits***

Spearman rank correlation – a non-parametric test for correlation between two variables expressed on a rank or ordinal scale. ***coefficient de corrélation de rang de Spearman***

corridor A band of vegetation, usually older forest, which serves to connect distinct patches on the landscape, thus permitting the movement of plant and animal species between what would otherwise be isolated patches. **corridor**

corrosion (1) The *chemical weathering* of rocks by chemical processes (e.g., solution, hydrolysis, hydration, oxidation); commonly used to describe the single chemical process of solution and electrochemical processes (2) The process by which metals are oxidized in the atmosphere. **corrosion**

cortex The outer primary tissue of a stem or root bounded externally by the epidermis and internally by the vascular system. **cortex**

cote (*geomorphology*) A prominent dissected escarpment forming the edge of a plateau. Also used in reference to hills or a hilly upland, an elevated rough plain, or a low ridge within a swampy area. **côte**

cotyledons The first leaves to develop in a seed. In some plants they remain within the seed; in others they emerge from the seed upon germination. **cotylédons**

coulee (*geomorphology*) (1) A glacial *meltwater channel* that is dry or has an intermittent stream. (2) A tongue-like mass of debris moved by *solifluction*. **coulée**

C

counteradaptation The evolution of characteristics of two or more species to their mutual disadvantage. **contradaptation**

counterevolution The development of traits in a population in response to exploitation, competition, or other detrimental interaction with another population. **contrévolution**

counting chamber (*microbiology*) A device used to facilitate the counting of cells by microscopic techniques. **cellule de comptage**

country-rock A general term for any type of rock penetrated by an igneous intrusion or invaded by a mineral vein. **roche encaissante**

covalent bond See *chemical bond*. **liaison covalente**

covariance (*statistics*) The relationship between two variables. The covariance of two independent random variables is zero, but a zero covariance does not imply independence. Covariance is also called the "first product-moment," and is defined for a set of paired data as the mean of the products obtained by pairwise multiplying the deviations from the respective means. **covariance**

cover crop A crop used primarily for the purpose of protecting and improving the soil between periods of regular crop production, or between rows of permanent standing crops (e.g., orchards and vineyards). **culture de couverture**

cover material Soil placed over solid waste at a landfill at the end of each day to control odors, flies, rodents, mosquitoes, blowing litter, and fire. **matériel de recouvrement**

cover soil (*land reclamation*) Unconsolidated materials including salvaged surface soil, salvaged regolith, or selected bedrock spoil used to top-dress spoils to build a better quality minesoil. See *topsoil.* **terre de recouvrement**

cover, percent The soil area covered by the aerial parts of plants and mulch; expressed as a percent of the total area. **pourcentage de couverture ou de recouvrement**

cracking (*chemistry)* The utilization of heat and/or a catalyst to reduce high-molecular weight hydrocarbons in crude oil to smaller molecules, which are the constituents of products such as gasoline, ethylene, or heating oil. **cracking, craquage**

cradle knoll A small knoll formed by earth that was raised and left by an uprooted tree. See *microrelief.* **butte de chablis**

crater (1) A basin-like, rimmed depression surrounding the vent at the summit or on the flanks of a volcano, formed a major eruption or by collapse of a volcanic cone. (2) A saucer-shaped pit or depression on the Earth's surface (also on the moon and planets) resulting from impact or explosion. **cratère**

creep Slow mass movement of soil and soil material down steep slopes primarily under the influence of gravity, but aided by saturation with water and alternate freezing and thawing. In engineering usage, creep is any general, slow displacement under load. See *mass movement.* **reptation**

crest The slope component commonly at the top of an erosional ridge, hill, or mountain. See *slope morphological classification.* **crête**

crevasse (1) A deep break or fissure in the Earth. (2) A crack or breach in the bank of a river. (3) A deep fissure or crack in a glacier, caused by differential movement within the ice resulting from shear stresses. **crevasse**

crevasse filling (*geomorphology*) A short, linear ridge or hummock of glacial sediments deposited by water in the cracks and crevasses of a stagnating icesheet. **remplissage de crevasse**

crib dam (*irrigation*) A barrier of timber built across a stream channel to form bays or cells that are filled with stone or heavy material. **barrage en encoffrement**

critical angle (*geomorphology*) The largest angle at which a slope remains stable. (*physics*) The smallest angle of incidence at which there is total reflection when an optic, acoustic, or electromagnetic wave passes from one medium to another less refractive medium. **angle critique**

critical area An area that, due to its size, location, condition, or value must be treated with special consideration because of inherent site factors or difficulty of management (e.g., a severely eroded, sediment-producing area; a polluted area that requires special management). **site fragile**

critical density The mass per unit volume of a saturated granular material below which it will lose strength, and above which it will gain strength when subjected to rapid deformation. **masse volumique critique**

critical depth (*hydrology*) Depth of flow in a channel of specified dimensions at which specific energy is a minimum for a given discharge. **profondeur critique**

critical load The maximum input or concentration of a pollutant or pollutants that will not cause changes leading to long-term harmful effects on the most sensitive components of ecosystems, according to present knowledge. Also called *threshold value* or critical concentration. **charge critique**

critical nutrient concentration The nutrient concentration in the plant, or specified plant part, below which the nutrient becomes deficient for optimum growth rate. **seuil critique**

critical path analysis (*geomorphology*) A type of geomorphological analysis that establishes the number of sequential steps that have occurred in formation of the present-day landforms. (*project management*) A system in which all aspects of the project are depicted in a sequence, and the data then translated into a schedule. **méthode du chemin critique**

critical pressure The pressure required to condense a gas at the critical temperature, above which, regardless of pressure, the gas cannot be liquefied. **pression critique**

critical velocity (*hydrology*) Velocity at which a given discharge changes from tranquil to rapid flow. **vitesse critique**

critical void ratio The *void ratio* corresponding to the *critical density*. **indice des vides critique**

critical wildlife habitat Part or all of a specific place occupied by a wildlife species or a population of such species and recognized as being essential for maintenance of the population. **habitat faunique critique**

crop land Land used for cultivated field crops, fruits, vegetables, and other crops, either alone or in rotation with crops or grassland. See *crop rotation*. **terre arable, terre en culture**

crop nutrient requirement The amount of nutrients needed to grow a specified yield of a crop plant per unit area. **besoin en éléments nutritifs**

crop residue Plant material (i.e., portion of plant or crop) remaining in the field after harvest (e.g., leaves, straw, seeds, stalks, and roots). See *mulch*. **résidus de cultures**

crop residue management See *tillage, crop residue management*. **gestion des résidus de cultures**

crop rotation The practice of growing different crops in a planned regular sequence or succession on the same land. Usually established for economic considerations, especially to aid in the control of insects and diseases, maintain soil fertility, and decrease soil erosion. **rotation culturale ou des cultures**

cropping intensity The share of farmland devoted to crops or cultivation. The portion of total crop land occupied by crops or a specific crop at any one time, or over a specific time period. **assolement**

cropping pattern The yearly sequence and spatial arrangement of crops or crops and fallow in a given area. **système de culture**

cropping system A system comprised of soil, crop, weeds, pathogen, and insect subsystems that transforms solar energy, water, nutrients, labor, and other inputs into food, feed, fuel, or fiber. The cropping system is a subsystem of a *farming system*. **système de culture**

cross cultivation See *tillage, cross cultivation*. **travail du sol en effectuant des passages croisés**

cross-section (1) A diagram showing the features transected by a vertical plane, such as a vertical section through a landform, ore body, fossil, plant, or other object. (2) An exposure or cut that shows transected geologic features. **(1) section transversale (2) coupe transversale**

cross-bedding An arrangement in which thin layers of stratified sediment are transverse or oblique to the main plane of stratification. **lits entrecroisés**

crossed nicols Two Nicol prisms or Polaroid plates oriented in a polarizing microscope such that the transmission planes of polarized light are at right angles. Light transmitted from one is intersected by the other unless there is an intervening substance. **nicols croisés, lumière polarisée analysée**

crosslinking (*chemistry*) The existence of bonds between adjacent chains in a *polymer*, thus adding strength to the material. **liaison transversale**

cross-pollination The transfer of pollen from the *anther* of the *stamen* of one

C

plant to the *stigma* of the *pistil* of another plant. **pollinisation croisée**

cross-slope farming Conducting field operations across the predominant slope rather than up- and down-slope, usually to prevent soil *erosion.* See *contour cultivation.* **culture en contre-pente**

cross-stratification See *cross-bedding.* **stratification entrecroisée**

crotovina A former animal burrow in one soil horizon that has become filled with organic matter or material from another horizon. Also spelled krotovina. **crotovina**

crown (*forestry*) The upper parts of a tree or other woody plant carrying the main branch system and foliage above a clean stem(s). **cime**

crown closure (*forestry*) The condition when the crowns of trees touch and effectively block sunlight from reaching the forest floor. **fermeture du couvert**

crucible furnace A furnace that raises the temperature of a small porcelain container (crucible), reducing the contents to *ash.* **fournaise à creusets**

crumb structure A soil structural condition in which soil particles or peds are in the form of crumbs. **structure grumeleuse ou granuleuse**

crushing See *tillage, crushing.* **écrasement, fragmentation**

crushing strength The force required to crush a mass of dry soil or, conversely, the resistance of a mass of dry soil to crushing, expressed in units of force per unit area (i.e., pressure). **résistance à l'écrasement**

crust (*agronomy*) (1) A surface layer of soil, from a few mm to several cm thick, that when wet forms a dispersed seal or barrier to water and air infiltration. On drying, the crust is much more compact, hard, and brittle than the soil material just beneath it. Crusts are generally formed by the breakdown, slaking, and resorting of *aggregates* by raindrops. (*geology*) (2) The outermost layer of the Earth, varying between 6 and 48 km in thickness, comprising all the material above the *Mohorovicic discontinuity.* It makes up less than 0.1% of the Earth's total volume. (*geology*) (3) A laminated, crinkled algal deposit, slightly arched to bulbous, formed on rocks or fossils by accretion or flocculation. **(1, 3) croûte, (2) écorce terrestre**

Cryands *Andisols* that have a *pergelic* soil temperature regime. A suborder in the U.S. system of soil taxonomy. **Cryands**

Cryerts A suborder in the U.S. system of soil taxonomy. Vertisols that have a cryic or pergelic soil temperature regime. **Cryerts**

cryic A soil temperature regime that has mean annual soil temperatures of >0°C but <8°C, >5°C difference between mean summer and mean winter soil temperatures at 50 cm, and cold summer temperatures. **cryique**

Cryids A suborder in the U.S. system of soil taxonomy. Aridisols that have a cryic soil temperature regime. **Cryids**

Cryods A suborder in the U.S. system of soil taxonomy. Spodosols that have a cryic or pergelic soil temperature regime. **Cryods**

cryogenic soil Soil that has formed under the influence of freezing soil temperatures. **sol cryogénique**

cryology The study of the properties of snow, ice, and frozen ground. **cryologie**

cryomorphology The part of *geomorphology* pertaining to the various processes and products of cold climates. **cryomorphologie**

cryopedology The study of the processes of intensive frost action and the occurrence of frozen ground, particularly permafrost, including the civil engineering methods used to overcome or minimize the difficulties involved. **cryopédologie**

cryophilous *(biogeography)* Relating to plant and animal responses to very low temperatures. See *cryophyte.* **cryophyle**

cryophyte A plant adapted to live in an environment of permanent ice or snow. See *cryophilous.* **cryophyte**

cryoplanation The slow denudation and reduction of a land surface by processes associated with intensive frost action supplemented by the actions of running water, moving ice, and other agents. **cryoplanation, géliplanation**

cryoplankton Microscopic plant and animal organisms which inhabit regions permanently blanketed by ice and snow. **cryoplancton**

Cryosolic An order in the Canadian system of soil classification consisting of mineral or organic soils that have perennially frozen material within 1 m of the surface in some part of the soil body, or pedon. The mean annual soil temperature is less than 0°C. They are the dominant soils of the zone of continuous permafrost and become less widespread to the south in the zone of discontinuous *permafrost*; their maximum development occurs in organic and poorly drained, fine textured materials. Vegetation associated with Cryosolic soils varies from sparse plant cover in the high arctic, through tundra, to subarctic and northern boreal forests. The active layer of these soils is frequently saturated with water, especially near the frozen layers, and colors associated with *gleyzation* are therefore common in mineral soils, even those that occur on well drained portions of the landscape. They may or may not be markedly affected by *cryoturbation.* The order has three great groups: *Turbic Cryosol*, comprising mineral soils that display marked cryoturbation and generally occur on patterned ground, *Static Cryosol*, mineral soils without marked *cryoturbation,* and *Organic Cryosols*, formed in organic materials. **cryosolique**

cryosphere The part of the Earth's surface that is perennially frozen. **cryosphère**

cryoturbation Frost action that causes churning, heaving, and considerable structural modification of the soil and subsoil. **cryoturbation**

cryovegetation Plant communities, especially algae, mosses, and lichens that have adapted to environments of permanent snow and ice. **cryovégétation**

cryptocrystalline A rock texture characterized by crystals so small that their features can only be seen with the aid of a powerful (i.e., electron) microscope; also describes a rock with such a texture. **cryptocristallin**

cryptogam A plant that reproduces by spores and not by flowers or seeds (e.g., mosses, lichens, and ferns). **cryptogame**

crystal Any homogeneous solid that has a regularly repeating atomic arrangement and is bounded by plane surfaces that form definite angles with each other to give the substance a regular geometrical form. It can be a chemical element, a compound, or an isomorphous mixture. **cristal**

crystal chemistry The study of the relationship among chemical composition, internal structure, and physical properties of crystalline matter. **chimie des cristaux**

crystal lattice See *lattice structure.* **réseau cristallin**

crystal morphology See *euhedral, subhedral, anhedral.* **morphologie des cristaux**

crystal structure The orderly and repeated arrangement of atoms in a crystal, the translational properties of which are described by the crystal lattice or space lattice. **structure des cristaux**

crystalline Of or pertaining to the nature of a crystal; having regular molecular structure. See *amorphous mineral.* **cristallin**

crystalline rock (1) A general term used in geology to distinguish igneous and metamorphic rock from sedimentary rock. (2) Rock consisting of minerals in a crystalline state. **roche cristalline**

crystalline solid A solid with a regular arrangement of its molecular structure. **solide cristallin**

crystallization The process by which matter becomes crystalline from a gaseous, fluid, or dispersed state. **cristallisation**

crystallography The scientific study of crystals including their growth, structure, physical properties, and classification by form. **cristallographie**

culm The jointed and usually hollow stem of various grasses and sedges. **tige**

cultipack See *tillage, cultipack.* **cultitassement, raffermissement du lit de semence**

cultivar (1) A variety, strain, or race that has originated and persisted under cultivation or was specifically developed for the purpose of cultivation. (2) For cultivated plants, the equivalent of botanical variety, in accordance with the International Code of Nomenclature of Cultivated Plants, 1980. **cultivar**

cultivated land Land tilled and used to produce crops; includes land left fallow. **terre cultivée**

cultivation See *tillage, cultivation.* **hersage, travail de sol**

cultural eutrophication (*limnology*) Acceleration of the natural process of enrichment (i.e., aging) of bodies of water by human activities. **eutrophisation**

cultural vegetation Any type of vegetation influenced by human activities either directly or indirectly. **plante cultivée**

culture (*microbiology*) A population of microorganisms growing in an artificial media. A pure culture is grown from a single cell; a mixed culture consists of more than one microorganism growing together. **culture**

culture dish A shallow device for the cultivation of microorganisms on a solid nutrient medium (*agar*) prepared in the laboratory. The most common device is a petri dish measuring 100 by 15 mm. **boîte de Pétri**

culture media Solid or liquid substances prepared and sterilized in the laboratory to provide the nutrients needed for the growth of bacteria or fungi. **milieu de culture**

cumulative distribution *(statistics)* A distribution for a data set which shows how many items are "less than" or "more than" given values. **distribution cumulée ou cumulative**

cumulative effects Effects on biota, or on the physical environment of biota, of stress imposed by more than one mechanism (e.g., stress in fish imposed by elevated suspended sediment concentrations in the water and high water temperature). See *additive effects.* **effets cumulatifs**

cumulative infiltration Total volume of water infiltrated per unit area of soil surface during a specified time period. See *infiltration rate.* **infiltration cumulée ou cumulative**

cumulization A process of soil formation whereby mineral particles are added to the surface of a soil solum by eolian, hydrologic, or human agents. **accumulation**

cumulose deposits Superficial deposits composed largely of organic materials (e.g., *peat*). **dépôts cumuliques**

cupriferous Copper-bearing. **cuprifère**

current meter (*hydrology*) Instrument used for measuring the velocity of flowing water. **moulinet**

curve fitting *(statistics)* The fitting of a mathematical curve to any statistical data that can be plotted against a space or time variable. **ajustement d'une courbe**

cut and fill (1) (*geology*) A process in which material eroded from one place by waves or streams is deposited nearby until the surfaces of erosion and deposition are continuous. (2) (*engineering*) The excavation of earth material from one place and its deposition as compacted fill in an adjacent place, as in road-building. (3) (*land reclamation*) The process of earthmoving by excavating part of an area and using the excavated material for adjacent embankments or fill areas. **(1) cut and fill (2, 3) déblai-remblai**

cutan (*soil micromorphology*) A modification of the texture, structure, or fabric at natural surfaces in soil materials due to concentration of particular soil constituents or *in situ* modification of the matrix. Cutans may be composed of any of the component substances of the soil material. See *alban, argillan, ferran, ferri-argillan, gypsan, mangan, matran, organan*. **cutane, revêtement, patine**

cuticle (1) The waxy layer on the surface of leaves that helps prevent desiccation. (2) The non-cellular outer layer of the body wall of an insect or other arthropod. **cuticule**

cutin Fatty substance deposited in many plant cell walls and on outer surface of epidermal cell walls, where it forms a layer known as the *cuticle*. **cutine**

cut-over forest A forest in which most or all of the merchantable timber has been cut. **friche**

cutting See *tillage, cutting*. **découpage**

cybernetic systems (*modeling*) Systems subject to regulation by feedback mechanisms due to interactions among system components, as characterized by non-linear relationships between causes and effects. **systèmes cybernétiques**

cybernetics The study of guidance and control problems in biological or mechanistic systems. See *general systems theory, system*. **cybernétique**

cycle (1) A sequence of events which returns to its starting point (e.g., hydrologic cycle). (2) A sequence or succession of events that runs to completion with the last stage different from the first (e.g., cycle of erosion). **cycle**

cycle of erosion (*geomorphology*) A concept describing the evolution and modification of the physical landscape, in which the various stages of erosion are considered parts of a cycle that follows an orderly sequence. **cycle d'érosion**

cyclic salt Salt derived from the sea or salt lakes, deposited on the landscape from wind or rainfall. **embrun salé**

cyclotron A type of particle accelerator in which an ion introduced at the center is accelerated in an expanding spiral path by use of alternating electrical fields in the presence of a magnetic field. **cyclotron**

cytochrome Iron-containing chemical species, composed of heme and a protein, that serve as the principal electron carriers utilized in respiratory processes. **cytochrome**

cytokinin Substituted adenines that are growth regulators in plants. **cytokinine**

cytoplasm The protoplasm of a cell exclusive of the nucleus. **cytoplasme**

D

D

Dalton's law For a mixture of gases in a container, the total pressure exerted is the sum of the pressures that each gas would exert if it were alone. See *partial pressure*. **loi de Dalton**

dam Any artificial barrier which impounds or diverts water. A dam is generally considered hydrologically significant if it is 0.4 to 2.0 m or more in height from the natural bed of the moving body of water (e.g., stream). **barrage**

damping The process by which an effect is diminished in intensity; occasionally used to describe the process of self-regulation or *negative feedback*. **amortissement**

Darcy's law A law describing the rate of flow of water through porous media. Water flow, or the Darcy velocity (v), through a permeable medium depends on the conductivity of the medium (K, hydraulic conductivity) and the driving force of the water (i, hydraulic potential gradient). The formula is: $v = -Ki$. It is named for Henri Darcy of Paris, who formulated it in 1856 from extensive work on the flow of water through sand filter beds. **loi de Darcy**

Dark Brown Chernozem A great group of soils in the *Chernozemic* order (Canadian system of soil classification). The soils occur in the cool to cold semiarid grassland regions, and have a dark brown surface (Ah or Ap) horizon on a lighter-colored brownish B (Bm, Btj, or Bt) horizon, which may be absent, over a highly base-saturated, usually calcareous, C horizon. **chernozem brun foncé**

Dark Gray Chernozem A great group of soils in the *Chernozemic* order (Canadian system of soil classification). The soils occur in the cool to cold subhumid grassland/forest transitional regions, and have a dark gray, partially eluviated surface (Ahe or Ap) horizon and a brownish B (Bm, Btj, or Bt) horizon, which may be absent, over a highly base-saturated, usually calcareous C horizon. **chernozem gris foncé**

datum A geographical or numerical quantity or fact which serves as a base or reference point. It is the starting point in any type of measurement or reasoning. **donnée**

datum elevation A level reference elevation used in mapping. **plan horizontal de référence**

datum level The zero level (usually sea level or a mean based on tidal levels) from which land elevations and ocean depths are measured. **niveau de référence**

DDT (dichloro-diphenyl-trichloro-ethane) 1,l,l-tricholoro-2,2-bis (4-chloriphenyl) ethane, the first modern chlorinated hydrocarbon insecticide. DDT has a *half-life* of 15 years and persists in the environment; it accumulates in fatty tissues and in food chains, and is now banned from use in many countries. **DDT (dichloro-diphényl-trichloréthane)**

dead furrow See *tillage, dead furrow*. **ados, sillon terminal, raie de curage**

dead-ice features (*geomorphology*) A hummocky terrain consisting of glaciofluvial sediments and ablation till, produced by the melting of a stagnant icesheet or glacier (dead ice) *in situ*. See *kame*. **modelé de glace morte**

dealkalization The leaching of sodium ions and salts from *natric horizons*. **déalcalinisation**

deamination The removal of an amino group ($-NH_2$), freeing ammonia. **désamination**

death phase (*microbiology*) The terminal stage of growth of bacteria. After the bacteria have exhausted the supply of available nutrients, growth (cell division) stops. When the cells can no longer maintain viability, cell death results and the number of viable cells decreases. See *growth phase*. **phase de déclin**

debris The loose material arising from disintegration of rocks and vegetative material, transportable by streams, ice, or floods. **débris**

debris basin A detention structure to retain soil materials and detritus transported by runoff, or to collect solids from livestock feedlot runoff. **bassin de décantation ou de déjection.**

debris cone (*geology*) An alluvial fan with steep slopes, generally composed of coarse fragments. Also called alluvial cone or *debris fan*. **cône de déjection**

debris dam (*irrigation*) A barrier built across a stream channel to retain rock, sand, gravel, silt, or other material. **barrage de retenue des débris**

debris fall (*geology*) The relatively free collapse of weathered mineral and rock material from a steep slope or cliff; especially common along the undercut banks of streams. **éboulis**

debris fan See *debris cone*. **cône de déjection, cône de débris**

debris flow A moving mass of rock fragments, soil, and mud, with more than half of the particles larger than sand size. Movement can be slow (less than 1 m per year) to rapid (up to 160 km per hour). **coulée de débris**

decalcification The removal of calcium ions from a soil by the process of *leaching*, in which calcium bicarbonate is carried away in solution. Decalcification proceeds downward from the surface and is thought to be an initial stage of *podzolization* and clay translocation. **décalcification**

decay product An element or isotope resulting from *radioactive decay*; it may be stable or radioactive and therefore undergo further decay. **produit de désintégration**

deciduous forest, temperate See *temperate deciduous forest*. **forêt tempérée à feuilles caduques**

deciduous plant A plant that sheds its leaves at the end of a defined growing season or during a period of temperature or moisture stress. In the temperate zone, leaf fall occurs during autumn, when broadleaf and some *coniferous* trees (e.g., larch) shed their leaves (or needles). In tropical forests, leaf fall may be at any time, since there are few restrictions on the growing season. **plante à feuilles caduques**

decision analysis model An organized system that policy makers and managers can use to select a course of action; often, but not necessarily, a formal model. **modèle d'analyse de décision**

decision region (*statistics*) A region in the measurement space corresponding to a specific class, defined by means of discriminant functions and used to classify data vectors of unknown class association. **région décisionnelle**

decision rule (*statistics*) The criterion used to establish discriminant functions for classification (e.g., nearest-neighbor rule, minimum-distance-to-means rule, maximum-likelihood rule). Also called classification rule. **critère de décision**

decision tree A logical branching arrangement of decision processes involving class features and decision rules, that allows class-membership decisions to be made in sequential steps and among relatively fewer alternatives at each node of the tree. **arbre de décision**

declination (magnetic) The angle between true (geographic) north and magnetic north (direction of the compass needle). Declination varies from place to place, and can be set on a compass for a particular location; also called deviation. **déclinaison magnétique**

decomposer A heterotrophic organism, chiefly a microorganism, that breaks down the bodies of dead animals or parts of dead plants and absorbs some of the decomposition products while releasing similar compounds usable by producers. **décomposeur**

decomposition The breakdown of a complex material into simpler materials. The complex material can be organic or inorganic, and heat, sunlight, water, chemicals, or metabolism can cause the decomposition. Metabolic decomposition is carried out by *decomposer* organisms. **décomposition**

decreaser (*range management*) For a given plant community, a species that decreases in number as a result of a specific abiotic/biotic influence or management. **espèce en évolution régressive**

deduction Reasoning from the general to the particular; inference of consequences from evidence; derivation of applications from general principles. See *induction.* **déduction**

deep ecology An ecological perspective of environmental problems that emphasizes the interrelatedness of the Earth and its biota, the equal importance of all species, and the need for radical social, economic, and political reforms to maintain and improve the quality of the environment. **écologisme radical**

deep percolation Water that percolates below the root zone and cannot be used by plants. **percolation profonde**

deep weathering (*geology*) A type of *chemical weathering* of rocks (controlled by temperature, precipitation, rock type, and vegetation cover) that extends to great depths (30 m) below the surface in the humid tropics, where the process reaches its maximum efficiency. **altération profonde**

deficiency symptom Discernible evidence of abnormal growth resulting from a nutrient shortage. The most common plant nutrient deficiency symptoms are stunted growth and *chlorosis.* **symptôme de carence**

deflation The removal of material from a beach, desert, or other land surface by wind action**. déflation**

deflation basin or hollow A large-scale basin or depression formed by the action of the wind (i.e., *deflation*) in arid or semi-arid lands. See *eolian.* **creux de déflation**

deflocculate (1) To separate the individual components of compound particles by chemical or physical means or both. (2) To cause the particles of the dispersed phase of a colloidal system to become suspended in the dispersion medium. **défloculer**

deforestation The removal of trees and other vegetation on a large scale, usually to expand agricultural or grazing lands. Deforestation is often cited as one of the major causes of the *enhanced greenhouse effect* because burning or decomposition of the wood releases carbon dioxide, and trees that once removed carbon dioxide from the *atmosphere* through *photosynthesis* are no longer present and contributing to carbon storage. **déboisement**

deformation Any change in the original volume or fabric of a material. (*geology*) A change in the original form or volume of rock masses produced by tectonic forces. **déformation**

degradation (1) (*biology, ecology*) The diminution of biological productivity or diversity. (2) (*soil science*) The process whereby a compound is transformed into simpler compounds. (3) (*geomorphology*) The gradual lowering of a land surface by *denudation,* in which erosive forces (e.g., rivers, ice) erode material, transport it, and deposit it elsewhere;

the opposite of aggradation. **(1, 2) dégradation (3) aplanissement**

degree of saturation An index that expresses the volume of water present in a soil relative to the volume of pores. Ranges from 0 to 100%. Also called water-filled pore space. **taux de saturation**

degree-day The difference between the average temperature each day and 0°C. Degree days are positive for daily averages above 0°C and negative for those below 0°C. **degré-jour**

degrees of freedom (*statistics*) The capability of variation within a system. The number of degrees of freedom in a particular system is the number of independent variables that can be freely assigned or changed to bring the system into a new state of equilibrium without causing the disappearance or appearance of a phase within the system. It has several uses in statistics. For example, a random sample of size n is said to have $n–1$ degrees of freedom for estimating the population variance, because there are $n–1$ independent deviations from the mean on which to base such an estimate. **degrés de liberté**

dehydrogenase An enzyme that accelerates oxidation of a substrate by removing hydrogen. **déshydrogénase**

dehydrogenation A chemical reaction in which hydrogen is removed from a molecule. **déshydrogénation**

deionization The removal of all charged atoms or molecules from some material such as water. The process usually involves one resin that attracts all positive ions and another resin to capture all negative ions. **désionisation**

deionized water Water that has been passed through resins that remove all ions. **eau déionisée**

delayed runoff Water from precipitation that sinks into the ground before discharging into streams through seeps and springs; also, runoff delayed by any means (e.g., temporary storage). **écoulement retardé**

delineation A portion of a landscape shown by a closed boundary on a *soil map* that defines the area, shape, and location of one or more component soils plus inclusions and/or miscellaneous area. See *map unit*. **délimitation**

deliquescence The absorption of water from the air by a hygroscopic solid. **déliquescence**

delta (*geomorphology*) A fan-shaped alluvial deposit at the mouth of a river formed by deposition of successive layers of sediments brought down from the land and spread out on the bottom of a basin. Where the stream current reaches quiet water, the bulk of the coarser load is dropped and the finer material is carried farther out. Deltas are recognized by nearly horizontal beds overlain by more steeply inclined and coarser-textured beds. **delta**

deltaic Pertaining to or characterized by a *delta* (e.g., deltaic sedimentation). **deltaïque**

demonstration area An area of land with definite boundaries on which practices are demonstrated to show their application and benefit to agriculture and/or resource management (e.g., improved soil and water conservation and land use practices). **site de démonstration, site d'essais**

denaturation The structural breakdown of a protein resulting in the loss of its function. **dénaturation**

denature To change the natural configuration of a substance (e.g., heating an enzyme which causes a change in conformation and usually in biologic activity). **dénaturer**

dendritic Branched like a tree. **dendriforme, dendritique**

dendritic drainage (*hydrology*) A type of drainage pattern which develops as an entirely random network because of the absence of structural controls. **drainage dendritique**

dendrochronology The study of the growth ring patterns in trees to determine age and the prevailing environmental conditions that affected the tree during its lifetime. Used by archeologists to date sites, and by climatologists to analyze past climates. The methodology is based on the premise that the width of the growth ring will reflect the amount of precipitation and the temperature of the year in which it was formed. The reliability of the technique can be limited because other factors (e.g., soil type, forest fires) may also affect the width of the growth ring. **dendrochronologie**

dendrology The study, identification, and systematic classification of trees and shrubs. **dendrologie**

denitrification Reduction of nitrate, nitrite to molecular nitrogen, or nitrogen oxides by microbial activity (*dissimilatory reduction of nitrate*) or chemical reactions involving nitrite (chemical denitrification). Microbial denitrification is an anaerobic respiratory process characteristic of facultative aerobic bacteria growing under oxygen-depleted conditions (denitrifying bacteria). **dénitrification**

density The mass of a substance per unit volume at a specified temperature and pressure. The accepted SI units are kilograms per cubic meter (kg m^{-3}) and megagrams per cubic meter (Mg m^{-3}), which are equivalent to grams per cubic centimeter (g cm^{-3}). **masse volumique**

density slicing (*remote sensing*) A technique used to assign image points or data vectors to particular classes based on the density or level of the response in a single image or channel; classification by thresholds; also called level slicing. **découpage en plages de densité**

density stratification (*limnology*) Arrangement of water masses into separate, distinct horizontal layers of different density. May be caused by differences in temperature, dissolved salts, or dissolved and suspended solids. See *thermal stratification, turnover.* **stratification**

denudation (*geology*) The wearing away or progressive lowering of the land surface by weathering and erosion, leveling mountains and hills to flat or gently undulating plains. **dégradation**

deoxyribonucleic acid (DNA) The macromolecule containing the genetic information governing the properties of individual cells and organisms; the genes of a cell. DNA is composed of chains of phosphate, sugar molecules (deoxyribose), and purines and pyrimidines; it is capable of self-replication and determines RNA synthesis. Exact copies of the genes of each cell are transferred to each daughter cell during cell division. Changes in the structure of DNA (mutations) can occur naturally or as a result of exposure to certain chemicals or radiation. Some changes result in death of the cell, and others are passed on to future generations as altered genes. **acide désoxyribonucléique (ADN)**

dependent variable A characteristic or condition of an item (e.g., object, plant, animal, population, system) that changes its value or degree with changes in another *independent variable.* **variable dépendante**

depleted soil Soil that has lost most of its plant-available nutrients. **sol épuisé, sol appauvri**

deposit Material left in a new position by a natural transporting agent (e.g., water, wind, ice, or gravity) or by the activity of humans. **dépôt**

deposition (1) (*geology*) The laying down of material after having been eroded and transported by various physical processes (e.g., wind, ice, running water, marine waves, and ocean currents), as the remains of former organisms (coal, peat, coral), or by evaporation. See *evaporite.* (2) (*meteorology*) The transition of a substance from the vapor phase directly to the solid phase, without passing through an intermediate liquid phase. (3) (*air quality*) The washout or settling of material from the atmosphere to the ground, vegetation, or surface waters. **(1) sédimentation, accumulation (2, 3) dépôt**

deposition velocity The rate at which a specific air pollutant is deposited on the Earth divided by the ambient concentration of that pollutant. **vitesse de dépôt**

depression storage (*hydrology*) Water stored in surface depressions and therefore not contributing to surface runoff. **emmagasinement dans les dépressions**

depth, effective soil The depth of soil material that plant roots can penetrate readily to obtain water and plant nutrients. **profondeur de la couche arable**

deranged drainage pattern (*hydrology*) A disordered arrangement of drainage in a recently glaciated area, characterized by only a few short tributaries, wandering streams, and extensive swampy areas between streams. **tracé de drainage dérangé**

derelict land (*agriculture*) Land voluntarily abandoned due to poor crop growing capability. (*non-agricultural*) Land voluntarily abandoned due to industrial pollution (e.g., abandoned mine) or within urban areas mainly due to socio-economic considerations. **terre abandonnée, terre en friche**

desalination The removal of salts form water to allow use for drinking, irrigation, or industrial processing. The two main desalination techniques are distillation and osmosis. **desalinisation**

desalinization (1) The removal of salts from saline soils, usually by leaching. (2) See *desalination*. **désalinisation**

desert An arid region with sparse vegetative cover and a mean annual precipitation of less than 250 mm. Four kinds are distinguished: polar deserts having continual snow cover and intense cold; middle-latitude deserts in the interiors of continents, characterized by low precipitation and high summer temperatures; trade-wind deserts, with negligible precipitation and a large range in daily temperature; coastal deserts, caused by the effects of a cold current of water on the western coast of a large land mass. One of the four main subdivisions of world vegetation. See *biochore*. **désert**

desert biome Any region with scant precipitation and extreme temperatures, and having plants and animals that have adapted to those conditions; typically, but not exclusively, found in hot climates. See *biome*. **biome de désert**

desert crust A hard layer containing calcium carbonate, gypsum, or other binding material, exposed at the surface in desert regions. **croûte désertique**

desert pavement The layer of gravel or stones left on the land surface in desert regions after removal of the fine material by wind erosion and deflation. **pavé désertique, reg**

desert polish A smooth, shiny surface imparted to rocks of desert regions by windblown sand and dust. **poli désertique**

desertification The progressive destruction or degradation of existing vegetative cover to form desert, which occurs as a result of overgrazing, *deforestation*, drought, and the burning of extensive areas. Once formed, deserts can only support a sparse range of vegetation. Climatic effects associated with this phenomenon include increased *albedo*, reduced atmospheric humidity, and greater atmospheric dust (aerosol) loading. Also called desertization. See *desert*. **désertification**

desiccant A chemical agent used to remove moisture from a material or object. **desséchant, dessicant**

desiccation The removal of moisture from a material. **dessication**

desiccation crack A crack formed in clays and muds caused by shrinkage during drying. **fente ou fissure de retrait**

desiccation, soil A progressive increase of aridity generally accompanied by a falling *water table,* drying out of the soil and deterioration in the quality and amount of vegetation cover. **dessèchement du sol**

desilication (1) Process of soil formation involving the chemical migration of silica out of the soil solum, leaving an accumulation of iron and aluminum oxides. Also called *ferralitization*, ferritization, allitization. (2) The removal of silica from rocks by *chemical weathering* or by reaction between a body of magma and the surrounding wall rock (e.g., formation of lime silicates). **désilicification**

de-silting area An area of grass, shrubs, or other vegetation used for inducing deposition of silt and other debris from flowing water. Also called *filter strip*. **zone de dessablement**

desorption The displacement of ions from the solid phase of the soil into solution by a displacing ion. **désorption**

desulfuration Biological oxidation of a sulfur-containing organic compound, resulting in incorporation of molecular oxygen into the molecule and the concurrent elimination of sulfur. **désulfuration**

detachment The removal of fragments of material from a soil mass by an erosive agent such as raindrops, running water, or wind. **détachement**

detection limit The smallest amount of a particular chemical that can be detected by a specific analytical instrument or method. **limite de détection**

detention basin (*hydrology*) A relatively small storage lagoon for slowing storm water runoff by temporary storage. **bassin de retenue**

detention dam (*irrigation*) A dam constructed for the purpose of temporary storage of streamflow or surface runoff and for releasing the stored water at controlled rates. **barrage de retenue du sol**

detergent A cleansing agent; a substance which allows hydrophobic particles such as oil and grease to be removed using an aqueous washing medium. The classical structure of a detergent, such as soap, consists of a long hydrophobic tail topped off by a charged terminal group (e.g., a long-chain fatty acid). The affinity of different parts of the molecule for both hydrophobic and hydrophilic regions aids the process of the hydrophobic material being taken into colloidal aqueous suspension. **détergent**

deterministic model A *model* that contains no random elements and for which the future course of the system is determined by its position or character at some fixed point in time. See *general systems theory.* **modèle déterministe**

deterministic process A time series or other sequence of numbers whose future values can be predicted with certainty. **processus déterministe**

detrital *Clastic*; Pertaining to rock and minerals occurring in sedimentary rocks that were derived from preexisting igneous, sedimentary, or metamorphic rocks. **détritique**

detritivore (*zoology*) Animal that feeds on particulate material derived from the remains of plants or animals, including large scavengers, smaller animals such as earthworms and some insects, and *decomposers*. **détritivore**

detritus (1) Material produced by the disintegration and weathering of rocks that has been moved from its site of origin. (2) A deposit of this material. (3) Any fine particulate debris, usually of organic origin but sometimes defined as organic and inorganic debris. (4) Freshly dead or partially decomposed organic matter. **détritus**

detritus food web A community of organisms in which one group feeds on another, involving interlinked food chains, and where the primary consumers feed on plant or animal detritus. **réseau trophique détritivore**

deuterium An isotope of hydrogen that contains one proton and one neutron

D

in the nucleus of the atom. Also called heavy hydrogen. **deutérium**

deviation, standard *(statistics)* A measure of the average variation of a series of observations or items of the mean of a population; in normally distributed sets of moderate size, the interval of the mean plus or minus the standard deviation includes about two-thirds of the items. The standard deviation is given by the square root of the sum of squared deviations from the mean divided by *n–1*. This is the most widely used measure of dispersion of a frequency distribution, and is denoted by *s*. Division by *n* (rather than *n–1*) is sometimes applied, in which case it is called the root-mean-square deviation. The square of the sample standard deviation is called the *variance*. **écart type, déviation standard**

dew Moisture deposited in the form of water droplets on the surface of vegetation and other objects located near ground level. **rosée**

dew point The temperature at which air becomes saturated with water. **point de rosée**

dewatering Removal of water from sludge by pressing, centrifuging, or air drying and other methods. The resulting product can be composted, landfilled, or burned. **déshydratation, assèchement**

diagenesis (*geology*) The post-depositional processes of compaction and cementation of sediments when they are at or near to the surface, at relatively low temperatures and pressures, and exclusive of weathering and metamorphism. When diagenesis leads to the formation of massive rock layers it is called *lithification*. **diagénèse**

diagnostic horizon (*Canadian system of soil classification*) A horizon which is unique to a soil order (e.g., *podzolic B horizon*). (soil classification, *U.S. soil taxonomy*) Combinations of specific soil characteristics indicative of certain classes of soils. Those that occur at the surface are called epipedons; those below the surface are called diagnostic subsurface horizons. **horizon diagnostique**

dialysis A method in which large molecules (e.g., protein) and small molecules (e.g., glucose) in solution may be separated by selective diffusion through a semipermeable membrane. **dialyse**

diamagnetism A type of magnetism, associated with paired electrons, that causes a substance to be repelled from the inducing magnetic field. **diamagnétisme**

diatom Alga having siliceous cell walls that persist as a skeleton after death. Any of the microscopic unicellular or colonial algae constituting the class Bacillariaceae. They occur abundantly in fresh and salt waters, and their remains are widely distributed in soils. **diatomée**

diatomaceous earth Fine, grayish siliceous material composed chiefly or wholly of the remains of *diatoms*. It may occur as a powder or a porous, rigid material. **terre de diatomées**

diatomite See *diatomaceous earth*. **diatomite**

diazotroph An organism or association of organisms that can reduce molecular nitrogen (N_2) to ammonia. **diazotrophe**

dibble A tool for opening holes for planting seeds or small seedlings; also called planting bar, spud, or planting iron. **plantoir**

dicotyledon A plant whose embryo has two *cotyledons*; one of the two classes of *angiosperms*. **dicotylédone**

dieback (*ecology*) A dramatic decline in the number of individuals within a population of organisms. (*botany*) A condition in woody plants in which the ends of the branches die, often progressively toward the base, because of disease, insect injury, winter injury, or excessive or insufficient moisture. **dépérissement**

differential erosion See *erosion*. **érosion différentielle**

differential rate law An expression that gives the rate of a reaction as a function of concentrations; often called rate law. **loi de vitesse différentielle**

differential thermal analysis Thermal analysis carried out by uniformly heating a sample that undergoes chemical and physical changes while simultaneously heating a reference material that undergoes no changes. The temperature difference between the sample and the reference material is measured as a function of the temperature of the reference material. **analyse thermale différentielle**

differential water capacity The change in water content in a unit volume of soil per unit change in *matric potential.* The water capacity at a given water content will depend on the particular desorption or adsorption curve used. Distinction should be made between volumetric and specific water capacity. Also called specific water capacity. **capacité différentielle de rétention de l'eau**

differential weathering (*geology*) *Chemical weathering* that occurs at different rates in a rock mass as a result of variations in porosity, grain size, mineral composition, or degree of joint development. Often results in an uneven surface. **altération différentielle ou sélective**

diffraction A phenomenon observed in the propagation of waves, in which the propagation direction is changed due to a change in the amplitude or phase of waves that strike an object. **diffraction**

diffraction pattern The experimentally measured intensities, diffracting angle (direction), and order of diffraction for each diffracted beam obtained when a crystal is placed in a narrow beam of x-rays or neutrons (usually monochromatic). Each substance has a characteristic diffraction pattern. **patron de diffraction**

diffraction spacing See *d-spacing*. **distance interfeuillet**

diffuse double layer A heterogeneous system that consists of a solid surface (e.g., clay) having a net electrical charge together with an ionic swarm in solution containing ions of opposite charge, neutralizing the surface charge. **double couche diffuse**

diffuse radiation Radiation that does not reach the subject from a single direction. **rayonnement diffus**

diffused air A type of sewage aeration; air is pumped into the sewage through a perforated pipe. **aération par diffusion d'air**

diffusion The independent or random movement of ions or molecules that tends to bring about their uniform distribution within a continuous system. **diffusion**

diffusion (nutrient) The movement of nutrients in soil that results from a concentration gradient. See *figure*. **diffusion d'un élément nutritif**

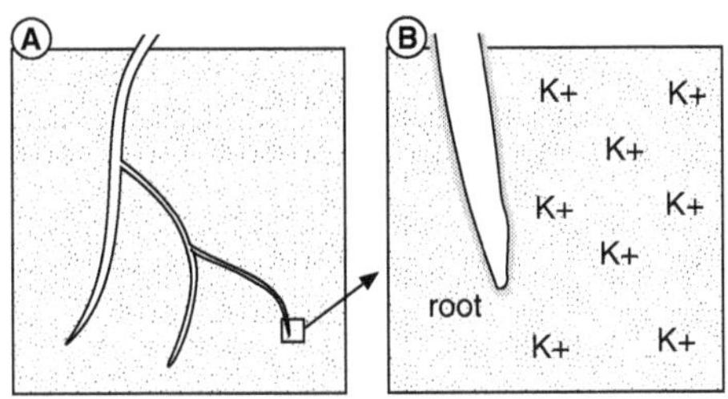

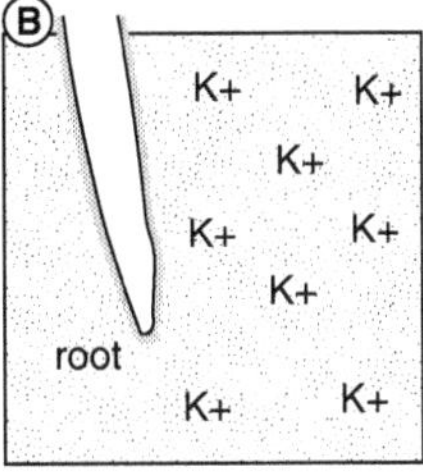

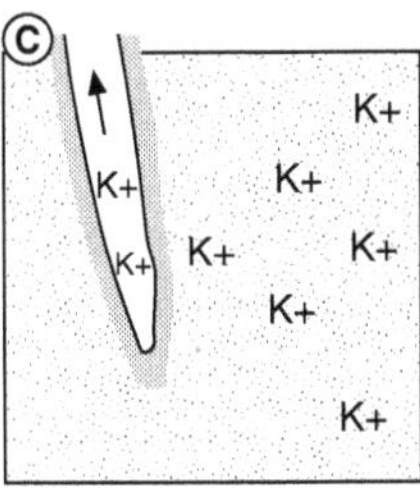

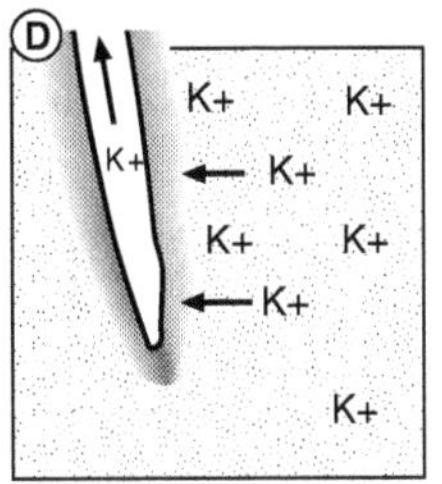

Roots draw nutrients from near their surface.

Even distribution of nutrients.

Plant uptake decreases nutrient concentration near root.

To equalize concentration, nutrients move from areas of higher concentration to lower.

Nutrient diffusion (adapted from Ontario Ministry of Agriculture, Food, and Rural Affairs, 1998).

D

diffusion coefficient A statistical parameter used in the calculation of rates of *diffusion*. It varies with the temperature, the nature of the particles being diffused, and the nature of the diffusion medium. Defined as length squared divided by time (i.e., I^2/t, where I = length and t = time). **coefficient de diffusion**

diffusion pressure gradient (*botany*) The process by which water is drawn upwards through tree trunks or plant stems, stimulated by the differences in vapor pressure between root hairs and leaf surfaces. **gradient de pression de diffusion, tension sur la colonne d'eau**

diffusivity The ratio of the *hydraulic conductivity* to the *differential water capacity*. Sometimes called hydraulic diffusivity. **diffusivité**

dig See *tillage, dig*. **creuser**

digester A closed tank in a wastewater treatment plant, that decreases the volume of solids and stabilizes raw sludge by bacterial action. **digesteur**

digestion (*waste management*) The mineralization or biochemical decomposition of particulate organic material to lessen the impact of adding domestic waste to streams. **digestion**

dike (1) (*engineering*) An embankment constructed to confine or control water, especially one built along the banks of a river to prevent overflow of lowlands; a levee. (2) (*geology*) A tabular body of igneous rock that cuts across the structure of adjacent rocks or cuts massive rocks. **(1) digue (2) dyke, filon intrusif**

diluent Substance used to dilute a solution or suspension. **diluant**

dilution The process of adding solvent to lower the concentration of solute in a solution. **dilution**

dilution factor The extent to which the concentration of some solution or suspension has been lowered through the addition of a *diluent*. **facteur de dilution**

dimer A molecule formed by the joining of two identical monomers. **dimère**

dimictic lake A stratified lake that undergoes two overturns each year. The water in dimictic lakes becomes layered in response to differences in the temperatures of surface and deep waters. See *overturn*. **lac dimictique**

dimorphism (1) The crystallization of the same chemical compound in two crystal forms (e.g., pyrite and marcasite). (2) The characteristic of having two distinct forms in the same species (e.g., male and female, sterile and fertile leaves in ferns). **bimorphisme**

dinitroaniline Any of a group of soil-applied *herbicides* that curtails the lateral root formation in weeds. **dinitroaniline**

dinitrogen fixation Conversion of molecular nitrogen (N_2) to ammonia and subsequently to various organic combinations or forms utilizable in biological processes. **fixation de l'azote atmosphérique**

dinitrophenol Any of a group of chemical compounds toxic to humans, used for a variety of purposes ranging from weed control to a weight-control agent. **dinitrophénol**

dioctahedral An octahedral sheet, or a mineral containing such a sheet, that has two-thirds of the octahedral sites filled by trivalent ions such as aluminum or ferric iron. **dioctaédrique**

dioecious (*botany*) Plants having staminate (male) flowers on one plant and pistillate (female) flowers on another plant. See *monoecious*. **dioïque**

dip The angle that a stratum or any planar feature makes with the horizontal, measured perpendicular to the *strike* of the strata and in the vertical plane. **pendage**

diphenyl ethers Any of a group of *contact herbicides* that affects broadleaf weeds more than grasses, and also inhibits photosynthesis. **éthers diphényles**

diploid Pertaining to cells or organisms having two sets of chromosomes. **diploïde**

dipole A pair of separated, opposite electrical charges. **dipôle**

dipole moment A property of a molecule whose charge distribution can be represented by a center of positive charge and a center of negative charge. **moment dipolaire**

dipole-dipole attraction The attractive force resulting when polar molecules line up so that the positive and negative ends are close to each other. **attraction dipolaire**

direct competition The exclusion of individuals from resources by aggressive behavior or the repelling effect of toxins. **compétition directe**

direct count In soil microbiology, any one of several methods for estimating, by direct microscopic examination, the total number of microorganisms in a given mass of soil. **numération directe, décompte direct**

direct photolysis The transformation or decomposition of a material caused by the absorption of light energy. **photolyse directe**

direct plating Method used to identify fungi present in soil or litter. A known weight of soil or litter is distributed over the surface of an appropriate medium and incubated to facilitate the growth of fungi. Following development of the fungal colonies, the organisms are counted and identified. The method is used in a similar manner to quantify specific bacteria present in water or wastewater. **ensemensement direct**

direct shear test A shear test in which soil under an applied normal load is stressed to failure by moving one section of the sample or sample container relative to the other section. **essai de cisaillement direct**

disaccharide A sugar formed from two *monosaccharides* joined by a *glycosidic linkage*. **disaccharide**

disc trencher (*silviculture*) A machine designed for mechanical site preparation in reforestation activities. It provides continuous rows of planting spots rather than intermittent patches as provided by patch scarifiers. Consists of scarifying steel discs equipped with teeth. **trancheuse à disque**

discharge (*hydrology*) The volume of water flowing past a point in a stream or pipe for a specific time interval. **débit**

discharge curve (*irrigation*) (1) A rating curve showing the relation between stage and rate of flow of a stream. (2) A curve showing the relation of discharge of a pump and the speed, power, and head. **courbe des débits, courbe d'étalonnage**

disconformity (*geology*) A type of *unconformity* in which there is a considerable time gap in the sedimentary sequence but in which the strata above and below the plane of unconformity are parallel, having similar *dip* and *strike* characteristics. **discordance parallèle**

discontinuity (*geology*) (1) Any interruption in sedimentation; an *unconformity* (2) A surface separating two unrelated groups of rocks (e.g., a fault). (*geophysics*) A surface at which seismic wave velocities abruptly change; a boundary between seismic layers of the Earth. (*meteorology*) A sharp change in temperature, humidity, wind speed, and direction at a marked boundary surface, usually a frontal surface between air masses. **discontinuité**

discordance Any phenomenon which does not conform to the normal order of things. (*geomorphology*) Any topographical feature that bears little or no relationship to the geological structure of the region (e.g., discordant drainage). (*geology*) An igneous intrusion cutting through the bedding or foliation of the country-rock. **discordance**

D

discrete variable A quantitative variable limited to a finite or countable set of values, regardless of the accuracy of the measurement method. There can be no intermediate values between the numbers representing a discrete variable. For example, a count of the number of trees in a stand must be represented by a whole number, and exclude decimal fractions. See *continuous variable*. **variable discrète ou discontinue**

discriminant function A function used in conjunction with a set of threshold values in a classification procedure. For example, in remote sensing it is one of a set of mathematical functions commonly derived from training samples and a decision rule, used to divide the measurement space into decision regions. **fonction discriminante**

disinfection The effective killing by chemical or physical processes of all organisms capable of causing infectious disease. (e.g., chlorination). **désinfection**

disintegration (*radiochemistry)* A spontaneous change in the nucleus of a radioactive element that results in the emission of some type of radiation and the conversion of the element into a different element. (*geology*) The breakdown of rock and mineral particles into smaller particles by physical or mechanical forces such as wetting and drying, temperature changes, frost action, ice formation, or root penetration and expansion. **désintégration**

dispersal The movement of organisms away from the place of birth or from centers of population density. **dispersion**

dispersant A chemical agent used to break up concentrations of organic material. In cleaning oil spills, used to disperse oil from the water surface. **agent dispersant**

disperse (1) To break up compound particles, such as aggregates, into the individual component particles. (2) To distribute or suspend fine particles (e.g., clay) in or throughout a medium (e.g., water). **disperser**

dispersion (*soils*) The process of disrupting and destroying the structure or aggregation of the soil so that each particle is separate. (*ecology*) The local distribution of organisms in space. (*chemistry*) A two-phase system consisting of finely divided particles (the disperse phase) distributed throughout a bulk substance (the continuous phase), e.g., clay particles dispersed in an aqueous solution; liquid particles dispersed in a gas (e.g., fog). (*seismology*) A division of *seismic waves* of different wavelengths owing to a variation of speed with wavelength. (*statistics*) The spread of a set of observations or objects about some central point or place. **dispersion**

dispersion diagram A diagram to illustrate graphically the distribution of a set of data, usually over time (e.g., monthly temperatures at a given station). Data plotted in space or on a surface are generally referred to as a *distribution*. **diagramme de dispersion**

dispersive clay A soil that has a high percentage of sodium in the pore water salts and is subject to rapid colloidal erosion from concentrated flow through cracks or channels in the soil. **argile dispersée ou peptisée, collosol**

dispersivity The ratio of the *hydrodynamic dispersion coefficient* (d) divided by the pore water velocity (v); thus $D=d/v$. **dispersibilité**

disposal field The area used for spreading liquid effluent to separate wastes from water, degrade impurities, and improve drainage waters. **champ d'épuration**

disposal pond See *lagoon*. **lagune, étang d'épuration**

dissected (*geomorphology*) Cut by erosion, especially by streams; commonly applied to plains in the process of erosion after an uplift. **disséqué**

dissection (*geomorphology*) The process of fluvial erosion resulting in disruption of a relatively even land surface by cutting ravines or valleys into it, generally as a result of regional uplift. **dissection**

dissimilation The release from cells of inorganic or organic substances formed by metabolism. **désassimilation**

dissimilatory reduction of nitrate The use of nitrate by organisms as an alternate electron acceptor in the absence of O_2, causing reduction of nitrate to ammonium. See *denitrification*. **réduction non-assimilatoire des nitrates**

dissipation capacity The ability of a soil to adsorb, retain, and degrade chemical compounds. **capacité de dissipation**

dissociation The breakdown of a substance (e.g., molecule, ion) into smaller substances. **dissociation**

dissociation constant At equilibrium, the ratio of the molar concentrations of the ions produced by the dissociation of a molecule to the molar concentration of the undissociated molecule. For acids, the ratio is abbreviated *Ka* and usually expressed as p*Ka*, i.e., the negative logarithm of *Ka*. For bases the abbreviations are *Kb* and *pKb*. For example, carbonic acid dissociates into hydrogen ions and bicarbonate ions. At 25°C, the *Ka*, the ratio of the molar concentrations of hydrogen ions and bicarbonate ions to the molar concentration of carbonic acid, is 4.47×10^{-7} and the p*Ka* is 6.35. **constante de dissociation**

dissolved gases Gases that are in solution homogeneously mixed in water. **gaz dissous**

dissolved load (*hydrology*) The portion of a river's load that is carried in solution, in contrast to *bed load*. **charge en solution**

dissolved organic carbon (DOC) A measure of the organic carbon that is dissolved in water or soil water extracts. **carbone organique dissous (COD)**

dissolved organic matter (DOM) Water soluble organic compounds found in soil solution (generally these compounds are considered < 0.45 µm in size). Dissolved organic matter consists of simple compounds of biological origin (e.g., metabolites of microbial and plant processes) including sugars, amino acids, and low molecular weight organic acids, but may also include large molecules. **matière organique dissoute (MOD)**

dissolved oxygen (DO) The amount of gaseous oxygen dissolved in a liquid — usually water. An important measure of the quality of a water body for aquatic organisms. **oxygène dissous (OD)**

dissolved solid The total amount of dissolved material, organic and inorganic, contained in water or wastes. Excessive dissolved solids make water unpalatable for drinking and unsuitable for industrial uses. **matière solide en solution**

distillation The process of separation and/or purification of the components of some substance based on differences in their boiling points. The components within a mixture of two liquids can be separated if the two liquids have different boiling points. The mixture is heated, and the material with the lowest boiling point is converted to a vapor first. That vapor can then be condensed to produce a liquid that consists primarily of the single liquid. **distillation**

distilled water Water purified by *distillation*; free of dissolved salts and other compounds. **eau distillée**

distribution (1) A graphic representation that illustrates the interrelationships between a set of objects or observations, usually in a spatial rather than time dimension. (2) (*statistics*) See

D

normal distribution, skewness. (3) The extent of the geographic area (continuous or discontinuous) over which a species occurs at a given time. **distribution**

disturbed area See *disturbed land.* **zone perturbée**

D

disturbed land (*land reclamation*) Area where vegetation, topsoil, or overburden is removed, or where topsoil, spoil, and processed waste is placed (as in mining). Also called *disturbed area.* **terrain perturbé**

disulfide linkage An S–S bond that stabilizes the tertiary structure of many proteins. **liaison disulfide**

ditch A shallow graded channel for collecting excess water within a field, usually constructed with flat side slopes for ease of crossing. **fossé**

diurnal (1) During daytime. (2) At daily intervals. **(1) diurne (2) journalier, quotidien**

diurnal temperature variations Daily variations in temperature at a particular location, related to the local radiation budget. **variations journalières de température**

divergence (*hydrology*) The flow of water in different directions away from a particular area or zone. The taking of water from a stream or other body of water into a canal, pipe, or other conduit. **divergence**

diversion dam A structure or barrier that diverts part of or all the water of a stream to a different course. **digue, barrage de dérivation**

diversity (*ecology*) (1) The physical or biological complexity of a system. (2) A measure of the number of different species along with the number of individuals in each representative species in a given area. Diversity is made up of *richness* and *evenness.* **diversité**

diversity index A mathematical expression that depicts species diversity in quantitative terms. The *Shannon-Weaver index* is a widely used measure. **indice de diversité**

divide (*geomorphology*) The area of high ground that separates two different drainage systems. See *interfluve.* **ligne de partage des eaux, interfluve**

doline (*geomorphology*) A closed depression in *karst* terrain, usually rounded or elliptical in shape, formed by the solution and subsidence of the limestone near the surface. The bottom may lead down into a vertical shaft descending into the limestone, or there may be a sink hole into which surface water flows and disappears underground. **doline**

dolomite (1) A natural mineral consisting of calcium magnesium carbonate, $CaMg(CO_3)_{(2)}$. Dolomite is white to light-colored and has perfect rhombohedral coverage. (2) A sedimentary rock in which dolomite is the dominant mineral; also called magnesian limestone. **(1) dolomite (2) dolomie**

dolomitic lime A naturally occurring liming material composed chiefly of carbonates of Mg and Ca in approximately equimolar proportions. **chaux dolomitique**

domestic sewage Wastewater and solid waste characteristic of the flow from toilets, sinks, showers, and tubs in a household. **eaux usées domestiques**

domestic waste Waste from a household. See *domestic sewage.* **eaux usées domestiques**

domestic water Water used within a household. **eau de consommation**

dominance (*ecology*) A community property reflected in the patterns of relative abundance of species that may functionally describe the competitive relations between species. **dominance**

dominance-diversity curve (*ecology*) A plot of the logarithmic measure of species abundance versus species rank from the most common to the rarest species in the community. **relation abondance-diversité**

dopplerite An amorphous, brownish black, gel-like calcium salt of a *humic acid* that is found at depths within or beneath *peat* deposits. **dopplerite**

dormancy (*botany*) A special condition of arrested growth in which the plant and plant parts such as buds and seeds do not begin to grow without special environmental cues. The requirement for such cues, which include cold exposure and a suitable photoperiod, prevents the breaking of dormancy during superficially favorable growing conditions. (*biology*) A period of reduced biologic activity in animals (e.g., hibernation). **dormance**

double bond A bond in which two pairs of electrons are shared by two atoms. **liaison double**

double refraction See *birefringence*. **biréfringence**

double sample A sample of a sample; used, for example, when attributes from remote sensing or cartographic materials can be measured on a larger sample than those attributes requiring field measurements. Attributes on the former sample can be used to guide selection of the latter sample. **échantillon double**

dough stage A stage of seed development at which endosperm development is pliable, like dough (e.g., soft, medium, hard); usually used when 50% of seeds on an inflorescence are in this stage of development. **stade pâteux**

down-scaling (*modeling*) The reductionist approach where an empirically defined system is dismantled to explain how it works according to its constituent processes. **réduction ou diminution d'échelle**

drag (1) See *tillage, drag*. (2) The force retarding the flow of water or wind over the surface of the ground. **(1) niveler avec une traîne (2) entrave, résistance à l'écoulement**

drag scarification (*silviculture*) In reforestation, a method of site preparation that disturbs the forest floor and prepares logged areas for regeneration. Often carried out by dragging chains or drums behind a skidder or tractor. **scarifiage par traînage**

drain (1) A buried pipe or conduit (closed drain). (2) A ditch (open drain) for carrying off surplus surface water or groundwater. (3) Any other channel (open or closed) for removing surplus water. **drain, canal de drainage**

drain tile Concrete, ceramic, or plastic pipe used to conduct water from the soil. **tuyau de drainage**

drainage (1) Removal or outflow of excess surface water or groundwater from land by surface or subsurface drains. (2) Soils are characterized according to their natural drainage on the basis of the frequency and duration of periods when the soil is not saturated. Terms used to describe drainage are: excessively, well, moderately, imperfectly, and poorly drained soil. **drainage**

drainage basin A part of the Earth's surface occupied by a drainage system which consists of a surface stream with all its tributaries and impounded bodies of water. Also known as *watershed*, *catchment* area, and drainage area. The geographic area bounded peripherally by a drainage divide within which all surface water flows into a single river or stream via its tributaries. **bassin versant, bassin hydrographique**

drainage density Ratio of the total length of all streams within a drainage basin to the area of that basin. **densité de drainage**

drainage divide The boundary line, along a topographic ridge or a subsurface formation, separating two adjacent *drainage basins*. **ligne de faîte, ligne de partage des eaux**

drainage pattern The spatial configuration or arrangement of stream courses in an area as related to local geologic and geomorphic features and history. The spatial relationships of all streams within a drainage system. **tracé de drainage**

drainage structures Include metal and wooden culverts, open-faced culverts, bridges, and ditches. **ouvrages de drainage**

draw (*geomorphology*) (1) A natural linear depression followed by surface drainage. (2) A sag or depression leading from a valley to a gap between two hills. **(1) vallon (2) ravine**

drawdown (*hydrology*) (1) The lowering of the surface elevation of a body of water, the water surface of a well, the water table, or the *piezometric surface* adjacent to the well, resulting from the withdrawal of water therefrom. (2) The difference between the height of the water table and that of the water in a well. **rabattement**

drift See *glacial drift.* **sédiment, dépôt**

drill cuttings Rock or other materials forced out of the borehole as a well is drilled. **déblais de forage**

drill seeding Planting seed with a drill in relatively narrow rows, generally less than 30 cm apart. See *seed drilling.* **semis en ligne à espacement rapproché**

drilling fluid A dense fluid material, often containing bentonite clay and barite, used to cool and lubricate a well drilling bit, seal openings in the wall of the borehole, transport drill cuttings to the surface, reduce drill pipe friction, and control well pressure. **fluide de forage**

drilling mud A carefully formulated heavy suspension, usually in water but sometimes in oil, used in rotary drilling. It commonly consists of bentonitic clays, chemical additives, and weighting materials such as barite. It is pumped continuously down the drill pipe, out through openings in the drill bit, and back up in the annulus between the pipe and the walls of the hole to a surface pit, where it is screened and reintroduced through the mud pump. The mud lubricates and cools the bit, carries the cuttings up from the bottom; and prevents blowouts and cave-ins by plastering friable or porous formations and maintaining a hydrostatic pressure in the borehole, offsetting pressures of fluids that may exist in the formation. Also called *mud* and *drilling fluid.* **boue de forage**

drilling waste fluid Heterogeneous mixture of water, drilling muds, borehole cuttings, additives, and various other wastes that are specifically related to the drilling activity. Components of the drilling waste that contain less than or equal to 5% solids. **résidus liquides de forage**

drilling waste solid The bottom layer of sump material composed of drill cuttings, flocculated bentonite, weighting materials, and other additives. Drilling waste solids can be stacked with minimal or no overflow of liquid. Components of the drilling waste that contain more than 5% solids. **résidus solides de forage**

driving force An input that interacts with other components and generates the flow of materials and energy in an ecosystem. Also called driving function. **force motrice**

drop spillway (*irrigation*) A structure in which the water drops over a vertical wall onto an apron at a lower elevation. **déversoir vertical**

drop structure (*irrigation*) A structure used to convey water to a lower level and dissipate its surplus energy. The structure may be vertical or inclined. **chute**

drought A continued period of lack of moisture, so serious that crops fail to develop and mature properly. **sécheresse**

drumlin (*geomorphology*) A low, smoothly rounded, elongate hill of glacial drift, commonly glacial till, built under the margin of the ice and shaped by its flow, or carved out of an older moraine by re-advancing ice. Its long axis is parallel to the direction of ice movement. **drumlin**

dry aggregate *Aggregates* or secondary soil particles that are not broken down (i.e., stable) and can be separated from a soil by dry sieving. **agrégat sec**

dry bulk density The ratio of the mass of dry soil to its bulk volume. See *bulk density.* **masse volumique apparente sèche, densité apparente sèche**

dry deposition The introduction of acidic material to the soil or surface waters by particles containing sulfate or nitrate salts. See *wet deposition.* **dépôt sec**

dry mass The mass of solid soil particles after all water has been vaporized by heating to 105°C. **masse sèche**

dry matter (DM) The substance in a plant remaining after oven drying to a constant weight. **matière sèche (MS)**

dry permafrost Loose, crumbly permafrost containing little or no ice or moisture. **pergélisol sec**

dry scrubber (1) An air pollution abatement device that removes sulfur dioxide from stack gases by injecting a fine dry chemical reagent (e.g., limestone) into the flue gas. The sulfur dioxide reacts with the injected chemical to form a solid particulate, which is captured in a fabric bag. (2) A gravel bed filter used to collect particulate matter from gas streams. **épurateur à sec**

dry weight basis Expressing the composition of a soil or plant material based on the weight of the material without moisture. Dry weight is more accurate, because moisture content can be variable. **à base de poids sec**

dry-bulb temperature The temperature reading from an ordinary thermometer; the reading is used with the *wet-bulb temperature* to compute. See humidity, relative. **température du thermomètre sec**

dryland farming A method of farming in arid and semiarid areas (e.g., drier regions of India, the former Soviet Union, Canada, and Australia) without using irrigation (i.e., dry or rainfed agriculture), the land treated so as to conserve moisture by use of mulches and removal of weeds. **aridoculture, culture en région sèche**

d-spacing In diffraction of x-rays by a crystal, the distance or separation between the successive and identical parallel planes in the crystal lattice. It is expressed as *d* in *Bragg's equation.* Also called diffraction spacing. **espace interfeuillet**

duckfoot An implement with horizontally spreading, V-shaped tillage blades or sweeps that are normally adjusted to provide shallow cultivation without turning over the surface soil or burying surface crop residues. A type of point or foot used on a *chisel cultivator.* **en patte d'oie**

duff The layer of partially and fully decomposed organic materials lying below the litter and immediately above the mineral soil. **matière organique, humus**

duff mull A type of forest humus transitional between *mull* and *mor.* **humus intermédiaire**

dugout pond A pond constructed by excavation rather than by constructing a dam. **fosse-réservoir**

Dumas method (1) A method of determining the amount of nitrogen in an organic compound. The sample is combusted in the presence of copper, and the nitrogen present is converted to nitrogen oxides and reduced to nitrogen gas. The gas is collected and the mass of nitrogen in a known mass of sample is determined. (2) A method of finding the relative molecular masses of volatile liquids by weighing. The air in an empty flask of known volume is weighed at a known temperature. A liquid sample is vaporized at a known temperature into the flask. The mass difference between an empty and gas-filled flask is the mass of the gas. **méthode de Dumas**

dune (*geomorphology*) Ridge or hill of sand built up by wind action in areas of sandy glacial drift on sea coasts, in deserts, and elsewhere. Dunes are started by some obstruction (e.g., a bush, boulder, or fence) that causes an eddy or otherwise thwarts the sand-laden wind. Once begun, the dunes themselves offer further resistance and they grow to form various shapes. **dune**

dune complex (*geomorphology*) An area of moving and fixed sand dunes, together with sand plains and the ponds, lakes, and swamps produced by the blocking of streams by the sand. **champ de dunes**

dune, mobile A dune in the process of being built up and reshaped, and that may be continuously advancing or shifting position. See *dune, stabilized.* **dune active**

dune, stabilized A dune that has become fixed or anchored by vegetation and is therefore protected from further eolian action. The stabilization may occur naturally through a gradual colonization, or artificially by planting suitable grass species and trees. See *dune, mobile.* **dune stabilisée**

duric A diagnostic B horizon (Canadian system of soil classification) that is strongly cemented and usually has an abrupt upper boundary and a diffuse lower boundary. Cementation is usually strongest near the upper boundary. Ordinarily the color is similar to that of the parent material, and the structure is amorphous or coarse platy. Air dried clods do not slake when immersed in water. The duric horizon does not meet the requirements of a *podzolic B* horizon but may meet those of a Bt horizon. **durique**

duricrust A hard crust on the surface, or a layer in the upper horizons of a soil in a semiarid climate. It is formed by the accumulation of soluble minerals deposited by mineral-bearing waters that move upward by capillary action and evaporate during the dry season. **croûte pédologique concrétionné**

Durids A suborder in the U.S. system of soil taxonomy. Aridisols which have a duripan that has its upper boundary within 100 cm of the soil surface. **Durid**

durinodes Weakly cemented to indurated soil nodules cemented with SiO_2. Durinodes break down in concentrated KOH after treatment with HCl to remove carbonates, but do not break down on treatment with concentrated HCl alone. **durinodes**

duripan A mineral soil horizon cemented by silica, usually opal or microcrystalline forms of silica, to the point that air-dry fragments will not slake in water or HCl. A duripan may also have other cementing agents such as iron oxide or calcium carbonate. A subsurface diagnostic horizon in the U.S. system of soil taxonomy, and analogous to the *duric* horizon in the Canadian system of soil classification. **duripan**

dust Particles of finely divided, dry, solid matter of silt- and clay-sized earthy particles (i.e., less than 0.0625 mm in diameter), occurring anywhere in the atmosphere, and light enough to be carried in suspension by the wind (e.g., dust from human industrial and domestic activity, volcanic eruptions, wind erosion, and cosmic sources). On average, 1 km^3 of air contains 600 Mg of dust. **poussière**

dust mulch A loose, finely granular, or powdery layer on the soil surface, usually produced by excessive shallow cultivation under dry soil conditions, and also by deposition. **sol pulvérisé**

Dutch elm disease Fatal infectious disease of elm trees (genus *Ulmus*) caused by a fungus (genus *Ceratocystis*), and spread from tree to tree by the elm bark beetle (genus *Scolytus*). **maladie hollandaise de l'orme**

dy Finely divided, partly decomposed organic material accumulated in peat soils in the transition zone between the peat and the underlying mineral material. Dy peat also refers to amorphous material formed from humus soils that have settled in lake waters. Dy is poorer in nutrients than *gyttja* and is characterized by a high *C:N ratio*. (Swedish). **bourbe dystrophe**

dyke See *dike*. **dyke, filon intrusif**

dynamic equilibrium state The condition of a system in which the inflow of materials or energy equals the outflow. **état d'équilibre dynamique**

dynamic viscosity A measure of the resistance of a fluid to flow, expressed as mass per length-time (kg m^{-1} s^{-1}). In soil, stresses and intermolecular attractions cause a restriction to water flow. For liquids, viscosity decreases with increasing temperature. For gases, viscosity increases with increasing temperature. **viscosité dynamique**

dynamometer An instrument for measuring the draft of tillage implements and the resistance of soil to penetration by tillage implements. **dynamomètre**

dynecology A subdivision of ecology that deals with the study of the processes of change in related communities. See *autecology, synecology*. **dynécologie**

Dystric Brunisol A great group of soils in the *Brunisolic* order (Canadian system of soil classification). The soils may have mull Ah horizons less than 10 cm thick. Bm horizons are generally low in base status, and pH ($CaCl_2$) is usually 5.5 or lower. **brunisol dystrique**

dystrophic (*limnology*) Lake water high in humic substances and plant degradation products, usually brown in color. Plant and animal life are typically sparse, and the water has a high *BOD*. **dystrophe**

dystrophic lake A lake that is between the *eutrophic* and swamp stages of aging, and has a shallow and brownish or yellowish water with high levels of organic matter. See *dystrophic*. **lac dystrophe**

E

E horizon A horizon formed at or near the mineral surface by *eluviation* of silicate clay, iron, or aluminum, alone or in combination, and normally consisting mainly of silt or sand. Used in the U.S. system of soil taxonomy. See *Appendix D* for equivalent Canadian and FAO/UNESCO terms. **horizon E**

Earth (1) The solid material of the globe that constitutes the land surface, in contrast to the water surface. (2) The loose surface material (including soil) as distinct from solid rock. (3) The planet of the solar system which is third in order of distance from the sun, and fifth in size of the nine major planets. It has the shape of a slightly flattened spheroid, with the North Pole 45 m farther from the equatorial plane than the South Pole. Planet Earth has the following constant characteristics: mean density = 5517 $\times 10^3$ kg m^{-3}; mass 5976 $\times 10^{24}$ kg; volume = 1083 $\times 10^{21}$ m^3; gravity acceleration 9.812 m sec^{-2}; total surface area = 510 million km^2; land area (29.22%) = 149 million km^2; ocean area (70.78%) = 361 million km^2; mean radius = 6371 km; equatorial radius = 6378.5 km; polar radius = 6357 km; equatorial circumference = 40,067 km; meridional circumference = 39,999.7 km; average height = 875 m above sea level; greatest height = 8850 m (Mt. Everest); lowest point of land surface = 396.2 m below sea level (shores of Dead Sea); greatest ocean depth = 10,430 m below sea level (the Marianas trench). **terre**

earth dam Dam constructed of compacted soil materials. **barrage en terre**

earthflow A rapid movement of soil and loose surface material downslope when saturated and buoyed up by water. The mass of saturated material slumps downwards and outwards, and the contained water generates a flow, in the form of a tongue, that may extend to the foot of the slope or beyond, until the velocity of the flow is insufficient to move the plastic mass. **coulée de boue**

earth hummock A type of patterned ground due to frost heaving in the soil under *periglacial* conditions. **butte gazonnée, thufur**

earth pillar A column or pinnacle of clay (often till) or relatively soft earthy material, capped by a boulder that serves to protect it from erosion by rain. Once the boulder falls from the pinnacle, the pillar will rapidly be destroyed. It is typical of the *badlands* terrain. **cheminée de fée, demoiselle coiffée**

Earth science All studies concerned with physical characteristics of the Earth, in contrast to its biological characteristics. It includes the disciplines of cartography, climatology, geochemistry, geodesy, geomorphology, geophysics, hydrology, meteorology, mineralogy, oceanography, palaeontology, petrology, remote sensing, sedimentology, soil science, stratigraphy, structural geology, and surveying. See *geography, physical*. **science de la terre, géosciences**

earth slide A downhill movement of a mass of superficial material due to slope failure, usually as a result of water reducing the friction along a shear plane in the soil mantle. See *mass movement.* **glissement de terrain**

earth tremor A slight *earthquake.* **faible secousse sismique**

earthquake A sudden motion of the Earth's crust caused by faulting or volcanic activity. Earthquakes can occur in near-surface rocks or down to as deep as 700 km below the surface. The actual area of the earthquake is called the focus, and the point on the Earth's surface above the focus is called the epicenter. **tremblement de terre, séisme**

easting The first half of a grid reference, always preceding the northing when map coordinates are being quoted. Represents the distance measured eastwards from the origin of the grid. **abscisse**

EC See *electrical conductivity.* **EC**

ECe See *electrical conductivity, extract.* **ECe**

ecoclimate Climate in relation to *flora* and *fauna.* See *ecosystem.* **écoclimat**

ecodistrict A detailed mapping unit in Canada's ecological classification system, two or more of which comprise an ecoregion. See *ecological land classification system, Canadian.* **écodistrict**

ecoelement See *ecological land classification system, Canadian.* **écoélément**

eco-fallow See *tillage, chemical fallow.* **jachère écologique**

ecologic Pertaining to the living environment. **écologique**

ecologic counterparts Species of different phylogenetic origin that occur in different areas and have converged phenotypically to fulfill similar ecologic roles. **contreparties écologiques**

ecologic efficiency The percentage of energy in biomass produced by one trophic level that is incorporated into biomass by the next highest trophic level. **efficacité écologique**

ecological approach Resource management and planning based on consideration of the relationships among all organisms, including humans, and their physical environment. **approche écologique**

ecological balance A state of dynamic equilibrium within a community of organisms in which genetic, species, and ecosystem diversity remain relatively stable, subject to gradual changes through natural succession. **équilibre écologique**

ecological classification An approach used to categorize and delineate, at different levels of resolution, areas of land and water having similar characteristic combinations of features of the physical environment (e.g., climate, topography, geology, soil, and hydrologic function), biological communities (e.g., plants, animals, and microorganisms), and human influence (e.g., economic, cultural, and infrastructure). See also *biogeoclimatic classification system.* **classification écologique**

ecological density The number of organisms from a specific part of an ecosystem or habitat expressed as individuals weight of biomass per unit area, or per unit volume. **densité écologique**

ecological efficiency The effective transfer of useful energy from a food to the animal consuming it; normally applied to the feeding structure of animal communities in the natural environment; the transfer is usually expressed as a percent. **efficacité écologique**

ecological equivalents Species that occupy the same or similar ecological niches in similar ecosystems located in different geographical locations (e.g., the zebra of Africa and the buffalo of North America both occupy grasslands and are grazers). **équivalents écologiques**

ecological factor Any part or condition of the environment that influences the life of one or more organisms. See *biotic factors.* **facteur écologique**

ecological health The occurrence of certain attributes deemed to be present in a healthy, sustainable resource, with the absence of conditions that result from known stresses or problems affecting the resource. **santé écologique**

ecological integrity The quality of a natural unmanaged or managed ecosystem in which the natural ecological processes are sustained with genetic species, and ecosystem diversity is ensured for the future. **intégrité écologique**

ecological land classification An integrated approach to classification, delineation, and description of ecologically distinct areas of the Earth's surface at different levels of generalization using various abiotic and biotic factors at each of the levels. Ecological classification integrates information about geologic, landform, soil, vegetative, climatic, wildlife, water, and human factors that may be present. The dominance of any one or more of these factors varies with the given ecological land unit. This approach to land classification can be applied incrementally on a scale-related basis from site-specific ecosystems to very broad ecosystems. See *ecological land classification system, Canadian.* **classification écologique du territoire**

ecological land classification system, Canadian A hierarchical *ecological land classification* system developed for sustainable resource management and planning purposes in Canada. This national ecological framework, consisting of a map and attribute database, has been developed at the ecozone, ecoregion, and ecodistrict levels of the hierarchy. **système canadien de classification écologique du territoire**

Components of the system include:

ecodistrict – a part of an ecoregion characterized by distinctive assemblages of relief, geology, landforms and soils, vegetation, water, fauna, and land use. ***écodistrict***

ecoelement – a part of an ecosite displaying uniform soil, topography, vegetation, and hydrology. ***écoélément***

ecoprovince – a part of an ecozone characterized by major assemblages of structural or surface forms, faunal realms, and vegetation, hydrological, soil, and climatic zones. ***écoprovince***

ecoregion – a part of a province characterized by distinctive regional ecological factors including climatic, physiography, vegetation, soil, water, fauna, and land use. ***écorégion***

ecosection – a part of an ecodistrict throughout which there is a recurring assemblage of terrain, soils, vegetation, water bodies and fauna. ***écosection***

ecosite – a part of an ecosection in which there is relative uniformity of parent material, soil, hydrology, and vegetation. ***écosite***

ecozone – an area of the Earth's surface representative of large and very generalized ecological units characterized by interactive and adjusting abiotic and biotic factors. The ecozone defines, on a subcontinental scale, the broad mosaics formed by the interaction of macroscale climate, human activity, vegetation, soils, and geological and physiographic features of the country. Fifteen ecozones are recognized in Canada: Tundra Cordillera, Boreal Cordillera, Montane Cordillera, Boreal Plains, Taiga Plains, Prairie, Taiga Shield, Boreal Shield, Hudson Bay Plains, Mixed Wood Plains, Pacific Maritime, Atlantic Maritime, Southern Arctic, Northern Arctic, and Northern Cordillera. See *Appendix E, Figure 1.* ***écozone***

ecological niche The physical space in a habitat occupied by an organism; its functional role in the community (e.g., its trophic position). **niche écologique**

E

E

ecological process The actions or events that link organisms (including humans) and their environment (e.g., *carbon sequestration*, disturbance, nutrient cycling, successional development, and productivity). **processus écologique**

ecological pyramid A visual representation, resembling a pyramid, that depicts the total mass of organisms residing in each *trophic level* in a given area. The bottom bar of the pyramid represents the mass of plants, the top bar represents the mass of carnivores, and intermediate bars represent other forms of biota in the feeding structure of the community. **pyramide écologique**

ecological release The expansion of habitat and food preferences by populations in regions of low species diversity, resulting from reduced interspecific competition. **expansion écologique**

ecological reserve Areas representative of natural ecosystems in which rare or endangered native plants or animals may be preserved in their natural habitat. **réserve écologique**

ecological risk assessment A process that evaluates the likelihood that adverse ecological effects may occur, or are occurring, as a result of exposure to one or more stressors. **évaluation du risque écologique**

ecological succession The slow change in an ecosystem during which one community of organisms is gradually replaced by a different kind of community. **succession écologique**

ecology The study of the relationships between organisms and their environment. **écologie**

ecology, soil The branch of soil science that deals with interrelations among soil organisms, and between soil organisms and their environment. **écologie du sol**

ecoprovince See *ecological land classification system, Canadian.* **écoprovince**

ecoregion See *ecological land classification system, Canadian.* **écorégion**

ecoregion classification A system used to stratify terrestrial and marine ecosystem complexity into areas of similar climate, physiography, oceanography, hydrology, vegetation, and wildlife potential. **classification des écorégions**

ecosection See *ecological land classification system, Canadian.* **écosection**

ecosite See *ecological land classification system, Canadian.* **écosite**

ecospecies A biological group comprising organisms fully fertile among themselves but only weakly fertile with members of allied groups. **écoespèce**

ecosphere The mantle of earth and troposphere inhabited by living organisms. **écosphère**

ecosystem A functional unit consisting of all the living organisms (plants, animals, and microbes) in a given area, and all the non-living physical and chemical factors of their environment, linked together through nutrient cycling and energy flow. **écosystème**

ecosystem composition The constituent elements of an ecosystem. **composition écosystémique**

ecosystem function The processes through which the constituent living and non-living elements of ecosystems change and interact, including biogeochemical processes and succession. **fonction écosystémique**

ecosystem health The degree to which an *ecosystem* maintains its organization, function (e.g., energy and matter transformations), structure (e.g., information linkages, food webs, and biodiversity), and autonomy over time, and retains its resilience to stress. **santé des écosystèmes**

ecosystem management An attempt to set explicit goals for an *ecosystem* based on the available understanding of ecological interactions and processes necessary to sustain ecosystem composition, structure, and function. **gestion écosystémique**

ecosystem perspective The emphasis of ecosystem structure and functioning at large scales in space and time, rather than single points. **vision écosystémique**

ecosystem productivity The ability of an ecosystem to produce, grow, or yield products (e.g., trees, shrubs, or other organisms). **productivité de l'écosystème**

ecosystem structure The spatial arrangement of the living and nonliving elements of an ecosystem. **structure écosystémique**

ecosystem sustainability The ability to sustain diversity, productivity, resilience to stress, health, renewability, and/or yields of desired values, resource uses, products, or services of an ecosystem while maintaining the integrity of the ecosystem over time. **durabilité écosystémique**

ecotone A transition area between two adjacent ecological communities usually exhibiting competition between organisms common to both. **écotone**

ecotype A locally adapted population of a species with a distinctive limit of tolerance to environmental factors. See *biotype*. **écotype**

ecozone See *ecological land classification system, Canadian*. **écozone**

ectomycorrhiza A mycorrhizal association in which the fungal mycelia extend inward, between root cortical cells, to form a network, and outward into the surrounding soil. See *Hartig net*. Usually the fungal hyphae also form a mantle on the surface of the roots. See *figure*. **ectomycorhize**

ectotrophic mycorrhiza See *ectomycorrhiza*. **mycorhize ectotrophe**

edaphic Pertaining to the soil, and particularly the influence of soil on organisms. **édaphique**

edaphic climax The climax stage of an ecological succession in which the community is in equilibrium with localized soil conditions. **climax édaphique**

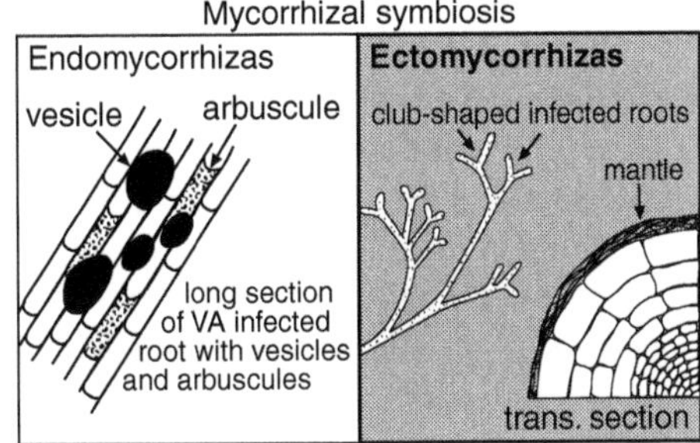

Ectomycorrhizas (adapted from Killham, 1994).

edaphology The study of the influence of soil on living things, particularly plants, including human use of land for plant growth. **édaphologie**

edatope Refers to a specific combination of soil moisture regime and soil nutrient regime. **édatope**

edge The outer band of a patch with a significantly different environment from the interior of the patch. **bordure**

edge effect Habitat conditions (such as degree of humidity and exposure to light or wind) created at or near the well-defined boundary between ecosystems (i.e., between open areas and adjacent forest). **effet de bordure**

EDTA The sodium salt of ethylenediaminetetraacetic acid $(CH_2COO^-Na^+)_2NCH_2\text{-}CH_2N\text{-}(CH_2COOO^-Na^+)_2$, a strong chelating agent, used as a soil test extractant for Zn and other micronutrients. **EDTA**

effect (*statistics*) In *analysis of variance*, the influence exerted by each separate controlling factor over the average values assumed by the variable. **effet**

effective calcium carbonate equivalent See *calcium carbonate equivalent*. **équivalent effectif de carbonate de calcium**

effective cation exchange capacity The sum of cations that a soil can adsorb at its native pH value. **capacité d'échange cationique effective**

effective precipitation The portion of total precipitation that becomes available for plant growth. **précipitation effective**

effective soil depth See *depth, effective soil.* **profondeur effective de sol**

effective stress The stress transmitted through a soil by intergranular pressures. **contrainte effective**

efflorescence A white powder produced on the surface of a rock or soil in an arid region by evaporation of water, or by loss of water of crystallization on exposure to the air. It commonly consists of soluble salts such as gypsum, calcite, natron, or halite. **efflorescence**

effluent (*waste water management*) (1) The discharge of solid, liquid, or gaseous wastes that enter the environment as a byproduct of human processes. (2) The discharge or outflow of water from ground or subsurface storage. **effluent**

E_H The potential generated between an oxidation or reduction half-reaction and the standard hydrogen electrode. In soils, it is the potential created by oxidation–reduction reactions that take place on the surface of a platinum electrode measured against a reference electrode minus the E_H of the reference electrode. **E_H (potentiel redox)**

eigenvalues (*statistics*) The characteristic or latent roots of the correlation matrix in factor analysis. Eigenvalues determine the value of the factors, and there are as many eigenvalues as variables in the matrix. **valeurs propres**

El Niño A climatic phenomenon involving seasonal changes in the direction of the tropical winds over the Pacific Ocean and abnormally warm surface ocean temperatures. It occurs irregularly, but generally every three to five years, often first becoming evident during the Christmas season (El Niño means Christ child) in the surface waters of the eastern tropical Pacific Ocean. These changes can disrupt weather patterns throughout the tropics and can extend to higher latitudes, especially in Central and North America. **El Niño**

elastic Said of a body in which strains are instantly and totally recoverable, and in which deformation is independent of time. **élastique**

elastic deformation A temporary deformation of a material or body, after which it returns to its former character and shape once the stress has been released. **déformation élastique**

elastic limit The maximum stress that a material can withstand without undergoing permanent deformation. Also called yield point, yield limit. **limite d'élasticité**

elasticity The property or quality of being *elastic*. **élasticité**

electrical conductivity (EC) The reciprocal of electrical resistivity. The conductivity of electricity through water or an extract of soil; expressed in decisiemens or siemens per meter ($dS\ m^{-1}$) at 25°C. It is a measure of soluble salt content in solution. **conductivité électrique**

electrical conductivity, extract (Ece) The electrical conductance of an extract from a soil saturated with distilled water, normally expressed in units of decisiemens per meter ($dS\ m^{-1}$) at 25°C. **conductivité électrique d'un extrait saturé**

electricity The form of energy arising from the movement or accumulation of electrons. The movement of electrons (called electric current) produces a magnetic field, a phenomenon used to convert electrical energy to mechanical energy in electric motors. Conversely, generators use mechanical energy to move a magnetic field, producing an electric current in a conductor. **électricité**

electrochemistry The study of the interchange of chemical and electrical energy. **électrochimie**

electrodialysis A process that uses electric current and an arrangement of semipermeable membranes to separate soluble minerals from water. Often used in desalination. **électrodialyse**

electrokinetic (zeta) potential The electrical potential at the surface of shear plane between immobile liquid attached to a charged particle and mobile liquid further from the particle. **potentiel électrocinétique, potential zêta**

electrolysis The passage of an electric current through an *electrolyte*, causing migration of the positively charged ions to the negative electrode (i.e., the cathode) and the negatively charged ions to the positive electrode (i.e., the anode). **électrolyse**

electrolyte Any compound (e.g., NaCl) that dissociates into ions when dissolved in water. The resulting solution will conduct an electric current. **électrolyte**

electrolytic Pertaining to *electrolysis* or an *electrolyte*. **électrolytique**

electrolytic cell A cell in which *electrolysis* occurs. It uses electrical energy to produce a chemical change that would otherwise not occur spontaneously. **cellule électrolytique**

electromagnetic radiation Radiation consisting of electric and magnetic waves that travel at the speed of light (e.g., light, radio waves, gamma rays, x-rays); also called electromagnetic energy. **radiation électromagnétique**

electron A negatively charged particle that moves around the nucleus of an atom. **électron**

electron capture detector (ECD) A sensitive detector used in conjunction with *gas chromatography*. The chromatograph uses a stream of gas to carry and separate the compounds of interest. As the carrier gas flows through the detector, the gas is ionized by a radioactive substance. The electrons resulting from this ionization migrate to an anode, creating a voltage. When the compounds in question are encountered in the carrier gas, the electric current is interrupted. The decrease in current is proportional to the concentration of solute in the gas. **détecteur à conduction**

electron diffraction pattern The interference pattern seen when a beam of electrons is transmitted through a substance. Each substance has a characteristic pattern that conveys basic crystallographic information as well as information about orientation, defects, crystal size, and additional phases. See *x-ray diffraction*. **patron de diffraction des rayons-x**

E

electron microprobe An analytical instrument that uses a finely focused beam of electrons to excite x-ray emission from selected portions of a sample. The emitted x-ray spectrum provides data about the composition of the sample at the point of excitation. **microsonde électronique**

electron microscope An electron-optical instrument that uses electrons focused by systems of electrical or magnetic lenses to magnify very small objects onto a fluorescent screen or photographic plate. It is capable of resolving much finer structures than an optical microscope (up to 1,000,000 times actual size without loss of definition). See *scanning electron microscope*. **microscope électronique**

electron-transport chain Final sequence of reactions in biological oxidations composed of a series of oxidizing agents arranged in order of increasing strength, and terminating in oxygen. **chaîne de transport d'électrons**

electrophoresis A technique used to separate, identify, and quantify proteins and similar macromolecules as these molecules migrate within gel or cellulosic substrates under the influence of an electric current. **électrophorèse**

electrostatic precipitator An air pollution abatement device that removes particulates from chimney emissions by imparting an electrical charge to particles in a gas stream for mechanical collection on an electrode. **précipitateur électrostatique**

element A substance that cannot be decomposed into simpler substances. In an element, all the atoms have the same number of protons and electrons; only the number of neutrons varies. There are 92 naturally occurring elements. **élément**

E

elevation (1) (*geodesy*) The vertical distance from mean sea level to a point or object on the Earth's surface (i.e., height above sea level). (2) (*surveying*) The angle between the horizontal and a point at a higher level. (3) (*civil engineering*) The drawing of a building made in projection on a vertical plane. **(1) élévation, altitude (2, 3) élévation**

elevation head (*hydrology*) The potential energy in a hydraulic system, represented by the vertical distance between the hydraulic system (e.g., pipe, channel) and a reference level. **charge d'eau d'élévation**

elutriation (*soil*) Mechanical analysis of a sediment in which the finer, lightweight particles are separated from the coarser, heavy particles by means of a slowly rising current of air or water of known and controlled velocity, carrying the lighter particles upward and allowing the heavier ones to sink. (*chemistry*) Purification, or removal of material from a mixture or in suspension in water, by washing and decanting, leaving the heavier particles behind. **élutriation**

eluvial horizon A soil horizon that has been formed by the process of *eluviation*. See *illuvial horizon*. **horizon éluvial**

eluviation The transportation of soil material in suspension or in solution within the solum by the downward or lateral movement of water. See *figure*. **éluviation**

embankment An artificial deposit of material that rises above the natural surface of the land and is used to contain, divert, or store water, to support roads or railways, or for other similar purposes. **endiguement**

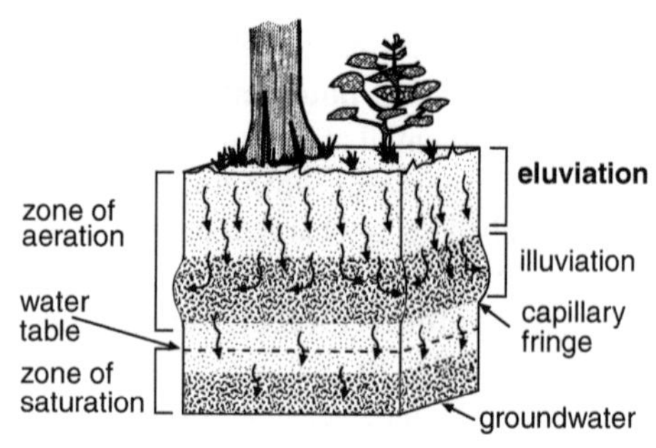

Eluviation (adapted from Dunster and Dunster, 1996).

emergency spillway (*irrigation*) A spillway used to carry runoff exceeding a given design flood. **évacuateur de secours**

emergent shoreline A former coastline that is now either raised above present high-water mark (i.e., *raised beach*) or located at some distance inland from the present coastline. The emergence of the land/sea margin may be due to either the fall of sea level (*eustasy*) or uplift of the land by Earth movements (*isostasy*). **ancienne ligne de rivage, ligne de rivage perchée**

emission Release of a substance (usually a gas) into the *atmosphere*. **émission**

emission factor The average amount of a specific pollutant emitted from each type of polluting source in relation to the amount of source material handled, processed, or burned (e.g., number of kilograms of particulates per tonne of raw material). **coefficient d'émission**

emission inventory The amount of each primary air pollutant released daily into a community's atmosphere. **inventaire des émissions**

emission rate The rate at which a pollutant is released into the *atmosphere*. **taux d'émission**

emissivity The amount of energy given off by an object relative to the amount given off by a blackbody at the same temperature; normally expressed as a positive number between 0 and 1. **émissivité**

empirical Based on experience or observation, as opposed to theory. **empirique**

emulsion A colloidal dispersion of one liquid in another. **émulsion**

en echelon (*geology*) Pertaining to geological structures that, though remaining parallel, are offset or overlapping (similar to tiles on a roof viewed from the side). **en échelons**

enantiomers Molecules whose relationship to each other is as non-superposable mirror-image partners; they show *optical activity* or exhibit *chirality*. **énantiomères**

encapsulation (*land reclamation*) The total containment of decanted sump contents (drilling wastes) by using low permeability liners. These can be clay liners with a permeability of 1 $\times 10^{-7}$ cm s^{-1} or less, or an approved synthetic liner. After treatment and decanting of the liquid layer, the remaining solids are capped by 1 m of compacted material. Finally, subsoil and topsoil are spread over the encapsulated area. **encapsulation**

end moraine A ridge-like accumulation of drift built chiefly along the terminal margin of a valley glacier or the margin of an ice sheet. It is mainly the result of deposition by ice, deformation by ice thrust, or both. Synonym: terminal moraine. **moraine terminale**

endemic Characteristically found in a particular region; originally occurring only in a given area; continually present in a community. **endémique**

endemic species A species originating in, or belonging to, a particular region. See *species*. **espèce endémique**

endoenzyme, or intracellular enzyme An enzyme formed within the cell and not excreted into the medium. **enzyme endogène**

endogenous Produced from within; originating from or due to internal causes. **endogène**

endogenous variable (*modeling*) Output variable (response) of the model, generated by the effect of exogenous variables on the system's state variables. **variable endogène**

endolithic Growing within stone (e.g., thalli of some lichens). **endolithique**

endomycorrhiza A mycorrhizal association with intracellular penetration of the host root cortical cells by the fungus as well as outward extension into the surrounding soil. See *arbuscule*, *vesicle*, and *vesicular arbuscular*. See *figure*. **endomycorhize**

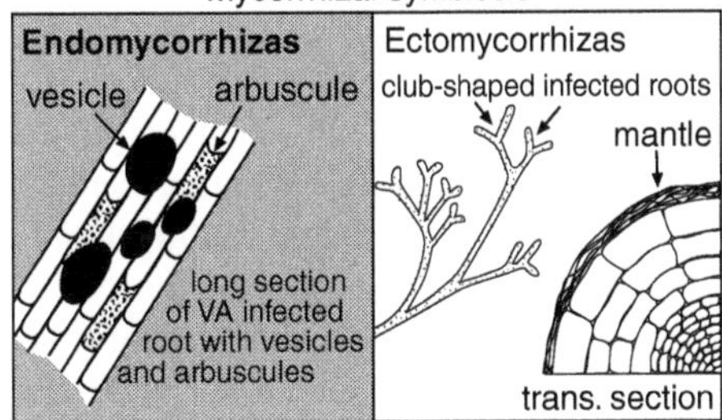

Endomycorrhizas (adapted from Killham, 1994).

endophyte An organism that lives at least part of its life cycle within a host plant as a parasite or symbiont. **endophyte**

endoreic drainage (*hydrology*) An inward flowing pattern of drainage in the world's semi-arid zones. **écoulement endoréique**

endotherm An organism that has the ability to maintain a constant body temperature through physiological mechanisms. Mammals are endotherms. Also called homeotherm. **endotherme**

endothermic Pertaining to a chemical reaction that occurs with absorption of heat. See *exothermic*. **endothermique**

endotoxin A poison produced by a microbial cell, released following cell lysis. **endotoxine**

E

endotrophic Nourished or receiving nourishment from within (e.g., fungi or their hyphae receiving nourishment from plant roots in a mycorrhizal association). **endotrophe**

endotrophic mycorrhiza See *endomycorrhiza.* **mycorhize endotrophe**

energy The ability or capacity of a system to do work. Different kinds of energy include: kinetic energy, the "free" energy which is dissipated continually as heat friction; solar radiant energy, the energy transferred by *electromagnetic radiation* from the sun; potential energy, the energy stored in any object prior to its release as "free" energy; geothermal energy, kinetic energy derived from the Earth's interior and which enters into numerous geological phenomena (e.g., geysers); elastic energy, potential energy associated with a condition of mechanical strain; heat energy, kinetic energy generated by the internal random motion of molecules; nuclear energy, potential energy contained in the nucleus of an atom; and chemical energy, potential energy stored in the molecules of compounds. **énergie**

energy budget An estimate of the energy added to and lost from a system. **bilan énergétique**

energy flow The path of energy as it moves through the various components of a community of organisms, including the input of solar energy, energy captured by photosynthesis, the utilization of energy by various animal groups, and the loss of heat from the community. **flux d'énergie**

englacial Contained, embedded, or carried within a glacier or ice sheet; refers to meltwater streams, till, drift, or moraine. Also called intraglacial. **intraglaciaire**

enhanced greenhouse effect An increase in the natural *greenhouse effect* through emission of anthropogenic *greenhouse gases* into the *atmosphere.* These gases trap more *infrared radiation*, thereby exerting a warming influence on the *climate.* See *climate change* and *global warming.* **effet de serre amplifié**

enrichment (*limnology*) The addition of nutrients into surface water to the point that the trophic state is greatly increased because of stimulation of the growth of algae and other aquatic plants. See *trophicgenic region.* (*radiochemistry*) The process of increasing the abundance of a specific isotope in a mixture of isotopes. **enrichissement**

enrichment culture A technique in which environmental, including nutritional, conditions are controlled to favor the development of a specific organism or group of organisms. **façon culturale d'enrichissement**

ensilage See *silage.* **ensilage**

enteric bacteria A group of bacteria that inhabit the intestinal tracts of humans and other animals. This group includes pathogenic bacteria such as *Salmonella* and *Shigella.* **entébactérie, entérobactérie**

enthalpy A thermodynamic quantity defined as the sum of a body's internal energy plus the product of its volume multiplied by the pressure. **enthalpie**

Entisols An order in the U.S. system of soil taxonomy. Mineral soils that have no distinct subsurface diagnostic horizons within 1 m of the soil surface. **Entisols**

entropy The availability of energy to do work. Often used as a thermodynamic measure of the randomness or disorder within a system (i.e., the higher the entropy, the more disordered the system). **entropie**

environment The sum total of the conditions within which an organism lives. **environnement**

environment management system A system providing a framework to monitor and report on an organization's environmental performance. **système de gestion de l'environnement**

environmental resistance The forces of nature (e.g., predators, drought) that tend to maintain populations of organisms at stable levels. **résistance environnementale**

environmentally sensitive areas Areas requiring special management attention to protect important scenic values, fish and wildlife resources, historical and cultural values, and other natural systems or processes. **zones écologiquement vulnérables**

environmental sustainability Using environmental resources in a manner that maintains their potential to meet the needs of present and future generations. **durabilité de l'environnement**

environmentally sustainable agriculture Agriculture that can be carried out indefinitely without significantly harming the environment. **agriculture durable du point de vue de l'environnement, environnement durable en agriculture**

enzyme A protein compound produced within an organism that acts as a catalyst for biochemical reactions. Enzymes are sensitive to changes of temperature, pH, and other substances in the environment. **enzyme**

eolian Pertaining to materials deposited by wind. **éolien**

ephemeral data (*remote sensing*) Data that help to characterize the conditions under which remote sensing data were collected. Ephemeral data may be used to calibrate the sensor data prior to the analysis, and include such information as the positioning and spectral stability of sensors, sun angle, and platform attitude. **données accessoires ou auxiliaires**

ephemeral stream A stream or portion of a stream that flows only in direct response to precipitation. **cours d'eau intermittent**

epicenter The point on the Earth's surface directly above the focus of an earthquake. See *earthquake*. **épicentre**

epidermal cells The cells forming the outer layer of an animal's skin or the thin outer covering of plant leaves, stems, and roots. **cellules épidermiques**

epidermis The outermost layer of cells in plants and animals. **épiderme**

epifluorescence A type of microscopy in which the specimen under examination is illuminated by the projection of ultraviolet light through the microscope from above the microscope stage. It differs from standard microscopy, in which the specimen is illuminated from below the slide with white light. The observer sees the object as light emitted from the specimen by fluorescence mechanisms (glowing) stimulated by the ultraviolet light. **épifluorescence**

epigenetic (epigenic) process (*geology*) Any geological or rock-forming process that takes place upon the surface of the Earth. Its usage has come to include rocks formed at or near the surface of the Earth, and therefore includes both sedimentary and volcanic rocks. **processus épigénétique**

epilimnion The warmer, less dense topmost layer in the water body of a lake or ocean. See *hypolimnion, thermocline*. **épilimnion**

epipedon A *diagnostic horizon*, in the U.S. system of soil taxonomy, occurring at the surface of the soil. **épipédon**

epiphyte A plant that grows nonparasitically on another plant, obtaining its nutrients from the atmosphere. **épiphyte**

epoch The third rank order in a subdivision of geological time. Several epochs form a *period* (e.g., the Pleistocene and the Holocene are epochs and together form the Quaternary period, which is part of the Cenozoic *era*). See *Appendix C*. **époque**

equation of state An equation derived from a combination of *Boyle's law* and *Charles' law* that describes the relationships among the pressure, density, and temperature of a gas:

E

pressure = gas constant × density × absolute temperature. $P= RrT,$ where R is the universal gas constant, r is density, and T is the *absolute temperature*. **équation caractéristique**

equifinality A concept of general systems theory indicating that the same final state may be achieved from differing initial conditions and in different ways. The concept is applicable to landforms, where the overall shape of a particular terrain may be achieved by different processes but the initial landforms may have been very different. **équifinalité**

equilibrium A steady-state condition in which flow in equals flow out. **équilibre**

equilibrium constant (*chemistry*) The value obtained when equilibrium concentrations of the chemical species are substituted in the equilibrium expression. **constante d'équilibre**

equiplanation The reduction of topography to a plain-like surface. **équiplanation**

equi-potential line (*hydrology*) A line, in a field of flow, such that the total head is the same for all points on the line, and therefore the direction of flow is perpendicular to the line at all points. A contour line indicating equal pressure head of ground water in an aquifer. **ligne équipotentielle**

equivalent depth The amount of water in the soil expressed as its equivalent depth (mm) when standing as free water on the surface of the soil. Calculated from the product of *volume wetness* and soil depth (mm). **profondeur équivalente**

equivalent diameter In sedimentation analysis, the diameter assigned to a nonspherical particle. It is numerically equal to the diameter of a spherical particle having the same density and velocity of fall. **diamètre équivalent**

equivalent land capability The ability of land to support various uses after reclamation similar to that which existed prior to any activity being conducted on the land. However, the ability to support individual land uses will not necessarily be equal after reclamation. **possibilité d'utilisation équivalente des terres**

equivalent or equivalent weight The weight (grams) of an ion or compound that combines with or replaces 1 gram of hydrogen. The atomic weight or formula weight divided by its valence. **équivalent-gramme**

equivalent radius In sedimentation analysis, a measure of sedimentary particle size equal to the radius of a spherical particle of the same density that would have the same settling rate as the sedimentary particle. **rayon équivalent**

era A geologic time unit next in order of magnitude below an eon, and larger than a period (e.g., the Paleozoic era is in the Phanerozoic eon and includes, among others, the Devonian period). The four principal eras are Precambrain, Paleozoic, Mesozoic, and Cenozoic. See *Appendix C*. **ère**

ergot A disease of cereals and other grasses caused by the fungus *Claviceps pupurea* in which the hard, black sclerotia of the fungus replace the grain in an infected plant. These sclerotia contain alkaloids which can cause severe poisoning if ingested by animals or humans; also the fungus itself, or the sclerotia of the fungus. **ergot**

ericaceous Of or relating to the heath family, *Ericaceae* (e.g., blueberry). **éricacé**

erode To deplete or remove the land surface by wind, water, or other agents. See *erosion*. **éroder**

erodibility (1) The degree or intensity of a soil's state or condition of, or susceptibility to, being eroded. (2) The K factor in the *universal soil loss equation*. See *erosion*. **érodabilité ou érodibilité**

erodible Susceptible to *erosion*. **érodable**

erosion The wearing away of the land surface by water, wind, ice, or other agents, including such processes as weathering, solution, transportation, and gravitational creep. Often called *geological* or *natural erosion*, as it occurs without human intervention. (2) Detachment and movement of soil from one location to another mainly by wind and water, and also by tillage. See *accelerated erosion*, *gully erosion*, *rill erosion*, *sheet erosion*, *splash erosion,* and *tillage erosion*. **érosion**

Types of erosion include:

accelerated erosion – erosion that occurs at a faster rate than normal, natural, or geological erosion, primarily as a result of human activities. ***érosion accélérée***

differential erosion – erosion that occurs at irregular or varying rates, caused by differences in the resistance of surface geological material. Relative rates of erosion are associated with climatic differences, contrasting rock hardness, terrain contrasts, or different tectonic histories. On a large scale, softer or weak rocks are rapidly worn away, whereas more hard or resistant rocks remain to form ridges, hills, or mountains. In local situations, the above is evident along rivers, glaciers, and coastlines. ***érosion différentielle***

geological erosion – the normal or natural erosion caused by geological processes acting over long periods and resulting in the wearing away of mountains, the dissection of plains, and the buildup of floodplains and coastal plains. Often called natural or normal erosion. ***érosion géologique, érosion naturelle***

surficial erosion – A process of soil formation whereby material is removed from the surface layer of a soil. ***érosion de surface***

erosion classes A grouping of *erosion* conditions based on the degree or the characteristic patterns of erosion. The classes apply to *accelerated erosion*, not to normal, natural, or geological erosion. Four erosion classes are recognized for water erosion (slightly, moderately, severely, and gullied) and three for wind erosion (slightly, severely, and blown-out). **classes d'érosion**

erosion control Methods used to reduce soil losses resulting from *erosion*. **méthodes de lutte contre l'érosion**

erosion control structures Structures (e.g., diversion terraces, grassed waterways, surface inlets, hedgerows, and *buffer strips*) that are used to curtail soil *erosion*. **structures de lutte contre l'érosion**

erosion enrichment ratio (ER) The ratio of a compound's concentration in eroded soil to non-eroded soil; the same for eroded water flow to normal water flow. **rapport d'enrichissement érosif**

erosion pavement A layer of coarse fragments (e.g., sand, gravel, cobbles, or stones) that remains on the surface of the ground after erosion removes fine particles. **dallage d'érosion**

erosion potential A numerical value expressing the inherent *erodibility* of a soil or maximum potential erosion. In the *universal soil loss equation* (under clean tillage, up- and downslope) EI = *RKLS/T*. **érosion potentielle**

erosion risk, actual Risk of soil *erosion*, a measure of the potential for erosion that leads to the need for preventative measures such as soil cover, and improved land use and management practices. **risque ou potentiel d'érosion**

erosion risk, inherent Risk of *erosion* on bare soil. See *erodibility*. **risque inhérent d'érosion**

erosion surface (*geomorphology*) A near-level, regional land surface shaped and degraded by the erosive action

E

of ice, wind, or water, especially by running water. Often called planation surface at the global scale, and erosion platform for a limited area. **surface érodée**

erosive velocity Velocity of the erosive agent necessary to cause erosion. **vitesse érosive**

E

erosivity The measured or predicted ability of water, wind, gravity, or other agents to cause *erosion*. The R factor in the *universal soil loss equation* is a measure of the erosivity of rainfall. **érosivité**

erratic A transported rock fragment different from the bedrock where it lies. Generally applied to fragments transported by glacier ice or floating ice. **erratique**

error (1) In a statistical model, the deviation of experiment from theory, possibly from two sources: measurement (or observation) error, in which the elements of the model were wrongly measured; and equation error, in which the wrong formula was used. (2) In hypothesis testing, the type of error associated with the acceptance or rejection of the null hypothesis. See *Type I error, Type II error, standard error*. **erreur**

escarpment A long, more or less continuous cliff or relatively steep slope facing in one general direction, separating two level or gently sloping surfaces, and produced by erosion or faulting. **escarpement**

Escherichia The genus name of a type of bacteria whose normal habitat is the colon in humans and other warm-blooded animals. The organism is *Gram-positive*, ferments lactose at 37°C, and can grow with or without molecular oxygen. If members of this genus, referred to as *fecal coliforms*, are found in water, the water is considered to be contaminated with fecal material. ***Escherichia***

esker (*geomorphology*) A serpentine ridge of roughly stratified gravel and sand that was deposited by a stream flowing in or beneath the ice of a stagnant or retreating glacier and was left behind when the ice melted. Length ranges from less than 100 m to more than 500 km (including gaps), and in height from 3 m to more than 300 m. **esker**

esker, beaded (*geomorphology*) Sinuous ridge of sand and gravel in which wider parts on the ridge are linked by narrower segments, creating the overall impression of a necklace of beads. See *esker*. **esker en chapelet**

essential element A chemical element required for the normal growth of plants. **élément essentiel ou constitutif**

ester An organic compound produced by the reaction between a *carboxylic acid* and an *alcohol*. Triesters, molecules which contain three ester groups, occur in nature as oils and fats. **ester**

estivation The reduction of biologic activity by an organism during the summer or, more generally, during periods that are hot, dry, or both. **estivation**

estuarine Of, pertaining to, or formed in an *estuary*. **estuarien**

estuary An arm of the sea at the mouth of a river, or the drowned mouth of a river, occurring where the tide meets the river current. **estuaire**

euhedral A crystal with good to perfect crystallographic form. See *anhedral* and *subhedral*. **euèdre**

euic High level of bases in soil material, specified at the family level of classification. **euique**

eukaryote A cell that has a membrane-bound nucleus, membrane-bound organelles, and chromosomes in which the DNA is associated with proteins; an organism composed of such cells. **eucaryote**

eukaryotic Of a cell type in which the nuclear material is bounded by a membrane. **eucaryote**

euphotic zone (*limnology*) The surface or upper layer of any body of water through which light can penetrate, thereby leading to photosynthesis. **région euphotique**

euryhaline Tolerant of considerable difference in osmotic pressure and salinity. **euryhalin**

eustasy, eustatism World-wide changes of sea level due to an actual fall or rise of the ocean as opposed to a movement of the land. See *isostasy*. **eustatisme, eustasie**

eustatic See *eustasy*. **eustatique**

Eutric Brunisol A great group of soils in the *Brunisolic* order (Canadian system of soil classification). The soils may have mull Ah horizons less than 10 cm thick, and they have Bm horizons with a high degree of base saturation. **brunisol eutrique**

eutroph A species that grows at high nutrient concentrations. **eutrophe**

eutrophic Habitats, particularly soils and water, that are rich in nutrients. See *mesotrophic, oligotrophic*. **eutrophe**

eutrophic lake A lake characterized by large amounts of plant nutrients and algae, low water transparency, low dissolved oxygen, and high biological oxygen demand. Eutrophic lakes are often shallow and seasonally deficient in oxygen in the *hypolimnion*. See *oligotrophic lake, hypolimnion*. **lac eutrophe**

eutrophication The normally gradual process of aging whereby a lake evolves into a marsh, due to nutrient enrichment leading to an increase in algae and other aquatic plants and depletion of dissolved oxygen. The process can be accelerated by human activities that promote enhanced surface runoff and the enrichment of waters with nitrogen and phosphorus. **eutrophisation**

evaporation The process in which a liquid is changed into a gas by molecular transfer. Evaporation rates depend on *insolation*, temperature differences between air and water, humidity of the air, wind velocity, and the nature of the ground surface. Rates of evaporation are increased by large amounts of insolation, dry air, high wind speeds, and bare ground; they are decreased by small amounts of insolation, low wind speeds, a weak vapor flux, high humidity, and vegetation-covered ground. **évaporation**

evaporation pond Shallow artificial pond in which sewage sludge is placed to dry and then removed for further treatment or disposal. **étang d'évaporation, lagune**

evaporimeter A device used to measure rates and amounts of *evaporation*. **évaporimètre**

evaporite Sediment deposited from an aqueous (water) solution as a result of extensive or local evaporation (e.g., anhydrite, rock salt, and borates). **évaporite**

evapotranspiration The loss of water from a given area during a specified time by evaporation from the soil surface and transpiration from the plants. Potential or reference evapotranspiration is the maximum transpiration that can occur in a given weather situation with a low-growing crop (e.g., grass or alfalfa) that is not short of water. **évapotranspiration**

evenness A measure of the the distribution of individuals across species. The more equitable the distribution, the greater the evenness. Evenness is a component of species *diversity*. **régularité**

evergreen Vegetation that is never entirely without green foliage, with leaves persisting until a new set has appeared. **végétation à feuillage persistant**

evergreen plant A plant that retains its leaves throughout the year. The term is most commonly applied to trees such as spruces and firs. See *deciduous plant*. **plante à feuillage persistant**

evolution The development of progressively more complex forms of life from simple ancestors; the phenomenon of common ancestry, appearance, diversification, change, and extinction of organisms throughout the history of Earth. Darwin proposed natural selection as the mechanism by which evolution occurs. See *natural selection*. **évolution**

E

evolutionary opportunism The principle that adaptations are based on whatever genetic variation is available in a population. **opportunisme évolutif**

evolutionary time Periods of time extending across generations. **période d'évolution**

exchange complex The surface-reactive particles in soils that exchange *cations*. **complexe d'échange**

exchangeable anion A negatively charged *ion* held on or near the surface of a solid particle by a positive surface charge and which may be replaced by other negatively charged ions. **anion échangeable**

exchangeable base *Cation* or base adsorbed onto the surface of soil particles; in most soils Ca^{+}, Mg^{2+}, and K^{+} predominate. This term is often considered incorrect by soil chemists because these cations are not bases by any modern definition. **base échangeable**

exchangeable cation A positively charged ion held on or near the surface of a solid particle by a negative surface charge and which may be replaced by other positively charged ions in the soil solution. Often determined as the salt-extractable minus water-soluble cations in a saturation extract, and expressed in centimoles or millimoles of charge per kilogram. **cation échangeable**

exchangeable cation percentage The extent to which the adsorption complex (i.e., *cation exchange capacity)* of a soil is occupied by a particular cation. **pourcentage de cations échangeables**

exchangeable nutrient A plant nutrient that is held by the adsorption complex of the soil and is easily exchanged with the *anion* or *cation* of normal salt solution. **élément nutritif échangeable**

exchangeable sodium fraction The fraction of the cation exchange capacity of a soil occupied by sodium ions. **fraction de sodium échangeable**

exchangeable sodium percentage (ESP) The *exchangeable sodium fraction* expressed as a percentage. **pourcentage de sodium échangeable, sodivité**

exchangeable sodium ratio (ESR) The ratio of exchangeable sodium to all other exchangeable cations. **rapport de sodium échangeable**

exclosure An area fenced to exclude animals. **exclos**

exfoliation (*geology*) The process by which concentric scales, plates, or shells of rock, from less than 1 cm to several m in thickness, are successively stripped from the bare surface of a large rock mass by physical or chemical forces. **desquamation (mm-cm), exfoliation (m)**

exhumed paleosol See *paleosol, exhumed*. **paléosol exhumé**

exoenzyme An enzyme excreted by a microorganism into the environment. An enzyme that acts outside the cell; also called extracellular enzyme. **exoenzyme**

exogenetic Pertaining to processes originating at or near the surface of the Earth, (e.g., weathering and denudation) and to rocks, ore deposits, and landforms that owe their origin to such processes. Also called exogenic, exogenous. **exogène**

exogenous Originating outside an organism. See *endogenous*. **exogène**

exogenous variable (*modeling*) An input variable, independent of the internal state of the system. It represents an external force that is imposed on the system and, acting on it, induces changes within it. Also called forcing function or driving function. **variable exogène**

exothermic Pertaining to a chemical reaction that occurs with a liberation of heat. See *endothermic*. **exothermique**

exotic Not indigenous to the area in which it is found, referring to an organism or species. **exotique**

exotoxin A protein produced and released by the bacteria that cause certain diseases in humans (e.g., *botulism* and tetanus). These proteins are extremely toxic in microgram quantities. **exotoxine**

experimental design *(statistics)* The design or planning of an experiment having the following statistical aspects: (1) selection of treatments (factors, and levels of factors) whose effects are to be studied; (2) specifying a layout for the experimental units (e.g., plots) to which the treatments are to be applied; (3) identifying rules according to which the treatments are to be distributed among the experimental units, and (4) specifying the measurements to be made for each experimental unit. These various aspects must be completed in a manner such that the techniques to be used in analysis of the results are clear prior to the conduct of the experiment. **dispositif ou plan expérimental plan d'experience**

experimental unit The subject, object, area, grouping, or subdivision to which a treatment is applied in an experimental design. **unité expérimentale**

exploit To make use of; to derive benefit from. **exploiter**

exploitation competition Reduction of the availability of a resource, and thus the efficiency of exploitation of that resource by other individuals. **compétition d'exploitation**

exploitation equilibrium Stability between predator and prey populations in which the counter-evolutionary responses of one just balances those of the other. **équilibre d'exploitation**

exponential decay The decline in the number of a population, amount of a pollutant, or level of radioactivity according to the exponential function $N=Noe^{-kx}$, where *N* is the amount left after decay, *No* is the initial amount, *k* is a constant, and *x* is a variable such as time, altitude, or water depth. **décroissance exponentielle**

exponential growth Growth in the size of a population in which the number of individuals increases by a constant percentage of the total population each time period; represented by the equation $N(t) = N_0e^{kt}$, where $N(t)$ is the population size at time *t*, *e* is the base of the natural logarithm, and *k* is a constant. **croissance exponentielle**

exponential growth phase (*microbiology*) The phase in the growth, usually of bacteria, in which the statistical population is growing at an exponential rate. Same as *log growth phase*. See *growth phase*. **phase de croissance exponentielle**

exposure Contact between a chemical, physical, or biological agent and the outer surfaces of an organism; exposure to an agent does not imply that it will be absorbed or that it will have an effect. **exposition**

exposure characterization Identification of the conditions of contact between a substance and an individual or population; may involve identification of concentration, routes of uptake, target sources, environmental pathways, and the population at risk. **condition d'exposition**

extended rotation A cropping sequence that is lengthened by adding additional crop species to the sequence. **rotation longue ou allongée**

extension services Assistance provided, usually by government agencies, to individual operators in farming, agroforestry, and woodland production. **vulgarisation**

E

E

external drainage The natural elimination or accumulation of precipitation water on the soil surface. **drainage superficiel**

extinction (*ecology*) The total disappearance of a species or higher taxon, so that it no longer exists anywhere. The disappearance of a lake by drying up or by destruction of the lake basin. **extinction**

extinction angle The angle at which a crystal goes black in crossed polarized light. **angle d'extinction**

extirpation The elimination of a plant or animal species from a particular area, but not from its entire range. **extirpation**

extracellular enzyme See *exoenzyme*. **enzyme extracellulaire**

extract, soil The solution separated from a soil suspension or from a soil by filtration, centrifugation, suction, or pressure. **extrait de sol**

extractant The solvent or chemical used to remove a substance (e.g., nutrients) into solution out of the soil. Extractants are formulated to dissolve and remove specific substances. **solution d'extraction**

extrapolation Estimating values which lie beyond the range of values on the basis of which a predicting equation was obtained. **extrapolation**

extrinsic factor An environmental agent independent of an ecosystem that influences organisms and their environments without itself being modified. **facteur extrinsèque**

exudate Material (fluids or cells) that has been released or sloughed off from microbes or plant roots (e.g., the products of root growth in soil). **exsudat**

F

F

F horizon An organic soil horizon containing more than 17% organic carbon, normally associated with upland forest soils, which is characterized by an accumulation of partly decomposed leaves, twigs, woody materials, and mosses. The material may be partly comminuted by soil fauna as in *moder*, or it may be a partly decomposed mat permeated by fungal hyphae as in *mor*. An organic horizon in the Canadian system of soil classification. See *Appendix D* for equivalent terms in the U.S. and FAO/UNESCO systems of soil classification. See also *horizon, soil; L horizon*; *H horizon*. **horizon F**

fabric, soil The physical constitution of a soil material expressed by the spatial arrangement of the solid particles and associated voids. **organisation du sol, état structural du sol**

facet (1) A face on a crystal. (2) Any plane surface abraded on a fragment of rock by ice or wind. **facette**

facies (*geology*) (1) The aspect, appearance, and characteristics of a rock unit, especially as compared with adjacent or associated units. (2) A rock unit or group of units that exhibits lithological, sedimentological, and faunal (fossil) characteristics that enable them to be classified as distinct from another rock unit or group. (3) A lateral change of character within a stratigraphic unit, especially in its *lithology*. **faciès**

factor Any variable or quantity under investigation in an experiment. **facteur**

factor analysis *(statistics)* A type of multivariate analysis whereby a set of data are linearly expressed so as to identify the minimum number of influences necessary to account for the maximum observed variation, and for indicating the extent to which each influence accounts for the variation observed. **analyse factorielle**

factors of soil formation See *soil formation factors*. **facteurs de formation du sol**

factorial experiment An experiment used in conjunction with variance analysis to determine the effect of more than one factor upon a variate by examining all the possible combinations of their different levels. See *variance*. **plan d'expérience factoriel**

facultative Capable of adaptive response to varying environment. **facultatif**

facultative organism An organism that is able to carry out both options of a mutually exclusive process (e.g., aerobic and anaerobic metabolism). May also be used in reference to other processes such as photosynthesis (e.g., a facultative photosynthetic organism is one that can use either light or the oxidation of organic or inorganic compounds as a source of energy). **organisme facultatif**

failure The deformation of a rock or soil marked by the sudden formation of fractures and loss of cohesion, and the inability to resist continued stress. **rupture**

Fairfield-Hardy digester A machine that decomposes garbage, sewage sludge, and industrial and other organic wastes by a controlled continuous aerobic-thermophilic process. **digesteur ou bioréacteur Fairfield-Hardy**

F

fall cone penetrometer A variety of *cone penetrometer* that utilizes dropping weights to provide known increments of force applied to the cone, resulting in measured increments of soil penetration. **pénétromètre à cône tombant**

fall overturn (*limnology*) Circulation of lake water in autumn. The *epilimnion* cools and the *thermocline* disappears, resulting in the whole lake being circulated by wind friction. **renversement automnal**

fallout (1) Fragmental material ejected from an impact or explosion crater and eventually redeposited in and around it. (2) The descent of usually radioactive particles through the atmosphere following a nuclear explosion. Also refers to the particles themselves. **(1) dépôt atmosphérique (2) retombée radioactive**

fallow See *tillage, fallow.* **jachère**

fallow land Land not being used to grow a crop, but cultivated or left untilled during the whole or greater portion of the growing season to preserve water, kill weeds, and increase soil nutrients. In arid and semi-arid regions, a fallow year is commonly used in crop rotations. See *summerfallow.* **terre en jachère**

false negative An erroneous test result that labels a chemical or individual as not having a certain property or condition when in fact the property or condition is present. For example, the false determination that a chemical is a non-carcinogen when it actually is a carcinogen. See *false positive.* **faux négatif**

false positive An erroneous test result that labels a chemical or individual as having a certain property or condition when in fact the property or condition is not present. For example, a false test result stating that a person is infected with a virus when in fact the virus is not present. See *false negative.* **faux positif**

family A unit or group in the taxonomic classification of organisms composed of one or more genera; a subdivision of an *order.* **famille**

family, soil A category in the Canadian, U.S., and other systems of soil classification. Soil families are differentiated primarily by texture, drainage, thickness of horizons, permeability, mineralogy, consistence, and reaction. **famille de sols**

fan (*geomorphology*) Landform with a perceptible gradient from the apex to the toe. Deposited by a stream when it emerges from an upslope position onto a lowland with a marked decrease in gradient. **éventail, glacis**

FAO/UNESCO Soil Classification system See *Soil Classification, FAO/UNESCO soil units.* **système de classification des sols FAO/UNESCO**

farm forestry The practice of forestry on farm or ranch lands generally integrated with other farm or ranch operations. See *agroforestry.* **foresterie paysane ou rurale**

farm pond A water impoundment made by constructing a dam or embankment or by excavating a pit. **étang fermier**

farming system A decision-making unit comprised of a farming household, livestock, and cropping systems, that produces crop and animal products for consumption. **système de production**

fat (glyceride) An ester composed of glycerol and fatty acids. **graisse (glycéride)**

fat clay Clay of relatively high plasticity. **argile plastique**

fats Organic compounds containing carbon, hydrogen, and oxygen, (see *fat glyceride*); the proportion of oxygen to carbon is much less in fats than it is in carbohydrates; fats in the liquid state are called oils. **gras**

fatty acid An organic compound consisting of a hydrocarbon chain and a terminal carboxyl group. See *carboxylic acid.* **acide gras**

fault (*geology*) A fracture or fracture zone in the Earth's crust, along which there has been slippage and displacement of sides relative to one another parallel to the fracture. **faille**

faulting The process of fracturing and displacement that produces a *fault*. **formation de faille.**

fauna The animals in a particular region or geologic period. **faune**

fecal bacteria Any type of bacteria whose normal habitat is the colon of warm-blooded mammals, such as humans. These organisms are usually divided into groups, such as *fecal coliform* and *fecal streptococcus*. **bactéries fécales**

fecal coliform A type of bacteria whose natural habitat is the colon of warm-blooded mammals, such as humans. Specifically, the group includes all of the rod-shaped bacteria that are non-spore forming, *Gram-negative*, lactose-fermentating in 24 hours at 44.5°C, and which can grow with or without oxygen. The presence of this type of bacteria in water, beverages, or food is usually taken to mean that the material is contaminated with solid human waste. Bacteria included in this classification represent a subgroup of the larger group called coliform. **coliforme, colibacille fécal**

fecal material Solid waste produced by humans and other animals and discharged from the gastrointestinal tract; a component of domestic sewage. **matières fécales, excréments**

fecal pellets The excreta of fauna. **turricules**

fecal streptococcus A group of bacteria normally present in large numbers in the intestinal tracts of warm-blooded animals other than humans. By assessing the ratio of coliforms to streptococci in a water sample, a rough estimate can be made of the relative contribution of fecal contamination from the two sources. **streptocoque fécal**

feed Harvested forage (e.g., hay, fodder, or grain, grain products, and other foodstuffs processed for feeding livestock). **aliment des animaux**

feedback A process by which the result of a controlling operation is used as part of the data on which the next controlling operation is based, enabling a system to make a response. Thus, when a change is introduced via one of the system variables, its transmission through the structure leads the effect of the change back to the initial variable, giving a circularity of action. There are two types of feedback: negative feedback, in which the effect of the change is to counteract the impact of the initial alteration, and positive feedback, when the effect of the change is to cause the system to continue changing in the same direction (i.e., a snowballing effect which may cause the system to go out of control). A feedback process in which chance does not operate is known as deterministic feedback. A feedback loop is the path by which the feedback process is accomplished. See *general systems theory*. **feedback, rétroaction**

feedback control system A system that contains one or more feedback loops and is capable of exerting different orders of self-control. **système à rétrocontrôle, système asservi**

feedback inhibition Inhibition of an enzyme caused by a product of that enzyme's activity on a substrate. **inhibition rétroactive**

feedback mechanism A mechanism that connects one aspect of a system to another. The connection can be either amplifying (positive feedback) or moderating (negative feedback). See *climate feedback*. **mécanisme de rétroaction**

feedlot A confined area in which cattle are held and fed to promote maximum weight gain prior to marketing. The large quantities of animal waste produced can sometimes lead to pollution problems in nearby water bodies. **parc d'engraissement**

feldspar (1) A group of minerals of the general formula, M-Al(Al, Si)$_3$O$_8$, where M can be K, Na, Ca, Ba, Rb, Sr, or Fe. Feldspars are the most abundant of any mineral group, constituting about 60% of the Earth's crust and occurring in all types of rock. Feldspars are white and gray to pink, have a hardness of 6 (*Mohs scale*), are commonly twinned, have monoclinic or triclinic symmetry, and show good cleavage in two directions. (2) A mineral of the feldspar group, (e.g., microcline). **feldspath**

feldspathic Said of a rock or other mineral aggregate containing *feldspar.* **feldspathique**

felsic Descriptive of light-colored rocks containing an abundance of feldspar and quartz. **felsique**

fen A *peatland* with the water table usually at or just above the surface. The waters are mainly nutrient-rich and *minerotrophic* due to contact with mineral soils. The dominant materials are moderately to well-decomposed sedge and/or brown moss peats of variable thickness. The soils are mainly *Mesisols*, *Humisols*, and Organic *Cryosols* (*Hemists*, *Saprists*, and *Histels*). The vegetation consists predominantly of sedges, grasses, reeds, and brown mosses, with some shrubs and, at times, a sparse tree layer. (A *wetland class* in the Canadian wetland classification system). **tourbière minérotrophe**

fen peat Peat of sedge and sedge-brown moss origin. It may contain wood, and it is generally neutral in pH but ranges from weakly acidic to alkaline. (Fen or sedge peat is an organic genetic material in the *Canadian system of soil classification*). **tourbe minérotrophe**

fen soils A general term pertaining to organic soils formed in fen peat in an environment of base-rich groundwater. Thus, fen soils have an alkaline or neutral reaction, and when drained give rise to extremely fertile and productive agricultural soils. See *peat soils.* **sols de fen**

fermentation A type of bacterial or yeast metabolism (chemical reaction) characterized by the conversion of carbohydrates to acids and alcohols, usually occurring in the absence of molecular oxygen. **fermentation**

ferallitic soil A soil of hot and humid climates that has undergone intense and prolonged weathering. Most or much of the silica and virtually all base cations have been removed, leaving iron and aluminum. The profile consists largely of quartz, kaolinite, gibbsite, and hematite or goethite. **sol ferrallitique**

ferrallitization See *desilication.* **ferrallisation**

ferran (soil micromorphology) A *cutan* composed of a concentration of iron oxides. **ferrane**

ferri-argillan (*soil micromorphology*) A *cutan* composed of intimately mixed clay minerals and iron oxides. **ferri-argilane**

ferricrete An accumulation of iron oxides and hydroxides in the soil, usually as a zone of iron oxide cementation near the surface, resulting from weathering or soil-forming processes such as *laterization.* **conglomérat à ciment ferrugineux**

ferriferous Iron-bearing; refers to minerals containing iron, or to materials (e.g., sedimentary rock) that are richer in iron than usual. **ferrifère**

ferrihydrite A dark reddish-brown, poorly crystalline iron oxide mineral ($Fe_5O_7(OH).4H_2O$) that forms in wet soils. It occurs in concretions and placic horizons and often can be found in ditches and pipes that drain wet soils. **ferrihydrite**

Ferrods A suborder in the U.S. system of soil taxonomy. Spodosols that have more than six times as much free iron (elemental) than organic carbon in the spodic horizon. Ferrods are rarely saturated with water or do not have characteristics associated with wetness. **ferrods**

Ferro-Humic Podzol A great group of soils in the *Podzolic* order (Canadian system of soil classification). The upper 10 cm of the Bhf horizon contain 5% or more organic carbon, 0.6% or more pyrophosphate-extractable Al and Fe, and either a ratio of organic carbon to pyrophosphate-extractable Fe of less than 20 or a percentage of pyrophosphate-extractable Fe greater than 0.3, or both. The *B horizon* is usually overlaid with a light-colored, eluviated horizon (Ae) and a light-colored humus layer. **Podzol ferro-humique**

ferrolysis Destruction of clay involving its disintegration and solution in water by a process based upon the alternate reduction and oxidation of iron. **ferrolyse**

ferromagnesian Containing iron and magnesium; applied to mafic minerals, especially amphibole, pyroxene, biotite, and olivine. **ferromagnésien**

ferruginous Pertaining to or containing iron; used in reference to rock having a red or rusty color due to the presence of ferric oxide, the quantity of which may be very small (e.g., red-colored sandstone cemented by ferric oxides). **ferrugineux**

fersiallic soil Soil characterized by a red-colored *B horizon* containing illuvial clay, a lack of acidification, and the redisposition of calcium carbonate within the *C horizon*. It is associated with Mediterranean climates which have a cool, wet season followed by a hot, dry summer. **sol fersiallique**

fersiallitization A weathering process typical of subtropical climates with strong seasonal contrasts. It involves the *inheritance* and *neoformation* of smectitic clays and the immobilization of iron oxides owing to alkaline conditions. **fersiallitisation**

fertigation Application of plant nutrients through irrigation water. **irrigation fertilisante, fertigation**

fertility, soil The ability of a soil to supply nutrients necessary for plant growth. **fertilité du sol**

fertilization Application of plant nutrients to the soil in the form of commercial fertilizers, animal manure, green manure, compost, and other *amendments*. **fertilisation**

F

fertilization, foliar Application of a dilute solution of fertilizer nutrients to plant foliage, usually to supplement nutrients absorbed by plant roots. **fertilisation foliaire**

fertilizer Any organic or inorganic material (other than liming material) of natural or synthetic origin that is added to a soil to supply elements essential to plant growth. **engrais**.

Types of fertilizer include:

acid-forming – at fertilizer that increases the residual acidity of the soil and eventually decreases its pH following application. ***engrais acidifiant***

bulk-blend – two or more granular fertilizers of similar size mixed together to form a compound fertilizer (also called blended fertilizer). ***engrais de mélange***

chemical – a manufactured fertilizer product containing a substantial amount of one or more of the primary nutrients unless otherwise noted; the manufacturing process usually involves chemical reactions, but may consist of refining or physically processing naturally occurring materials (e.g., potassium salts or sodium nitrate). ***engrais minéral***

coated – a granular fertilizer that has been coated with a thin layer of some substance (e.g., clay) to prevent caking or to control dissolution rate. ***engrais enrobé***

complete – a fertilizer containing significant quantities of the primary plant nutrients (nitrogen, phosphorus, and

F

potassium) in appropriate forms for increasing soil fertility. ***engrais ternaire***

compound – a fertilizer formulated with more than one plant nutrient. ***engrais composé***

conditioned – a fertilizer treated with an additive to improve physical condition or prevent caking; the conditioning agent may be applied as a coating or incorporated into the product. ***engrais conditionné***

controlled-release – a fertilizer that dissolves in a controlled fashion at a lower rate than conventional water-soluble fertilizers (also called delayed release, slow release, controlled availability, slow acting, and metered release fertilizer). Delayed dissolution may result from coatings on water-soluble fertilizers or from low dissolution and/or mineralization rates of fertilizer materials in soil. ***engrais à libération lente***

granular – a fertilizer processed to achieve uniform size, stability, and shape, sized between an upper and lower limit or between two screen sizes, usually within the range of 1 to 4 millimeters, often more closely sized. ***engrais granulaire***

injected – placement of fluid fertilizer or anhydrous ammonia into the soil either through use of pressure or nonpressure systems. ***engrais injecté***

inorganic – a fertilizer material in which carbon is not an essential component of its basic chemical structure. Urea is often considered an inorganic fertilizer because of its rapid hydrolysis to form ammonium ions in soil. ***engrais minéral***

liquid – a fertilizer wholly or partially in solution that can be handled as a fluid; includes clear liquids, liquids containing solids in suspension, and usually anhydrous ammonia; however, anhydrous ammonia sometimes is referred to as a gaseous fertilizer even though it is applied as a liquid (also called fluid fertilizer). ***engrais liquide, ammoniac anhydre***

mixed –two or more fertilizer materials blended or granulated together into individual mixes; includes dry mix powders, granulated, clear liquid, suspension, and slurry mixtures. ***engrais composé***

nongranular – a fertilizer containing fine particles, usually with some upper limit, such as 3 millimeters, but no lower limit. ***engrais non granulaire***

organic – a material containing carbon and one or more plant nutrients in addition to hydrogen and/or oxygen. ***engrais organique***

pop-up – fertilizer placed in small amounts in direct contact with the seed. ***engrais de démarrage***

prilled – a type of granular fertilizer of near-spherical form made by solidification of free-falling droplets in air or other fluid medium. ***engrais perlé***

salt index – the ratio of the decrease in osmotic potential of a solution containing a fertilizer compound or mixture to that produced by the same weight of $NaNO_3 \times 100$. ***indice de salinité de l'engrais***

sidedressed – fertilizer application made to the side of crop rows after plant emergence. ***engrais appliqué en bande en post-levée***

slow-release – Controlled release fertilizer. ***engrais à libération lente***

solution – an aqueous liquid fertilizer free from solids. ***engrais liquide***

starter – a fertilizer applied in small amounts with or near the seed for the purpose of accelerating early growth of the plant. ***engrais de démarrage***

straight – a fertilizer containing only one nutrient (e.g., urea or superphosphate). ***engrais simple***

suspension – a liquid (fluid) fertilizer containing solids held in suspension (e.g., by the addition of a small amount of clay); the solids may be water-soluble materials in a saturated solution, insoluble, or both). ***engrais en suspension***

top-dressed – the surface application of fertilizer to a soil after the crop has been established. ***engrais en post-levée***

fertilizer analysis The percentage composition of fertilizer, expressed in terms of nitrogen (N), phosphorus pentoxide (P_2O_5), and potassium oxide (K_2O) and sometimes minor elements (e.g., a fertilizer with a 6–12–6 analysis contains 6% nitrogen (N), 12% P_2O_5, and 6% K_2O. **composition élémentaire d'un engrais**

fertilizer fixation The process by which available plant nutrients are rendered less available or unavailable in soil. **rétrogradation d'engrais**

fertilizer formula The quantity and grade of the crude stock materials used in making a fertilizer mixture. **formule d'engrais**

fertilizer grade The guaranteed minimum analysis, expressed in percent, of the major plant nutrient elements contained in a fertilizer material or in a mixed fertilizer; usually designated as N–P_2O_5–K_2O, but possibly N–P–K in areas where permitted or required by law. See *fertilizer analysis*. **formule d'engrais, teneur d'un engrais**

fertilizer ratio The relative proportion of each nutrient supplied by the fertilizer, obtained by dividing the fertilizer grade by a factor that produces the smallest possible whole numbers (e.g., 6–24–24 fertilizer grade thereby becomes a 1–4–4 ratio). **rapport d'éléments fertilisants**

fertilizer requirement The quantity of certain plant nutrient elements needed, in addition to the amount supplied by the soil, to increase plant growth to a designated optimum. **besoin en engrais**

fertilizer value A monetary value assigned to a quantity of organic wastes representing the cost of obtaining the same plant nutrients in their commercial form and in the amounts found in the waste. **valeur fertilisante**

fertilizer, manufactured Manufactured fertilizers are the traditional method of marketing mixed fertilizers. The manufacturer chooses a fertilizer ratio that fits the needs of a market area, selects appropriate ingredients, mixes them in the right proportions, and forms them into granules. Micronutrients are sometimes included, especially if the fertilizer is designated as a premium grade. Manufactured fertilizers are mass-produced to minimize production costs, have uniform composition, and are sold in bags labeled with descriptive information. **engrais de synthèse**

F

fiber The organic material retained on a 100-mesh sieve (0.15 mm) either with or without rubbing, except for wood fragments that cannot be crushed in the hand and are larger than 2 mm in the smallest dimension. Rubbed fiber refers to materials rubbed between the fingers ten times, or processed in a blender. This characteristic is an indicator of the degree of humification and is used in soil classification to distinguish *fibric, humic,* and *mesic* types of organic materials. **fibre**

fibric (peat) (1) Organic material containing large amounts of weakly decomposed fibers whose botanical origins are readily identifiable; fibric material has 40% or more of rubbed fiber by volume (or weight of rubbed fiber retained on a 100 mesh sieve) and is classified in the *von Post* scale of decomposition as class 1 to class 4. (2) In the U.S. system of soil taxonomy and the Canadian system of soil classification, descriptive of the least decomposed of the *organic soils*. See *Fibrisol, Histosol*. **fibrique**

fibric layer A layer of organic soil material containing large amounts of weakly decomposed fiber whose botanical origin is readily identifiable. **couche fibrique**

fibric materials See *organic soil materials*. **matériaux fibriques**

Fibrisol A great group of soils in the *Organic* order (Canadian system of soil classification) that are saturated for most of the year. The soils have a dominantly fibric middle tier or middle and surface tiers if a terric, lithic, hydric, or cryic contact occurs in the middle tier. **fibrisol**

Fibrists A suborder in the U.S. system of soil taxonomy. *Histosols* that have a high content of undecomposed plant fibers and a bulk density less than about 0.1 g cm^{-3}. Fibrists are saturated with water for periods long enough to limit their use for most crops unless they are artificially drained. **Fibrist**

fibrous Containing, consisting of, or like fibers. **fibreux**

fibrous root system A plant root system having a large number of small, finely divided, widely spreading roots but no large individual roots (e.g., grass root system). See *taproot system*. **système racinaire fasciculé**

fiducial mark (*surveying*) An index line or point; a line or point used as a basis of reference. (*photogrammetry*) (1) Index markers, usually four in number, rigidly connected to the camera lens through the camera body, that form images on the negative defining the principal point of a photograph. (2) The markers in any instrument that define the axes whose intersection fixes the principal point of a photograph and fulfills the requirements of interior orientation. **repère de cadre**

field (1) An area of open land, especially one used for pasture or crops. (2) A region that is known for a particular mineral resource (e.g., coal field, gold field). (3) That space in which an electric, gravitational, or magnetic effect occurs and is measurable. It has continuity in that there is a value associated with every location within the space. **champ**

field of view (*remote sensing*) The solid angle through which an instrument is sensitive to electromagnetic radiation. The field of view controls the area of the Earth's surface sensed by the sensor. For example, the *Landsat* satellite has a field of view of 11 to 56°, compared to 90 to 120° for most *multispectral scanners*. **angle de champ, champ de visée**

field blanks Samples of *analyte*-free media that are transferred from one vessel to another, or exposed to the same sampling environment. They are used to measure incidental or accidental contamination of a sample during the whole process (sampling, transport, sample preparation, and analysis). **blancs d'essai**

field capacity For agronomic purposes, the *in situ* content of water, on a mass or volume basis, remaining in a soil two or three days after having been wetted with water and after free drainage becomes negligible. Often called field water capacity. See *available water*. See *figure*. **capacité au champ**

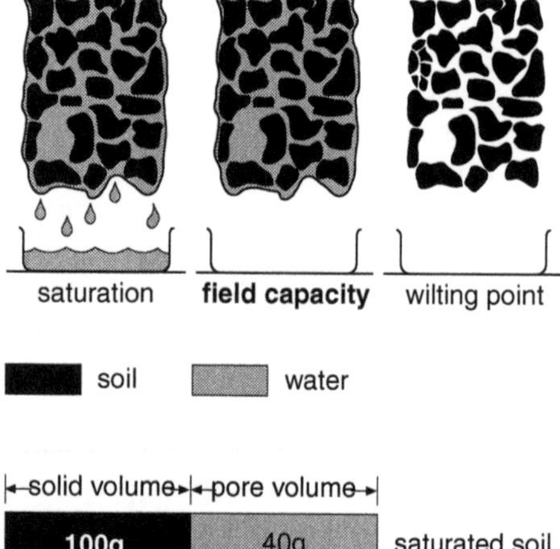

Volume of solids, water, and air in a loam soil at saturation, **field capacity,** and wilting point (adapted from Brady and Weil, 1996).

field trial An experiment conducted under field conditions. Ordinarily less subject to control than a formal experiment, and maybe less precise; also called field test. **essai en champ**

filamentous algae Aggregations of one-celled plants that grow in long strings or mats in water and are either attached or free floating; tend to plug canals, weirs, and other structures; also provide habitat for invertebrate animals. **algue filamenteuse**

fill (1) (*geology*) Any sediment deposited by any agent so as to fill or partly fill a channel, valley, sink, or other depression. (2) (*waste management*) A site on which solid waste is disposed of using sanitary landfilling techniques. (3) (*civil engineering*) A site on which earth is moved to raise elevation. (4) (*land reclamation*) Depth of material which is to be placed (filled) to bring the surface to a predetermined grade. Also, the material itself. See *backfill*. **(1, 2) décharge (3, 4) remblai**

filling (*land reclamation*) The process of depositing dirt and mud in marshy areas to create more land for real estate development. Filling can disturb natural ecological cycles. **remblaiement**

filling in See *terrestrialization*. **comblement**

film water A layer of water (i.e., hydration envelope) that is bound to and surrounds soil particles due to adsorptive forces, and varies in thickness from 1 or 2 to 100 or more molecular layers. **eau pelliculaire**

filter cloth (*land reclamation*) Synthetic fabric used as a filter, usually beneath rock riprap or between materials with significant differences in size, to prevent movement of fine material through coarser material. **toile filtrante**

filter sand (*wastewater management*) Sand suitable for use in filtering suspended matter from water in a water treatment facility. **sable filtrant**

filter strip A strip of vegetation along a watercourse for removing sediment, organic material, organisms, nutrients, and chemicals from runoff or wastewater. See *buffer strip*. **bande filtrante**

filtration (*wastewater management*) The mechanical process that removes particulate matter by separating water from solid material, usually by passing it through sand. **filtration**

final clarifier (*wastewater management*) A gravitational settling tank placed after the *biological wastewater treatment step*. The tank functions to remove *suspended solids*. Also called a secondary clarifier. **clarificateur terminal**

F

fine clay A clay fraction of specified size less than 2 μm, usually less than 0.2 or 0.08 μm. **argile fine**

fine earth The fraction of mineral soil consisting of particles less than 2 mm in *equivalent diameter*. **terre fine**

fine grained (*geology*) (1) Descriptive of igneous rock and its texture, in which particles have an average diameter less than 1 mm. (2) Descriptive of sedimentary rock and its texture, in which particles have an average diameter of silt and smaller. **à grain fin**

fine sand (1) A soil separate. (2) A soil textural class. See *soil separate, texture*. **sable fin**

fine tailings (fine tails, sludge) The material accumulating at the bottom of oil sands tailings ponds. It is a matrix of dispersed clays, fine minerals, residual hydrocarbons, and various contaminants. Whole tailings (plant tailings) includes tailings sand, which settles rapidly and is used to form tailings dykes. See *tailings*. **résidus fins**

fine texture Consisting of or containing large quantities of the fine soil fractions, particularly of silt and clay. It includes all the textural classes of clay loams and clays: clay loam, sandy clay loam, silty clay loam, sandy clay, silty clay, and clay. Sometimes it is subdivided into clayey texture and moderately fine texture. See *texture*. **texture fine**

fines A term used in soil mechanics for the portion of a soil finer than a No. 200 (75-μm opening) U.S. standard sieve. **fraction fine**

finished water (*wastewater management*) Water that has completed a purification or treatment process. **eau traitée**

fire blight A disease of apples, pears, hawthorns, pyracantha, and related members of the rose family caused by a bacteria (genus *Erwinia*); results in death of twigs and branches; tips of shoots turn black and die back, even to main limbs. **brûlure bactérienne**

fire management (*forestry*) The activities concerned with the protection of people, property, and forest areas from wildfire, and the use of prescribed burning for the attainment of forest management and other land use objectives, all conducted in a manner that considers environmental, social, and economic criteria. **gestion des incendies ou des feux**

fire point The lowest temperature at which a liquid will vaporize at a sufficient rate to support continuous combustion. See *flash point*. **point d'inflammation spontanée**

fire suppressant An agent directly applied to burning fuels to extinguish the flaming and smoldering or glowing stages of combustion. **agent extincteur**

firebreak A natural or artificial barrier usually created by the removal of vegetation and used to prevent or retard the spread of a fire. **pare-feu**

fireclay A sedimentary clay containing large amounts of hydrous aluminum silicates, which can resist high temperatures without disintegrating or deforming and is therefore useful in manufacturing brick and crucibles. **argile réfractaire**

firm The *consistence* of a moist soil that noticeably resists crushing, but can be crushed with moderate pressure between the thumb and forefinger. **ferme**

firming See *tillage, firming*. **tassement, raffermissement**

first law of thermodynamics A law stating that during any chemical or physical change in a closed system, energy is not destroyed or created but is changed from one form to another. Its expression for a closed system is $Q = AU + W$, where Q is the net heat absorbed by the system, W is the work performed, and AU is the change in internal energy. For example, chemical energy may be changed to heat energy during the burning of fossil fuels. Also referred to as the law of conservation of energy. **première loi de la thermodynamique**

first-order reaction A chemical reaction in which the rate of reaction is proportional to the concentration of one of the reactants, and not to any other chemical within the reaction mixture. See *rate constant, zero-order reaction*. **réaction de premier ordre**

fishway A passageway designed to enable fish to ascend a dam, cataract, or velocity barrier; also called a fish ladder. **passe à poissons**

fissile Capable of being easily split along closely spaced planes. **fissile**

fissile bedding The bedding of sedimentary rocks in which the individual sheets or laminae are less than 2 μm in thickness. **litage fissile, en feuillet**

fission (*radiochemistry*) The splitting of an atomic nucleus into at least two or more nuclei, with the concurrent release of neutrons and a large amount of energy. (*biology*) Asexual reproduction occurring when a single cell divides into two equal parts. **fission**

fissure An extensive crack, break, or fracture in rocks. **fissure**

fixation The process or processes in a soil by which certain chemical elements essential for plant growth are converted from a plant-available, soluble, or exchangeable form to a much less plant-available, soluble, or exchangeable form (e.g., phosphate fixation). **fixation**

fixed ammonium The ammonium in soil that cannot be replaced by a neutral potassium salt solution (e.g., 1*M* KCl). **ammonium fixé**

fixed carbon The ash-free carbonaceous material remaining after volatile matter is driven off during the proximate analysis of a dry solid waste sample. **charbon calciné**

fixed phosphorus (1) Phosphorus that has been changed to a less soluble form as a result of reaction with the soil; slowly available phosphorus. Specifically, it is the quantity of soluble phosphorus compounds that, when added to soil, becomes chemically or biologically attached to the solid phase of soil and cannot be recovered by extracting the soil with a specified extractant under specified conditions. Some of these extractants are water, carbonated water, and dilute solutions of strong mineral acids with or without fluoride or other exchangeable anion. (2) Applied phosphorus that is not absorbed by plants during the first cropping year. (3) Soluble phosphorus that has become attached to the solid phase of the soil, unavailable phosphorus, or phosphorus in other than readily or moderately available forms. **phosphore fixé**

flag leaf The uppermost leaf on a fruiting (fertile) grass culm; the leaf immediately below the *inflorescence* of the seed head. **feuille étendard, dernière feuille, feuille de l'épi**

flagellate Protozoan that moves by means of one to several flagella. **flagellé**

flagellum (plural flagella) A flexible, whiplike appendage on cells, used as an organ of locomotion. See *figure*. **flagelle**

flaggy See *coarse fragments*. **en dalles**

flagstone A thin fragment of sandstone, limestone, slate, shale, or (rarely) of schist, 15 to 37 cm long. See *coarse fragments*. **dalle**

flash flood A short-lived but rapid (usually less than six hours) rise of water in a river due to heavy rainfall, the collapse of an ice dam, log jam, or artificial dam. **crue éclair**

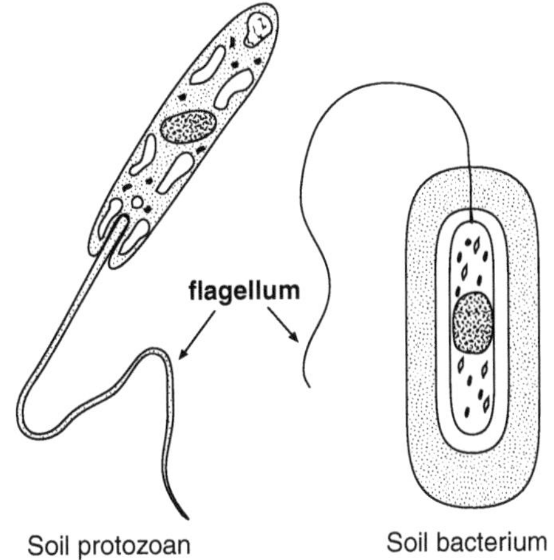

Flagellum

flash point The lowest temperature at which a flammable liquid produces a sufficient amount of vapor to ignite with a spark. **point éclair**

flat planting See *tillage, flat planting*. **semis superficiel ou à plat**

flexible cropping A nonsystematic sequence of growing adapted crops with cropping and fallow decisions at each prospective date of planting based on available water in the soil plus expected growing season precipitation. **assollement facultatif ou souple**

floc (*waste water management*) The clumped solids or precipitates formed by the coagulation of smaller particles, or colloids. See *flocculation*. **floc**

flocculation The process by which small particles or colloids coagulate due to the presence of ions in solution. In most soils the clay and humic materials remain flocculated due the presence of divalent and trivalent cations. In wastewater management, rapid mixing of chemicals into wastewater enhances the formation of *floc*, which must be of sufficient size for removal, filtration, or sedimentation. **floculation**

flood Any relatively high streamflow that overtops the stream banks and covers land that is not normally under water.

F

Inundation of the land surface by water following the rapid rise of the water level of a water body. **crue**

flood control Methods or facilities for reducing flood flows. **lutte contre l'inondation**

flood frequency The average occurrence of flooding of a given magnitude, over a period of years. **fréquence de crue**

flood peak Flood crest. The maximum rate of flow attained at a given point during a flood. The highest value of the stage or stream-flow attained by a flood; the top of the flood wave. **débit de pointe de crue**

flood plain The land bordering a stream, built up of sediments from overflow of the stream and subject to flooding. The lowland which borders a river, usually dry but subject to flooding. Also the portion of a river valley which has been inundated by the river during historic floods. **plaine alluviale**

flood-plain meander scar A crescentic mark indicating the former position of a river meander on a flood plain. **échancrure de méandre de plaine alluviale**

floodway (1) A large-capacity channel constructed to divert floodwaters safely through or around populated areas. (2) Part of a flood plain, otherwise leveed, reserved for emergency diversion of water during floods. (3) The channel of a river or stream and those parts of the flood plains adjoining the channel, which are reasonably required to carry and discharge the flood water of any river or stream. **canal de crue ou de dérivation**

flora The plants of a particular region, environment, or geologic period. **flore**

flow velocity The volume of water transferred per unit of time and per unit of area in the direction of the net flow of water in soil. See *Darcy's law*. **vitesse d'écoulement, débit**

flower The reproductive organ of most seed-bearing plants. Flowers carry out multiple roles in plants including sexual reproduction, seed development, and fruit production. **fleur**

flue dust Particles of solid matter smaller than 100 microns in diameter carried in the products of combustion. **poussière de combustion**

flue gas A mixture of gases resulting from combustion and coming out of a chimney. Includes nitrogen oxides, carbon oxides, sulfur oxides, particulates, and water vapor. **gaz de combustion**

flue gas scrubber A device that removes fly ash and other solid matter from flue gas by the use of sprays, wet baffles, or other techniques that require water as the primary mechanism for separation. Also called flue gas washer. **épurateur de gaz de combustion**

flume A channel, either natural or artificial, which carries water. **canal**

fluorapatite A mineral of the apatite group ($Ca_5(PO_4)_3F$). It is a varicolored, hexagonal mineral occurring in granular and fibrous forms and as prismatic crystals, having a specific gravity of 3.1 to 3.2 and a hardness of 5 (*Mohs scale*); found in igneous rocks, hydrothermal veins, and bedded deposits. See *apatite*. **fluoroapatite**

fluorescein A fluorescing pigment produced by many *pseudomonads* and observed by illumination with ultraviolet light. **fluorescéine**

fluorescence The phenomenon in which absorption of light of a given wavelength by a fluorescent molecule is followed by the emission of light at longer wavelengths. **fluorescence**

fluorescent antibody An antiserum conjugated with a fluorescent dye (e.g., *fluorescein* or rhodamine). Fluorescent-labeled antiserum can be used to stain burred slides or other preparations and visualize the specific microorganism (antigen) of interest by fluorescence microscopy. **anticorps fluorescent**

fluorocarbons Carbon-fluorine compounds that often contain other elements such as hydrogen, chlorine, or bromine. Common fluorocarbons include *chlorofluorocarbons* and *related compounds* (also known as ozone-depleting substances), *hydrofluorocarbons* (HFCs), and perfluorcarbons (PFCs). **fluorocarbures**

flushing Feeding female animals a concentrate feed shortly before and during the breeding period to stimulate ovulation. **alimentation intensive, flushing**

fluted moraine (*geomorphology*) Plain of glacial till having long, smooth, gutter-like depressions and/or low, elongated *drumlins* molded beneath moving ice and oriented parallel to the direction of ice flow. **moraine cannelée**

Fluvents A suborder in the U.S. system of soil taxonomy. *Entisols* that form in recent loamy or clayey alluvial deposits, are usually stratified, and have an organic carbon content that decreases irregularly with depth. Fluvents are not saturated with water for periods long enough to limit their use for most crops. **Fluvents**

fluvial Descriptive of materials transported and deposited by flowing water; growing or living in a stream or river; produced by the action of a stream or river. **fluvial, fluviatile**

fluvial cycle of erosion The lowering of a region to base level largely by running water, especially the action of rivers. **cycle d'érosion fluviale**

fluvial deposits All sediments, past and present, deposited by flowing water, including glaciofluvial deposits. Wave-worked deposits and deposits resulting from sheet erosion and mass wasting are not included. **dépôts fluviatiles**

fluvial processes All processes and events by which the configuration of a stream channel is changed; especially processes by which sediment is transferred along the stream channel by the force of flowing water. **processus fluviatiles**

fluvioeolian Descriptive of materials transported and deposited by the combined action of streams and wind. **fluvio-éolien**

fluvioglacial See *glaciofluvial deposits*. **fluvio-glaciaire**

fluviolacustrine Descriptive of materials pertaining to sedimentation partly in lake water and partly in streams, or to sediments deposited under alternating or overlapping lacustrine and fluvial conditions. **fluvio-lacustre**

flux (1) The rate of movement of a quantity (e.g., mass or volume of fluid, electromagnetic energy, number of particles, or energy) across a given area. (2) A state of change. (3) A substance that reduces the melting point of a mixture (e.g., limestone used in iron smelting to lower the fusion temperature of the ore). **(1, 2) flux (3) fondant**

F

flux density The time rate of transport of a quantity (e.g., mass or volume of fluid, electromagnetic energy, number of particles, or energy) across a unit area perpendicular to the direction of flow. **densité de flux**

fly ash The finely divided residue resulting from the combustion of ground or powdered coal and that is transported from the firebox through the boiler by flue gases. **cendres volantes**

fodder The dried, cured plants of tall, coarse grain crops such as maize and soybeans, including the grain, stems, and leaves; grain parts not snapped off or threshed. See *stover*. **fourrage**

fogging Pesticide application in which material in a solution is atomized by mechanical or other physical means to form very small droplets, giving the appearance of fog or smoke. **nébulisation, atomisation, brumisage**

fold (*geology*) A bend or flexure in a layer or layers of rock. See *figure*. **pli**

foliage The green or live leaves of growing plants; plant leaves collectively. Often used in reference to aboveground development of forage plants. **feuillage**

foliar analysis Chemical evaluation of the status of plant nutrients or the plant-nutrient requirements of a soil by the analysis of leaves or needles. **analyse foliaire**

foliar diagnosis An estimation of plant mineral nutrient status, based on the chemical composition of selected plant parts and the color and the growth characteristics of the plant foliage. **diagnostic foliaire**

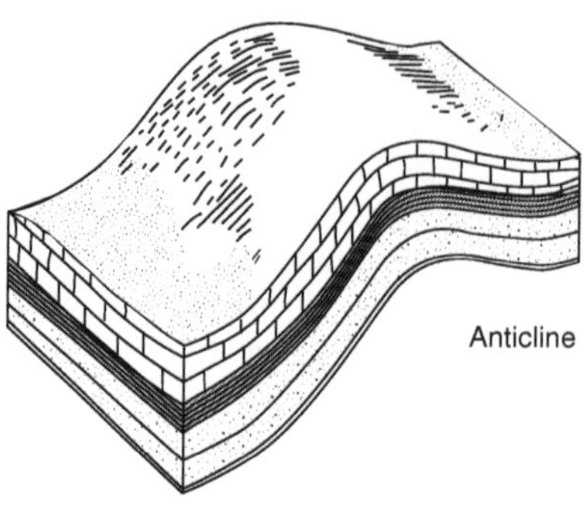

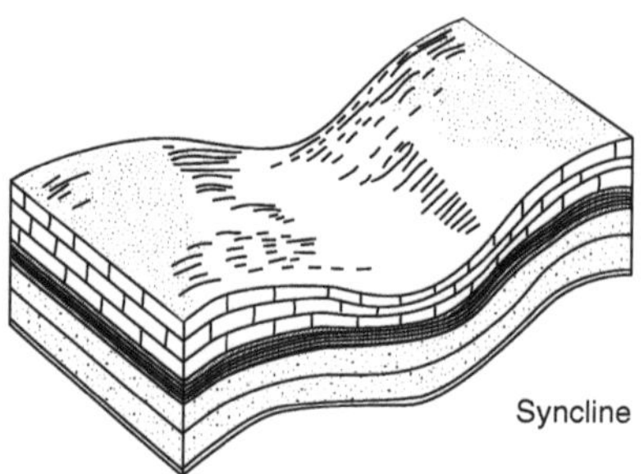

Types of **folds** (adapted from Dunster and Dunster, 1996).

folic Organic materials developed primarily from accumulation of leaves, twigs, and woody materials with or without a minor component of mosses. They are normally associated with upland forested soils with imperfect drainage or drier. See *LFH horizon, L horizon*, *F horizon* and *H horizon.* **folique**

Folisol A great group of soils in the *Organic* order (Canadian system of soil classification). The soils are not usually saturated for more than a few days a year, and consist of 10 cm or more of *LFH horizons* derived from leaf litter, twigs, branches, and mosses. A lithic contact or fragmented material occurs at a depth of less than 160 cm. Mineral layers less than 10 cm thick may lie above the lithic contact. **folisol**

folistic epipedon A surface layer of soil (U.S. system of soil taxonomy) generally consisting of organic soil material that is saturated for less than 30 days (cumulative) in normal years, and is not artificially drained. **épipédon folistique**

Folists A suborder in the U.S. system of soil taxonomy. Histosols that have an accumulation of organic soil materials mainly as forest litter that is <1 m deep to rock or to fragmental materials with interstices filled with organic materials. Folists are not saturated with water for periods long enough to limit their use if cropped. **Folists**

food chain The flow of carbon and energy in the form of food from organisms in one trophic level to those in another. The trophic levels depend on how an organism obtains its food. The first link in the chain is green plants, called primary producers; those animals that consume the plants are called herbivores or consumers, and animals that eat other animals are called carnivores or secondary consumers and are placed in the highest feeding level. The transfer of materials or mass from one trophic level to the next is approximately 10% efficient. See *ecological pyramid*, *trophic level.* **chaîne alimentaire**

food cycle All the interconnecting food chains in a community; also called *food web.* **cycle alimentaire**

food web The interrelationship among the organisms in a community according to the transfer of useful energy from food resources to organisms eating those resources. **chaîne alimentaire ou de prédation**

foot (*geomorphology*) The bottom of a slope, grade, or declivity. The lower part of any elevated landform. **pied**

foothills The lower line of hills that lie parallel at the foot of a higher mountain range. **piedmont, contrefort**

footing The portion of the foundation of a structure that transmits loads directly to the soil. **semelle**

footslope See *slope morphological classification.* **base de pente**

forage (1) Edible parts of plants, other than separated grain, that can provide feed for grazing animals or can be harvested for feeding, including browse, and herbage. (2) To search for or to consume forage (of animals). **(1) fourrage (2) fourrager**

forage quality Characteristics that make *forage* valuable to animals as a source of nutrients; the combination of chemical and biological characteristics of forage that determines its potential to be used by animals to produce meat, milk, wool, or work. **qualité des fourrages**

forb Any herbaceous (non-woody) plant that is not a *grass*, *sedge*, or rush. **plante herbacée dicotylédone**

forcing functions Significant factors that determine the composition of natural ecosystems (e.g., temperature, rainfall, relative humidity, and solar radiation). **fonctions motrices**

forcing mechanism A process that alters the energy balance of the climate by changing the relative balance between incoming *solar radiation* and outgoing *infrared radiation* from the Earth (e.g., changes in solar irradiance, volcanic eruptions, and enhancement of the natural *greenhouse effect* by emission of carbon dioxide). See *radiative forcing*. **mécanisme de forçage**

forcing variable A constant or variable acting on a system from the outside; a driving variable. **variable de forçage**

forest A continuous tract of trees over a large area; land with at least 10% of its surface area stocked by trees of any size. One of the four main subdivisions of world vegetation. See *biochore*. **forêt**

forest cover Forest stands or cover types consisting of a plant community made up of trees and other woody vegetation growing more or less closely together. **couvert forestier**

forest cover map A map showing relatively homogeneous forest stands or cover types, produced from the interpretation of aerial photos and information collected in field surveys. Commonly includes information on species, age class, height class, site, and stocking level. **carte forestière**

forest cover type A descriptive term used to group stands of similar characteristics and species composition (due to given ecological factors) by which they may be differentiated from other groups of stands. **strate forestière**

forest ecology The relationships between forest organisms and their environment. **écologie forestière**

forest fire Any wildfire or prescribed fire that burns in forest, grass, alpine, or tundra vegetation types. **incendie forestier**

forest fire weather index system, Canadian A part of the Canadian forest fire danger rating system. The components of the system provide numerical ratings of relative fire potential in a standard fuel type (i.e., a mature pine stand) on level terrain, based solely on consecutive observations of four fire weather elements measured daily at noon at a suitable fire weather station; the elements are dry bulb temperature, *relative humidity*, wind speed, and precipitation. The system provides a uniform method of rating fire danger across Canada. **méthode canadienne de l'indice forêt-météo**

forest floor All dead vegetation and organic matter, including fresh leaf and needle litter, moderately decomposed organic matter, and humus or well-decomposed organic residue, on the mineral soil surface under forest vegetation. **couche holorganique, couverture morte, litière**

forest health A forest condition that is naturally resilient to damage; characterized by biodiversity, it contains sustained habitat for timber, fish, wildlife, and humans, and meets present and future resource management objectives. **santé des forêts**

forest health agents Biotic and abiotic influences on the forest that are usually a naturally occurring component of forest ecosystems. Biotic influences include fungi, insects, plants, animals, bacteria, and nematodes. Abiotic influences include frost, snow, fire, wind, sun, drought, nutrients, and human-caused injury. **facteurs de santé des forêts**

forest health treatments The application of techniques to influence pest or beneficial organism populations, mitigate damage, or reduce the risk of future damage to forest stands. Treatments can be either proactive (for example, spacing trees to reduce risk of attack by bark beetles) or reactive (for example, spraying insecticides to treat outbreaks of gypsy moth). **traitements pour améliorer la santé des forêts**

forest hydrology The study of hydrologic processes as influenced by forest and associated vegetation. **hydrologie forestière**

forest influences The effects of forests on soil, water supply, climate, and environment. **influences forestières**

forest interior conditions Conditions found deep within forests, away from the effects of open areas. Forest interior conditions include particular microclimates found within large forested areas. **conditions intérieures des forêts**

forest inventory An assessment of forest resources, including digitized maps and a database which describes the location and nature of forest cover (including tree size, age, volume, and species composition) as well as a description of other forest values (e.g., soils, vegetation, and wildlife features). **inventaire forestier**

forest land Land bearing a stand of trees at any age or stature, including seedlings, and of species attaining a minimum average height of 1.8 meters at maturity, or land from which such a stand has been removed but on which no other use has been substituted. **terrain forestier**

forest management The practical application of scientific, economic, and social principles to the administration and working of a forest for specified objectives. Particularly, that branch of forestry concerned with the overall administrative, economic, legal, social, scientific, and technical aspects, especially silviculture, protection, and forest regulation. **aménagement forestier**

forest peat See *swamp peat*, *woody peat*. **tourbe forestière**

forest practice Any activity carried out on forest land to facilitate the use of forest resources, including but not limited to timber harvesting, road construction, silviculture, grazing, recreation, pest control, and wildfire suppression. **pratique forestière**

forest profile The range of forest conditions that exists across the landscape, including such factors as timber species, quality, condition and age, location, elevation, topography, accessibility, and economic viability. **profil des forêts**

forest regeneration, forest renewal The renewal of a tree crop through either natural means (e.g., seeded on-site from adjacent stands or deposited by wind, birds, or animals) or artificial means (e.g., by planting seedlings, or direct seeding). **régénération forestière**

forest resources Resources and values associated with forests and range including timber, water, wildlife, fisheries, recreation, botanical forest products, forage, and biological diversity. **ressources forestières**

forest service road A road used to provide access to managed forest land. **chemin forestier**

forest soils Soils developed under forest vegetation. **sols forestiers**

forested fen or treed fen A fen with a sparse, open canopy tree cover of about 20 to 60% of the surface area. Typical tree species in Canada are *Larix laricina* or *Picea mariana*. **fen arboré**

forestry The theory and practice of managing and using the natural resources that occur on and in association with forest lands. **foresterie**

formation (*geology*) A body of rock strata that consists predominantly of a certain lithologic type or combination of types. It is the fundamental lithostratigraphic unit. Formations may be combined into groups or subdivided into members. (*geomorphology*) A topographic feature differing conspicuously from adjacent features (e.g., *hoodoo* formation). (*biogeography*) A group of plant or animal associations that exist together because of closely similar life patterns, habits, and climatic requirements. **formation**

formation-class A unit of vegetation defined on a geographic basis with reference to its external features; a major subdivision of a biochore (e.g., the tropical forest and the taiga are formation-classes of the forest biochore). **formation végétale**

formula weight The sum of the atomic weights of the elements in a compound. **masse moléculaire**

formulation The form in which a fertilizer or agrichemical is sold, e.g., solid or liquid. **forme**

fossil The remains of the whole or part of any formerly living organism, preserved by natural causes in crustal rocks. Fossils can be chemically altered or replaced; they can be recognized by the mold or impression that an organism had made. Fossilized remnants of the effects of the organism, rather than the fossil itself, are called trace-fossils (e.g., tracks, worm burrows). **fossile**

fossil fuel Combustible geologic deposits of carbon in reduced (organic) form and of biological origin (e.g., coal, oil, natural gas, oil shales, and tar sands). They emit *carbon dioxide* into the *atmosphere* when burnt, thus significantly contributing to the *enhanced greenhouse effect*. **combustible fossile**

fossil soil See *paleosol*. **sol fossile**

founder principle The principle that a population started by a small number of colonists contains only a small fraction of the genetic variation of the parent population. **principe fondateur**

Fourier analysis (*statistics*) Mathematical representation of functions of a variable as the sum of a series of sine and cosine terms. It is used to analyze a harmonic data series (a complex curve that exhibits periodic variations), e.g., analysis of changes in seasonal rainfall. **analyse de Fourier**

Fourier transform A mathematical technique for converting time domain data to frequency domain data, and vice versa. **transformation de Fourier**

fracture (1) The characteristic break pattern of a mineral. (2) A break in a rock due to extension or compression, as in faulting or folding. **fracture**

fragic A loamy subsurface soil horizon of high bulk density and low organic matter content. When dry, it has a hard consistence and seems to be cemented. When moist, it has moderate to weak brittleness. It frequently has bleached fracture planes, and it is overlaid with a friable *B horizon*. Air-dried clods slake when immersed in water. Fragic soil materials show evidence of pedogenesis, including one or more of the following: oriented clay within the matrix or on faces of peds, redoximorphic features within the matrix or on faces of peds, strong or moderate soil structure, and coatings of albic materials or uncoated silt and sand grains on faces of peds or in seams. The term is used in the Canadian, U.S., and other soil taxonomies. See *fragipan*. **fragique**

fragile land Land areas that are sensitive to degradation due to slope, soil type,

F

exposure, and degree of cover. Areas especially subject to soil *erosion* and rapid deterioration. **terre vulnérable ou fragile**

fragipan A natural subsurface horizon having a higher bulk density than the solum above; seemingly cemented when dry, but showing moderate to weak brittleness when moist. The layer is low in organic matter, mottled, and slowly or very slowly permeable to water; it usually has some polygon-shaped bleached cracks. It is found in profiles of either cultivated or virgin soils but not in *calcareous* material. A subsurface diagnostic horizon in the U.S. system of soil taxonomy, and analogous to the *fragic* horizon in the Canadian system of soil classification. **fragipan**

fragmentation The process of transforming large continuous forest patches into one or more smaller patches surrounded by disturbed areas. This occurs naturally through such agents as fire, landslides, windthrow, and insect attack. In managed forests, timber harvesting and related activities have been the dominant disturbance agents. **fragmentation**

frame A representation of a population, used to implement a sampling strategy. There are two types of frames: a list frame that lists the identifying units in the population (e.g., a list of all the lakes in a state or province), and an area frame that consists of explicit descriptions of a partition of the areal extent of a zone or map unit (e.g., a soil zone or ecological zone). **base de sondage**

Frasch process A method of recovery of sulfur from underground deposits by melting it with hot water and forcing it to the surface by air pressure. **procédé de Frasch**

free energy The capacity of a system to perform work, with a change in free energy measured by the maximum work obtainable from a given process. See *Gibbs free energy*. **énergie libre**

free iron oxides A general term for those iron oxides that can be reduced and dissolved by a dithionite treatment (e.g., goethite, hematite, ferrihydrite, lepidocrocite, and maghemite, but not magnetite). See *iron oxides*. **oxydes de fer libres**

free lime test A simple test using acid to determine if a soil has excess carbonates. The test is positive if effervescence (fizzing) occurs when acid is added to the soil sample. **test de carbonates libres**

free liquids (*waste management*) Liquids capable of migrating from waste and contaminating groundwater. Hazardous waste containing free liquids may not be disposed of in landfills. **lixiviat**

free oxides Oxides and hydroxides of iron, aluminum, silicon, manganese, and titanium, usually of fine particle size, that occur uncombined with other elements and often as coatings on primary and secondary minerals. **oxydes libres**

free radical An atom or group of atoms with an unpaired valence electron which makes them highly reactive and thus short-lived. Free radicals can be produced by *photolysis* or pyrolysis in which a bond is broken without forming ions. **radical libre**

free water (*soil*) Water in soil (or rock) that is free to move in response to the pull of gravity. Also called *gravitational water*. (*geology*) Water that can be removed from a substance (e.g., ore) without changing the structure or composition of the substance. **eau libre**

free-living Capable of living, reproducing, or carrying out a specific function without the direct assistance of a plant or animal of a different species. See *symbiosis*. **libre**

freeze-thaw action (*geology*) A type of weathering process related to the freezing and thawing of water in cracks and fissures, which leads to frost-splitting (*congelifraction*). (*soil*) A mechanical process that leads to the breakdown of soil *aggregates*. **action gel-dégel**

freezing point The temperature at which a liquid freezes or solidifies at normal pressure. For fresh water this temperature is 0°C. **point de congélation**

frequency (*statistics*) (1) The number of occurrences of a given type or event, or the number of members of a population falling into a specified class. (2) A statistical expression of the presence or absence of an individual item in a series of subsamples (i.e., the ratio between the number of sample areas that contain an individual item and the total number of sample areas). (*electromagnetic radiation*) The number of waves (cycles) per second that pass a given point in space. Frequency is given in units of inverse time (e.g., second^{-1}). (*remote sensing* or *physics*) See *frequency band*. **fréquence**

frequency band The number of oscillations per unit time or number of wavelengths that pass a point per unit time. The frequency bands used by radar (radar frequency bands) were first designated by letters for military secrecy. Those designations include:

Frequency Band	Approximate Frequency Range (gigahertz)	Approximate Wavelenth Range (centimeter)
P-band	0.225 – 0.39	140 – 76.9
L-band	0.39 – 1.55	76.9 – 19.3
S-band*	1.55 – 5.20	19.3 – 5.77
X-band	5.20 – 10.90	5.77 – 2.75
K-band	10.90 – 36.00	2.75 – 0.834
Q-band	36.00 – 46.00	0.834 – 0.652
V-band	46.00 – 56.00	0.652 – 0.536

*The C-band, 3.9 to 6.2 gigahertz, overlaps the S- and X-bands; these letter designations have no official sanction but still are widely used. **bande de fréquences**

frequency curve (*statistics*) A curve that graphically represents a *frequency distribution* e.g., a smooth line drawn on a histogram if the class interval is made smaller and the steps between several bars grow smaller. **courbe de fréquence**

frequency distribution (*statistics*) A statistical method of plotting numerical data to show the frequency with which different values of a variable occur within a sample. An ungrouped frequency distribution is simply a list of figures occurring in the raw data, together with the frequency of each figure. Once the figures are grouped into classes, it becomes a grouped frequency distribution, in which class intervals and class limits can be identified. When the data are plotted graphically on a histogram the graph is called a *frequency curve*. See *normal distribution*, *Poisson distribution*. **distribution de fréquence, distribution statistique**

F

fresh mulch The primary layer of bulky, coarse, largely undecayed herbage residue. **paillis frais**

fresh water Water with less than 0.2% salinity. **eau douce**

fresh water biome The aquatic *biome* consisting of water containing fewer salts than the waters in the marine biome; consists of two major types: running waters (rivers, streams) and standing waters (lakes, ponds). **biome d'eau douce**

Freundlich isotherm An *adsorption isotherm* which assumes a logarithmic fall in the *enthalpy* of adsorption with surface coverage. The adsorption is described mathematically by the Freundlich equation: $x/m = KC^{1/n}$ where x/m is the quantity of solute adsorbed per unit weight of adsorbent, C is the equilibrium concentration of the adsorbing compounds, and K and n are constants. **formule de Freundlich**

friable A descriptor of *consistence,* pertaining to the ease of crumbling of soils. **friable**

friction cone penetrometer A cone penetrometer with the additional capacity of measuring the local side-friction component of *penetration resistance*. Penetration resistance developed by the friction sleeve equals the vertical force applied to the sleeve divided by its surface area. See *cone penetrometer*, *cone index*. **pénétromètre à cône à friction latérale**

F

frigid A soil temperature regime that has mean annual soil temperatures of >0°C but <8°C, >5°C difference between mean summer and mean winter soil temperatures at 50 cm below the surface, and warm summer temperatures. Isofrigid is the same except the summer and winter temperatures differ by <5°C. **frigide**

frits A type of low solubility fertilizer produced by mixing a compound containing the desired nutrient with other materials that can be fused to make glass. The glass is ground to a powder that can be added to soil to serve as a slow-release fertilizer. Frits are most appropriate for acid soils where their weathering rates are sufficiently rapid to supply the included micronutrient (or micronutrients) at adequate rates. **forme frittée**

fritted trace element Sintered silicate having total guaranteed analyses of *micronutrients* with controlled (relatively slow) release characteristics. **oligo-éléments frittés**

frost (1) The state of freezing or of becoming frozen, when the air temperature at screen level (i.e., at 1 meter) falls to or below the freezing point. Qualifying adjectives generally accompany the term in order to indicate its intensity; i.e., hard, sharp, killing. (2) A weathering agent which breaks up rocks and soil owing to freezing of the interstitial water. (3) Frozen dew, fog, or water vapor are sometimes called white frost. See *ground frost*. **gel**

frost boil An area of bare soil which is sufficiently disturbed by frost action to prevent plant colonization. **ventre de boeuf**

frost crack A fissure opened up in the soil by the development of an incipient ice wedge. **fissure de gel, fente de gel**

frost creep Soil creep resulting from frost action. **reptation des sols causée par le gel**

frost free period This is considered to be the number of days between the last spring *frost* and the first autumn frost. Frost is assumed to occur when the minimum daily temperature is 0°C or less. **période sans gel**

frost heave The lifting or lateral movement of soil as caused by freezing processes in association with the formation of ice lenses or ice needles. **soulèvement par le gel**

frost line The maximum depth of frozen ground in areas where there is no *permafrost*; it may be expressed for a given winter, as the average of several winters, or as the greatest depth on record. **profondeur maximum du gel, seuil du gel**

frost weathering The mechanical weathering process caused by alternating cycles of freezing and thawing of moisture in soil or rock pores, fissures, and cracks. See *freeze-thaw action*, *congelifraction*. **gélivation, gélifraction, cryoclastie**

frost, concrete Ice in the soil in such quantity as to constitute a virtually solid block. **masse cryoconsolidée**

frost, honeycomb Ice in the soil in insufficient quantity to be continuous, thus giving the soil an open, porous structure permitting the ready entrance of water. **gel alvéolaire**

frozen ground Ground having a temperature below freezing and generally containing a variable amount of water (as ice). **gélisol**

fruiting body A specialized structure in some bacteria (e.g., *Myxobacteria*) and fungi that produce spores. **organe de fructification**

F*-test** *(statistics)* A parametric statistical test based on a statistic that has an *F* distribution, most commonly applied in tests concerning the equality of population variances, as in an *analysis of variance*. **test *F

fuel cell A cell in which the chemical energy of a fuel is converted directly into electrical energy. **pile à combustible**

fuller's earth A clay possessing a high adsorptive capacity for water and other fluids, consisting largely of montmorillonite or palygorskite. It is extensively used as an adsorbent in refining and decolorizing oils and fats, and is a natural bleaching agent. **argile de fuller**

fulvic acid Organic materials that are soluble in alkaline solution and remain soluble on acidification of the alkaline extracts. **acide fulvique**

fumigant A chemical compound in the form of a gas, particulate, vapor, or smoke, usually used to kill pests (e.g., fungi, insects, or rodents). **fumigant**

fumigation, soil Treatment of the soil with volatile or gaseous substances that penetrate the soil mass and kill one or more forms of soil organisms. **fumigation du sol**

functional group An atom or group of atoms responsible for the characteristic reactions of a compound (e.g., –OH for alcohols, –CHO for aldehydes, –COOH for carboxylic acids). **groupement fonctionnel**

fungicide A chemical compound (i.e., *pesticide*), usually synthetic, that kills fungi or prevents them from causing diseases on plants. **fongicide**

fungistat A chemical compound that inhibits or prevents the growth of fungi. **fongistatique**

fungus (plural fungi) Nonphototrophic *eukaryotic* microorganisms that contain rigid cell walls. Fungi range in size from microscopic (but larger than either *bacteria* or *actinomycetes)* to such large and readily visible specimens as mushrooms. Microscopic fungi have larger *hyphae* in their *mycelia* than actinomycetes, and have visible cell walls in the hyphae. Fungi generally tolerate acid conditions better than other *microbes* and therefore dominate the microbial population of most forest soils. **champignon**

furans See *polychlorinated dibenzofurans.* **furannes**

furrow See *tillage, furrow.* **sillon, raie**

furrow dams Small earth ridges or rows used to impound water in furrows. **barrages de rigoles**

furrow erosion The erosion that occurs with the process of furrow irrigation. **érosion des rigoles d'irrigation**

furrow mulching The practice of placing straw or other mulch materials in irrigation furrows to increase infiltration and reduce erosion. **paillage des rigoles d'irrigation**

furrow slice The layer of soil that is plowed or otherwise tilled. The dry mass of a hectare furrow slice is often assumed to be 2,000,000 kg for purposes of calculating storage capacity or required amounts of soil amendments. **couche de labour, couche travaillée**

fusion (1) The combination of two light nuclei to form a heavier nucleus. The reaction is accompanied by the release of a large amount of energy. See *fission.* (2) The process of melting a solid by addition of heat. (3) The unification of two or more substances, as by melting together. **fusion**

G

gabion A mesh container used to confine rocks or stones, and used to construct dams and groins or line stream channels and slopes to prevent erosion. **gabion**

gage or gauge (1) A device for indicating magnitude or position in specific units, when such magnitude or position undergoes change (e.g., elevation of a water surface, amount or intensity of precipitation, depth of snowfall). (*hydrology*) (2) The operation of measuring the discharge of a stream of water in a waterway. **jauge**

gaging station (*hydrology*) A particular site on a stream, canal, lake, or reservoir where systematic observations of gage height, discharge, or water quality (or any combination of these) are obtained. **station de jaugeage**

Gaia hypothesis The proposition that the composition and temperature of the atmosphere is a product of interrelated activities in the biosphere, especially those of microorganisms, and that the biosphere behaves as a single self-regulating organism. Gaia was the ancient Greek goddess of the Earth. **hypothèse de Gaia**

galvanic cell A device in which chemical energy from a spontaneous redox reaction is changed to electrical energy that can be used to do work. **pile galvanique**

galvanizing See *cathodic protection*. **galvanisation**

gametangium A differentiated structure in which *gametes* are produced, or the contents of which may function as a gamete. **gamétange**

gamete A sexually reproductive cell or unisexual protoplasmic body incapable of giving rise to another individual until after *conjugation* with another gamete and the joint production of a zygote. **gamète**

gamma diversity Diversity across a range of habitats in a landscape or geographic area. See *alpha diversity, beta diversity*. **diversité gamma**

gamma probe An instrument for measuring total soil density by relating the fraction of radiation emitted through a soil layer to that received by the detector. See also *neutron probe*. **sonde à rayons gamma**

gamma ray A highly penetrating type of nuclear radiation, similar to x-rays and light, except that it comes from within the nucleus of an atom, and in general has a shorter wavelength. Also referred to as gamma radiation. **rayon gamma**

gamma-ray attenuation Soil density in the field can be measured by means of gamma-ray attenuation. The gamma rays can pass from the source to the detector by following a curved path through the soil. Soil particles absorb the gamma rays, and the count can therefore be taken as an indication of soil *bulk density* after suitable correction for water content. **atténuation du rayonnement gamma**

gangue The impurities (e.g., clay or sand) in an ore. **gangue**

gas One of the three states of matter. In a gaseous state, there is little attraction between the particles, which have continual, random motion. The gas has no fixed shape or volume, can

G

expand indefinitely, and can assume the shape of the space in which it is held. It is also easily compressed, with the random collisions between particles exerting pressure on the walls of the container. **gaz**

gas chromatography (GC) An analytical technique that can yield both qualitative and quantitative evaluations of sample mixtures of volatile substances. The compounds of interest are separated by using an inert gas to flush a sample preparation through a column packed with a substance that selectively absorbs and releases the volatile constituents. See *chromatography.* **chromatographie en phase gazeuse**

gas pressure potential A pneumatic potential in soil usually only considered when the external gas pressure differs from atmospheric pressure, as in a pressure membrane apparatus. **potentiel pneumatique**

gas, natural A hydrocarbon, consisting mainly of ethane and methane, usually associated with crude oil accumulations. **gaz naturel**

gasification See *coal gasification.* **gazéification**

gas-phase absorption spectrum The pattern of radiation absorbance by chemical compounds in a gaseous mixture. Particular gases will absorb radiation in certain frequency ranges. The spectrum is used to monitor some air pollutants. **spectre d'absorption de la phase gazeuse**

gastropod (*zoology*) Any mollusk belonging to the class Gastropoda, characterized by a distinct head with eyes and tentacles and, in most, by a single calcareous shell that is closed at the apex, sometimes spiralled, not chambered, and generally asymmetrical (e.g., snail). **gastéropode, gastropode**

Gaussian curve *(statistics)* A type of *frequency curve*, possessing perfect symmetry about the central value and fitting many of the *frequency distributions* which occur most often. It is synonymous with a normal curve. See *normal distribution.* **courbe de Gauss, courbe gaussienne**

Gaussian distribution See *Gaussian curve.* **distribution de Gauss, distribution gaussienne**

Geiger-Muller counter (Geiger counter) An instrument that measures the rate of radioactive decay based on the ions and electrons produced as a radioactive particle passes through a gas-filled chamber. **compteur à pointe de Geiger**

gelic material Mineral or organic materials that have evidence of cryoturbation and/or ice segregation in the active layer (seasonal thaw layer) and/or upper part of the permafrost; used in the U.S. system of soil taxonomy. **matériau gélique**

gelifluction A type of *solifluction* occurring in frozen ground under both seasonal freezing and permafrost conditions. It differs from *congelifluction*, which refers only to permanently frozen ground (i.e., *permafrost*). **gélifluxion**

gelisol See *frozen ground.* **gélisol**

Gelisols An order in the U.S. system of soil taxonomy. Soils of very cold climates that contain permafrost within 100 cm of the soil surface, or *gelic materials* within 100 cm of the soil surface and permafrost within 200 cm of the soil surface. **Gélisols**

general circulation model (GCM) A global, three-dimensional computer model of the climate system that can be used to simulate human-induced climate change. GCMs are highly complex, representing the effects of such factors as reflective and absorptive properties of atmospheric water vapor, greenhouse gas concentrations, clouds, annual and daily solar heating, ocean temperatures, and ice boundaries. The most recent GCMs include global representations of the atmosphere, oceans, and land surface. **modèle de circulation générale**

general systems theory A logico-mathematical framework developed in biological science and widely adopted in Earth and physical sciences to show the interrelationships of objects and ideas with functions and processes as part of an integral *system*. It outlines the general principles relating to systems irrespective of the nature of the component elements and the relations or forces between them. The principles are defined in mathematical language, and introduce notions such as wholeness and sum, progressive mechanization, centralization, leading parts, hierarchical order, individuality, finality, and equifinality. **théorie du système général**

genesis, soil The mode of origin of the soil, especially the processes or *soil formation factors* responsible for the development of the *solum* from unconsolidated parent material. **genèse ou formation des sols**

genetic diversity Variation among and within species that is attributable to differences in hereditary material. **diversité génétique**

genetics The science that deals with the storage, replication, transfer, and expression of information that governs the transmission of traits from parents to offspring. **génétique**

genophore In *prokaryotes*, the circular DNA molecule containing a set of the genetic instructions for the cell. **génophore**

genotype The total genetic makeup of an organism. All of the genetic traits contained by a specific organism are not usually expressed. Some genes are recessive, and their presence is masked by the overriding influence of dominant traits, determining the organism's actual appearance, or *phenotype*. **génotype**

genus A group in the taxonomic classification of organisms, composed of one or more species. The name of the genus is the first word in *binomial nomenclature*. The plural form of genus is genera, several of which make up a *family*. **genre**

geo- A prefix meaning Earth. **géo**

geobotanical prospecting The study of plants and their distribution as indicators of soil composition and depth, bedrock lithology, the possibility of ore bodies, and groundwater conditions. **prospection géobotanique**

geochemical cycle The sequence of stages in the migration of elements during geologic changes. A major cycle proceeds from magma to igneous rock to sediments to sedimentary rocks to metamorphic rocks. A minor or exogenic cycle proceeds from sediments to sedimentary rocks to weathered material. **cycle géochimique**

geochemistry The study of the distribution, abundance, and circulation of the chemical elements (and isotopes) in minerals, ores, rocks, soils, water, and the atmosphere on the basis of the properties of their atoms and ions. **géochimie**

geocryology The study of *frozen ground* (both seasonally frozen and *permafrost*), excluding the study of glaciers. **géocryologie**

geodesy (1) The science related to determining the size and shape of the Earth and the precise location of points on its surface. (2) The science of determining the gravitational field of the Earth, and the study of temporal variations such as Earth tides, polar motion, and rotation. **géodésie**

geodetic surveying The applied science of *geodesy*; surveying that takes into account the shape and size of the Earth and makes corrections for Earth curvature. **relevé géodésique**

geographic information system (GIS) A computer system for capturing, storing, checking, integrating, manipulating, analyzing, and displaying data related to positions on the Earth's surface. Typically, a GIS (sometimes called Spatial Information System) is used for handling maps of one kind

G

or another. These might be represented as several different layers, where each layer holds data about a particular kind of feature. Each feature is linked to a position on the graphical image of a map. Layers of data are organized to be studied, to perform statistical analysis, and to support decision making. **système d'information géographique**

geographic province A large region in which all parts are characterized by similar geographic features. **province géographique**

G

geography The science dealing with the distribution and interaction of phenomena on the Earth's surface, including physical environment and behavioral environment. (See *geography, physical*) **géographie**

geography, physical The science related to the spatial and the temporal characteristics and relationships of all phenomena within the Earth's physical environment (i.e., *atmosphere, biosphere, hydrosphere,* and *lithosphere*). **géographie physique**

geography, soil Geography dealing with the areal distribution of soil types. **géographie des sols**

geologic erosion See *erosion*. **érosion géologique**

geologic hazard A geologic condition, either natural or human-made, that poses a potential danger to life and property (e.g., earthquakes, landslides, slumping, flooding, faulting, beach erosion, land subsidence, and pollution, and geotechnical as related to phenomena such as foundation and footing failures). **risque géologique**

geological map A map showing surface distribution of rock varieties, age relationships, and structural features. **carte géologique**

geological time See *chronostratigraphy*. **temps géologique**

geology The study of the origin, structure, composition, and history of the Earth, together with the processes, products formed, and history of the planet and its life forms since its origin. It comprises *crystallography, geochemistry, geomorphology, geophysics, mineralogy, petrology, sedimentology, stratigraphy,* and *structural geology*. **géologie**

geomagnetic field The magnetic field of the Earth that causes a compass needle to align north–south. **champ géomagnétique**

geomagnetic poles The poles of the Earth's magnetic field, located some 6400 km above the surface (and not, therefore, corresponding to the surface magnetic poles). They are situated above $78^1/_2$° N, 69°W and $78^1/_2$° S, 11° E, respectively. **pôles géomagnétiques**

geomagnetism The study of the Earth's magnetic field. **géomagnétisme**

geometric growth A doubling of the number of individuals constituting a population, with each time interval corresponding to a generation. Also called *exponential growth*, this is the characteristic growth pattern of bacteria. **croissance géométrique**

geometric mean *(statistics)* The mean of a *log-normal distribution* of values, calculated as the nth root of the product of n values. **moyenne géométrique**

geometric rate of increase Factor by which the size of a population changes over a specified period. **taux géométrique d'augmentation**

geometric series A sequence of numbers in which the ratio of successive numbers is constant. **progression géométrique**

geometric standard deviation *(statistics)* An expression of the dispersion of a set of measurements about a *geometric mean*. Normally distributed measurements have a mean (average), with the dispersion indicated by standard deviations added to or subtracted from the mean. For the *log-normal distribution*, however, the geometric mean is multiplied and divided by the geometric standard deviation to calculate the dispersion of the data set. **écart type géométrique**

geometrical *(cis-trans)* isomerism Isomerism in which atoms or groups of atoms can assume different positions around a rigid ring or bond. **isomérie géométrique cis-trans**

geomorphic (1) Pertaining to the form of the Earth or its surface features (e.g., a geomorphic province). (2) Pertaining to geomorphology; geomorphologic. **géomorphologique**

geomorphic cycle See *cycle of erosion.* **cycle géomorphologique, cycle d'érosion**

geomorphic surface A portion of the landscape specifically defined in space and time that has determinable geographic boundaries and is formed by one or more agencies during a given time period. **unité géomorphologique**

geomorphological map A map depicting terrain features (e.g., maps of river terraces, moraines, or kame fields). **carte géomorphologique**

geomorphology The science that deals with the form of the Earth, the general configuration of its surface, the systematic examination of landforms, and the history of geologic changes as recorded by these surface features. **géomorphologie**

geophysics The study of the Earth by quantitative physical methods, including specialties such as *seismology*, tectonophysics, and engineering geophysics. **géophysique**

geosphere The soils, sediments, and rock layers of the Earth's crust, both continental and beneath the ocean floors. **géosphère**

geosyncline (*geoology*) A long, linear basin that contains large (1000 or more meters) accumulations of sediments, and which may be uplifted, folded, and faulted, eventually forming a mountain range. **géosynclinal**

geothermal Describing hot water, steam, or energy that is produced by the transfer of heat from the interior of the Earth to geological deposits close to the surface. Hot springs are an example of geothermal activity. **géothermique**

geothermal energy Energy that can be extracted from the Earth's internal heat. **énergie géothermique**

geotropism A directional movement of a plant in response to the stimulus of gravity. **géotropisme**

germicide An agent capable of killing germs, usually pathogenic microorganisms. **germicide**

germination (*botany*) The resumption of the growth of the seed embryo after a period of dormancy. Germination does not take place unless the seed has been transported to a favorable environment (e.g., adequate water and oxygen, and suitable temperature) by one of the agents of seed dispersal. (*microbiology*) The process of vegetative cell formation following a period of dormancy (i.e., spore formation). **germination**

gibberellins A group of plant hormones that stimulate the growth of leaves and shoots. **gibbérellines**

Gibbs free energy The thermodynamic potential for a system whose independent variables are the absolute temperature, applied pressure, mass variables, and other independent, extensive variables. The change in Gibbs free energy, as a system passes reversibly from one state to another at constant temperature and pressure, is a measure of the work available in that change of state. **énergie libre de Gibbs**

gibbsite A mineral, $Al(OH)_3$, with a platy habit that occurs in highly weathered soils and in laterite. It may also be prominent in the subsoil and *saprolite* of soils formed on crystalline rock high in feldspar. **gibbsite**

gilgai The microrelief of soils produced by expansion and contraction caused by changes in moisture. Gilgai is found in soils that contain large amounts of clay, which swells and shrinks noticeably with wetting and drying. It usually occurs as a succession of microbasins and microknolls in nearly level areas, or as microvalleys and microridges parallel to the direction of the slope. See *microrelief.* **gilgai**

girdling An incision made completely around the stem of a living tree preventing the transport of nutrients and water across the cut; often carried out well into the outer sapwood, for the purpose of killing the tree in horticulture, sometimes used to promote growth of larger fruits or to reduce rooting. **annélation, incision annulaire**

G

glacial Pertaining to, characteristic of, produced or deposited by, or derived from a glacier. **glaciaire**

glacial boulder A large rock fragment that has been transported by a glacier. See *erratic.* **bloc glaciaire, erratique**

glacial drainage channel Stream channel associated with a glacier or icesheet. See *meltwater channel.* **chenal d'eau de fonte**

glacial drift All rock material carried by glacier ice and glacial meltwater, or rafted by icebergs (e.g., till, stratified drift, and scattered rock fragments). **sédiment glaciaire, matériau de transport glaciaire, dépôt glaciaire**

glacial epoch The Pleistocene epoch, the earlier of the two epochs in the Quaternary period, characterized by the extensive glaciation of regions now free from ice. **époque glaciaire**

glacial erosion Wearing away of the Earth's surface as a result of grinding and scouring by passage of a glacier or an ice-sheet and the removal of the eroded material, along with the erosive action of meltwater streams. **érosion glaciaire**

glacial erratic See *erratic.* **bloc erratique**

glacial geology The study of the geologic features and effects resulting from erosion and deposition by glaciers and ice-sheets; also, the features of a glaciated region. Also called glaciogeology. **glaciogéologie**

glacial lake (1) A lake partly or entirely fed by meltwater and lying against or on a glacier. (2) A lake formed by a morainal dam. (3) A lake occupying a bedrock basin produced by glacial erosion. **lac glaciaire**

glacial maximum The time or position of the greatest advance of a glacier, or the largest extent achieved by ice-sheets during the Pleistocene ice age. **maximum glaciaire, pléniglaciaire**

glacial minimum The time or position of the greatest retreat of a glacier. **minimum glaciaire**

glacial recession A decrease in the area, thickness, and/or length of a glacier resulting from melting exceeding the rate of glacier flow. It is characterized by downwasting of the ice mass and probably a retreat of the ice front, although the glacier itself may still be in a state of forward movement. Also called *glacial retreat.* **retrait, recul glaciaire**

glacial retreat See *glacial recession.* **retrait glaciaire**

glacial scour The eroding action of a glacier, including the removal of surficial material and the abrasion and polishing of the bedrock surface by rock fragments dragged along by the ice. **affouillement glaciaire**

glacial till See *till.* **till glaciaire**

glacial spillway (*geology*) A channel that has been cut in solid rock or in drift by the overflow of an ice-damned lake. It differs from a *meltwater channel*, being shallower, with more subtle features, and is most easily recognized where it is associated with deltas, lake shorelines, and lake bottom sediments. Also called overflow channel. **déversoir de lac glaciaire**

glaciated A land surface that has been covered and modified by the action of a glacier or an ice-sheet. **glacié**

glaciation (1) The formation, movement, and recession of *glaciers* or ice-sheets. (2) A collective term for the geological processes of glacial activity and the resulting effects on the Earth's surface. (3) A glacial phase during an ice age in which extensive glaciers developed, attained a maximum, and receded. **glaciation**

glacier A large mass of ice formed on land by the compaction and recrystallization of snow, which exhibits movement downslope or outward in different directions under the influence of gravity, or may be retreating or stagnant. The term is usually restricted to descriptions of ice masses confined by topographic features (e.g., cirque glaciers, valley glaciers) and is distinguished from an ice-sheet or an ice-cap, both of which have larger dimensions. **glacier**

glacier milk Meltwater from a glacier, characterized by a high percentage of light-colored, fine clay- or silt-sized particles suspended in the water. **lait de glacier**

glacier retreat The receding of the front, or snout, of a glacier, occurring as the body of the ice is still moving forward. This condition is reached when the rate of ablation exceeds the rate of accumulation, leading to a negative mass balance. **retrait de glacier**

glacio-eustasy World-wise changes to sea level caused by the depletion of the oceanic reservoir of water at the expense of *ice-sheets* and *glaciers*. **glacio-eustasie**

glaciofluvial Descriptive of material moved by glaciers and subsequently sorted and deposited by streams flowing from the melting ice. The deposits are stratified and may occur in the form of *outwash plains, deltas, kames, eskers,* and *kame terraces*. **fluvio-glaciaire**

glaciolacustrine Pertaining to, derived from, or deposited in glacial lakes; especially said of the deposits and landforms composed of suspended material brought by meltwater streams flowing into lakes bordering the glacier, such as *deltas,* kame deltas, and varved sediments. **glacio-lacustre**

glaciology The study of all processes associated with existing solid water, including snow and ice, and their distribution on the Earth. **glaciologie**

glass electrode An electrode for measuring pH from the potential difference that develops when it is dipped into an aqueous solution containing H^+ ions. **électrode de verre**

glauconite An Fe-rich dioctahedral mica with tetrahedral Al (or Fe^{3+}) usually greater than 0.2 atoms per formula unit and octahedral R^{3+} correspondingly greater than 1.2 atoms. A generalized formula is $K(R_{1.33}^{3+}R_{0.67}^{2+})(Si_{3.67}Al_{0.33})O_{10}(OH)_2$ with $Fe^{3+} >> Al$ and Mg>Fe(II) (unless altered). Further characteristics are d(060) >0.151 nm and (usually) broader infrared spectra than celadonite. Generally green in color and common in sedimentary rocks from Cambrian to the present; it is an indicator of very slow sedimentation. Mixtures containing this mineral as a major component can be called glauconitic. **glauconie, glauconite**

gleyed soil Soil affected by *gleyzation*. **sol gleyifié**

Gleysol A great group of soils in the *Gleysolic* order (Canadian system of soil classification). A thin (less than 8 cm) Ah horizon is underlaid with mottled gray or brownish gleyed material, or the soil has no Ah horizon. Up to 40 cm of mixed peat (bulk density 0.1 or more) or 60 cm of fibric moss peat (bulk density less than 0.1) may occur on the surface. **gleysol**

Gleysolic An order of soils in the Canadian system of soil classification developed under wet conditions and permanent or periodic reduction. These soils have low *chromas*, or prominent *mottling*, or both, in some horizons. The great groups *Gleysol, Humic Gleysol*, and *Luvic Gleysol* are included in the order. **gleysolique**

gleyzation, gleysation A soil-forming process, operating under poor drainage conditions, which results in the

G

reduction of iron and other elements and in gray colors and mottles. See *Gleysolic* and *Gleysol.* **gleyification**

global warming An increase in the near surface temperature of the Earth. Global warming has occurred in the distant past as the result of natural influences, but the term is most often used to refer to the warming predicted as a result of increased emissions of greenhouse gases. See *climate change* and *enhanced greenhouse effect.* **réchauffement planétaire**

global warming potential (GWP) The index used to translate the level of *emissions* of various gases into a common measure in order to compare the relative *radiative forcing* of different gases without directly calculating the changes in atmospheric concentrations. GWPs are calculated as the ratio of the radiative forcing that would result from the emissions of one kilogram of a *greenhouse gas* to that from emission of one kilogram of carbon dioxide over a period of time (usually 100 years). Gases involved in complex atmospheric chemical processes have not been assigned GWPs due to complications that arise. Greenhouse gases are expressed in terms of *carbon dioxide equivalent.* The chart below shows the GWPs (assigned in 1996) for the most important greenhouse gases. **potentiel de réchauffement planétaire**

Gas	GWP
Carbon dioxide	1
Methane	21
Nitrous Oxide	310
HFC-134a	1,300
HFC-23	11,700
HFC-152a	140
HCF-125	2,800
PFCs	7,850*
SF_6	23,900

*This value is an average GWP for the two PFCs, CF_4 and C_2F_6.

globally stable Referring to a system that returns to its initial conditions after all perturbations. **système résilient**

glossic horizon A degraded *argillic*, *kandic*, or *natric horizon* from which clay and free iron oxides are removed. A subsurface diagnostic horizon in the U.S. system of soil taxonomy. **horizon glossique**

glycogen A carbohydrate similar to starch; the reserve food of bacteria, including cyanobacteria; also known as cyanophycean starch. **glycogène**

glycophyte A plant that does not grow well when the osmotic pressure of the soil solution rises above 200 kPa; a nonhalophytic plant. **glycophyte**

glycosidic linkage A C–O–C bond formed between the rings of two cyclic monosaccharides by the elimination of water. **liaison glycosidique**

gneiss A coarse-grained, banded crystalline rock resulting from high-grade regional metamorphism. Most gneisses comprise bands or lenticles of granular quartz and feldspar, which alternate fairly regularly with thinner schistose bands or lenticles comprising micas and amphiboles. Varieties are distinguished by texture (e.g., augen gneiss), characteristic minerals (e.g., hornblende gneiss), or general composition and/or origin (e.g., granite gneiss). **gneiss**

gob (*mining*) Waste material resulting from coal mining, usually high in carbon content and dark in color, and generally stored in piles. **remblai**

goethite A yellow-brown iron oxide mineral, FeOOH. Goethite occurs in almost every soil type and climatic region, and is responsible for the yellowish-brown color in many soils and weathered materials. **goethite**

Golgi complex Membrane complex found in *eukaryotes* that packages proteins and lipids, which are transferred to selected areas of a cell. **appareil de Golgi**

goniometer (1) An instrument used for measuring angles or finding directions (e.g., the angles between crystal faces). (2) A device used to determine the direction of maximum response of a received radio signal or the direction of maximum radiation for a transmitted signal (e.g., an x-ray goniometer, which measures the angle of diffraction of crystals). **goniomètre**

grab sample A single soil, air, water, or biological sample drawn over a short time period. As a result, the sample is not representative of long-term conditions at the sampling site. This type of sampling yields data that provide a snapshot of conditions or chemical concentration at a particular point in time. **échantillon instantané**

gradation (1) (*geology*) The bringing of a land surface to a state of uniform grade or slope through erosion or deposition. (2) (*civil engineering*) The frequency distribution of the various sized grains that constitute a sediment, soil, or other material. **(1) aplanissement (2) granulométrie**

grade (1) The slope of a road, channel, or natural ground. (2) The finished surface of a canal bed, roadbed, top of embankment, or bottom of excavation; any surface prepared for the support of construction such as paving or laying a conduit. **(1) pente (2) niveau du sol**

grade stabilization structure A structure for the purpose of stabilizing the grade of a gully or other watercourse, thereby preventing further head-cutting or lowering of the channel grade. **ouvrage de stabilisation de pente**

graded (1) (*geomorphology*) A land surface on which erosion and deposition are balanced, maintaining a general slope of equilibrium. Also referred to as "at grade". (2) (*geology*) A sediment or rock containing particles of essentially uniform size. Also called sorted. (3) (*engineering*) A soil or sediment consisting of particles of many sizes, or having a uniform distribution of particles from coarse to fine. (This definition is opposite that for geology.) **(1) à l'équilibre (2) classé (3) gradué**

graded terrace See *terrace*. **terrasse régularisée**

gradient Change of elevation, velocity, pressure, or other characteristics per unit length. See *slope*. **gradient, inclinaison**

gradient analysis A means of relating species distributions to environmental changes by sampling a series of communities along a habitat gradient. **analyse des gradients**

grading curve A curve on which the *particle size* of a sample of soil or sediment is plotted on a horizontal, logarithmic scale and percentages are plotted on a vertical, arithmetic scale, during *mechanical analysis*. **courbe granulométrique**

graft A relatively small part (e.g., tissue) of a plant or animal to be transplanted (i.e., grafted) to a larger part. In plants, grafting involves implanting a scion (any unrooted portion of a plant) into a growing plant, called a stock, resulting in contact of their cambium layers, thus enabling the scion to derive water and nutrients from the stock, eventually resulting in a union of the two parts. **greffe**

Graham's law The rate at which a gas diffuses is inversely proportional to the square root of its density. **loi de Graham**

grain (1) Mineral or rock particle with a diameter of less than a few millimeters (e.g., sand grain). (2) General term for particles of all sizes (e.g., fine-grained, coarse-grained). (3) Single crystal or a separate particle of ice in snow or ice. (4) The linear arrangement of topographic features in a region (e.g., parallel ridges and valleys). **grain**

grain density See *particle density*. **densité particulaire**

grain-size analysis See *particle-size analysis*. **analyse granulométrique, analyse mécanique**

grain-size distribution See *particle-size distribution*. **granulométrie**

G

Gram negative/Gram positive The response of bacteria to a procedure called Gram staining. When treated with crystal violet, Gram's iodine, 95% ethanol, and safranin, bacteria usually retain either a red color (Gram negative) or a purple/blue color (Gram positive). In addition to the differences in color upon staining, these two types of bacteria represent bacteria that are fundamentally different in terms of structure, physiology, ecology, and pathogenicity. Common Gram-negative species are *Escherichia, Salmonella,* and *Pseudomonas*; common Gram-positive species are *Bacillus, Staphylococcus,* and *Streptococcus.* **Gram négative, Gram positive**

Gram stain A technique used to differentiate bacteria. Gram-positive bacteria stain violet, and Gram-negative bacteria stain red. **coloration de Gram**

graminoid Grasslike; pertaining to narrow-leaved grasses, sedges, rushes, or similar plants. See *grass, sedge.* **graminiforme**

granite A coarse-grained, light-colored igneous rock of plutonic origin, consisting mainly of quartz (20 to 40%) together with alkaline and plagioclase feldspars, and very commonly a mica. It is one of the most common igneous rocks, but there are many varieties, their classification depending upon grain size and mineral composition. One of the finest grained is called aplite, while the coarsest is a pegmatite. Where mica predominates, it becomes a biotite or muscovite granite. Granite that includes hornblende or pyroxene is called a hornblende or pyroxene granite. When alkali and plagioclase feldspars are about equal, the rock is called an adamellite, but if plagioclase dominates, it is a quartz diorite or granodiorite. **granite**

granular soil structure A shape of soil structure. See *Appendix B, Figure 1* and *Table 1*. **granulaire**

granular disintegration (*geology*) A main type of mechanical weathering caused by *freeze thaw action* and temperature changes resulting in the grain-by-grain breakdown of rock masses, especially coarse-grained rocks (e.g., granite, gneiss, sandstone, and conglomerate). **désagrégation granulaire**

granular fertilizer See *fertilizer, granular.* **engrais granulaire**

granulation The process of producing granular materials. **granulation**

granule Spheroidal soil aggregate. **granule**

graph A plane surface depicting points and lines in their correct relative positions in relation to coordinates. A graph is a visual representation of data in the form of a continuous curve, usually on squared paper (graph paper). A line on a graph is always referred to as a curve, even though it may be straight. **graphique**

graphic scale A graduated line by means of which distances on the map or chart may be measured in terms of ground distances; also called bar scale. **échelle graphique**

grass Plant in the family Gramineae; characterized by bladelike leaves arranged on the *culm* or stem, and *inflorescence* surrounded by glumes. Grass roots may be *fibrous, rhizomatous,* or *stoloniferous.* Many grasses have basal meristems, unlike other plants that have *apical meristems.* **graminée**

grass barrier A strip of grass placed across a field for snow management or control of wind *erosion.* **bande enherbée**

grass tetany (hypomagnaesemia) A condition of cattle and sheep marked by tetanic staggers, convulsions, coma, and frequently death; characterized by a low level of blood magnesium. **tétanie d'herbage (hypomagnésiémie)**

grassed waterway A natural or constructed waterway, usually broad and shallow, covered with erosion-resistant grasses, used to convey surface water from or across cropland along natural depressions, to prevent *gully erosion.* **voie d'eau engazonnée**

grassland An ecosystem with limited precipitation (about 250 to 750 mm annually) in which grasses and shrubs constitute the dominant vegetation. Several types of natural grasslands are recognized, including tropical, temperate (e.g., pampas in South America, prairie in North America, steppe in Asia, or veldt in South Africa), and montane. Grassland has been introduced in other parts of the world (e.g., Britain and New Zealand) in areas that were originally forested as parts of pastoral farming systems. One of the four main subdivisions of world vegetation. See *biochore.* **prairie**

grassland biome See *grassland.* **biome de prairie**

grasslike plant A plant that resembles a true grass but is taxonomically different (e.g., sedges and rushes). **plante graminiforme**

gravel Rock fragments 2 mm to 7.5 cm in diameter. **gravier**

gravel envelope Selected aggregate placed around the screened pipe section of well casing, or a subsurface drain to facilitate the entry of water into the well or drain. **enveloppe de gravier**

gravel filter Graded sand and gravel aggregate placed around a drain or well screen to prevent the movement of fine materials from the aquifer into the drain or well. **filtre de gravier**

gravelly Material modifier descriptive of an accumulation of rounded to subrounded particles ranging in size from pebbles to boulders (2 mm to greater than 60 cm). For soil classification purposes, limits are placed on the volumetric amount of gravel (e.g., gravelly sandy loam has a sandy loam texture with 20 to 50%). See *coarse fragments.* **graveleux**

gravimeter A device used to measure differences in the Earth's gravitational field. **gravimètre**

gravimetric analysis Quantitative analysis that depends on weighing. **analyse gravimétrique**

gravitational potential The amount of work that must be done per unit quantity of soil water in order to transport it reversibly and isothermally, from a pool at a specified elevation and at atmospheric pressure to a similar pool at the elevation of the point under consideration. See *soil water.* **potentiel d'eau libre, potentiel gravitaire**

gravitational water Water that moves into, through, or out of the soil under the influence of gravity. See *soil water, soil water potential.* **eau libre, eau gravitaire**

gravity (1) The effect on any body in the universe due to attractive forces between it and all other bodies, and to any centrifugal force that may act on the body because of its motion in an orbit. (2) The force exerted by the Earth and its rotation on a unit mass, or the acceleration of a body falling freely in a vacuum in the gravitational field of the Earth. **gravité**

gravity flow Water flow that is not pumped but flows due to the acceleration of gravity. Used in irrigation, drainage, inlets, and outlets. **écoulement par gravité**

Gray Brown Luvisol A great group of soils in the *Luvisolic* order (Canadian system of soil classification) occurring in a moderate climate, higher than 8.0 °C mean annual soil temperature, and developed under deciduous and coniferous forest cover. These soils have a dark-colored mull-like surface (Ah) horizon, a light-colored eluviated (Ae) horizon, a brownish illuvial B (Bt) horizon, and a basic or calcareous C horizon. The solum is highly base saturated (with neutral salt extraction). This group includes soils formerly called Gray Brown Podzolic. **luvisol brun gris**

Gray Luvisol A great group of soils in the *Luvisolic* order (Canadian system of soil classification) occurring in moderately cool climates, where the mean annual soil temperature is usually lower than 8.0 °C. The soils have developed under deciduous and coniferous forest cover, and have an eluviated light-colored surface (Ae) horizon, a brownish illuvial B (Bt) horizon, and usually a calcareous C horizon. The solum is base saturated (with neutral salt extraction). The Ahe horizon, if present, is less than 5 cm thick. This group includes soils formerly called Gray Wooded. **luvisol gris**

G

graywacke A dark-gray, firmly indurated, coarse-grained sedimentary rock that consists of poorly sorted angular to subangular grains of quartz and feldspar, with a variety of dark rock and mineral fragments, embedded in a compact clayey matrix having the general composition of slate and containing an abundance of very fine-grained illite, sericite, and chloritic minerals. It commonly exhibits graded bedding and is believed to have been deposited by turbid submarine currents. **grauwacke**

graze (*range management*) The consumption of forage *in situ* by animals. **pâturer**

grazing capacity The maximum stocking rate possible without inducing damage to vegetation or related resources; may vary from year to year. **charge limite d'un pâturage**

grazing distribution Dispersion of livestock grazing within a management unit or area. **distribution du pâturage**

grazing management The manipulation of animal grazing in pursuit of a defined objective. **conduite du pâturage**

grazing period The length of time that grazing livestock or wildlife occupy a specific land area. **temps de pâturage**

grazing pressure The relationship between the number of animal units or forage intake units and the weight of forage dry matter per unit area at any one point in time. **taux de charge des pâturages**

grazing system A defined, integrated combination of animal, plant, soil, and other environmental components and the grazing methods by which the system is managed to achieve specific results or goals. **système de pâturage**

great group A category in the Canadian system of soil classification, the U.S. Soil Taxonomy, and other classification systems. It is a taxonomic group of soils having certain morphological features in common and a similar pedogenic environment. Examples are Black, Solonetz, Gray Brown Luvisol, Humic Podzol, Melanic Brunisol, Regosol, Gleysol, and Fibrisol. **grand groupe**

green belt An area restricted from use for buildings and houses, frequently planned around the periphery of urban settlements to break up the continuous pattern of urban development. It often serves as a separating buffer between a pollution source and a concentration of population. **ceinture verte, ceinture de verdure**

green manure Any crop or plant grown and plowed under when green or soon after maturity to improve the soil by adding organic matter and subsequently releasing plant nutrients, especially nitrogen. **engrais vert**

green-chopped Mechanically harvested forage fed to animals while it is fresh and succulent. **affourragement vert**

greenhouse effect The heating effect of the atmosphere upon the Earth. Short wave solar radiation passes through the atmosphere and is absorbed by the Earth's surface. The Earth reradiates long wave radiation, which is absorbed by the air (mainly by carbon dioxide and water vapor). Thus, the atmosphere traps heat much like the glass in a greenhouse. See *figure*. **effet de serre**

greenhouse gas An atmospheric gas or vapor that absorbs outgoing infrared energy emitted from the Earth,

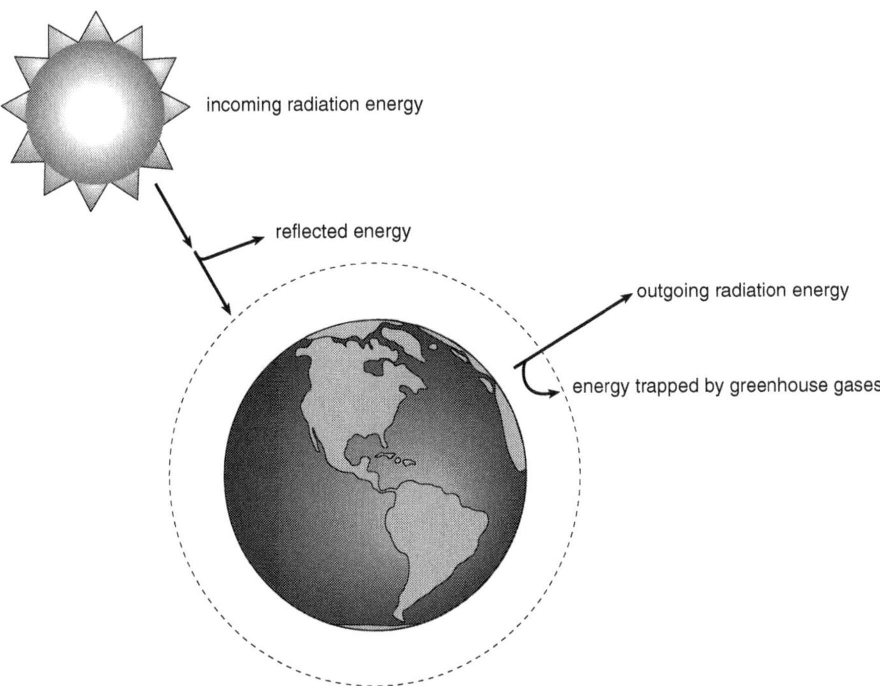

The greenhouse effect. Short-wavelength radiation emitted from the sun is absorbed by the Earth and re-radiated at longer wavelengths. Carbon dioxide, CH_4, and N_20 account for 90% of this greenhouse effect (adapted from Janzen et al., 1998).

G

contributing to the *greenhouse effect.* Important greenhouse gases are *carbon dioxide,* water vapor, *methane, nitrous oxide,* and the *chlorofluorocarbons.* **gaz à effet de serre**

gray box system A system whose structure is only partly known. See *black box system, white box system.* **système de type boîte grise**

grid A systematic array of points or lines (e.g., a rectangular pattern of pits or cores). **grille, maille**

groin A structure to protect and improve a shoreline (built usually to trap littoral drift or retard erosion of the shore). **épi**

gross primary productivity (GPP) The total fixation of energy by photosynthesis. **productivité primaire brute**

gross productivity The rate at which energy or nutrients are assimilated by an organism, a population, or an entire community. **productivité brute**

ground control (*remote sensing*) A point or system of points on the Earth's surface whose position has been established by ground survey(s), also called field control. **canevas d'appui, amer**

ground cover (*soil conservation*) Grasses or other plants grown to keep soil from being blown or washed away. (*horticulture*) Low-growing plants (e.g., vinca and ginger) that do not require mowing. **couvert végétal**

ground data Supporting data collected on the ground, and information derived therefrom, as an aid to the interpretation of remotely recorded surveys (e.g., airborne imagery). Generally, this should be performed concurrently with the airborne surveys. Also called *reference data.* **réalité de terrain, données de terrain**

ground fire A fire that consumes all the organic material of the forest floor and also burns into the underlying soil. It differs from a surface fire by being invulnerable to wind. In a surface fire, the flames are visible and burning is accelerated by wind, whereas in a ground fire, wind is not generally a serious factor. **feu de surface**

ground frost A condition when the minimum temperature at the ground surface is below 0°C, although the air temperature may remain above the freezing point. **gel au sol**

ground moraine (*geology*) An unsorted mixture of rocks, boulders, sand, silt, and clay deposited by glacial ice. The predominant material is till, but some *stratified drift* is present. Most of the till is thought to have accumulated under the ice by lodgment, but some till has been let down from the upper surface of the ice by ablation. Ground moraine is usually in the form of undulating plains having gently sloping swells, sags, and enclosed depressions. **moraine de fond**

ground state The lowest possible energy state of an atom or molecule. **état fondamental**

ground truthing The use of a ground survey to confirm the findings of an aerial survey or to calibrate quantitative aerial observations. **vérification au sol, vérification de terrain**

groundwater That portion of the water below the surface of the ground at a pressure equal to or greater than atmospheric (i.e., always saturated or below the water table). Water that is passing through or standing in the soil and the underlying strata. It is free to move by gravity. See *water table*. **eau souterraine**

groundwater flow The movement of water in the zone of saturation, whether naturally or artificially produced. **écoulement souterrain**

groundwater hydrology The science dealing with the movement of the soil solution in the saturated zone of the soil profile. **hydrogéologie, hydrologie des eaux souterraines**

groundwater level See *water table*. **niveau de l'eau souterraine**

groundwater reservoir See *aquifer*. **réservoir aquifère**

groundwater runoff Groundwater that is discharged into a stream channel as spring or seepage water. That part of runoff which has passed into the ground as precipitation or snow melt, has become groundwater, and has been discharged into a stream channel as spring or seepage water. **ruissellement souterrain**

groundwater table The upper limit of the groundwater. **nappe phréatique**

groundwater velocity The rate of water movement through openings in rock or sediment. Estimated using *Darcy's law*. **vitesse de l'eau souterraine**

group (of the periodic table) A vertical column of elements having the same valence electron configuration and showing similar properties. **groupement**

grove A small group of trees, usually without understory, planted or natural. **bocage**

growing season The length of time that cultivated crops can be grown in a particular location. The number of days between the last freeze in the spring and the first frost in autumn. In equatorial latitudes the growing season is usually continuous, whereas in northern latitudes it may be reduced to 8 to 10 weeks. **saison de croissance**

growth phase (*microbiology*) Pattern of change in growth (cell division) that occurs when bacteria are placed in an environment suitable for their growth. The various phases of growth are *lag growth phase* (a pause in cell division), *log growth phase* (cell divison is at maximum rate), *stationary growth phase* (cells stop dividing), and death phase (cells are dying). See *figure*. **phase de croissance**

growth regulator A substance that, when applied to plants in small amounts, inhibits, stimulates, or otherwise

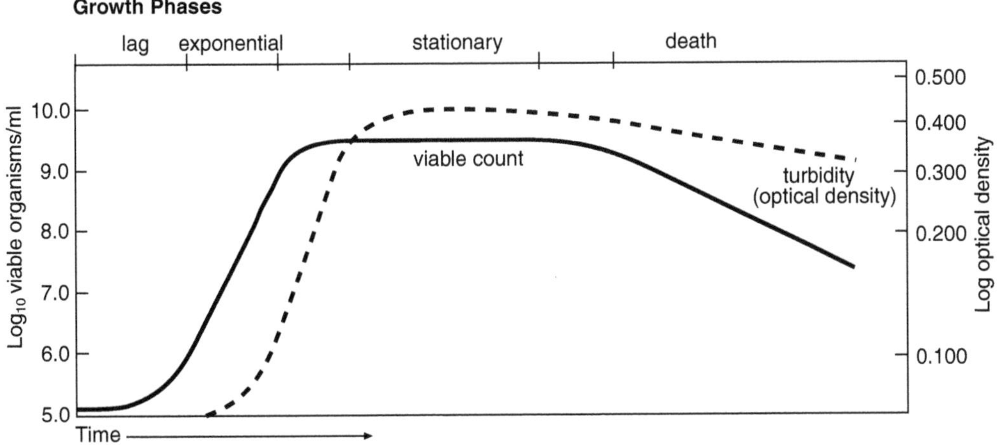

Typical **growth phases** of a bacterial population (adapted from Brock et al., 1984).

G

modifies the growth process. **régulateur de croissance**

growth ring The layer of wood produced in a tree during its annual growth period. Growth rings can be analyzed for chronologic and climatic data based on number and relative size. See *dendrochronology*. **anneau de croissance**

guano The dried, decomposed excrement of birds and bats, rich in phosphates and nitrogenous compounds, used as a fertilizer. **guano**

guard cells Specialized kidney-shaped epidermal cells surrounding a *stoma* in a leaf or green stem; changes in turgor of a pair of guard cells cause the pore to open or close, thereby regulating the opposing priorities of permitting CO_2 uptake and avoiding H_2O-vapor loss. **cellules de garde, cellules stomatiques**

guess row See *tillage, guess row*. **ligne de guidage**

guild A group of populations that exploit the same class of resources in a similar way. **guilde**

gullied land An area where all diagnostic soil horizons have been removed by water, resulting in a network of V-shaped or U-shaped channels. Some areas resemble miniature badlands. Generally, gullies are so deep that extensive reshaping is necessary for most uses. **terrain raviné**

gully An *erosion* channel caused by the concentrated but intermittent flow of water, usually across bare soil during and immediately following heavy rains; may consist of small channels in the early stage, but with time or intense rainfall it can become deep enough to interfere with, and not to be repaired by, normal tillage operations. **ravin**

gully control plantings The planting of forage, legume, or woody plant seeds, seedlings, cuttings, or transplants in gullies to establish or re-establish a vegetative cover adequate to control runoff and erosion. **stabilisatiation de ravins, végétalisation de ravins**

gully erosion The *erosion* process whereby gullies are formed on the land surface, owing to the effects of heavy rainstorms. The surface runoff becomes concentrated into shallow, narrow channels (see *rill*), which then combine to form deep gullies. Over short periods, gully erosion can remove the soil from these narrow areas to various depths, from about 0.3 m to as much as 30 m. **érosion par ravinement**

gully reclamation Management procedures designed to restore and prevent further erosion in gullies (e.g., filling them in or planting vegetation to stabilize the banks). **remise en état de ravins**

gullying Removal of soil by running water, with formation of channels that cannot be smoothed out (and often cannot be crossed) during normal tilling by wheeled implements. **ravinement**

Gutenberg discontinuity The discontinuity that marks the boundary between the mantle and core of the Earth, located about 2900 kilometers below the surface. **discontinuité de Gutenberg**

guttation The extrusion of water (and sometimes salts) from the aerial parts of plants, usually at night when transpiration rates are low. **guttation**

gymnosperm Plants that do not have their seeds protected by an outer covering (e.g., pine, fir, and spruce). See *angiosperm*. **gymnosperme**

gypsan (*soil micromorphology*) A *cutan* composed of gypsum. **gypsane**

gypsic horizon A diagnostic mineral subsurface horizon (U.S. system of soil taxonomy) of secondary $CaSO_4$ enrichment that is > 15 cm thick, has at least 50 g kg^{-1} more *gypsum* than the *C horizon*, and in which the product of the thickness in centimeters and the amount of $CaSO_4$ is equal to or greater than 1500 g kg^{-1}. **horizon gypsique**

Gypsids A suborder in the U.S. system of soil taxonomy. Aridisols which have a gypsic or petrogypsic horizon that has its upper boundary within 100 cm of the soil surface and lack a petrocalcic horizon overlying any of these horizons. **Gypsids**

gypsum The common name for calcium sulfate ($CaSO_4 \cdot 2H_2O$) used to supply calcium to ameliorate soils with a high exchangeable sodium fraction. **gypse**

gypsum requirement The quantity of gypsum (e.g., Mg ha^{-1}) or its equivalent required to reduce the exchangeable sodium content of a given amount of soil to an acceptable level. **besoin en gypse**

gyttja A nutrient-rich, sedimentary peat consisting mainly of plankton, other plant and animal residues, and mud. It is deposited in water in a finely divided condition. (Swedish) **bourbe eutrophe**

H

H horizon An organic soil horizon containing more than 17% organic carbon, normally associated with upland forest soils, which is characterized by an accumulation of decomposed leaves, twigs, woody materials, and mosses in which the structures are indiscernible. This horizon differs from an *F horizon* by having greater *humification* due chiefly to the action of organisms. It is frequently intermixed with mineral grains, especially near the junction with a mineral horizon. An organic horizon in the Canadian system of soil classification. See *Appendix D* for equivalent terms in the U.S. and FAO/UNESCO systems of soil classification. See *horizon, soil; L horizon*; *F horizon*. **horizon H**

Haber process The industrial method used for the fixation of atmospheric nitrogen for use as fertilizer. The process forms ammonia by the direct combination of atmospheric nitrogen and molecular hydrogen from natural gas. The reaction requires a pressure between about 15 and 100 MPa and a temperature of about 500°C in the presence of an iron catalyst. The ammonia can be chemically combined with carbon dioxide to produce urea, or further reacted with oxygen to make nitric acid. Nitric acid added to ammonia will form ammonium nitrate, which, along with urea, is a widely used nitrogen fertilizer. **procédé de Haber**

habitat The place where an organism lives and/or the conditions of that environment including the soil, vegetation, water, and food. **habitat**

habitat breadth The distribution of a population among types of *habitats*. **amplitude de l'habitat**

habitat enhancement Any manipulation of *habitat* that improves its value and ability to meet specified requirements of one or more species. **amélioration de l'habitat, aménagement de l'habitat**

habitat function The purpose(s) served by a particular *habitat*, in the life cycle of a species including breeding, denning, feeding (seasonal or year-round), molting, over-wintering, shelter, staging, and young-rearing. **fonction de l'habitat**

habitat management Management of the forest to create environments that provide *habitats* (food, shelter) to meet the needs of particular organisms. **gestion de l'habitat**

habitat needs The various requirements for survival and prosperity of a species that must be met by the *habitat*, including food, water, shelter, suitable breeding, and denning sites. See *habitat function*. **besoins en habitat**

habitat niche A *habitat* that supplies the factors necessary for the existence of a specific organism or species. **niche d'habitat**

habitat selection Preference for certain *habitats*. **sélection d'habitat**

habitat type The kind of *habitat* that exists in an area, based on the dominant vegetation (e.g., tundra; grassland; coniferous, deciduous, or mixed forest). **type d'habitat**

habitat, critical A *habitat* that is necessary for the survival of a species in an area (the function served by this habitat cannot be accomplished by another habitat). **habitat critique**

habitat, preferred *Habitats* for which species have a natural inclination over other, often similar habitats within the same general area (e.g., beaver have a natural inclination for using stands of *Populus* spp. even though stands of other tree species may be present and, in many cases, better situated for use). The biological or traditional reason for the preference is often unknown. **habitat préféré**

half-life (1) The time it takes for one-half the atoms of a radioactive isotope to decay into another isotope. The half-life is different for each radioactive element. (2) The time required for a reactant to reach half of its original concentration. **(1) période (2) demi-vie**

half-reactions The two parts of an oxidation-reduction reaction, one representing oxidation, the other reduction. **demi-réactions**

halite Sodium chloride (NaCl) or common salt, formed from sea water or in a salt lake as an evaporite. **chlorure de sodium, halite**

halloysite A member of the kaolin subgroup of clay minerals. It is similar to kaolinite in structure and composition, except that hydrated varieties occur that have interlayer water molecules. Halloysite usually occurs as tubular or spheroid particles and is most common in soils formed from volcanic ash. **halloysite**

halocarbon A chemical compound consisting of carbon, sometimes hydrogen, and either chlorine, fluorine, bromine, or iodine. **halocarbone**

halogen One of the reactive nonmetals: fluorine, chlorine, bromine, iodine, and astatine. These elements are known for their reactivity and are only found combined with other chemicals in nature. At room temperature, fluorine and chlorine are gases, bromine is a liquid, and iodine and astatine are solids. Chlorine is used extensively as an industrial chemical and is a component in many compounds of environmental interest. **halogène**

halogenation The chemical reaction in which a *halogen* atom is introduced into a compound. **halogénation**

halomorphic soil A general term for *saline* and *alkali soils*. **sol halomorphe**

halophile A microorganism that lives at high salinity. **halophile**

halophyte Plant that is tolerant of salt in the soil or in the air (i.e., sea spray); common in salt-pan and sea-shore environments. **halophyte**

hamada An accumulation of stones at the surfaces of deserts, formed by the washing or blowing away of the finer material. **hamada**

haploid A cell or organism that contains a single set of chromosomes. **haploïde**

hard rock (1) A term used loosely for igneous or metamorphic rock, as distinguished from sedimentary rock. (2) Rock that requires drilling and blasting for its economical removal. **roche dure**

hard water See *hardness*. **eau dure**

hardening In soil formation, the decrease in volume of voids by collapse and compaction, and by filling some voids with fine earth, carbonates, silica, and other materials. **durcissement**

hardiness The ability of a plant or other organism to withstand severe climates, especially frost, during the growing season. **rusticité**

hardness A measure of the amount of calcium and magnesium salts dissolved in water. **dureté**

hardness scale (*geology*) A measure of the resistance of a material to scratching. The most common measure is the *Mohs scale,* in which minerals are compared to ten reference minerals arranged so that each one can scratch all of those that precede it on the scale, which from softest to hardest are: talc, 1; gypsum, 2; calcite, 3; fluorite, 4; apatite, 5; feldspar, 6; quartz, 7; topaz, 8; corundum, 9;

diamond, 10. (*limnology*) The quality of water produced by soluble salts of calcium, magnesium, and other substances, generally expressed as mg calcium carbonate per liter of water. Water hardness classes are: <60 mg L^{-1}, soft; 61 to 120 mg L^{-1}, moderately hard; 121 to 180 mg L^{-1}, hard, >180 mg L^{-1}, very hard. **échelle de dureté**

hardpan A hardened soil layer, in the lower *A* or in the *B horizon*, caused by cementation of soil particles with organic matter or with materials such as silica, sesquioxides, or calcium carbonate. The hardness does not change appreciably with changes in moisture content, and pieces of the hard layer do not slake in water. **carapace, horizon induré**

hardwood Wood produced from *deciduous*, broad-leafed tree species (e.g., oak, alder, maple). **feuillus**

harrowing See *tillage, harrowing*. **hersage**

Hartig net The intercellular net of *hyphae* between the cortical cells of a root in an *ectomycorrhiza*. **filet de Hartig**

harvest index The quantity of harvestable biomass produced per unit total biomass produced; the quantity of biomass produced per unit input of plant nutrient. **indice de récolte**

haul road Road from a mine pit, timber cutting, or other resource extraction area to a loading dock, tipple, ramp, or preparation plant used for transporting mined material by truck. **chemin d'exploitation**

hay Entire herbage of forage plants that is harvested and dried for feed; sometimes including seed of grasses and legumes. **foin**

haylage Product resulting from ensiling forage with about 45% moisture in the absence of oxygen. See *silage*. **ensilage mi-fané, ensilage préfané**

hazardous waste Any waste that poses a significant threat to human health or safety because it is toxic, ignitable, corrosive, or reactive. **déchets dangereux**

head (1) (*hydrology*) The pressure exerted by a liquid upon a unit area dependent upon the height of the surface of the liquid above the point where the pressure is determined. Units of head are force per unit area (kg m^{-2}). (2) (*limnology*) The upper part of an aquatic system such as the source of a stream, the upper part of a catchment, the apex of a delta, or the end of a lake opposite the outlet. (3) (*geomorphology*) A high coastal promontory or cape. **(1) pression hydrostatique, hauteur (2) tête de réseau (3) promontoire**

head ditch (*irrigation*) An open ditch used to convey and distribute water in a field for surface irrigation. **canal de distribution**

headwater The beginning of a stream or river, its source, or its upstream portion. **cours supérieur d'une rivière**

healthy ecosystem An ecosystem in which structure and functions allow the maintenance of biodiversity, biotic integrity, and ecological processes over time. **écosystème sain**

heartwood Nonliving and commonly dark-colored wood in which no water transport occurs; surrounded by sapwood. **bois de coeur**

heat The kinetic energy of a material arising from the molecular motion of the material. **chaleur**

heat capacity The amount of heat required to raise the temperature of a unit volume of soil by one degree. It may also be expressed in terms of mass of soil, in which case it is called specific heat. **capacité calorifique**

heat flux density The quantity of heat flowing per unit of time across a unit area. **densité du flux thermique**

heat of combustion The energy liberated when one mole of given substance is completely oxidized (e.g., the chemical reaction between an organic fuel and oxygen to form carbon dioxide and water). **chaleur de combustion**

heat of condensation The heat released when a vapor changes state to a liquid. See *heat of vaporization.* **chaleur latente de condensation**

heat of hydration The *enthalpy* change associated with placing gaseous molecules or ions in water; the sum of the energy needed to expand the solvent and the energy released from the solvent–solute interactions. **chaleur d'hydratation**

heat of immersion The heat evolved on immersing a soil with a known initial water content (usually oven dry) in a large volume of water. **chaleur d'immersion**

H

heat of solution The *enthalpy* change associated with dissolving a solute in a solvent; the sum of the energies needed to expand both solvent and solute in a solution and the energy released from the solvent–solute interactions. **chaleur de dissolution**

heat of vaporization The energy required to vaporize one mole of a liquid at a pressure of one atmosphere. **chaleur latente de vaporisation**

heat sink Any material used to absorb heat. **puits thermique**

heath (1) (*general*) A type of moorland landscape at low altitude, especially on the coast. (2) (*botany*) Originally used narrowly to describe a shrubby vegetation dominated by plants of the genus *Calluna* (heather), but now extended to include a vegetation comprising *Erica*, gorse, broom, bilberry, and other small, close-growing woody shrubs. **(1) lande (2) plante éricacée**

heave (1) An upward movement of a surface caused by expansion due to swelling clay, removal of overburden, frost action, or other agent (e.g., frost heaving). (2) The horizontal component of separation or displacement on a fault. **(1) soulèvement (2) rejet horizontal**

heavy atom An *isotope* of an element that contains more neutrons than in the most frequently occurring form of that element. **atome lourd**

heavy clay A soil textural class. See *texture*. **argile lourde**

heavy fraction (organic matter) The material that settles in a heavy liquid; consists of mineral particles and organic matter adsorbed to mineral surfaces or contained in aggregates. See *light fraction*. **fraction lourde ou dense de la matière organique**

heavy hydrogen The *isotope* of hydrogen that contains one proton and one neutron in the nucleus rather than the more common form having one proton and no neutron. Also called deuterium. See *heavy water*. **hydrogène lourd (deutérium)**

heavy liquid In analysis of minerals, a liquid of high density (e.g., bromoform, methylene iodide) in which specific-gravity tests can be made, or in which mechanically mixed minerals can be separated. Also used to separate soil constituents (i.e., mineral and organic particles) on the basis of density. **liqueur lourde**

heavy media separation Separation of solid material into heavy and light fractions in a fluid medium whose density is between theirs. **flottation ou séparation gravimétrique**

heavy metal A metal with a high relative atomic mass; usually applied to common transition metals such as Cu, Fe, Mn, Mo, Zn, Cd, and Hg. Excess levels of these metals are a cause of environmental pollution. **métal lourd**

heavy mineral A rock-forming mineral having a specific gravity greater than about 2.85 (e.g., magnetite, ilmenite, rutile, garnet). See *light mineral*. **minéral lourd**

heavy soil A soil having a high content of the fine separates, particularly clay, or a soil having a high drawbar pull and therefore hard to cultivate. See *fine texture*. **sol lourd**

heavy water Water in which hydrogen atoms are replaced by the heavier isotope, deuterium. See *heavy hydrogen*. **eau lourde**

heliophyte A plant adapted to life in full sunlight. In shade, a heliophyte may germinate but would not complete its full life cycle of flowering and fruiting. See *sciophyte*. **héliophyte**

heliotropism The action by which plants turn towards the light. **héliotropisme**

hematite An iron mineral of the hematite group (α–Fe_2O_3), dimorphous with maghemite, gray to black, trigonal, usually occurring in massive, fibrous, micaceous, or granular form, having a specific gravity of 5.26 and a hardness of 5 to 6 (*Mohs scale*). Hematite is found in igneous, sedimentary, and metamorphic rocks, both as a primary constituent and as an alteration product. It is the most abundant iron ore and is also one of the main cementing agents of sandstone. **hématite**

hemic material Organic soil material at an intermediate degree of decomposition; up to two-thirds of the material cannot be recognized. Its bulk density is very low and water-holding capacity is very high. **matériau hémique**

hemicellulose (*biochemistry*) Polysaccharides that are associated with cellulose and lignin in the cell walls of green plants. It differs from cellulose in that it is soluble in alkali and, with acid hydrolysis, gives rise to glucose, uronic acid, xylose, galactose, and other carbohydrates. **hémicellulose**

Hemists A suborder in the U.S. system of soil taxonomy. *Histosols* that have an intermediate degree of plant fiber decomposition and a bulk density between about 0.1 and 0.2 g cm^{-3}. Hemists are saturated with water for periods long enough to limit their use for most crops unless they are artificially drained. **Hémists**

hemoglobin A biomolecule composed of four myoglobin-like units (proteins plus heme) that can bind and transport oxygen molecules in the blood to other body tissues. **hémoglobine**

Henry's law For a gas in contact with water at equilibrium, the ratio of the atmospheric concentration of the gas and its aqueous concentration is equal to a constant (i.e., *Henry's law constant*), which varies only with temperature. One form of Henry's law is $K_H = P/C_w$, where K_H is Henry's law constant, P is the equilibrium partial pressure of a gas in the air above the water, and C_w is the aqueous concentration of the gas dissolved in the water. **loi de Henry**

Henry's law constant The ratio, at equilibrium and for a given temperature, of the atmospheric concentration of a gas to its aqueous concentration; typical units are MPa per mole. Henry's law constant is sometimes given as the equilibrium ratio of the aqueous concentration of a gas to its atmospheric concentration; in this case the units will be inverted, in mole fraction/atmosphere. **constante de la loi de Henry**

herb Any flowering plant except those developing persistent woody bases and stems above ground. **herbe**

herbaceous Nonwoody vegetation. **herbacé**

herbage The biomass of herbaceous plants, other than separated grain, generally above ground but including edible roots and tubers. **herbage**

herbicide A chemical compound (i.e., *pesticide*), usually synthetic, that kills or causes damage to specific plants (usually weeds) without significant disruption of other plant or animal communities. Types of herbicides include contact, selective (designed to kill specific weeds or plants), non-selective (destroys or prevents all plant growth), post-emergence (designed to be applied after the crop is above ground), and pre-emergence (designed to be applied before the crop emerges through the soil surface). See *contact herbicide*, *amide*, *benzoic*, *dinitroanaline*, *diphenyl ethers*, *phenoxy*, and *triazine*. **herbicide**

herbivore A consumer that eats plants or other photosynthetic organisms to obtain its food and energy. **herbivore**

heredity The transmission of characteristics from parent to offspring through genes. **hérédité**

heterocyst A specialized vegetative cell in certain filamentous blue-green *algae*; larger, clearer, and thicker walled than the regular vegetative cells. **hétérocyste**

heterogeneous Diverse in character; having unlike qualities; varied in content. See *homogeneous*. **hétérogène**

heterogeneous equilibrium An equilibrium involving reactants and/or products in more than one phase. **équilibre hétérogène**

heterotroph An organism able to derive carbon and energy for growth and cell synthesis by utilizing organic compounds. See *autotroph*. **hétérotrophe**

heterotrophic Capable of deriving energy for life processes only from the decomposition of organic compounds, and incapable of using inorganic compounds as sole sources of energy or for organic synthesis. **hétérotrophe**

heterotrophic ecosystem An ecosystem that obtains its energy principally from preformed organic energy sources. **écosystème hétérotrophe**

heterotrophic nitrification Biochemical oxidation of ammonium to nitrate and nitrite by heterotrophic microorganisms. **nitrification hétérotrophe**

heterotrophy The nutritional mode of organisms that gain both carbon and energy from organic compounds. **hétérotrophie**

heterozygous Containing two forms (alleles) of a gene, one derived from each parent. **hétérozygote**

heuristic approach An exploratory approach to a problem using successive evaluations of trial and error to arrive at a final approach. **approche heuristique**

hiatus (*geology*) (1) A break or interruption in the continuity of a stratigraphic sequence of rocks, where the missing strata either were never deposited or were destroyed by erosion prior to deposition of overlying strata. (2) The time period of an interval of nondeposition. **hiatus, lacune**

hierarchical Based on a hierarchy, whereby objects or features are placed in a systematic ordering or ranking of factors or attributes based on a list of criteria. Higher levels in the hierarchy provide broader information as compared to lower levels in the hierarchy. **hiérarchique**

hierarchy A graded system that ranks any number of objects one above another by classes or orders. **hiérarchie**

high performance liquid chromatography (HPLC) A chromatographic technique in which the mobile phase is a solvent and the stationary phase is a liquid on a solid support, a solid, or an ion-exchange resin. The sample is forced through the column under pressure. See *chromatography*. **chromatographie liquide à haute performance**

hill (1) A natural elevation or an area of upland smaller than a *mountain* but with no specific definition of absolute elevation. (2) See *tillage, hill*. **(1) colline (2) buttage**

hill creep Slow downhill movement, on a steep hillside and under the influence of gravity, of soil and loose rock; it is an important factor in the wasting of hillsides during dissection. See *creep*. **reptation**

Histels A suborder in the U.S. system of soil taxonomy. Histels are organic soils similar to *Histosols* except that they have permafrost within 2 meters below ground surface. They have 80% or more organic materials from the soil surface to a depth of 50 cm or to a glacic layer or densic, lithic, or paralithic contact, whichever is shallowest. These soils occur predominantly in Subarctic and Low Arctic regions of continuous or widespread permafrost. Less than one-third of the active layer (the soil between the ground surface and permafrost table) or an ice layer, which is at least 30 cm thick, has been *cryoturbated.* **Histels**

histic epipedon A thin organic soil horizon (U.S. system of soil taxonomy) that is saturated with water at some period of the year unless artificially drained, and that is at or near the surface of a mineral soil. The histic epipedon's maximum thickness depends on the types of materials in the horizon, and the lower limit of organic carbon is the upper limit for the *mollic epipedon.* **épipédon histique**

histogram *(statistics)* The graphical display of a set of data that shows the frequency of occurrence (along the vertical axis) of individual measurements or values (along the horizontal axis); a frequency distribution. **histogramme**

histones A class of positively charged chromosomal proteins that bind to DNA and comprise chromatin; they are rich in lysine and arginine amino acid residues. **histones**

Histosols An order in the U.S. system of soil taxonomy. Organic soils that have organic soil materials in more than half of the upper 80 cm, or of any thickness if overlying rock or fragmental materials have interstices filled with organic soil materials. **Histosols**

hoe See *tillage, hoe.* **biner, passer la houe**

hogback, hog's-back ridge A sharp-crested ridge formed by a hard bed of rock that dips steeply downward. **hogback, dos d'âne**

Holarctic Of or relating to the geographical distribution of fauna in the northern or Arctic region. **holarctique**

holding pond A reservoir, pit, or pond, usually made of earth, used to retain polluted runoff water for disposal on land. **étang de retenue**

holistic Of or related to a view of the natural environment that encompasses an understanding of the functioning of the complete array of organisms and chemical–physical factors acting in concert rather than the properties of the individual parts. **holistique**

holistic ecology The study of populations and communities as a whole in consideration of relationships between organisms and the total environment. **écologie holistique**

hollow (1) A low tract of land surrounded by hills or mountains; a small, sheltered valley or basin, especially a rugged area. (2) A landform represented by a depression (e.g., a cirque, cave, large sink, or blowout). (3) A low area of any shape and size in *peatland* (e.g., wet depressions, mud bottoms, pools, or ponds). **creux**

holocrystalline Consisting entirely of crystallized minerals with no glass present; generally pertains to igneous rock. **holocristallin**

holomictic lake A lake where bottom and surface waters are mixed periodically. **lac holomictique**

holophytic Obtaining food after the manner of a green plant. **autotrophe**

holoplankton Organisms which spend their entire life cycles as *plankton.* **holoplancton**

holozoic Obtaining food after the manner of most animals by ingesting complex organic matter. **holozoïque**

homeostasis (1) The tendency towards a relatively stable equilibrium between interdependent elements, especially as maintained by physiological processes. The maintenance of constant level conditions in the face of a varying external environment. (2) *(ecology)* The buffering capacity of an ecosystem that allows it to resist perturbations; the tendency of a biologic system to remain at or return to normal after or during an outside stress. **homéostasie**

homogeneous Of the same kind; consisting of similar parts; uniform. See *heterogeneous.* **homogène**

homogeneous equilibrium An equilibrium system where all reactants and products are in the same phase. **équilibre homogène**

homologous chromosomes Corresponding chromosomes in the male and female gametes, which pair during meiosis. **chromosomes homologues**

homologous series A series of related molecules with the same functional group or groups but whose structures differ by a fixed group of atoms. Each member of a homologous series is known as a homologue. The empirical formulae of the homologues can be represented by a simple formula (e.g., the straight-chain *alkanes* form a homologous series, with the general formula C_nH_{2n+2}). **série homologue**

H

homozygous Containing two identical alleles at a gene locus. **homozygote**

honeycomb frost Ice in the soil in insufficient quantity to be continuous, thereby giving the soil an open, porous structure that readily permits water to enter. **gel à structure alvéolaire**

hoodoo An unusually shaped column, pinnacle, or pillar of rock produced in a semi-arid environment by differential weathering or erosion of horizontal strata. **cheminée de fée, demoiselle coiffée**

horizon, soil A layer of soil or soil material approximately parallel to the land surface; it differs from adjacent layers by physical (e.g., color, structure, texture, consistence) and/or chemical (e.g., pH) properties and mineralogical composition. Soil horizons in various soil classification systems are designated by a capital letter, with or without a lower case letter and/or number indicating a subdivision, or a numerical annotation (e.g., Ah horizon, A1 horizon, Bt horizon). See *Appendix D* for soil horizon nomenclature in the Canadian, U.S., and FAO/UNESCO systems of soil classification. **horizon du sol, horizon pédologique**

hormogonium A segment of cells, marked off by heterocysts in a cyanobacterial filament, capable of detachment and development into a new filament. **hormogonie**

hormone A chemical substance produced usually in minute amounts in one part of an organism, from which it is transported to another part of that organism on which it has a specific effect. **hormone**

horneblende A mineral of the amphibole group having the general formula $Ca_2(Mg,Fe^{+2})_4Al(Si_7Al)O_{22}(OH,F)_2$, occurring as dark green to black crystals and in granular masses, found in many different types of igneous and metamorphic rocks. **hornblende**

horticulture The art and science of growing fruits, vegetables, and ornamental plants. **horticulture**

host A plant or animal having a parasite. **hôte**

hot spring A thermal spring with water temperature above 37°C. It can occur in non-volcanic areas but is more common in areas of current or recently active vulcanicity. **source thermale**

hue The aspect of color determined by the wavelengths of light, and changes with the wavelength. *Munsell color system* hue notations indicate the visual relationship of a color to red, yellow, green, blue, or purple, or an intermediate of these hues. See *chroma, value,* and *color.* **teinte, tonalité, gamme**

humic Pertaining to or derived from *humus.* **humique**

humic peat Peat material that is at an advanced stage of decomposition. It has the lowest amount of fiber, the highest bulk density, and the lowest saturated water-holding capacity of the organic materials; it is physically and chemically stable over time, unless it is drained; the rubbed fiber content is <10% by volume, and the material usually is classified in the *von Post humification scale* of decomposition as class 7 or higher. **tourbe humique**

humic acid Organic materials that are soluble in alkaline solution but precipitate on acidification of the alkaline extracts. **acide humique**

Humic Gleysol A great group of soils in the *Gleysolic* order (Canadian system of soil classification). The soils have a dark-colored *A* (Ah or Ap) *horizon* more than 8 cm thick underlaid with mottled gray or brownish gleyed mineral material. It may have up to 40 cm of mixed peat (bulk density 0.1 or more) or up to 60 cm of fibric moss peat (bulk density less than 0.1) on the surface. **gleysol humique**

humic layer A layer of highly decomposed organic soil material containing little fiber. **couche humique**

Humic Podzol A great group of soils in the *Podzolic* order (Canadian system of soil classification) occurring in cool humid coastal regions, cool humid inland locations at higher altitudes, and some peaty depressions. The soils have a dark brown to black Bh horizon at least 10 cm thick, having more than 1% organic carbon, less than 0.3% pyrophosphate-extractable Fe, a ratio of organic carbon to pyrophosphate-extractable Fe of 20 or more, and a very low base saturation (neutral salt extraction). A thin iron pan or a series of very thin (totaling less than 2.5 cm) iron pans may be present. **podzol humique**

Humic Regosol A great group of soils in the Regosolic order (Canadian system of soil classification). They have an Ah or dark colored Ap horizon at least 10 cm thick at the mineral surface. They may have organic surface horizons and buried mineral-organic horizons. They do not have a *B horizon* at least 5 cm thick. **régosol humique**

humic substance A relatively high molecular-weight, brown- to black-colored soil organic substance, obtained on the basis of solubility characteristics in alkali and acid. See *humic acid, fulvic acid.* **substance humique**

Humic Vertisol A great group of soils in the *Vertisolic* order (Canadian system of soil classification). The soils have a mineral-organic (Ah) horizon more than 10 cm in thickness, or, if cultivated, have at least 2% organic C, and an Ap *color* value of ≤5 (dry) and a chroma ≤1.5 (dry). The *A horizon* is easily distinguishable from the rest of the solum. **vertisol humique**

humid Pertaining to a climate in which the lower limit of annual precipitation is 50 cm in cool regions and the upper limit is 150 cm in hot regions, with vegetation dominated by forest. When distributed normally throughout the year, moisture does not limit the production of most crops. See *arid, semi-arid, subhumid.* **humide**

humidity, absolute The actual quantity or mass of water vapor present in a given volume of air; expressed as g m^{-3} grams per cubic meter. **humidité absolue**

humidity, relative The ratio of the actual amount of water vapor present in a portion of the atmosphere at a given temperature to the quantity that would be required to saturate it at the same temperature. **humidité relative**

humification The processes by which organic matter transforms to form humus such that the initial structures or shapes can no longer be recognized. See *humus.* **humification**

humin The fraction of the soil organic matter not dissolved when the soil is treated with dilute alkali. **humine**

Humisol A great group of soils in the *Organic* order (Canadian system of soil classification) that are saturated for most of the year. The soils have a dominantly humic middle tier, or middle and surface tiers if a terric, lithic, hydric, or cryic contact occurs in the middle tier. **humisol**

H

hummock (1) (*pedology*) In general, a knoll or mound above a level surface. (2) A mound of earth in a periglacial environment having either a core of mineral soil (*earth hummock*) or a core of stones (*turf hummock*), up to 70 cm in height and 3 m in diameter. (3) (*peatland*) A mounded type of microhabitat in a *peatland*. A mound of moss that usually rises 20 to 30 cm above the low (hollow) portion of the peatland surface. **(1) button (2) thufur (3) butte, massif**

hummocky Uneven, hilly terrain; also, peatland characterized by *hummocks*. **en bosses et creux**

hummocky moraine (*geomorphology*) A type of moraine characterized by a very complex surface of low-relief (generally less than 10 meters local relief) knolls or mounds of glacial sediment separated by irregular depressions, all of which lack linear or lobate forms. The depressions commonly lack external drainage outlets and are not part of an integrated fluvial system. Hummocky moraine is thought to form by the deposition of supraglacial glacial sediment as a stagnating ice mass melts out *in situ*. Although glacial till is the most common sediment type, areas of hummocky glaciofluvial and glaciolacustrine moraine also occur. Often called knob-and-kettle terrain. **moraine bosselée**

Humods A suborder in the U.S. system of soil taxonomy. *Spodosols* that have accumulated organic carbon and aluminum, but not iron, in the upper part of the spodic horizon. Humods are rarely saturated with water or do not have characteristics associated with wetness. **Humods**

Humo-Ferric Podzol A great group of soils in the *Podzolic* order (Canadian system of soil classification). In these soils the upper 10 cm of the *B horizon* (Bf) contains between 0.5 and 5% organic carbon and 0.6% or more pyrophosphate-extractable Al and Fe (>0.4% for sands). The ratio of organic carbon to pyrophosphate extractable Fe is less than 20. **podzol humo-ferrique**

Humox A suborder in the U.S. system of soil taxonomy. *Oxisols* that are moist all or most of the time and have a high organic carbon content within the upper 1 m. Humox have a mean annual soil temperature of <22°C and a base saturation within the oxic horizon of <35%, measured at pH 7. **Humox**

Humults A suborder in the U.S. system of soil taxonomy. *Ultisols* that have a high content of organic carbon. Humults are not saturated with water for periods long enough to limit their use for most crops. **Humults**

humus (1) The fraction of the soil organic matter that remains after removal of *macroorganic matter* and *dissolved organic matter*. It is usually dark-colored. See *humification*. (2) Also used in a broader sense to designate the *humus forms* referred to as forest humus. They include principally *mor*, *moder*, and *mull*. (3) All the dead organic material on and in the soil that undergoes continuous breakdown, change, and synthesis. **humus**

humus form A group of soil horizons located at or near the surface of a *pedon*, formed from organic residues, either separately from or intermixed with mineral particles. **forme d'humus**

hyaline Transparent, like glass. **hyalin**

hybrid Offspring that is a cross between parents of different species, subspecies, or cultivars. **hybride**

hybridization The binding or annealing of two complementary single strands of nucleic acid. **hybridation**

hydrate (1) A mineral compound produced by hydration, or one in which water is part of the chemical composition. (2) To cause the incorporation of water into the chemical composition of a substance (e.g., mineral). **hydrate**

hydrated lime A liming material composed chiefly of calcium and magnesium hydroxides. It reacts quickly to neutralize acid soils. **chaux hydratée**

hydration Chemical combination of water with another substance. **hydratation**

hydraulic conductivity The rate at which water can pass through a soil material under unit gradient; it is the proportionality factor (*K*) in *Darcy's law* as applied to the viscous flow of water in soil (i.e., the flux of water per unit gradient of the *hydraulic potential*). It depends on the *intrinsic permeability* of the medium (*k*) and the fluid properties (mass density *r*, viscosity *h*), and can be defined as $K = kgr/h$, where *g* is the gravitational acceleration. If conditions require that the viscosity of the fluid be separated from the conductivity of the medium, it is convenient to define the permeability (or *intrinsic permeability*) of the soil. **conductivité hydraulique**

hydraulic dredging Removal of underwater deposits by pumping them as a slurry through a pipe. **dragage hydraulique**

hydraulic gradient (*hydrology*) The slope of the hydraulic grade line; the slope of the free surface of water flowing in an open channel. A vector (macroscopic) point function equal to the decrease in the hydraulic head per unit distance through the soil in the direction of the greatest rate of decrease. In isotropic soils, this will be in the direction of the water flux. **gradient hydraulique**

hydraulic head Total water potential expressed in terms of energy per unit weight of water. An expression of water pressure in length units (e.g., level at which water stands in a piezometer or tensiometer); the height of a liquid column corresponding to a given pressure. Hydraulic head *(h)* is given by $h = z + y$, where *z* is the elevation head and *y* is the pressure head. It can be identified as the sum of *gravitational, pressure,* and *matric potentials,* and may be called the hydraulic potential. See *soil water.* **charge hydraulique**

hydraulic mining See *mining.* **abattage hydraulique**

hydraulic potential The sum of gravitational, hydrostatic, and matric *soil water potential*, expressed in units of head, pressure, or potential. **potentiel hydraulique**

hydraulic pressure The pressure within water produced by a combination of forces, such as capillary and gravitational, acting on the water. **pression hydraulique**

hydraulic radius The flow area of any conduit, open or closed, divided by its wetted perimeter. The cross-sectional area of a stream of water divided by the length of that part of its periphery in contact with its containing conduit; the ratio of area to wetted perimeter. Also called hydraulic mean depth. **rayon hydraulique**

hydraulics The science concerned with water and other fluids at rest or in motion. **hydraulique**

hydric An extremely wet, often submerged environment. Plants that characterize such an environment are called hydrophytic. **hydrique**

hydric layer A layer of water in the control section of *Organic soils*, extending from a depth of not less than 40 cm to more than 160 cm. **couche hydrique**

hydric soil A soil that is wet long enough to periodically produce anaerobic conditions, influencing the growth of plants. **sol hydromorphe, sol hydrique**

hydrocarbon Any organic compound, gaseous, liquid, or solid, consisting solely of carbon and hydrogen. Fossil fuels are complex mixtures of hydrocarbons. **hydrocarbure**

hydrocarbon derivative An organic molecule that contains one or more elements in addition to carbon and hydrogen. **dérivé d'hydrocarbure**

hydrodynamic dispersion The process wherein the solute concentration in flowing solution changes in response to the interaction of solution movement with the pore geometry of the soil, similar to diffusion but taking place when solution moves. **dispersion hydrodynamique**

hydrodynamic dispersion coefficient The coefficient in the solute convection equations that accounts for *hydrodynamic dispersion*; usually determined by solving an inverse problem. **coefficient de dispersion hydrodynamique**

hydrodynamics The branch of *hydraulics* which deals with the pressures created by water turbulence and its flow through channels, conduits, and pipes, and over weirs. **hydrodynamique**

hydrofluorocarbons (HFCs) These chemicals (along with perfluorocarbons) were introduced as alternatives to ozone-depleting substances in serving many industrial, commercial, and personal needs. HFCs are emitted as byproducts of industrial processes and are also used in manufacturing. They do not significantly deplete the stratospheric ozone layer, but are powerful greenhouse gases with a *global warming potential* ranging from 140 (HFC-152a) to 12,100 (HFC-23). **hydrofluorocarbures**

hydrogen bond The chemical bond between a hydrogen atom of one molecule and two unshared electrons of another molecule. **liaison hydrogène**

hydrogen bonding A type of electrostatic interaction between molecules occurring in molecules that have hydrogen atoms bound to electronegative atoms (e.g., nitrogen, oxygen). An example of this type of bond is the weak attraction between separate water molecules by way of the oxygen atom of one molecule and the hydrogen atom of an adjacent molecule. This weak attraction constantly forms and breaks among the water molecules in liquid water. Evaporation of water occurs when individual water molecules at the air-water interface break free of all hydrogen bonds and enter the gas phase. Hydrogen bonds are also found between many organic molecules (e.g., between the two strands of deoxyribonucleic acid (DNA) that constitute individual genes). **liaison hydrogène**

hydrogen sulfide (H_2S) A malodorous gas made up of hydrogen and sulfur, with the characteristic odor of rotten eggs. It is usually emitted during anaerobic decomposition of organic matter and in advanced stages of eutrophication. It is also a byproduct of refinery activity and the combustion of oil. In heavy concentrations, it can cause illness. **sulfure d'hydrogène**

hydrogenation reaction A chemical reaction with hydrogen; in particular, a reaction in which hydrogen is added, with a catalyst (e.g., nickel) present, to an unsaturated compound. **réaction d'hydrogénation**

hydrogenic soil Soil developed under the influence of water standing within the profile for prolonged periods. It is formed mainly in cold, humid regions. **sol hydrogénique**

hydrogen-ion concentration See *pH scale*. **concentration en ions hydrogène**

hydrogeology The science that deals with subsurface water and with related geologic aspects of surface water. **hydrogéologie**

hydrograph (*hydrology*) A graph showing variation in stage (i.e., water level, depth) or discharge of a moving body of water (e.g., stream, river) over time. **hydrogramme**

hydrologic Relating to water. **hydrologique**

hydrologic budget Accounting of the inflow, outflow, and storage in a hydrologic unit (e.g., drainage basin, aquifer, soil zone, lake, reservoir); relationship between evaporation, precipitation, runoff, and the change in water storage. Also called water balance; water budget; hydrologic balance. **bilan hydrologique**

hydrologic cycle The distribution and movement of water from the time of precipitation until the water has been returned to the atmosphere by evaporation and is again ready to be precipitated. **cycle hydrologique**

hydrologic equation The water inventory equation (inflow = outflow + change in storage) which expresses the basic principle that during a given time interval, the total inflow to an area must equal the total outflow plus the net change in storage. **équation hydrologique**

hydrologic model Mathematical formulations that simulate hydrologic phenomena considered as processes or systems. A conceptual or physically-based procedure for numerically simulating a process or processes which occur in a watershed. **modèle hydrologique**

hydrology The applied science concerned with the waters of the Earth, their properties, occurrences, distribution, circulation, and movement through the *hydrologic cycle* (*precipitation*, consequent runoff, *infiltration*, and storage; *evaporation*; *condensation*). Hydrology can be differentiated based on water movement and distribution in specific situations including agrohydrology (root zone, *irrigation,* and surface water in conveyance systems on agricultural land), groundwater hydrology (saturated zone of the soil profile), soil hydrology (soil solution in the soil profile), and surface hydrology (water on the soil surface). **hydrologie**

hydrolysis (1) A chemical reaction of a compound with water. (2) The process by which a substrate is split to form two end products by the intervention of a molecule of water. **hydrolyse**

hydrometallurgy A process for extracting metals from ores by use of aqueous chemical solutions. Two steps are involved: selective leaching and selective precipitation. **hydrométallurgie**

hydrometeorology The science of the application of meteorology to hydrologic problems. **hydrométéorologie**

hydrometer A tubular device for measuring the *specific gravity* of a liquid by the depth to which the hydrometer sinks when immersed. It is made of glass, with the lower end weighted, and graduated in specific gravity or other units. **hydromètre**

hydrometry The measurement and analysis of the flow of water, especially by the use of a *hydrometer.* **hydrométrie**

hydromica See *illite*. **hydromica**

hydromorphic Developed under conditions of excess moisture; hydromorphic soils are found in areas of poor drainage. **hydromorphe**

hydromorphic soil Soil that develops under conditions of poor drainage in marshes, swamps, seepage areas, or flats. See *Gleysolic*. **sol hydromorphe**

hydromuscovite Pertaining to any fine-grained, muscovite-like clay mineral, commonly but not always high in water content and deficient in potassium. It is probably an *illite*. **muscovite hydratée, illite**

hydronium ion The H_3O^+ ion; a hydrated proton. **ion hydronium**

hydrophilic Having strong affinity for water. **hydrophile**

hydrophobic Having little or no affinity for water molecules. Hydrophobic substances have more affinity for other hydrophobic substances than for water. **hydrophobe**

hydrophobic soil Soil that is water repellent, often due to dense fungal mycelial mats or hydrophobic substances vaporized and reprecipitated during fire. **sol hydrophobe**

hydrophobicity The tendency for a soil particle or soil mass to resist hydration, usually quantified using the water drop penetration time test. **hydrophobicité**

hydrophyte Plant that grows in water, or in wet or saturated soils. **hydrophyte**

hydroseeding The application of seed in a water slurry that contains fertilizer, a soil binder and/or mulch. **ensemencement hydraulique**

hydrosphere The part of the Earth composed of water, including clouds, oceans, seas, ice caps, glaciers, lakes, rivers, and underground water. **hydrosphère**

hydrostatic pressure Pressure exerted by nonmoving water due to depth alone. **pression hydrostatique**

hydrostatics The branch of *hydraulics* concerned with the pressures of fluids at rest. **hydrostatique**

hydrous Said of a mineral compound containing water. **hydraté**

hydrous mica See *illite*. **mica hydraté**

hydroxide A compound of an element with the radical or ion OH^-, (e.g., sodium hydroxide, NaOH). **hydroxyde**

hydroxy-aluminum interlayers Polymers of general composition $[Al(OH)_{3-x}]_m^{mx+}$ that are adsorbed on interlayer cation exchange sites. Although not exchangeable by unbuffered salt solutions, they are responsible for a considerable portion of the titratable acidity (and pH-dependent charge) in soils. **polymères d'hydoxyaluminium en position inter-feuillet**

hydroxy-interlayered vermiculite A vermiculite with partially filled interlayers of hydroxy-aluminum groups, normally dioctahedral in both the interlayer and the vermiculite layer. It is common in the coarse clay fraction of acid surface soil horizons, and has intermediate cation exchange properties between vermiculite and chlorite. Also called chlorite-vermiculite intergrade or vermiculite-chlorite intergrade. See *hydroxy-aluminum interlayers*. **vermiculite chloritisée**

hydroxylapatite (1) A rare mineral of the apatite group, $Ca_5(PO_4)_3(OH)$. (2) An apatite mineral in which hydroxyl predominates over fluorine and chlorine. Also called hydroxyapatite. **hydroxyapatite**

hygrometer An instrument for measuring humidity of the air. **hygromètre**

hygroscopic Absorption of moisture vapor from the atmosphere. **hygroscopique**

hygroscopic water (1) Water adsorbed by a dry soil from an atmosphere of high relative humidity. (2) Water lost from an air-dry soil when it is heated to 105°C. (3) Water held by the soil when it is at equilibrium with an atmosphere of a specified relative humidity at a specified temperature, usually 98% relative humidity at 25°C. This is considered an obsolete term. **eau hygroscopique**

hyperbolic reaction A reaction having a rate constant for which the plot of the product formed versus time yields a curve approaching some maximum value, best described by a hyperbolic equation. See *Monod equation*. **réaction hyperbolique**

hyperosmotic Having a salt concentration greater than that of the surrounding medium. **hyperosmotique**

hypersaline Excessively saline (e.g., salinity substantially greater than that of normal sea water). **hypersalin**

hyperthermic A soil temperature regime that has mean annual soil temperatures of 22°C or more and >5°C difference between mean summer and mean winter soil temperatures at 50 cm below the surface. Isohyperthermic is the same except the summer and winter temperatures differ by <5°C. **hyperthermique**

hypha (plural hyphae) Filament of fungal cells. A large number of hyphal filaments (hyphae) constitute a *mycelium*. Bacteria of the order *Actinomycetes* also produce branched mycelium. See *figure*. **hyphe**

hypogene Formed at great depths within the Earth; pertaining to *plutonic rocks* and some of the rocks that are metamorphosed at depth. **hypogénique**

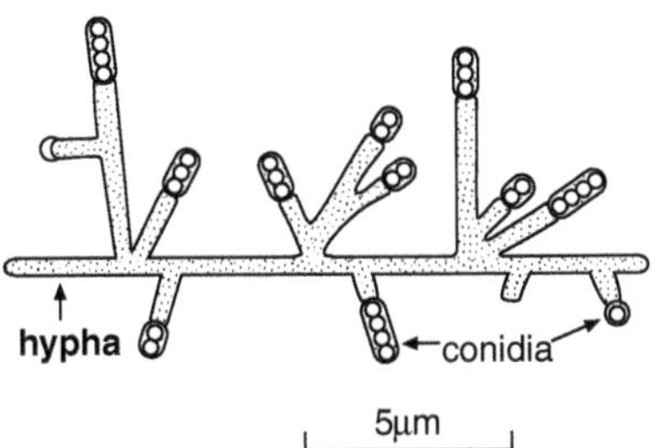

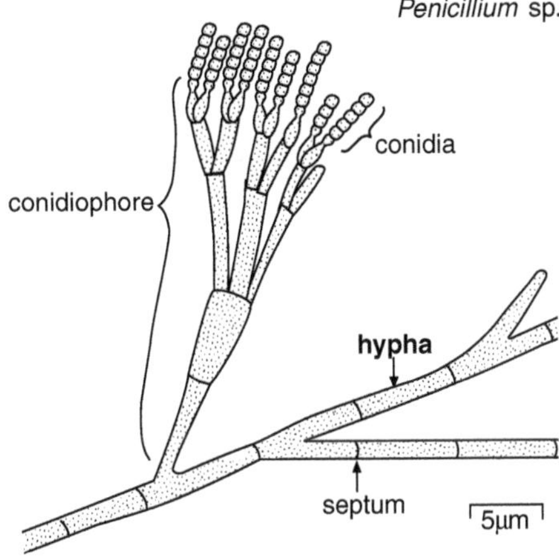

Hypha (adapted from Killham, 1994).

hypolimnion The lower, cooler, oxygen-poor water layer found in stratified lakes, that extends from the thermocline to the bottom of the lake. See *epilimnion*. **hypolimnion**

hypoosmotic Having a salt concentration less than that of the surrounding medium. **hypoosmotique**

hypothesis In general, a theory or proposition based on certain assumptions and that can be evaluated scientifically, as by experimentation. In statistics, an assertion about the functional form of a population (e.g., the hypothesis that a sample comes from a normal population). **hypothèse**

hypotonic A solution having a lower salt concentration than an *isotonic solution*. **hypotonique**

hypoxia (*limnology*) A condition in which natural waters have a low concentration of dissolved oxygen (DO). Most game and commercial species of fish avoid waters that are hypoxic. **hypoxie**

hypsithermal A period of relatively warm climatic conditions during the Holocene epoch, between about 9000 and 2600 BP, (the dating varies with latitude). In Britain it is sometimes referred to as the Climatic Optimum but is more specifically confined to the period between 7500 and 5200 BP. **hypsithermal, altithermal**

hypsography Theories and techniques used in surveying and map making for describing elevations of land surfaces with reference to a datum, usually sea level. **hypsométrie**

hypsometer A simple instrument, often a stick or other straight edge, used to measure tree heights on the basis of similar angles. **hypsomètre**

hysteresis A non-unique relationship between two variables, wherein the curves representing the relationship depend on the sequences or starting point used to observe the variables. **hystérèse**

I

ice age A glacial period or part of a glacial period; most frequently refers to the last glacial period, namely the latest of the glacial epochs, also known as the Pleistocene Epoch. **âge glaciaire**

ice rafting Transport of rocks and other glacially eroded debris by floating ice, either ice floes or icebergs. See *erratic*. **transport glaciel**

ice-sheet A glacier of considerable thickness, more than 50,000 sq km in area, forming a continuous cover of ice and snow over a land surface, spreading outward in all directions and not confined by the underlying topography; a *continental glacier*. During the Pleistocene epoch, ice-sheets covered large parts of North America and northern Europe and Asia, but are now confined to polar regions (as in Greenland and Antarctica). **inlandsis**

ice wedge (*geology*) Wedge-shaped, foliated ground ice produced in permafrost, occurring as a vertical or inclined sheet, dike, or vein tapering downward, and measuring from a few millimeters to as much as 6 m wide and from 1 to 30 m high. It originates by the growth of hoar frost or by the freezing of water in a narrow crack or fissure produced by thermal contraction of the permafrost. **fente de remplissage de glace**

icecap A covering of ice over a tract of land, smaller in size than an ice-sheet, such as a dome-shaped cover in a mountain glacier. Usage also includes reference to the Greenland icecap and the Polar icecaps that are of much greater dimensions. See *glacier, ice-sheet*. **calotte glaciaire**

ice-contact deposit (*geology*) Stratified drift deposited in contact with melting glacier ice, such as an *esker, kame, kame terrace*, or deposits marked by numerous kettles. **sédiment de contact glaciaire**

ice-dammed lake See *glacial lake*. **lac de barrage glaciaire**

ice-thrust (*geology*) Descriptive of materials with deformation structures, including folds and faults, and landforms, including moraines and arcuate ridges of sediment and/or bedrock, created by glacial thrusting at or near the termini of glaciers. **de chevauchement glaciaire**

ice-thrust moraine (*geology*) Broadly arcuate, subparallel ridges, commonly high (up 60 m), large, and long (up to several km), resembling moraines but composed mostly of detached blocks of unconsolidated bedrock, and/or Quaternary deposits that have been folded and thrusted by glacial pressure. Ice-thrust moraine is intimately associated with lodgement till, and in many places it is covered by ablation till, forming hummocky knobs and knolls on the surface of large areas of subparallel arcuate ridges. **moraine de chevauchement**

ice-wedge polygon (*geology*) A large, nonsorted polygon characterized by borders of intersecting ice wedges, found only in permafrost regions and formed by contraction of frozen ground. The fissured borders may be ridges or shallow troughs, and are underlaid with ice wedges. The diameter is up to 150 m, averaging 10 to 40 m. In plan, the pattern tends to be three- to six-sided. **polygone de fente de gel**

iconic model A model of a real world situation reduced in scale but presenting the same properties. See *simulation model.* **modèle iconique**

ideal gas A gas that would perfectly obey the *ideal gas law.* A gas that would consist of molecules that occupy negligible space and have negligible forces between them. All collisions made between molecules and the walls of the container or between molecules would be perfectly elastic because the molecules would have no means of storing energy. At normal temperatures and pressures, most gases behave similarly to an ideal gas, and the ideal gas law can be routinely applied as in air pollution calculations. **gaz parfait**

ideal gas law (universal gas equation) An equation of state for a gas, where the state of the gas is its condition at a given time; expressed by $PV= nRT$, where P = pressure, V= volume, n = moles of the gas, R = the universal gas constant, and T = absolute temperature. This equation expresses behavior approached by real gases at high T and low P. **loi des gaz parfaits**

ideal solution A solution whose vapor pressure is directly proportional to the mole fraction of solvent present. **solution idéale**

igneous Rock formed from the cooling and solidification of magma, and that has not been changed appreciably since its formation. Igneous rocks constitute one of the three main classes into which rocks are divided, the others being *metamorphic rock* and *sedimentary rock.* **ignée**

illite (1) A discrete, non-expansible mica of detrital or authigenic origin or the micaceous component of interstratified systems (e.g., illite-smectite). If used to refer to the species, it should meet the following requirements: the micaceous layers ideally are non-expansible, the octahedral sheet is dioctahedral and aluminous, the interlayer cation is primarily potassium, and the composition deviates from that of muscovite in two main ways: (i) phengitic component is present in which substitution of R^{2+} cations for octahedral A1 is balanced by addition of tetrahedral Si beyond the ideal Si:Al ratio of 3:1 for muscovite. This substitution gives the octahedral sheet an overall negative charge of about 0.2 to 0.3 per formula unit. (ii) Interlayer vacancies or water molecules amounting to about 0.2 to 0.4 atoms per formula unit are compensated by additional tetrahedral Si cations beyond those required by the phengitic component. Where reference is made to the species illite, a clear statement should be made to that effect in order to avoid confusion with the general usage. (2) In soil taxonomy, the presence of a 1 nm x-ray diffraction peak and greater than or equal to 4% K_2O is used to denote the presence of illite. **illite**

illuvial horizon A soil horizon in which material carried from an overlying layer has been precipitated from solution or deposited from suspension as a layer of accumulation. See *eluvial horizon.* **horizon illuvial**

illuviation The process of deposition of soil material removed from one horizon in the soil to another, usually from an upper to a lower horizon in the soil profile. Illuviated substances include silicate clay, hydrous oxides of iron and aluminum, and organic matter. See *figure.* **illuviation**

immigrant Any animal or plant newly introduced into an environment. **immigrant**

immiscible Inability of two phases to mix together to form one phase. See *miscible.* **immiscible**

immobilization The conversion of an element from the inorganic to the organic form in microbial or plant tissue, thus rendering the element not readily

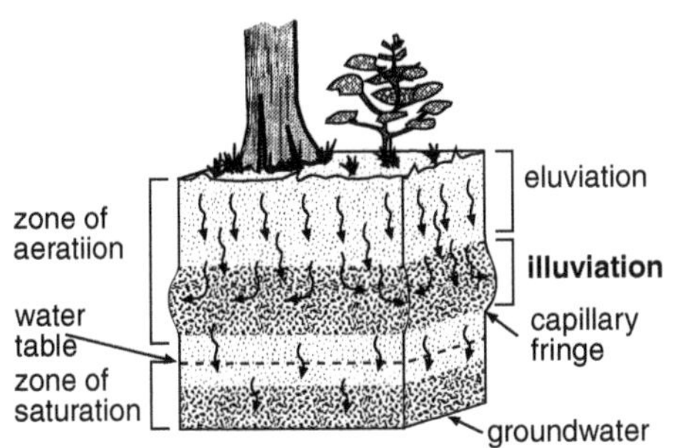

Illuviation (adapted from Dunster and Dunster, 1996).

available to other organisms or plants. **immobilisation, organisation**

imogolite A poorly crystalline aluminosilicate mineral with an ideal composition $SiO_2\,Al_2O_3\,2.5H_2O(^+)$. It appears as threads consisting of assemblies of a tube unit with inner and outer diameters of 1.0 and 2.0 nm, respectively. Imogolite is commonly found in association with allophane, and is similar to allophane in chemical properties. It is mostly found in soils derived from volcanic ash, and in weathered *pumice* and *spodosols*. **imogolite**

impact assessment A study of the effect of resource development on other resources. **étude d'impacts**

impedance The resistance to an energy flow calculated as the magnitude of the cause of the flow (e.g., force, pressure, voltage) divided by the energy flow. **impédance**

impeded drainage A condition in which the movement of water through soils under the influence of gravity is impeded. **drainage entravé**

impeding horizon A horizon that hinders the movement of water by the influence of gravity through soils. **couche ou horizon d'impédance**

imperfect flower A flower containing only the male (stamen) or the female (pistil) parts. See *perfect flower*. **fleur incomplète**

impermeability A characteristic of a material that prevents the passage of a fluid through that material. **imperméabilité**

impervious Resistant to penetration by fluids or by roots. **imperméable**

impoverishment, soil The process or the result of making the soil less productive. **appauvrissement du sol**

improved pasture Area improved by seeding, draining, irrigating, fertilizing, brush or weed control, not including areas where hay, silage, or seeds are harvested. **pâturage amélioré**

improvement cutting The silvicultural practice of removing trees of undesirable species, form or condition from the main canopy of the stand to improve the health, composition and value of the stand. **coupe d'amélioration**

in situ — In the natural or original position. A measurement taken in the field, without removal of a sample to the laboratory (e.g., rocks, soil, and fossils situated in the place where they were originally formed or deposited). ***in situ***

in vitro — In a test tube or other laboratory environment. Experiments performed outside an intact living organism (e.g., in Petri dishes or test tubes). ***in vitro***

in vivo — In a living organism. Experiments performed inside living organisms (e.g., to assess chemical toxicity). ***in vivo***

inactivation A reaction resulting in a substance (e.g., herbicide) no longer chemically active. **inactivation**

Inceptisols An order in the U.S. system of soil taxonomy. Mineral soils that have one or more pedogenic horizons in which mineral materials other than carbonates or amorphous silica have been altered or removed but not accumulated to a significant degree. Under certain conditions, Inceptisols may have an *ochric, umbric, histic, plaggen,* or *mollic epipedon*. Water is available to plants more than half of the year or

I

more than 90 consecutive days during a warm season. **Inceptisols**

inclination (*geology*) The angle of *dip* of a rock stratum, fault, or mineral vein. **pendage**

inclined (*geomorphology*) A sloping, unidirectional surface of at least 300 meters length and not broken by marked irregularities. Slopes can be 2 to 70%. **incliné**

inclinometer See *clinometer.* **clinomètre**

incomplete block design *(statistics)* An experimental design in which all treatments are not represented in each block. **plan d'expérience à blocs incomplets**

I

incongruent solution Dissolution accompanied by decomposition or by reaction with liquid, so that one solid phase is converted into another; dissolution to give dissolved material in different proportions from those in the original solid. **dissolution non congruente**

incorporation (into soil) Mixing one material (e.g., fertilizer, crop residue, or manure) into the soil, usually by tillage. See *tillage, incorporation.* **incorporation**

increaser (*range management*) For a given plant community, those species that increase in number as a result of a specific abiotic/biotic influence or management practice. **espèce en évolution progressive**

incubation Growth of a microbial culture under specific environmental conditions. **incubation**

incubation period The average time between exposure to an infectious, disease-causing agent and manifestation of the signs and/or symptoms of the disease. **période d'incubation**

independent variable A measurable quantity that, as it takes different values, can be used to predict the value of a *dependent variable*; the variable manipulated by the experimenter in order to determine changes in the dependent variable. **variable indépendante**

index map A small-scale map portraying the numbers and/or names of larger-scale maps in the same or related series. It is generally in the form of a simple outline map showing the location of the more detailed maps. **carte-index**

index of refraction (*crystallography*) A number that expresses the ratio of the velocity of light in a vacuum to the velocity of light within the crystal. Its conventional symbol is *n*. Modifying factors include wavelength, temperature, and pressure. *Birefringent* crystals have more than one index of refraction. **indice de réfraction**

indicator (1) (*soil*) An indirect measure of a soil property; a related or associated property or that can serve to indicate change in the original property. (2) A measure of a *soil quality* attribute that is a related or associated property (e.g., pedotransfer function, surrogate, or proxy). Indicators should be subject to standardization, specific to a quality attribute, sensitive to change, and easily measured or collected. (3) (*chemistry*) A chemical that changes color and is used to mark the end point of a titration. **indicateur**

Types of indicators include:

indicator bacteria – nonpathogenic bacteria whose presence in water indicate the possibility of pathogenic species in the water. ***bactéries indicatrices***

indicator microorganisms – microorganisms that, if present above certain levels in water, indicate contamination by human sewage. The *coliform* bacteria are used commonly as indicators because the test for them is reliable, relatively inexpensive, and produces timely results. ***micro-organismes indicateurs***

indicator plants – plants characteristically associated with specific soil or site conditions. ***plantes indicatrices***

indicator species – animal or plant species used to predict site quality and characteristics. ***espèce indicatrice, bio-indicateur***

indigenous Native; naturally present in an area. See *autochthonous.* **indigène**

indirect competition Exploitation of a resource by one individual that reduces the availability of that resource to others. See *direct competition.* **compétition indirecte**

indoor air pollution The presence of excessive levels of air contaminants inside a home or building. Sources include cigarette smoke, fuel combustion for heating or cooking, wallboards, carpets, insulation, and the geology of the area (e.g., radon in soil or rocks beneath the structure). **pollution intérieure des locaux**

induction Reasoning from the particular to the general, or from the individual to the universal; a derivation of general principles from the examination of separate facts. See *deduction.* **induction**

indurated Rock or soil hardened or consolidated by pressure, *cementation*, or heat. **induré**

indurated layer A soil layer that has become hardened, generally by *cementation* of soil particles. **couche indurée**

inert gas A gas that does not react with other substances under ordinary conditions; also called noble or rare gas. **gaz inerte**

inert organic matter Soil organic matter that is highly carbonized. Inert organic matter consists of charcoal, charred plant materials, graphite, and coal with long turnover times. **matière organique inerte**

infection Contamination with disease-causing microorganisms (e.g., bacteria, fungi, viruses). **infection**

infectious Describing a virus, bacterium, fungus, or protozoan that can invade a host to produce disease. The term is also applied to a disease caused by some pathogenic microbe (e.g., typhoid fever) as opposed to a disease that is not (e.g., coronary artery disease). **infectieux**

infestation Inhabitation by insects, rodents, or weeds, in contrast to *infection* by microorganisms. **infestation**

infiltrability The flux (or rate) of water *infiltration* into soil when water at atmospheric pressure is maintained on the atmosphere-soil boundary, with the flow direction being one-dimensionally downward. **taux maximal d'infiltration**

infiltration The entry of water into soil. **infiltration**

infiltration flux (or rate) The volume of water entering a specified cross-sectional area of soil per unit time (e.g., L s^{-1}). **taux d'infiltration**

infiltrometer A device for measuring the volume or *flux* (or rate) of liquid (usually water) entry downward into the soil. **infiltromètre**

inflorescence (1) The complete flower head of a plant, including stems, stalks, and flowers. (2) The process of flowering. **(1) inflorescence (2) inflorescence, exertion**

information theory The science of the transmission of messages or information by communication systems. Information is a property not intrinsic to any one message but to a set of messages. **théorie de l'information**

infrared (IR) The portion of the electromagnetic spectrum with wavelengths just beyond the red end of the visible spectrum in the wavelength interval from about 0.75 μm to 1 mm. Consists of far IR, the longer wavelengths of the infrared region, from 25 μm to 1 mm; middle IR, with wavelengths from around 2 or 3 μm to around 25 μm; and near IR (visible red) consisting of shorter wavelengths in the IR region extending from about 0.75 μm to around 2 or 3 μm. **infrarouge**

infrared (IR) absorption spectroscopy A technique that measures the wavelength and intensity of the absorption of infrared light by a sample. The infrared light used is energetic enough to excite molecular vibrations to higher energy levels. The wavelengths of IR absorption bands are characteristic of specific types of chemical bonds. It is most often used for identification of organic and organometallic molecules. **spectroscopie d'absorption dans l'infrarouge**

infrared radiation The heat energy emitted from all solids, liquids, and gases. (*meteorology*) The heat energy emitted by the Earth's surface and its atmosphere. *Greenhouse gases* strongly absorb this radiation in the Earth's atmosphere and re-radiate some back towards the surface, creating the greenhouse effect. See *greenhouse effect.* **rayonnement infrarouge**

ingesta Nutritive materials consumed by animals. **ingesta**

inheritance (*weathering*) The formation of secondary clays by the very slight transformation of primary minerals; associated mainly with temperate and subtropical climates. **héritage**

inhibition The prevention of growth or multiplication of organisms. **inhibition**

initial storage (*hydrology*) That portion of precipitation required to satisfy interception by vegetation, the wetting of the soil surface, and depression storage. **rétention initiale**

innate capacity for increase (r_0) The intrinsic growth of a population under ideal conditions without the restraining effects of competition. **capacité de croissance innée**

inoculant A seed or soil additive, especially for legume seed, composed of specific nitrogen-fixing bacteria that facilitate nitrogen fixation. **inoculant**

inoculate To treat, usually seeds, with microorganisms for the purpose of creating a favorable response. Most often refers to the treatment of legume seeds with *Rhizobium* to stimulate dinitrogen fixation. **inoculer**

inoculation The artificial introduction of microorganisms into a habitat or their introduction into a *culture medium.* **inoculation**

inoculum The organisms used to *inoculate* a *culture medium.* **inoculum**

inorganic Pertaining or relating to a compound that is not *organic*, usually of mineral origin. **inorganique**

inorganic soil A soil made mainly of mineral particles. See *mineral soil.* **sol minéral**

inputs and outputs (*modeling*) Two interrelationships which link an open system with its external environment. Inputs are stimuli that energize a system and provide it with the external material to make it operative; they link an open system with its environment. The variable whose value is given as input to a system is called the independent variable. Outputs result from a system's processes, representing the effects of the system itself on the environment. The variable whose value is determined by the system is the dependent variable. See *variable.* **entrées et sorties**

in-row subsoiling See *tillage, in-row subsoiling.* **sous-solage**

insecticide A chemical compound, usually synthetic, that kills insects or prevents them from causing diseases on plants. **insecticide**

inselberg (*geology*) An isolated residual knob or hill, rising abruptly from a lowland erosion surface. It is characteristic of a late stage of the erosion cycle. See *monadnock.* **inselberg**

insolation (1) The combined solar and sky radiation reaching the Earth; also the rate at which it is received, per unit of horizontal surface. (2) The geologic effect of the sun's rays on the Earth's surficial materials, specifically, the effect of changes of temperature on the mechanical weathering of rocks. **rayonnement solaire direct incident**

instar (*zoology*) An insect or other arthropod in any of the forms between successive molts. Insects with complete metamorphosis usually have several larval instars, and both the pupa and adult (or imago) are also considered instars. **stade larvaire**

instrument blank A solvent or reagent blank (minus the *analytes* of interest) used to measure interference or contamination from an analytical instrument. See *laboratory blank*. **blanc d'instrument**

intake area An area of *recharge*. **région d'alimentation**

intake rate The rate of entry of water into soil. See *infiltration flux*. **taux d'infiltration**

integrated pest management (IPM) The combination of biological, cultural, and/or chemical control methods to limit pest damage to agricultural crops. IPM attempts to maintain pests at manageable levels rather than eradicate them. **lutte intégrée**

integrated resource management The identification and consideration of all resource values, including social, economic, and environmental needs, in land use and development decision making. **gestion intégrée des ressources**

intensity, map The number of delineations demarcated per unit area on a soil map. **densité de délimitations cartographiques**

intensive cropping Maximum use of crop land by means of frequent succession of harvested crops; usually implies reliance on management inputs (e.g., fertilizer, pesticides). **culture intensive**

intensive grazing management Grazing management that attempts to increase production or utilization per unit area or production per animal through a relative increase in stocking rates, forage utilization, labor, resources, or capital. **conduite de pâturage intensive**

interaction (*ecology*) Reciprocal action or influence between organisms, between organisms and the environment, or between environmental factors. (*statistics*) In analysis of variance, the effects of two or more controlling factors upon each other. **interaction**

interbed A *bed*, typically thin, of one kind of rock material occurring between or alternating with beds of another kind. **couche interstratifiée**

intercalated (*geology*) Layered material that exists or is inserted between layers of a different character, especially relatively thin strata of one kind of material that alternate with thicker strata of some other kind (e.g., beds of shale intercalated in a body of sandstone). **intercalé**

interception (*hydrology*) The process by which precipitation is caught and held by foliage, twigs, and branches of trees, shrubs, and other vegetation, and lost by evaporation, never reaching the surface of the ground. Interception equals the precipitation on the vegetation minus stream-flow and through fall. **interception**

interception channel A channel excavated at the top of earth cuts, at the foot of slopes, or at other critical places to intercept surface flow; a catch drain; also called interception ditch. **fossé de crête, fossé d'interception**

interceptor drain Surface or subsurface drain, or a combination of both, designed and installed to intercept flowing water. **drain d'interception**

intercropping Sowing two or more crops simultaneously in alternate rows. **cultures intercalaires**

interflow (*hydrology*) That portion of rainfall that infiltrates the soil and moves laterally through the upper soil horizons until intercepted by a stream channel or until it returns to the surface. Interflow usually takes longer to reach stream channels than runoff. **écoulement divergent**

interfluve The relatively undissected upland separating adjacent streams flowing into the same drainage system. See *slope morphological classification*. **interfluve**

interglacial A relatively mild period occurring between two glacial stages. **interglaciaire**

intergrade minerals A series of minerals, mainly nonexpanding, ranging in structure from expanding 2:1 types (e.g., montmorillonite and vermiculite) to nonexpanding chlorite, resulting from partial filling of the interlayer space by octahedral material, mainly hydroxy-Al or hydroxy-Fe substances. **intergrades**

I

intergrade, soil A soil that possesses moderately well-developed distinguishing characteristics of two or more genetically related taxa. **sol intergrade, sol de transition**

interlayer See *phyllosilicate*. **position inter-feuillet**

intermediate rock Any igneous rock containing between 52 and 66% silica by weight (e.g., diorite). See *basic rock*, *ultrabasic rock*. **roche mafique**

intermediate scale map See *map scale*. **carte à moyenne échelle ou à échelle intermédiaire**

intermittent grazing A method that imposes grazing for indefinite periods at irregular intervals. **pâturage intermittent**

intermittent stream A stream that flows periodically, at certain times of the year, as when it receives water from springs or from a surface source. In contrast to *perennial stream*. **cours d'eau temporaire ou intermittent**

intermolecular forces Relatively weak interactions that occur between molecules. See *van der Waals' forces*. **forces de van der Waals**

internal friction The portion of the shearing strength of a soil due to the interlocking of soil grains and the resistance to sliding between the grains. Indicated by the term $d \tan q$ in Coulomb's equation $t = c + d \tan q$, where t is shear stress, d is normal stress, c is cohesion, and q is friction angle. **friction interne**

internal soil drainage The downward movement of water through the soil profile. **drainage interne du sol**

interplanting (1) Planting several crops together on the same land (e.g., planting one crop between the rows of another). (2) Planting of crops between rows of trees, especially while the trees are too small to occupy the land completely. (3) Planting young trees among existing trees or brushy growth. See *interseeding*. **(1, 2) cultures intercalaires (3) plantation intercalaire**

interpolation The process of determining the value of a function between two known values without using the equation of the function itself. See *extrapolation*. **interpolation**

interpretation, soil survey The organization and presentation of knowledge about characteristics, qualities, and behavior of soils as they are classified and outlined on maps, in relation to soil use. In general, interpretation is using soil information to develop such things as productivity ratings of soils, land use evaluations, tax assessments, soil and farm management recommendations, forest management recommendations, environmental quality protection, reclamation, recreation suitability, and construction material suitability. **interprétation des données pédologiques**

interpretive forest site A designated forest site and ancillary facilities developed to interpret, demonstrate, or facilitate the discussion of the natural environment, forest practices, and integrated resource management. **site d'interprétation forestier**

interpretive map A map that presents the interpretation of soil survey information for specific uses. **carte interprétative**

inter-rill erosion The removal of a relatively even, thin layer of surface soil by *runoff* water flowing over land between rills. Also called *sheet erosion*. **érosion en nappe ou en couche**

interseeding Sowing two or more crops simultaneously in the same rows. See *interplanting*. **semis en intercalaire ou en plante-abri**

interstade A warmer substage of a glacial stage marked by a temporary retreat of the ice, and of shorter duration than an interglacial. **interstade**

interstitial water Water in the pore spaces in soil or rock. **eau intersticielle**

interstratified See *interbedded*. **interstratifié**

interstratified clay mineral An aggregation composed of random or regular intergrowths of two or more clay minerals. **minéral argileux interstratifié**

intertidal zone The area of coastal land that is covered by water at high tide and uncovered at low tide. **zone intertidale**

interval The range between two extremes within which a variable can take any value. In statistics, interval data contrast with ordinal data in having real number values rather than integer placings. **intervalle**

intracellular Inside the cell. **intracellulaire**

intraspecific Referring to interactions that occur between individuals of the same species. **intraspécifique**

intrazonal soil A soil having a morphology that shows the influence of some local or factor of relief, parent material, or age, rather than that of climate and vegetation. **sol intrazonal**

intrinsic permeability The property of a porous material (e.g., soil) that expresses the ease with which gases or liquids flow through it. Soil **hydraulic conductivity (K)** is composed of the fluidity of the liquid or gas and the intrinsic permeability *(k)* of the soil. Calculated by $k = K\eta/\rho g$, where η is the viscosity of the fluid, ρ is the density of the fluid, and g is the acceleration of gravity. **perméabilité intrinsèque**

intrinsic rate of increase (r_m) Exponential growth rate of a population with a stable age distribution. **taux de croissance intrinsèque**

introduced species See *exotic*. **espèce introduite**

intrusive Denoting igneous rocks in a molten state that have been forced into older rocks or between rock layers and solidified below the surface of the Earth. **intrusive**

inverse square law The relationship describing the reduction in a physical quantity (e.g., radiation) with increasing distance from a source. The emitted quantity decreases by a factor equal to one divided by the square of the distance from the source. **loi de l'inverse des carrés**

inversion See *tillage, inversion*. **inversion, bouleversement**

invertebrate (*zoology*) An animal belonging to the Invertebrata (i.e., without a backbone) (e.g., mollusks, arthropods, and earthworms). **invertébré**

ion Atom, group of atoms, or compound that is electrically charged as a result of the loss of electrons (i.e., *cation*) or the gain of electrons (i.e., *anion*). **ion**

ion activity The effective concentration of a particular ion in a solution or soil-water system. Expressed analogously to pH (e.g., pCa, pNa). **activité ionique**

ion exchange The exchange of ions of the same charge between an aqueous solution and a solid in contact with it. An example of cation exchange involves the conversion of hard water to soft water. Hard water contains the divalent ions of calcium (Ca^{+2}) and magnesium (Mg^{+2}), which cause soap and detergents to form precipitates in water. A water softener consists of a resin that is saturated with sodium ions (Na^{+}). As hard water percolates through the resin, the ions of calcium or magnesium are removed as they attach to the resin,

I

thus releasing (being exchanged for) sodium ions. **échange d'ions**

ion pairing A phenomenon occurring in solution when oppositely charged ions aggregate and behave as a single particle. **formation d'une paire ionique**

ion selective electrode An electrochemical sensor, used in conjunction with a *reference electrode*, in which the potential depends on the logarithm of the activity of a given ion in aqueous solution (e.g., pH, nitrate, sodium). **électrode à ion spécifique**

ion selectivity (1) The relative adsorption of an ion by the solid phase in relation to the adsorption of other ions. (2) The relative absorption of an ion by a plant root in relation to absorption of other ions. **sélectivité ionique**

ionic bond An attractive force that draws electrons from one atom to another, thus transforming neutral atoms into electrically charged ions. Also, electrovalent bond. **liaison ionique**

ionic bonding The electrostatic attraction between oppositely charged ions. **liaison ionique**

ionic compound (binary) A compound that results when a metal reacts with a nonmetal to form a *cation* and an *anion*. **combinaison ionique**

ionic diffusion The movement of charged ions by the mechanism of *diffusion*. **diffusion ionique**

ionic solid A solid containing *cations* and *anions* that dissolves in water to give a solution containing the separated ions which are mobile and thus free to conduct electrical current. **solide ionique**

ionic strength A parameter that estimates the interaction between ions in solution. It is calculated as one-half the sum of the products of ionic concentration and the square of ionic charge for all the charged species in a solution. **force ionique**

ionic substitution The partial or complete proxying of one or more types of ions for one or more other types of ions in a given structural size in a crystal lattice. **substitution ionique**

ionization The process by which an atom or molecule acquires an electric charge (positive or negative) through the loss or gain of electrons in a chemical reaction, in solution, or by *ionizing radiation*. **ionisation**

ionization energy The quantity of energy required to remove an electron from a gaseous atom or ion. **énergie d'ionisation**

ionizing radiation Electromagnetic radiation or atomic particles capable of displacing electrons from around atoms or molecules, thereby producing charged atoms, molecules, or ions. Common types of ionizing radiation are *x-rays, gamma rays, alpha particles,* and *beta particles*. **radiation ionisante**

ionosphere See *atmosphere*. **ionosphère**

iridescence The display of prismatic colors in the interior or on the surface of a mineral, caused by interference of light from thin films or layers of different refractive index. **iridescence, irisation**

iron bacteria Usually refers to bacteria that cause the precipitation of iron oxide through their metabolic processes. However, in certain acid environments with relatively low oxidation potential, there also are bacteria that reduce iron. **bactéries ferrugineuses, ferrobactéries**

iron oxides A group name for the oxides and hydroxides of iron (e.g., minerals goethite, hematite, lepidocrocite, ferrihydrite, maghemite, and magnetite). Also called *sesquioxides*, or iron hydrous oxides. **oxydes de fer**

iron pan A thin, indurated soil horizon in which iron oxides are the major constituent of the cementing material. Several kinds of cementing materials may occur, including iron-organic matter complexes, hydrous oxides of manganese and iron, and hydrous iron oxides. **alios**

iron pyrite See *pyrite*. **pyrite**

ironstone An in-place concentration of iron oxides that is at least weakly cemented. **terre de fer, roche ferrugineuse**

irradiance The measure of radiant flux incident on a surface; it has the dimensions of energy per unit time (e.g., watts). **irradiance**

irradiated Exposed to radiation. **irradié**

irreversible process Any process that proceeds in one direction spontaneously, without external interference. **processus irréversible**

irrigable land Land having soil, topographic, drainage, and climatic conditions favorable for irrigation, and located in a position where a water supply is or can be made available. **terre irrigable**

irrigation Application of water to lands for agricultural purposes by various methods. **irrigation**

Methods of irrigation include:

basin – irrigation in which a level or nearly level area, surrounded by an earth ridge or dike, is flooded with water. ***irrigation par bassin***

border-strip – irrigation in which water is applied at the upper end of an area of land bounded by two border ridges or dikes to guide the irrigation stream from the point of application. Also called border irrigation. ***irrigation par calants ou à la planche***

center-pivot – sprinkler irrigation in which a rotating sprinkler pipe or boom supplies water to the sprinkler heads from the center or pivot point of the system. ***irrigation par pivot central***

check-basin – irrigation in which water is applied rapidly to relatively level area that is practically or entirely surrounded by earth ridges. ***irrigation par bassin de retenue***

contour-furrow – application of water in furrows that run across the slope with a forward grade in the furrows. ***irrigation par rigoles d'infiltration suivant les courbes de niveau***

corrugation – irrigation in which water is applied to small, closely spaced furrows to confine the flow of irrigation water to one direction. ***irrigation par infiltration ou par billons***

drip – irrigation in which water is applied directly on or below the ground surface by means of applicators (e.g., perforated pipe) operated under low pressure. ***irrigation au goutte-à-goutte***

flood – irrigation in which water is applied rapidly over the entire surface of the soil creating a sheet of water; called "controlled flooding" when water is impounded or the flow directed by border dikes, ridges, or ditches. ***irrigation par submersion***

furrow – irrigation in which water is applied to row crops in ditches between the rows. ***irrigation par rigoles***

wild flooding – irrigation in which water is released at high points in a field and water distribution is uncontrolled. ***irrigation par submersion non-contrôlée***

winter – irrigation of lands between growing seasons in order to store water in the soil for subsequent use by plants. ***irrigation d'hiver***

irrigation efficiency The ratio of the water actually consumed by crops on an irrigated area to the amount of water applied to the area. **efficacité d'irrigation**

irrigation flume (1) Open conduit for conveying water across obstructions. (2) An entire canal elevated above natural ground (e.g., aqueduct). (3) A specially calibrated structure for measuring open channel flows. **(1) canal sur appuis (2) canal surélevé (3) canal de jaugeage**

irrigation requirement, consumptive The quantity of irrigation water per unit area (e.g., cm per ha) needed for crop production, and which is exclusive of precipitation, stored soil moisture, or groundwater. **besoin des cultures en eau d'irrigation**

isobath An imaginary line on a land surface along which all points are the same vertical distance above the upper or lower surface of an aquifer or above the water table. **isobathe**

isodyne (1) The points of a cultivating implement registering equal pull on a *dynamometer.* (2) A line on a map of a cultivated field connecting points of equal pull on a *dynamometer.* **isodyne**

isoelectric point The pH value of a solution in equilibrium with a constant potential surface whose net electrical charge is zero and is dependent only on the presence of H^+, OH^- or H_2O to form species on the surface. **point isoélectrique**

I

isoelectronic ions Ions containing the same number of electrons. **ions isoélectriques**

isomer A chemical compound (usually organic) that has the same molecular formula as another compound but a different molecular structure and therefore different chemical and physical properties (e.g., butane and isobutane both have the molecular formula C_4H_{10}, but they have a different arrangement of the carbon and hydrogen atoms). **isomère**

isomorphism (*chemistry*) The characteristic in which two or more chemical compounds have essentially the same crystalline structure. Such substances are so closely similar that they can generally crystallize together to form a continuous series of solid solutions, forming an isomorphous series. (*evolution*) The similarity that develops in organisms of different ancestry as a result of evolutionary convergence. **isomorphisme**

isomorphous substitution The replacement of one atom by another of similar size in a crystal structure without disrupting or seriously changing the structure. When a substituting cation is of a smaller valence than the cation it is replacing, there is a net negative charge on the structure. **substitution isomorphe**

isopach A line drawn on a map through points of equal true thickness of a designated stratigraphic unit or group of stratigraphic units. **isopaque**

isopleth A general term for a line on a map that connects points of equal value (e.g., elevation), or of any quantity that can be numerically measured and plotted on a map. **isoplète**

isostasy The state of balance which the Earth's crust tends to maintain or return to through the equalizing of pressures at depth in the Earth's crust. The granitic rocks prevalent in continental areas are less dense than the basaltic rocks of the ocean basins. The density difference causes continents to rise above sea level. **isostasie**

isotactic chain A polymer chain in which the substituent groups such as CH_3 are all arranged on the same side of the chain. **chaîne isotactique**

isotonic solutions Solutions having identical osmotic pressures. **solutions isotoniques**

isotope dilution An analytical method in which a known quantity of an element with an isotopic composition different from that of the natural element (called a spike) is mixed with the sample being analyzed. Measurement of the mixture allows calculation of the amount of the natural element in the sample. **dilution isotopique**

isotope One of two or more atoms of the same element that have the same number of protons in their nucleus but different numbers of neutrons. Hydrogen (one proton, no neutrons), deuterium (one proton, one neutron), and tritium (one proton, two neutrons) are isotopes of hydrogen. Most elements in nature consist of a mixture of isotopes. **isotope**

isotopic fractionation The relative enrichment of one isotope of an element over another, owing to slight variations in their physical and chemical properties. It is proportional to differences in their masses. **fractionnement isotopique**

isotopically exchangeable ion An ion, bonded to a soluble or solid substance, that can exchange with similar isotopically labeled ions in solution in a specified period of time. As equilibrium is approached, the ratio of native to isotopically labeled ions approaches the same value for each reactive form in which the ion occurs. **ion isotopiquement échangeable**

isotropic shrinkage Soil shrinkage that occurs equally in all directions. **retrait isotrope**

isotropy Having the same properties in all directions. **isotropie**

isthmus A narrow strip of land separating two bodies of water and connecting two larger bodies of land. **isthme**

iteration A repetitive process of successive approximations in a procedure, used, for example, in solving complex equations numerically. **itération**

J

Jackson turbidity unit (JTU) (*limnology*) A unit that expresses the cloudiness (turbidity) of water; the measurement is related to the distance through the water that light can be seen by the unaided eye. **unité de turbidité Jackson (UTJ)**

jarosite A pale yellow potassium iron sulfate mineral, $KFe_3(OH)_6(SO_4)_2$. **jarosite**

joint (*geology*) A fracture or parting that abruptly interrupts the physical continuity of a rock mass but which exhibits no differential movement, in contrast to a fault. **joint, diaclase**

joint planes Planes of parallel orientation. **plans de séparation ou de diaclase**

judgement sample A form of non-probability sample in which the sample is chosen according to the judgment of the sampler. **échantillon choisi à dessein, échantillon au jugé**

juvenile water Water contained in rock or of magmatic origin derived from the Earth's interior and not from atmospheric or surface water. **eau juvénile**

K

K factor See *universal soil loss equation.* **facteur K**

kame (*geomorphology*) A mound, knob, or short irregular ridge, composed of stratified sand and gravel deposited by a subglacial stream as a fan or delta at the margin of a melting glacier, by a supraglacial stream in a low place or hole on the surface of the glacier, or as a ponded deposit on the surface or at the margin of stagnant ice. **kame**

kame complex, kame field (*geomorphology*) An assemblage of closely spaced kames, interspersed in places with *kettles* and *eskers*, and having a characteristic hummocky topography. Much the same as *kame-and-kettle terrain.* **complexe juxtaglaciaire**

kame moraine (*geomorphology*) (1) An *end moraine* that contains numerous *kames.* (2) A group of kames along the front of a stagnant glacier, commonly developed from the slumped remnants of an *outwash plain* built up over the foot of rapidly wasting or stagnant ice. **moraine de kame**

kame terrace (*geomorphology*) A flat-topped ridge or terrace feature occurring between a valley glacier and the valley slopes. It is composed of bedded glaciofluvial or glaciolacustrine materials deposited by meltwater streams flowing laterally to the glacier. The kame-terrace surface is sometimes carved into by lateral meltwater channels and pitted with hollows (i.e., *kettle holes*), features that distinguish the kame terrace from a pro-glacial lake shoreline. The terrace is subsequently dissected by post-glacial tributary streams or cliffed by the meandering river after disappearance of the glacier. **terrasse juxtaglaciaire, terrasse de kame**

kame-and-kettle terrain (*geomorphology*) An undulating landscape composed of groups of *kames* and/or *kame terraces* interspersed or pitted with *kettle holes*. This type of landform is sometimes called a *kame complex* and is created when supraglacial, englacial glaciofluvial, or glaciolacustrine sediments are lowered onto the surface as the ice-sheet decays. **topographie en bosses et creux**

kandic horizon A subsoil diagnostic horizon (U.S. system of soil taxonomy) having a clay increase relative to overlying horizons, and has low activity clays (i.e., <160 $cmol_c$ kg^{-1} clay). **horizon kandique**

kaolin group See *kaolin.* **groupe du kaolin**

kaolin, kaolinite (1) An aluminosilicate mineral with a 1:1 layer structure; that is, consisting of one silicon tetrahedral layer and one aluminum octahedral layer with very little *isomorphous substitution.* More commonly called kaolinite. (2) A soft, usually white rock composed largely of kaolinite. **kaolin, kaolinite**

kaolinitic A *soil mineralogy* class for family groupings in the Canadian system of soil classification and the U.S. soil taxonomy. **kaolinitique**

karst (*geomorphology*) A type of topography that is formed over limestone, dolomite, or gypsum by dissolution, characterized by sinkholes, caves, and underground drainage. **karst**

karst plain (*geomorphology*) A plain, usually of limestone, on which karst features are developed. **plaine karstique**

karst topography (*geomorphology*) An irregular land surface in a limestone region. The principal features are depressions (e.g., *dolines*) which sometimes contain soils which have been washed off the rest of the surfaces, leaving them bare and rocky. Drainage is usually by underground streams. **topographie superficielle de karst**

karst valley A closed depression formed by the coalescence of several sinkholes. Its drainage is subsurface and may be hundreds of meters to a few kilometers in size, and it usually has an irregular floor and a scalloped margin inherited from the sinkholes. **ouvala**

karst, covered Limestone terrain in which the characteristic karstic features are buried beneath a cover of superficial materials (e.g., alluvium, blown sand, or glacial till). See *karst*. **karst couvert**

K-bentonite See *potassium bentonite*. **bentonite potassique**

kernel A mature ovule of a grass plant that has the ovary wall fused to it. See *caryopsis*. **grain**

kerogen A solid, mineraloid hydrocarbon embedded in subsurface rock formations. It is not the same as oil shale, but yields oil when subjected to destructive distillation. **kérogène, kérobitume**

ketone An organic compound containing the carbonyl group (>C=O) bonded to two carbon atoms. **cétone**

kettle lake (*geomorphology*) A water body occupying a *kettle*. **lac de kettle**

kettle moraine (*geomorphology*) A morainal area characterized by an extremely undulating terrain of *kames* and *kettle holes*. **moraine à kettles**

kettle, kettle hole (*geomorphology*) A depression in glacial drift, especially in outwash, and a *kame complex* or *kame field*, formed by the melting of a detached block of stagnant ice that was buried in the drift. A kettle is usually 10 to 15 m deep and 30 to 150 m in diameter and often contains a wetland. **kettle**

keystone species A species that plays an important ecological role in determining the overall structure and dynamic relationships within a biotic community. A keystone species' presence is essential to the integrity and stability of a particular ecosystem. **espèce clé**

K-feldspar See *potassium feldspar*. **feldspath potassique**

kinematic viscosity The *dynamic viscosity* of a fluid divided by its density (mass per unit volume). Units are length squared divided by time (m^2 s^{-1}). **viscosité cinématique**

kinetic energy The energy inherent in a substance because of its motion, expressed as a function of its velocity (v) and mass (m) (i.e., $mv^2/2$). **énergie cinétique**

kingdom The highest category in the hierarchy of classification of animals and plants. **règne**

Kjeldahl method A method for measuring the amount of nitrogen in an organic compound. The compound is boiled with concentrated sulfuric acid and a catalyst (e.g., copper sulfate) to convert the nitrogen to ammonium sulfate. Alkali is added and the mixture heated to distill off ammonia. This is passed into a standard acid solution, and the amount of ammonia can then be found by estimating the amount of unreacted acid by titration. Kjeldahl nitrogen is the amount of nitrogen contained in an organic material as determined by this method. **méthode Kjeldahl**

k_m The Michaelis constant, the substrate concentration of half the maximum velocity of an enzyme reaction. **k_m, constante de Michaelis-Menten**

knickpoint (*geomorphology*) Any interruption or break of slope, especially a point of abrupt change or inflection in the long profile of a stream or its valley. **rupture de pente, brisure**

knob (*geomorphology*) A rounded hill or mountain, especially an isolated one. **mamelon**

knob-and-kettle topography (*geomorphology*) An undulating landscape in which a disordered assemblage of knolls, mounds, or ridges of glacial drift is interspersed with irregular depressions, pits, or *kettles* that are commonly undrained and may contain wetlands. See *hummocky moraine*. **paysage en bosses et creux**

knoll A rounded hill with little elevation. **colline**

Koppen's climatic classification An empirical climatic classification devised in 1918 by German biologist W. Koppen based on the climatic requirements of certain vegetation types, with six major categories, each of which is subdivided according to different rainfall and temperature characteristics. **classification climatique de Koppen**

1. Tropical zone climate – wet, hot, equatorial regions; monthly average temperature above 18°C. **climat de zone tropicale**
2. Dry climate – arid and semi-arid deserts and steppes; evaporation exceeds precipitation; 4 to 11 months above 18°C; 1 to 8 months between 10°C and 20°C. **climat sec**
3. Mesothermal (temperate) climate – humid, sub-tropical; may have dry summers; warmest month above 10°C; coldest month above 0°C but below 18°C. **climat mésothermique**
4. Microthermal climate – humid climate with long winters and mild summers; warmest month above 10°C; coldest month below 0°C. **climat microthermique**
5. Polar climate – no true summer, with warmest month average temperatures below 10°C. **climat polaire**
6. Highland (mountain) climate – relatively low temperatures and high precipitation. **climat montagneux**

kriging (*statistics*) A weighted, moving-average estimation technique based on geostatistics that uses the spatial correlation of point measurements to estimate values at adjacent, unmeasured points. The kriging routine preserves known data values, estimates missing data values, and estimates the variance at every missing data location. After kriging, the filled matrix contains the best possible estimate of the missing data values, in the sense that the variance has been minimized. **krigeage**

krotovina See *crotovina*. **krotovina**

k-selected An individual for which selection is strong for relatively few well-developed offspring that have a good chance of surviving and reproducing; presumed to be important when competition is intense. See *r-selected*. **individu à strategie k**

Kubiena box or tin A container (usually made of metal with two sides open and separate lids) in which intact soil samples are taken for preparation for micromorphological analysis. Named in honor of W.L. Kubiena, who in 1938 first presented the concepts and techniques for microscopic examination of the distribution of soil particles and voids. **échantillonneur de Kubiena, boîte d'échantillonnage de Kubiena**

kurtosis (*statistics*) The peakedness of a distribution curve. A high peak is called leptokurtic, a flat-topped peak is called platykurtic, and a curve intermediate between the two is called mesokurtic. **aplatissement, kurtosis**

L

L factor See *universal soil loss equation.* **facteur L**

L horizon An organic soil horizon containing more than 17% organic carbon, normally associated with upland forest soils, characterized by an accumulation of leaves, twigs, woody materials, and mosses, in which the original plant structures are easily discernable. An organic horizon in the Canadian system of soil classification. See *Appendix D* for equivalent terms in the U.S. and FAO/UNESCO systems of soil classification. See *horizon, soil; F horizon; H horizon.* **horizon L**

labile (*biochemistry*) An organic substance readily transformed by microorganisms or readily available to plants. (*geology*) Referring to rocks and minerals that are easily decomposed. **labile**

laboratory blank An artificial sample, usually distilled water, introduced to a chemical analyzer to observe the response of the instrument to a sample that does not contain the material being measured, and also to detect any contamination occurring during laboratory processing of the sample. **blanc de laboratoire**

lacustrine (1) Pertaining to lakes. See *limnetic.* (2) Descriptive of materials that either have settled from suspension in bodies of standing fresh water or have accumulated at their margins through wave action. **lacustre**

lacustrine deposit Material deposited in lake water and later exposed either by lowering of the water level or by uplifting of the land. These sediments range in texture from sands to clays. **dépôt lacustre**

lacustrine plain (1) A low-lying, nearly level tract of land formed in a lake by well-sorted sediments from inflowing streams. (2) A flat lowland or a former lake bed bordering an existing lake. **plaine lacustre**

lag (1) (*geology*) Coarse-grained material that is rolled or dragged along the bottom of a stream at a slower rate than the finer material, or is left behind after currents have washed away the finer material (e.g., lag deposit). (2) (*geology*) The residual accumulation of coarse material left on a desert surface after wind has blown away the finer material (e.g., lag gravel). (3) (*meterology*) Any time delay between the arrival of signal in a measuring instrument and the response of that instrument, especially in meteorological instruments (e.g., time lag). **(1) dépôt résiduel (2) résidu de déflation (3) retard de réponse**

lag growth phase (*microbiology*) The period following the inoculation of a growth medium with a culture of bacteria. Some time is required for the organism to adjust to the new environment, and during this interval there is no increase in the number of bacteria per unit volume of culture fluid. See *growth phase.* **phase de latence**

lagoon (*waste management*) An artificial pond in which the manure from some livestock operations is collected. Bacteria in the lagoon decompose much of the material and release plant nutrients; the effluent from a lagoon can make an effective fertilizer. **lagune, étang**

lake Any inland body of standing water, larger and deeper than a pond. **lac**

lake plain See *lacustrine plain.* **plaine lacustre**

lake rampart (*geomorphology*) A ridge of material along a lake shore, produced by ice pushed shoreward by winds, waves, currents, or by expansion of ice against yielding lake-shore deposits. **bourrelet lacustre, bourrelet glaciel**

lake terrace (*geomorphology*) A narrow shelf, partly cut and partly built along a lake shore in front of a nip or line of low cliffs, and later exposed when the water level falls. **terrasse lacustre**

lamella (*mineralogy*) A thin scale, *lamina*, or layer (e.g., one of the units of a polysynthetically twinned mineral, such as plagioclase). **lamelle**

lamellar Composed of or arranged in *lamellae*, like the leaves of a book. **lamellaire**

lamina (plural laminae) (1) (*geology*) A thin layer, plate, or scale. (2) (*soil structure*) The thinnest recognizable layer in a layered sequence of soils or sedimentary rocks, differing from other layers in color, composition, or particle size; commonly 0.05 to 1.00 mm thick. (3) (*botany*) A thin, plate-like or scale-like structure in an organism. **(1, 2) dépôt laminique (3) lamina**

laminar Consisting of, arranged in, or resembling laminae. See *lamina.* **laminaire**

laminar flow A type of flow in which particles in a fluid are flowing smoothly in the same direction; fluid flow without turbulence. (*hydrology*) Streamline flow in which successive flow particles follow similar path lines and head loss varies with velocity to the first power. **écoulement laminaire**

land The solid part of the Earth's surface or any part thereof. A tract of land is defined geographically as a specific area of the Earth's surface. Its characteristics include all stable or predictably cyclic attributes of the biosphere vertically above and below this area, including those of the atmosphere, soil and underlying geology, hydrology, plant and animal populations, and results of past and present human activity, to the extent that these attributes exert a significant influence on the present and future uses of land by humans. **terre, terrain**

land capability The ability of the land to support a given use, based on an evaluation of its physical, chemical, and biological characteristics including topography, drainage, climate, hydrology, soils, and vegetation. **possibilités d'utilisation des terres, capabilité des terres**

land classification The arrangement of land units into various categories based on the properties of the land or its suitability for some particular purpose. **classification des terres**

land ecosystem Any ecologically defined unit of land identified for any hierarchical level of an ecological land classification system (e.g., ecoprovince, ecoregion, ecodistrict, ecosection, ecosite, and ecoelement). **écosystème terrestre**

land farming (*waste management*) A technique for the controlled biodegradation of organic waste that involves mixing waste sludge with soil. Tilling the mixture ensures adequate oxygen and controls the moisture content, nutrient levels, and soil pH. **épandage sur le sol**

land leveling Formation of a horizontal or near-horizontal landscape by removal or redistribution of earth, for purposes such as flood irrigation or development of a pad for equipment or buildings. **profilage**

land planing See *tillage, land planing.* **nivellement, planage**

land quality The value placed on land with respect to its fitness for a specific use; derived from a larger environmental system than soil, and includes the natural integration of soil, water, climate, landscape (i.e., topographic and hydrologic factors), and vegetation characteristics. **qualité des terres**

land reclamation See *reclamation.* **mise en valeur restauration des terres**

land resting Temporary discontinuance of cultivation of a piece of land. See *fallow land.* **jachère, friche temporaire**

land system A land area recognized and separated by differences in one or more of: general pattern of land surface form, surficial geological materials, amount of lakes or wetlands, or general soil pattern. The concept is applied in some soil mapping systems, wherein a land system is a subdivision of an *ecodistrict* in the hierarchical *ecological land classification* system. All land systems within one ecodistrict have the same general climate for agriculture, but differences in microclimate patterns can be recognized. **système paysager**

land system inventory An inventory compiled at small (generally 1:250,000) scale where the differentiating land unit is the *land system.* **inventaire ou cartographie des systèmes paysagers**

land use The way in which land is used (e.g., pasture, orchards, producing field crops, or recreation). **utilisation des terres ou des sols**

land use capability An evaluation of land according to its limitations for use; employs three categorical levels known as classes, subclasses, and units. The land use capability system has had the widest use of any interpretive system of soil classification. **possibilités d'utilisation des terres**

land use planning The process by which decisions are made on future land uses over extended periods of time that are deemed to best serve the general welfare. This process includes evaluating the status, potential, and limitation of the land and its resources, and interacting with the populations associated and/or concerned with the area to determine their needs, wants, and aspirations for the future. **aménegement du territoire, planification de l'utilisation des terres**

landfill A site where the layers of waste materials from municipal or industrial sources are buried and covered regularly (usually daily) with soil. **site d'enfouissement, décharge**

landfill trench method A technique for placing municipal solid waste in a landfill; the waste is spread and compacted in 60 cm layers within trenches about 75m long, 6m wide, and 2m deep. Cover material for a full trench is obtained by excavating an adjacent trench. See *landfill.* **enfouissement par la méthode de la tranchée**

landfill, secure A ground location for the deposit of hazardous wastes. The material is placed above natural and synthetic liners that prevent or restrict the leaching of dangerous substances into groundwater. **site d'enfouissement à accès contrôlé**

landform A three-dimensional part of the land surface, formed of soil, sediment, or rock that is distinctive because of its shape, is significant for land use or to landscape genesis, repeats in various landscapes, and also has a fairly consistent position relative to surrounding landforms. **forme de terrain, unité de relief**

landform classification The arrangement of landforms into various categories based on their various properties (e.g., structure, composition, configuration, genesis, age). **classification des formes de terrain**

landforming See *tillage, landforming.* **modelage**

Landsat An unmanned Earth-orbiting NASA satellite that transmits *multispectral* images in the 0.4 to 1.1 μm region to Earth receiving stations. It was formerly called Earth Resource Technology Satellite, or ERTS. **Landsat**

landscape (*general*) All the natural features (e.g., fields, hills, forests, and water) that distinguish one part of the Earth's surface from another. Usually it is the portion of land or territory that the eye can see in a single view, including all its natural characteristics. (*physical geography*) An area composed of interacting ecosystems that are repeated because of geology, landform, soils, climate, biota, and human influences throughout the area. Landscapes are generally of a size, shape, and pattern determined by interacting ecosystems. **paysage**

L

landscape ecology The study of the distribution patterns of communities and ecosystems, the ecological processes that affect those patterns, and changes in pattern and process over time. **écologie du paysage**

landscape model A conceptualization of a typical or modal landscape, or a range of landscape types, that includes elements of slope class, surface form, a surface form modifier, and various other attributes of a landscape. **modèle de paysage**

landscape unit A planning area based on topographic or geographic features (e.g., a watershed or series of watersheds). **unité de paysage**

landslide, landslip (1) A mass of material that has slipped downhill under the influence of gravity, often assisted by water when the material is saturated. (2) Rapid movement of a mass of soil, rock, or debris down a slope. **glissement de terrain**

Langmuir isotherm An *adsorption isotherm* that describes the case where the solid has a high affinity for the solute. The adsorption is described mathematically by the Langmuir equation: $x/m = KbC/1+KC$; where x/m is the quantity of solute adsorbed per unit weight of adsorbent, C is the equilibrium concentration of the adsorbing compounds, K is a constant related to the bond energy, and b is the adsorption maximum or total amount of solute capable of being adsorbed. **isotherme de Langmuir**

Lantz process A destructive distillation technique in which the combustible components of solid waste are converted into combustible gases, charcoal, and a variety of distillates. **procédé de distillation de Lantz**

larva (*zoology*) The immature form of animal, chiefly insects, that must pass through metamorphosis before reaching the adult form. **larve**

latent heat The amount of heat absorbed or released by a substance during a change of state, under conditions of constant temperature and pressure. **chaleur latente**

lateral (1) (*drainage*) Secondary or side channel, ditch or conduit; sometimes called branch line or drain, spur, lateral ditch, or group lateral. (2) (*horticulture*) Plant growth from the side of the root, stem, or other plant part; not terminal or central. **(1) canalisation secondaire (2) racine latérale ou secondaire**

lateral moraine (1) (*geomorphology*) A low ridge-like moraine carried on, or deposited at, the side of a mountain glacier. It is composed chiefly of rock fragments loosened from the valley walls by glacial abrasion and plucking, or fallen onto the ice from the bordering slopes. (2) An end moraine built along the side margin of a glacial lobe occupying a valley. **moraine latérale**

lateral planation (*geology*) The reduction of the land in an interstream area to a plain or a nearly flat surface by the lateral erosion of a meandering stream. **aplanissement des interfluves**

laterite A highly weathered red subsoil rich in secondary oxides of iron, aluminum, or both, nearly devoid of bases and primary silicates, and commonly with quartz and kaolinite. It develops in a tropical or forested warm to temperate climate, and is a residual product of weathering. **latérite**

laterization A soil-forming process in the humid tropics leading to soils with a low silica/sesquioxide ratio in the clay fractions, low clay activity, low content of most primary minerals and soluble constituents, a high degree of *aggregate stability*, and usually red in color. **latérisation**

latitude Angular measurement in degrees north or south of the equator; lines denoting latitude are also called parallels. (One minute of arc or meridian is one nautical mile.) **latitude**

latitude zonation Zonation of soils arising from climate and vegetation that changes with latitude, of the land (from tropical to arctic). See *zonation, vertical zonation.* **zone de latitude, zonalité horizontale**

latosol Soils characterized by deep weathering and abundant hydrous oxides; virtually synonimous with *laterite.* **latosol**

latosolization see *desilication.* **latosolisation**

lattice A regular geometric arrangement of points in a plane or in space. Lattice is a mathematical concept, used to represent the distribution of repeating atoms or groups of atoms in a crystalline substance. Atomic substitutions take place in a structure and not in a lattice. See *phyllosilicate mineral terminology.* **réseau**

lattice energy The energy required to separate the ions of a crystal by an infinite distance. **énergie de réseau ou réticulaire**

lattice structure The orderly arrangement of atoms in a crystalline material. **structure maillée, réticulée, ou en treillis**

lava flow Area covered by lava. Most lava flows have a sharp, jagged surface, crevices, and angular blocks characteristic of lava. Others are relatively smooth and have a ropy glazed surface. Some soil material may occur in cracks and pockets, but the flows are virtually devoid of plants except for lichens. **coulée de lave**

law In science, a formal statement of the invariable and regular manner in which natural phenomena occur under given conditions. **loi**

law of conservation of energy See *first law of thermodynamics.* **loi de la conservation de l'énergie**

layer charge The magnitude of charge per formula unit of a clay that is balanced by ions of opposite charge external to the unit layer. See *phyllosilicate mineral terminology.* **charge électrique de la couche**

layer silicate mineral See *phyllosilicate.* **phyllosilicates**

Le Chatelier's principle If a change is imposed on a system at equilibrium, the position of the equilibrium will shift in a direction that tends to reduce the effect of that change. **principe de Le Chatelier**

leaching The removal of soluble materials (e.g., humus, bases, and sesquioxides) from one horizon or zone in soil to another by water movement in the profile. Over time, the upper layer of a leached soil can become increasingly acidic and mineral-deficient. See *illuvial horizon.* **lessivage, lixiviation**

leaching field (*waste management*) The area of land into which a septic tank drains. Wastewater exiting the tank is dispersed over, and percolates through, a defined area of land. **champ d'épuration ou de percolation**

leaching fraction The fraction of infiltrated irrigation water that percolates below the root zone. **facteur de perte par lessivage**

leaching requirement The fraction of irrigation water that must pass through the root zone to prevent soil salinity from exceeding a specific value; applies to steady-state or long term average conditions. **besoin en lessivage**

leaf curl The rolling back of leaf edges as a result of damage caused by disease, insects, or environmental factors. **enroulement de la feuille**

leaf mold A substance formed by decayed and partially decayed leaves, found on the surface of some forested areas, or produced by composting leaves. **terreau de feuilles**

leaf spot Fungus and bacterial diseases that cause distinct discolored spots on plant foliage, thus weakening the plant's ability to produce food. **helminthosporiose**

least squares method (*statistics*) A method of curve fitting, and therefore a method of estimating the parameters appearing in the corresponding equations. It consists of minimizing the sums of squares of the differences between observed values and the corresponding values calculated by means of a model equation. It may also be used to determine the most probable value of a single quantity from a number of measurements of that quantity, and to discover the probable error of the mean value of a number of observations. **méthode des moindres carrés**

lee The side of a hill, dune, or other prominent object that is sheltered or turned away from the wind. Also, the side or slope of a hill or knob that faces away from an advancing glacier or ice sheet and is relatively protected from its abrasive action. **côté sous le vent**

legend (*cartography*) A brief explanatory list of the symbols, cartographic units, patterns (shading and color hues), and other cartographic conventions appearing on a map, chart, or diagram. **légende**

leghemoglobin A red pigmented protein containing iron, produced in root nodules formed by symbiotic *Rhizobium* bacteria in leguminous plants. Leghemoglobin is similar to, but not identical to, mammalian hemoglobin. **leghémoglobine**

legume A plant of the botanical family Leguminosae, (e.g., pea or bean), which has the ability to fix atmospheric nitrogen through a symbiotic association with nodule-forming bacteria of the genus *Rhizobium*. **légumineuse**

Lemna gibba The genus and species name of a small, stemless, free-floating plant used in experiments to determine the toxicity of pollutants to aquatic plant life; called duckweed. ***Lemna gibba***

lens, lensing A body of ore, rock, sand, or water that is thick in the middle and thins at the edges, similar to a double convex lens. **stratification lenticulaire**

lentic Standing (non-flowing) water biome such as lakes, ponds, and swamps. **lentique**

lepidocrocite An orange iron oxide mineral (FeOOH) found in mottles and concretions of wet soils. **lépidocrocite**

lessivage The process by which *clay minerals* are moved mechanically downward through a soil by percolating water, leading to *illuviation* in a lower horizon. Evidence of lessivage is the gradual appearance with increasing depth of clay skins on the soil peds. **lessivage, éluviation**

leucinization In soil formation, the paling of soil horizons by disappearance of organic materials either through transformation to lighter colored ones or through removal from the horizons. **leucinisation**

levee A natural or artificial embankment confining a stream to its channel or, if artificial, limiting the area of flooding; also a landing place, pier, or quay along a river or stream. **levée, digue**

level A flat or very gently sloping, unidirectional surface with a generally constant slope not broken by marked elevations and depressions. Slopes generally less than 2%. **plat, horizontal, subhorizontal**

level rod A straight rod or bar, with a flat face graduated in plainly visible linear units with zero at the bottom, used in measuring the vertical distance between a point on the Earth's surface and the line of sight of a leveling instrument that has been adjusted to a horizontal position. Also called rod, leveling rod, surveyor's rod. **mire**

leveling A surveying method for determining height differences between points in relation to a point of known height (i.e., *datum*), usually by sighting through a leveling instrument at one point to a level rod at another point. Also, the finding of a horizontal line or the establishing of grades by means of a level. See *land leveling*. **nivellement**

Lewis structure A diagram of a molecule showing how the valence electrons are arranged among the atoms in the molecule. **structure de Lewis**

ley A biennial or perennial hay or pasture portion of a rotation, including cultivated crops. **prairie temporaire**

LFH horizon A designation commonly applied to upland organic soil horizons in the Canadian system of soil classification. LF, LH, and FH may also be applied. See *horizon, soil; L horizon*; *F horizon*; *H horizon*. **horizon LFH**

lichen Organism resulting from the symbiotic relationship between a fungi and algae or cyanobacterium. They live in a wide range of habitats, from desert to polar regions and the tropics, but are commonly found in barren environments, and are dominant in tundra regions and mountains. Lichens are very sensitive to air pollution and serve as an indicator species. They occur in one of four basic growth forms: crustose – crustlike, growing tight against the substrate; squamulose – tightly clustered and slightly flattened pebble-like units; foliose – leaflike, with flat sheets of tissue not tightly bound; fruticose – free-standing branching tubes. **lichen**

Liebig's law The growth and reproduction of an organism is dependent on the nutrient substance (e.g., nitrogen, phosphorus, potassium, calcium, and others) that is available in minimum quantity. See *limiting nutrient*. **loi de Liebig, loi du minimum**

life cycle (*biology*) The phases, changes, or stages that an organism passes through during its lifetime. See *ontogeny*. (*environmental marketing*) The various stages through which a product, process, or activity passes, during which environmental effects may result that affect the acceptance of these things by environmentally conscious consumers. **cycle de vie**

L

lift See *tillage, lift*. **arracher**

ligand A molecule or ion that donates a pair of electrons to a metal atom or ion to form a *complex*. Molecules that function as ligands act as a *Lewis bases*. **ligand (coordinat)**

light fraction (organic matter) Organic matter isolated from mineral soils by flotation of dispersed soil suspensions on water or heavy liquids with densities (1.5 to 2.0 Mg m^{-3}). Light fraction organic matter consists primarily of plant debris but also may contain fungal hyphae, spores, seeds, charcoal, and animal remains. **fraction légère de la matière organique**

light mineral A rock-forming mineral having a specific gravity lower than a standard, usually 2.85 (e.g., quartz, feldspar, calcite, dolomite, muscovite, feldspathoids). See *heavy mineral*. **minéral léger**

lignin (*biochemistry*) A macromolecular polymer that is the structural constituent of woody fibers in plant tissue. The polymer is a random arrangement of phenylpropane units. Biodegradation of lignin is slow. **lignine**

lignite A brownish-black coal of low rank with high inherent moisture and volatile matter; alteration of vegetal matter has proceeded further than peat but not so far as subbituminous coal. **lignite**

lime (*chemistry)* Calcium oxide (CaO). (*agriculture*) A soil amendment containing calcium carbonate, magnesium carbonate, and other materials, used to neutralize soil and furnish calcium and magnesium for plant growth. **chaux**

lime concretion An aggregate of precipitated calcium carbonate or other material cemented by precipitated calcium carbonate. **concrétion calcaire**

lime potential The negative logarithm of the ratio of the hydrogen ion activity to the square root of the sum of the activities of calcium and magnesium in the soil solution. It is generally written as pH – p(Ca + Mg). It can also be written as 1/2 log [(Ca + Mg)$(OH)_2$] + pK_w, where K_w is the ionic product of water. Lime potential is an expression of the sum of the activities of calcium and magnesium hydroxides in the soil solution. **potentiel de chaux, potentiel calcaire**

lime requirement The amount of agricultural limestone, or the equivalent of another liming material, required to raise the pH of a volume of soil to a specific value under field conditions. **besoin en chaux**

lime-soda process A water-softening method in which lime and soda ash are added to water to remove calcium and magnesium ions by precipitation. **procédé chaux-soude**

limestone A sedimentary rock consisting chiefly of the mineral calcite (calcium carbonate, $CaCO_3$), with or without magnesium carbonate (dolomite). Common impurities include *chert* and clay. Limestone is the most commonly occurring carbonate rock and is the consolidated equivalent of limy mud, calcareous sand, and/or shell fragments. It yields lime on heating (or calcination). **calcaire**

liming The application of lime to land, primarily to reduce soil acidity (see *reaction, soil*) and to supply calcium for plant growth, and also to maintain a high degree of availability of most nutrient elements required by plants. *Dolomitic lime* supplies both calcium and magnesium. **chaulage**

limited water-soluble substance A chemical causing water pollution that is soluble in water at less than one milligram of substance per liter of water. **substance peu soluble à l'eau**

limiting factor A factor whose presence or absence restricts the continued reproduction or spread of a particular species; the constraint may be physical (e.g., light), chemical (e.g., an essential nutrient), ecological (e.g., competition), or a combination of factors. **facteur limitant**

limiting nutrient Any nutrient limiting plant growth. See *nutrient*. **élément déficient ou limitant**

limits of tolerance Minimum and maximum amounts of a required physical or chemical factor in the environment within which a particular species can exist; an optimum level, supporting the maximum number of organisms, lies within the tolerance limits. **limite de tolérance**

limnetic zone The open-water region of a lake, especially in areas too deep to support rooted aquatic plants; supports plankton and fish as the principal plants and animals. See *littoral zone*. **région limnétique**

limnetic, limnic (*biology*) Pertaining to body of fresh water and organisms that inhabit fresh water. (*geology*) Pertaining to coal deposits formed in fresh water basins or peat formation. **limnétique**

limnic A component of organic soils that includes both organic and inorganic materials deposited in water by

L

precipitation or through the action of aquatic organisms, or derived from underwater and floating aquatic plants and aquatic animals (e.g., *marl*, *diatomaceous earth*, and *coprogenous earth* or *sedimentary peat*). **limnique**

limno layer In organic soil, a layer at least 5 cm thick composed of *limnic* materials (e.g., *marl*, *diatomaceous earth*, or *coprogenous earth* or *sedimentary peat*). **couche limnique**

limnology The study of fresh waters, especially lakes and ponds, dealing with the physical, chemical meterological, and especially biological and ecological conditions pertaining to such bodies of water. **limnologie**

limnophyte Any vegetation species that grows in an environment of constant submergence in fresh water. **limnophyte**

limonite Yellowish-brown hydrous oxides of iron ($2Fe_2O_3.3H_2O$). Limonite is a common secondary material formed by weathering (oxidation) of iron-bearing minerals; it may also occur as a precipitate in bogs or lakes. It occurs as coatings, earthy masses, and in a variety of other forms, and is the coloring material of yellow clays and soils. Limonite is a minor ore of iron. See *bog iron ore*. **limonite**

lindane A commercial *chlorinated hydrocarbon* insecticide that consists of several isomers of hexachlorocyclohexane (HCH). Lindane is a non-aromatic hydrocarbon, persistent in the environment. **lindane**

line plot survey A survey using plots as sampling units; plots of specified size are laid out, usually at regular intervals along parallel survey lines. **inventaire par échantillonnage en ligne**

lineament (*geology*) A linear topographic feature of regional extent that appears to reflect the crustal structure of a region (e.g., fault lines, straight stream courses). **linéament**

linear accelerator A type of particle accelerator in which a changing electrical field is used to accelerate a positive ion along a linear path. **accélérateur linéaire**

linear alkyl sulfonate (LAS) A common surfactive agent used in detergents. **alkylsulfonate linéaire**

linear programming A mathematical method of systematically budgeting enterprises to use available resources efficiently. **programmation linéaire**

lineation (*geology*) Any linear structure in a rock (e.g., flow lines, ripple marks). **linéation**

line-intercept method A method to sample vegetation by recording the plants intercepted by a measured line set close to the ground or by vertical projection from the line; in regeneration assessments, this method is called linear regeneration sampling. **méthode d'intersection**

liner A low-permeability material that retards the escape of leachate from the landfill to underlying groundwater. (e.g., clay or high density polyethylene). **revêtement**

lipid Any of a diverse group of organic compounds that contain long-chain aliphatic hydrocarbons and their derivatives (e.g., fatty acids, alcohols, amines, amino alcohols, and aldehydes); includes waxes, fats, and derived compounds. Lipids are insoluble in water but soluble in organic solvents such as *chloroform*. See *aliphatic compound*. **lipide**

liquefaction Generally, a change from the gaseous phase to the liquid phase. (*soil mechanics*) The sudden large decrease of the shearing resistance of a cohesion-less soil. It is caused by a collapse of the structure by shock or other strain, and is associated with a sudden, temporary increase of the interstitial water pressure. It involves a temporary transformation of the material into a fluid mass. Often called spontaneous liquefaction. **liquéfaction spontanée**

liquid chromatography A technique for the separation and analysis of higher-molecular-weight organic compounds. See *chromatography*. **chromatographie en phase liquide**

liquid fertilizer See *fertilizer, liquid*. **engrais liquide**

liquid limit One of the *Atterberg limits*. (1) The water content corresponding to an arbitrary limit between the liquid and plastic states of *consistence* of a soil; the upper *plastic limit*. (2) The water content at which a pat of soil, cut by a standard-sized groove, will flow together for a distance of 12 mm under the impact of 25 blows in a standard liquid-limit apparatus. **limite de liquidité**

liquid manure Manure obtained from agricultural operations that has a moisture level greater than 88% and contains less than 12% dry matter. Liquid manure can be pumped or moved by gravity. **lisier**

liquid waste Pollutants such as soap, chemicals, or other substances in liquid form. **effluent**

lister planting See *tillage, lister planting*. **semis au creux du sillon**

listing (middle breaking) See *tillage, listing (middle breaking)*. **forme de buttage ou billonnage**

lithic contact A boundary between soil and continuous, coherent, underlying material. The underlying material is sufficiently coherent to make digging with a spade impractical. If the underlying material is composed of a single mineral, its *hardness* is at least 3 (*Mohs scale*) and gravel-size chunks will not disperse in water or sodium hexametaphosphate solution with 15 hours shaking. See *paralithic contact*. **contact lithique**

lithic layer Bedrock under the control section of a soil. In *Organic* soils, bedrock occurring within a depth of 10 to 160 cm from the surface. A diagnostic layer in the Canadian system of soil classification. **couche lithique**

lithification, lithifaction The process of converting a sedimentary deposit into a solid rock by compaction and cementation. See *diagenesis*. **lithification**

lithiophorite A black manganese oxide, $(Al, Li)MnO_2(OH)_2$, common in iron-manganese nodules of acid soils. It has a layer structure. **lithiophorite**

litho- A prefix meaning rock or stone. **litho**

lithology The study of rock characteristics, particularly color, mineralogical composition, and grain size, by megascopic examination of samples or outcrop material. Also, the physical character of a rock. **lithologie**

lithophile element An element that is concentrated in the Earth's silicate crust rather than in its mantle or core. **élément lithophile**

lithophyte Vegetation able to flourish on a bare rock surface (e.g., lichens). See *lithosere*. **lithophyte**

lithosequence A group of related soils that differ from each other because of differences in the parent rock. **lithoséquence**

lithosere The successional development of a group of plant communities on bare rock faces or blocky scree slopes (i.e., relatively dry, hard, and exposed surfaces); a type of *xerosere*; the first colonizing plants are usually lichens and terrestrial algae. **lithosère**

lithosphere The solid portion or crust of the Earth's surface, comprising of a number of mobile plates. The plates from which the continents emerge are composed of primarily basaltic, granitic, and various sedimentary rocks. The lithosphere is distinguished from the *atmosphere, hydrosphere,* and the *biosphere*, as well as from the Earth's interior, i.e., its mantle and core. **lithosphère**

lithostratigraphic unit A body of rock consisting dominantly of a certain lithologic type or combination of types, or has other unifying lithologic features. It may be igneous, sedimentary, or metamorphic, and it may or may not be consolidated. **division lithostratigraphique**

lithostratigraphy The component of *stratigraphy* that deals with the lithology of strata, their organization into units based on lithologic character, and their correlation. **lithostratigraphie**

lithotroph A type of bacteria capable of obtaining metabolically useful energy from the oxidation of inorganic chemicals, chiefly ammonium, nitrite, iron, and various forms of sulfur. These bacteria obtain their carbon from carbon dioxide as do the green plants. **lithotrophe**

lithotrophy Metabolic mode in which oxidation of inorganic compounds provides the source of energy for metabolism. See *chemoautotrophy*. **lithotrophie**

litter The uppermost layer of organic materials on the soil surface containing freshly fallen or slightly decomposed plant material. **litière**

littering The process of soil formation whereby organic litter and associated humus accumulates to a depth of less than 30 cm on a mineral soil surface. **litiérage**

Little Ice Age A period of climatic cooling in the Northern Hemisphere between the mid-16th century and the mid-19th century, during which mountain glaciers re-advanced and the climate was characterized by cold winters. **petit âge glaciaire**

littoral zone An interface region or zone at the edge of aquatic ecosystems, between the land and a lake or the sea. The area of a lake or pond close to the shore; includes rooted plants. See *profundal zone*. **région ou zone littorale**

livestock Domestic animals produced or kept primarily for farm, ranch, or market purposes, including beef and dairy cattle, hogs, sheep, goats, and horses. **bétail**

load The quantity of a material that enters a water body over a given time interval. **charge**

loading (1) (*soil mechanics*) The increased weight brought to bear on the ground surface, an increase that might lead to failure or flowage of the stressed surface material. (2) (*waste management*) The addition of organic wastes to soils at a rate that enhances plant growth and helps meet fertility requirements of a soil. (3) (*statistics*) A term used in factor analysis denoting the degree to which a given variable is attached to a factor. **(1, 2) chargement (3) saturation**

loading capacity (*waste water management*) The greatest amount of chemical materials or thermal energy that can be added to a stream without exceeding water quality standards established for that stream. **capacité de charge**

loading schedule (*waste management*) A timetable indicating an agreed-upon mass or volume of a material to a water body for a given time interval. **calendrier d'épandage**

loam See *texture, soil*. **loam**

loamy Intermediate in texture and properties between fine-textured and coarse-textured soils. It includes all textural classes having loam or loamy as a part of the class name, such as clay loam or loamy sand. See *texture, soil*. **loameux**

loamy coarse sand See *texture, soil*. **sable grossier loameux**

loamy fine sand See *texture, soil*. **sable fin loameux**

loamy sand See *texture, soil*. **sable loameux**

loamy very fine sand See *texture, soil*. **sable très fin loameux**

lodgment till, lodgement till A *basal till* commonly characterized by compact *fissile* structure and containing stones oriented with their long axes generally parallel to the direction of ice movement. **till de fond**

L

loess Material transported and deposited by wind and consisting of a homogeneous, non-stratified, unindurated deposit made up predominantly of silt, with smaller amounts of very fine sand and/or clay. **loess**

log (1) (*geology*) A continuous record as a function of depth, usually graphic and plotted to scale on a narrow paper strip, of observations made on the rocks and fluids encountered in a *borehole*. (2) (*modeling*) An abbreviation of logarithm. **(1) diagraphie (2) log**

log growth phase (*microbiology*) That time period during which growth (cell division) of bacteria is occurring at a maximum rate. Same as *exponential phase*. See *growth phase*. **phase logarithmique de croissance**

logistic curve (*statistics*) A graph resembling a flattened S (i.e., sigmoid curve), obtained when certain phenomena (e.g., population size for a species) are plotted against time. The growth rate is high at the beginning, reaches a maximum during the early stages, starts to slow near the center, and declines to zero at the end. It may also be described as a modified exponential curve. **courbe logistique**

log-normal distribution (*statistics*) The distribution of a variable if the logarithms of the values of that variable have a normal distribution. **distribution log-normale**

lone pair (1) An electron pair that is localized on a given atom. (2) An electron pair not involved in bonding. **paire ionique libre**

longitude The angle at the pole between the meridian of the place and some standard meridian; in North America, the standard meridian is the one passing through Greenwich, England, and longitude is measured east or west from Greenwich. **longitude**

longitudinal dune A long, narrow sand dune, usually symmetrical in cross profile, oriented parallel to the direction of the prevailing wind; it commonly forms under conditions of strong and constant winds. Such dunes may be a few meters high, and up to 100 km long. **dune longitudinale**

longitudinal fault (*geology*) A fault whose strike is parallel with that of the general structural trend of the region. **faille longitudinale**

loosening (1) The increase in volume of voids in soils by plant, animal, and human activity, by freeze-thaw or other physical processes, and by removal of materials by leaching. (2) See *tillage, loosening*. **ameublissement**

lotic Pertaining to flowing water, such as streams and rivers. See *lentic*. **lotique**

lower plastic limit See *plastic limit*. **limite de plasticité inférieure**

lowland (1) Land that is lower and generally flatter than adjacent higher country. (2) A general term for extensive plains lying just above sea level. (3) Bottomland along a stream. **basse terre**

Luvic Gleysol A great group of soils in the *Gleysolic* order (Canadian system of soil classification) developed under wet conditions, under grass or forest or both. The soils have Aeg and Btg horizons. **Gleysol luvique**

Luvisolic An order of soils in the Canadian system of soil classification that have eluvial (Ae) horizons, and illuvial (Bt) horizons in which silicate clay is the main accumulation product. The soils developed under forest or forest-grassland transition in a moderate to cool climate. **Luvisolique**

luxury uptake The absorption by plants of nutrients in excess of their requirements for optimum growth; a tendency that occurs to some degree with any nutrient but is especially evident with potassium and nitrogen. Luxury concentrations during early growth may be used in later growth. **consommation de luxe**

lyotropic series The sequence of relative attractive forces between cations and *micelles,* reflecting both the number of charges and the hydrated sizes, which is $Al^{3+} > Ca^{2+} > Mg^{2+} > K^+ = NH_4^+ > Na^+$. This series may be considered as a preferential adsorption sequence. *Cation-exchange capacity* sites adsorb more of an ion early in the sequence than later in the sequence if the two are present in equal amounts. **série lyotropique**

lysimeter (1) A device for measuring percolation and leaching losses of water and solutes from a column of soil under controlled conditions. (2) A device for measuring gains (i.e., precipitation and condensation) and losses (i.e., evapotranspiration) of water by a column of soil. **lysimètre**

lysis Rupture or dissolution of a cell. **lyse**

lysosome A membrane-enclosed intracellular vesicle containing hydrolytic enzymes; the primary intracellular constituent involved in digestion in animal and some plant cells. Collectively, lysosomes can hydrolyze all classes of macromolecules. **lysosome**

M

macro-aggregate Soil *aggregates* that are greater in size than 250 μm, generally composed of clay, silt, and sand particles, *clay-organo complexes*, *micro-aggregates*, and organic matter. Temporary stabilization is provided for these large aggregates by plant roots and activity of soil fungi and fauna. **macroagrégat**

macroconsumer A class of heterotrophic organisms that ingests other organisms or particulate organic matter (e.g., *herbivores*, carnivores, and *omnivores*). **macroconsommateur**

macroelement See *macronutrient.* **élément majeur**

macrofauna, soil Soil animals that are >1000 μm in length (e.g., vertebrates, earthworms, and large arthropods). **macrofaune du sol**

macroflora See *macrophyte.* **macroflore**

macrofossil A *fossil* large enough to be studied without the aid of a microscope. **macrofossile**

macronutrient A chemical element needed in large amounts, usually greater than 1 mg kg^{-1} in the plant, for the growth of plants, usually applied artificially in fertilizer or liming materials. The prefix macro refers to the quantity and not to the essentiality of the element to the plant. See *micronutrient.* **élément majeur**

macroorganic matter Fragments of soil organic matter greater than the lower size limit of sand (i.e., >20 m or >53m). Macroorganic matter is isolated by sieving a dispersed soil, and consists primarily of partially decomposed plant residues but also may contain fungal *hyphae*, spores, seeds, charcoal, and animal remains. Often called particulate organic matter. See *particulate organic matter.* **matière organique particulaire ou grossière**

macroorganism A plant, animal, or fungal organism visible to the unaided eye without the aid of magnification. See *microorganisms.* **macroorganisme**

macrophyte Any plant visible to the unaided eye (e.g., aquatic mosses, ferns, liverworts, and rooted plants). Also called *macroflora.* **macrophyte**

macropore See *pore-size distribution.* **macropore**

macropore flow The tendency for water applied to the soil surface at rates exceeding the upper limit of *unsaturated hydraulic conductivity* to move into the soil profile mainly via *saturated flow* through macropores. See *preferential flow.* **écoulement préférentiel**

macroporosity The volume of large pores in the soil. **macroporosité**

macroscopic Capable of being seen with the naked eye. **macroscopique**

made land Areas filled with earth, or earth and trash mixed, usually by or under the control of humans. A miscellaneous land type. **terrain anthropique, terre rapportée**

mafic Descriptive term for dark-colored minerals, commonly found in basic igneous rocks. It is a term derived from magnesian and iron-rich (i.e., *ferromagnesian*) minerals. **mafique**

maghemite A dark reddish-brown, magnetic iron oxide mineral (Fe_2O_3) chemically similar to hematite but structurally similar to magnetite. Often found in well-drained, highly weathered soils of tropical regions. **maghémite**

magma Molten rock found in the *mantle*, beneath the crust of the Earth, which when forced toward the surface cools and solidifies to become igneous rock. **magma**

magnesian limestone A limestone containing a small proportion (<10%) of magnesium carbonate ($MgCO_3$). See *dolomite*. **calcaire magnésien, dolomite**

magnesite Magnesium carbonate ($MgCO_3$), one of the minerals of the carbonate group. **magnésite**

magnetic anomaly A positive or negative departure from the predicted value of the Earth's magnetic field, measured at a particular point on the ground surface and resulting from a local disturbance such as a concentration of magnetic materials. **anomalie magnétique**

magnetic declination The acute angle between magnetic north and *true north*, expressed in degrees W and E of true north. Also called magnetic variation. **déclinaison magnétique**

magnetic inclination The acute angle between the vertical and the direction of the Earth's magnetic field in the plane of the magnetic meridian. It is often called the magnetic dip. **inclinaison magnétique**

magnetic poles The two specific points of the Earth's magnetic field at which the lines of magnetic force are vertical. The magnetic poles slowly change positions and do not coincide with the geographical poles. **pôles magnétiques**

magnetite A black, magnetic iron oxide mineral (Fe_3O_4), usually inherited from igneous rocks; often found in soils as black magnetic sand grains. **magnétite**

magnetometer An instrument used to measure the direction and intensity of the Earth's magnetic field. **magnétomètre**

management plan Detailed long-term plan for resource management in an area; contains inventory and other resource data. **plan directeur**

mangan (*soil micromorphology*) A *cutan* containing enough manganese oxide or hydroxide to effervesce upon application of hydrogen peroxide. **mangane**

manganese oxides Oxides of manganese, typically black and frequently occurring in soils as nodules and coatings on ped faces, usually in association with *iron oxides*. Birnessite and lithiophorite are common manganese oxide minerals in soils. **oxydes de manganèse**

mangrove (1) A vegetation type characterizing a coastal swamp of brackish or saline water in which specially adapted trees form a dense swamp forest. (2) Tropical tree species adapted to live on saline muds in the tidal zone of creeks and estuaries (e.g., *Rhizophora*, *Avcennia*, *Laguncularia* spp.). **mangrove**

manometer An instrument for measuring pressure, consisting of a transparent U-shaped tube filled with a liquid such as water or mercury. **manomètre**

mantle The layer of the Earth between the *crust* and the *core*, with its upper boundary marked by the *Mohorovicic discontinuity* and its lower boundary by the *Gutenberg discontinuity*. It consists of ultrabasic rocks with densities of 3.0 to 3.3 g cm^{-3}. **manteau**

manure The excreta of animals, with or without the admixture of bedding or litter, in varying stages of decomposition. It is also called barnyard manure or stable manure. This is the usual meaning in North America. In some countries manure is used to refer to any fertilizer. **fumier, fumure, engrais**

map, electronic A collection of geographically referenced data stored in a file and capable of being displayed on a graphics screen or plotted in hard copy form. **carte numérique**

map projection A presentation on a planar surface of the system of lines representing parallels and meridians on the surface of the Earth or a section of the surface of the Earth; determined by means of geometrical construction or mathematical computation. **projection cartographique**

map resolution The accuracy with which the location and shape of map features are depicted for a given map scale. **résolution cartographique**

map scale A statement of a measure on the map and the equivalent measure on the Earth, often expressed as a representative fraction of distance (e.g., 1:24,000). **échelle cartographique**

map series A set of maps, usually of uniform scale and style, referring to a specific area and identified by a series number. **série de cartes**

map unit, soil (1) A conceptual group of one to many delineations identified by the same name in a soil survey that represent similar landscape areas comprising the same kind of component soil, plus inclusions; two or more kinds of component soils, plus inclusions; component soils and miscellaneous area, plus inclusions; two or more kinds of component soils that may or may not occur together in various delineations but all have similar, special use and management, plus inclusions; or a miscellaneous area and included soils. (2) Sometimes called a delineation. See *delineation, soil complex, soil association, undifferentiated, miscellaneous land type*. **unité cartographique**

map, large-scale A map having a scale of 1:100,000 or larger. **carte à grande échelle**

map, medium-scale A map having a scale from 1:100,000, exclusive, to 1:1,000,000, inclusive. Also called intermediate-scale. **carte à moyenne échelle**

map, small-scale A map having a scale smaller than 1:1,000,000. **carte à petite échelle**

map, soil A map showing the distribution of soil types or other soil mapping units related to the prominent physical and cultural features of the Earth's surface. **carte pédologique**

Five kinds of soil maps are as follows:

generalized soil map – a small-scale soil map showing the general distribution of soils within a large area in less detail than on a detailed soil map. ***carte pédologique généralisée ou de synthèse***

schematic soil map – a soil map compiled from scant knowledge of the soils of new and undeveloped regions by applying available information about the soil-formation factors of the area. The scale is usually 1:500,000 or smaller. See *soil formation factors*. ***carte pédologique schématique***

exploratory soil map – a map, with less detail than the broad reconnaissance map, showing the general distribution of soils determined by traverses at intervals of 10 to 20 km. The scale is generally less than 1:100,000. ***carte pédologique exploratoire***

broad reconnaissance soil map – map, with less detail than the reconnaissance map, showing the general distribution of soils determined by traverses at intervals of 4 to 10 km. The scale is generally 1:50,000 to 1:300,000. ***carte pédologique de reconnaissance générale***

reconnaissance soil map – a map showing the distribution of soils and physical features as determined by traversing the area at intervals of 2 to 4 km. The scale, depending on the detail required, is 1:20,000 to 1:200,000. ***carte pédologique de reconnaissance ou semi-détaillée***

detailed soil map – a soil map showing the boundaries between soil types or complexes of intimately associated soil types. The scale of the map depends on the purpose of the map, the intensity of land use, the pattern of soils, and the scale of other cartographic materials available. Traverses are usually made at 0.5 to 2-km intervals, or more frequently. The scale commonly used for field mapping is 1:10,000 to 1:40,000. ***carte pédologique détaillée***

very detailed soil map – a soil map showing the boundaries between soil types, mainly simple units. This map is used for very intensive planning requiring appraisal of soil resources of very small areas. Traverses are usually made at less than 0.5-km intervals, or more frequently. The scale commonly used for field mapping is less than 1:10,000. ***carte pédologique très détaillée ou intensive***

mapping unit See *map unit, soil.* **unité cartographique**

marble A metamorphic limestone, consisting of *calcite* and/or *dolomite,* formed by the alteration of sedimentary limestone by regional or contact metamorphism. It is widely used as a decorative building stone and in sculpture. **marbre**

marginal land Land that has various limitations (e.g., poor drainage, low fertility) for crop growth or agriculture in general. In economic terms, it provides barely enough return to meet expenses related to a specific use. **terre marginale, terre à faible potentiel agricole**

marine (1) Pertaining to the sea. (2) Descriptive of materials that have settled from suspension in salt or brackish waters, or have accumulated at the margins of these waters through shoreline processes such as wave action and longshore drift. **marin**

marine biome The aquatic *biome* consisting of waters containing 3.5% salt on average; includes the *oceans* and covers more than 70% of the Earth's surface; divided into *benthic* and *pelagic* zones. **biome océanique**

marine terrace (1) A narrow coastal strip formed of deposited material, sloping gently seaward. (2) A wave-cut platform that has been exposed by uplift along a seacoast or by lowering of sea level. (3) A flat, narrow surface that drops off to a steep embankment, formed by the merging of a wave-cut platform and a wave-built terrace. **terrasse marine**

Markov chain A *system* comprising a sequence of possible states such that the conditional probability of transition from one state to another is independent of the way in which the former state was achieved. The statistical model of the Markov chain has been used to describe, but not explain, how in many parts of the world there is a tendency for dry weather to occur in spells, with the occurrence of a wet day often being independent of the previous weather conditions. **chaîne de Markov**

Markov process Any process that generates a *stochastic* series, which will have a statistical influence over any finite-length sequence. **processus de Markov**

marl A soft, unconsolidated earthy deposit consisting of calcium carbonate or magnesium carbonate, or both, and often shells, usually mixed with varying amounts of clay or other impurities. **marne**

marsh (1) Wetland periodically inundated by standing or slowly moving nutrient-rich water, with a substratum of mineral or peat material. Surface water levels may fluctuate seasonally, with declining levels exposing drawdown zones of matted vegetation or mud flats. The associated soils are dominantly *Gleysols* with some *Humisols* and *Mesisols*. Marshes characteristically show a zonal or mosaic surface pattern of vegetation consisting of unconsolidated grass and sedge sods, frequently interspersed with channels or pools of open water. Marshes may be bordered by peripheral bands of trees and shrubs, but the predominant vegetation consists of a variety of emergent non-woody plants (e.g., rushes, reeds, and sedges). Where open water areas occur, a variety of submerged and floating aquatic plants flourish. (2) One of five *wetland classes* in the *Canadian wetland classification system*. **marais**

marsh gas Gas produced during the decomposition of organic material buried in wetland soils. The primary gas is methane. **gaz des marais**

marsh, tidal A low, flat area traversed by interlacing channels and tidal sloughs and periodically inundated by high tides; vegetation is usually salt-tolerant plants. **marais tidal**

mass flow (nutrient) The movement of solutes associated with the net movement of water. See *figure*. **écoulement massique**

mass flow rate The rate at which a liquid or gas flows, calculated as m = Qρ, where m is the mass flow rate, Q is the volumetric flow rate, and ρ is the fluid density, expressed as mass per unit time. **débit massique**

mass movement Dislodgement and downslope transport of soil and rock material as a unit under direct gravitational stress. The process includes slow displacements (e.g., *surface creep*) and rapid movements (e.g., slides and slumps). Fluid transport (water, ice, air) may play an important role. See *figure*. **mouvement de masse**

mass number The number of protons plus the number of neutrons in the nucleus of an atom. This number is also the approximate atomic weight of an atom. See *atomic weight*. **nombre de masse**

mass spectrometer An instrument used to determine the relative masses of atoms by the deflection of their ions in a magnetic field. **spectromètre de masse**

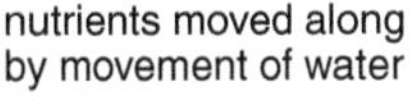

Mass flow.

mass transport The carrying of material in a moving medium such as water, air, or ice. See *mass wasting*. **transport en masse**

mass wasting The downslope movement of soil and rock material under the direct influence of gravity. The debris removed is not carried within, on, or under another medium. **mouvement en masse**

mass wetness A measure of *soil wetness*. The mass of water relative to the mass of dry soil particles (usually at 105°C). Sometimes called gravimetric water content. **teneur en eau gravimétrique ou massique**

massive (1) (*pedology*) A kind of soil structure within the structureless category

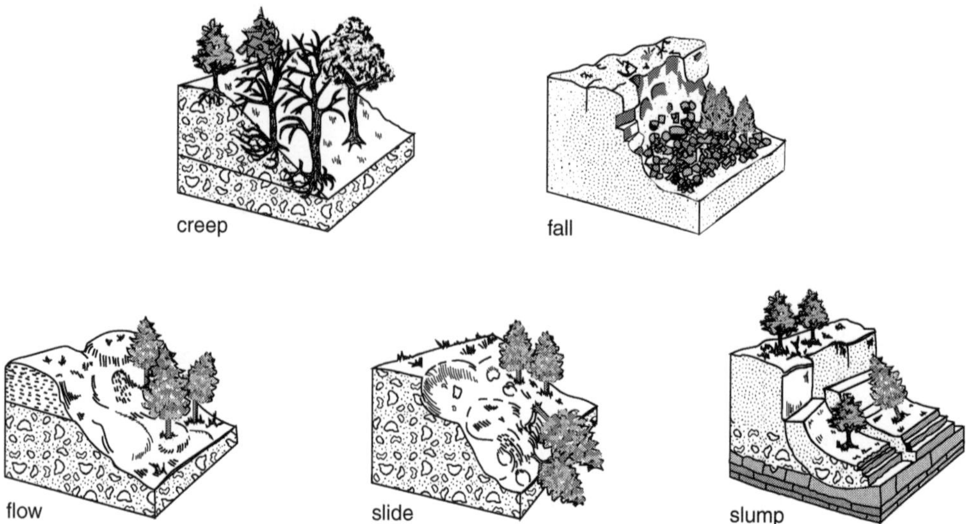

Mass movement of rock, soil, and organic matter initiated by creep, fall, flow, slide, and slump (adapted from Dunster and Dunster, 1996).

M

of soil structure classification. Massive soil material is a coherent mass showing no evidence of any distinct arrangement of soil particles. See *soil structure*. (2) (*geology*) Descriptive term for a very thick rock unit without stratification, jointing, foliation, cleavage, or flow-banding. **(1) structure massive (2) homophane**

mast Fruits and seeds of shrubs, woody vines, trees, cacti, and other non-herbaceous vegetation available for animal consumption. **paisson**

matching The act by which detail or information on the edge, or overlap area, of a map or chart is compared, adjusted, and corrected to agree with the existing overlapping chart. Also called edge matching. **raccordement**

material blank Samples of construction materials, such as those used in groundwater wells and pump and flow testing, to document decontamination (or measure artifacts) from use of these materials. **blanc de matériau**

matran (*soil micromorphology*) A *cutan* that contains s-matrix skeletal grains within the plasma concentration. **matrane**

matric potential The amount of work that must be done per unit quantity of soil water in order to transport it reversibly and isothermally from a pool at the elevation and the external gas pressure of the point under consideration to the soil water. The matric potential results from *capillarity* and adsorptive forces. See *soil water*. See *figure*. **potentiel matriciel**

matrix (*geology, pedology*) The natural material in which any metal, fossil, pebble, crystal, or other contrasting material is embedded. (*statistics*) An ordered array of numbers on which certain operations may be performed. **matrice**

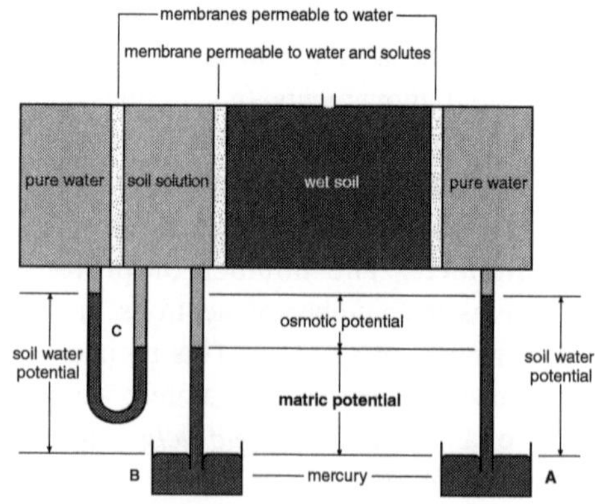

Relationship among osmotic (C), **matric** (B), and combined (A) soil water potential (adapted from Brady and Weil, 1996).

maturity stage (*botany*) The stage of development when a plant is fully developed or ripe. **stade de maturité**

maximum sustainable yield (*crops*) The maximum crop yield that can be sustained for a specific soil type and climate without undue harm to the environment. (*animals*) The maximum number of animals that may be harvested without initiating a long-term decline in population numbers. (*natural resources*) The greatest amount of a renewable natural resource (e.g., forests or wildlife) that can be removed without diminishing the continuing production and supply of the resource. **rendement optimal, rendement maximal durable**

M-discontinuity See *Mohorovicic discontinuity*. **discontinuité M ou de Mohorovicic**

meadow (*ecology*) An opening in a forest, generally at higher elevations, with high production of herbaceous plants usually resulting from high soil water content or a perched water table. (*agriculture*) An area of perennial, herbaceous vegetation, usually grass or grasslike, used primarily for hay production. **pré**

mean (*statistics*) See *arithmetic mean*. **moyenne**

mean deviation (*statistics*) The arithmetic mean of the absolute deviations from the mean of a set of data. **écart moyen**

mean velocity (*hydrology*) The average velocity (i.e., the flow rate or discharge) divided by the cross-sectional area. **vitesse moyenne**

meander (*geomorphology*) One of a series of somewhat regular loop-like bends, twists, or curves in the course of a stream, developed when the stream shifts its course laterally toward the convex sides of the original curves. **méandre**

meander scar (*geomorphology*) A crescentic cut in a bluff or valley wall, produced by sideward cutting of a meandering stream and indicating its former route. It is partly infilled by slumping of the channel walls and vegetation growth; called an oxbow lake if it contains water. **bras mort**

measure of central tendency A statistical parameter which indicates the degree of clustering within the distribution of a sample. See *central tendency, distribution, mean, median, mode*. **mesure de la tendance centrale**

measuring weir A shaped notch through which water flows are measured. **déversoir**

mechanical analysis See *particle-size analysis* and *particle-size distribution*. **analyse mécanique**

mechanical weathering (*geology*) The process of *physical weathering* of rocks by *exfoliation*, *freeze-thaw action*, *granular disintegration*, and other forces. All the processes of mechanical weathering are distinct from those of *chemical weathering*. **désagrégation mécanique**

mechanistic The viewpoint that the behavior of all biological phenomena may be explained in mechanico-chemical terms. **mécaniste**

medial moraine, median moraine (*geology*) Linear, often ice-cored, accumulation of rubbly material extending down the center of a glacier, varying in width from a narrow ridge to a broader spread of morainic material. It is caused by the merging of two *lateral moraines* from the point at which two glaciers unite. **moraine médiane**

median (*statistics*) The central value in an ordered series of values, with an equal number of values occurring above and below it. **médiane**

Mediterranean scrub biome See *chapparal*. **biome de maquis méditerranéen**

medium A material or substance that acts as a carrier for some compound or chemical (e.g., air, water, soil, food). (*microbiology*) A sterile mixture, either solid or liquid, of nutrients used for the cultivation of bacteria or fungi. Solid media are prepared using *agar* as the agent to solidify the liquid solution containing required nutrients. **milieu**

medium texture An intermediate class between fine-textured and coarse-textured soils that includes the following textural classes: very fine sandy loam, loam, silt loam, and silt. **texture moyenne**

megafauna Animals, living or fossil, large enough to be seen and studied with the naked eye. **mégafaune**

meiosis The process of nuclear division in a cell by which the chromosomes are reduced to half their original number; occurs during the formation of sex cells. **méiose**

Melanic Brunisol A great group of soils in the *Brunisolic* order (Canadian system of soil classification). The soils have mull Ah horizons thicker than 5 cm and base-saturated Bm horizons. **brunisol mélanique**

melanic epipedon A surface layer of soil (U.S. system of soil taxonomy) having *andic* soil properties. **épipédon mélanique**

melanization A soil forming process in which light-colored mineral materials are darkened by admixture of organic matter (as in Ah horizons). **mélanisation**

melting point The temperature at which a solid changes to a liquid. The temperature will vary, and is consistent at equal temperatures and pressures for each element or solid. **point de fusion**

meltwater Water resulting from the melting of snow or glacier ice. **eau de fonte**

meltwater channel (*geology*) A channel cut into solid rock or drift deposits by water flow in areas of former glaciation, but unrelated to the present drainage system. It may or may not currently carry a stream. It frequently cuts across current drainage divides, has steep sides, has fairly constant width, and is often associated with *deltas, kame terraces,* or *eskers* at its intake end, and with outwash fans, *eskers,* and *kames* at its distal end. **chenal d'eau de fonte**

membrane filter A flat modified cellulose disk, which has a small pore size (e.g., millipore filter), used to recover and count bacteria or other organisms and substances in samples of liquid substances, such as water. **filtre à membrane**

mercaptans Organic compounds that contain the –SH group. They are compounds that are sulfur analogs of alcohols and phenols and are named according to the parent hydrocarbon (e.g., the chemical formula of methyl alcohol is CH_3OH, and that of methyl mercaptan is CH_3SH). Mercaptans are noted for their odor and are used to impart odor to natural gas in public gas supplies; air emissions from chemical manufacturers and pulp and paper mills may contain mercaptans. **mercaptans**

meridian One of a group of abstract lines on the Earth's surface formed by the longitude and latitude coordinate system. Meridians represent lines of equal longitude and thus converge at the poles. All meridians are great circles. **méridien**

meristem The region of active cell division in plants located either at the stem and root tips (*apical meristem*), where they are responsible for the primary growth of plant, or laterally in stems and roots (lateral meristems). The cells so formed then become modified to form the various tissues such as the epidermis and cortex. **méristème**

meromictic lake A lake fed by water from mineral springs that supply dissolved substances; there is a permanent density gradient or stratification and little nutrient exchange between the upper and lower regions of the water. See *dimictic lake.* **lac méromictique**

meroplankton Temporary planktonic stage in the life cycle of many marine organisms. Organisms that spend only a portion of their life as plankton, ultimately becoming nekton or benthos. See *nekton, benthos.* **méroplancton**

mesa (*geomorphology*) A tableland or isolated flat-topped hill, bounded on at least one side by steep slopes or cliffs, usually composed of nearly horizontal strata of bedrock. **mesa**

mesic (1) A soil temperature regime that has mean annual soil temperature of 8°C or more but <15° C, and >5°C difference between mean summer and mean winter soil temperatures at 50 cm below the surface. (2) Mesic peat is organic material at a stage of decomposition between that of fibric and humic materials; peat soil material with >10% and <40% rubbed fibers. Mesic material usually is classified in the *von Post humification scale* of decomposition as class 5 or 6. **mésique**

mesic layer A layer of organic material at an intermediate stage of decomposition between that of the *fibric* and *humic layers*. **couche mésique**

Mesisol A great group of soils in the *Organic* order (Canadian system of soil classification) that are saturated for most of the year. The soils have a dominantly mesic middle tier, or middle and surface tiers, if a terric, lithic, hydric, or cryic contact occurs in the middle tier. **mésisol**

mesobiota See *mesofauna*. **mésobiote**

mesofauna, soil The soil-dwelling animals (200 to 1000 µm long) large enough to cause disturbance of the soil pores when moving through soil (e.g., nematodes, mites, collembola, enchytraeids, ants, insect larvae). **mésofaune**

mesophile An organism growing best at moderate temperatures of 25 to 40°C. **mésophile**

mesophyll (*botany*) The leaf cells that contain chloroplasts and are located between the upper and lower epidermis. The mesophyll is a *parenchyma* tissue, the main assimilation tissue of plants. **mésophylle**

mesophyte A plant that requires only moderate amounts of water for growth. **mésophyte**

mesosphere See *atmosphere*. **mésosphère**

mesotrophic Containing a moderate amount of plant nutrients. See *eutrophic, oligotrophic*. **mésotrophe**

metabolism The sum of all chemical processes occurring within an organism including the synthesis (anabolism) and breakdown (catabolism) of organic compounds. **métabolisme**

metabolite A chemical substance produced by the metabolic reactions of an organism. **métabolite**

metal An element that gives up electrons relatively easily and is lustrous, malleable, and a good conductor of heat and electricity. **métal**

metalloids (semimetals) Elements along the division line in the periodic table between metals and nonmetals. These elements exhibit both metallic and nonmetallic properties. **métalloïdes**

metallurgy The process of separating a metal from its ore and preparing it for use. **métallurgie**

metamorphic rock Rock derived from preexisting rocks, but differing from them in physical, chemical, and mineralogical properties as a result of natural geological processes, principally heat and pressure, originating within the Earth. The preexisting rocks may have been igneous, sedimentary, or another form of metamorphic rock. **roche métamorphique**

metamorphism The processes by which rocks are altered in their mineralogy, texture, and internal structure due to external sources of heat, pressure, and the introduction of new chemical substances, rather than by changes induced solely by burial, as in *diagenesis*. **métamorphisme**

meteorite A solid body that reaches the Earth's surface from an extraterrestrial source despite the frictional effect of the atmosphere, which consumes the majority of meteorites. It is usually composed of various proportions of a nickel-iron alloy (typically 10% nickel and 90% iron) and silicate minerals. **météorite**

meteorology The science of weather-related phenomena. **météorologie**

methane (CH_4) (*chemistry*) A hydrocarbon molecule produced through anaerobic decomposition of waste in landfills, animal digestion, decomposition of animal wastes, production and distribution of natural gas and oil, coal production, and incomplete fossil fuel combustion. (*meteorology*) A *greenhouse gas*; its atmospheric concentration is increasing by about 0.6% per year, but this increase may be stabilizing. **méthane**

methanotroph An organism capable of oxidizing *methane*. **méthanotrophe**

M

methemoglobinemia State of oxygen starvation produced, especially in babies, when nitrite is absorbed into the bloodstream from the digestive tract; it impairs the ability of hemoglobin to transport oxygen. **méthémoglobinémie**

methyl mercury One of the common alkyl mercury compounds in which an atom of mercury is bonded to a methyl (CH_4) group; produced by a variety of organisms (chiefly bacteria) in the natural environment following contamination by ionic forms of inorganic mercury. Methyl mercury is absorbed by a variety of animals and undergoes bioaccumulation within the food chain. **méthyle de mercure**

mica A layer-structured aluminosilicate mineral group of the 2:1 type characterized by its non-expandability and high layer charge, which is usually satisfied by potassium. The major types are muscovite, biotite, and phlogopite. **mica**

mica schist See *schist*. **schiste micacé, micaschiste**

micaceous (1) Consisting of, containing, or pertaining to mica (e.g., a micaceous sediment). (2) Resembling mica, especially in terms of the capability of another mineral or other substance to be split into thin sheets. **micacé**

micelle (1) An electrically charged particle of *clay* and *humus*. Micelles contain most of the negative charges in soils but only a small portion of the positive charges. Micelles can be considered polyanions, because they carry large numbers of negative charges. (2) An aggregate of fatty acid anions having their hydrophobic tails in the interior and their polar heads pointing outward to interact with the polar water molecules. **micelle**

Michaelis-Menton equation A hyperbolic equation used to describe the kinetics of enzymatic activity. See k_m. **équation de Michaelis-Menton**

microaerophile A microorganism growing best in the presence of small amounts of atmospheric oxygen. **micro-aérophile**

micro-aggregate Soil *aggregates* that range in size from 2 to 250 μm composed of clay and silt particles, *clay-organo complexes*, and organic matter stabilized by microbial materials (e.g., *polysaccharides*, hyphal fragments, and bacterial cells or colonies). See *macro-aggregate*. **microagrégat**

microbe Microscopic organisms are classed generally as microbes or microorganisms, in contrast to larger organisms called macrobes. Microbes can be divided into microfauna such as *protozoa* and microflora such as *bacteria, actinomycetes,* and *fungi*. **microbe**

microbial biomass The total mass of living microorganisms in a given volume or mass of a particular environment such as soil. **biomasse microbienne**

microbial load The total number of bacteria and fungi in a given quantity of water or soil, or on the surface of food. The presence of the bacteria and fungi may not be related to the presence of disease-causing organisms. **charge microbienne**

microbial mat A community of microorganisms forming flat, cohesive muddy, layered structures living precursors of stromatolites. **feutre ou mat microbien, natte microbienne**

microbial oxidation Process in which soil microorganisms "burn up" organic matter in the soil during their normal metabolism. **oxydation biologique**

microbial population Total number of living microorganisms in a given volume or mass of soil. **population microbienne**

microbiology The study of organisms that can be seen only with the aid of a microscope. The science deals with the structure and chemical composition of various microbes, the biochemical changes within the environment that are caused by members of this group, the diseases caused by microbes, and the reactions of animals, including humans, to their presence. **microbiologie**

microbiology, soil The branch of soil science that deals with soil-inhabiting microorganisms and their relationship to both plant and animal growth. **microbiologie du sol**

microbiota The plants, animals, and microorganisms that can only be seen with the aid of a microscope. **microbiote**

microclimate (1) The climate of a small area resulting from the modification of the general climate by local differences in elevation or exposure. (2) The sequence of atmospheric changes within a very small region. **microclimat**

microcline A white, gray, brick-red, or green mineral ($KAl–Si_3O_8$) of the alkali feldspar group. It is the fully ordered, triclinic modification of potassium feldspar and is dimorphous with orthoclase, being stable at lower temperatures; it usually contains sodium in minor amounts. Microcline is a common rock-forming mineral of granitic rocks and pegmatites, and is often secondary after orthoclase. **microcline**

microcosm (1) A laboratory model of an ecosystem in which environmental variables and physical, chemical, and biological factors can be manipulated to observe the response; the results of microcosm studies may not always apply to an actual ecosystem because of the simplification of system components. (2) A miniature habitat. **microcosme**

microcrystalline A rock texture characterized by crystals that are visible only under the microscope; also said of a rock with such a texture. See *cryptocrystalline*. **microcristallin**

microelement See *micronutrient*. **élément mineur**

microfauna, soil (*zoology*) The small animals (generally <200 μm long) that can only be identified with a microscope (e.g., protozoa, nematodes, and arthropods). **microfaune**

microflora Bacteria (including actinomycetes), fungi, algae, and viruses. **microflore**

microfossil A *fossil* too small to be identified without the aid of a microscope. **microfossile**

microgranite An acid *igneous* rock with the mineralogical and chemical properties of a *granite* but having a medium-grained texture. **microgranite**

microhabitat, soil (1) A small, usually distinctly specialized and effectively isolated habitat in soil. (2) Clusters of microaggregates with associated water within which microbes function. May be composed of several *microsites* that can be aerobic or anaerobic. **microhabitat du sol**

micrometeorology The study of small-scale atmospheric processes, particularly those operating near the ground surface, where large varieties of *microclimate* are found because of the different soils, crops, vegetation, surface textures, and terrain characteristics. **micrométéorologie**

micronutrient A chemical element needed in only small amounts by plants and animals, usually less than 1 mg kg^{-1} in the plant, for growth and health (e.g., boron, molybdenum, copper, iron, manganese, and zinc). The prefix "micro" refers to the amount, not the essentiality of the element to the organism. See *macronutrient.* **oligo-élément**

microorganism Living organism too small to be seen with the naked eye (<0.1 mm); includes bacteria, fungi, protozoans, microscopic algae, and viruses. **micro-organisme**

microplankton Plankton ranging in size from 60 μm to 1 mm, including most phytoplankton. **microplancton**

microrelief Small-scale, local differences in relief, including mounds, swales, or hollows, generally only a few meters in diameter and with elevation differences of up to 2 meters. **microrelief**

microscopic An object or organism visible only under the microscope. Objects that can be seen with the naked eye are called *macroscopic.* **microscopique**

microsere A series of successional ecologic stages that occur within a microhabitat (e.g., a tree stump). **microsère**

microsite, soil A small volume of soil where biological or chemical processes differ from those of the soil as a whole (e.g., an anaerobic microsite of a soil aggregate or the surface of decaying organic residues). **microsite du sol**

microwave A very short electromagnetic wave. Any wave between 1 meter and 1 millimeter in wavelength or 300 to 0.3 gigahertz in frequency; the portion of the electromagnetic spectrum in the millimeter and centimeter wavelengths, bounded on the short wavelength sides by the far infrared (at 1 mm) and on the long wavelength side by very high-frequency radio waves. Passive systems operating at these wavelengths sometimes are called microwave systems; active systems are called radar, although the literal definition of radar requires a distance measuring capability not always included in active systems. The exact limits of the microwave region are not defined. **micro-onde, hyperfréquence**

midden A refuse heap marking the site of previous habitation. **tertre, butte-témoin**

milk stage (*botany*) In grain (seed), the stage of development following pollination, in which the endosperm appears as a whitish liquid that is somewhat like milk. **stade laiteux**

mine Any opening in, excavation in, or working of the Earth's surface or subsurface for the purpose of recovering, opening up, or proving mineral ore, coal, a coal-bearing substance, oil sands, or an oil sands-bearing substance; it includes any associated infrastructure. See *pit, quarry.* **mine**

mine drainage Water pumped or flowing from a mine. **drainage de mine**

mine dump Area covered with overburden and other waste materials from ore and coal mines, quarries, and smelters, and usually having little or no vegetative cover prior to reclamation. See *spoil.* **terril**

mine wash Water-deposited accumulations of sandy, silty, or clayey material recently eroded in mining operations. It may clog streams and channels, and damage land on which it is deposited. **boue de mines**

mined land Land with new surface characteristics due to the removal of a mineable commodity. **terrain exploité**

mineral (1) A naturally occurring homogeneous solid, inorganically formed, with a definite chemical composition and an ordered atomic arrangement.

(2) An economic *mineral deposit* exploited by mining. **minéral**

mineral deposit A mass of naturally occurring mineral material, usually of economic value (e.g., metal ores, nonmetallic minerals). **dépôt, gisement**

mineral soil A soil consisting predominantly of, and having its properties determined by, mineral matter. Contains <17% organic carbon (30% organic matter) by weight, but may contain an organic surface layer up to 30 cm thick. **sol minéral**

mineralization (*biochemical*) The conversion of an organic substance to an inorganic form as a result of microbial decomposition. Gross mineralization is the total amount converted, and net mineralization, the gross minus the growth demand or immobilization by the decomposer organisms, usually measured in laboratory incubations. (*geology*) The process whereby minerals are introduced into preexisting rocks, in the form of veins, resulting in ore deposit. (*geology*) A process of fossilization by which organic components are replaced by inorganic material. **minéralisation**

mineralogical analysis The estimation or determination of the kinds or amounts of minerals present in a rock or a soil. **analyse minéralogique**

mineralogy The study of minerals in terms of their formation, occurrence, composition, classification, and properties (e.g., crystal form, color, cleavage, fracture, hardness, luster, specific gravity, streak, and transparency). **minéralogie**

mineralogy, soil The subspecialization of soil science dealing with the kinds and proportions of minerals present in a soil. **minéralogie des sols**

minerotrophic A supply of water to vegetation, originally derived from mineral soils or rocks but sometimes via lakes or rivers as intermediates; it may be *eutrophic*, *mesotrophic*, or *oligotrophic*. **minérotrophe**

minesoil Soil produced by mining and reclamation activities that is capable of supporting plant growth. See *reconstructed profile*. **sol minier**

minimum data set A collection of quantitative information on a minimal number of key properties or attributes that can be used to assess a soil process in *soil quality* assessment. **ensemble de données minimal**

minimum tillage See *tillage, minimum tillage*. **travail minimal du sol**

mining The process of obtaining useful minerals from the Earth's crust; includes both underground excavations and surface workings. Types include: area mine — surface mining performed on level to gently rolling topography; auger mine — a mining method in which holes are drilled into the side of an exposed mineral seam and the mineral transported along an auger bit to the surface; contour surface mine — the removal of overburden in steep or mountainous areas and mining from a mineral seam that outcrops or approaches the surface at about the same elevation; hydraulic mine — mining by washing sand and dirt away with water, leaving the desired mineral; open-pit mine — mining of metalliferous ores by surface mining methods; surface mine — mining method in which the overlying materials are removed to expose the mineral seam for extraction; underground mine — mining method in which the minerals are extracted without removing the overlying materials. **extraction minière, exploitation minière**

mining by-products Any of the spoils, tailings, and slag resulting from mining and processing of minerals. **résidus miniers**

minor element See *micronutrient*. **élément mineur**

mire (1) In general, all kinds of *peatlands* and peatland vegetation (*bog* and *fen*). (2) A section of wet, swampy ground; bog; marsh; wet, slimy soil of some depth; deep mud. **tourbière, milieu tourbeux, bourbier**

miscellaneous land type (*soil survey*) A mapping unit for areas of land that have little or no natural soil, or that are inaccessible for orderly examination, or where, for some reason, it is not feasible to classify the soil (e.g., rough mountainous land, eroded slopes, marshes). **type de terrain divers**

miscible The ability of two or more phases to mix and form one phase. See *immiscible*. **miscible**

miscible displacement The process that occurs when a fluid mixes with and displaces another fluid. For example, salts are leached from a soil because the added water mixes with and displaces the soil solution. **déplacement de fluides miscibles**

mite (*zoology*) A member of the Arachnida which include spiders; they occur in large numbers in many organic surface soils. **mite**

M

mitigation Actions taken to lessen the actual or foreseen adverse environmental impact of a project or activity. **atténuation**

mitochondrion An organelle in which the chemical energy in reduced organic compounds is transferred to ATP molecules by oxygen-respiring respiration. **mitochondrie**

mitosis The process of cell division involving the replication and division of the cell nucleus; results in two genetically identical cells. **mitose**

mix, bury, and cover (drilling wastes) (*land reclamation*) A method of drilling waste management whereby sump solids are stabilized and diluted by mixing with subsoil. The waste materials must be mixed at least 1:1 by volume with the subsoil. The stable waste is then placed into the original sump, or other sumps, and is covered with at least one meter of clean subsoil and then the original topsoil. **technique d'élimination des résidus de forage**

mixed farm A farm that raises a mixture of cash crops, feed crops, and livestock. **ferme ou entreprise agricole diversifiée**

mixed forest A forest composed of two or more species of trees; in practice, usually a forest in which at least 20% are trees of other than the dominant species. See *pure forest*. **forêt mixte**

mixed-layer mineral A mineral whose structure consists of alternating layers of clay minerals and/or mica minerals. For example, chlorite, which is made up of alternating biotite and brucite sheets. **minéral interstatifié**

mixing See *tillage, mixing*. **mélange**

mixing ratio The concentration of water vapor in the atmosphere, commonly expressed as grams of water vapor per kilogram of dry air. See *relative humidity*. **rapport de mélange**

mixotrophy Nutritional mode of bacteria and protoctists characteristic of organisms nourished by both autotrophic and heterotrophic mechanisms. **mixotrophe**

mixture A material of variable composition that contains two or more substances. **mélange**

modal profile The soil profile (pedon) with physical, chemical, and biophysical characteristics lying close to the center of the ranges of properties that define a soil series; the most frequently occurring profile within a soil series. **profil modal, typique ou représentatif**

mode *(statistics)* The value of a variate which is possessed by the largest number of members of a population. **mode, valeur dominante**

model A formalized expression of a theory, event, object, process, or system used for prediction or control; an experimental design based on a causal situation that generates observed data. Types of models developed and used in the Earth sciences include scale models, in which a prototype object is scaled up or down to a size convenient for study (e.g., to emphasize micro-relief); conceptual models, which are based on observation and express a segment of the real world in idealized form, retaining essential features but omitting extraneous details (see *process-response model*) (e.g., a volcanic eruption); mathematical models, which are abstractions of physical models in that they replace events, forces, and/or objects by using expressions that contain mathematical variables, constants, and parameters. These latter models can be further subdivided into: *deterministic, statistical,* and *stochastic-process models.* See *iconic model* and *simulation model.* **modèle**

model building The creation of a model to simplify a real world problem while maintaining a balance between accurate representation of reality and mathematical manageability. This aim may be achieved by omitting certain variables that have only a small effect on the performance of the system but add great mathematical complexity; changing the nature of the variables (e.g., by treating some as constants); changing the relationship between variables (e.g., by assuming that a function is continuous rather than discontinuous); or modifying the constraints by adding to or subtracting from them as appropriate. **modélisation**

modeling Development of a mathematical or physical representation of a theory, event, object, process, or system that accounts for all or some other known properties. Models are often used to test the extent of changes of components on the overall performance of the system. **modélisation**

moder A type of forest *humus* made up of plant remains partly disintegrated by the soil *fauna*, but not matted as in raw humus. It is a transitional form of humus between *mull* and *mor.* Also called *duff mull.* **moder**

moderately coarse texture Consisting predominantly of coarse particles. In soil textural classification, it includes all the *sandy loams* except very *fine sandy loam.* **texture modérément grossière**

moderately fine texture Consisting predominantly of intermediate-sized soil particles with or without small amounts of fine or coarse particles. In soil textural classification it includes *clay loam, sandy clay loam,* and *silty clay loam.* **texture modérément fine**

Mohorovicic discontinuity A seismic discontinuity occurring between the *crust* of the Earth and the underlying *mantle,* across which the velocities of P-waves and S-waves (i.e., *seismic waves*) are significantly modified owing to the different densities of the crust and mantle. The discontinuity, commonly called the moho, occurs at an average depth of 35 km below the continents and at about 10 km below the oceans. See *Gutenberg discontinuity.* **discontinuité de Mohorovicic**

Mohr circle A graphical representation of the components of stress in soil acting across the various planes at a given point, drawn with reference to axes of normal stress and shear stress. **cercle de Mohr**

Mohr envelope The envelope of a series of Mohr circles representing stress conditions at failure for a given material. **enveloppe de Mohr**

Mohs scale See *hardness scale.* **échelle de Mohs**

moisture release curve, moisture retention curve See *water retention curve.* **courbe de désorption, courbe de rétention en eau**

moisture tension (or suction) The ability of soil to attract and adsorb pure water, due to matric or osmotic potentials. The terms "tension" and "suction" are semantic devices to express "potential" (which is usually negative in soil) in positive rather than negative units. See *soil water*. **tension de l'eau du sol, potentiel hydrique**

molality The number of moles of solute per unit mass of solvent. **molalité**

molar concentration The number of moles (*M*) of a chemical substance per unit volume of a medium (e.g., 0.2 *M* [8 g] of sodium hydroxide per liter of water). **concentration molaire**

molar heat capacity The energy required to raise the temperature of one mole of a substance by 1°C. **chaleur molaire**

molar mass See *molecular weight*. **masse molaire**

molar volume The volume of one mole of an ideal gas; equal to 22.42 liters at STP. **volume molaire**

molarity The number of moles of a dissolved chemical substance per liter of solution. **molarité**

moldboard plowing See *tillage, moldboard plowing*. **labour avec charrue à versoirs**

mole drain Unlined drain formed by pulling a bullet-shaped cylinder through the soil. **galerie-taupe**

mole fraction The number of moles of a given component in a phase, divided by the total number of moles of all components in the phase. Mole fractions are useful in defining the composition of a phase. **fraction molaire**

mole ratio (stoichiometry) The ratio of moles of one substance to moles of another substance in a balanced chemical equation. **rapport molaire**

molecular equation An equation representing a reaction in solution, showing the reactants and products in undissociated form, whether they are strong or weak electrolytes. **équation moléculaire**

molecular formula The exact formula of a molecule, giving the types of atoms and the number of each type. **formule moléculaire**

molecular structure The three-dimensional arrangement of atoms in a molecule. **structure moléculaire**

molecular weight The mass in grams of one mole of molecules or formula units of a substance; the same as molar mass. **poids moléculaire**

molecule The smallest part of a substance that can exist separately and still retain its chemical properties and characteristic composition; the smallest part of a chemical compound that can take part in a chemical reaction. **molécule**

mollic epipedon A surface horizon of mineral soil (U.S. system of soil taxonomy) that is dark-colored and relatively thick, contains at least 5.8 g kg^{-1} organic carbon, is not massive and hard or very hard when dry, has a base saturation of >50% when measured at pH 7, has <110 mg P kg^{-1} soluble in 0.05 M citric acid, and is dominantly saturated with bivalent cations. **épipédon mollique**

Mollisols An order in the U.S. system of soil taxonomy. Mineral soils that have a *mollic epipedon* overlying mineral material with a base saturation of 50% or more when measured at a pH of 7. Mollisols may have an *argillic, natric, albic, cambic,* or *petrocalcic* horizon, a *histic epipedon*, or a *duripan*, but not an *oxic* or *spodic* horizon. **Mollisols**

molt (*zoology*) To periodically cast or shed the outer body covering which permits an increase in size; especially characteristic of invertebrates. See *instar*. **mue**

moment measure *(statistics)* A descriptor of the character of a distribution curve. Moment measures include: *mean, standard deviation, skewness,* and *kurtosis*. **mesure des moments**

monadnock (*geomorphology*) An isolated hill or type of residual resulting from denudation, which has left it rising conspicuously above a plain. See *peneplain*. It is usually related to an outcrop of more resistant rocks, although the term refers to the morphology, not the structure. **monadnock**

monitoring The process of checking, observing, or keeping track of something for a specified period of time or at specified intervals. **suivi, monitoring, surveillance**

monocotyledon A plant whose embryo has one *cotyledon* (e.g., grass); one of the two classes of *angiosperms*. **monocotylédone**

monoculture (1) The repetitive growing of the same crop on the same land, season after season. (2) The cultivation of a single crop. **monoculture**

Monod equation The hyperbolic equation commonly used to describe microbial growth rate; see *hyperbolic reaction*. **équation de Monod**

monodentate (unidentate) ligand A *ligand* that can form one bond to a metal ion. **coordinat monodenté**

monoecious (*botany*) Plants having reproductive organs of both sexes on the same individual, either in different flowers or in the same flowers. See *dioecious*. **monoïque**

monolith, soil A vertical section of a soil profile removed from the soil and mounted for display or study. **monolithe**

monomer A compound that under certain conditions will join to other compounds of the same type to form a molecular chain called a *polymer* (e.g., vinyl chloride monomers can polymerize to polyvinyl chloride [PVC]). **monomère**

monomictic lake A lake with only one *overturn* per year. There are two types: warm with overturn in winter, and cold with overturn in summer. **lac monomictique**

monoprotic acid An acid with one acidic proton. **monoacide**

monosaccharide (simple sugar) A carbohydrate that cannot be split into smaller units by the action of dilute acids. Monosaccharides are classified according to the number of carbon atoms they possess (e.g., trioses have three carbon atoms, tetroses have four, pentoses have five). Each of these is further divided into whether the molecule contains an aldehyde group (–CHO), called an aldose, or a ketone group (–CO–), called a ketose. Monosaccharides can exist as either straight-chain or ring-shaped molecules. **monosaccharide, sucre simple, ose**

Monte Carlo method *(statistics)* A procedure that produces a statistical estimate of a quantity by taking many random samples from an assumed probability distribution, such as a normal distribution. The method is typically used when experimentation is unfeasible or when the actual input values are difficult or impossible to obtain. **méthode de Monte-Carlo**

montmorillonite An aluminum silicate (smectite) of the general composition $Si_4Al_{1.5}Mg_{0.5}O_{10}(OH)_2Ca_{0.25}$, in which calcium is readily exchangeable with other cations. It has a 2:1 layer structure composed of two silica tetrahedral sheets and a shared aluminum and magnesium octahedral sheet. Montmorillonite has a permanent negative charge that attracts interlayer cations existing in various degrees of hydration, thus causing expansion and collapse of the structure (i.e., shrink swell). **montmorillonite**

montmorillonite-saponite group Replaced by *smectite*. **groupe des smectites**

mor (or raw humus) A type of forest humus distinguished by a matted F layer and a holorganic H layer with a sharp delineation from the *A horizon*. It is generally acid, and has high organic carbon content (52% or more) and a high C:N ratio (usually 25 to 35 but sometimes higher). Various subgroups of mor can be recognized by the morphology, as well as chemical and biological properties. **mor, humus brut**

morainal Relating to material deposited by glacial ice. **morainique**

moraine (*geology*) An accumulation of heterogeneous rubbly material, including angular blocks of rock, boulders, pebbles, and clay, that has been transported and deposited by a glacier or ice-sheet. **moraine**

moraine kame (*geomorphology*) A *kame* that forms one of a group having the characteristics of a terminal moraine. **moraine de kame**

morph A phenotypic variant of a species with a specific form, shape, or structure. **morphe**

morphochronology The dating of morphological features or landforms. **morphochronologie**

morphographic map A map illustrating the terrain by pictorial symbols as if the landscape was being viewed obliquely from the air. **carte physiographique**

morphologic unit (1) A rock-stratigraphic unit identified by its topographic features (e.g., an alluvial fan). (2) A surface, either depositional or erosional, that is recognized by its topographic character. **unité morphologique**

morphology (*biology*) The study of the form and structure of animals and plants or their fossil remains, especially the relations and development of organs apart from their functions. (*geomorphology*) The study of the landforms of the Earth's surface. **morphologie**

morphology, soil The physical constitution of a soil profile as exhibited by the kinds, thickness, and arrangement of the horizons in the profile, and by the texture, structure, consistence, and porosity of each horizon. **morphologie du sol**

morphometry The precise measurement of shape, such as that of landforms and water bodies. **morphométrie**

morphostasis The perpetuation of a *steady-state equilibrium* by countering any tendency to change in a system. See *general systems theory*. **morphostase**

mortality Death or destruction as a result of competition, disease, insect damage, drought, wind, fire, and other factors (excluding harvesting). **mortalité**

mosaic (*photogrammetry*) An overlapped grouping of vertical aerial photographs that produces an overall impression of an area too large to be depicted on a single air photograph. (*petrology*) A petrologic texture in which mineral grains form a regular pattern with relatively straight or slightly curved boundaries or contacts. **mosaïque**

moss A class of lower green plants of the phylum Bryophytes. Moss plants consist of small slender stalks and leaves, and lack vascular tissue and true roots. Underground support and conductance are carried out by filamentous structures called rhizoids. **mousse**

moss peat See *peat moss*, *sphagnum peat*, *brown-moss peat*. **mousse de tourbe**

most probable number (MPN) Method for estimating microbial numbers in soil based on extinction dilutions. **nombre le plus probable**

motile Capable of movement. **motile**

mottled zone A layer that is marked with spots or blotches of different color or shades of color. The pattern of mottling and the size, abundance, and color contrast of the *mottles* may vary markedly and should be specified in the soil description. See *mottles*. **zone marbrée, tachetée, marmorisée**

mottles Spots or blotches of different color or shades of color interspersed with the dominant color of the soil matrix. **marbrures, mouchetures, taches**

mottling Formation or presence of *mottles* in the soil. **marmorisation, marbrures**

mounded A type of microtopography consisting of small basins and knolls; commonly applied to *peatlands*. An example of mounded categories is as follows: level or micromounded; slightly mounded (mounds 0.3 to 1 m high and >7 m apart); moderately mounded (mounds 0.3 to 1 m high and 3 to 7 m apart); strongly mounded (mounds 0.3 to 1 m high and 1 to 3 m apart); severely mounded (mounds >1 m high and >3 m apart; ultramounded (mounds >1 m high and <3 m apart). **à buttes**

mountain A portion of the land surface rising considerably above the surrounding country either as a single eminence or in a range or chain. Generally, mountains are considered to project at least 300 m above the surrounding land, those less than this being referred to as *hills*. **montagne**

mountain range A single, large mass consisting of a succession of mountains or narrowly spaced mountain ridges, with or without peaks, similar in position, direction, formation, and age; a component part of a mountain system or a mountain chain. **chaîne de montagnes**

mountain-building See *orogeny*. **orogénèse, formation des montagnes**

mucigel The gelatinous material at the surface of roots, containing mostly water and also natural and modified plant *exudates* (mucilages), bacterial cells, and their metabolic products, as well as colloidal mineral and organic matter from the soil. **mucigel**

muck Highly decomposed organic material in which the original plant parts are not recognizable. Contains more mineral matter and is usually darker in color than peat. **terre tourbeuse**

muck soil An organic soil consisting of highly decomposed plant materials. In a muck soil the total thickness of all organic layers is greater than that of mineral layers. **terre tourbeuse, terre organique, terre noire**

mud (1) A mixture of clay and/or silt with water, ranging from semi-fluid to soft and plastic. (2) Artificial slurry created for use in drilling boreholes (see *drilling mud*). (3) The mixed material of a mudflow. **boue**

mud rock See *mudstone*. **argilite**

mudflow A general term for a mass-movement landform and process characterized by a flowing mass of fine-grained earth material with a high degree of fluidity. The water content may range up to 60%. See also *earthflow, debris flow*. **coulée boueuse**

mudslide A *landslide* of essentially homogeneous clay-silt material, but without the velocity or linear extent of a *mudflow*. See *slump*. **coulée de boue**

mudstone (1) An indurated sedimentary rock composed mainly of clay. (2) Any fine-grained, clastic sedimentary rock where doubt exists about the proportions of sand, silt, and clay, or in which the proportions vary in such a way that more precise terms cannot be used in a meaningful way. **mudstone**

muffle furnace An analytical oven used to determine the organic or mineral content of soil or plant material by dry ashing at temperatures (e.g., up to 500°C) sufficient to oxidize organic material. **fournaise à mouffles**

mulch (1) A layer of dead plant material on the soil surface. See *tillage, mulch*. (2) An artificial layer of material such as paper or plastic on the soil surface. (3) The cultural practice of placing rock, straw, asphalt, plastic, or other materials on the soil surface as a surface cover. **paillis, mulch**

mulch farming See *tillage, mulch farming*. **culture sur résidus ou sur paillis**

mulch tillage See *tillage, mulch tillage*. **travail du sol avec conservation des résidus, culture sur paillis**

mull A type of forest humus consisting of an intimate mixture of well-humified organic matter and mineral soil that makes a gradual transition to the horizon underneath. It is distinguished by its crumb or granular structure. Because of the activity of the burrowing microfauna (mostly earthworms) partly decomposed organic debris does not accumulate as a distinct layer (F layer) as in *mor* and *moder*. It is a kind of Ah horizon, wherein various subgroups can be distinguished by the morphology and chemical characteristics. Mull is a diagnostic layer in the Canadian system of soil classification, where it is defined as having an organic matter content of 5 to 25% and a *C:N ratio* of 12 to 18. **mull**

mull-like moder *Humus* forms characterized by the external appearance of *mull*; a relatively weak binding together of organic matter and the mineral soil particles; little or no clay and a relatively high percentage of silt-sized and sand-sized mineral grains. **moder à tendance mull**

multilevel sampling (*remote sensing*) Combining remotely sensed data from different types of platforms with ground data. **échantillonnage à altitudes multiples**

multiple correlation See *correlation, statistical*. **corrélation multiple**

multispectral Remote sensing in two or more spectral bands, such as visible and infrared. **multispectral**

multispectral scanner (MSS) A line-scanning sensor that uses an oscillating or rotating mirror, a wavelength-selective dispersive mechanism, and an array of detectors to simultaneously measure the energy available in several wavelength bands, often in several spectral regions; the movement of the platform usually provides for the along-track progression of the scanner. **scanneur multibande, balayeur multispectral**

multivariate analysis *(statistics)* A data-analysis approach that makes use of multidimensional interrelationships and correlations within the data for effective discrimination. Some types are factor and component analysis, classification, discriminatory analysis, cannonical correlation, Hotelling's T, Wishart's distribution, and Wilks' criterion. **analyse multivariable, analyse à plusieurs variables**

Munsell color system A color designation system specifying the relative degrees of the three simple variables of color: hue, value, and chroma. For example, 10YR 6/4 is the color of a soil having a hue of 10YR, value of 6, and chroma of 4. These notations can be translated into several different systems of color names. See *chroma*, *hue*, and *value*. **code de couleurs Munsell**

muscovite A clear, dioctahedral layer silicate of the mica group with Al^{3+} in the octahedral layer and Si and Al in a ratio of 3:1 in the tetrahedral layer. See *mica*. **muscovite**

muskeg A common term for *peatland* in Canada and parts of the U.S. The word is of Algonquin origin and is applied to natural and undisturbed areas partially covered with *Sphagnum* mosses, tussocky sedges, and an open growth of scrubby trees. **muskeg, plée, savane**

mutualism An interaction between two or more distinct biological species in which all members benefit from the *association*. Symbiotic mutualism requires an intimate association between species in which neither can carry out a requried function alone. Nonsymbiotic mutualism, also called protocooperation, is a beneficial but not obligatory relationship between organisms (i.e., the organisms are capable of independent existence). **mutualisme**

mycelium (plural mycelia) A mass of threadlike filaments, branched or composing a network, that constitutes the vegetative structure of a fungus. See *figure*. **mycélium**

myco Prefix designating an association or relationship with a fungus (e.g., mycotoxins are toxins produced by a fungus). **myco**

mycology The study of fungi. **mycologie**

mycorrhiza The association, usually symbiotic, of *fungi* with the roots of seed plants. Under favorable soil conditions, plants with mycorrhizae have been shown to produce up to four or five times as much growth as similar plants without mycorrhizae. The mycorrhizae absorb water and

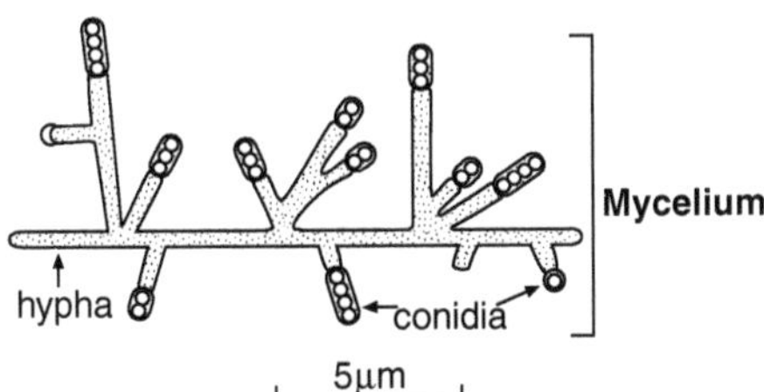

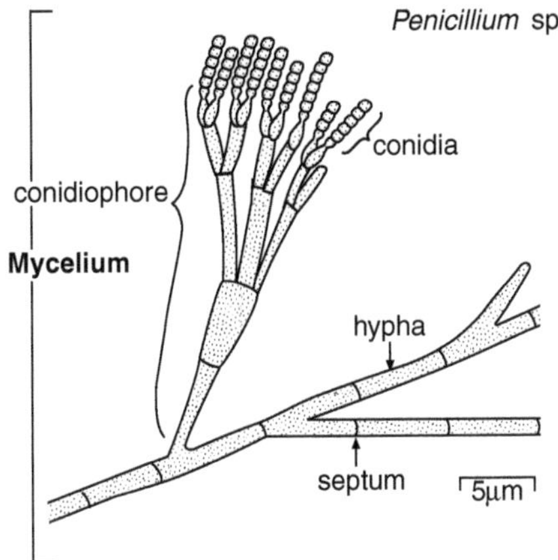

Mycelium (adapted from Killham, 1994).

nutrients and pass them on to the plant, secrete hormones that stimulate plant growth, and help protect the plant roots from disease organisms. In return, the fungi depend on the plant for carbohydrates. See *ectotrophic mycorrhiza* and *endotrophic mycorrhiza*. **mycorhize**

mycotoxin A toxin or toxic substance produced by a fungus. *Aflatoxin*, a natural carcinogen, is a mycotoxin commonly found in rice, peanuts, wheat, and maize. **mycotoxine**

myriapod (*zoology*) Any terrestrial arthropod belonging to the superclass Myriapoda (e.g., centipedes and millipedes). **myriapode**

myxamoeba The naked cell formed when a zoospore sheds its flagella and becomes ameboid. **myxamibe**

N

nadir That point on the celestial sphere vertically below the observer, or 180° from the zenith. **nadir**

nannoplankton Plankton in the size range 50 to 60 μm defined as uncatchable in standard plankton nets. **nannoplancton**

narrow row planting See *tillage, narrow row planting*. **plantation en rangs étroits ou serrés**

native prairie An area of unbroken grassland or parkland dominated by non-introduced species. **prairie naturelle**

native species A species that is a part of an area's original fauna or flora. **espèce indigène**

natric horizon A mineral soil horizon (in the U.S. system of soil taxonomy) that satisfies the requirements of an *argillic horizon* but that also has prismatic, columnar, or blocky structure and a subhorizon having >15% saturation with exchangeable Na^+. **horizon natrique**

natural abundance Percentage of an element occurring on Earth in a particular stable isotopic form. **abondance naturelle**

natural disturbance The impact of periodic natural events such as fire, severe drought, attack by insect or disease, or wind. **perturbation naturelle**

natural disturbance regime The historic pattern (frequency and extent) of fire, insects, wind, landslides, and other natural processes in an area. **régime de perturbations naturelles**

natural ecosystem An ecosystem minimally influenced by humans that is, in the larger sense, diverse, resilient, and sustainable. **écosystème naturel**

natural erosion See *erosion, geological erosion*. **érosion naturelle**

natural event system A *system* that incorporates the magnitude, frequency, duration, and temporal spacing of a natural event. **système à événement naturel**

natural gas A mixture of colorless, highly flammable *hydrocarbon* gases occurring within pore spaces of subterranean rock reservoirs, often in association with petroleum deposits. It is composed of 80 to 85% methane and 10% ethane, with the balance consisting of propane, butane, nitrogen, carbon dioxide, hydrogen sulfide, and occasionally small proportions of helium and some other minor hydrocarbons. **gaz naturel**

natural increase Growth in the number of individuals in a community as a result of more births than deaths. **croissance naturelle**

natural law (1) A statement that expresses generally observed behavior. (2) The laws of nature; regularity in nature. **loi naturelle**

natural radioactivity Ionizing radiation from sources not related to human activities; for example, cosmic rays and radiation emitted by *radioisotopes* found naturally in the Earth's crust. **radioactivité naturelle**

natural range barrier A river, rock face, dense timber, or any other naturally occurring feature that stops or significantly impedes livestock movement to and from an adjacent area. **barrière à la répartition naturelle**

natural regeneration The renewal of a plant by natural seeding, sprouting, suckering, or layering, or seeds that may be deposited by wind, birds, or mammals. **régénération naturelle**

natural resources (1) Resources needed by an organism, population, or ecosystem that support an increasing rate of energy conversion when their availability is increased up to an optimal or sufficient level. (2) Materials or conditions occurring in nature that can be exploited or used to satisfy the needs of humans, including air, soil, water, native vegetation, minerals, and wildlife. Renewable natural resources naturally replenish themselves within the limits of human time, whereas nonrenewable natural resources do not. **ressources naturelles**

natural revegetation Natural re-establishment of plants; propagation of new plants over an area by natural processes. **remise en végétation naturelle**

N

natural seeding Natural distribution of seed over an area; also called volunteer seeding. **ensemencement naturel**

natural selection The natural process by which organisms best adapted to their environment survive and those less well adapted are eliminated; proposed by Charles Darwin as the process by which evolution occurs in order to account for the origin and diversity of organisms, often referred to as survival of the fittest. See *evolution*. **sélection naturelle**

natural sink (1) A habitat that traps or immobilizes chemicals (e.g., plant nutrients, organic pollutants, or metal ions) through natural processes. (2) A natural process whereby pollutants are removed from the atmosphere. The sink process can be physical (e.g., removal of particulate matter by rain), chemical (e.g., reaction of ozone with nitric oxide to form nitrogen dioxide and oxygen), or biological (e.g., uptake of airborne hydrocarbons by soil microorganisms). **puits naturel**

natural vegetation (1) Any plant life that is not organized or influenced by humankind. (2) More strictly, the world's major vegetation climaxes. See *climax*. **végétation naturelle**

naturalized plant A plant introduced from other areas which has become established in and more or less adapted to a given region by long-continued growth there. **plante naturalisée**

nauplius Free-swimming microscopic larval stage characteristic of many crustaceans and barnacles. **nauplie**

near infrared reflectance spectroscopy (NIRS) A method of forage quality analysis based on spectrophotometry at wavelengths in the near infrared region. **spectroscopie de réflexion dans le proche infrarouge**

negative correlation See *correlation, statistical*. **corrélation négative**

negative feedback See *feedback*. **rétroaction négative, feedback négatif**

negligible residue A small amount of pesticide remaining in or on raw agricultural commodities that would result in a daily intake of the compound by consumers of that commodity; regarded as toxicologically insignificant. **résidu négligeable**

nekton Animals in aquatic systems that are free-swimming, independent of currents or waves. See *benthos, plankton*. **necton**

nematode (*zoology*) A small, threadlike worm common to most soils. Most species live on decaying organic matter, but some infect plant roots and live as parasites (Still others are parasites on animals, including human beings). **nématode**

neoformation The formation of clay minerals by materials freed from parent material by complete weathering; associated mainly with tropical climates. **néogenèse, néoformation**

nephelometer A device that measures the scattering of light by particles (or bacteria) suspended in air or water, compared to a reference suspension. **néphélomètre**

neritic zone Shallow regions (<200 m depth) of a lake or ocean that border the land. In oceans, relates to water overlying the continental shelf. **région néritique**

Nernst equation An equation relating the potential of an electrochemical cell to the concentrations of the cell components; $E_{cell} = E^{o}_{cell} - 0.0592/n \times \log(Q)$ at 25°C. (n, number of electrons, *Q*, the reactin quotient). **équation de Nernst**

nesosilicate One of a group of silicate minerals having a characteristic structure in which individual SiO_4 tetrahedra are linked only by ionic bonds from interstitial cations. **nésosilicate**

nested experiment An experiment in which a different set of levels of a second factor is used in conjunction with each level of a first factor; also called a hierarchical experiment. **expérience en tiroirs**

nested sampling (1) Sampling where certain units are embedded in larger units which form part of the whole sample. (2) Multi-stage sampling in which the higher-stage units are contained within or nested in the lower stage units. **échantillonnage hiérarchique**

net radiation The net exchange between all outgoing and incoming radiation. **rayonnement net, bilan radiatif**

net aboveground productivity (NAP) The accumulation of biomass in aboveground parts of plants (e.g., trunks, branches, leaves, flowers, and fruits) over a specified period; usually expressed on an annual basis. **productivité épigée nette**

net community productivity (NCP) The gain of biomass within a defined region over time; the total amount of carbon dioxide fixed by photosynthetic plants within the area (i.e., gross primary production) minus that lost through metabolism at all trophic levels within the same region. **productivité communautaire nette**

net primary productivity (NPP) The amount of carbon dioxide fixed during photosynthesis minus that amount lost during respiration of those plants per unit area of land or volume of water; represents the actual rate of production of new biomass. See *primary productivity.* **productivité primaire nette**

net production efficiency The percentage of assimilated energy incorporated during growth and reproduction. **efficacité de production nette**

network A system of lines (arcs, edges, links) that join a set of points. It may be planar (two-dimensional) or non-planar (three-dimensional), with most physical geography networks being the former. Network analysis is the technique that plans a number of steps or operations within a system. **réseau**

neurotoxin A substance that causes nervous dysfunction. **neurotoxine**

neuston Small particles or microorganisms found in the surface film that covers still bodies of water. **neuston**

neutral detergent fiber (NDF) That portion of a forage that is insoluble in neutral detergent: also called *cell well constituents.* **fibre au détergent neutre, fibre détergente neutre**

neutral soil A soil in which the surface layer, at least in the tillage zone, is in the pH 6.6 to 7.3 range. See *acid soil, alkaline soil, pH,* and *reaction, soil.* **sol neutre**

neutralization The addition of an alkaline material to an acid material to raise the pH and overcome an acid condition, or the addition of an acid material to an alkaline material to lower the pH. **neutralisation**

neutron One of the elementary particles in the nucleus of all atoms except hydrogen. A neutron does not have a charge, and the atomic mass is approximately 1. **neutron**

neutron probe A probe with a radioactive source that measures soil water content through the reflection of scattered neutrons by hydrogen atoms in soil water. **sonde à neutrons**

new forestry A philosophy or approach to forest management that has as its basic premise the protection and maintenance of ecological systems. In new forestry, the ecological processes of natural forests are used as a model to guide the design of the managed forest. **nouvelle foresterie**

niche The particular role that a species plays in an ecosystem; the physical space occupied by an organism. **niche**

nitrate (NO_3) The most highly oxidized phase in the nitrogen cycle, which normally reaches important concentrations in the final stages of biologic oxidation. **nitrate**

nitrate reduction The biochemical process whereby nitrate is reduced by plants and microorganisms to ammonium for cell synthesis (nitrate assimilation, assimilatory nitrate reduction) or to various lower oxidation states (N, N_2O, NO, NO_2) by bacteria using nitrate as the terminal electron acceptor in anaerobic respiration (respiratory nitrate reduction, dissimilatory nitrate reduction, denitrification). **réduction du nitrate**

nitric acid A strong mineral acid having the formula HNO_3; one of the constituents of *acid rain.* **acide nitrique**

nitric oxide (NO) A gas formed as an intermediate product during *denitrification* in soil, by heating air to high temperatures or by the oxidation of organic nitrogen contaminants in a fuel during combustion. It is not itself a pollutant, but it converts to *nitrogen dioxide,* which is a major constituent of photochemical smog. **oxyde nitrique**

nitric phosphates Fertilizers made by processes involving treatment of phosphate rock with nitric acid or a mixture of nitric, sulfuric, or phosphoric acids, usually followed by ammoniation. The water solubility of the phosphorus content may vary over a wide range. **nitrophosphates**

nitrification A two-stage biological process in which ammonia is oxidized to nitrite (NO_2) and the nitrite to nitrate (NO_3) by a discrete group of bacteria found in soil or water. **nitrification**

nitrite An oxidized nitrogen molecule, NO_2^- that is an intermediate product formed in the nitrogen cycle during *nitrification* and *denitrification* in soil. **nitrite**

nitrogen The most abundant gas of the *atmosphere,* essential element for plant growth and an important constituent of soil. See *nitrification, nitrogen cycle.* **azote**

nitrogen assimilation The incorporation of nitrogen into organic cellular constituents by living organisms. **assimilation de l'azote**

nitrogen cycle The continuous cycling of nitrogen from the atmosphere to the Earth and back again, during which nitrogen is converted from one form to another through a combination of biological, geological, and chemical processes. In soil, the nitrogen cycle is the sequence of biochemical changes by which nitrogen is used by a living organism, liberated upon the death and decomposition of the organism, and converted to its original state of oxidation. See *figure.* **cycle de l'azote**

nitrogen dioxide (NO_2) A gas readily produced in the atmosphere from nitric oxide by the addition of an oxygen atom (i.e., $NO + O \rightarrow NO_2$) or as an intermediate product during *denitrification* in soil; a major ingredient in photochemical smog. It can be converted by atmospheric reactions to *nitric acid*, an ingredient in acid rain. See *nitrogen oxides.* **dioxyde d'azote**

nitrogen fixation The conversion of nitrogen gas in the atmosphere (N_2) to a

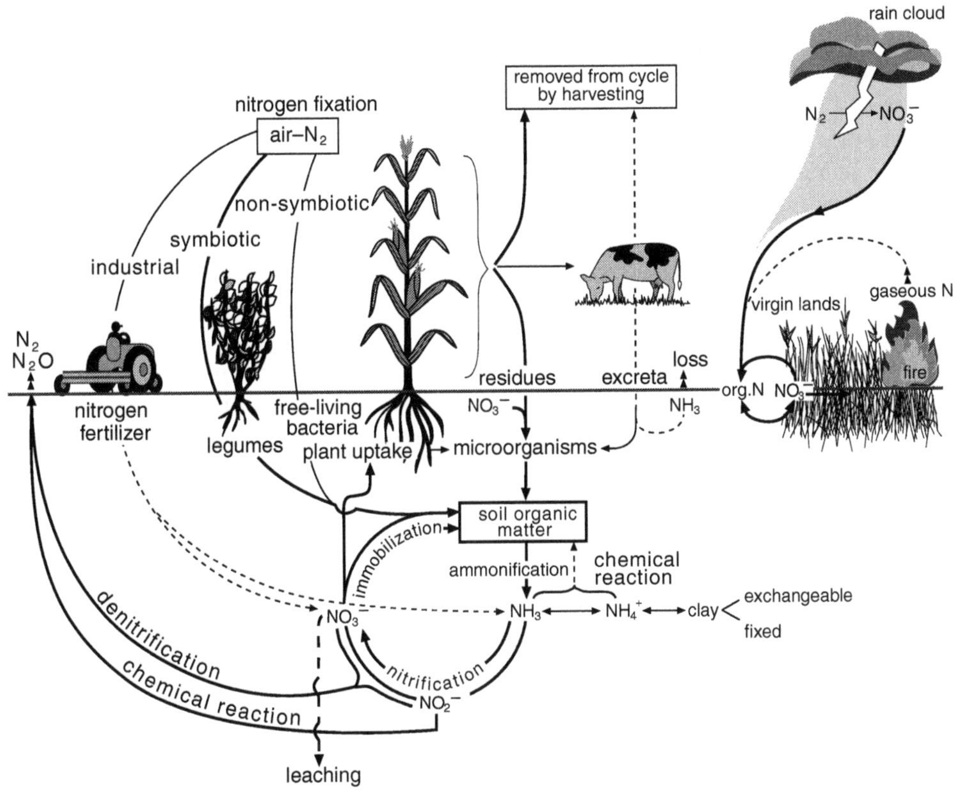

The **nitrogen cycle** in soil (adapted from Stevenson, 1982).

reduced organic form (e.g., ammonia and amino groups of amino acids) that can be used as a nitrogen source by organisms. Biological nitrogen fixation is carried out by a variety of organisms; however, those responsible for most of the fixation are certain species of blue-green algae, the soil bacterium *Azotobacter*, and the symbiotic association of leguminous plants and the bacterium *Rhizobium*. **fixation biologique de l'azote**

nitrogen oxides (NOx) (*chemistry*) Gases consisting of one atom of nitrogen and a varying number of oxygen atoms. (*meteorology*) Nitrogen oxides are produced in the emissions of vehicle exhausts and from power stations and are considered pollutants because they contribute to the formation of photochemical ozone (smog), impair visibility, and have health consequences. **oxydes d'azote (NOx)**

nitrogen requirement The actual amount of nitrogen needed by a plant for desired growth, determined by multiplying the yield goal (e.g., kg ha^{-1}) by the amount of nitrogen required for producing one kg. In contrast, the fertilizer recommendation is determined by taking the difference of the nitrogen requirement minus the soil nitrogen sources (e.g., manure, residual nitrate). **besoin en azote**

nitrogenase An enzyme complex containing iron and molybdenum; it reduces atmospheric nitrogen to organic nitrogen compounds. **nitrogénase**

nitrogen-fixing bacteria Bacteria in the root nodules of plants that can convert atmospheric nitrogen to ammonia and other nitrogen-containing compounds useful to plants. **bactéries fixatrices d'azote**

nitrogen-fixing plant A plant that can assimilate and fix the free nitrogen of the atmosphere with the aid of bacteria living in the root nodules. *Legumes* are the most important nitrogen-fixing plants. **plante fixatrice d'azote**

nitrosamines A large and diverse family of synthetic and naturally occurring compounds having the general formula (R)(R')N–N=O, where R and R' are side groups with a variety of possible structures. Several nitrosamines produced from nicotine by bacterial activity during the tobacco curing process are responsible, in part, for the cancer-causing properties of tobacco products. **nitrosamines**

nitrous oxide (N_2O) A powerful *greenhouse gas* with a *global warming potential* of 320. Major sources of nitrous oxide include soil cultivation practices, especially the use of commercial and organic fertilizers, fossil fuel combustion, nitric acid production, and biomass burning. **protoxyde d'azote**

N

nodule A structure developed on the roots of most legumes and some other plants in response to the stimulus of root nodule bacteria; legumes bearing these nodules are nitrogen-fixing plants, using atmospheric nitrogen instead of depending on nitrogen compounds in the soil. **nodule, nodosité**

nodule bacteria The bacteria that fix dinitrogen (N_2) within organized structures (nodules) on the roots, stems, or leaves of plants. Sometimes used as a synonym for "rhizobia." **bactéroïde**

nodule, soil A rounded unit within the soil matrix that differs from the surrounding material because of the concentration of some constituent or a change in fabric. **nodule de sol**

noise The *variance* after the effects of the known variables have been accounted for. Any unexplained variance exhibiting no discernible pattern in terms of any other variable, or in space and time, is called random noise. Unexplained spatial variance is called spatial noise. **bruit**

nominal variable A sample characteristic expressed as a class or descriptive category; e.g., slope class, parent genetic material, soil structure type. These are qualitative, not quantitative, data, and only nonparametric tests are appropriate for data collected as nominal variables. See *ordinal variable*, *parametric tests*. **variable qualitative**

nomograph A graphical solution to a multivariable equation, wherein parallel vertical scales, one for each variable, are arranged such that a straight line across the scales produces values for each variable that together solve the equation. **nomogramme**

nonconformity A type of *unconformity* in which the underlying older rocks are of igneous or metamorphic rather than sedimentary origin. **discordance majeure**

nondestructive testing (NDT) Geophysical methods used to detect subsurface water, subsurface containers, or the areal extent of soil and groundwater contamination without soil borings. Methods include acoustic sounding, infrared radiation, x-rays, magnetic field perturbation, and electrical resistivity. **essai non destructif**

nongranular fertilizer See *fertilizer, nongranular*. **engrais non granulaire**

non-inversive tillage See *tillage, non-inversive tillage*. **travail sans inversion de sol**

non-limiting water range The region bounded by the upper and lower soil water content over which water, oxygen, and mechanical resistance are not limiting to plant growth; determined using *soil water characteristic* and *soil strength*. Also called least-limiting water range. **étendue des teneurs en eau non-limitantes**

nonlinear function A function whose variables do not increase in a linear fashion. **fonction non linéaire**

nonlinear regression *(statistics)* Any type of *regression* equation in which the exponent of the independent variables is greater than unity. The associated regression curve cannot be precisely constructed, and predictions cannot be made precisely, using a least-squared straight regression line (i.e., linear regression). The regression line must therefore be manually constructed, or a best fit using polynomial equations can be attempted. **régression non linéaire, régression curvilinéaire**

non-native species A species introduced into an ecosystem through human activities. **espèce introduite**

non-parametric statistical test Statistics used in a distribution-free test that make no assumptions about the distribution of the background data population. The non-parametric tests become more important when the samples being tested are small. They can be applied to data on nominal and ordinal scales only and include: chi-square test, Mann-Whitney U-test, and Wilcoxon test. See *parametric statistical tests.* **test statistique non paramétrique**

nonpolar solvent A solvent with no positive polarity or negative polarity (e.g., benzene). This type of solvent is a good dissolver of other nonpolar materials. See *polar solvent.* **solvant non polaire**

nonpressure solution Usually nitrogen fertilizer solutions of such low free ammonia (NH_3) content that no vapor pressure develops and the solution can be applied without having to control vapor pressure. **solution azotée**

nonsaline-alkali soil See *sodic soil.* **sol alcalin non-salin**

non-silicate Rock-forming minerals that do not contain silicon. **non-silicaté**

nonsoil The collection of soil-like material that does not meet the definition of soil. It includes soil displaced by unnatural processes and unconsolidated material beyond the influence of soil-forming processes, except for the material that occurs within 25 cm below soil as defined. Nonsoil also includes unconsolidated mineral or organic material thinner than 10 cm overlying bedrock, organic material thinner than 40 cm overlying water, and soil covered by more than 60 cm of water in the driest part of the year. **non-sol**

non-symbiotic fixation Conversion of atmospheric nitrogen to protein by soil microbes. **fixation non symbiotique**

nontronite An iron-rich mineral of the smectite group, $Na_{0.33}Fe_2^{+3}(Al_{0.33}Si_{3.67})O_{10}(OH)_2 \bullet nH_2O$. **nontronite**

nonuniform flow *(hydrology)* Flow in which the mean velocity or cross-sectional area vary at successive channel cross-sections. **écoulement varié**

normal (1) A data distribution that corresponds to an accepted frequency distribution about a population mean. See *mean, normal distribution.* (2) At right angles to or perpendicular to (i.e., orthogonal). **normale**

normal atmospheric pressure Standard pressure, usually taken to be equal to that of a column of mercury 760 mm in height; approximately 1 kg per square cm. **pression atmosphérique normale**

normal distribution *(statistics)* A *frequency distribution* having a plot which is a continuous, infinite, bell-shaped curve that is symmetrical about its arithmetic *mean, mode,* and *median* (which in this distribution are numerically equivalent). Also called *Gaussian curve.* **distribution normale**

normal erosion See *erosion, geological erosion.* **érosion normale**

N

normal solution A solution that contains one gram equivalent of replaceable hydrogen per liter of solution (e.g., a normal solution of hydrochloric acid (HCl) contains 1 gram molecular weight of HCl per liter of solution, while a normal solution of sulfuric acid (H_2SO_4) contains half a gram molecular weight of H_2SO_4 per liter of solution). **solution normale**

no-tillage (zero tillage) See *tillage, no-tillage (zero tillage).* **systèmè zéro-labour, semis-direct, système de culture sans labour**

NOx See *nitrogen oxides.* **oxydes d'azote, NOx**

noxious species (*range management*) A plant that is undesirable because it conflicts with, restricts, or otherwise causes problems under the management objectives. **espèce nuisible**

nuclear energy Energy released as particulate or electromagnetic radiation and heat during reactions of atomic nuclei. **énergie nucléaire**

N

nuclear fission The splitting of a heavy nucleus into two approximately equal parts, which are radioactive nuclei of lighter elements, accompanied by the release of a large amount of energy and generally one or more *neutrons.* **fission nucléaire**

nuclear fusion The combining of atomic nuclei of very light elements by collision at high speed to form new and heavier elements, resulting in release of large amounts of energy. **fusion nucléaire**

nuclear magnetic resonance (NMR) spectroscopy A spectroscopic technique which measures the absorption of radiofrequency (RF) radiation by a nucleus in a strong magnetic field. Absorption of the radiation causes the nuclear spin to realign or flip in the higher-energy direction. After absorbing energy, the nuclei will re-emit RF radiation and return to the lower-energy state. The energy of an NMR transition depends on the magnetic field strength and a proportionality factor for each nucleus, called the magnetogyric ratio. (Of the elements useful in soil research, ^{1}H has the most favorable gyromagnetic ratio, followed by ^{31}P, ^{13}C, and ^{15}N). The local environment around a given nucleus in a molecule will slightly perturb the local magnetic field exerted on that nucleus and affect its exact transition energy. This dependence of the transition energy on the position of a particular atom in a molecule makes NMR spectroscopy extremely useful for determining the structure of molecules. **spectroscopie de résonance magnétique nucléaire**

nuclear reactor A device to promote and control *nuclear fission* for producing heat and then steam to generate electricity, or for producing certain *radioisotopes.* All reactors have a core containing nuclear fuel, which serves as the energy source, and control rods, which regulate the rate of fission. Other than their safe operation, one of the primary environmental concerns related to nuclear reactors is the accumulation of waste materials that are very radioactive and remain dangerous for hundreds of years. These radioactive wastes accumulate within the reactor structure itself. Consequently, used fuel rods and reactors that are decommissioned are sources of large amounts of dangerous radioactive wastes. **réacteur nucléaire**

nucleic acid An organic acid composed of purine and pyrimidine bases, sugars, and phosphoric acid joined into nucleotide complexes; the two types are *deoxyribonucleic acid (DNA)* and *ribonucleic acid (RNA).* **acide nucléique**

nucleon A particle in an atomic nucleus, either a *neutron* or a *proton.* **nucléon**

nucleus (*chemistry*) A central point or mass of an atom around which the orbital electrons revolve; composed of *protons* and *neutrons.* (*biology*)

The membrane-bound structure of a cell that carries genetic information and which regulates protein synthesis. The presence of a nucleus distinguishes the more complex *eukaryotic* cells of plants and animals from the simpler *prokaryotic* cells of bacteria and cyanobacteria that lack a nucleus. (*meteorology*) A particle which is suspended in the atmosphere. **noyau**

nuclide A type of atom characterized by its *atomic number* and its number of neutrons. The elements and their various isotopes number about 1000 nuclides. **nucléide**

null hypothesis *(statistics)* A hypothesis, H_0, that is to be tested against an alternative hypothesis, H_1, and whose erroneous rejection is considered a *Type I error.* **hypothèse nulle**

nurse crop See *companion crop, underseeding*. **plante-abri**

nutrient Any substance that an organism obtains from its environment to supply energy and growth, or contribute to biomass production; most often used to identify substances used for growth by plants. **élément nutritif**

nutrient antagonism The depressing effect caused by one or more plant nutrients on the uptake and availability of another nutrient in the plant. **antagonisme entre éléments nutritifs**

nutrient balance An undefined theoretical ratio of two or more plant nutrient concentrations necessary for maximum growth rate and yield. An imbalance results when one or more nutrients is present in either deficient or excess supply. **équilibre nutritif**

nutrient budget A quantitative determination of the major nutrients flowing to, retained within, and released from a system. **bilan nutritif**

nutrient concentration The weight of nutrients in a given weight of volume of material. The nutrient concentration of plants is usually expressed as grams per kilogram (g kg^{-1}) or milligrams per kilogram (mg kg^{-1}) of dry or fresh weight and should not be confused, or used interchangeably, with *nutrient content.* **concentration en éléments nutritifs**

nutrient content The amount of nutrients contained in a material. The nutrient content of plants is usually expressed as weight per unit area (kg ha^{-1}) and should not be confused, or used interchangeably, with *nutrient concentration*. **teneur en éléments nutritifs**

nutrient cycle The cyclic conversions of nutrients from one form to another within the biosphere. An example is the *nitrogen cycle* in which the nitrogen atom undergoes several changes in oxidation state (N_2, NO_3^-, R–NH_2, and NH_4^+, among others) during the cycling of this element through the biological community, and into the air, water, or soil, and back. **cycle des éléments nutritifs**

nutrient diffusion See *diffusion, nutrient.* **diffusion des éléments nutritifs**

nutrient interaction The resultant effect or response from two or more nutrients when applied together that differs from the sum of the additive individual responses when applied separately; often used to describe metabolic or ion-uptake phenomenon. **interaction entre éléments nutritifs**

nutrient retention Holding onto nutrients. **rétention des éléments nutritifs**

nutrient stress A condition occurring when the quantity of available nutrients, either too much or too little, reduces plant growth. **stress nutritif**

nutrient uptake The process by which roots absorb *plant nutrients* from the soil. Three processes facilitate the necessary contact between the root and the soil nutrients: *root interception, mass flow,* and *diffusion.* **prélèvement des éléments nutritifs**

nutrient-supplying power of soils The capacity of the soil to supply nutrients to growing plants from both the available and adsorbed forms. See *fertility, soil.* **capacité de libération des éléments nutritifs du sol**

nutrition Ingestion, digestion, or assimilation of food by plants and animals. **nutrition**

nymph (*zoology*) An immature developmental form characteristic of the preadult stage in insects that do not have a pupal stage (e.g., mayflies and stoneflies). See *larva.* **nymphe**

O

O horizon An organic soil horizon containing more than 17% organic carbon, normally associated with poorly drained peatland soils, and developed mainly from mosses, rushes, sedges, and woody materials (*peat* materials). The term is applied in the Canadian and U.S. systems of soil taxonomy, although sub-horizons are designated differently (See *Appendix D*). See *horizon, soil* and A, *B, C, L, F,* and *H horizons.* **horizon O**

obligate Necessary or required. **stricte**

occurrence The presence of a species within an ecosystem, regardless of the reason for, or the duration of, this presence. **occurrence, présence**

Ochrepts A suborder in the U.S. system of soil taxonomy. *Inceptisols* formed in cold or temperate climates and that commonly have an *ochric epipedon* and a *cambic horizon.* They may have an *umbric* or *mollic epipedon* <25 cm thick or a *fragipan* or *duripan* under certain conditions. These soils are not dominated by amorphous materials and are not saturated with water for periods long enough to limit their use for most crops. **Ochrepts**

ochric epipedon A surface horizon of mineral soil (U.S. system of soil taxonomy) that is too light in color, too high in chroma, too low in organic carbon, or too thin to be a *plaggen, mollic* and *umbric epipedon, anthropic soil,* or *histic epipedon*, or that is both hard and massive when dry. **épipédon ochrique**

octahedron (1) In the isometric system, a crystal form consisting of eight triangular faces, each having equal intercepts on all three crystallographic axes. (2) A lattice structure in phyllosilicates consisting of an arrangement of ions that can be visualized as four spheres in a square, with one additional sphere centered above and one below, thus forming a symmetrical figure with eight sides and six points (a double pyramid). The space inside an octahedron is the right size for medium-sized cations such as Al^{3+}, Mg^{2+},· or Fe^{2+} (as well as several other less abundant cations of similar size). **octaèdre**

off-site damage costs Costs borne by society as a result of agricultural practice (e.g., deposition of soil and soil nutrients in water bodies as a result of soil erosion). Such costs are often not considered by producers in decision making or agriculture. **coût des dommages hors-ferme**

Ohm's law Law that states that the potential difference (in volts) is equal to the product of the current (amperes) and resistance ohms); V=IR, where V is volts, I is the current, and R is the resistance. **loi d'Ohm**

oil pool An accumulation of *petroleum* in the pores of a sedimentary rock. **gisement de pétrole**

oil sand Porous sandstone or sand in which *petroleum* occurs. **sables bitumineux**

oil shale A fine-grained argillaceous rock containing *hydrocarbons* in the form of a waxy substance called *kerogen.* Crude *petroleum* is obtained by distillation of the hydrocarbons. **pyroschiste**

oil wasteland Land on which oily wastes have accumulated, including slush pits and adjacent areas affected by

O

oil waste. **terrain d'épandage de pétrole**

old age (*geography*) That state in the development of streams and landforms when the processes of erosion are decreasing in vigor and efficiency or the forms are tending toward simplicity and subdued relief. **vieillesse**

old growth forest Forest that contains live and dead trees of various sizes, species, composition, and age class structure. Old-growth forests, as part of a slowly changing but dynamic ecosystem, include climax forests but not sub-climax or mid-seral forests. **vieille forêt**

old-growth forest attributes Structural features and other characteristics of old-growth forests, including large trees for the species and site, wide variation in tree sizes and spacing, accumulations of large dead standing and fallen trees, multiple canopy layers, canopy gaps and understory patchiness, elements of decay such as broken or deformed tops or trunks and root decay, and the presence of species characteristic of old growth. **attributs des vieilles forêts**

O

oligoclase A mineral of the plagioclase feldspar group with albite-anorthite composition ranging from $Ab_{90}An_{10}$ to $Ab_{70}An_{30}$. It is common in igneous rocks of intermediate to high silica content. **oligoclase**

oligotrophic (1) Containing a small amount of nutrients; refers to waters low in nutrient loading with low primary production of organic material by algae and/or macrophytes. Growth in an oligotrophic water is often limited by low levels of phosphorus and nitrogen. (2) Designation for *peatlands* that are poor to extremely poor in nutrients. See *eutrophic, mesotrophic*. **oligotrophe**

oligotrophic lake A lake that has a low supply of nutrients and thus contains little organic matter. Oligotrophic lakes are characterized by high water transparency and high dissolved oxygen. **lac oligotrophe**

olivine A general term used for members of the isomorphous series forming the olivine group. A continuous series of solid solutions ranging from a silicate of magnesium, Mg_2SiO_4, to a silicate of iron, Fe_2SiO_4. The general formula is $M^{+2}SiO_4$, where M^{+2} is Fe, Mg, Mn, or Ni. Olivine is a component of basic rocks and ultrabasic rocks, and occasionally of acid rocks. It is usually dark green in color and exhibits no cleavage. **olivine**

ombrogenous Produced by rain. Commonly applied to *peatland* areas which receive nutrients from precipitation. See *ombrotrophic*. **ombrogène**

ombrotrophic Referring to a supply of nutrients exclusively from rain water (including snow and atmospheric fallout), therefore making nutrition extremely oligotrophic. Commonly applied to *peatland* areas where plants are dependent on nutrients from precipitation. **ombrotrophe**

omnivore (*zoology*) An organism which eats both plants and animals; also called diversivore. **omnivore**

once-over tillage See *tillage, once-over tillage*. **préparation du sol par un passage unique d'outils combinés**

one-way analysis of variance (*statistics*) An *analysis of variance* in which the total sum of squares is expressed as the sum of the treatment sum of squares, the error sum of squares, and no other. **analyse de variance à un critère de classification**

ontogeny The life history or development of an individual (as opposed to that of a race). See *life cycle, phylogeny*. **ontogénie**

onyx A banded variety of chalcedonic *silica,* used as a semiprecious stone in the manufacture of jewelry and objets d'art. **onyx**

oolite See *oolitic limestone*. **oolite**

oolith A spherical rock particle formed by the gradual accretion of material around an inorganic (e.g., a sand grain) or organic (e.g., a piece of shell) nucleus, possibly through the action of algae. Ooliths have a small diameter (0.25 to 2 mm); larger varieties (3 to 6 mm diameter) are called *pisoliths*. Many sedimentary iron ores are made up of ooliths, and are called oolitic ironstones, but ooliths are most common in the form of *oolitic limestone*. **oolithe, oolite**

oolitic limestone A sedimentary rock made up mostly of *ooliths*. Oolitic limestone is usually *calcareous*, but noncalcareous oolites are known (e.g., those formed from iron minerals that replaced calcareous ooliths). Also called *oolite*. **calcaire oolithique**

opacity The degree or quality of being impenetrable to light or other forms of radiation. See *opaque*. **opacité**

opaque (1) Not transmitting light. (2) Not transmitting the particular wavelengths (which may or may not be visible) that affect given photosensitive materials; a substance may be opaque to some colors and not others. (3) A material applied to areas of a negative to make it opaque in those areas. (4) To apply a material to block out light. **opaque**

open drain Natural watercourse or constructed open channel that conveys drainage water. **tranchée de drainage**

open range (*range management*) (1) An extensive grazing area on which the movement of livestock is unrestricted. (2) Range on which the livestock owner is not required to confine his livestock. **terrain de parcours libre**

open system A *system* in which energy and matter are exchanged between the system and its environment. **système ouvert**

open-cast mining See *strip mining*. **exploitation à ciel ouvert**

open-pit mine Mining facilities where the ratio of overburden to mineral is small. See *mining*. **mine ou exploitation à ciel ouvert**

opportunistic species A species that takes advantage of temporary or local conditions. (Populations of opportunistic species usually fluctuate markedly.) **espèce oportuniste**

optical activity The ability of a substance to rotate the plane of plane-polarized light as it passes through a crystal, liquid, or solution. It occurs when the molecules of the substance are asymmetric, so that they can exist in two different structural forms, each being a mirror image of the other. **activité optique**

optical coefficient The percentage of light energy striking a surface that is absorbed, reflected, or passes through the material. **coefficient optique**

optical density The amount of opacity of a translucent object. **densité optique**

optimum level The percentage of a nutrient element present in a crop when it is producing the maximum profit possible. It can be evaluated by an economic interpretation. The optimum level of an element can never exceed the critical percentage of that element in the same plant tissue. **niveau optimal**

orchard A group of crop trees usually arranged in a symmetrical pattern. **verger**

order A unit or group in the taxonomic classification of organisms composed of one or more families; a subdivision of a *class*. **ordre**

order, soil The broadest category in the Canadian, U.S., and other systems of soil classification. All the soils within an order have one or more characteristics in common. Orders are subdivided into suborders in the U.S. soil taxonomy. Ten soil orders are recognized in Canada: *Brunisolic, Chernozemic, Cryosolic, Gleysolic, Luvisolic, Organic, Podzolic, Regosolic, Solonetzic*, and *Vertisolic*. The twelve orders in the U.S. system of soil taxonomy are: *Alfisols, Andisols, Aridisols, Entisols, Gelisols, Histosols, Inceptisols, Mollisols, Oxisols, Spodosols, Ultisols,* and *Vertisols.* **ordre de sols**

ordinal variable A characteristic expressed by rank or order. For example, a sample group might be classified by shades of the color red, with the number 1 being the lightest red, and 10 the darkest red (see *Munsell color system*). Although the ranking is often in numerical order, the data do not represent actual quantities; there is no indication of the measured difference between the ranked samples. Used only in *nonparametric* statistical tests. See *nominal variable.* **variable ordinale ou ordonnée**

ordinate The vertical axis (or *y* axis) in a graph, in contrast to the horizontal axis (or *x* axis). See *abscissa.* **ordonnée**

ordination A method used in areas such as vegetation ecology whereby stands are ordered by the similarity of species composition and relative abundances or environmental gradients. **ordination**

ore A metalliferous mineral or aggregate of minerals in which the metal content is of sufficient economic value to justify extraction. **minerai**

organan (*soil micromorphology*) A *cutan* composed of a concentration of organic matter. **organane**

organelle A distinguishable part of a cell's intracellular structure composed of an organized complex of macromolecules and small molecules (e.g., nucleus, mitochondrion, plastid, ribosome, and mesosome). **organite cellulaire**

organic Derived from living organisms. (*chemistry*) Any compound containing carbon; a molecule that is of biologic origin and contains carbon. All living matter is organic. The original definition of the term related to the source of chemical compounds, with organic compounds being those carbon-containing compounds obtained from plant or animal sources, whereas *inorganic* compounds were obtained from mineral sources. **organique**

Organic An order of soils in the Canadian system of soil classification that have developed dominantly from organic deposits. The majority of Organic soils are saturated for most of the year unless artificially drained, but some are not usually saturated for more than a few days. They contain 17% or more organic carbon, and: (i) if the surface layer consists of fibric organic material and the bulk density is less than 0.1 (with or without a *mesic* or *humic* Op less than 15 cm thick), the organic material must extend to a depth of at least 60 cm; or (ii) if the surface layer consists of organic material with a bulk density of 0.1 or more, the organic material must extend to a depth of at least 40 cm; or (iii) if a *lithic* contact occurs at a depth shallower than stated in (i) or (ii), the organic material must extend to a depth of at least 10 cm. **organique**

organic acid An acid with a carbon-atom backbone; often contains the carboxyl group. **acide organique**

organic carbon (1) Carbon atoms in an organic compound that are linked to other carbon atoms by a covalent bond. Used to distinguish such compounds from *inorganic* forms of carbon, such as those found in the carbonates and cyanides. (2) The amount of organic material in soil or water, usually determined by the difference between total carbon (determined by dry combustion or wet chemistry) and inorganic carbon (determined by acid dissolution). **carbone organique**

organic chemistry The study of carbon-containing compounds (typically chains of carbon atoms) and their properties. **chimie organique**

Organic Cryosol A great group of soils in the Cryosolic order (Canadian system of soil classification). They have developed primarily from organic materials containing more than 17% organic carbon by weight and are underlaid with permafrost within 1 m of the surface. They are greater than 40 cm thick, or greater than 10 cm thick over either a lithic contact or an ice layer that is at least 30 cm thick. **cryosol organique**

organic deposit Materials that have accumulated by growth and death of plants, and that contain more than 17% organic carbon; includes *peat* and *litter* materials. See *chalk, coal, diatomaceous earth, organic soil, peat.* **dépôt organique**

organic farming Crop production systems that attempt to rely solely on organic matter or natural materials for crop production. They avoid or largely exclude the use of synthetic fertilizers, *pesticides*, growth regulators, and livestock feed additives. Organic farming relies instead on crop rotations, crop residues, animal manures, legumes, *green manures*, off-farm organic wastes, mechanical cultivation, mineral-bearing rocks, and biological pest control to maintain soil productivity and tilth, to supply plant nutrients, and to control insects, weeds, and other pests. **agriculture biologique, agriculture organique, agriculture écologique**

organic fertilizer A byproduct of the processing of animal or vegetable substances that contains sufficient plant nutrients to be of value as fertilizer. **engrais organique**

organic load (*limnology*) The amount of organic material added to a body of water. The amount of material, usually added by human activities, that must be mineralized or degraded within a particular environment. **charge organique**

organic matter, soil The organic fraction of the soil; includes plant and animal residues at various stages of decomposition, cells and tissues of soil organisms, and substances synthesized by the soil population. It is usually determined on soil that has passed through a 2.0-mm sieve. **matière organique du sol**

organic nitrogen Nitrogen that is bound to carbon-containing compounds. **azote organique**

organic phosphorus Phosphorus present as a constituent of an organic compound, or a group of organic compounds (e.g., glycerophosphoric acid, inositol phosphoric acid, and cytidylic acid). **phosphore organique**

organic reef (*geology*) A sedimentary rock structure composed mainly of the remains of marine organisms, especially corals and algae. **récif organique**

organic soil A soil composed predominantly of organic matter, in contrast to a mineral soil. See *bog, peat.* **sol organique**

organic soil materials *(Canadian system of soil classification)* Organic soil materials consisting of peat deposits that contain >17% organic carbon (>30% organic matter) by weight. These deposits may be as thin as 10 cm if they overlie bedrock, but are otherwise greater than 40 cm and generally greater than 60 cm thick. Four classes of organic materials are bog (sphagnum or forest peat), fen (fen peat or sedge peat), swamp (forest peat), and undifferentiated organic peat. *(U.S. system of soil taxonomy)* Soil materials that are saturated with water and have 174 g kg^{-1} or more organic carbon if the mineral fraction has 500 g kg^{-1} or more clay, or 116 g kg^{-1} organic carbon if the mineral fraction has no clay, or has proportional intermediate contents, or if never saturated with water, have 203 g kg^{-1} or more organic carbon. Organic soil materials are usually classified according to their degree of decomposition, where *fibric* materials are least decomposed and *mesic* materials are intermediate in

O

degree of decomposition between the less decomposed fibric and the more decomposed *humic* materials. **matériaux de sol organique**

organic solvent A liquid, usually a hydrocarbon such as benzene or toluene, used to dissolve materials such as paints, varnishes, grease, oil, or other hydrocarbons. **solvant organique**

organic weathering (*geology*) The breakdown of rocks by microbial and plant activity, especially by production of organic acids and carbonic acid from dissolved carbon dioxide. See *chemical weathering* and *mechanical weathering*. **altération biologique**

organism Any living or once-living thing. **organisme**

organochlorine See *chlorinated hydrocarbons*. **organochloré**

organomercurial Any of a group of organic molecules complexed with mercury. In general, these compounds are toxic to humans, with the nervous system the most sensitive. **composé organomercuriel**

organo-mineral complex An assemblage of primary particles (e.g., clay, silt, and sand) and organic compounds stabilized by organic and inorganic chemical bonds, electrostatic forces, and biological agents. Generally considered to be a basic unit of soil *aggregation*. See *clay-organo complex*. **complexe organo-minéral**

organophosphate Any of a group of organic insecticides (i.e., *pesticides*) that contain phosphorus. They interfere with nerve transmission through the inhibition of cholinesterase, and can be toxic to humans. Most of the organophosphates are relatively nonpersistent in the environment. **organophosphate**

organotroph An organism capable of obtaining carbon and energy for growth and cell development by utilizing organic compounds. **organotrophe**

organotrophic Obtaining growth, energy, and cell carbon from organic materials. **organotrophe**

organotrophy Nutritional mode of organisms that obtain their electrons from organic compounds. **organotrophie**

oriented tillage See *tillage, oriented tillage*. **travail du sol directionnel**

origin An arbitrary starting point on a scale. Also, the point defined by the intersection of two axes in a system of coordinates. **origine**

ornamental A plant grown for the beauty of its form, foliage, flowers, or fruit, rather than for food, fiber, or other uses. **plante ornementale**

orogenic Of or pertaining to mountain building. **orogénique**

orogeny The process of forming mountain chains, involving buckling, deformation, and compression of strata and resulting in folding, faulting, and thrusting features, often as a result of plate tectonics. **orogenèse**

orophyte A plant that grows only in a mountainous environment. **orophyte**

orphan lands (*land reclamation*) Disturbed surfaces resulting from mine operations that were inadequately reclaimed by the operator and for which the operator no longer has any fixed responsibility; usually refers to lands mined previous to the passage of comprehensive reclamation laws. **terrains orphelins**

Orthels A suborder in the U.S. system of soil taxonomy. *Gelisols* that show little or no *cryoturbation* (less than one-third of the pedon). Patterned ground (except for polygons) generally is lacking. These soils occur primarily within the zone of discontinuous *permafrost*, in alpine areas where precipitation is greater than 1400 mm per year. **Orthels**

Orthents A suborder in the U.S. system of soil taxonomy. *Entisols* that have either textures of very fine sand or finer in the fine earth fraction, or textures of loamy fine sand or coarser and a coarse fragment content of

35% or more, and that have an organic carbon content that decreases regularly with depth. Orthents are not saturated with water for periods long enough to limit their use for most crops. **Orthents**

Orthic A subgroup referring to the modal or central concept of various great groups in the *Brunisolic, Chernozemic, Cryosolic, Gleysolic, Luvisolic, Podzolic,* and *Regosolic* orders of the Canadian system of soil classification. **orthique**

Orthids A suborder in the U.S. system of soil taxonomy. *Aridisols* that have a *cambic, petrocalcic, gypsic,* or *salic* horizon or a *duripan* but that lack an *argillic* or *natric* horizon. **Orthids**

orthoclase A colorless, white, cream-yellow, reddish, or gray monoclinic mineral ($KAlSi_3O_8$) of the feldspar group, with partially ordered Al–Si arrangement, dimorphous with microcline, and having a specific gravity of 2.55 to 2.63 and a hardness of 6 to 6.5 (*Mohs scale*). Orthoclase is a common constituent of granitic rocks, and also occurs in some metamorphic and sedimentary rocks. **orthoclase**

Orthods A suborder in the U.S. system of soil taxonomy. *Spodosols* that have less than six times as much free iron (elemental) than organic carbon in the *spodic* horizon but the ratio of iron to carbon is 0.2 or more. Orthods are not saturated with water for periods long enough to limit their use for most crops. **Orthods**

orthogneiss See *gneiss*. **orthogneiss**

orthophosphate A salt of orthophosphoric acid (e.g., $(NH_4)_2HPO_4$, $CaHPO_4$, or K_2HPO_4). **orthophosphate**

Orthox A suborder in the U.S. system of soil taxonomy. *Oxisols* that are moist all or most of the time, and that have a low to moderate content of organic carbon within the upper 1 m or a mean annual soil temperature of 22°C or more. **Orthox**

ortstein A *diagnostic horizon* (Canadian and U.S. systems of soil taxonomy) which is strongly cemented with illuviated sesquioxides and organic matter, is at least 3 cm thick, and occurs in more than one-third of the exposed face of the *pedon*. **ortstein**

osmoregulation Regulation of the salt concentration in cells and body fluids. **régulation osmotique**

osmosis The tendency of a fluid (i.e., water) to pass through a semipermeable membrane separating solutions of different concentrations (e.g., a cell membrane) into the solution of higher concentration, equalizing the concentrations on both sides of the membrane. **osmose**

osmotic gradient The difference in salt concentrations between the plant and the soil; it may prevent water from being absorbed by the roots and, if great enough, causes water to be drawn out of the plant. **gradient osmotique**

osmotic lysis The rupture of a cell placed in a dilute solution, resulting from an increase in osmotic pressure generated as a result of the concentration of the materials inside the cell. **lyse osmotique**

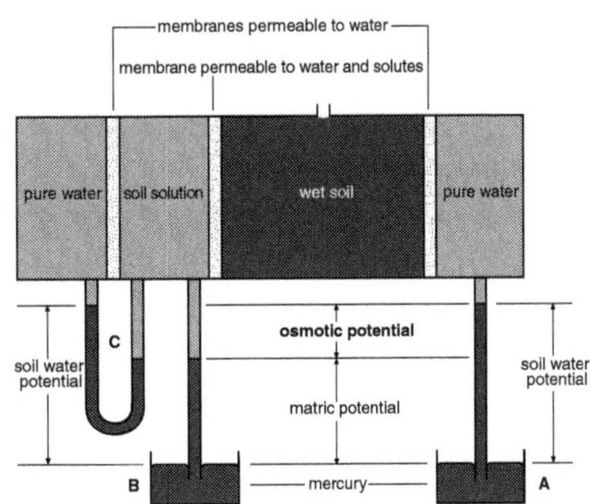

Relationship among **osmotic** (C), matric (B), and combined (A) soil water potential (adapted from Brady and Weil, 1996).

osmotic potential The amount of work that must be done per unit quantity

O

of pure water in order to transport it reversibly and isothermally from a pool of pure water, at atmospheric pressure, to a pool of water identical in composition with the soil water at the same elevation as the point under consideration. See *soil water.* See *figure.* **potentiel osmotique**

osmotic pressure The pressure to which a pool of water, identical in composition with the soil water, must be subjected in order to be in equilibrium, through a semipermeable membrane, with a pool of pure water. (Semipermeable means permeable only to water.) It may be identified with the *osmotic potential.* See *soil water.* **pression osmotique**

osmotroph An organism that obtains nutrients through the active uptake of soluble materials across the cell membrane. This class of organism, which includes the bacteria and fungi, cannot directly utilize particulate material as nutrients. **osmotrophe**

Ostwald process A commercial process for producing nitric acid by the oxidation of ammonia. **procédé d'Ostwald**

outcrop (*geomorphology*) An exposure of a geologic formation or structure at the Earth's surface. **affleurement**

outgassing The loss of vapors or gases by a material, usually as a result of raising the temperature of and/or reducing the pressure on the material. **dégazage**

outlet (*hydrology*) Point of water disposal from a stream, river, lake, tidewater, or artificial drain. An opening through which water can be freely discharged from a reservoir. **sortie d'eau**

outwash Sediments washed out by flowing water beyond active glacier ice and laid down as a sorted and stratified deposit. The deposits consist mostly of sand and gravel, although the particle size may vary from boulders to silt. See *glaciofluvial deposits.* **épandage fluvio-glaciaire**

outwash plain (*geomorphology*) A broad, level to gently sloping sheet of outwash deposited by meltwater streams flowing in front of or beyond a glacier; a broad body of outwash. **plaine d'épandage fluvio-glaciaire**

oven-dry soil Soil that has been dried at 105°C until it has reached constant weight. **sol séché à l'étuve**

overburden Material of any nature, consolidated or unconsolidated, that overlies a deposit of useful materials, ores, or coal, especially those deposits that are mined from the surface by open cuts. **morts-terrains, stérile**

over-consolidated soil deposit A soil deposit that has been subjected to an effective pressure greater than the present overburden pressure. **dépôt de sol surcompacté**

overdraft (*hydrology*) Extraction of groundwater from an aquifer at a rate greater than the aquifer is recharged. **surexploitation**

overgrazed range (*range management*) A range showing loss of plant cover and accelerated erosion as a result of heavy grazing or browsing pressure. **terrain de parcours surpâturé**

overgrazing (*range management*) Grazing so heavy and prolonged that it impairs future forage production and causes deterioration through damage to plants or soil, or both. **surpâturage**

overland flow (*hydrology*) The flow of rainwater or snow melt over the land surface, which is not intercepted by vegetation and which runs as a shallow unchannelled sheet across the soil toward stream channels. After it enters a watercourse, it becomes *runoff.* **ruissellement**

overpopulation A population density that exceeds the capacity of the environment to supply the health requirements of the individual organism. **surpopulation**

overseeding Sowing into an already established crop. **sursemis**

overstocking (*range management*) Placing more animals on a grazing area than that area can sustain. **surcharge pastorale**

overstory The portion of the trees in a forest stand forming the upper crown cover. See *understory*. **étage dominant**

overstory species Species that grow in the top stratum of a community, such as a forest's canopy or the top shrub stratum in a shrubland. **espèce de l'étage dominant**

overstripping (*land reclamation*) Soil salvage operation in pipeline trench reclamation involving the intentional stripping of the upper subsoil with the topsoil. This would only be done where incorporation of the upper subsoil would not significantly degrade the quality of the topsoil. This procedure may be suitable for areas with a shallow topsoil layer and good quality upper subsoil. **surexploitation du sol**

overtopped Trees with crowns entirely below the general level of the crown cover receiving little or no direct light from above or from the sides. **dominé**

overtopping Vegetation higher than the favored species, as in brush or deciduous species shading and suppressing more desirable coniferous trees. **domination**

overturn (*limnology*) The vertical mixing of layers of large bodies of water caused by seasonal changes in temperature. See *fall overturn, spring overturn*. **renversement**

oxbow lake A crescent-shaped lake formed in an abandoned river bend that has become separated from the main stream by a change in the course of the river. **lac en croissant**

oxic horizon A mineral soil horizon (U.S. system of soil taxonomy) that is at least 30 cm thick and characterized by the absence of weatherable primary minerals or 2:1 lattice clays, the presence of 1:1 lattice clays, and highly insoluble minerals such as quartz sand, the presence of hydrated oxides of iron and aluminum, the absence of water-dispersible clay, and the presence of low cation exchange capacity and small amounts of exchangeable bases. **horizon oxique**

oxidant The ability of some oxygen-containing compounds to oxidize other compounds (e.g., ozone, peroxyacetyl nitrates, and nitrogen dioxide). **oxydant**

oxidation (1) The loss of electrons by atoms or compounds. Oxidation reactions are always accompanied by simultaneous reduction reactions in which an atom or molecule gains electrons. (2) The mineralization or decomposition of organic compounds by microorganisms. See *reduction*. **oxydation**

oxidation state The number of electrons to be added (or subtracted) from an atom in a combined state to convert it to the elemental form. Also called oxidation number. **état d'oxydation**

oxidation-reduction (redox) reaction A reaction in which one or more electrons are transferred. **réaction d'oxydo-réduction**

oxidative phosphorylation Conversion of inorganic phosphate into the energy-rich phosphate of adenosine 5′–triphosphate. **phosphorylation oxydative**

oxide mineral Minerals dominated by iron, aluminum, and silicon oxides. The oxide minerals are most abundant in highly weathered materials in tropical areas. **oxyde**

oxides of nitrogen (NOx) See *nitrogen oxides*. **oxydes d'azote**

oxidizing agent (electron acceptor) A reactant that accepts electrons from another reactant. **agent oxydant**

Oxisols An order in the U.S. system of soil taxonomy. Mineral soils that have an *oxic horizon* within 2 m of the surface or *plinthite* as a continuous phase within 30 cm of the surface, and that do not have a *spodic* or *argillic horizon* above the oxic horizon. **Oxisols**

oxygen debt A temporary phenomenon that occurs in an organism when available oxygen is inadequate to supply the respiratory demand and often resulting in the accumulation of breakdown products that are not oxidized until sufficient oxygen becomes available. **déficit en oxygène**

oxygen demand The need or requirement for molecular oxygen (O_2) to meet the needs of biological and chemical processes in water. **demande en oxygène**

oxygen depletion The removal of dissolved oxygen from a body of water as a result of bacterial metabolism of degradable organic compounds added to the water, typically by human activities. See *biochemical oxygen demand*. **épuisement d'oxygène**

oxygen diffusion rate (ODR) The rate of diffusion of oxygen through soil as defined by Fick's law; often estimated in soil by use of platinum electrodes to diffusion-governed oxygen reduction rate (ODR). **taux de diffusion de l'oxygène**

oxyphyte A plant adapted to the excessively acid soils of cool, humid environments (e.g., peat bogs or the floor of a coniferous forest). **oxyphyte**

ozone (*chemistry*) A molecule made up of three atoms of oxygen. (*meteorology*) Ozone occurs naturally in the stratosphere and provides a protective layer shielding the Earth from ultraviolet radiation and the subsequent harmful health effects on humans and the environment. In the troposphere, it is a chemical oxidant and major component of photochemical smog. Ozone is an effective greenhouse gas, especially in the middle and upper troposphere and lower stratosphere. **ozone**

P

P factor See *universal soil loss equation.* **facteur P**

packing voids (1) Voids formed by the random packing of single skeletal grains. (2) Voids formed by the random packing of peds that do not accommodate each other. **vides d'entassement**

paddock A grazing area that is a subdivision of a grazing management unit and is enclosed and separated from other areas by a fence or barrier. **enclos**

palatability (*range management*) A food preference based on plant characteristics resulting in a choice between two or more forages or parts of the same forage, conditioned by the animal and environmental factors that stimulate a selective intake response. **palatabilité, appétibilité**

paleobotany The scientific study of plant life throughout geological time based on the fossil record. **paléobotanique**

paleoecology The scientific study of the relationship between fossil organisms and their environments, using such techniques as pollen analysis and radiocarbon dating to examine the fossil record. **paléoécologie**

paleomagnetism The study of variations in the Earth's magnetic field as recorded in ancient rocks. **paléomagnétisme**

paleosol An ancient soil or a buried soil horizon formed during the geologic past. See *buried soil.* **paléosol**

paleosol, buried A soil formed on a landscape during the geological past and subsequently buried by sedimentation. **paléosol enfoui**

paleosol, exhumed A formerly buried *paleosol* that has been exposed on the landscape by the erosive stripping of an overlying mantle of sediment. **paléosol exhumé**

paleozoology The scientific study of fossil fauna, both vertebrate and invertebrate. **paléozoologie**

palisade parenchyma *Chlorophyll*-containing cells that, along with the spongy *parenchyma*, are located between the upper and lower epidermis, or outer layers, of a leaf. **parenchyme palissadique**

pallid zone A whitish-colored horizon of decomposed kaolinitic clay and quartz sand, free of iron minerals, occurring above deeply weathered bedrock in the tropics. **lithomarge, zone pâle**

palsa A peat-covered mound with a permafrost core; usually *ombrotrophic* (although commonly surrounded by *minerotrophic fen*), generally much less than 50 m across, and from one to several meters high. Palsas may be treeless or may contain a few stunted trees. Palsa growth is due to the build-up of segregated ice mainly in the mineral soil. See *peat mound, peat plateau.* **palse**

paludal A marsh or swamp environment; the sediments deposited therein. **paludien**

paludification The process of *peat* accumulation leading to peatland formation over previously forested land, grassland, or even bare rock, due to climatic or autogenic processes. A characteristic feature is the development of anaerobic conditions due to waterlogging. **paludification**

paludization Accumulation of deep deposits (>30 cm) of organic matter, as in *peat* or *Organic soils.* **paludication**

palygorskite (1) A fibrous clay mineral ($Si_8Mg_2Al_2O_{20}(OH)_2(OH_2)\bullet 4H_2O$) composed of two silica tetrahedral sheets and one aluminum and magnesium octahedral sheet that make up the 2:1 layer that occurs in strips. The strips, which have an average width of two linked tetrahedral chains, are linked at the edges, forming tunnels where water molecules are held. Palygorskite is most common in soils of arid regions. Also called attapulgite. (2) A magnesium aluminum silicate clay used in fertilizer production, including conditioning of fertilizer products, and as a suspending agent in suspension fertilizers. **palygorskite**

palynology The study of pollen and spores. See *pollen analysis*. **palynologie**

pan (1) A horizon or layer, in soils, that is strongly compacted, indurated, or very high in clay content. See *caliche, claypan, fragipan,* and *hardpan*. (2) See *tillage, pans*. **(1) pan (2) couche indurée**

pan, genetic A natural subsurface soil layer of low or very low permeability, with a high concentration of small particles, and differing in certain physical and chemical properties from the soil immediately above or below. See *claypan, fragipan*, and *hardpan*, all of which are genetic pans. **pan d'origine naturelle**

panicle An *inflorescence*, the main axis of which is branched, and whose branches bear loose flower clusters. **panicule**

parabola A mathematical curve ($y = x^2$) that passes through the origin of a graph and increases steadily in gradient away from the axis on both sides, with the value on the *y*-axis increasing as the square of the value on the *x*-axis. **parabole**

parabolic dune (*geomorphology*) A type of curved sand dune, similar in shape to a *barchan* but with the horns pointing upwind instead of downwind. It is usually formed by a *blow-out* process in which the center of a dune is partly removed and carried downwind, leaving the horns behind and drawn out in an elongated form. **dune parabolique**

paradigm A pattern, example, or *model* constructed to test a hypothesis or to formalize a set of data so that they may be systematically analyzed and synthesized. **paradigme**

paraffins Straight- or branched-chain hydrocarbons with carbon-carbon single bonds; the members of this group are relatively nonreactive. A large component of crude oil. Also called *aliphatic* hydrocarbons. When the carbon atoms are arranged in a ring structure, they are called cycloparaffins. **paraffines**

paragneiss See *gneiss*. **paragneiss**

paralic A *littoral* environment where shallow waters predominate, such as a lagoonal environment. See *littoral zone*. **paralique**

paralithic Weathered bedrock that is permeable and penetrable by plant roots. The material has a hardness which is low on *Mohs scale*, and it can be dug manually with a spade or auger. The coarse fragment content of this boundary layer between soil and solid bedrock increases with depth until consolidated rock is encountered. **paralithique**

paralithic contact A boundary between soil and continuous coherent underlying material that has a *hardness* of less than 3 (*Mohs scale*). When moist, the underlying material can be dug with a spade, and chunks will disperse in water or sodium hexametaphosphate solution with 15 hours' shaking. Similar to *lithic contact,* except that the material is softer. **contact paralithique**

parallax (1) The apparent displacement of the position of a body, with respect to a reference point or system, caused by a shift in the point of observation. (2) The apparent displacement between objects on the

P

Earth's surface due to their difference in elevations. **parallaxe**

parameter The true value of a population characteristic. The estimate of a parameter, called a statistic, is a measurement of a sample of the population (e.g., *mean, standard deviation*). It is used more loosely as being synonymous with a *variable* or measurable quantity (e.g., volume, height). **paramètre**

parametric statistical tests Statistical tests applied where certain assumptions are made about the distribution of the background population data from which the samples are drawn, assumptions which are not always warranted (e.g., that the data are drawn from a *normal distribution*). In general, the smaller the sample to be tested, the less likely the validity of a parametric test. Thus, in the latter case, and where normality cannot be assumed, *non-parametric statistical tests* should be used. The *t-test* is the best known example of a parametric test. The latter type of test can only be applied to data on an interval scale. **tests paramétriques**

paraplow See *tillage, paraplow.* **décompacteur de type paraplow**

parasite An organism (the parasite) whose natural habitat is either on (i.e., ectoparasites) or inside (i.e., endoparasites) the host. The parasite consumes part of the blood or tissues of its host, usually without killing the host. **parasite**

paratill See *tillage, paratill.* **décompacteur de type paratill**

parenchyma (*botany*) A tissue of higher plants consisting of living cells with thin walls that are agents of photosynthesis and storage. The *pith* of the shoots, the storage tissue of the fruits, the seeds, the roots, and other underground organs are all parenchyma tissues, as is the *mesophyll* (the assimilation tissue of leaves). **parenchyme**

parent material, genetic material The unconsolidated and more or less chemically weathered mineral or organic matter from which the solum of a soil has developed by *pedogenic* processes. The classes of genetic materials for unconsolidated mineral materials used in soil surveys in Canada are *anthropogenic, colluvial, eolian, fluvial, lacustrine, marine, morainal, saprolite, volcanic ash,* and *undifferentiated*; and for the organic materials there are *bog, fen, swamp,* and *undifferentiated organic material.* **matériau parental ou originel**

parent rock The rock from which the parent materials of soils are formed. **roche-mère**

partial correlation See *correlation, statistical.* **corrélation partielle**

partial cutting (*silviculture*) A general term referring to silvicultural systems other than clearcutting, in which only selected trees are harvested. Partial cutting systems include seed tree, shelterwood, selection, and *clearcutting with reserves.* **coupe partielle**

partial pressure The pressure exerted by an individual gas in a gaseous mixture. If the mixture of gases is assumed to behave as an *ideal gas*, the sum of the individual partial pressures in a mixture is equal to the total pressure of the mixture of gases. For example, the atmosphere contains roughly 78% nitrogen and 21% oxygen, and many other gases comprise the remaining 1%. Of a total atmospheric pressure of 760 millimeters of mercury (mm Hg), the partial pressure of the nitrogen is about 593 mm Hg (i.e., 78% of 760 mm Hg), the oxygen partial pressure is about 160 mm Hg (21%), and the other gases total about 7 mm Hg. **pression partielle**

partial sterilization The elimination of a portion of a population of microorganisms, usually by heat or chemical treatment. The process is selective, and certain organisms or groups of organisms are destroyed to a greater extent than others. **stérilisation partielle**

particle accelerator A device used to accelerate nuclear particles to very high speeds. **accélérateur de particules**

particle density The mass per unit volume of the soil; usually expressed in Mg m^{-3}. Sometimes called grain density. See also *bulk density*. **densité particulaire**

particle size The effective diameter of a particle measured by sedimentation, sieving, or micrometric methods. Sometimes called grain size. **calibre**

particle-size analysis The determination of the various amounts of the different separates in a soil sample, usually by sedimentation, sieving, micrometry, or a combination of these methods. Sometimes called grain-size analysis or mechanical analysis. **analyse mécanique ou granulométrique**

particle-size distribution The fractions of the various soil separates (i.e., clay, silt, and sand) in a soil sample, expressed as mass percentages. Sometimes called grain-size distribution. **distribution granulométrique**

particulate matter (*water*) Particles of living or dead organic material of plant or animal origin, and in some cases mineral (e.g., silt, clay) material, that are suspended in water. (*pollution*) Solid material in water in either the solid or dissolved state. Solid particles or liquid droplets carried by a stream of air or other gases. **matière particulaire**

particulate organic matter (POM) (*soil*) Relatively large particles of organic matter mainly consisting of plant residues (e.g., fine roots and partially decomposed organic residue) but also possibly contain fungal hyphae, spores, seeds, charcoal, and animal remains. Particulate organic matter is often operationally defined as the organic matter that is retained on a sand-size sieve (i.e., $>53\mu m$ or $>20\mu m$) after dispersing a soil. See *macroorganic matter*. (*water*) Material of plant or animal origin that is suspended in water. **matière organique particulaire (MOP)**

particulate phosphate The portion of the total amount of phosphate (PO_4^{-3}) suspended in water that is attached to particles and will not pass through a filter in either inorganic or organic forms; must be solubilized before it can be used as a plant nutrient. **phosphate particulaire**

partition coefficient A measure of the distribution of some chemical between two immiscible solvents. **coefficient de partage ou de répartition**

pass band (*remote sensing*) A wavelength range over which a sensor is capable of receiving and measuring electromagnetic energy. **bande passante**

pasteurization A process in which fluids are heated at temperatures below boiling to kill pathogenic microorganisms in the vegetative state. **pasteurisation**

pasture An area devoted to the production of forage, introduced or native, and harvested by grazing. **pâturage**

pasture improvement Any practice of grazing, mowing, fertilizing, liming, seeding, or other methods of management designed to improve vegetation for grazing purposes. **amélioration du pâturage**

pasture, tame Grazing lands, planted to introduced or domesticated forage species, that may receive periodic cultural treatments such as renovation, fertilization, and weed control. **pâturage cultivé ou amélioré**

pasture, temporary A field of crop or forage plants grazed for only a short period, usually not more than one crop season. **pâturage temporaire**

patch (*landscape ecology*) A particular unit with identifiable boundaries that differs from its surroundings in one or more ways. These can be a function of vegetative composition, structure, age, or some combination of the three. **polygone**

pathogen A microorganism, such as a bacterium or fungus, that has the capacity to cause disease under normal conditions. **agent pathogène**

pathogenic Causing or capable of causing disease. Having the potential to cause disease. **pathogène**

patina (*geology*) A discoloring or surface film found on rock exposures due to *chemical weathering*. **patine**

patterned (ribbed) (*geomorphology*) A type of surface expression associated with *fen peatlands* and consisting of a pattern of parallel or reticulate low ridges. **côtelé, strié**

PCBs See *polychlorinated biphenyls*. **BPC**

peak discharge (*hydrology*) Rate of discharge of a volume of water passing a given location. See *flood peak*. **débit maximal**

peat Material constituting *peatland*, exclusive of live plant cover, consisting largely of organic residues accumulated as a result of incomplete decomposition of dead plant constituents under conditions of excessive moisture (e.g., submergence in water and/or water-logging). Peatland may contain a variable proportion of transported mineral material; may form in both base-poor and base-rich conditions, and either as *autochthonous peat* or *allochthonous peat*; and may contain basal layers of coprogenic elements and comminuted plant remains (such as *gyttja*) or *humus* gels (such as *dy*). The physical and chemical properties of peat are influenced by the nature of the plants from which it has originated, by the moisture relations during and following its formation and accumulation, by geomorphological position and climatic factors. The moisture content of peat is usually high; the maximum water-holding capacity occurs in *Sphagnum* peat, being over 10 times its dry weight and over 95% of its volume. Most peats have a high organic content (85% and more). However, in general, peat must have an organic matter content of not less than 30% of the dry weight (about 17% carbon content). **tourbe**

peat bog See *bog*. Also, a term frequently used to describe any *peatland*. **tourbière**

peat moss *Peat* sold for horticultural uses, generally composed of weakly decomposed *Sphagnum* spp. Also called moss peat, raw moss, Sphagnum moss. **mousse de tourbe**

peat mound Localized accumulation of *peat*, larger than a *hummock*, usually *ombrotrophic*. A peat mound with a permafrost core is called a *palsa*. **butte tourbeuse**

peat plateau A low, generally flat-topped expanse of *peat*, rising one or more meters about the surface of a *peatland*. A layer of permafrost exists in the peat plateau, which may extend into the peat below the peatland surface and into the underlying mineral soil. See *palsa*. Some controversy exists about whether peat plateaus and palsas are only morphological variations of the same feature or are genetically different. **plateau tourbeux**

peat soil A soil developed in peat. **sol tourbeux**

peat stratigraphy A vertical sequence of layers of different materials within a peat deposit. Differences may be due to floristic composition, state of decomposition, or incidence of extraneous materials. **stratigraphie de tourbière**

peatland All types of peat-covered terrain. See *muskeg*. **milieu tourbeux, tourbière**

ped A unit of *soil structure* used in pedology and for soil classification purposes to describe *aggregates* formed by natural processes, in contrast to a *clod* or aggregate formed in cultivated soils. **ped**

pedality (*soil micromorphology*) The degree to which peds are surrounded by voids. **pédalité, pédicité**

pedestal A column of soil supporting stones and plant debris, indicative of water erosion. Also, a rock pinnacle with a narrow base and stem and a large head or cap rock, thought to have been

P

formed in an arid climate by the action of sand blasting upon less-resistant rock strata. **pilier d'érosion**

pediment (*geomorphology*) A gently inclined erosion surface of low relief, occurring below a markedly steeper slope and extending at a low gradient down towards a river or alluvial plain; typically developed in arid or semiarid regions at the foot of a receding mountain slope; may be bare or mantled by a thin layer of alluvium in transit to the adjoining plain or basin. **pédiment**

pediplain (*geomorphology*) A surface of low relief, broken by occasional residual hills, thought to be produced by the coalescence of several *pediments*. It is a relatively smooth surface with a gently concave profile and with relatively few streams. **pédiplaine**

pedogenesis The natural process of soil formation. **pédogenèse**

pedogenic Pertaining to the processes responsible for the development of the solum. **pédogénétique**

pedological feature Recognizable entities observed microscopically in the soil material that are unique from the surrounding adjacent materials due to any reason, i.e., genesis, deposition, chemical concentration, or morphology (e.g., *cutans, concretions*). **trait ou caractéristique pédologique**

pedology The study of soils that integrates their distribution, formation, morphology, and classification as natural landscape bodies. **pédologie**

pedon The smallest, three-dimensional unit at the surface of the Earth that is considered as a soil. Its lateral dimensions are large enough to permit the study of horizon shapes and relations, with area ranging from 1 to 10 m^2. Where horizons are intermittent or cyclic, and recur at linear intervals of 2 to 7 m, the pedon includes one-half of the cycle. Where the cycle is <2 m, or all horizons are continuous and of uniform thickness, the pedon has an area of approximately 1 m^2. If the horizons are cyclic but recur at intervals >7 m, the pedon reverts to the 1 m^2 size, and more than one soil will usually be represented in each cycle. **pédon**

pedotransfer function A function that predicts difficult-to-measure soil properties from readily obtained, or basic, properties of the same soil. An interpolative technique based on regression equations often used in *soil quality* assessment. There are two types: continuous pedotransfer functions use soil properties as regressed variables; class pedotransfer functions use soil type or horizon as the regressed variable. **fonction de pédotransfert**

pedoturbation A process of soil formation in which soil materials undergo mixing by biological and physical (freeze-thaw and wet-dry cycles) churning and cycling of soil materials, thereby homogenizing the solum to various degrees. **pédoturbation**

pegmatite A very coarse-grained (>3 cm diameter) igneous rock, with some crystals reaching lengths of more than 1 m. It occurs in the form of granite veins and dykes, and other coarse-grained rocks (e.g., gabbro-pegmatite), which are called pegmatitic rocks. The pegmatites contain abundant accessory minerals (e.g., tourmaline, topaz, fluorite, and apatite) and are important economically because they contain many rare elements. **pegmatite**

pelagic zone The environment of the open ocean, away from the shore; a marine biome type comparable with the limnetic zone of lakes. See *limnetic zone*. **région ou milieu pélagique**

peneplain (*geomorphology*) An undulating surface of low relief, interspersed with occasional residual hills known as *monadnocks,* thought to have been formed by the widening of floodplains and the wearing down of *interfluves* by *subaerial* denudation. **pénéplaine**

P

peneplanation The lowering of a land surface by *subaerial* denudation to form a *peneplain*. **pénéplanation**

penetrability The ease with which a probe can be pushed into the soil. It may be expressed in units of distance, speed, force, or work, depending on the type of penetrometer used. See *cone penetrometer*. **pénétrabilité**

penetration resistance The force per unit area on a standard cone necessary for penetration by the cone into soil. The load required to maintain a constant rate of penetration into soil of a probe or instrument. **résistance à la pénétration**

penetrometer A device for measuring *penetration resistence*. **pénétromètre**

pentachlorophenol (PCP) A white organic solid with needle-like crystals and a phenolic odor. Its most common use is as a wood preservative (fungicide). Though once widely used as a herbicide and insecticide, it has been banned in most countries for these and other uses, as well as for any over-the-counter sales. When released to soil or water, PCP will be slowly broken down by microbes and may gradually leach into ground water. If released in water, it adsorbs to sediment or is degraded by sunlight. Its accumulation in fish will is moderate. **pentachlorophénol (PCP)**

peptide Any of a group of organic compounds comprising two or more amino acids linked by *peptide bonds*. **peptide**

peptide bond The bond resulting from the condensation reaction between amino acids; formed by the reaction between adjacent carboxyl (–COOH) and amino (–NH_2) groups, with the elimination of water. **liaison peptidique**

peptidoglycan The rigid acetylglucosamine, acetylmuramic acid-amino acid layer of bacterial cell walls. **peptidoglycane**

percent error *(statistics)* The error of a measurement expressed as a fraction of the measurement itself. For example, a measurement expressed as 20 cm, plus or minus 5%, means that the true value is considered to be between 19 and 21. **pourcentage d'erreur**

percentile *(statistics)* A value of a ranked data series expressed as a percentage division. For example, the upper ten percentile indicates the value exceeded by 10% of the data. **percentile, centile**

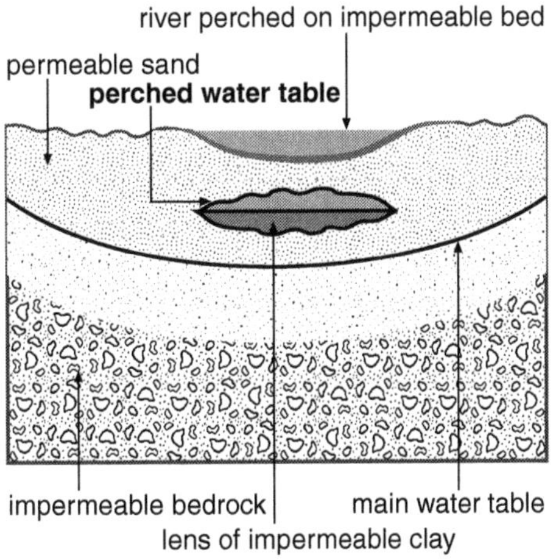

Perched water table (adapted from Dunster and Dunster, 1996).

perched water table A water table that is elevated above an impermeable layer at some depth within the soil. The soil within or below the impermeable layer is not saturated with water. See *figure*. **nappe d'eau suspendue ou perchée**

percolation (of soil water) The downward movement of water through soil; specifically, the downward flow of water in saturated or nearly saturated soil at hydraulic potential gradients of 1.0 or less. **percolation**

perennial plant A plant that persists from year to year and usually produces reproductive structures in two or more different years. See *annual plant, biennial plant*. **plante pérenne ou vivace**

perennial stream A stream that flows at all times of the year, in contrast to *intermittent stream*. **cours d'eau pérenne**

perfect elasticity A property possessed by a material that is able to return to its original form after the removal of an applied force. See *elastic limit.* **élasticité parfaite**

perfect flower A flower containing both male (stamen) and female (pistil) parts. See *imperfect flower.* **fleur complète**

perfect plasticity A property possessed by a material that is unable to return to its original form after an applied force has been removed, but retains a new form instead. **plasticité parfaite**

pergelic A soil temperature regime that has mean annual soil temperatures of <0°C. Permafrost is present in soils with a pergelic temperature regime. **pergélique**

periglacial Non-glacial processes and features of cold climates regardless of age and any proximity to glaciers. **périglaciaire**

period (*physics*) For *electromagnetic radiation*, the time required for one wave cycle to pass a given point; the inverse of *frequency*. A period is given in units of time (e.g., seconds). (*geology*) A unit of geological time in the standard *chronostratigraphic* classification. A period comprises a number of *epochs,* and several periods make up an *era.* See *Appendix C*. The body of rock formed in an era is known as a *system.* **période**

periodic drift (*limnology*) The drift of bottom organisms in water bodies at regular or predictable intervals, such as diurnal and seasonal. **dérive périodique**

periodic table An arrangement of the chemical elements by *atomic number* (increasing number of protons) to show the relationship between elements. Horizontal rows are called periods; vertical rows are called groups. **tableau périodique**

periphyton Microscopic organisms attached to and growing on the bottom of a waterway or on submerged objects; also called *aufwuchs.* **périphyton**

perlite A lightweight, granular material made of a volcanic mineral treated by heat and water so that it expands like popcorn; used as, or a component in, growing media. **perlite**

permafrost A condition existing below the ground surface, irrespective of its texture, water content, or geological character, in which the temperature in the material has remained below 0°C continuously for more than a year and, if pore water is present in the material, a sufficiently high percentage is frozen to cement the mineral and organic particles. The term describes permanently frozen ground, but permafrost has been subdivided into continuous and discontinuous permafrost, while sporadic permafrost is confined to alpine environments. **pergélisol**

permafrost table The upper boundary of permafrost, usually coincident with the lower limit of seasonal thaw (active layer). See *permafrost.* **limite du pergélisol**

permanent charge The net negative or positive charge of clay particles inherent in the crystal lattice of the particle. It is not affected by changes in pH or by ion exchange reactions. **charge permanente**

permanent crop cover A perennial crop, such as a forage, that protects the soil throughout the year. **couverture végétale permanente**

permanent hardness of water Water hardness that cannot be removed by boiling because of the very high concentration of calcium and magnesium sulfate. **dureté permanente de l'eau**

permanent pasture Grazing land occupied by perennial pasture plants or by self-seeding annuals, usually both, that remains unplowed for many years. See *rotation pasture.* **pâturage permanent**

permanent wilting point For agronomic purposes, the soil water content associated with wilting of test plants (that

P

do not recover under humid conditions) as a measure of the lower end of the plant *available water*. Often estimated by the water content at a soil matric potential of -1.5 MPa. See *field capacity*, *soil water*. See *figure*. **point de flétrissement permanent**

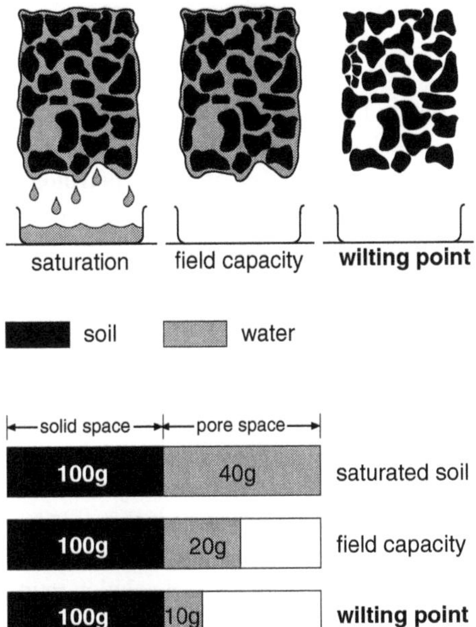

Volume of solids, water, and air in a loam soil at saturation, field capacity, and **wilting point** (adapted from Brady and Weil, 1996).

permeability The ease with which a porous medium (e.g., bulk soil or a layer of soil) transmits gases, such as air, or liquids, such as water. In soil, because different layers or horizons vary in permeability, the specific horizon should be designated. Used synonymously with *hydraulic conductivity*. See *intrinsic permeability*. **perméabilité**

permeameter A device for confining a sample of soil or porous medium and subjecting it to fluid flow, in order to measure the *hydraulic conductivity* or *intrinsic permeability* of the soil or porous medium. **perméamètre**

permease An enzyme that increases the rate of transfer of a substance across a membrane. **perméase**

Perox A suborder in the U.S. system of soil taxonomy. Oxisols that have a perudic soil moisture regime. **Perox**

persistence The relative ability of a chemical to remain stable (e.g., resist decomposition) and effective following its release into the environment. **persistance, stabilité**

persistent pesticide A *pesticide* that is slowly decomposed, often by physical rather than biological processes, after its application, and remains active in the environment for more than one year. **pesticide rémanent**

perudic A *udic* moisture regime in which the soil is wet throughout the year. **perudique**

pest Any organism regarded as harmful, irritating, or offensive to humans, either directly or indirectly through its effect on animals and plants; the term is typically applied to rats and other rodents, insects that transmit disease or destroy crops, and pathogenic fungi and bacteria. In agriculture and forestry, a pest is any agent designated as detrimental to the health of vegetation or animals and which impedes effective resource management. **organisme nuisible, ennemi des cultures, ravageur**

pest incidence A measurement of the presence and magnitude of pests within a given area. **incidence des ravageurs**

pest resistant varieties Plant varieties of economic crops that have been improved so they are less likely to be affected by plant diseases and/or other pests. **variétés ou cultivars résistants aux organismes nuisibles**

pesticide Chemical or biological substances used to kill organisms (e.g., insects, microorganisms, rodents, weeds) that are harmful to plants, animals, or humans. Pesticides include *insecticides*, *fungicides*, *herbicides*, and *rodenticides*. **pesticide**

pesticide residue A small amount of *insecticide*, *fungicide*, or *herbicide* remaining in or on a harvested food or feed crop. **résidu de pesticide**

pesticide tolerance A scientifically and legally determined limit for the amount of chemical (i.e., *pesticide*) residue that can be permitted to remain in or on a harvested food or feed crop as a result of the application of a chemical for pest control purposes; such tolerances or safety levels are set well below the point at which residues may be harmful to consumers. **tolérance aux pesticides**

Petri dish A shallow covered dish used for the isolation and/or culture of bacteria. The standard dish is 100 × 15 millimeters and consists of two overlapping halves. A solid media (agar) is placed in the bottom portion of the dish for inoculation of bacterial cultures. **boîte de Pétri**

petrocalcic horizon A continuous, indurated *calcic horizon* that is cemented by calcium carbonate and, in some places, with magnesium carbonate. It is difficult to penetrate with a spade or auger when dry. Dry fragments do not slake in water, and it is impenetrable to roots. A diagnostic subsoil horizon in the U.S. system of soil taxonomy. **horizon pétrocalcique**

P

petrochemicals Organic chemicals obtained from petroleum or natural gas either by steam cracking or as byproducts of refinery operations. **dérivés du pétrole**

petrofabric analysis The study and measurement of the spatial relationships of rock fragments, particles, and mineral grains in a rock, till, or other sediment. **analyse de la fabrique**

petrogenesis The branch of *petrology* which examines the origins of rocks, especially igneous rocks. **pétrogenèse**

petrogypsic horizon A continuous, strongly cemented, massive *gypsic horizon* that is cemented by calcium sulfate. It can be chipped with a spade when dry. Dry fragments do not slake in water, and this horizon is impenetrable to roots. A diagnostic subsoil horizon in the U.S. system of soil taxonomy. **horizon pétrogypsique**

petroleum A complex mixture of hydrocarbons, derived from crude oil, with a carbon content of 83 to 87%, a hydrogen content of 11 to 14%, and minor amounts of oxygen, nitrogen, sulfur, and traces of metals (e.g., lead). The mixture is found in various geological deposits and can be refined to produce such products as gasoline, fuel oil, kerosene, and asphalt. **pétrole**

petrology The scientific study of rocks. It includes *geochemistry, lithology, mineralogy,* and *petrogenesis.* **pétrologie**

pH curve (titration curve) A plot showing the pH of a solution being analyzed as a function of the amount of titrant added. **courbe de titration**

pH scale A logarithmic scale for expressing the acidity or alkalinity of a solution, i.e., equal to $-\log[H^+]$. **échelle pH**

pH, soil The pH of a solution in equilibrium with soil. It is determined by means of a glass, quinhydrone, or other suitable electrode or indicator at a specified moisture content or soil-solution ratio, in a specified solution such as distilled water or 0.01 M $CaCl_2$. **pH du sol**

phagocyte A cell capable of engulfing particles including living organisms (e.g., bacteria) and particulate organic matter. **phagocyte**

phagotroph (*zoology*) An organism that obtains nutrients through the ingestion of solid organic matter. This class of organism includes all animals, from the simplest single-celled protozoa to the higher forms. **phagotrophe**

phase (*pedology*) A subdivision of a soil type of other unit of classification having characteristics that affect the use and management of the soil, but which do not vary sufficiently to differentiate it as a separate soil type. A variation in a property or characteristic (e.g., degree of slope, degree of erosion, stone content). (*geology*) A variety differing slightly from the normal type; a facies. (*geophysics*) An event on a seismogram marking the arrival of a group of *seismic waves* and indicated by a change of amplitude and/or period. (*general systems theory*) A distinct type, region, or economy of systems operation, commonly demarcated by thresholds. (*chemistry*) The distinct, physically separable conditions of a substance, as a solid, liquid, or gas. **phase**

pH_C . The calculated pH that a water would have if it were in equilibrium with calcium carbonate. Numerically, pH_C is equal to $(pK_2 - pK_C) + p(Ca) + pAlk$, where p(Ca) and pAlk are the negative logarithms of the molar concentrations of Ca and of the equivalent concentration of $(CO_3 + HCO_3)$, respectively, and pK_2 and pK_C are the negative logarithms of the second dissociation constant of H_2CO_3 and the solubility constant of $CaCO_3$, respectively, both corrected for ionic strength. It is used in conjunction with the measured pH of a water to determine if $CaCO_3$ will precipitate from the water, or if the water will dissolve $CaCO_3$ as it passes through a calcareous soil. **pH_C**

pH-dependent cation exchange capacity The difference between the *effective cation exchange capacity* and the *cation exchange capacity* of a soil measured at a pH higher than that of its natural value. **capacité d'échange cationique dépendante du pH**

pH-dependent charge The portion of the cation or anion exchange capacity which varies with pH. See *acidity, residual*. **charge dépendante du pH**

phenocryst A relatively large, prominent crystal embedded in a finer-grained matrix or groundmass of igneous rock. **phénocristal**

phenology The study of periodic biological phenomena that recur (e.g., flowering and seeding), especially as related to climate. **phénologie**

phenols A group of organic compounds that contain a hydroxyl group (–OH) bound directly to a carbon atom in a benzene ring. Unlike normal alcohols, phenols are acidic because of the influence of the aromatic ring, thus phenol itself ionizes in water. In very low concentrations phenols produce a taste and odor problem in water, and at higher levels are toxic to aquatic life; they are byproducts of petroleum refining, tanning, and other manufacture. **phénols**

phenotype An organism as observed by its visible characteristics, resulting from the interaction of its genotype with the environment. See *genotype*. **phénotype**

phenoxy Any of a group of broadleaf *herbicides* that cause abnormal growth by upsetting a plant's hormone balance. They have both pre-emergence and post-emergence activity. **herbicide de type hormonal, phénoxy**

phi scale A scale, based on the negative logarithm to the base 2, used in sedimentary *petrology* to describe and delimit the range of particle sizes into a number of size classes to indicate degrees of *sorting*. See *particle size*. **échelle granulométrique logarithmique des unités phi**

phloem Vascular tissue in plants that conducts sugars and other synthesized food materials from the regions of manufacture to those of consumption and storage. Phloem is found in the vascular bundles, the longitudinal strands of conductive tissue, in association with the water-conducting tissue, or *xylem*. See *figure*. **phloème**

P

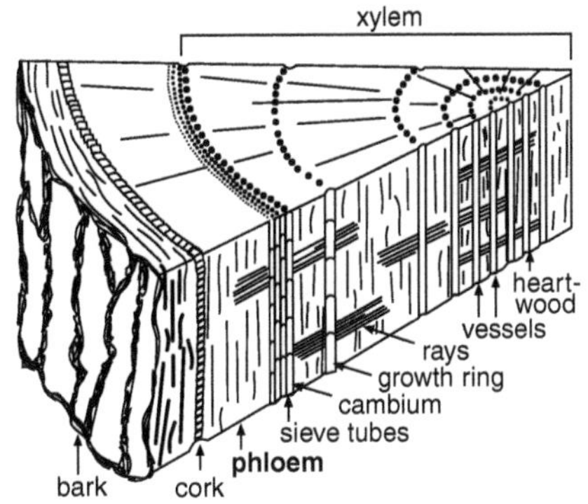

Phloem (adapted from Dunster and Dunster, 1996).

phosphate A compound salt of phosphoric acid with a complex mineralogy. Phosphatic deposits include: marine phosphates, which are rare as sediments but occasionally form phosphatic limestones; bone beds and *coprolites*; and *guano*. Phosphate is one of the three main elements needed for successful plant growth. A super-phosphate is one that has been treated with sulfuric acid to form an important agricultural fertilizer. In fertilizer terminology, phosphate is used to express the sum of the water-soluble and citrate-soluble phosphoric acid (P_2O_5); also referred to as available phosphoric acid (P_2O_5). **phosphate**

phosphate rock A porous, lower-density, microcrystalline, calcium fluorophosphate of sedimentary or igneous origin, often containing impurities of iron, aluminum, and magnesium. It is usually concentrated and solubilized to be used directly or concentrated in the manufacture of commercial phosphate fertilizers. **roche phosphatée**

phosphobacteria Bacteria capable of converting *organic phosphorus* into orthophosphate. **phosphobactéries**

phospholipids Esters of glycerol containing two fatty acids and a phosphate group. Having nonpolar tails and polar heads, they tend to form bilayers in aqueous solution. **phospholipides**

phosphorescence Emission of light that continues after the exciting mechanism has ceased. **phosphorescence**

phosphoric acid In commercial fertilizer manufacturing, it is used to designate orthosphorphoric acid (H_3PO_4). In fertilizer labeling, it denotes the phosphate content in terms of available phosphorus, expressed as percent P_2O_5. **acide phosphorique**

phosphorite Rock phosphate containing various calcium phosphates, most of which are derived from apatite. See *phosphate*. **phosphorite**

phosphorus An element essential to life and important in many plant metabolic processes including photosynthesis and respiration, energy storage and transfer, protein and carbohydrate metabolism, cell division and enlargement; also contributes to the eutrophication of lakes and other bodies of water. **phosphore**

phosphorus cycle The conversion of phosphorus from one form to another via biological, geological, and chemical processes. The sequence of biological, chemical, and other changes undergone by phosphorus wherein it is used by living organisms, liberated upon the death and decomposition of the organism, and converted to its original state of oxidation. In soil, the phosphorus cycle is mainly controlled by microbial processes in which plant *organic phosphorus* or insoluble *inorganic phosphorus* are transformed via biochemical and chemical processes. **cycle du phosphore**

phosphorus, total Soluble and insoluble orthophosphates, condensed phosphates, and organic and inorganic species. **phosphore total**

photic zone (*limnology*) See *euphotic*. **zone photique**

photoautotroph An organism capable of utilizing light energy for growth; an organism that utilizes sunlight as its primary energy source for the synthesis of organic compounds. **photoautotrophe**

photochemical Describing a chemical reaction that is driven by sunlight. **photochimique**

photoelectric cell A transducer that converts *electromagnetic radiation* in the infrared, visible, or ultraviolet regions into electrical quantities, such as voltage, current, or resistance; also called photocell. **cellule photoélectrique**

photoelectric effect The emission of an electron from a substance exposed to *electromagnetic radiation*. **effet photoélectrique**

photoelectron An electron emitted from a substance as a result of the *photoelectric effect*. **photoélectron**

photogrammetry The preparation of charts and maps from aerial photographs using stereoscopic equipment and

P

methods; also called aerial photogrammetry. **photogrammétrie**

photolithotroph An organism that uses light as a source of energy and CO_2 as the source of carbon for cell biosynthesis. **photolithotrophe**

photolysis The breakdown of a material by sunlight; an important mechanism for the degradation of contaminants in surface water and the terrestrial environment. **photolyse**

photomap A mosaic map made from aerial photographs showing physical and cultural features as on a *planimetric map*. **photo-carte**

photometer An instrument for measuring the intensity of light or the relative intensity of a pair of lights; also called illuminometer. If the instrument is designed to measure the intensity of light as a function of wavelength, it is called a spectrophotometer. **photomètre**

photomicrograph A photograph of an enlarged or macroscopic view of a microscopic object, taken by attaching a camera to a microscope. Also referred to as a microphotograph. **photographie microscopique**

photon A *quantum* of electromagnetic radiation. A unit of intensity of electromagnetic radiation, including light. A photon has properties that relate to both particles and waves; it has no charge or mass; however, it does have momentum. **photon**

photoperiod (*botany*) A period of a plant's daily exposure to light. **photopériode**

photoperiodism The growth and flowering response of flora in relation to latitudinal and seasonal changes in the length of daylight hours. **photopériodisme**

photorespiration (*botany*) Respiratory activity due to O_2 reaction instead of CO_2 in the photosynthetic pathway that takes place in cool-season plants during the light period; no useful form of energy is derived. **photorespiration**

photosensitization (*range management*) A noncontagious disease resulting from the abnormal reaction of light-colored skin to sunlight after a photodynamic agent has been absorbed through the animal's system. Grazing on certain kinds of vegetation or ingesting certain molds under specific conditions causes photosensitization. **photosensibilisation**

photosynthate Carbohydrates and other organic molecules produced by plants during the process of photosynthesis. **photosynthat**

photosynthesis A process in green plants and some bacteria whereby light energy is absorbed by chlorophyll-containing molecules and converted to chemical energy (the light reaction). During the process, carbon dioxide is reduced and combined with other chemical elements to provide the organic intermediates that form plant biomass (the dark reaction). Green plants release molecular oxygen, which they derive from water during the light reaction. **photosynthèse**

photosynthetic quotient The volume of oxygen produced during photosynthesis as a proportion of the volume of carbon dioxide used. **quotient chlorophyllien**

phototriangulation The process for the extension of horizontal and/or vertical control whereby the measurements of angles and/or distances on overlapping photographs are related into a spatial solution using the perspective principles of the photographs; generally, this process involves using aerial photographs and is called aero-triangulation, aerial triangulation, or photogrammetric triangulation. **phototriangulation**

phototrophic Obtaining growth energy from light by means of photosynthetic processes. **phototrophe**

phototropism The response of a plant or animal to a source of bright light

P

(e.g., the orientation of a flowering plant towards the sun, or the flight of an insect towards a light at night). See *heliotropism*. **phototropisme**

phreatic water See *groundwater*. **eau souterraine, nappe phréatique**

phthalates Agents added to plastics to improve their flexibility. Over 25 different compounds are produced for commercial use, with di(2-ethylhexyl) phthalate (DEHP) and di-n-butyl phthalate (DBP) the most common. Because they are used in every major category of consumer products and are only slowly degradable, phthalates are distributed throughout the environment. The level of acute toxicity of the compounds is very low; however, some evidence indicates a possible capability to induce tumors. **phtalates**

phycosphere The zone of water influenced by algal exudates. **phycosphère**

phyllite A *clay* sediment altered by low-grade regional metamorphism into a metamorphic rock midway between a slate and a *schist*. It is coarser-grained and less perfectly cleaved than a slate. **phyllite**

P

phyllosilicate A class or structural type of *silicate* in which the SiO_4 tetrahedra are lined together in infinite two-dimensional sheets and are condensed with layers of AlO or MgO octahedra in the ratio 2:1 or 1:1. Isomorphous substitution of certain elements often occurs. Also called layer silicate mineral. **phyllosilicate**

Terms used to describe phyllosilicate minerals include:

interlayer – materials between structural layers of minerals, including cations, hydrated cations, organic molecules, and hydroxide groups or sheets. ***position inter-couche***

layer – a combination of sheets in a 1:1 or 2:1 assemblage. ***couche***

plane of atoms – a flat (planar) array of one atomic thickness (e.g., a plane of basal oxygen atoms within a tetrahedral sheet). ***plan atomique***

sheet of polyhedra – a flat array more than one atom thick and composed of one level of linked coordination polyhedra. A sheet is thicker than a plane and thinner than a layer (e.g., tetrahedral sheet, octahedral sheet). ***couche polyédrique (tétraédrique ou octaédrique)***

phyllosphere The surface of above-ground living plant parts. **phylloplan**

phylogeny The evolutionary development of a group or species of organisms. See *ontogeny*. **phylogénie, phylogenèse**

phylum One of the primary divisions of the animal and plant kingdom; a group of closely related classes of animals or plants. **phylum**

physical geology The study of *geology* combined with aspects of *geomorphology*. **géologie physique**

physical properties of soils The characteristics, processes, or reactions of a soil that are caused by physical forces and are described by, or expressed in, physical terms or equations. Sometimes physical properties are confused with, and/or difficult to separate from, chemical properties; hence, terms *physical-chemical* or physico-chemical properties are often used. Examples of physical properties are *bulk density, water-holding capacity, hydraulic conductivity, porosity, pore size distribution*. **propriétés physiques des sols**

physical rooting conditions Physical soil characteristics, such as waterholding capacity and porosity, that control root growth. **conditions physiques d'enracinement**

physical system A research methodology based on the scientific method and analogous to *general systems theory*, but differing from it in being characterized by a greater dissection of the specific problem into its component parts, such that the operation of each part and the interactions between the parts can be conveniently examined. In this way it becomes possible to synthesize the components into a working whole. See *systems analysis*. **système physique**

physical weathering The breakdown of rock and mineral particles into smaller particles by physical forces such as frost action and wind. See *mechanical weathering*. **désagrégation ou altération physique**

physical-chemical environment The nonbiological factors that characterize the environment of an organism. **environnement physico-chimique**

physiography (1) An outdated term for the study of landforms, now replaced by *geomorphology*. (2) The scientific study of *geomorphology, pedology*, and *biogeography*. **physiographie**

physiological drought A temporary state of drought that affects plants during daytime, often causing them to wilt because water losses by transpiration are more rapid than the uptake by roots, although the soil may have an adequate supply of water. During the night, when transpiration slows down, the plants normally recover. **sécheresse physiologique**

physisorption See *adsorption*. **physisorption**

phytoecology The branch of ecology concerned with the relationship between plants and their environment. **phytoécologie**

phytogeography A study of the geographical distributions of plants on the Earth's surface. **phytogéographie**

phytolith An inorganic body derived from replacement of plant cells; literally plant stones. Phytoliths are usually opaline. **phytolithe**

phytometer A plant or group of plants used to measure the physical factors of the habitat in terms of physiological activities. **phytomètre, plante indicatrice**

phytomorphic soils Well-drained soils of an *association* that have developed under the dominant influence of the natural vegetation characteristic of a region. The *zonal soils* of an area. **sols phytomorphes**

phytoplankton The plant portion of plankton, the plantlike unicellular organisms in water. See *zooplankton*. **phytoplancton**

phytotoxic Injurious to plants. **phytotoxique**

piedmont (*geomorphology*) Lying or formed at the base of mountains, generally describing the gentle slope leading down from steep mountain slopes to the plains, including both the *pediment* and the accumulation of *colluvial* and *alluvial* material which forms a low-angle slope beyond the pediment. **piémont**

piezometer An instrument used to measure the pressure head in a pipe, tank, or soil. It usually consists of a small pipe or tube connected or tapped into the side or wall of a pipe or tank and connected to a manometer pressure gage, water or mercury column, or other device for indicating pressure head. **piézomètre**

piezometric surface The imaginary surface to which groundwater rises under hydrostatic pressure in wells or springs. The surface at which water will stand in a series of *piezometers*. **surface piézométrique**

pilot project A research project requiring field work to meet a stated objective providing preliminary estimates of a resource condition; used to evaluate indicators, sampling strategy, methods, and logistics. Research activities on indicators should be described as individual pilot projects, usually during a single index period. **projet pilote**

pingo An Inuit term for a conical, asymmetrical mound or hill, with a circular or oval base and commonly fissured summit, occurring in the continuous and discontinuous permafrost zone. It has a core of massive ground ice, is covered with soil (including *peat*) and vegetation, and exists for at least two winters. **pingo**

pioneer plant A plant or a plant community capable of occupying a bare or newly exposed site (e.g., a freshly deposited flood sediment, or new sand dune) and persisting there until supplanted by invader or other succesion species. **végétation ou plante pionnière**

pisolith See *oolith.* **pisolite, pisolithe**

pistil The female part of a flower, comprising the ovary, stigma, and style. **pistil**

pistillate A flower that has a *pistil* but no *stamen*, or a plant or variety whose flowers have this characteristic. **pistillé**

pit An excavation in the surface made for the purpose of removing, opening up, or proving sand, gravel, clay, marl, peat, or any other substance it includes any associated infrastructure, but does not include a mine or quarry. See *mine, quarry*. **fosse, puits**

pit and mound topography (*geomorphology*) Complex *microrelief* created by numerous cradle knolls and their attendant pits. Usually associated with forested sites or cleared sites that have not been plowed. **topographie en bosses et creux**

pitch (1) An alternative name for *asphalt* or *bitumen.* (2) The amount of *plunge* in a fold (i.e., the angle formed by the dipping axis of a fold away from the horizontal plane). **(1) bitume (2) inclinaison**

pitchblende A black, lustrous oxide of uranium (uraninite); the chief ore of uranium, occurring in sulfide-bearing veins. **pechblende**

pith The ground tissue occupying the center of the stem or root within the *vascular bundle*; usually consists of *parenchyma.* **coeur**

pitted outwash (*geomorphology*) Terrain consisting of outwash with pits or *kettles*, produced by the partial or complete burial of stagnant glacial ice by outwash and the subsequent thaw of the ice and collapse of the surficial materials. **plaine d'épandage piquée**

pK_a The negative logarithm of the acid *dissociation constant.* Lower pK_a values indicate stronger acids. **pK_a**

pK_b The negative logarithm of the base *dissociation constant.* Lower pK_b values indicate stronger bases. **pK_b**

placer deposit A mass of sand, gravel, or similar sedimentary detrital materials deposited near to the site where they have been weathered from solid rock and laid down as alluvial or beach placers. Such deposits frequently contain valuable minerals such as gold or rutile. **dépôt placérien**

placic A black to dark reddish mineral soil horizon that is usually thin (< 5 mm) but that may range from 1 to 25 mm in thickness. It is cemented with Fe, Al-organic complexes, hydrated Fe oxides, or a mixture of Fe and Mn oxides. It is slowly permeable or impenetrable to water and roots. The term is used in both the Canadian and U.S. systems of soil taxonomy. **placique**

plaggen epipedon An anthropogenic surface horizon (U.S. system of soil taxonomy) more than 50 cm thick that is formed by long-term manuring and mixing. **épipédon de plaggen**

Plaggepts A suborder in the U.S. system of soil taxonomy. Inceptisols that have a plaggen epipedon. **Plaggepts**

plagioclase A series of rock-forming, triclinic silicates of the *feldspar* group with the general formula (Na,Ca) Al $(Al,Si)Si_2O_8$. At high temperatures it forms a complete solid-solution series ranging from albite, $NaAlSi_3O_8$, through the intermediate members oligoclase, andesine, labradorite, and bytownite, to anorthite, $CaAl_2Si_2O_8$. **plagioclase**

plagiosere (plagioclimax) The development of a vegetation community where the plant succession has been temporarily arrested by non-climatic controls in the form of human interference such as burning or grazing. If the human interference becomes permanent, the plagiosere is a plagioclimax. See *sere*. **plagiosère**

plain (*geomorphology*) A region of generally uniform slope that is comparatively level, of considerable extent, and not broken by marked elevations and depressions. **plaine**

plane surface system A type of *process-response system* in which flows of mass and energy possess only spatially variable horizontal components. **système à surface plane**

planimetric map A map that represents only the relative horizontal positions of cultural or natural features, using lines and symbols. It is distinguished from a topographic map by the omission of relief or elevation contours. **carte planimétrique**

planimetry The determination of horizontal distances, angles, and areas by measurements on a map. **planimétrie**

planisaic A photomap in which the *planimetry* detail is shown by overprints in color. See *photomap* and *toposaic*. **photo-carte planimétrique**

plankton Very small marine animals (*zooplankton*) and plants (*phytoplankton*) that drift along with ocean and lake currents, living at or near the surface. Plankton are the first link in the marine food chain. Zooplankton is composed of small crustaceans, worms, and mollusks; larvae and eggs from many animals; Protozoans, including foraminifera and dinoflagellates (which, in overabundance, cause toxic red tides). Phytoplankton is composed of *algae* (including green algae, golden algae, blue-green algae, and diatoms), *bacteria, fungi*. **plancton**

plankton bloom Large quantity of plankton, usually related to rapid growth of the organisms, that give the water a definite color. Water usually appears green, but blooms may also be black, yellow, red, brown, or blue-green. **fleur d'eau planctonique**

plant Member of the kingdom Plantae. *Eukaryotes* composed of cells containing plastids (e.g., chloroplasts, their precursors, or derivatives) that develop from nonblastular embryos. The kingdom includes mosses, ferns, conifers, and flowering plants. **plante**

plant analysis Analytical procedures to determine the nutrient or mineral content of plants or plant parts. **analyse végétale**

plant competition The struggle among plants for available light, moisture, and soil nutrients. **compétition végétale**

plant food The inorganic compounds used by a plant to nourish its cells; often used as a synonym for *plant nutrients*, particularly in the fertilizer trade. **élément nutritif minéral**

plant growth promoting rhizobacteria (PGPR) Rhizobacteria (bacteria from the *rhizosphere* and *rhizoplane*) that colonize plant roots and stimulate plant growth. **rhizobactéries favorisant la croissance des plantes**

plant nutrient An element essential for plant growth (e.g., carbon, hydrogen, oxygen, nitrogen, phosphorus, potassium, calcium, magnesium, sulfur, boron, copper, iron, manganese, molybdenum, zinc, and chlorine). **élément nutritif**

plant nutrition The biochemical process of absorption, assimilation, and utilization of elements and compounds necessary for plant growth and reproduction. **nutrition des plantes**

plant residue See *crop residue* and *mulch*. **résidus végétaux**

plant succession The development of vegetation in an area over time, with a trend toward occupation by plant communities of successively higher ecological order. **succession végétale**

plantation, forest A stand of trees established by planting young trees or by sowing seed. **plantation forestière**

planting Establishing a crop, orchard, or forest by setting out seedlings, transplants, or cuttings in an area. **plantation**

plasma (*soil micromorphology*) That part of the soil material capable of being moved or that has been moved, reorganized, and/or concentrated by the processes of soil formation. It includes all the material, mineral or organic, of colloidal size and relatively soluble material that is not contained in the *skeleton grains*. **plasma**

plasmid (episome) Extrachromosomal DNA. **plasmide (épisome)**

plasmodium A body of naked, multinucleated protoplasm exhibiting amoeboid motion. **plasmodium**

plastic Capable of being deformed permanently without rupture; usually involves some recrystallization or realignment of grains. See *elastic*. **plastique**

plastic limit One of the *Atterberg limits*. The minimum water mass content at which a small sample of soil material can be deformed without rupture; the lower plastic limit. **limite de plasticité**

plastic soil A soil capable of being molded or deformed continuously and permanently into various shapes by moderate pressure. See *consistence*. **sol plastique**

plasticity index The numerical difference between the *liquid limit* and the *plastic limit*. Also called plasticity number. **indice de plasticité**

plasticity range The range of water mass content within which a small sample of soil exhibits plastic properties. See *plastic limit*. **étendue du comportement plastique**

plastid A body in a plant cell that contains photosynthetic pigments. **plaste**

plate See *plate tectonics*. **plaque**

plate count (*microbiology*) The number of colonies formed on a culture medium that has been inoculated with a small amount of soil in order to estimate the number of certain organisms present in the soil sample. **numération sur plaque**

plate tectonics (*geology*) A geological theory of global-scale dynamics involving the movement of seven major plates and several minor plates in the Earth's crust. Considerable tectonic activity occurs along the margins of the plates as their slow independent movements relative to one another cause deformation, volcanism, earthquakes, and faulting along their margins. **tectonique des plaques**

plateau An extensive, relatively elevated area of comparatively flat land, commonly limited on at least one side by an abrupt descent to lower land. **plateau**

platy soil structure A shape of soil structure. See *Appendix B, Table 1* and *Figure 1*. **structure lamellaire**

playa An ephemerally flooded, vegetatively barren shallow central basin of a plain where water gathers after a rain and then evaporates. It is generally veneered with fine-textured sediment and usually contains large amounts of soluble salts. **playa**

plinthite A non-indurated mixture of iron and aluminum oxides, clay, quartz, and other diluents that commonly occurs as red soil mottles usually arranged in platy, polygonal, or reticulate patterns. It can change irreversibly to ironstone hardpans or irregular aggregates on exposure to repeated wetting and drying. **plinthite**

ploidy Referring to the number of chromosome sets contained by a cell; one = haploid; two = diploid; three = triploid; and so on. **ploïdie**

plot A carefully measured area laid out for experimentation or measurement. **parcelle**

plottable error The smallest distance on the ground that can be depicted on a map, according to the scale. This is due to the minimum thickness (about 0.25 mm) attainable when drawing lines on a map. **erreur du tracé cartographique**

plow layer See *tillage, plow layer*. **couche de labour, couche de sol arable**

plow pan A subsurface soil layer, of varying thickness, which has a higher bulk density and lower total porosity than the soil material immediately above or below it. Normally occurs just below the maximum depth of tillage and is related to pressure from tractor tires and tillage implements. Also called plow sole. See *tillage, plow pan*. **semelle de labour**

plowing See *tillage, plowing*. **labour**

plowless farming See *tillage, plowless farming*. **culture sans labour**

plow-planting See *tillage, plow-planting*. **labour et semis en un passage (à l'aide d'un train d'outils)**

plug seedling A seedling grown in a small container under carefully controlled (nursery) conditions. When seedlings are removed from containers for planting, the nursery soil remains bound up in their roots. See *bareroot seedling* and *container seedling*. **semis en contenant**

plunge (*geology*) The departure of an axis of a fold from the horizontal the fold; the amount of plunge is the *pitch* of the fold. **plongement, inclinaison**

plutonic rock (1) In general, igneous rock originating from deep within the Earth. (2) Rock with coarse *particle size* that has resulted from formation at considerable depths where cooling and crystallization occurred slowly. **roche plutonique**

Podzolic An order of soils in the Canadian system of soil classification having *podzolic B horizons* (Bh, Bhf, or Bf) in which amorphous combinations of organic matter (dominantly fulvic acid), Al, and usually Fe are accumulated. The sola are acid and the B horizons have a high pH-dependent charge. The great groups in the order are *Humic Podzol, Ferro-Humic Podzol,* and *Humo-Ferric Podzol.* **podzolique**

podzolic B A *diagnostic horizon* (Canadian system of soil classification) that consists of a *B horizon* at least 10 cm thick having accumulations of organic matter and/or Fe and Al oxides, in specified minimum amounts. (A Bf, Bhf, or Bh horizon at least 10 cm thick. *See horizon, soil.* **B podzolique**

podzolization A process of soil formation resulting in the genesis of *Podzolic soils*; it involves the translocation of Fe and/or Al organic matter complexes from the *A horizon* to the *B horizon*, resulting in the concentration of silica in the layer eluviated. **podzolisation**

point bar (*geomorphology*) A depositional feature composed of sand and gravel that accumulates on the inside of a river *meander,* from which it is usually separated by a trough. The trough (or swale) is eventually infilled by the deposition of finer alluvial sediments. **banc arqué**

point distribution *(statistics)* A statistical method of summarizing the location characteristics of data on a map, whereby each item is allocated to a discrete point on the map, e.g., a dot map. **distribution ponctuelle**

point gage A sharp point attached to graduate scale, staff, or vernier for accurate measurement of the surface elevation of soil or water. **pointe linimétrique droite**

point of zero net charge The pH value of a solution in equilibrium with a particle whose net charge from all sources is zero. It is often determined for soils low in permanent charge minerals and high in oxides, and hydrous oxides of iron and aluminum. **point de charge nette nulle**

poised equilibrium The condition occurring when opposing forces are balanced. See *equilibrium, steady state equilibrium*. **équilibre indifférent**

P

Poisson distribution *(statistics)* A type of statistical data *distribution,* used to describe the case in which the *probability* of an event occurring (p) is very small compared with the probability that it will not occur *(q).* Whereas in a *normal distribution* $p = q = {}^1/_2$, in the Poisson distribution p is very much smaller than q. **distribution de Poisson**

polar covalent bond A covalent bond in which the electrons are not shared equally because one atom attracts them more strongly than the other. See *chemical bond.* **liaison covalente polaire**

polar molecule A molecule that has a permanent dipole moment, i.e., a molecule in which there is some separation of charge in the chemical bonds, so that one part of the molecule has a positive charge and the other part a negative charge. **molécule polaire**

polar solvent A solvent, with a slight negative charge on one part of the molecule and a slight positive charge at another position, that dissolves other polar materials. The most common polar solvent is water. See *nonpolar solvent.* **solvant polaire**

P

polarity A property of a molecule that results in one part of the molecule having a slight negative charge while the other has a slight positive charge. A water molecule exhibits polarity because it is composed of one oxygen atom joined to two hydrogen atoms by covalent bonds. The oxygen atom exerts a slightly stronger attraction for the shared electrons than do the hydrogen atoms. Thus, the oxygen atom tends to be slightly negative, while the hydrogen atoms tend to be slightly positive in character. This creates a molecule with a positive pole and a negative pole. **polarité**

polarized light microscopy Use of a light microscope equipped with polarizing filters that screen out all light waves not vibrating or moving in the same plane, used to observe specific optical characteristics of geological materials. This type of microscopy is useful in the identification of asbestos. **microscopie à lumière polarisée**

polished section A fragment of rock, mineral, or soil that is mounted on a glass slide and mechanically ground to about 0.03 mm in thickness using fine grinding powders. No cover glass is applied, allowing for observations with several microscopic techniques other than the light microscope (e.g., electron microscopes). See *thin section.* **lame polie**

pollen The fine dust-like grains discharged from the male part of a flower. Since it is very resistant to decay, pollen assists in the deciphering of paleoenvironments, especially those of the quaternary. See *pollen analysis.* **pollen**

pollen analysis A technique of identifying and counting the different pollen types that have been preserved in peat, organic soils, and lake muds; used by paleontologists, botanists, and biogeographers to assist in the reconstruction of paleoenvironments. Careful analysis of each horizon of the organic sample allows the periodic changes of the former vegetational assemblages to be understood, thereby enabling reconstructions of changing climatic conditions to be made. Although pollen analysis cannot give an absolute age for a deposit, it can assist in relative dating. **analyse pollinique**

pollination The transfer of pollen from the *stamen* or *staminate* flower to the *pistil* or *pistillate* flower. **pollinisation**

pollution The addition to, and accumulation in, the environment of harmful or objectionable material into the environment in sufficient quantities that will adversely influence the functioning, quality, or aesthetic value of that environment. **pollution**

polychlorinated biphenyls (PCBs) A derivative of biphenyl ($C_6H_5C_6H_5$) in which some of the hydrogen atoms on the benzene rings have been replaced by chlorine atoms. PCBs have been used in a variety of applications, including as heat transfer fluids in large transformers and as dielectric fluids in capacitors. Though their use has now ceased, they are still present in many older electrical installations. The compounds are very stable, are widely distributed in the environment, and bioaccumulate in mammals; they have been shown to induce cancer development in mammals. **biphényles polychlorés (BPC)**

polychlorinated dibenzofurans (PCDFs, furans) A contaminant found in commercial preparations of *polychlorinated biphenyls* and *pentachlorophenol* that may be responsible for some of the physiological effects ascribed to those compounds. **dibenzofurannes polychlorés (DFCP, furannes)**

polycyclic Denoting a chemical compound composed of two or more units of the six-carbon aromatic nucleus, *benzene*. Polycyclic compounds may contain single rings (e.g., phenylbenzene) or fused rings (e.g., naphthalene). **polycyclique**

polycyclic aromatic hydrocarbons (PAH) A group of aromatic ring compounds that are derivatives of anthracene, which consists of three benzene rings in a row. Other aromatic rings or organic groups are attached to the anthracene. They are found in coal, tar, and petroleum, and are emitted by combustion-related activities. Many different compounds can be formed through metabolic conversions involving the basic aromatic nucleus within biological systems, as well as during chemical syntheses, because of the reactive nature of anthracene and its derivatives. **hydrocarbures aromatiques polycycliques (HAP)**

polycyclic landscape (*geomorphology*) A landscape having geomorphic features that developed under several whole or partial *cycles of erosion,* possibly initiated by changes of *base-level* or by climatic change. Different landscapes may exhibit elements from several different cycles. **paysage polycyclique**

polygenetic landscape (*geomorphology*) A landscape having geomorphic features that developed under different types of climates and processes at various times during its formation. A landscape modified by *glaciation* is an example. A polygenetic landscape is also likely a *polycyclic landscape.* **paysage polygénique**

polygenic (polygenetic) soil A soil formed by two or more different processes so that none of the horizons is genetically related to another. **sol polygénique**

polygon A closed geometric figure used to graphically represent area features with associated attributes. **polygone**

polygonal ground A form of patterned ground marked by the polygonal arrangements of rock, soil, and vegetation, produced on a level or gently sloping surface by frost action. **sol polygonal**

polymers Macromolecules composed of repeating units of some smaller molecule (*monomer*). Many natural materials (e.g., proteins, starches, nucleic acids) and synthetic materials (e.g., nylon, plastic, rubber) are made from linking smaller units together. *Cellulose* is a natural polymer composed of glucose molecules; polyethylene plastic is an artificial polymer composed of ethylene molecules. **polymères**

polymerase chain reaction (PCR) An *in vitro* method for amplifying defined segments of DNA. PCR involves a repeated cycle of oligonucleotide hybridization and extension on single-stranded DNA templates. **réaction en chaîne de la polymérase**

P

polymerization A chemical reaction in which many small molecules (monomers) are joined together to form a large molecule. **polymérisation**

polymictic lake A lake that is continually mixing or that has very short stagnation periods. **lac polymictique**

polynuclear aromatic hydrocarbons (PAH) See *polycyclic aromatic hydrocarbons*. **hydrocarbures aromatiques polynucléaires (HAP)**

polypedon A group of contiguous similar *pedons*. The limits of a polypedon are reached at a place where there is no soil or where the pedons have characteristics that differ significantly. **polypédon**

polypeptide A *polymer* formed from amino acids joined together by *peptide bonds*. The properties of a polypeptide are determined by the type and sequence of its constituent amino acids. **polypeptide**

polyploidy The production of more than two sets of chromosomes. **polyploïdie**

polyprotic acid An acid with more than one acidic *proton*. It dissociates in a stepwise manner, one proton at a time. **acide polyprotique**

P

polysaccharide A polymeric carbohydrate formed by synthesis from *monosaccharides* or *disaccharides* (sugars) (e.g., starch, cellulose, and glycogen). **polysaccharide**

polyvinyl chloride (PVC) Polychloroethene; produced from vinyl chloride monomers. A strong synthetic *polymer* plastic, resistant to fire, chemicals, and weather; used in pipe, toys, electrical coverings, and other products. **chlorure de polyvinyle**

pomology The science of cultivating plants for their fruit. **pomologie**

pond A natural body of standing fresh water occupying a small surface depression, usually smaller than a lake and larger than a pool. **étang**

poorly graded soil (*engineering*) A soil material consisting mainly of particles nearly the same size. Compaction of poorly graded soil material can increase its density only slightly. **matériau de granulométrie médiocre**

population (*general*) People. (*biology*) All individuals of a species living in defined geographical area at the same time in *association* with each other. (*statistics*) All objects or values making up a related group. **population**

population dynamics The numerical changes in a population that occur within a stated period of time. **dynamique de population**

pore A void or space in a soil or rock not occupied by solid mineral material. The part of the bulk volume of soil not occupied by soil particles. Also called interstices or voids. **pore**

pore ice Frozen water in the interstitial pores of a porous medium. **glace de remplissage des pores, glace interstitielle**

pore space The total space not occupied by soil particles in a bulk volume of soil. **espace poral**

pore water pressure The pressure exerted by water contained in the voids and interstices of a soil or rock, which, under saturated conditions, will force particles apart and possibly cause failure. It is measured by inserting tubes into the soil linked to a Bourdon pressure-gauge. **pression de l'eau porale**

pore water velocity The velocity at which water travels in pores relative to a given axis. It is equal to the flux density divided by the soil water content. **vitesse d'écoulement de l'eau porale**

pore-size distribution The volume fractions of the various size ranges of pores in a soil, expressed as percentages of the soil bulk volume (soil particles plus pores). **distribution volumétrique des pores**

porosity The volume percentage of the total soil bulk not occupied by solid particles. **porosité**

porphyrin An organic pigment characterized by a cyclic group of four linked nitrogen-containing rings, the nitrogen atoms of which are usually coordinated to metal ions. They differ in the nature of their side-chain groups. They include *chlorophyll*, which contains magnesium, and heme, which contains iron. See *chlorophyll* for a diagram of the structure. **porphyrine**

porphyritic An igneous rock texture in which large crystals are set in a finer-grained ground mass that may be crystalline and/or glassy. See *porphyroblastic*. **porphyrique**

porphyroblastic A rock texture that looks like a *porphyritic* rock but in which the large grains or crystals are pseudo-*phenocrysts* produced by thermodynamic metamorphism. Thus, it refers to metamorphic rather than igneous rocks. **porphyroblastique**

positional probability A type of probability that depends on the number of arrangements in space that yields a particular state. **probabilité de position**

positive association The direct relationship between two variables, the values of which fluctuate together in the same direction (e.g., as the strength of incoming solar radiation increases seasonally, the atmospheric temperature increases). **association positive**

positive feedback See *feedback*. **rétroaction positive**

potable water Water suitable for drinking or cooking purposes by both health and esthetic standards. Used for both water and beverages. **eau potable**

potamology The scientific study of rivers. **potamologie**

potamon zone The lower reaches of stream where velocities over the riverbed are lower and the bed is mostly sand and mud. **potamon**

potash Potassium or potassium fertilizers, usually designated as K_2O. **potasse**

potassium fixation The process of converting exchangeable or water-soluble potassium to that occupying the position of K^+ in the micas. They are counter-ions entrapped in the ditrigonal voids in the plane of basal oxygen atoms of some phyllosilicates as a result of contraction of the interlayer space. The fixation may occur spontaneously with some minerals in aqueous suspensions or as a result of heating to remove interlayer water in others. Fixed K^+ ions are exchangeable only after expansion of the interlayer space. See *ammonium fixation*. **fixation du potassium**

potassium-supplying power of soils The capacity of soil to supply potassium to growing plants from both the exchangeable and the moderately available forms. **capacité de libération du potassium du sol**

pot-bound The condition of a potted plant whose roots have become densely matted; also called root-bound. **feutrage radiculaire**

potential evapotranspiration See *evapotranspiration*. **évapotranspiration potentielle**

potential model A *model* that measures the force exerted by a given object on a spatial point, by reference to the same object located at all other points on the spatial field under investigation. **modèle de potentiel**

potential natural plant community The biotic community that would become established if all successional sequences were completed without interference by humans under the present environmental conditions. **végétation naturelle potentielle**

potentially mineralizable organic matter Soil organic matter (most often organic carbon and/or organic nitrogen) that is mineralized during a laboratory incubation, usually at optimum temperature and moisture conditions for soil microbial activity. **matière organique potentiellement minéralisable**

pothole (1) A more or less circular hole in the rocky bed of a stream, carved by the scouring and grinding effect of pebbles rotated in an eddy in a stretch of rapids. (2)A small, shallow pond formed by glaciation in prairie grasslands. **marmite torrentielle**

power of test *(statistics)* The ability of a statistical test to discriminate correctly between a true and false hypothesis. **puissance d'un test**

prairie A tract of level to hilly land dominated by grasses and forbs, and with few shrubs and no trees; the natural plant community consists of various mixtures of tall-, mid-, and short-growing native species, also known as true prairie, mixed prairie, and short-grass prairie, respectively. **prairie**

pre-(post) emergence tillage See *tillage, pre-(post-)emergence tillage.* **travail du sol en pré- ou post-émergence**

pre-(post) harvest tillage See *tillage, pre-(post-)harvest tillage.* **travail du sol en pré- ou post-récolte**

pre-(post) planting tillage See *tillage, pre-(post-)planting tillage.* **travail du sol en pré- ou post-semis**

precipitant An agent added to a liquid mixture to facilitate the formation of solid materials that will settle from the mixture (e.g., alum [aluminum sulfate] is added to sewage to promote the formation of flocculent, which facilitates the removal of organic materials from the wastewater). **précipitant**

precipitate The solid that settles from a liquid suspension. The solid produced by a chemical reaction involving chemicals that are in solution. **précipité**

precipitation All forms of falling moisture, including rain, snow, hail, and sleet. **précipitations**

precipitation efficiency (effectiveness) A technique whereby the efficiency or usefulness of rainfall for crop growth, water supplies, or industrial potential can be measured. **efficacité des précipitations**

precipitation interception The stopping, interrupting, or temporary holding of precipitation in any form by mulch, a vegetative canopy, vegetation residue, or any other physical barrier. **interception des précipitations**

precision The degree of agreement among several measurements of the same quantity; the reproducibility of a measurement. **précision**

preconsolidation pressure (or prestress) The greatest effective pressure to which a soil has been subjected. **pression de contrainte**

predictive Referring to what will happen, especially in relation to *model building*. See *prescriptive.* **prédictif**

preferential flow The process whereby free water and its constituents move by preferred pathways through a porous medium. Also called *bypass flow.* **écoulement préférentiel**

prescriptive Referring to what ought to happen, especially in relation to *model building.* See *predictive.* **prescriptif**

preservation The action of reserving, protecting, or safeguarding a portion of the natural environment from unnatural disturbance. It does not imply preserving an area in its present state, as natural events and natural ecological processes are expected to continue. Preservation is part of, and not opposed to, conservation. **préservation**

pressure filter A device used to remove fine particulate matter from water; the filter consists of a filter medium, such as sand or anthracite coal, packed in a watertight vessel. **filtre à pression**

pressure head Energy contained by fluid because of its pressure, usually expressed in a depth of fluid. See *head.* **hauteur piézométrique**

pressure melting point The temperature at which materials may be induced to melt by the application of pressure. Pertinent, for example, to glaciers in which the melting point diminished with depth due to the increased pressure (weight) of the ice. **pression de fusion *in situ***

pressure membrane A membrane, permeable to water and only very slightly permeable to gas when wet, through which water can escape from a soil sample as a result of a pressure gradient. **membrane de tensiomètre**

pressure pan See *tillage, pressure pan.* **couche indurée d'origine anthropique, semelle de labour**

pressure wave See *compressional wave.* **onde de pression**

prilled fertilizer See *fertilizer, prilled.* **engrais perlé**

primary treatment The first phase in the treatment of wastewater in which different components of wastewater are physically separated. **traitement primaire**

primary (p) wave See *seismic waves.* **onde primaire**

primary consumer In a food chain, a heterotrophic organism that feeds on plants. An animal that eats other animals is a secondary consumer. **consommateur primaire**

primary energy Energy contained in fossil fuels such as *coal* and *petroleum,* and energy derived from renewable sources such as the sun, wind, and ocean waves. **énergie primaire**

primary growth (*botany*) Growth originating in the *apical meristems* of shoots and roots. **croissance primaire**

primary mineral A mineral that has not been altered chemically since deposition and crystallization from molten lava. See *secondary mineral.* **minéral primaire**

primary nutrient A nutrient applied most often and in the largest amounts as fertilizers (e.g., nitrogen, phosphorus, and potassium). See *macronutrient.* **élément majeur, élément nutrif primaire**

primary particles Individual soil particles resulting from a standard dispersion treatment. **particules primaires ou élémentaires**

primary productivity The amount of plant biomass accumulated per unit area of land or volume of water as a result of photosynthesis over a specific time interval, expressed either in units of energy (e.g., joules m^{-2} day^{-1}) or dry organic matter (e.g., kg ha^{-1} $year^{-1}$); it can be expressed as gross primary productivity or net primary productivity. See *secondary productivity.* **productivité primaire**

primary succession The ecological succession that first appears on a surface that has not been previously occupied by a community of organisms (e.g., a newly exposed glaciated rock surface or a lava-flow). See *pioneer, secondary succession.* **succession primaire**

primary tillage See *tillage, primary tillage.* **préparation primaire du sol**

prime agricultural land An economic term to loosely describe land of high value and/or good quality for crop production. **terre agricole de première qualité**

prime meridian The meridian of longitude 0°, from which longitude is measured; the meridian of Greenwich, England, is almost universally used for this purpose. **méridien origine**

principal meridian The line extending north and south along the astronomical meridian that passes through the initial point of a survey, along which areas of land measurement are established. **méridien principal**

principal stresses The maximum, minimum, and intermediate intensities of *stress* along each of three mutually perpendicular axes in a given material. **contraintes principales**

principle of allocation The balance between advantages and costs reflected in the division of a limited resource among various uses by an organism. **principe d'allocation**

prismatic soil structure A shape of soil structure having prism-like aggregates that have vertical axes much longer than the horizontal axes. See

Appendix B, Table 1, and *Figure 1.* **structure prismatique**

probability *(statistics)* The measure of the uncertainty associated with events of unknown outcome. Several statistical tests have been designed to predict the probability of many types of relationships occurring or not occurring. Probability is assessed by dividing the number of occurrences by the total number of cases. **probabilité**

probability distribution *(statistics)* A distribution of values of a variable indicating the odds (i.e., probability) of encountering each of the values. **distribution de probabilité**

probability sample *(statistics)* A sample obtained by a method in which every element of a finite population has a known, but not necessarily equal, chance of being included in the sample. **échantillon probabiliste**

process-response system A combination of a *cascading system* and a morphological system, with the linkages between the two provided by morphological components that are the same as, or closely correlated with, storages or *regulators,* which are fundamental parts of the cascading system. When a process-response system is modified by human intervention (e.g., a river authority controlling a drainage basin) it becomes a *control system,* because its *inputs* and *outputs* have been regulated. The process-response system shows the manner in which form is related to process. See *systems analysis.* **système de régulation des procédés**

process water Any water that comes in contact with raw materials during the processing or fabrication of a product. The water is often released as wastewater following use. **eau de traitement**

process-response model A representation of the *process-response system* to illustrate the ways in which forms are related to processes. See *inputs and outputs, model.* **modèle de régulation des procédés**

proclimax A stable plant community thought to have been originally established under climatic conditions different from those of today. See *climax.* **proclimax**

procumbent A prostrate plant whose branches usually do not root. **procombant**

producer (*biology*) An organism capable of fixing carbon dioxide, usually by photosynthesis, to produce organic biomass and thus contribute to primary productivity and provide food for consumers; usually green plants, but also some bacteria. (*agriculture*) A farmer or rancher. **producteur**

production forest A forest used for production of various commodities (e.g., timber). **forêt de production**

productive forest land Forest land capable of producing a merchantable stand within a defined period of time. **forêt productive**

productivity The capacity to produce. (*ecology*) The rate at which organic matter is stored in any organism. (*land*) The physical yield expected from a land unit assuming specific management practices and input levels. (*economics*) A measure of technical efficiency, which may be expressed as the added output for an additional unit of input or the average output per unit input. **productivité**

productivity, soil (*soil*) The capacity of a soil, in its normal environment, to produce a particular plant or sequence of plants under a defined management system; usually expressed in terms of yield. **productivité du sol**

product-moment correlation See *correlation, statistical.* **corrélation du moment des produits**

profile, soil A vertical section of the soil through all its horizons and extending into the parent material. **profil de sol**

profundal zone (*limnology*) In deep lakes, the deep, bottom-water area beyond the depth of effective light penetration; all of the lake floor beneath the *hypolimnion.* See *hypolimnion, limnetic zone.* **zone profonde**

progeny (*biology*) The offspring of animals or plants. (*chemistry*) Products resulting from the decay of radioactive elements. **progénie**

proglacial (*geology*) Features located in front of or just beyond the outer limits of a glacier or ice-sheet, generally at or near its lower end (e.g., lakes, streams, deposits, and other features produced by or derived from the glacier ice). **proglaciaire**

projection (*geometry*) The extension of lines or planes to intersect a given surface; the transfer of a point from one surface to a corresponding position on another surface by graphical or analytical methods. (*surveying*) The extension of a line beyond the points that determine its character and position. The transfer of a series of survey lines to a single theoretical line by a series of lines perpendicular to the theoretical line. In surveying a traverse, a series of measured short lines may be projected onto a single long line, connecting two main survey stations, and the long line is then treated as a measured line of the traverse. **projection**

prokaryotic An organism composed of cells that do not contain membrane-bound organelles such as nuclei, mitochondria, or chloroplasts. The genetic material of this type of cell is not associated with large chromosomes within the nucleus, as is characteristic of higher plants or animals. Bacteria and bluegreen algae comprise this group. See *eukaryotic*. **procaryotique**

propagation The increase or multiplication of plants by sexual (e.g., seeds) or asexual methods (e.g., cuttings, grafts, layers, or meristem culture). **multiplication**

propagule Any cell unit capable of developing into a complete organism. For fungi, the unit may be a single spore, a cluster of spores, hyphae, or a hyphal fragment. **propagule**

prospecting The removal of overburden, core drilling, construction of roads, or any other disturbances of the surface for the purpose of determining the location, quantity, or quality of a natural mineral deposit. **prospection**

protease An enzyme that hydrolyzes peptide bonds, thereby converting proteins into polypeptides and eventually amino acids. **protéase**

protected areas Areas, such as provincial parks, federal parks, wilderness areas, ecological reserves, and recreation areas, that have protected designations according to federal and provincial/state statutes. Protected areas are land and freshwater or marine areas set aside to protect diverse natural and cultural heritage. **territoires protégés**

protection forest An area wholly or partly covered with woody growth, managed primarily for its beneficial effects on soil and water conservation rather than for wood or forage production. **forêt de protection**

protein A macromolecule made up of a long chain of amino acids linked by peptide bonds and containing carbon, hydrogen, nitrogen, oxygen, and sometimes sulfur. Protein occurs in all animal and vegetable matter and is essential in the diet of animals. Some proteins are enzymes, which hasten chemical reactions in living organisms; others play a structural role (e.g., tubulin, actin, and myosin). **protéine**

protein, crude An estimate of protein content based on a determination of total nitrogen (N) content multiplied by 6.25, because proteins average about 16% N. **protéine brute, matière azotée totale**

proteolytic enzyme A type of enzyme that degrades proteins. **enzyme protéolytique**

protist A microscopic, usually single-celled, member of the kingdom Protoctista; an informal name for heterotrophic (*protozoa*) or autotrophic (*algae*) *eukaryotic* microorganisms. **protiste**

protoctist A member of the kingdom Protoctista. *Eukaryotic*, heterotrophic, and autotrophic microorganisms and their larger descendants, exclusive of animals, plants, and fungi. None forms embryos. The kingdom includes diatoms, dinoflagellates, brown seaweeds and other algae, ciliates, amoebas, malarial parasites, slime molds, slime nets, and many other groups. **protoctiste**

proton A positively charged particle in an atomic nucleus. **proton**

protoplasm The material inside the living cells of plants and animals. **protoplasme**

protoplast An intact bacterial cell or plant cell from which the cell wall has been removed. Protoplasts are produced in the laboratory, and they must be protected from osmotic lysis. **protoplaste**

P

protozoan (plural protozoa) Unicellular *eukaryotic* microorganisms that move by protoplasmic flow (*amoebae*), flagella (*flagellates*), or cilia (*ciliates*). Most species feed on bacteria, fungi, or detrital particles. See *figure*. **protozoaire**

provenance The geographic source or place of origin of seed (or pollen); the native geographic location of parent plant(s) within which their genetic makeup has developed through natural selection. **provenance**

provenance test A progeny test of populations of the same species but of different provenances; objectives may include studying their performance under a range of site and climatic conditions, identifying the most desirable provenances for seed or seedling increase, and establishing a collection of biotypes of direct and potential plant breeding value. **test de provenance**

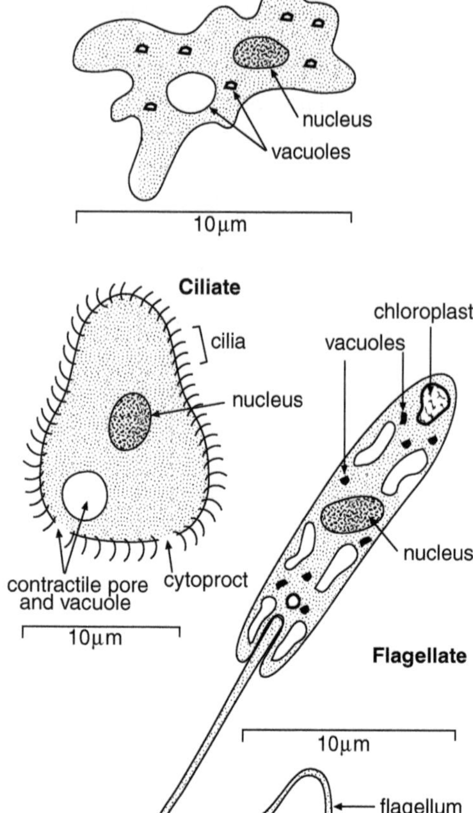

Representatives of the soil **protozoa** (adapted from Killham, 1994).

province A geographic area uniform in some or all of its characteristics, and large in extent. A faunal province, for example, has a particular assemblage of animal species which differs from assemblages in different contemporaneous environments elsewhere. A physiographic province exhibits similar geologic structures and climate, and has a pattern of topographic relief that differs considerably from adjacent areas, indicating a uniform geomorphic history. **province**

proximate analysis An analytical system for feedstuffs that includes the determination of ash, crude fiber, crude protein, ether extract, moisture (dry matter), and nitrogen-free extract. **analyse immédiate**

proximate factors Aspects of the environment that organisms use as cues for behavior (e.g., day length); proximate factors are often not directly important to the organism's wellbeing. **facteurs proximaux**

proxy property See *surrogate property*, *pedotransfer function*. **propriété subrogative**

pruning (*forestry*) The removal of live or dead branches from standing trees, usually the lower branches of young trees, and the removal of multiple leaders in plantation trees, for the improvement of the tree or its timber; the cutting away of superfluous growth, including roots, from any plant to improve its development. (*horticulture*) The cutting back of woody plants, especially fruit and ornamental trees, bushes, and shrubs to train them while young by selecting several main scaffold limbs able to support heavy loads of fruit; to cut off unnecessary plant parts so that energies are concentrated where needed, or kept in bounds; to control the quality and quantity of flowers and fruit; to remove dead, damaged or diseased parts. See *thinning*. **élagage, taille**

prussic acid A poison produced as a glucoside by several plant species, especially sorghums. Also called hydrocyanic acid. **acide prussique, ou cyanhydrique**

Psamments A suborder in the U.S. soil taxonomy. *Entisols* that have textures of loamy fine sand or coarser in all parts, have < 35% coarse fragments, and that are not saturated with water for periods long enough to limit their use for most crops. **Psamments**

psammon Interstitial organisms found between sand grains. **psammon**

psammophyte A plant that grows best in, or tolerates, sand, particularly fine to medium sand. **psammophyte**

psammosere The succession of vegetation that develops in an environment of moving sand dunes. See *sere*. **psammosère**

pseudomonad A member of the genus *Pseudomonas*, a large group of *Gram-negative,* obligately respiratory (i.e., never fermentative) bacteria. **pseudomonade**

pseudomorph (1) A mineral that has replaced another mineral and assumed its external form (e.g., *gypsum* replacing anhydrite). (2) A fossil that has gradually been replaced by secondary material to form a cast that has preserved its former shape. **pseudomorphe**

pseudopodium (plural pseudopodia) A protrusion of an amoeboid cell formed by the extrusion of the cytoplasm (but still enclosed in the membrane) for the purpose of movement or feeding. **pseudopode**

psychrometer An instrument for determining atmospheric humidity by the reading of two thermometers, the bulb of one being kept moist and ventilated. **psychromètre**

psychrophile A microorganism capable of thriving at temperatures as low as 0°C. **psychrophile**

Pteridophyta A phylum of vascular plants that reproduces from spores instead of seeds (e.g., ferns, club mosses, and horsetails). ***Pteridophyta***

puddled soil A dense soil with a degraded soil structure; dominated by massive or single-grain structure, resulting from handling the soil when it is in a wet, plastic condition so that when it dries it becomes hard and cloddy. **sol sans stucture ou à structure massive**

pulp Fiber material produced by chemical or mechanical means, or a combination of the two, from fibrous cellulose raw material and from which, after suitable treatment, paper and paperboard are made. **pâte**

pulpwood Roundwood cut from trees and prepared primarily for manufacture into wood pulp or wood fiber; does not include chips and sawdust produced as residues of lumber and plywood operations, but does include chips manufactured from roundwood in the forest or at chip mills remote from a pulp mill. **bois à pâte**

pulse The edible seeds of any leguminous plant. **graines de légumineuses**

pulse labeling A technique in which radioisotopes are used for measuring the rate of synthesis of compounds within living cells. The cells are exposed to a small quantity of an isotope for a brief period, or "pulse". **marquage bref, marquage de courte durée**

pumice A vesicular igneous rock with a sponge-like structure, so light in weight that it floats in water. It is formed from the bubbly, glassy scum on the surface of a lava flow or lake. Fragments of pumice are blown out of a volcanic vent during an eruption to form an accumulation on the slopes of the volcano. **pierre ponce**

P

pumping station A location and apparatus whereby water, sewage, or other liquids are pumped from a lower level to a higher level. **poste de pompage**

pupa (*zoology*) An intermediate, usually *quiescent*, form assumed by insects after the larval stage and maintained until the beginning of the adult stage. **pupe**

pure culture A population of microorganisms composed of a single strain. Such cultures are obtained through selective laboratory procedures and are rarely found in a natural environment. **culture pure**

pure forest A forest composed essentially of trees of one species; in practice, a forest in which at least 80% of the trees are of one species; contrast with *mixed forest*; also called pure timber stand. **peuplement pur**

purine An organic nitrogenous base with a double-ring structure, (e.g., adenine or guanine) that occurs in nucleic acids. **purine**

push moraine (*geomorphology*) Landform produced by the bulldozing effect of an ice-sheet advancing across the *glacial drift* from an earlier glaciation. **moraine de poussée**

putrefaction The partial degradation of organic materials under conditions of insufficient oxygen supply. The result is the release of noxious oxidation products and gases. **putréfaction**

putrescible Describing any substance capable of rapid decomposition by microbes and that likely results in the production of a rotten, foul-smelling odor. **putrescible**

pycnocline (*limnology*) A zone where the water density rapidly increases due to changes in temperature or salinity. **pycnocline**

pyramid of biomass See *ecological pyramid*. **pyramide de biomasse**

pyramid of energy A drawing similar to an *ecological pyramid* that depicts the total caloric content of the organisms at each trophic level. **pyramide d'énergie**

pyrethroid Any of a group of botanical insecticides developed from pyrethrum extracted from plants belonging to the genus *Chrysanthemum*. Pyrethroids are non-persistent in the environment. **pyréthrinoïde**

pyrimidine An organic nitrogenous base with a single-ring structure, (e.g., cytosine, thymine, or uracil) that occurs in nucleic acids. **pyrimidine**

pyrite A yellowish mineral (FeS_2), generally metallic appearing; also known as fool's gold. Pyrites are often found in coal and coal mine spoil. Burning coal with sulfides results in the release of oxides of sulfur (SOx) into the atmosphere, which contributes to *acid deposition*. When pyrites in mine spoil are exposed to oxidizing conditions, sulfur-oxidizing bacteria produce sulfuric acid that makes the material so acid (near pH 2) that nothing will grow in it. **pyrite**

pyroclastics Detrital volcanic materials that have been explosively or aerially ejected from a volcanic vent. **pyroclastique**

pyrolysis The thermal destruction of some material (e.g., coal, oil, wood, or other organic substance) in the absence of molecular oxygen. Also called destructive distillation. **pyrolise**

pyrophoric Describing a material that can ignite spontaneously in air. **pyrophorique**

pyrophosphate Any of a class of phosphorus compounds produced by the reaction of either anhydrous ammonia or potassium hydroxide with pyrophosphoric acid ($H_4P_2O_7$). Pyrophosphoric acid is a condensation product of two molecules of orthophosphoric acid (H_3PO_4). The main polyphosphate species in polyphosphate fertilizers. **pyrophosphate**

pyrophyllite An aluminosilicate mineral, $Si_4Al_2O_{10}(OH)_2$, with a 2:1 layer structure but without isomorphous substitution. It is dioctahedral. **pyrophyllite**

pyroxene A group of common rock-forming minerals with the general formula $ABSi_2O_6$, where A is chiefly Mg, Fe^{+2}, Ca, or Na, and B is Mg, Fe^{+2}, or Al. Colors include white, yellow, green, brown, and greenish black; specific gravity ranges from 3.2 to 4.0; and hardness ranges from 5 to 7 (*Mohs scale*). Pyroxene occurs as stout prismatic crystals and in massive form in igneous and high-temperature metamorphic rocks rich in magnesium and iron. **pyroxène**

Q

Q_{10} See *temperature coefficient.* **Q_{10}**

quadrat A measured area of any shape and size used as a sample plot or sample area in a study survey. **quadrat**

qualitative analysis The examination of a substance or sample to determine what chemical compounds or elements are present irrespective of the amounts of those compounds or elements. **analyse qualitative**

qualitative variable See *categorical variable.* **variable qualitative**

quality assurance An integrated system of activities that include *quality control,* quality assessment, reporting, and quality improvement to ensure that a product or service meets defined standards of quality with a stated level of confidence. **assurance de qualité**

quality control A system of procedures, checks, audits, and corrective actions to ensure that the quality of a product or service is satisfactory, adequate, dependable, and economical, and meets defined standards of quality with a stated level of confidence. **contrôle de qualité**

quantitative analysis The examination of a substance or sample to determine the precise amounts of certain chemicals or elements that are present. **analyse quantitative**

quantitative variable Data having the form of numerical quantities such as measurements or counts, as distinguished from *categorical variables.* **variable quantitative**

quantum A unit of energy conveyed by an electromagnetic wave. The magnitude of the unit is proportional to the frequency of the wave; for example, a unit of ultraviolet light (relatively high frequency) conveys more energy than a unit of infrared light (relatively low frequency). Also referred to as a *photon.* **quantum**

quarry Any opening in, excavation in, or working of the surface or subsurface for the purpose of working, recovering, opening up, or proving any mineral other than coal, a coal-bearing substance, oil sands, or an oil sands-bearing substance; includes any associated infrastructure. See *pit, mine.* **carrière**

quartile *(statistics)* One of four equal values into which a data distribution may be divided around the *median* value. **quartile**

quartz The crystalline form of silicon dioxide (SiO_2). It is lustrous and sufficiently hard to scratch glass (see *hardness scale).* In its most common form it is transparent and uncolored, but there are several varieties, including amethyst, yellow quartz (citrine), rose quartz, rock crystal (watery quartz), and smoky quartz (cairngorm). **quartz**

quartzitic schist See *schist.* **schiste quartzique**

quasi-equilibrium A state of near *equilibrium,* reached when a system moves towards a *steady-state equilibrium,* but where absolute equilibrium is never actually achieved in the face of a constantly changing gross energy environment. **quasi-équilibre**

quench Fertilizer application sufficient to saturate the soil's fixation capacity; the amount required is often too large to be economical. Fertilizer needs on soils with quenched fixation capacities

Q

are thereafter comparable with those on soils that fix little or none of that particular nutrient. **surfertilisation**

quick clay Clayey material having the tendency to change from a relatively stiff condition to a liquid mass when disturbed. See *liquefaction.* **argile sensible**

quick sand Sand of low bearing capacity caused by the upward flow of water and the resultant decrease in *intergranular pressure.* **sable mouvant**

quicksilver Mercury where it occurs as a native mineral. **mercure**

quiescent (*zoology*) The temporary cessation of development, movement, or other activity (e.g., the pupal stage of a life cycle). **quiescent**

Q-wave See *seismic waves.* **onde de Love, onde Q**

R

R factor See *universal soil loss equation.* **facteur R**

R layer Underlying consolidated bedrock. See *horizon, soil.* **couche R**

radar A method, system, or technique, including equipment components, for using beamed, reflected, and timed electromagnetic radiation to detect, locate, and/or track objects, to measure altitude, and to acquire a terrain image. Acronym derived from radio detection and ranging. **radar**

radar, synthetic aperture (SAR) *Radar* in which a synthetically long apparent or effective aperture is constructed by integrating multiple returns from the same ground cell, taking advantage of the Doppler effect to produce a phase history record that may be processed to reproduce an image. Abbr. SAR. **radar à ouverture synthétique, radar à antenne synthétique, radar à synthèse d'ouverture**

radial drainage A pattern of outflowing rivers away from a central point. **drainage radial**

radiance The accepted term for radiant flux in power units (i.e., watts). **irradiance**

radiant energy The electromagnetic radiation transmitted by waves through space or other media. **énergie de rayonnement, énergie radiante**

radiant flux Amount of radiation impinging on a given surface per unit time. Also called radiant power. **flux énergétique**

radiant heat Infrared energy emitted by a surface. The amount of heat is proportional to the fourth power of the absolute temperature of the radiating body. See *infrared.* **chaleur rayonnante**

radiation Energy emitted in the form of electromagnetic waves, having differing characteristics depending upon the wavelength. The radiation from the sun is relatively energetic and has a short wavelength (ultraviolet, visible, and near infrared), whereas energy re-radiated from the Earth's surface and the *atmosphere* has a longer wavelength (*infrared radiation*) because the Earth is cooler than the sun. **rayonnement**

radiation sterilization The use of ionizing radiation (e.g., *gamma rays*) either to render a plant or animal incapable of reproduction or to kill all microorganisms associated with some material or product. **radappertisation**

radiative forcing A change in the balance between incoming *solar radiation* and outgoing *infrared radiation.* Without any radiative forcing, solar radiation coming to the Earth would continue to be about equal to the infrared radiation emitted from the Earth. The addition of *greenhouse gases* to the atmosphere traps an increased fraction of the infrared radiation, re-radiating it back toward the surface and creating a warming influence (i.e., positive radiative forcing, because incoming solar radiation will exceed outgoing infrared radiation). See *greenhouse effect.* **forçage radiatif**

radical A group of atoms, either in a compound or existing alone. See *free radical.* In soil most anions are radicals. Polyvalent cations like Al^{3+} and Fe^{3+} may also become radicals by

R

attracting one or two hydroxyls and forming cations of lower valence (lesser charge) such as $Al(OH)_2^+$ and $Fe(OH)_2^+$. **radical**

radicle The first root of a plant that elongates during germination of a seed and forms the primary root. **radicule**

radioactive decay See *radioactivity*. **dégradation radioactive**

radioactive isotope An *isotope* of an element that is subject to radioactive decay. Also called radionuclide or radioisotope. See *radioactivity*. **isotope radioactif**

radioactive series A series of elements produced by radioactive decay of unstable atoms, with one decay product following another until a stable element is reached. For example, the uranium series begins with uranium-238 and ends with lead-206. The other two naturally occurring radioactive element series are the thorium and actinium series. **série radioactive**

radioactivity The spontaneous disintegration of an atom accompanied by the emission of *gamma rays, alpha particles,* or *beta particles*. Natural radioactivity is the result of the spontaneous disintegration of naturally occurring radioactive isotopes. Radioactivity can be induced in many nuclides by bombarding them with neutrons or other particles. **radioactivité**

radiocarbon See *carbon-14*. **radiocarbone**

radiocarbon dating See *carbon dating*. **datation au radiocarbone**

radiochemistry The branch of chemistry dealing with the study of the properties and use of radioactive compounds and *ionization*. See *radioactivity*. **radiochimie**

radioecology The study of the effects of radiation on plants and animals in natural communities. **radioécologie**

radiography The production of images using radiation other than visible light. **radiographie**

radioisotope See *radioactive isotope*. **radio-isotope**

radiometer An instrument for quantitatively measuring the intensity of electromagnetic radiation in some band of wavelengths in any part of the electromagnetic spectrum. **radiomètre**

radionuclide See *radioactive isotope*. **radionucléide**

radiotracer A radioactive isotope used to enable its path through a biological or mechanical system to be traced by the radiation it emits. See *tagged molecule, tracer*. **radiotraceur**

radon (Rn) A radioactive element that is a gas produced directly from the radioactive decay of radium, a member of the uranium decay series. The chemically inert gas enters homes through soil, water, and building materials. An important source of personal exposure to radon can be drinking water obtained from wells. The threat comes from the inhalation of the gas released from water during showering, bathing, cooking, and other water uses. Ingestion of water does not appear to present a threat. **radon**

rafting The transport of material (e.g., seeds, fauna, soil, *erratics*) by floating ice or other floating material. **transport glaciel, transport par corps flottant**

rain Condensed water vapor in the atmosphere occurring in drops large enough to fall under the influence of gravity. See also *precipitation*. **pluie**

rain gauge An instrument used to measure *rainfall* at a given point on the ground. **pluviomètre**

rain splash The process of raindrop impact that contributes to *splash erosion*. On level ground it merely rearranges the soil particles, but on a slope there is a net transport of material owing to a longer downslope flight path for rebounded droplets and to the downslope component of the impact force. The latter is related to raindrop size

R

and rainfall intensity. Once *overland flow* begins, raindrop impact is cushioned by the layer of water. Thus, rain splash is probably only significant in short, sharp storms during which little runoff is generated, or close to an *interfluve* where there is insufficient catchment to generate flow. See *rain wash.* **éclaboussement**

rain wash A type of surface wash characterized by the movement of soil particles and loose materials down a hillside as a result of *rain splash* and *runoff.* It can ultimately lead to *soil erosion*, especially *sheet erosion.* **délavage par les pluies**

raindrop erosion See *splash erosion, rain splash.* **érosion par éclaboussement gouttes de pluie**

rainfall The water equivalent of all forms of atmospheric *precipitation* received in a *rain-gauge,* assuming that there is no loss by *evaporation*, percolation, or *runoff.* The term includes rain, dew, hoar frost, and rime; parts of the two latter will, on melting, find their way into the rain gauge. **hauteur d'eau, hauteur pluviométrique**

rainfall duration The period of time during which rainfall occurs, exceeds a given intensity, or maintains a given intensity. **durée de la pluie**

rainfall erosivity index A measure of the erosive potential of a specific rainfall event. In the *Universal Soil Loss Equation* it is defined as the product of total kinetic energy of the storm times its maximum 30-minute intensity. Sometimes called rainfall erosion index. See *erosivity.* **indice d'érosivité par la pluie**

rainfall excess (*hydrology*) The volume of rainfall in excess of interception, infiltration, and storage. **pluie nette**

rainfall frequency The frequency at which a given rainfall intensity and duration can be expected to be equaled or exceeded; usually expressed in years. **fréquence de la pluie**

rainfall intensity The rate at which rain is falling at any given instant, usually expressed in units of centimeters per hour. **intensité pluviale**

rainfall interception See *precipitation interception.* **interception de la pluie**

raised beach A shoreline and its littoral deposits elevated above present sea level either by positive *isostasy* or by a fall in sea level (*eustasy*). There may be raised beaches at different levels resulting from repeated movements of sea level. **plage soulevée**

random error *(statistics)* An error that has an equal probability of being high or low, as though it were chosen at random from a probability distribution of such errors. **erreur aléatoire**

random sample A data *sample* in which any one individual measurement or count in the data population is as likely to be selected as any other (i.e., selected without bias). If, however, the data exhibit a marked clustering, there may be a tendency for a biased sample to be taken. This can be avoided by adopting the technique known as stratified random sampling, in which the data are divided into classes (strata) before taking a random sample within each stratum. See *sampling frame.* **échantillon aléatoire**

randomization *(statistics)* The process of imposing an element of chance on the selection of a sample. Randomization is a step in the design protocol and may take many forms; it is the basis for determining the design-based properties of the resulting probability sample. **randomisation**

randomized blocks *(statistics)* An experimental design in which each block contains a complete replication of the treatments, which are allocated to the various units within the blocks in a random manner and therefore allow unbiased estimates of error to be constructed. **blocs aléatoires**

random-walk model A type of *simulation model* that represents the randomness in the spatial progression of a physical process; used especially in geomorphology to denote the degrees of randomness exhibited in the growth of a drainage pattern, whereby no geological or other controls are allowed to dominate the development of the network. **modèle de cheminement aléatoire**

range *(statistics)* (1) The difference between the lowest and highest values in a set of observations. It is one of the measures of dispersion (see *standard deviation*). (*geology*) (2) A chain of mountains. (*ecology*) (3) The spatial area occupied by a particular species of flora or fauna. The ecological range is the actual habitat that it currently occupies, and the tolerance range is the area in which it could continue to survive and reproduce. (*stratigraphy*) (4) The stratigraphic range is the distribution of any taxonomic group of organisms throughout geologic time. (*ecology, range management*) (5) Land supporting indigenous vegetation that either is grazed or has the potential to be grazed, and is managed as a natural ecosystem. **(1) étendue, plage (2) chaîne (3) aire de distribution géographique (4) répartition (5) parcours, parcours naturel**

R

range improvement Any practice designed to improve range condition or facilitate more efficient use of range. **amélioration du parcours**

range management The planning and direction of range use to obtain sustained, maximum animal production, consistent with conservation of the natural resources. **conduite des parcours**

range science The organized body of knowledge upon which the practice of range management is based. **science des parcours**

range seeding Establishing adapted plant species on ranges by means other than natural revegetation. **semis des parcours**

rangeland Land on which the indigenous vegetation (climax or natural potential) is predominantly grasses, grasslike plants, forbs, or shrubs and is managed as a natural ecosystem. If plants are introduced, they are managed similarly. Rangelands include natural grasslands, savannas, shrub lands, many deserts, tundras, alpine communities, marshes, and meadows. **terrain de parcours**

rangeland health The degree to which the ecological components (e.g., soil, vegetation, water, air) and processes of the rangeland ecosystem are balanced and their integrity sustained. Integrity is defined as maintenance of the structure and functional attribute characteristics of a particular locale, including normal variability. **santé du terrain de parcours**

ranking The method by which any data are arranged in order according to a given criterion or set of criteria (e.g., size, hardness) to produce an ordinal scale. See *scales of measurement.* **classement**

Raoult's law The law that states that the vapor pressure of a solution is directly proportional to the mole fraction of solvent present. **loi de Raoult**

rapid sand filtration (*wastewater management*) A water treatment method that removes suspended or colloidal particles as drinking water passes through a sand filter. **filtration rapide sur sable**

rare Species that occur in very low numbers throughout its area of distribution but is currently in little danger of extinction. See *endangered and threatened.* **rare**

rate constant The proportionality constant appropriate to the rate of a chemical reaction. In the differential equation $dX/dt = -kX$, the concentration of X is decreasing at a rate proportional to the remaining concentration of X.

Such a reaction is called a *first-order reaction* and the *k* in the equation is the rate constant. **constante de vitesse**

rate variables Variables that control the rate at which various responses are generated by various changes in the system's state. **variables de vitesse**

rate-determining step The slowest step in a reaction mechanism, the one determining the overall rate. **étape déterminante de la vitesse**

ration The amount of feed allotted to a given animal for a 24-hour day; may be fed at one time or in portions at different times during the day. **ration fourragère, ration**

ration, balanced A ration that furnishes the various essential nutrients in such proportion and amounts that it will properly nourish an animal, calculated on a per-day basis. **ration équilibrée**

raw humus See *mor*. **humus brut**

raw sewage Domestic or commercial wastewater that has not undergone any treatment for the removal of pollutants. **eaux usées**

raw water Groundwater or surface water before it is treated for use as a public water supply. **eau brute**

reach (*hydrology*) A specified length of a stream or channel. The length of a river between two gaging stations. More generally, any length of a river. **tronçon**

reactant A substance that reacts with another substance in a chemical reaction. **réactif**

reaction rate The change in concentration of a reactant or product per unit time. **vitesse de réaction**

reaction, soil The degree of acidity or alkalinity of a soil, usually expressed as a pH value. Descriptive terms commonly used for specific ranges in pH are: extremely acid, less than 4.5; very strongly acid, 4.5 to 5.0; strongly acid, 5.1 to 5.5; moderately acid, 5.6 to 6.0; slightly acid, 6.1 to 6.5; neutral, 6.6 to 7.3; slightly alkaline, 7.4 to 7.8; moderately alkaline, 7.9 to 8.4; strongly alkaline, 8.5 to 9.0; and very strongly alkaline, greater than 9.0. **réaction du sol**

reactive waste (*waste management*) Solid waste exhibiting the characteristic of interacting chemically with other substances, typical of hazardous waste. **déchets réactifs**

reagent A chemical used in laboratory analysis for the purpose of promoting a specific reaction (e.g., digestion, oxidation, colorimetric). **réactif**

reagent blank An aliquot of *analyte-* free water or solvent analyzed with the analytical batch. **blanc de réactif**

receiving waters Rivers, lakes, oceans, or other bodies that receive treated or untreated waste waters. **eaux réceptrices**

recessional moraine (*geomorphology*) A moraine marking a recessional phase or still-stand in the overall decline of an ice-sheet or *glacier*. Also called stadial moraine. **moraine de retrait**

recharge The addition of water to the zone of saturation. **alimentation**

recharge area Area of ground over which water is absorbed and added to the zone of saturation. **zone d'alimentation en eau**

reclamation The process of reconverting disturbed land to its former or other productive uses. All practicable and reasonable methods of designing and conducting an activity to ensure stable, non-hazardous, non-erodible, favorably drained soil conditions, and a land capability equivalent to or better than prior land use. Reclamation may also involve removal of equipment or buildings or other structures and appurtenances; investigations to determine the presence of substances; decontamination of buildings or other structures or other appurtenances, or land or water; the stabilization, contouring, maintenance, conditioning, or reconstruction of the surface of land; and other procedures, operations, or requirements as specified by regulatory bodies. **bonification ou restauration des terres**

reconnaissance A general, exploratory examination or survey of the main features of a region, usually preliminary to a more detailed survey. **reconnaissance**

reconstructed profile (*land reclamation*) The result of selective placement of suitable overburden material on reshaped spoils. See *minesoil*. **profil reconstitué**

recruitment The addition of new individuals to a population by reproduction. **recrutement**

rectification The process of converting a tilted or oblique photograph to the plane of the vertical. **redressement, superposition**

rectilinear slope A straight slope, or part of a slope (i.e., having a constant angle). Rectilinear slopes occur most commonly below an upper convex slope element and above a lower concave slope element. **pente rectiligne**

recycling A resource recovery method that involves collecting and reprocessing waste materials for reuse. **valorisation, recyclage**

red tide A visible red-to-orange coloration of the sea caused by the presence of a bloom of certain plankton; often the cause of major fish kills. **marée rouge**

R

redistribution In reference to soil water, the process of soil-water movement to achieve an equilibrium energy state of water throughout the soil. **redistribution**

redox potential An expression of the oxidizing or reducing power of a solution. **potentiel d'oxydoréduction**

reduced tillage See *tillage, reduced tillage*. **travail du sol réduit ou simplifié**

reducers Organisms, usually bacteria or fungi, that break down complex organic material into simpler compounds. Also called decomposers. **décomposeurs**

reducing agent (electron donor) A reactant that donates electrons to another substance to reduce the oxidation state of one of its atoms. **agent réducteur (donneur d'électron)**

reducing environment An environment conducive to the removal of oxygen; also expressed by showing an increase in negative valence, representing the addition of electrons to an atom or ion. **environnement réducteur**

reduction A chemical reaction during which electrons are added to an atom or molecule. With organic compounds, the addition of electrons is frequently accompanied by the addition of hydrogen atoms. Also called hydrogenation of organic compounds. **réduction**

reed peat *Peat* consisting mainly of reeds (*Phragmites*, *Scirpus*, *Typha* spp.). Also called *telmatic peat*. **tourbe de roseaux**

reference data (*remote sensing*) Data about the physical state of the Earth obtained from sources other than the primary remote sensing data source and used in support of remote sensing data analysis. They may typically include maps and aerial photographs, topographic information, temperature measurements, and other types of ancillary and ephemeral data. Also called ground truth, ground data, ground-based measurements. **réalité de terrain, données de terrain**

reference electrode An electrode that maintains an invariant potential under the conditions prevailing in an electrochemical measurement and thereby permits measurement of the potential of an *ion-selective electrode* or other sensing electrode. **électrode de référence**

reference site One of a group of *benchmark sites* or control-sampling locations that represent an *ecoregion* or other large biogeographic area. The sites, as a whole, represent the best ecological conditions that can be reasonably attained, given the prevailing topography, soil, geology, potential vegetation, and general land use of the region. **site de référence, site témoin, site repère**

reflectance The ratio of the radiant energy reflected by a surface to that incident upon it. Reflectance is affected by the nature of the surface itself, the angle of incidence, and the viewing angle. **réflectance**

reflected radiation The amount of incident radiation that is reflected by a surface. **rayonnement réfléchi**

reflux A laboratory technique in which a liquid is boiled in a container attached to a condenser, so that the vapor condenses and continuously flows back into the container. **reflux**

reforestation The natural or artificial restocking (i.e., planting, seeding) of an area with forest trees. Also called *forest regeneration*. See *afforestation*. **reboisement**

refraction The deflection of the direction of wave propagation when waves pass obliquely from one region of velocity to another. **réfraction**

refractory (1) Descriptive of a material or substance that resists biological metabolism or chemical reaction. (2) A highly heat-resistant material that can be used as a lining for incinerators or furnaces. **réfractaire**

refuge (*ecology*) The aggregation of large numbers of organisms in a small area with nearby food supplies (e.g., bird colonies). **refuge**

refuse reclamation The process of converting solid waste to saleable products (e.g., composting organic solid waste to produce a soil conditioner that can be sold). **récupération d'ordures**

regional metamorphism The action of pressure and heat over a very large area of the Earth, in association with an *orogeny*, whereby a wide range of new rocks and minerals is formed. See *metamorphism*. **métamorphisme régional**

registration (*remote sensing, cartography*) The process of geometrically aligning two or more sets of image data such that resolution cells for a single ground area can be digitally or visually superimposed. Data being registered may be of the same type, from very different kinds of sensors, or collected at different times. **calage**

regolith The unconsolidated mantle of weathered rock and soil material overlying solid rock. **régolithe**

Regosol The only great group in the *Regosolic order* (Canadian system of soil classification). The soils in the group have insufficient horizon development to meet the requirements of the other soil orders. **régosol**

Regosolic An order of soils in the Canadian system of soil classification having no horizon development, or development of the *A* and *B horizons* insufficient to meet the requirements of the other orders. **régosolique**

regrading The movement of soil material over a surface or depression that results in changing the shape of the land surface. **remodelage de terrain**

regression A statistical method for studying and expressing the relationship between the mean of a random variable and one or more independent variables. **régression**

regulated population A population that regularly tends towards a density approximating the ability of the environment to support individuals. **population régulée**

regulator Any component that tends to stabilize a *system* from within. When the negative feedback process operates within a system, it is known as *self-regulation*. See *feedback*. **régulateur**

regulatory response A rapid, reversible physiologic or behavioral response by an organism to change in its environment. **réponse de régulation**

rehabilitation (*land reclamation*) Return of land to a form and productivity in conformity with a prior land use plan, including a stable ecological state that does not contribute substantially to environmental deterioration and is consistent with surrounding esthetic values. See *reclamation, restoration*. **remise en valeur**

reinforced manure Manure supplemented with a phosphorus fertilizer. **fumier enrichi**

relative compaction A measure of soil *compaction* that expresses the *bulk density* at any one time as a percentage of the maximum bulk density for the same soil. An index that usually ranges from 60 to 100%. Also called state of compactness. **compaction relative, compactage relatif**

relic geomorph (*geomorphology*) A landform that has survived the forces of erosion and decay (e.g., an erosional remnant). **lambeau ou résidu d'érosion**

relic sediments Sediments having characteristics that represent a past environment rather than present-day conditions. **sédiments reliques**

relict A species or community that is unchanged from some earlier period of time. **relique**

relict landform (*geomorphology*) A landform that created by processes which are no longer operative or that play only a minor role in its present fashioning (e.g., glacial landforms are relict because the landform-producing agent is no longer present). **forme relique ou héritée**

R

relief The difference in elevation between the high and low points of a land surface. Land having no unevenness or differences of elevation is called level; gentle relief is called undulating; strong relief, rolling; and very strong relief, hilly. Other expressions such as *microrelief*, mesorelief, and macrorelief are also applied in describing degrees of relief. **relief**

relief drain A drain designed to remove water from the soil in order to lower the *water table* and reduce hydrostatic pressure. **drain de décharge**

relief map A type of map showing the variations in elevation of the land. The different shapes, heights, and degrees of slope are depicted by using contours, form lines, hachures (parallel lines), hill shading, layer tinting, and spot heights. **carte de relief, carte topographique**

remote automatic weather station A weather station that measures selected weather elements automatically and is equipped with telemetry apparatus for transmitting the electronically recorded data via radio, satellite, or landline communication system at predetermined times on a user-requested basis. **station météorologique automatique**

remote sensing The measurement or acquisition of information of some property of an object or phenomenon, by a recording device that is not in physical or intimate contact with the object or phenomenon under study. The technique uses such devices as the camera, lasers, radar systems, seismographs, gravimeters, magnetometers, and scintillation counters. **télédétection**

rendering A process of recovering fatty substances from animal parts by heat treatment, extraction, and distillation. **équarrissage**

Rendolls A suborder in the U.S. system of soil taxonomy. *Mollisols* that have no argillic or calcic horizon but that contain material with a $CaCO_3$ equivalent >400 g kg^{-1} within or immediately below the mollic epipedon. Rendolls are not saturated with water for periods long enough to limit their use for most crops. **Rendolls**

rendzina A group of soils having brown to black surface horizons that have developed on parent material that contains more than 40% *calcium carbonate equivalent*. Used in FAO and other classification systems; not used in Canadian and U.S. soil taxonomies. **rendzine**

renewable energy Those energy sources that do not rely on finite reserves of fossil or nuclear fuels. They comprise solar energy, wind energy, wave energy, tidal energy, hydroelectric power, and geothermal energy, and also include timber. Conversely, fossil fuels are a nonrenewable source of energy. **énergie renouvelable**

renewable resource A natural substance or object that has the capacity to replace itself over a relatively short time frame, under specific conditions, so that the quantity of the substance or object is not readily depleted, involves biological organisms that can be replaced after harvesting by regrowth or reproduction of the removed species, such as seafood or timber. Resources such as natural gas or mineral ores are non-renewable. Soil is formed at a relatively slow rate, so it would be considered non-renewable. **ressource renouvelable**

replicate See *replication.* **répétition**

replication The performance of an experiment or survey more than once to increase precision and obtain a closer estimation of sampling error. **répétition**

repose See *angle of repose.* **repos**

reserves (*forestry*) The retention of live or standing dead trees, pole size or larger, on site following harvest for purposes other than regeneration. Reserves can be uniformly distributed as single trees or left in small groups, and they can be used with any silvicultural system. **arbres en réserve**

reservoir A facility for the storage, regulation, and controlled release of water. Types of reservoirs include ponds, lakes, tanks, or basins that are either natural or constructed for flood control, water supply, and power generation. **réservoir**

reservoir rock Porous rock containing oil or natural gas. **roche-magasin, roche-réservoir**

reservoir tillage See *tillage, reservoir tillage.* **modelage superficiel du sol pour former des petits bassins de captage**

residence time The average amount of time that a substance spends in a stock, pool, or reservoir. **temps de résidence**

resident species A species that remains in an area thoughout the year. **espèce résidente**

residual fertility The available nutrient content of a soil carried over to succeeding crops. **fertilité résiduelle**

residual insecticide An insecticide that remains active on or in a treated location, even after a long time. **insecticide à effet rémanent**

residual material Unconsolidated and partly weathered mineral materials formed by the disintegration of consolidated rock in place. **matériau residuel**

residual nitrate The nitrate-nitrogen remaining in the soil after harvest. **nitrate résiduel**

residual shrinkage Shrinkage that is less than the volume of water lost during the final stages of drying. **retrait résiduel**

residual soil Soil formed from, or resting on, consolidated rock of the same material as that from which it was formed and in the same location. **sol résiduel**

residue See *crop residue.* **résidu**

residue management See *tillage, crop residue management.* **gestion des résidus de cultures**

residue processing See *tillage, residue processing.* **broyage et déchiquetage des résidus de cultures**

residuum (*geology*) An unconsolidated rock mass, residual at a site, derived from weathering and disintegration. **matériau résiduel**

resilience The ability of an ecosystem to maintain diversity, integrity, and ecological processes following disturbance. **résilience**

resistance The power or capacity of an object or a material to withstand the force imposed upon it. (*physics*) The ratio of applied electromotive force to the resulting current in a circuit, as described by *Ohm's law*. The unit of resistance is the ohm. Resistance opposes the flow of current, generates heat, controls electron flow, and helps supply the correct voltage to a device, depending on factors such as the material, length, and cross-sectional area of the conductor, and the temperature. (*ecology*) The ability of a biological community or ecosystem to withstand a stress or perturbation without adverse change to its structure or function, thereby maintaining an equilibrium state. **résistance**

resistant rock (*geology*) Any rock that withstands weathering forces, relative to adjacent rocks, because of differences in jointing, hardness, compactness, cementation, and mineral composition. See *differential weathering*. **roche résistante**

resistivity An electrical property of materials, representing the inherent ability of a material to resist current flow; expressed as $r = R(A/l)$, where r is resistivity, A is the area of a conductor, l is its length, and R is its resistance. **résistivité**

R

resistivity survey A type of noninvasive survey of soil for the detection of buried metal objects and chemical contaminants capable of conducting an electric current; useful for mapping groundwater depth, contaminant plumes, and site geology. **levé de résistivité**

resolution (*remote sensing*) (1) The ability of an entire remote sensor system to render a sharply defined image. Quantitatively, it is the minimum distance between two adjacent features, or the minimum size of a feature, that can be detected; for photography, the distance is usually expressed in lines per millimeter recorded on a particular film under specific conditions; as displayed by radar, lines per millimeter; if expressed in size of objects or distances, on the ground, the distance is called ground resolution. (2) A measure of the ability of an optical system to distinguish between signals that are spatially near or spectrally similar. Types of resolution include spatial and spectral. Spatial resolution is a measure of the smallest angular or linear separation between two objects. It is usually expressed in radians or meters, with a smaller resolution parameter denoting greater resolving power. Spectral resolution is a measure of both the discreteness of the bandwidths and the sensitivity of the sensor to distinguish between gray levels. It is a function of the spectral contrast between objects in the scene and their background, of the shape of the objects, and of the signal-to-noise ratio of the system. **résolution**

resource (natural) Any biological or physical feature of an area that is useful to humans or wildlife for any purpose. Natural resources are dynamic and become available through a combination of increased human knowledge and expanding technology in addition to changing individual and societal objectives. Two types of natural resources are recognized: *renewable resources* are parts of functioning natural systems that maintain a sustained yield, in which they are formed at rates approximately comparable to those of their consumption; non-renewable resources are those that have a finite distribution or are formed at rates considerably slower than those of their use. **ressource naturelle**

resource management system A combination of *conservation* practices suited to the primary use of land or water; protects the resource base by maintaining tolerable soil losses, acceptable soil and water quality, and acceptable ecological and management levels for the selected resource use. The system may include *conservation practices* that provide for quality in the environment and quality in the standard of living. **système de gestion des ressources**

resource monitoring The act of continually or periodically observing resources to determine changes and trends in their condition. **suivi de l'évolution des ressources**

resource recovery (*waste management*) The processing of solid waste to extract recyclable paper, glass, metals, or combustible material, and energy. **tri, tri à la source**

respiration A metabolic process in an individual cell, tissue, or organism resulting in the release of chemical energy derived from organic nutrients. Specifically, a series of reactions in a cell during which electrons removed during the oxidation of organic nutrients (substrates) are transferred to a terminal acceptor. When the terminal acceptor is molecular oxygen, the process is called aerobic respiration. When the terminal acceptor is an inorganic substitute for molecular oxygen, such as sulfate or nitrate, the process proceeds without oxygen and is called anaerobic respiration. In either case, potential energy present in the organic nutrients is converted to a chemical form useful to the cell. **respiration**

respiration/biomass ratio An expression of ecological turnover; the ratio of total community respiration (often expressed in grams per square meter per day) to community biomass (in grams per square meter). The ratio indicates the rate of energy flowing through the community compared to the mass of the community. **respiration spécifique**

respiratory quotient (RQ) The number of molecules of carbon dioxide liberated for each molecule of oxygen consumed. **quotient respiratoire**

respiratory reduction of nitrate See *dissimilatory reduction of nitrate.* **réduction respiratoire des nitrates**

response An action or feeling that answers to a stimulus. The features produced by the operation of the stimulus are called response elements. See *process-response model.* **réponse**

response time The delayed response of a *system* to a change of input (e.g., the delayed reaction of glaciers and icesheets to an alteration in their regime as a result of *climatic change).* The response time varies according to the size of the input. **temps de réponse**

restoration The process of restoring site conditions (e.g., establishment of natural land contours and vegetative cover) as they were before a land disturbance, such as surface mining. See *reclamation, rehabilitation.* **restauration**

restriction enzyme A class of highly specific enzymes that make double-stranded breaks in DNA at specific sites near where they combine. **enzyme de restriction**

retardation factor The capability of a soil for slowing or retarding the movement of a solute; defined for solutes subject to equilibrium reactions with the soil matrix. **facteur de ralentissement**

retention The state of being retained or to keep in possession. (*hydrology*) The amount of precipitation on a drainage area that does not escape as runoff; the difference between total precipitation and total runoff. **rétention**

retention time (*waste management*) The period of time that some waste, fluid, or other material is in a treatment facility or process unit. **temps de rétention**

reticulate mottling A network of streaks of different color; most commonly found in the deeper profiles of Lateritic soils containing *plinthite*. See *laterite*. **marbrures réticulées**

return flow (*hydrology*) That portion of the water diverted from a stream that finds its way back to the stream channel either as surface or underground flow. **écoulement restitué**

reuse The reintroduction of a commodity into the economic stream without any change. **valorisation, recyclage**

revegetation (*land reclamation*) The establishment of vegetation that replaces original ground cover following land disturbance. **revégétation, remise en végétation, végétalisation**

reverse osmosis A method of obtaining pure water from water containing a salt (e.g., used in desalination). Pure water and salt water are separated by a semipermeable membrane and the pressure of the salt water is raised above the osmotic pressure, causing water from the salt solution to pass through the membrane into the pure water. **osmose inversée**

reversible process A cyclic process, carried out by a hypothetical pathway, that leaves the universe exactly the same as it was before the process. No real process is reversible. **processus réversible**

reversion The change of essential plant nutrient elements from soluble to less soluble forms by reactions in the soil; usually refers to the conversion of monocalcium phosphate to the less soluble phosphate forms. **rétrogradation**

revetment A facing of stone or other material, either permanent or temporary, placed along the edge of a stream to stabilize the bank and protect it from the erosive action of the stream. **revêtement**

reworked Descriptive of material modified after its preliminary deposition, commonly by water or wind. **remanié**

rheology The scientific study of the flowage of materials, especially that of plastic solids. **rhéologie**

rheophilous Descriptive of *peat* formed from plants grown under the influence of mobile groundwater. Also called rheotrophic. **rhéophile**

rheotrophic See *rheophilous*. **rhéotrophe**

rhizobium (plural rhizobia) Small heterotrophic bacteria of the genus *Rhizobium* capable of forming symbiotic nodules on the roots of leguminous plants. In the nodules the bacteria fix atmospheric nitrogen that is used by the plants. The bacteria receive their energy from the plants. **rhizobium**

rhizocylinder The plant root plus the adjacent soil that is influenced by the root. See *rhizosphere*. **rhizocylindre**

rhizomatous Having *rhizomes*. **rhizomateux**

rhizome A horizontal underground stem, usually sending out roots and aboveground shoots at the nodes. **rhizome**

rhizoplane The external surface of roots and of the soil particles and debris adhering to them. **rhizoplan**

rhizosphere The area of soil immediately surrounding plant roots in which the kinds, numbers, and activities of microorganisms differ from that of the bulk soil. See *figure*. **rhizosphère**

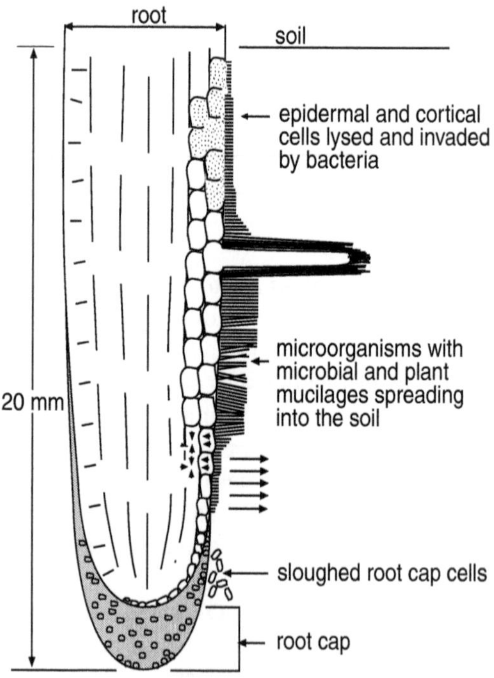

The **rhizosphere** (adapted from Bolton et al., 1993).

R

rhyolite A light-colored, silica-rich volcanic rock, or rock group, normally having small *phenocrysts* of quartz and alkali *feldspar* in a glassy to *cryptocrystalline* matrix. Rhyolite is the volcanic equivalent of granite. **rhyolite**

ribonucleic acid (RNA) A molecule composed of a linear sequence of nucleotides that are attached by 3′5′-phosphodiester linkages to nitrogenous bases such as adenine, guanine, uracil, or cytosine. RNA can store genetic information; as a component of ribosomes, it takes part in protein synthesis. **acide ribonucléique (ARN)**

ribosome A spherical organelle in all cells, composed of protein and *ribonucleic acid*; the site of protein synthesis. **ribosome**

richness The total number of species in an area, usually expressed as the number of species divided by the total number of individuals or the number of species per unit area. **richesse**

Richter scale An open-ended logarithmic scale that expresses the magnitude of a seismic disturbance, such as an earthquake, with respect to the energy dissipated in it. A value of 1.5 indicates the smallest earthquake that can be felt, 4.5 an earthquake causing slight damage, and 8.5 a very devastating earthquake. In theory, there is no upper limit to the magnitude of an earthquake. However, the strength of earth materials produces an actual upper limit of slightly less than 9. **échelle de Richter**

ridge (1) (*geomorphology*) A linear, narrow elevation of land, usually sharp-crested and with steep sides; a narrow spur of a mountain; a long, steep-sided rise on the ocean floor. (2) (*agriculture*) The raised area of soil thrown when a plow or other implement creates a furrow. See *tillage, ridge*. (3) (*meteorology*) An elongated area of relatively high atmospheric pressure. **(1, 3) dorsale (2) billon**

ridge planting See *tillage, ridge planting.* **semis sur billons**

ridge tillage See *tillage, ridge tillage.* **billonnage**

ridged (*geomorphology*) A type of surface expression of mineral landforms, characterized by a long, narrow elevation of the surface, usually sharp crested with steep sides. Ridges may be parallel, subparallel, or intersecting. **dorsal**

right-of-way The strip of land over which a power line, railway line, or road extends. **droit de passage**

rill A narrow, very shallow, intermittent water channel having steep sides usually only several centimeters deep. It presents no obstacle to normal tillage operations. **rigole**

rill erosion An erosion process on sloping fields in which numerous and randomly occurring small channels of only several centimeters in depth are formed; occurs mainly on recently cultivated soils. **érosion en rigoles, en ruisselets ou en filets**

rilled Channeled landscape on inclined slopes. Rills are ephemeral channels formed by runoff, and can be destroyed by plowing or by frost action. They can be seasonal in nature, and are caused by runoff following preferential flow lines. Rilling requires unidirectional and uniform (i.e., simple) slopes that are typically greater than 400 m in length. The term is used where there are four or more channels or rills per 800 m of cross-sectional distance. Rills are typically 10 to 50 cm deep and 50 to 150 cm wide, with lengths sometimes up to 3 kms. **en rigoles, raviné**

ring compound An organic compound containing an aromatic structure (e.g., *benzene)* in which six carbon atoms are joined together in a closed circle or ring. See *aromatic compound.* **composé cyclique**

riparian An area of land adjacent to a stream, river, lake, or wetland, that contains vegetation that, due to the presence of water, is distinctly different from the vegetation of adjacent upland areas. **riparien, riverain**

ripening (*pedogenesis*) Chemical, biological, and physical changes occurring in organic soils upon aeration of previously waterlogged materials. **maturation, vieillissement**

ripping A tillage operation that uses a deep chisel or other similar implement to break up plow pans or other impermeable layers to depths of 60 cm. See *chiseling* and *subsoiling*. **soussolage**

riprap A relatively lightweight covering of stones used to protect soil or surface bedrock from water erosion. **perré, enrochement de protection**

risk rating, forestry (assessment) (*forestry*) The process of identifying the degree of risk that timber harvesting imposes on adjacent and downslope social, economic, and forest resource values. The severity of each potential hazard and the magnitude of the potential consequences that correspond to each hazard provide the overall risk associated with harvesting a site. **évaluation du risque**

rithron zone The part of stream reach at higher elevations, characterized by rapid flow, low temperature, and high dissolved oxygen levels. **rithron**

R

river basin The total area drained by a river and its tributaries. **bassin versant**

river bed The channel that contains a river's waters. **lit**

river system All of the streams and channels draining a river basin. **réseau hydrographique**

river terrace (*geomorphology*) A portion of the former floodplain of a river, now abandoned and left at a higher level as the stream downcuts into its former floodplain. The term is used to describe both the bench-like form and the alluvial deposits of the former floodplain, and is applied to the *scarp* and to the flat tread above and behind it. **terrasse fluviatile ou fluviale**

river valley The depression made by a stream and the erosive processes that precede and accompany the development of the stream. **vallée fluviale**

river wash Barren, usually coarse-textured, alluvial soil in and along waterways, exposed at low water levels and subject to shifting during flood periods. A *miscellaneous land type*. **batture**

rock (1) A coherent, consolidated, and compact mass of mineral matter. It can be an aggregate of one or more minerals (e.g., granite, shale, marble), a body of undifferentiated mineral matter (e.g., obsidian), or solid organic material (e.g., coal). (2) Any prominent peak, cliff, or promontory, usually bare, when considered as a mass (e.g., Ayres Rock). **(1) roche (2) rocher**

rock creep The slow, downslope flow of rock fragments by intermittent slip along a plane between the fragment and the ground surface. The movement may be initiated by heating and cooling, or by the growth of ice crystals beneath the rock fragment. **reptation**

rock crystal See *quartz*. **cristal de roche**

rock drumlin (*geomorphology*) An elongated hill having a veneer of glacial drift over a rock core. **drumlin à noyau rocheux**

rock flour Fine-grained rock material with a *particle size* similar to that of *silt* or *clay*, produced by the grinding action of boulders. **farine de roche**

rock nodules Nodules with recognizable rock fabrics. **nodules rocheux**

rock phosphate (1) Deposits of the mineral apatite from which most phosphorus for fertilizer and other purposes comes. (2) Finely ground apatite applied as fertilizer without being treated with acid. **phosphate naturel**

rock weathering (*geology*) The chemical, physical, and organic processes that cause rocks to weather and decompose. **altération des roches**

rock-fill dam A dam fabricated of loose rock, usually dumped in place, with the upstream part constructed of hand or derrick-placed rock and faced with rolled earth or with an impervious surface of concrete, timber, or steel. **barrage en enrochement**

rockland An area of which usually 25 to 90% is occupied by rock outcrops and most of the remainder by shallow soils. A *miscellaneous land type*. **terrain rocheux**

rod weeding See *tillage, rod weeding*. **désherbage mécanique à l'aide d'un cultivateur à tige rotative**

rodenticide A chemical compound or agent that kills rats and other rodents or prevents damage by them. **rodenticide**

roguing (*silviculture*) The removal and destruction of plants considered off-type or diseased. **rebut de sélection**

rolling (*geomorphology*) (1) Descriptive of land surfaces with a very regular sequence of moderate slopes merging from rounded, sometimes confined concave depressions to broad, rounded convexities producing a wavelike pattern of moderate relief. Slope length is often 1.5 km or greater and gradients are greater than 5%. (*agriculture*) (2) See *tillage, rolling*. **(1) vallonné (2) roulage**

root (*botany*) The organ (usually below ground) of higher plants whose functions include the absorption and conduction of water and dissolved minerals, food storage, and anchorage of the plant in the soil. The root is distinguished from the stem by its structure, the manner in which it is formed, and the lack of such appendages as buds and leaves. See *figure*. **racine**

root bed See *tillage, root bed*. **profil racinaire**

root cap A mass of cells, shaped like a cap or thimble, that covers and pro-

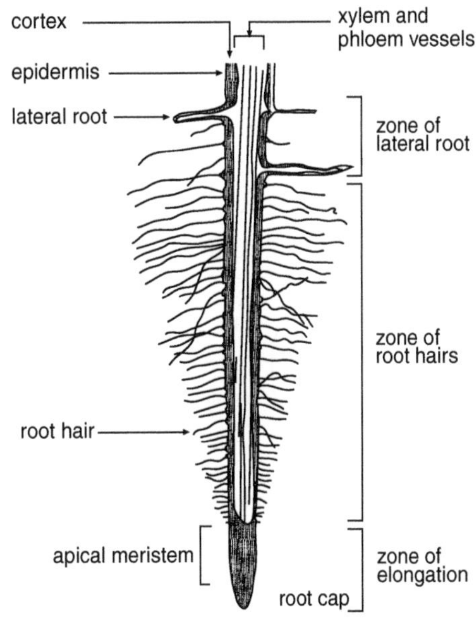

Parts of a **root** (adapted from Dunster and Dunster, 1996).

tects the growing tip of a root. **coiffe**

root hairs Tubular outgrowths of epidermal cells of the root in the zone of maturation, through which plants take in water, nutrients, and gases. **poils absorbants**

root hardy Descriptive of plants that regularly survive the winter, although the tops are killed to the ground. **plante vivace**

root interception A form of nutrient absorption by plant roots, based on direct contact between the nutrient and the root. Also called contact absorption. See *figure*. **absorption racinaire directe**

root mean square roughness (RMS) The standard deviation of (soil) surface height used as a measure of soil micro-relief; used to characterize the roughness of surfaces. **rugosité efficace, rugosité moyenne quadratique**

root nodule A hypertrophy formed on the roots of leguminous plants, caused by symbiotic nitrogen-fixing bacteria. See *figure*. **nodule racinaire**

R

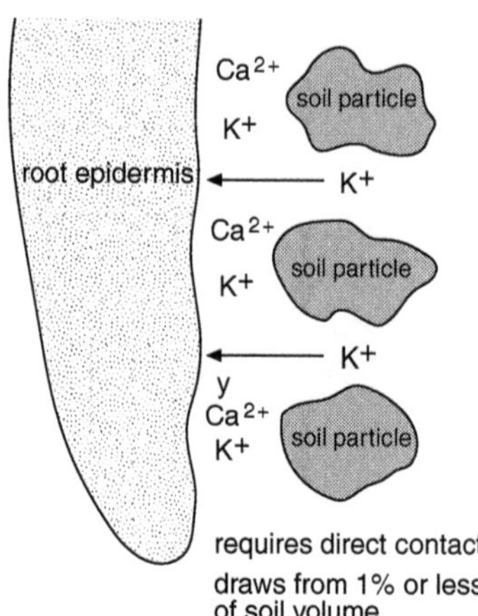

Root interception (adapted from Ontario Ministry of Agriculture, Food, and Rural Affairs, 1998).

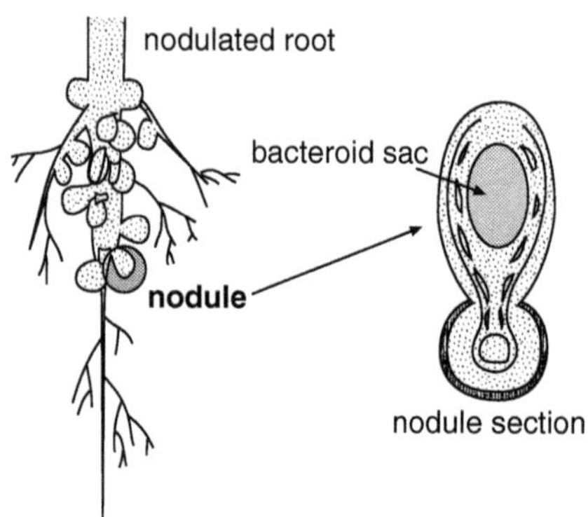

Root nodule

root pruning A technique to stimulate the development of a branched root system to facilitate transplanting, induce flowering or fruiting, control size, and rejuvenate plants grown in containers. **élagage ou taille des racines**

root rot Plant disease characterized by decay of the roots. **pourriture des racines**

root zone That part of the soil that is, or can be, penetrated by plant roots. **zone ou profil racinaire**

rootstock That part of a plant, including the roots, on which another variety has been budded or grafted. **porte-greffe**

rotary hoeing See *tillage, rotary hoeing.* **sarclage, hersage à la houe rotative**

rotary tilling See *tillage, rotary tilling.* **travail du sol avec bêche, charrue rotative**

rotation (1) (*agriculture*) A system of growing different crops on the same area of land each season. See *crop rotation.* (2) (*forestry*) The planned number of years between the formation or regeneration of a tree crop or stand and its final cutting at a specified stage of maturity. **(1) rotation (2) révolution**

rotation age (*forestry*) The age at which a forest stand is considered mature and ready for harvesting. **âge d'exploitabilité**

rotation grazing A system of pasture usage involving short periods of heavy stocking followed by periods of rest for herbage recovery during the same season; generally used on tame pasture or cropland pasture. **pâturage en rotation**

rotation pasture A cultivated area used as a pasture one or more years as part of crop rotation. **pâturage en rotation**

rotational landslide A *landslide* in which shearing takes place on a well defined, curved surface, concave upward, producing a backward rotation in the displaced mass. **glissement rotationnel**

rotenone An organic botanical compound extracted from the roots of tropical legumes, used as an *insecticide* and a fish poison. **roténone**

rough broken Descriptive of an area having steep slopes and many intermittent drainage channels, but usually covered with vegetation. A *miscellaneous land type* used in soil mapping. **escarpement, ravin**

roundness (*geology*) The degree of abrasion of a *clastic* particle as shown by the sharpness of its edges and corners. Quantitatively defined as the ratio of the average radius of curvature of the corners of the particle image to the radius of the maximum inscribed circle. **arrondi, émoussé**

row crop A crop planted in rows, normally to allow cultivation between rows during the growing season. **culture en rangées ou en lignes, grande culture**

r-selected Individuals that have been selected for high reproductive rates; presumed to be important in colonization or when competition is not intense. See *k-selected.* **individu à stratégie r**

rubbed fiber See *fiber.* **fibre frottée**

rubble An accumulation of loose, angular fragments, not water-worn or rounded like gravel. **blocaille**

rubifaction The development of a red color in soil resulting from the release of iron from primary minerals in soils. Also called braunification, ferrugination. **rubéfaction**

ruderal vegetation Any floral species that recolonize habitats that have suffered human disturbance, or any vegetation that flourishes on waste places (e.g., road verges, embankments, demolition sites). This vegetation often consists of weeds that have high demands for nutrients and/or are intolerant of competition. **végétation rudérale**

rumen The first compartment of the stomach of a ruminant or cud-chewing animal, in which food is partly digested with the help of bacteria. **rumen**

ruminant A suborder of mammals having a complex multichambered stomach; uses forages primarily as feedstuffs. **ruminant**

rumination Regurgitation and remastication of food in preparation for true digestion in ruminants. **rumination**

runoff (*soil*) (1) That portion of the total precipitation (rain water or snowmelt) on an area that does not infiltrate the soil but flows over it (i.e., surface flow). (*hydrology*) (2) That portion of the precipitation on a drainage area that is discharged from the area in stream channels; types include surface runoff and subsurface flow (e.g., groundwater runoff or seepage). See *erosion.* **(1) ruissellement (2) écoulement**

runoff plot An area of land, usually small, designed to measure the portion of rainfall or other precipitation that occurs as runoff or surface flow, including possible soluble nutrients and other materials dissolved or carried in the runoff. **parcelle de mesure d'érosion**

S

S factor See universal soil loss equation. **facteur S**

safe yield (*hydrology*) The rate at which water can be withdrawn from an aquifer without depleting the supply to such an extent that undesirable effects result. **débit de sécurité**

salic horizon A mineral soil horizon (in the U.S. system of soil taxonomy) enriched with secondary salts more soluble in cold water than *gypsum*. A salic horizon is 15 cm or more in thickness, contains at least 20 g kg^{-1} of salt, and the product of the thickness in centimeters and amount of salt by weight is >600 g kg^{-1}. **horizon salique**

Salids A suborder in the U.S. system of soil taxonomy. Aridisols which have a salic horizon that has its upper boundary within 100 cm of the soil surface. **Salids**

salination The process by which soluble salts accumulate in the soil. **salinisation**

saline (1) Salty; water containing sodium chloride (e.g., seawater). (2) Having a salinity greater than that of seawater (e.g., brine). (3) Hypersaline. (4) Said of a taste resembling that of common salt, especially in describing the properties of a mineral. **salin**

saline seep Intermittent or continuous saline discharge at or near the soil surface under dryland conditions that reduces or eliminates crop growth. **suintement salin**

saline soil A nonsodic soil containing enough soluble salts to interfere with the growth of most crop plants. The conductivity of the saturation extract is greater than 4 dS m^{-1} at 25°C, the *exchangeable-sodium percentage* is less than 15, and the pH is usually less than 8.5. **sol salin**

saline-alkali soil (1) A soil containing appreciable quantities of soluble salts with enough exchangeable sodium to interfere with the growth of most crop plants, as well as an *exchangeable-sodium percentage* greater than 15, a conductivity of a saturation extract greater than 4 dS m^{-1} at 25°C, and a pH 8.5 or less in the saturated soil. (2) A soil that has a combination of harmful quantities of salts and either a high alkalinity or high content of exchangeable sodium, or both, so distributed in the profile that the growth of most crop plants is reduced. Also called *saline-sodic soil*. **sol salin à alcalis**

saline-sodic soil See *saline-alkali soil*. **sol salin sodique**

salinity The amount of dissolved salts in a medium. See *salinity, soil*. **salinité**

salinity control The physical control, management, and use of water and related land resources in such a way as to maintain or reduce salt loading and concentrations of salt in water supplies. **lutte contre la salinisation**

salinity risk index A measure of the probability that an area has a certain level of salinity. **indice du risque de salinité**

salinity, soil The amount of soluble salts in a soil; usually determined by measuring the *electrical conductivity* of a saturation extract. See *saline soil*. **salinité du sol**

salinization The process of accumulation of salts in soil. **salinisation**

Salmonella A genus of bacteria that includes several species pathogenic to humans. Primary diseases are typhoid fever and food poisoning. ***Salmonella***

salt A compound formed of cations other than H^+ and anions other than OH^-. When an acid and a base react, they form water and a salt. **sel**

salt balance (*irrigation*) The quantity of soluble salts removed from an irrigated area in the drainage water minus that delivered in the irrigation water. **bilan de sels**

salt bridge A U-shaped tube containing an electrolyte that connects the two compartments of a galvanic cell, allowing ion flow without extensive mixing of the different solutions. **pont de sel**

salt flat The dried-up bed of a former salt lake, sometimes called a salt prairie. **fond salin**

salt lick An area containing an unusually large quantity of salt in the soil; animals consume the soil to obtain the salt. **salignon**

salt marsh A marsh periodically flooded with salt water. See *marsh*. **marais salant**

salt pan A shallow salt lake occurring in a small enclosed basin, smaller in its dimensions than a true salt lake. **étang salé**

salt-affected soil A soil that contains enough soluble salts and/or exchangeable sodium to interfere with normal plant growth. Salt-affected soils are usually categorized into three groups: *saline*, *sodic,* and *saline-sodic soil*. **sol altéré par le sel, sol salsodique**

saltation The rolling, bouncing, or jumping action of (1) soil particles (between 0.1 and 0.5 mm in diameter) caused by wind, usually <15 cm above the soil surface, for relatively short distances. The movement of soil particles by wind transport can impact and disrupt other soil particles; (2) solid particles (e.g., gravel, stones, or soil aggregates) caused by the action of flowing water. The size of the particle moved is determined by the velocity and density of the water current; (3) material downslope in response to gravity. **saltation**

saltation flux The rate of *saltation* per unit area. **flux de saltation**

salt-water intrusion (*limnology*) The phenomenon occurring when a body of salt water (lake or ocean), because of its greater density, invades or encroaches upon a body of fresh water. It can occur in both surface and groundwater bodies. Examples are the flooding of freshwater marshes, aquifers, and rivers by seawater. **invasion d'eau salée**

salvage harvesting (*forestry*) Logging operations specifically designed to remove damaged timber (dead or in poor condition) and yield a wood product; often carried out following fire, insect attack, or windthrow. **coupe de récupération**

sample A part of a population taken to estimate a parameter of the whole population. A sample drawn in such a way that it gives a true value for the population from which it was drawn. Types include *random sample*, *judgement sample*, *probability sample,* and representative sample. **échantillon**

sample plot An area of land, usually small, used for measuring or observing performance under existing or applied treatments; may be temporary or permanent. A sample plot can also be an *experimental unit*. **placette d'échantillonnage, parcelle d'échantillon**

sampling The technique of obtaining a *sample* to provide adequate information about a particular population. In order that every member of the population has an equal chance of being selected, bias must be excluded from the procedure. This may be done by obtaining a *systematic sample*, *random sample,* or by *nested sampling*. **échantillonnage**

sampling frame The entire statistical population from which a *sample is* taken. **base de sondage**

sampling unit One of the units into which an aggregate is divided or regarded as divided for the purposes of sampling, each unit being regarded as individual and indivisible when the selection is made. The definition of a unit may be made on some natural basis (e.g., tree stands, soil map units) or upon some arbitrary basis (e.g., areas defined by grid coordinates on a map). A sample consists of a set of sampling units or sites that will be characterized. Sampling units are defined by the frame; they may correspond to resource units, or they may be artificial units constructed for the sole purpose of the sampling design. **unité d'échantillonnage**

sand (1) A mineral particle between 0.05 and 2.0 mm in diameter. (2) Any one of five fractions within the sand size class. The names and size limits of sand separates recognized by pedologists in Canada and the U.S. are: very coarse sand, 2.0 to 1.0 mm; coarse sand, 1.0 to 0.5 mm; medium sand, 0.5 to 0.25 mm; fine sand, 0.25 to 0.10 mm; and very fine sand 0.10 to 0.05 mm. (3) A soil texture class. See *soil separates*, *texture*. **sable**

sand bank An accumulation of sand in an *estuary* or in a coastal environment, having a linear or parabolic form created by the tidal flow. **banc de sable**

sand dune See *dune, barchan, longitudinal dune*. **dune de sable**

sand filter See *filter sand*. **filtre à sable**

sand lens A lenticular band of sand in distinctly sedimentary banded material. **lentille sableuse**

sand wedge The infilling by sand of a fissure in the ground surface formerly occupied by an *ice wedge*. The sand takes on the shape of the ice-wedge cast and is regarded as evidence of formerly frozen ground in a *periglacial* environment. **fente de remplissage de sable**

sand-bearing method A method of testing the crushing strength of drain tile in which the tile is bedded in sand according to particular specifications. **essai de chargement à la boîte de sable**

sandstone A clastic sedimentary rock in which the particles are dominantly of sand size, from 0.062 to 2 mm in diameter. Its particles are mainly quartz, although mica, feldspar, and other minerals also occur. Sandstones can be subdivided according to their *particle size* into coarse, medium, and fine varieties. They can also be classified according to the agent that cements the particles: calcareous sandstones have a dolomitic type of cement, siliceous sandstones are bonded by silica, ferruginous sandstones are cemented by limonite. **grès**

sandur (*geomorphology*) An outwash plain formed from *glaciofluvial* material carried out from the front of an ice-sheet by melt-streams, and consisting of extensive accumulations of gravel, sand, and silt, crossed by braided streams. Also spelled sandr, sandar. **sandr, plaine d'épandage fluvio-glaciaire**

sandy Containing a large amount of sand. It may be applied to any one of the soil classes that contains a large percentage of sand. See *texture*. **sableux**

sandy clay A soil textural class. See *texture*. **argile sableuse**

sandy clay loam A soil textural class. See *texture*. **loam sablo-argileux**

sandy loam A soil textural class. See *texture*. **loam sableux**

sanitary landfill See *landfill*. **site d'enfouissement sanitaire**

sap (1) The fluid contents of the *xylem* or *phloem*. (2) The fluid contents of a cell's *vacuole* (cell sap). **sève**

sap test A test used to determine whether a particular ion is dissolved in the plant sap. **essai ou test de sève**

sapling A young tree, no longer a seedling but not yet a pole, about 1 to 2 m high and 2 to 4 cm in diameter at breast height, typically growing vigorously and without dead bark or more than an occasional dead branch. **gaule**

sapric material One of the components of organic soils with highly decomposed plant remains. Sapric material is not recognizable, and its bulk density is low. **matériau saprique**

Saprists A suborder in the U.S. soil taxonomy. *Histosols* that have a high content of plant materials so decomposed that original plant structures cannot be determined, and have a bulk density of about 0.2 Mg m^{-3} or more. Saprists are saturated with water for periods long enough to limit their use for most crops unless they are artificially drained. **Saprists**

saprobe An organism that excretes extracellular digestive enzymes and absorbs dead organic matter. **saprobionte**

saprobic Surviving on dead or decaying organic matter. See *saprophyte*. **saprobe**

saprolite Soft, friable, iso-volumetrically weathered bedrock that retains the fabric and structure of the parent rock, exhibiting extensive inter- and intra-crystal weathering. Saprolite usually contains a high proportion of residual silts and clays formed by alteration, chiefly by chemical weathering. In pedology, the term saprolite was formerly applied to any unconsolidated residual material underlying the soil and grading to hard bedrock below. **saprolithe**

saprophyte Any organism that derives its nutrition from dead or decayed organic material, in contrast to a parasitic organism that obtains its nutrition at the expense of a living organism. **saprophyte**

saprophytic competence The ability of a nodule symbiont or pathogenic microorganism to establish itself and live in soil as a saprophyte. **capacité saprophyte**

sapwood The light-colored wood that appears on the outer portion of a cross-section of a tree. See *cambium*. **aubier**

saturate (*soil water*) To fill all the voids between soil particles with a liquid. (*chemistry*) To form the most concentrated solution possible under a given set of physical conditions in the presence of an excess of the solute. (*chemistry*) To fill to capacity, as the adsorption complex with a cation species (e.g., H-saturated). **saturer**

saturated flow The flow of water in a soil under a gradient of positive pressure potential. **écoulement saturé ou à saturation**

saturated hydraulic conductivity The *hydraulic conductivity* of a soil at *saturation*, when all the soil pores are filled with water, where the moving force is the gradient of a positive pressure potential. **conductivité hydraulique saturée ou à saturation**

saturated hydrocarbon A hydrocarbon in which there are no double or triple bond linkages. Therefore, the hydrocarbon is more stable. **hydrocarbure saturé**

saturated soil paste A mixture of soil and water that forms a paste. At saturation the soil paste glistens as it reflects light, flows slightly when the container is tipped, and slides freely and cleanly from a spatula. **pâte de sol à saturation**

saturated zone (*geology*) A subsurface zone below which all rock pore space is filled with water. **zone de saturation**

saturation The wettest possible condition of a soil, when all the soil pores are filled with water. See *figure*. **saturation**

saturation content The mass water content of a saturated soil paste, expressed as a fraction of the dry soil mass. **teneur en eau de l'extrait de saturation**

S

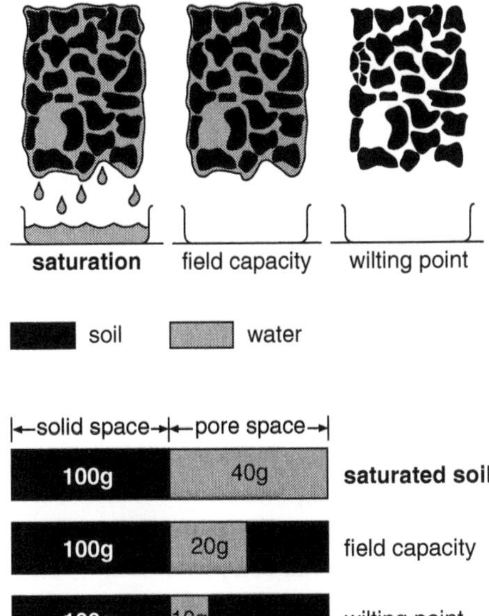

Volume of solids, water, and air in a loam soil at **saturation**, field capacity, and wilting point (adapted from Brady and Weil, 1996).

saturation mixing ratio The maximum water vapor concentration in the atmosphere for a given air temperature. The higher the air temperature, the higher the saturation mixing ratio. **rapport de mélange de saturation**

saturation percentage The water content of a saturated soil paste, expressed on a dry-weight percentage basis. **pourcentage de saturation**

saturation response A response of an organism to an environmental factor that increases to a plateau or asymptote beyond which further increases produce no response. **réaction de saturation**

saturation-paste extract The extract obtained (under vacuum) from a soil sample that has been saturated with water (i.e., *saturated soil paste*); Often used in soil solution and salinity assessment. **extrait de pâte à saturation**

sausage dam A dam composed of loose rock that has been wrapped with wire into cylindrical bundles and laid in a horizontal or vertical position. **barrage en gahions**

savanna (savannah, savana) A Spanish term usually used to describe the world's tropical grasslands. The vegetation is comprised of combinations of scattered trees, shrubs, and various grasses and sedges according to the length of the dry season. One of the four main subdivisions of world vegetation. See *biochore*. **savanne**

sawtimber Trees with logs suitable in size and quality for the production of lumber. **bois de sciage**

scale (*remote sensing*) The ratio of a distance on a photograph or map to its corresponding distance on the ground. The scale varies from point to point because of displacements caused by tilt and relief, but is usually taken as f/H, where f is the principal distance (i.e., focal length) of the camera and H is the height of the camera above mean ground elevation. It may be expressed as a ratio (e.g., 1:20,000), a representative fraction (e.g., 1/20,000), or an equivalence (e.g., 1 cm = 2,000 m.). (*forestry*) To estimate the content of sound wood in a log or bolt or group of logs or bolts, using a given unit of measure or weight. Also, the estimated content of a log or group of logs or bolts. **échelle**

scalping (*silviculture*) A method of preparing forest soils for tree planting that consists of removing the ground vegetation and root mat to expose mineral soil. **scalpage**

scanning electron microscope (SEM) An instrument that enlarges the image of a specimen by simultaneously and systematically sweeping an electron beam over the specimen, which is usually coated with some metal such as gold. The electrons that are reflected from the surface form an image on a photographic plate. The instrument can also be used to determine the composition of metal ions in a specimen (*electron microprobe*

S

analysis). **microscope électronique à balayage**

scarification (1) The process of scratching or abrading the seed coat of the seed of certain species to allow the uptake of water and gases as an aid to seed germination. (2) A method of seed-bed preparation that consists of exposing patches of mineral soil through mechanical action. **scarification**

scarifying See *tillage, scarifying.* **scarifiage léger, écroûtage**

scarp (*geomorphology*) A steep slope extending over a considerable distance and marking the edge of a terrace, plateau, or bench. **escarpement**

scattergram A graphic representation of the degree of correlation between two sets of statistical data. One set is plotted along the *x* axis, the other along the *y* axis, and a "best fit" line is drawn (see *regression*). If there are no residuals or anomalies the correlation is perfect, but if the scatter of points does not lie on a straight line there is imperfect correlation. The deviations of the residuals from the regression line can be shown by drawing residual lines. It is also called a scatter diagram. **diagramme de dispersion**

scattering (*remote sensing*) The reflection and refraction of electromagnetic energy by particles in the atmosphere; usually wavelength dependent. **diffusion**

S

schist (*geology*) A strongly foliated, coarsely crystalline metamorphic rock that can readily be split into slabs or flakes because most of the mineral grains are parallel to each other. **schiste**

schistosity The foliation in a schist, due largely to the parallel orientation of micas. **schistosité**

scientific method The lines of reasoning that scientists follow in attempting to explain natural phenomena. It includes observation, analysis, synthesis, classification, and inductive inference, in order to arrive at a *hypothesis* that seems to explain the problem. Hypothesis becomes *theory* if it withstands repeated testing and application. Deductive use of the theory may then explain additional problems. **méthode scientifique**

scintillation counter An instrument that detects and measures radioactivity by counting flashes of light (scintillations) produced when radiation strikes certain chemicals. **compteur à scintillation**

sciophilous Adapted to live entirely in the shade. **sciaphile**

sciophyte A plant that thrives in a shaded environment. **sciophyte**

sclerenchyma Plant tissue made up of cells with heavy lignified cell walls; it supports and protects the softer tissue of the plant. **sclérenchyme**

sclerophyllous Species of evergreen shrubs and trees that have adapted to lengthy seasonal drought by producing tough, leathery leaves to cut down moisture loss by transpiration; commonly found in regions with a Mediterranean climate. **sclérophylle**

scoria A type of *pyroclastic* material ejected from a volcano, characterized by a dark, vesicular structure of partly glassy and partly cindery material known as volcanic slag. It is formed by the rapid cooling of the blisters and bubbles of lava fragmented by a volcanic eruption. **scorie**

scoria land Areas of slag-like clinkers, burned shale, and fine-grained sandstone, characteristic of burned-out coal beds; usually sparsely vegetated. **champs de scories**

scour (*geology*) (1) Concentrated erosive action in a river channel or in estuaries and narrow straits. (*civil engineering*) (2) An artificial flow of water intended to remove mud from a stream bed; also, the structure built to produce such current. **(1) affouillement (2) curage, décapage**

scree An accumulation of fragmented rock waste below a cliff or rock face, formed as a result of disintegration, largely by mechanical weathering of a rock exposure. It is characterized by generally coarse *particle size,* varying degrees of soil or vegetation cover, and a concave slope profile. Scree is called *talus* in North America, but in some countries talus refers to loose material accumulated only beneath a free face. **talus d'éboulis**

screefing (*silviculture*) A method of preparing forest soils for tree planting that consists of mechanically pushing aside the humus layer to expose mineral soil. **scalpage**

screen A perforated plate or meshed fabric used to separate coarser from finer parts, as of sand or other particulate materials. **crible**

scrub A type of vegetation associated with poor soils, exposed locations, or semi-arid environments, in which the species are stunted, gnarled, or adapted to seasonal drought. **broussaille**

second bottom (*geomorphology*) The first terrace above the normal floodplain of a stream. **terrasse inférieure, basse terrasse**

second growth A forest or stand that has grown up naturally after removal of a previous stand by fire, harvesting, insect attack, or other cause. **forêt de seconde venue**

second law of thermodynamics One of the laws that describe the movement of energy within the environment. Two basic approaches can be used to interpret the second law. One involves the observation that whenever any form of energy is employed to do useful work, the conversion of energy to work is not 100% efficient. A portion of the energy source is always lost from the system as heat. Another way to describe the second law relates to the level of organization within a system. Any system tends to become more disorganized. When energy is converted from one form to another, randomness tends to increase. The measure of randomness is *entropy.* **deuxième loi de la thermodynamique**

secondary clarifier See *final clarifier.* **décanteur secondaire**

secondary consumer An animal that obtains its nourishment by eating other animals; a carnivore. **consommateur secondaire**

secondary metabolite A product of intermediary metabolism released from a cell. **métabolite secondaire**

secondary mineral A mineral resulting from the decomposition of a primary mineral or from the re-precipitation of the products of decomposition of a primary mineral. See *primary mineral.* **minéral secondaire**

secondary nutrient Plant nutrients required by the plant in moderate amounts (e.g., calcium, magnesium, sulfur) and usually not considered fertilizer elements, though they may be present in fertilizers as components of the materials used to provide desired fertilizer elements. **élément nutritif secondaire**

secondary productivity The annual rate at which biomass is stored by consumers at one trophic level, equal to the amount of biomass ingested from lower trophic levels, less predation and respiratory losses. See *primary productivity.* **productivité secondaire**

secondary response A change in a *system* that is not immediately evident because of either a pronounced operational time lag or the very small magnitude of the change. **réaction ou réponse secondaire**

secondary rocks or minerals Rocks or minerals formed as a result of alteration of existing minerals, or from material derived from the disintegration of rocks. See *clastic sediment.* **roches ou minéraux secondaires**

secondary succession The orderly and predictable changes that occur over time in the plant and animal communities of an area that has been subjected to the removal of naturally occurring plant cover. **succession secondaire**

secondary tillage See *tillage, secondary tillage*. **travail secondaire ou superficiel du sol, reprise de labour**

secondary treatment A phase in the treatment of wastewater in which bacteria from activated sludge or trickling filters consume the organic parts of the wastes. **traitement secondaire**

section (*geology, pedology*) (1) A vertical cut through rock, soil, or landform that reveals the nature of the subsurface material. For example, a geological section shows the details of the rock strata. A section may be artificial or natural, as in a cliff. (*petrography, pedology*) (2) A thin slice of a soil, sediment, or rock prepared for examination under a microscope. See *thin section*. (*biology*) (3) A faunal or floral division of a *genus*. **(1) coupe (2) lame mince (3) section**

sedge A grass-like or rush-like plant of the family Cyperaceae found worldwide but especially in subarctic and temperate marshes. Sedges differ from true grasses in having solid, usually triangular stems; most are perennial and reproduce by *rhizomes*. Stems and leaves of many genera are used for weaving mats, baskets, and hats, as well as in papermaking. **carex**

sedge peat See *carex peat*. **tourbe de Carex**

S

sedge-reed peat Peat composed of sedges, reeds, grasses; often laminated, stringy, matted, or felty; usually not too decomposed; may contain stems and roots of woody shrubs as inclusions, but generally no large logs or pieces of wood from trees. It originates in sedge and reed *fens*, reed sedge and cattail *marshes*, and sedge-shrub *carr*. It may occur as a single layer from surface to substratum, as a second layer overlying aquatic peats, as a basal layer, or in multi-layer sequences. See *carex peat, sphagnum peat, sedimentary peat*. **tourbe d'herbacées**

sediment Solid material, both mineral and organic, that is in suspension, is being transported, or has been moved from its site of origin by air, water, or ice and has come to rest on the Earth's surface. **sédiment**

sediment basin A reservoir for the confinement and retention of silt, gravel, rock, or other debris from a sediment-producing area. Commonly constructed at the upper end of a channel or reservoir to store sediment-laden water for a sufficient length of time for the sediment to be deposited. Also called sedimentation basin. **bassin de sédimentation**

sediment delivery ratio A measure of the *sediment* actually reaching a stream or lake; equal to the quantity of material reaching a defined point in a water system divided by the quantity of material eroded. **rapport de production de sédiments**

sediment discharge See *sediment load*. **débit solide**

sediment grade sizes Measurements of sediment and soil particles that can be separated by screening. **granulométrie des matériaux**

sediment load The quantity of *sediment*, measured in dry weight or by volume, transported through a stream cross-section in a given time. *Sediment discharge* consists of both suspended load and bedload. **charge solide**

sediment oxygen demand (SOD) The amount of dissolved oxygen removed from the water covering the sediment in a lake or stream because of microbial activity. **demande d'oxygène des sédiments**

sedimentary Pertaining to or containing *sediment*, or formed by its deposition. **sédimentaire**

sedimentary peat A material composed of plant debris and fecal pellets less than a few tenths of a millimeter in diameter and having brown or gray-brown colors when dry. It has slightly viscous water suspensions, is slightly plastic but not sticky, and shrinks upon drying to form clods that are difficult to rewet. It has few or no plant fragments recognizable to the naked eye. See *sedge peat, sphagnum peat.* **terre coprogène**

sedimentary rock A rock formed from materials deposited from suspension or precipitated from solution and usually more or less consolidated. The principal sedimentary rocks are sandstones, shales, limestones, and conglomerates. **roche sédimentaire**

sedimentation The process of subsidence and deposition by gravity of suspended matter carried in water; usually the result of the reduction of water velocity below the point at which it can transport the material in suspended form. **sédimentation, décantation**

sedimentation tank (*waste water management*) A tank in which waste solids are allowed to settle or to float as scum. **décanteur statique**

sedimentology The science dealing with the study of *sediments* and *sedimentary rocks* and the processes by which they were formed. **sédimentologie**

seed (*botany*) (1) A ripened (mature) ovule consisting of an embryo, a seedcoat, and a supply of food that, in some species, is stored in the endosperm. Seeds of *angiosperms* (flowering plants) are enclosed in the ovary that later forms a fruit. *Gymnosperm* seeds are exposed on the scales of cones. (*agriculture*) (2) To sow, as to broadcast or drill small-seeded crops. **(1) semence (2) semer**

seed coat The outer layer of the seed, developed from the integuments of the ovule. **tégument**

seed orchard A plantation of specially selected trees that is managed for the production of genetically improved seed. **verger à graines**

seed source The locality where a *seedlot* was collected. If the stand from which collections were made was exotic, the place where its seed originated is the original seed source. **origine des graines**

seed tree Trees selected to be left standing to provide seed sources for natural regeneration. Selection is usually on the basis of good form and vigor, the absence of serious damage by disease, evidence of the ability to produce seed, and wind firmness. **semencier**

seed tree silvicultural system An even-aged silvicultural system in which selected trees (seed trees) are left standing after the initial harvest to provide a seed source for natural regeneration. Seed trees can be left uniformly distributed or in small groups. Although regeneration is generally secured naturally, planting may augment it. Seed trees are often removed once regeneration is established, or may be left as reserves. **mode de régénération par coupe avec réserve de semenciers**

seedbed A prepared area in which seed is sown. In natural regeneration, the soil or forest floor on which seed falls. See *tillage, seedbed.* **lit de semence**

seed-drilling Seeding by means of an implement that plants seeds in holes or narrow furrows and covers them with soil. **semis à l'aide d'un planteur**

seedling A young plant grown from seed. **semis, plantule**

seedlot (*silviculture*) Seed from a particular collection event, either from a single tree collection or a pooling of seed from many trees. **lot de semences**

seep Wet areas, normally not flowing, that arise from an underground source. **suintement**

seepage (1) The escape of water downward through the soil. (2) The emergence of water from the soil along an extensive line of surface, in contrast to a spring where the water emerges from a local spot. (3) The interstitial movement of water that may take place through a dam, its foundation, or abutments. **suintement**

seepage pit A covered pit with a lining that permits treated sewage to seep into the surrounding soil. **fosse de rétention**

segetal vegetation A type of semi-natural vegetation growing where humans have interfered with the natural growth. **végétation ségétale**

segregation (1) The concentration of minerals within a *sedimentary rock* to form a nodule. (2) The separation of minerals into discrete masses during crystallization of *igneous rocks* (magmatic differentiation). (3) The concentration of minerals into bands in *metamorphic rocks*. (4) The formation of ice lenses in soil. **ségrégation**

seiche (*limnology*) Periodic oscillations in the water level of a lake or inland sea that occur with temporary local depressions or elevations of the water level. **seiche**

seismic focus, seismic origin See *earthquake*. **foyer sismique**

seismic survey An exploration technique utilizing seismic shooting to delineate subsurface geologic structures of economic importance. **relevé sismique**

seismic waves Earthquake shock waves generated from the focus within the Earth's crust. Three major types, referred to as primary (P) waves, secondary (S) waves, and L-waves, have been recognized. **ondes sismiques**

S

seismic zone See *earthquake*. **zone sismique**

seismograph A scientific instrument designed to record *seismic waves*. Early instruments were based on the principle of the pendulum, while modern instruments are highly sophisticated and sensitive enough to record the smallest tremors. **sismographe**

seismology The scientific study of earthquakes and the elastic properties of the Earth. **sismologie, séismologie**

selective cutting (*forestry*) A system of forest cutting in which single trees, usually the largest, or small groups of such trees are removed for commercial production or to encourage reproduction under the remaining stand in the openings. See harvest cutting, *improvement cutting, clear cutting*. **coupe d'écrémage ou sélective**

selective enrichment A technique for encouraging the growth of a particular organism or group of organisms. **enrichissement sélectif**

selective precipitation A method of separating metal ions from an aqueous mixture by using a reagent whose anion forms a precipitate with only one or a few of the ions in the mixture. **précipitation sélective**

self regulation The damping down of the effects made by external changes on the operation of a system. Also called negative feedback. **autorégulation**

self-mulching soil A soil in which the surface layer becomes so well aggregated that it does not crust and seal under the impact of rain, but instead serves as a surface mulch when it dries. **sol à autogranulation ou à autofoisonnement**

self-pollinated Pollinated by the anthers of the same flower. **autopollinisé**

self-pruning The natural death and fall of branches from live trees due to causes such as light and food deficiencies, decay, insect attack, snow, and ice; also called natural pruning. **élagage naturel**

self-purification (*limnology*) The removal of organic material, plant nutrients, or other biodegradable pollutants from a lake or stream by the activity of the resident biological community. **auto-épuration**

semi-arid Pertaining to a climate in which there is slightly more precipitation than in an arid climate, or to a region in which such a climate prevails and sparse grasses are the characteristic vegetation; annual precipitation is about 25 to 50 cm. Dryland farming methods or irrigation are usually required for crop production in these regions. See *arid, humid.* **semi-aride**

semi-desert The transition zone between true desert and more thickly vegetated zones. **région semi-désertique**

semipermeable membrane A membrane that is permeable to molecules of solvent but not to molecules of a solute. **membrane semi-perméable**

semivariogram See *variogram.* **semivariogramme**

senescence (1) The process of growing old; aging. Gradual deterioration of function in an organism leading to an increased probability of death. (2) (*botany*) The gradual degradation of a plant or plant organ, usually due to old age, accompanied or preceded by a loss of protein and some minerals. **(1) vieillissement (2) sénescence**

sensitive slope Any slope identified as prone to *mass wasting*. **pente sensible ou fragile**

sensitive watershed A watershed used for domestic purposes or that has significant downstream fisheries values, and in which the quality of the water resource is highly responsive to changes in the environment. Typically, such watersheds lack settlement ponds, are relatively small, are located on steep slopes, and have special concerns such as extreme risk of erosion. **bassin versant sensible ou fragile**

sensitivity The effect of remolding on the shear strength of an undrained cohesive soil. **sensibilité**

sepiolite A fibrous clay mineral, $Si_{12}Mg_8O_{30}(OH)_4(OH_2)\cdot 8H_2O$. It is composed of two silica tetrahedral sheets and one magnesium octahedral sheet that make up the 2:1 layer. The 2:1 layers occur in strips with an average width of three linked tetrahedral chains joined at the edges to form tunnels where water molecules are held. **sépiolite**

septage Septic tank sludge combining raw primary sludge and an anaerobically produced raw sludge. **boues activées**

septate Separated by cross walls. **septé**

septic tank A buried tank used to treat domestic wastes. The sanitary waste from a household is deposited and retained in a covered tank to allow solids to settle to the bottom and provide an environment for the decomposition of organic components by anaerobic bacteria. The liquid effluent that flows from the tank is allowed to seep into the soil. See *leaching field.* **fosse septique**

septic tank absorption field A soil absorption system for sewage disposal consisting of a subsurface tile system laid so that effluent from the septic tank is uniformly distributed into the soil. **champ d'épuration**

sequential cropping A type of multiple cropping, used where the season is long enough to permit a second (or even a third) crop to be grown in the same year. Also known as double cropping. **cultures successives**

sequestration To separate, isolate, or withdraw. See *carbon sequestration.* **séquestration**

sequum A sequence of an eluvial horizon and its related illuvial horizon in a soil. **séquum**

seral community One of the temporary development stages in the sequence of plant colonization leading up to a vegetation climax. **groupement végétal préclimatique**

seral stage Transitions in the orderly and predictable changes in a biological community from the *pioneer* stage to the climax (i.e., *serclimax*) stage. **stade séral**

serclimax The terminal phase in a *sere* in which all further progress towards a true climax is inhibited by a repeated natural intervention (e.g., regular flooding or wetting by salt spray). **serclimax**

sere A series of ecologic communities that succeed one another as an area's

S

vegetation develops from pioneer stage to climax community. See *succession.* **sère**

serial dilution Successive dilution of a specimen or solution (e.g., 1:10 dilution equals 1 ml of a specimen plus 9 ml of diluent) for analytical purposes. **suspension-dilution, dilution en série**

serial samples Samples collected according to some predetermined plan, such as along the intersections of grid lines or at stated distances or times. The method is used to ensure *random sampling.* **prélèvements sériés**

series, soil A category in the Canadian, U.S., and other systems of soil classification. It is the basic unit of soil classification, is a subdivision of a *soil family*, and consists of soils that are alike in all major profile characteristics except the texture of the *A horizon.* **série de sols**

serpentine A rock-forming mineral with a complex chemical composition, part of the kaolinite-serpentine group, with the general formula $A_3Si_2O_5(OH)_4$, where A = Mg, Fe^{+2}, Ni. It forms from the breakdown of *olivine* and *pyroxene* when basic and ultrabasic rocks are altered by metamorphism. It is commonly variegated green in color, with a greasy or silky luster. It is formed by the alteration of magnesium-rich silicates, and occurs as compact-to-fibrous masses in igneous and metamorphic rocks. Massive translucent varieties are used as an ornamental stone, and fibrous varieties are used as asbestos. **serpentine**

servomechanism A device that automatically monitors the progressive operation of a *system,* correcting any deviation by means of *feedback* and thereby maintaining a satisfactory performance within the system (e.g., a thermostat). **servomécanisme**

sesquan (*soil micromorphology*) A *cutan* composed of a concentration of *sesquioxides.* **sesquane**

sesquioxide Any of the oxides and hydroxides of iron and aluminum. **sesquioxyde, oxyde de fer et d'aluminium**

sessile (*biology*) Organisms that are closely attached to other objects or to a substrate. See *aufwuchs, periphyton.* (*botany*) Leaves or flowers that are attached to the stem directly by the base, not by a satlk or peduncle. **sessile**

seston All material, organic and inorganic, including the living and non-living bodies of plants or animals, suspended in a waterway. **seston**

settleable solids Suspended debris, sediment, or other solids heavy enough to sink when a liquid waste is allowed to stand in a pond or tank. See *settling chamber* and *settling pond.* **matières décantables**

settlement pond, settling pond Excavation larger than a catchment basin and preferably with lower velocity waterflows that enables suspended sediment to settle before the flow is discharged into a creek. (*wastewater management*) An open lagoon in which wastewater contaminated with solid pollutants is collected and allowed to stand. The solid pollutants suspended in the water sink to the bottom of the lagoon, and the liquid is allowed to overflow out of the enclosure. **étang de décantation**

settling basin (*hydrology*) An enlargement in the channel of a stream to permit the settling of materials carried in suspension. **bassin de décantation**

settling chamber (*wastewater management*) An enclosed container used to settle solid pollutants present in the wastewater. Solids that sink to the bottom are removed. **chambre de sédimentation**

settling tank See *settling chamber* and *settling pond.* **bac de décantation**

settling velocity The terminal speed attained by a particle as it descends through a fluid (e.g., a dust particle through the atmosphere, or a particle

S

of clay through water). See *Stokes' law*. **vitesse de sédimentation ou de décantation**

sewage All waterborne organic waste and wastewater generated by residential and commercial establishments. **eaux usées**

sewage lagoon See *lagoon*. **bassin de stabilisation des eaux usées**

sewage sludge (*wastewater management*) Solid and semisolid waste resulting from the treatment of domestic and industrial effluents released into sewers. **boues de station d'épuration**

sewage treatment plant A facility that receives wastewater from domestic sources and removes materials that reduce water quality and threaten public health when the water is discharged into receiving streams. **usine d'épuration des eaux usées**

sewer Any pipe or conduit used to collect and carry away sewage or storm water runoff from the generating source to treatment plants or receiving streams. **égout**

sewerage The entire system of sewage collection, treatment, and disposal. Also applies to all effluent carried by sewers whether it is sanitary sewage, industrial waste, or storm water runoff. **assainissement**

shade temperature The temperature normally taken in meteorological observations (i.e., that obtained in a *Stevenson screen),* away from direct insolation or radiation from nearby objects. **température sous abri**

shale A fine-grained detrital sedimentary rock, formed by the compaction of clay, silt, or mud. It has a finely laminated structure, which gives it fissures along which the rock splits readily, especially on weathered surfaces. Shale is well indurated, but not as hard as argillite or slate. It may be red, brown, black, or gray. Some shales are oil-bearing. **shale, phyllade**

shaly (1) Containing a large amount of shale fragments. (2) A *soil phase* (e.g., shaly phase). (3) A kind of fragment. See *coarse fragments*. **schisteux, en feuillet**

Shannon-Weaver index An expression of species diversity, equal to $-\Sigma P_i \log P_i$, where P_i is the number of individuals in species i divided by the total number of individuals in a community, and log P_i is the (natural or base 2) logarithm of P_i; the sum is taken for all i species in the community. The Shannon-Weaver index incorporates aspects of species diversity, evenness, and richness. **indice de Shannon-Weaver**

shatter See *tillage, shatter*. **fragmentation, écrasement**

shatter belt (*geology*) A narrow tract of broken crustal rocks where one or more faults have fractured the rocks to such a degree that it has become a zone of relative weakness, less resistant to *erosion* than neighboring rocks. Rivers sometimes pick out the shatter belts, leading to a marked linearity of their valleys in this particular tract. **zone de faille**

shear A distortion, strain, or failure producing a change in form, usually without change in volume, in which parallel layers of a body (e.g., soil) are displaced in the direction of their line of contact. See *shear stress*. **cisaillement**

shear force A force directed parallel to the surface element across which it acts. **force de cisaillement**

shear plane The surface along which *shearing* occurs during the application of tangential stress to a material. **surface de cisaillement**

shear strength The maximum resistance of a soil or rock to *shear stress*. **résistance au cisaillement**

shear stress The force per unit area acting tangentially to a given plane within a soil mass. **contrainte de cisaillement**

shearing The action caused by the application of tangential stress to a solid

S

material. See *tillage, shearing.* **cisaillement**

sheath (*botany*) A long, tubular structure surrounding another structure. **gaine**

sheet erosion The removal of a fairly uniform (usually thin) layer of soil from the land surface by *runoff* water on generally low-gradient slopes, at relatively slow speeds, and over long time periods. It comprises the two processes of *rain splash* impact and *runoff*, which carries away the particles dislodged by the falling rain. Often called *inter-rill erosion.* **érosion en nappe ou inter-rigoles**

sheet flow See *overland flow.* **écoulement en nappe**

sheet piling Material, often a meshing or interlocking wood, concrete, steel, or other material, placed vertically in the ground to contain erosion or obstruct water percolation (the lateral movement of groundwater) and to stabilize foundations. **structure de confinement**

Shelford's law The law that states that when one environmental factor approaches the limits of tolerance, either minimum or maximum, that factor, or another factor, will become the controlling one and determine whether or not a species is able to maintain itself. See *limiting factor.* **loi de Shelford**

shelterbelt See *windbreak.* **haie brise-vent**

shifting agriculture (slash-and-burn agriculture) A traditional agriculture system of semi-nomadic people in which a small area of the forest is cleared by burning and is cultivated for 1 to 5 years and then abandoned, as soil fertility and crop yields decline. After abandonment, vegetation succession returns the area to climax forest and soil fertility is restored. This system was once practiced worldwide but is now primarily used in tropical rain forest areas. **agriculture itinérante (agriculture sur brûlis)**

shingle Beach gravel that is coarser than ordinary gravel, especially if consisting of flat or flattish pebbles and cobbles. **galet de plage**

shoulder See *slope morphological classification.* **épaulement**

shrinkage coefficient, soil The change in soil bulk volume with change in mass water content at a constant stress; also equivalent to the rate of change in void ratio with moisture ratio at a constant stress. **coefficient de retrait**

shrinkage index The numerical difference between the *plastic* and *shrinkage limits.* **indice de retrait**

shrinkage limit The maximum water content at which a reduction in the water content will not cause a decrease in the volume of the soil mass; this defines the arbitrary limit between the solid and semi-solid states. See also *liquid limit, plastic limit.* **limite de retrait**

shrinkage, soil structural Shrinkage that is less than volume water loss due to water drainage from *macropores* at high soil water content. **retrait structural**

shrink-swell potential The susceptibility of soils, especially those with a relatively high clay content, to a change in volume due to loss or gain in moisture content. **potentiel de retrait et de gonflement**

shrub (*botany*) A perennial woody plant, less than 10 m tall, and having several stems arising at a point near the ground. It may be *deciduous* or *evergreen.* At the end of the growing season there is no die back of the aerial parts. **arbuste, arbrisseau**

sial That part of the Earth's crust composed of granitic material rich in silicon (Si) and aluminum (Al). It has a lower density (2.65 to 2.7 Mg m^{-3}) than the simatic rocks. See *sima.* The sial forms the continental crust. **sial**

side slope (1) A hill slope, as seen in plan view, with fairly straight boundaries above and below, situated on the side of an interfluve. See *slope morphological classification.* (2) The slope on the side of any cut or fill section, as in canals, dams, ditches, diversions, terraces, and channels; usually given as a ratio of horizontal distance to vertical distance (e.g., 2:1 slope means a horizontal distance of 2 meters for every 1 meter in vertical distance). **(1) pente latérale (2) pente de talus**

sidelooking radar (*remote sensing*) An all-weather, day/night remote sensor particularly effective in imaging large areas of terrain. It is an active sensor because it generates its own energy that is transmitted and received to produce a photo-like picture of the ground. Also called side-looking airborne radar. **radar à visée latérale**

siderophore A nonporphyrin metabolite secreted by certain microorganisms that forms a highly stable compound capable of chelating with iron. There are two major types: catecholate and hydroxamate. **sidérophore**

sieve analysis Determination of the percentage distribution of *particle size* by passing a measured sample of soil or sediment through standard sieves of various sizes. **analyse granulométrique par tamisage**

sigmoid growth A pattern of population growth that traces out an S-shaped or sigmoid curve. A population undergoing sigmoid growth begins at an exponential rate, but the rate slows as limiting factors are encountered and an equilibrium level of population is reached. See *exponential growth, r-selected, k-selected.* **croissance sigmoïde**

signal-to-noise ratio The ratio of the level of the information-bearing signal power to the level of the noise power. **rapport signal-bruit**

signature Any characteristic or series of characteristics by which a material may be recognized. **signature**

significance, statistical A statistical calculation to demonstrate to what degree a hypothesis is acceptable. A hypothesis will be acceptable if the calculated *probability* exceeds a given value (α, called the *level of significance*). If α is <0.05 (95% limit of confidence) the result is said to be significant; if α is <0.01 (99% limit of confidence) the result is highly significant. See *confidence limits.* **signification statistique**

silage *Forage* preserved in a succulent condition by organic acids produced by partial anerobic fermentation of sugars in the forage. **ensilage**

silcrete A conglomerate of surficial sand and gravel cemented together by silica. **silcrète, croûte siliceuse**

silica Silicon dioxide (SiO_2), occurring in crystalline, amorphous, and impure forms. **silice**

silica gel A nontoxic, water-absorbing activated silica used as a drying agent for comfort air conditioning and compressed air, in chromatography, and as a catalyst. **gel de silice**

silica, crystalline A mineral form of silica, or silicon dioxide, found in three types: quartz, tridymite, and cristobalite. Prolonged exposure to crystalline silica can cause silicosis in humans; exposure to noncrystalline silica, or amorphous silica, does not cause silicosis. Also called free silica. **silice cristalline**

silica-alumina ratio The number of molecules of silicon dioxide (SiO_2) per molecule of aluminum oxide (Al_2O_3) in clay minerals or in soils. **rapport silice-alumine**

silica-sesquioxide ratio The number of molecules of silicon dioxide (SiO_2) per molecule of aluminum oxide (Al_2O_3) plus ferric oxide (Fe_2O_3) in clay minerals or in soils. **rapport silice-sesquioxyde**

silicate A compound whose crystal structure contains SiO_4 tetrahedra, either isolated or joined through one or more of the oxygen atoms to form groups, chains, sheets, or three-dimensional structures with metallic elements. Silicates are classified according to crystal structure. See *nesosilicate, phyllosilicate, tectosilicate*. **silicate, minéral silicaté**

siliceous (1) Said of a rock or other substance containing abundant silica, especially as free silica rather than silicates. (2) A *soil mineralogy* class for family groupings in the U.S. soil taxonomy and the Canadian system of soil classification. **siliceux**

silicification The replacement of existing structures or minerals within an organic or inorganic body with silica, most commonly quartz, chalcedony, or opal. The silica is introduced either by groundwater solutions or from igneous sources. Fossil wood is an example. **silicification**

sill (*geology*) (1) An intrusive body of solidified magma, which is *concordant* with the bedding planes of crustal rocks (as opposed to a *dike,* which is a *discordance*). (*geomorphology*) (2) A ridge separating one partially closed ocean basin from the ocean or another basin. (*statistics*) (3) The distance in the *semi-variogram* at which the *semi-variance* no longer increases and is equal to the *variance*. The spherical semi-variogram develops a flat region at this point. **(1) filon-couche (2, 3) seuil**

silt (1) A *soil separate* consisting of particles between 0.05 and 0.002 mm in equivalent diameter. (2) A soil textural class. See *texture*. **limon**

silt loam A soil textural class. See *texture.* **loam limoneux**

silting (*hydrology*) The deposition of waterborne sediments in stream channels, lakes, reservoirs, or on flood plains, usually resulting from a decrease in the velocity of the water. **envasement**

siltstone A very fine-grained, consolidated, clastic rock composed predominantly of particles of silt grade. **grès fin, siltstone**

silty clay A soil textural class. See *texture.* **argile limoneuse**

silty clay loam A soil textural class. See *texture.* **loam limono-argileux**

silvic Pertaining to organic soils developed in forest peat; used in soil classification to describe organic soil families. **silvique**

silvics The study of the life history, requirements, and general characteristics of forest trees and stands in relation to the environment and the practice of *silviculture*. **écologie forestière appliquée**

silvicultural system A process that applies silviculture practices, including the tending, harvest, and replacement of a stand, to produce a crop of timber and other forest products. The system is named by the cutting method with which regeneration is established. The six classical systems are even-aged, uneven-aged, seed tree, shelterwood, selection, and clearcut. **pratique ou régime sylvicole**

silviculture The management or cultivation of forest trees. **sylviculture**

sima That part of the Earth's *crust* that is composed of material rich in silicon (Si) and magnesium (Mg). It is denser (2.9 to 3.3 Mg m^{-3}) than the *sial,* which covers it in some places, but where sial is missing, sima constitutes most of the ocean floor. **sima**

simulation A means of representing or imitating a real world situation or real *system* in an abstract form, usually by constructing a *simulation model.* **simulation**

simulation model A physical construction (e.g., a scale model of an object) or a mathematical *model* constructed for the purposes of experimentation into the workings of a *system*. A computer is normally used to assist in the calculations, hence the term *computer simulation model*. See *figure*. **modèle de simulation**

single bond A bond in which one pair of electrons is shared by two atoms. **liaison simple**

single-grain structure A broad category of *soil structure*, used to describe completely unattached soil particles that occur almost as individual

S

primary particles; secondary particles or aggregates are seldom present. It is usually found only in extremely coarse-textured soils. Sometimes called structureless. See *Appendix B, Table 1*. **structure particulaire**

sink (1) A reservoir that takes up a substance from another part of its cycle (e.g., soils and trees tend to act as natural sinks for carbon). (2) A depression in the land surface; a negative potential area. **(1) puits (2) cuvette**

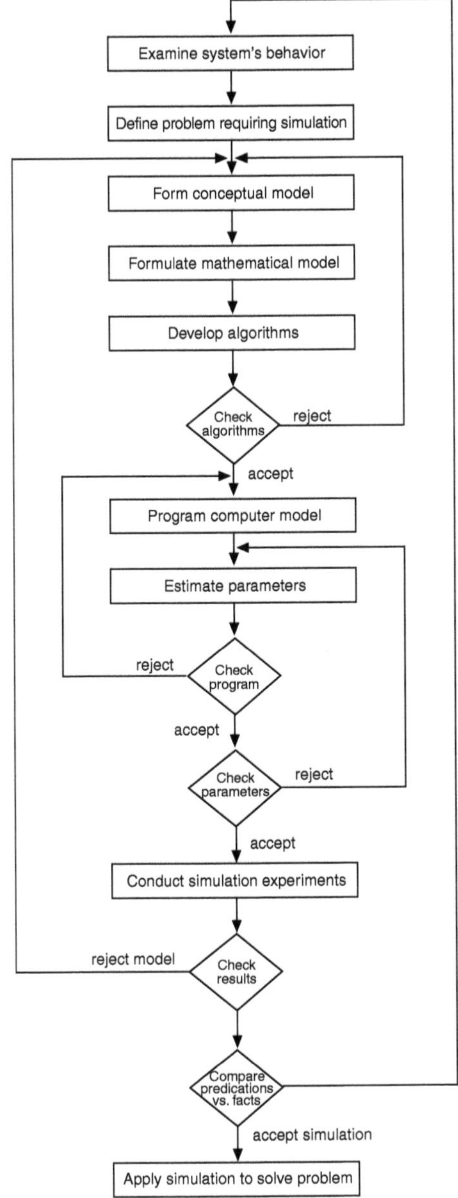

Phases in the construction of a **simulation model.**

sinkhole (*geomorphology*) A depression in the Earth's surface caused by the dissolution of underlying limestone, salt, or gypsum. Its drainage is through underground channels, its size is measured in meters or tens of meters, and it is commonly funnel-shaped. Sinkholes are common in *karst* terrain. **aven**

site (*ecology*) An area described or defined by its biotic, climatic, and soil conditions in relation to its capacity to produce vegetation. **site, station**

site factors Environmental factors present at a particular location. **facteurs de station**

site index (*forestry*) (1) A quantitative evaluation of the productivity of a soil for forest growth under the existing or specified environment. (2) The height (in meters) of the dominant forest vegetation taken at or calculated to an index age, usually 50 or 100 years. **indice de qualité de station**

site productivity The inherent capabilities of a site to produce or provide specific products or values over time. See also *net primary production*. **productivité ou potentiel du site**

site quality The capacity of a site to produce for a given purpose. **qualité de station**

size-exclusion chromatography (SEC) A chromatographic technique in which the mobile phase is a solvent and the stationary phase is a packing of porous particles. Also called gel-permeation chromatography (GPC). See *chromatography*. **chromatographie sur gel ou par exclusion**

skeletan (*soil micromorphology*) A *cutan* composed of skeleton grains. **squelettane**

skeleton grains (*soil micromorphology*) Individual grains that are relatively stable and not readily translocated,

S

concentrated, or reorganized by soil-forming processes; they include mineral grains and resistant siliceous and organic bodies larger than colloidal size. **grains du squelette**

skew planes (*soil micromorphology*) Planar *voids* that traverse the soil material in an irregular manner and are formed mostly by soil desiccation. **fentes de retrait irrégulières**

skewness The condition of being disordered or lacking symmetry; especially the state of asymmetry shown by a *frequency distribution* that is bunched on one side of the average and tails off on the other side. See *kurtosis*. **asymétrie**

skid trail (*forestry*) In timber harvesting, a random pathway traveled by ground skidding equipment while moving trees or logs to a landing. A skid trail differs from a skid road in that stumps are not removed, and the ground surface is untouched by the blades of earth-moving machines. **sentier de débardage**

skidder (*forestry*) In timber harvesting, a wheeled or tracked vehicle used for sliding and dragging logs from the stump to a landing. **débusqueuse**

skidding (*forestry*) In timber harvesting, the process of sliding and dragging logs from the stump to a landing, usually applied to ground-based as opposed to highlead operations. **débardage**

S

slaking The breakdown of *aggregates* into their primary particle constituents due to the addition of water to soil. **éclatement**

slash-and-burn agriculture See *shifting agriculture*. **agriculture sur brûlis**

slate A fine-grained rock produced from *argillaceous* rocks by metamorphism of a low grade. It is characterized by a well-developed cleavage, along which it splits quite easily. **ardoise**

slaty Containing a considerable quantity of slate fragments. It is used to modify soil texture class names (e.g., slaty clay loam). See *coarse fragments*. **ardoisier**

slick spots Small areas in a field that are slick when wet because of a high content of alkali or exchangeable sodium. **taches lisses**

slickens (*land reclamation*) Fine-textured materials separated in placer mining and in ore-mill operations; the materials may be detrimental to plant growth and so should be confined in specially constructed basins. A miscellaneous land type. **boue de mines fine**

slickensides Polished and striated surfaces that occur along planes of weakness resulting from the movement of one mass of soil against another in soils dominated by swelling clays; commonly observed in *Vertisols*. **surfaces de glissement, miroirs de faille**

slide A mass movement process in which slope failure occurs along one or more slip surfaces and in which the unit generally disintegrates into a jumbled mass en route to its depositional site. A debris flow or torrent flow may occur if enough water is present in the mass. **glissement**

slip (1) The downslope movement of a soil mass under wet or saturated conditions. (2) The movement of rock bodies in relation to each other on opposite sides of a *fault*. (3) The sliding movement of a glacier over its rock floor. **(1) glissement (2) rejet net (3) glissement basal**

slit planting, slot planting See *tillage, slit planting (slot planting)*. **semis-direct**

slit tillage See *tillage, slit tillage*. **travail du sol en bandes étroites**

slope An inclined surface, the gradient of which is determined by the amount of its inclination from the horizontal, and the length of which is determined by the inclined distance between the summit and the toe-slope. See *slope gradient*. **pente**

slope catchment area The area of a slope from which flow lines, following the direction of true slope, would cross a unit length of contour located anywhere in the slope complex; the area contributing flow to a unit length of contour. It is expressed as square meters of catchment area per meter of contour length. **bassin hydrographique, bassin versant**

slope classes The description of an area or region in terms of the steepness of slopes. **classes de pente**

Slope classes, class limits (in percent slope), and descriptive terminology

Class	Slope (%)	*Terminology*	*Terminologie*
1	0–0.5	*level*	*horizontal*
2	>0.5–2	*nearly level*	*subhorizontal*
3	>2–5	*very gentle slopes*	*pentes très faibles*
4	>5–10	*gentle slopes*	*pentes faibles*
5	>10–15	*moderate slopes*	*pentes modérées*
6	>15–30	*strong slopes*	*pentes fortes*
7	>30–45	*very strong slopes*	*pentes très fortes*
8	>45–70	*extreme slopes*	*pentes extrêmes*
9	>70–100	*steep slopes*	*pentes abruptes*
10	>100	*very steep slopes*	*pentes trés abruptes*

slope failure See *slide*. **glissement de talus**

slope gradient (angle) The degree of deviation of a surface from horizontal, measured in a numerical ratio, percent, or degrees. Expressed as a ratio or percentage, the first number is the vertical distance (i.e., rise) and the second is the horizontal distance (i.e., run), as 2:1 or 200%. Expressed in degrees, it is the angle of the slope from the horizontal plane, with a 90° slope being vertical (maximum) and 45° being a 1:1 slope. **inclinaison de la pente**

slope morphological classification The grouping of slopes into segments that have a defined range of slope gradient and plan and profile curvature (also called landform elements). The occurrence of a given slope morphological unit depends on the mode of formation of the landscape and the current geomorphic processes affecting the surface. In fluvially dissected landscapes, five basic slope morphological units may occur. See *figure – Basic slope morphological units.* In profile (downslope view), the units are:

summit or interfluve – area with a flat-to-gently-sloping surface at the drainage divide between two slopes. ***sommet ou interfluve***

shoulder – area with a convex profile curvature. ***épaulement***

backslope – area with a near-linear profile curvature and significant slope gradient. ***revers***

footslope – area with a concave profile curvature. ***base de pente***

toeslope (also called ***alluvial toeslope***) – area that is nearly level located at the base of the slope. ***pied ou base du talus***. In glaciated terrain that has not been fluvially dissected, large areas of near-level terrain without significant profile curvature but that function as neither summits nor toeslopes occur. See *figure – Shoulder, backslope, and footslope landform elements*. These are classified as level units. Each of the five profile units can be further divided on the basis of slope plan curvature into convergent units (those

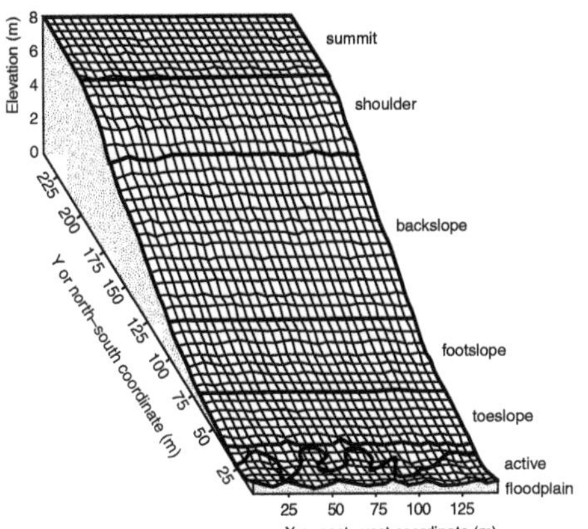

Basic **slope** morphological units.

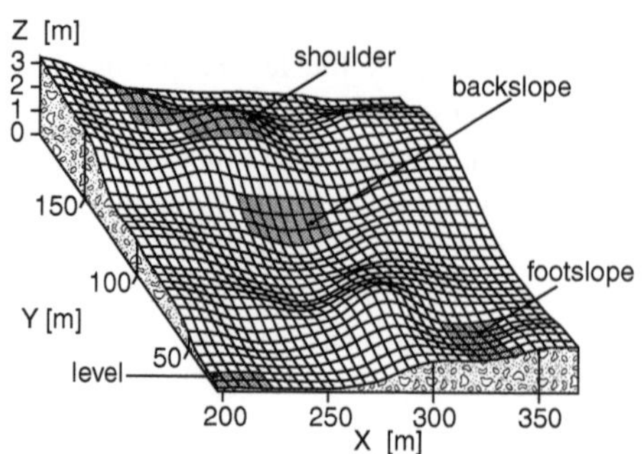

Shoulder, backslope, and **footslope** landform elements (D.J. Pennock, 2001).

with concave plan curvatures), divergent units (those with convex plan curvatures), and linear elements, which lack significant plan curvature. See *figure – Flowlines in areas of profile plan curvature*. **classification morphologique des versants**

slope morphology The study of the two- or three-dimensional shape of a slope. **morphologie des versants**

slope plan curvature The two-dimensional shape of the slope across the slope (i.e., parallel with the contour lines). It is measured as the rate of change in slope aspect along a given length of a contour line and reported in degrees per meter of contour length ($^{\circ}$ m^{-1}). Viewed from above (i.e., in plan view), the curvature is concave if the contour lines bow in, convex if they bow out, and linear if no significant curvature occurs. Flow lines (i.e., lines which are perpendicular to the contour lines) converge in areas of concave plan curvature and diverge in areas of profile plan curvature. See *figure – Flowlines in areas of profile plan curvature*. **courbure de niveau**

slope profile curvature The two-dimensional shape of the slope down the slope (i.e., perpendicular to the contour lines). It is measured as the rate of change in slope gradient along a given slope length measured perpendicular to the contour lines (i.e., along the true slope of the surface) and is reported in degrees per meter. Convex profile curvatures occur where the slope gradient increases downslope, linear curvatures where no significant change in gradient occurs downslope, and concave profile curvatures where the slope gradient decreases downslope. **courbure dans le sens de la pente**

slope stability The susceptibility of a slope to erosion and slides. **stabilité de la pente**

slope wash Soil and rock material moved down a slope predominantly by the action of gravity, assisted by running water that is not confined to channels (i.e., by *sheet erosion*). **érosion de pente**

sloping A type of surface expression associated with peatlands, consisting of a peat surface with a generally constant slope not broken by marked irregularities. **en pente**

slough (1) (*limnology*) A marsh or shallow undrained depression. A soft, muddy waterlogged ground or a sluggish body of water in a tidal flat or river area. In the prairie areas of Canada and the U.S. the terms slough and pothole refer to a pond in an enclosed basin, particularly the type occurring in hummocky moraines. (2) (*geology*) Rock material that has crumbled from the sides of a borehole. **(1) marécage, mare vaseuse, bourbier (2) éboulis**

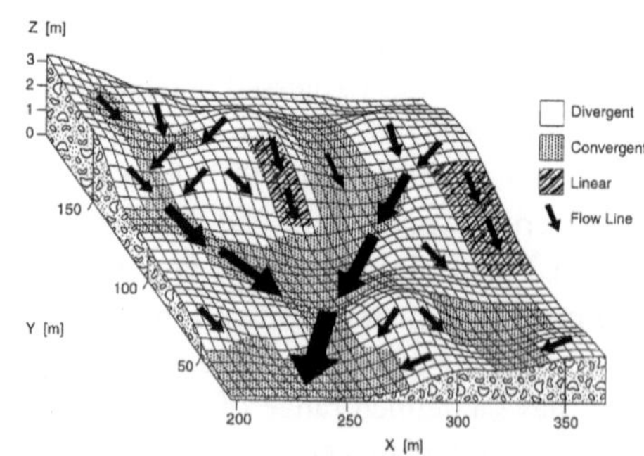

Flowlines in areas of **profile plan curvature** (D.J. Pennock, 2001).

S

slow release fertilizer See *fertilizer, controlled-release.* **engrais à libération lente**

sludge See *effluent* and *sewage sludge.* **boues**

sludge digestion The biological decomposition of *sewage sludge.* **digestion de boues**

sludge disposal The removal and discard of thick watery suspensions of particulate waste matter. Final disposal may involve the removal of excess water and the subsequent burning of the solids, or delivering the dewatered material to a landfill. **élimination de boue, mise en décharge**

sludge loading The maximum amount of *sewage sludge* that can be safely applied to soil. **charge de boues autorisée**

sluice (1) A channel constructed to carry water at a high velocity. (2) A vertically sliding gate used to control the flow of water in a stream channel. (3) To wash away earth or silt by directing a high velocity jet of water. **canal à écoulement rapide**

slump A mass of unconsolidated material that becomes detached from a hillside along a slip plane (see *slip*) and moves downslope, often as a *rotational landslide.* See *landslide.* **glissement**

s-matrix (of a soil material) (*soil micromorphology*) The material within the simplest peds or composing *apedal* soil materials in which the *pedological features* occur; it consists of the *plasma* (mineral and organic material of clay size), *skeleton grains* (individual mineral and organic particles larger than clay size), and *voids* that do not occur as pedological features other than those expressed by specific extinction (orientation) patterns. Pedological features also have an internal s-matrix. **fond matriciel**

smectite One of a group of 2:1 layer structured silicates with a high cation exchange capacity and variable interlayer spacing. Formerly called the montmorillonite group. The group includes di- and trioctahedral members. **smectite**

smelting A metallurgical process that involves reducing metal ions to the free metal. **fusion**

snow One of the solid forms of precipitation, consisting of snowflakes and minute spicules of ice. **neige**

snow density The water content of snow expressed as a percentage by volume. **densité de la neige**

snow hedge A planting of shrubs or other plants to intercept drifting snow; also called snow break or snow catch. See *snow management.* **trappe à neige (haie)**

snow management The management of snow to increase soil moisture for crop production; generally accomplished through the use of wind barriers, including grass and grass stubble barriers, as well as trees and shrubs. **gestion de la neige**

snowpack The snow mantle or accumulation of snow on the ground at any specified place and time. **couverture de neige**

soapstone A massive impure variety of *talc* that lends itself to carving. It is used by the Inuit to produce figurines for the commercial market. Also called steatite. **stéatite, pierre à savon**

sod Vegetation that grows so as to form a mat of soil and vegetation. **gazon**

sod grasses *Stoloniferous* or *rhizomatous* grasses that form a *sod* or turf. See *stolon.* **graminées à gazon**

sod planting See *tillage, sod planting.* **semis-direct sur prairie ou sur gazon**

sodic soil (1) A nonsaline soil containing sufficient exchangeable sodium to interfere with the growth of most crop plants. (2) A soil having an *exchangeable-sodium percentage* of 15 or more. Also called nonsaline-alkali soil. **sol sodique**

sodication The process whereby the exchangeable sodium content of a soil is increased. **sodisation**

sodicity The level of exchangeable sodium and its influence on a soil. **sodicité**

sodium adsorption ratio, adjusted The *sodium adsorption ratio* of a water adjusted for the precipitation or dissolution of Ca^{2+} and Mg^{2+} that is expected to occur where a water reacts with alkaline-earth carbonates within a soil. **rapport corrigé d'adsorption du sodium**

sodium estimation (*irrigation*) The estimation of sodium in irrigation water from known electrical conductivity (EC), calcium, and magnesium values as follows: sodium (mmol L^{-1}) = (EC (mS cm^{-1}) × 10) – (Ca^{2+} + Mg^{2+} (mmol L^{-1})). This equation is applied in some analytical laboratories as a screening tool for accuracy of determination of electrical conductivity and extractable cations in soils. **estimation de la teneur en sodium**

sodium adsorption ratio (SAR) The relationship of soluble sodium to soluble calcium plus magnesium in water or the soil solution, expressed by the equation:

$$SAR = [sodium] / [calcium + magnesium]^{1/2}$$

where the concentrations of ions, denoted by square brackets, are in millimoles per liter. This relationship can be used to predict the *exchangeable sodium fraction* of a soil. **rapport d'adsorption du sodium**

soft pesticide. An easily biodegradable pesticide that is non-persistent in the environment. See *persistent pesticide*. **pesticide non-persistant**

softrock. A colloquial term used by geologists and soil scientists to designate soft (i.e., paralithic) sedimentary rocks and to distinguish them from igneous and metamorphic rocks (i.e., hard rock). **roche tendre**

softwood. Wood produced from conifers such as pines, spruces, and firs, which dominate the forests of the temperate regions. The term does not relate to the hardness of the wood. Compare *hardwood*. **résineux**

soil. (1) The unconsolidated mineral or organic material on the immediate surface of the Earth that serves as a natural medium for the growth of land plants. (2) The natural, unconsolidated mineral or organic matter on the surface of the Earth that has been influenced by parent material, climate, macro- and micro-organisms, and relief, all acting over a period of time to produce a material different from which it was derived in many physical, chemical, biological, morphological properties. **sol**

soil absorption system. (*wastewater management*) Any system that uses soil to absorb treated sewage. (e.g., *seepage pit*). **site d'épandage sur sol**

soil acidity . See *acid soil; reaction, soil*. **acidité du sol**

soil aeration. (1) The gaseous composition of the soil *pore space*, particularly with respect to the amount and availability of oxygen for use by soil biota and/or soil chemical oxidation reactions. (2) The processes affecting this condition. **aération du sol**

soil air. The soil atmosphere; the gaseous phase of the soil (i.e. that volume not occupied by solid or liquid). **air du sol**

soil amendment Any material added to soil that enhances plant growth, via the nutrients they contain, or improves the soil condition (e.g., lime, gypsum, compost, animal manures, plant residues, certain industrial wastes, and synthetic soil conditioners). **amendement du sol**

soil association Groups of soils that occur in repeating patterns on the landscape. Often the hilltops are one soil series, the hillsides another, and the valley floors another. The separate areas may be large enough to be mapped on a detailed map but not on a generalized map. Soil associations are therefore used for making small-scale maps. **association de sols**

S

soil auger A tool for boring into the soil and withdrawing a small sample for field or laboratory observation. Soil augers may be classified into several types: those with worm-type bits, unenclosed; those with worm-type bits enclosed in a hollow cylinder; and those having a hollow cylinder with a cutting edge at the lower end. **tarière**

soil biological quality That component of soil quality derived from biological properties, including living organisms and material, and processes derived from living organisms (e.g., organic matter cycling, organic matter fractions, soil biota, microbial processes, and plants). **qualité biologique des sols**

soil boundary A line, or band, composed of points on the landscape at which soil characteristics are transitional between those of adjacent soil bodies, within which gradients of change are spatially less rapid than at the boundary. Three kinds of soil boundaries can be distinguished: sharp (i.e., abrupt on the land), distinct (i.e., neither abrupt nor gradual), and gradual (the cartographer usually draws a line in the middle of the zone of transition without indicating the true width of the boundary). Proportionate lengths of the three kinds of boundaries may be reported for a given soil body. Quantification of distinctness of soil boundaries has been suggested as: very sharp, < 0.3 m; sharp, 0.3 to 3 m; distinct, 3 to 5 m; gradual, 5 to 10 m; diffuse, > 10 m. Very broad diffuse boundaries are commonly subdivided arbitrarily into soil body delineations on maps. The form of a boundary may vary depending upon the specific property, or properties, considered. **limite de sol**

soil, buried See *buried soil.* **sol enfoui**

soil capability map See *capability class (soil), land capability.* **carte de possibilités des sols**

soil chemical quality That component of soil quality derived from chemical properties (e.g., mineralogy, organic matter, pH, pE, electrical conductivity, exchangeable sodium, and cation exchange capacity). **qualité chimique des sols**

soil classification The systematic arrangement of soils into categories on the basis of their characteristics. Broad groupings are made on the basis of general characteristics, and subdivisions on the basis of more detailed differences in specific properties. **classification des sols**

soil classification, Canadian system The Canadian national system of soil classification initially accepted for use in 1960, and most recently published as *The Canadian System of Soil Classification, Third Edition* in 1998. The system describes soils hierarchically, using five categories: *order*, *great group*, *subgroup*, *family,* and *series*. The system of soil classification for Canada at the order, great group, and subgroup levels is presented in *Appendix D, Table 1.* **système canadien de classification des sols**

Soil Classification, FAO/UNESCO soil units The FAO (Food and Agriculture Organization) and UNESCO jointly prepared the UNESCO Soil Map of the World (scale 1:5 million), first published in 1974. Map units developed for the map consist of classes at two levels based primarily on pedogenic processes. The terminology and conceptual basis for the higher soil categories are equivalent to the great group levels of systems such as for the U.S. Soil Taxonomy, as well as soil type for the Russian soil classification system. Class separations are based on *diagnostic horizons.* The lower class separations are based on special pedological features (e.g., gleying, salinization, lessivage). The soil units are mapped as *soil associations*, designated by

S

the dominant soil type, with soil phases differentiating characteristics important in use and management of soils. These consist of modifying properties (e.g., petric, saline, lithic, fragipan, stony), three textural classes (coarse, medium, and fine), and three slope classes (level to gently undulating, rolling to hilly, and steeply dissected to mountainous). The FAO soil map consists of very broad map units using a basic classification system, but it is the only truly international system in which soils all over the world can be accommodated. The system is intended for mapping soils at national and continental scales, but is not sufficiently detailed for application at local scales. The main categories of the system, along with equivalent U.S. Soil Taxonomy taxa, are indicated in the table. **système de classification des sols FAO-UNESCO, Carte mondiale des sols**

Soil Map Units for the FAO/UNESCO World Soil Map

FAO Soil Unit	U.S. Soil Taxonomy Equivalent	Description
Acrisols	Ultisols	Low base status soils with argillic horizons
Andosols	Andepts	Soils formed in volcanic ash that have dark surfaces
Arenosols	Psamments	Soils formed from sand
Cambisols	Inceptisols	Soils with slight color, structure, or consistency change due to weathering
Chernozems	Borolls	Soils with black surface color and high humus content, under prairie vegetation
Ferralsols	Oxisols	Highly weathered soils with sesquioxide-rich clays
Fluvisols	Fluvents	Water-deposited soils with little alteration
Gleysols	Aquic Suborders	Soils with mottled or reduced horizons due to wetness

Soil Map Units for the FAO/UNESCO World Soil Map (continued)

FAO Soil Unit	U.S. Soil Taxonomy Equivalent	Description
Greyzems	Boroll	Soils with dark surface, bleached E horizon, and textural B horizon
Histosols	Histosols	Organic soils
Kastanozems	Ustolls	Soils with chestnut surface color, steppe vegetation
Lithosols	Lithic Subgroups	Shallow soils over hard rock
Luvisols	Alfisols	Medium to high base status soils with argillic horizons
Nitosols	Ultisols and Alfisols	Soils with low CEC clay in argillic horizons
Phaeozems	Udolls	Soils with dark surface, more leached than Kastanozem or Chernozem
Planosols	–	Soils with abrupt A-B horizon contact
Podzols	Spodosols	Soils with light-colored eluvial horizons and subsoil accumulation of iron, aluminum, and humus
Podzoluvisols	Glossic Great Groups of Alfisols	Soils with leached horizons tonguing into argillic B horizons
Rankers	Lithic Haplumbrepts	Thin soils over siliceous materials
Regosols	Orthents, Psamments	Thin soil over consolidated material
Rendzinas	Rendolls	Shallow soils over limestone
Solonchaks	Salic Great Group	Soils with soluble salt accumulation
Solonetz	Natric Great Group	Soils with high sodium content
Vertisols	Vertisols	Self-mulching, inverting soils, rich in smectitic clays
Xerosols	Mollic Aridisols	Dry soils of semi-arid regions
Yermosols	Typic Aridisols	Desert soils

Soil Classification, U.S. Soil Taxonomy The system of soil classification described in *Soil Taxonomy: A Basic System of Soil Classification for*

Making and Interpreting Soil Surveys, prepared by the Soil Survey Staff, Soil Conservation Service, U.S. Department of Agriculture, originally published in 1975, and most recently published as *Soil Taxonomy* in 1999. Soil Taxonomy is a comprehensive soil classification system that provides a hierarchical grouping of soils based on observable or measurable properties. The system has six different levels of generalization, each level a category of classification. The soil order is the highest or most generalized level, followed by the suborder, great group, subgroup, family, and series. **système américain de classification des sols, Soil Taxonomy**

soil complex See *complex, soil.* **complexe de sols**

soil conditioner Any material added to a soil for the purpose of improving its physical properties, usually *aggregation.* **conditionneur de sol**

soil conservation (1) Protection of the soil against physical loss by erosion, or against chemical deterioration (i.e., excessive loss of fertility by either natural or artificial means). (2) A combination of all methods of management and land use that safeguard the soil against depletion, deterioration, or degradation by natural or human activities. (3) A division of soil science dealing with soil conservation and (1) and (2). See *conservation.* **conservation du sol**

soil core A volume of soil obtained by forcing a cylindrical device into the ground (e.g., soil, sediment), usually perpendicular to the horizontal. **carotte de sol**

soil correlation The process of defining, mapping, naming, and classifying the kinds of soils in a specific soil survey area for the purpose of ensuring that soils are adequately defined, accurately mapped, and uniformly named in all soil surveys made in a region. Soil correlation also deals with the techniques for describing soils and with the application and development of soil classification. **corrélation des sols**

soil correlation area A unit within a conceptual mapping framework described as an area with similar agroclimate and landscape ecology such that it defines the geographic limits for usage of soil series names; similar to *agroecological resource area.* See *zone, soil.* **aire de corrélation des sols**

soil creep The gradual, steady downhill movement of soil and loose rock material on a slope. **reptation de sol**

soil degradation The general process by which soil gradually declines in quality and is thus made less fit for a specific purpose, such as crop production. See *erosion.* **dégradation des sols**

soil drainage classes Seven classes that describe the overall natural drainage of soils, taking into account factors of external (i.e., surface runoff) and internal (i.e, perviousness) soil drainage in relation to supply of water. The classes from driest to wettest are: very rapidly, rapidly, well, moderately well, imperfectly, poorly, and very poorly drained. Each class describes water removal from the soil in relation to supply, and can be equated with a range in available water storage capacity. **classes de drainage du sol**

soil family See *family, soil.* **famille de sols**

soil fertility See *fertility, soil.* **fertilité du sol**

soil flushing A technique used to clean soil contaminated with inorganic or organic hazardous waste; the process involves flooding the soil with a flushing solution, which may be acidic or basic, or may contain surfactants, and the subsequent removal of the leachate via shallow wells or subsurface drains. The recovered leachate is then purified. **curage ou nettoyage du sol**

S

soil formation factors The interrelated natural agencies responsible for the formation of soil. The factors are grouped into five categories: parent material (physical and chemical attributes), climate (precipitation, temperature), organisms (flora and fauna), topography (elevation, slope, depth to water table) and time. **facteurs de formation des sols**

soil function The various roles that soil performs, or the tasks that are placed upon soil, that underpin the concept of *soil quality*. Soil functions in three main ways: as a medium for plant growth, a regulator or partitioner of water and energy, and an environmental buffer or filter. **rôle ou fonction du sol**

soil genesis (1) The mode of origin of the soil with special reference to the processes or soil-forming factors responsible for development of the solum, or true soil, from unconsolidated parent material. (2) The branch of soil science that deals with soil genesis. **genèse des sols, pédogenèse**

soil geography The branch of physical geography that deals with the areal distributions of soils. **géographie des sols**

soil health An approach to soil condition analogous to human or community health, by which the condition of a soil's properties and morphology are assessed against some optimum condition (i.e., soil-as-an-organism), or a soil's functions assessed against the goals placed upon them (i.e., soil-as-a-community), or against an optimum functional state. Often soil health is used synonymously with *soil quality*, except that a soil may have poor *inherent soil quality* but still have good health. **santé des sols**

soil improvement The process of making a soil more productive for growing plants (e.g., through drainage, irrigation, addition of fertilizers and soil amendments, and other management activities); the results of this process. **amélioration du sol**

soil individual A *polypedon* bounded by other soil polypedons or by nonsoil materials (e.g., water bodies). **individu-sol**

soil interpretations Predictions of soil behavior in response to specific uses or management based on inferences from soil characteristics and qualities (e.g., trafficability, erodibility, productivity). They are either qualitative or quantitative estimates or ratings of soil productivities, potentials, or limitations. **interprétation des sols**

soil inventory metadata Data that explains soil inventory data structures, terminology, classification systems, models, and procedures. **métadonnées d'inventaire pédologique**

soil landscape model A conceptual description of recurring soil and land patterns, and the relationships between them. It provides the basis in soil surveys for *soil map units* or concepts such as the *land system*. **modèle de pédopaysage**

soil loss See *soil erosion*. **perte de sol**

soil loss equation See *universal soil loss equation*. **équation de perte de sol**

soil loss tolerance (1) The maximum average annual soil loss that will allow continuous cropping and maintain soil productivity without requiring additional management inputs. (2) The amount of soil lost through *erosion* that is offset by soil development, thus maintaining an equilibrium between soil losses and gains. Soil loss tolerance can be expressed by a function (i.e., T = soil loss tolerance value generally ranges from 4 to 11 Mg ha^{-1} $year^{-1}$), which can be assigned to any soil type based on soil depth, previous erosion, and other factors. **perte de sol tolérable**

soil management The total of all tillage operations; cropping practices; and application of fertilizer, lime, and other treatments to a soil to produce plants and improve soil condition.

S

See *tillage, soil management.* **gestion du sol**

soil management group A group of soil units having similar adaptations or management requirements for one or more specific purposes (e.g., adapted crops or crop rotations, drainage practices, fertilization, forestry, and highway engineering). **groupe de gestion des sols**

soil map A map showing the distribution of soil types or other soil mapping units in relation to the prominent physical and cultural features of the Earth's surface. **carte de sols, carte pédologique**

soil map delineation A single soil area or polygon on a *soil map* that is differentiated from other areas on the basis of soil and landscape features. **délimitation cartographique des sols**

soil map unit A defined and named repetitive grouping of soil bodies occurring together in an individual and natural characteristic pattern over the soil landscape. The attributes of a map unit vary within more or less narrow limits that are determined by the intensity of the survey. A map unit comprises all the map delineations that have the same name. A map unit is conceptual; a map delineation is real. **unité cartographique de sols**

soil mechanics The application of the laws and principles of mechanics and hydraulics to engineering problems dealing with soil as an engineering material. **mécanique des sols**

soil micromorphology A branch of soil science dealing with the description, interpretation, and measurement of the spatial arrangement and identification of the mineral and organic constituents and associated pore space that makes up the morphology or microstructure of a soil. **micromorphologie du sol**

soil mineral (1) Any mineral that occurs as a part of, or in, the soil. (2) A natural inorganic compound with definite physical, chemical, and crystalline properties (within the limits of isomorphism) that occurs in the soil. See *clay mineral.* **minéral du sol**

soil mineralogy The branch of soil science that deals with the homogeneous inorganic materials found in the Earth's crust to the depth of weathering or of sedimentation. **minéralogie des sols**

soil model A conceptualization of the suite of soils in a landscape and their interrelationships; generally considered in terms of a dominant or co-dominant soil series and any significant soils or nonsoils. **modèle pédologique**

soil moisture See *soil water.* **humidité du sol**

soil order See *order, soil.* **ordre de sols**

soil organic matter quality The properties that characterize the labile or decomposable portion of organic matter; that component of *soil quality* derived from the composition of organic matter. **qualité de la matière organique du sol**

soil pedological quality That component of *soil quality* derived mainly from inherent soil properties or indicators that reflect the processes related to the formation of soils (e.g., soil spatial variability, soil landscape position, and surface features) and characterize changes in these processes that can ultimately influence *soil quality.* **qualité pédologique des sols**

soil physical quality That component of *soil quality* derived from physical properties used to assess soil quality, including those used to characterize the soil pore space, soil strength, and structure. **qualité physique des sols**

soil physics The branch of soil science that deals with the *physical properties of soils*, with emphasis on the state and transport of matter (especially water) and energy in the soil. **physique du sol**

soil pores That part of the bulk volume of soil not occupied by soil particles.

S

Also called interstices or *voids*. **pores du sol**

soil probe A soil sampler. A tool having a hollow cylinder with a cutting edge at the lower end, used for probing into the soil and withdrawing a small sample for field or laboratory observation. **sonde pédologique**

soil productivity See *productivity, soil*. **productivité des sols**

soil quality The value placed on a soil with respect to its fitness for a specific use; categorization of the fitness of a soil for a certain use based on ecological aspects, such as *soil functions*, that involve evaluating the capacity of a soil to function within specific ecosystem boundaries. See *soil health*. **qualité des sols**

soil quality attribute Properties that reflect or characterize a soil process, or processes that support a specific *soil function*. **attribut de la qualité des sols**

soil quality control The process of sustaining or improving *soil quality* by influencing or regulating soil management. **gestion de la qualité des sols**

soil quality framework An ecological approach that uses the sequence of function (i.e., what a soil does), process (i.e., those that support the function), properties or indicators (i.e., those critical to the process), and methods to evaluate *soil quality*. **cadre d'évaluation de la qualité des sols**

soil quality indicator An indirect measure of a soil quality attribute that is a related or an associated property (e.g., *pedotransfer function*, and *surrogate* and *proxy property*). Indicators should be subject to standardization (e.g., *standard methods*), specific to a quality attribute, sensitive to change, and easily measured or collected. **indicateur de la qualité des sols**

soil quality monitoring program Regular surveillance to assess the magnitude and direction of change in soil properties over time, often involving the establishment of monitoring or *benchmark sites*. The process helps identify and track *soil quality attributes*. **programme de surveillance de la qualité des sols**

soil quality susceptibility, index of A tool to identify agricultural areas at risk of declining *soil quality* because of various land use and management practices. **indice de vulnérabilité des sols**

soil quality, assessment of Characterization of *soil quality* by comparative (i.e., assessment of one system against another) or dynamic (i.e., continuously over time) means. The latter can involve the use of various approaches such as monitoring, statistics, and computer modelling. **évaluation de la qualité des sols**

soil quality, dynamic That aspect of *soil quality* relating to those soil properties subject to change over relatively short time periods, and responsive to management. **propriété dynamique du sol**

soil quality, inherent That aspect of *soil quality* relating to a soil's natural composition and properties as affected by parent material or development; mainly includes properties that show little change over time (e.g., mineralogy, particle size distribution). **qualité inhérente ou intrinsèque du sol**

soil redistribution The physical movement of soil components at the Earth's surface involving the detachment, movement (i.e., erosion), and deposition of soil. **redéposition de sol**

soil resiliency The capacity of a soil to recover its qualitative functions and dynamic properties, generally in a relatively short time frame, after some disturbance; an aspect of *soil quality*. **résilience du sol**

soil sample A representative sample taken from an area, a field, or portion of a field, from which the physical and chemical properties can be determined. **échantillon de sol**

soil separates Mineral particles, less than 2.0 mm in equivalent diameter, ranging between specified size limits. The names and size limits of separates recognized in Canada and the United States are: very coarse sand, 2.0 to 1.0 mm; coarse sand, 1.0 to 0.5 mm; medium sand, 0.5 to 0.25 mm; fine sand, 0.25 to 0.10 mm; very fine sand, 0.10 to 0.05 mm; silt, 0.05 to 0.002 mm; and clay, less than 0.002 mm. The separates recognized by the International Union of Soil Sciences are: coarse sand, 2.0 to 0.2 mm; fine sand, 0.2 to 0.02 mm; silt, 0.02 to 0.002 mm; and clay, less than 0.002 mm. **fractions du sol**

soil series See *series, soil.* **série de sols**

soil stabilization Chemical and/or mechanical treatment to increase or maintain the stability of a mass of soil or otherwise improve its physical properties. **stabilisation du sol**

soil sterilization A procedure whereby weed seeds in soil or soil organisms (total or specific organisms, or organisms at specific growth stages) are destroyed in soil by use of heat, chemicals, or irradiation. **stérilisation du sol**

soil strength A transient measure of a given soil's solid phase adhesive and cohesive status. This property is most easily affected by changes in soil water content and bulk density, although other factors including texture, mineralogy, cementation, cation composition, and organic matter content also affect it. See *cone index, penetration resistance.* **résistance du sol**

soil structural form The arrangement and size of the soil pore space. An approach used to characterize soil structure in cultivated soils, based on both *aggregates* and *pores*. See *soil structure.* **état structural du sol**

soil structural resiliency The ability of a cultivated soil to recover its *pore* space arrangement after removal of a specific stress (e.g., *compaction*). See *structure, soil.* **résilience structurale du sol**

soil structural stability The ability of a cultivated soil to retain the distribution and size of *aggregates* after exposure to various stresses (e.g., impact, shear, and *slaking* forces). See *structure, soil.* **stabilité structurale du sol**

soil structure (*pedology*) The combination or arrangement of primary soil particles into secondary particles (i.e., *aggregates*), units, or *peds*. These peds may be, but usually are not, arranged in the profile in such a manner as to give a distinctive characteristic pattern. Peds are characterized and classified on the basis of size, shape, and grade (i.e., degree of distinctness) into classes, types, and grades. See *Appendix B, Table 1* and *Figure 1*. (*agriculture*) The arrangement of soil aggregates in cultivated soils as modified by agricultural or human activity, mainly classified on the basis of size rather than shape. Generally classified by *soil structural form, soil structural stability,* and *soil structural resiliency*. See *microaggregate*, *macroaggregate*, and *organo-mineral complex.* **structure du sol**

soil structure grade A grouping or classification of soil structure on the basis of inter- and intra-aggregate adhesion, cohesion, or stability within the profile. The degree of distinctness of *aggregation*. It varies with the moisture content of the soil. Three grades of structure are designated: weak, moderate, and strong. **grade ou netteté de la structure du sol**

soil structure index Any measurement of a soil physical property, such as *aggregation*, porosity, permeability to air or water, or bulk density, that characterizes or indicates the structural condition of a soil. **mesure ou indicateur de la structure du sol**

soil structure type A classification of soil structure based on the shape of the *aggregates* or *peds* and their arrangement in the profile. See *Appendix B, Table 1* and *Figure 1.* **type de structure du sol**

soil suction See *moisture tension.* **succion du sol**

soil survey The systematic examination, description, classification, and mapping of soils in an area. Soil surveys are classified according to the kind and intensity of field examination. **prospection pédologique, levé des sols**

soil temperature The temperature obtained by a soil thermometer at any level within the soil. **température du sol**

soil test A chemical, physical, or biological procedure that estimates a property of soil pertinent to suitability of the soil to support plant growth. **analyse de sol**

soil transportation The movement of detached soil material across the land surface or through the air; may be accomplished by running water, wind, or gravity. See *erosion, detachment.* **transport de particules de sol**

soil variability Differences in soil properties due to spatial, temporal, and analytical changes; must be considered in *soil quality* evaluations. **variabilité (spatio-temporelle) des sols**

soil water The water in soil. It contains energy in different quantities and forms. The most dominant and important form of energy influencing the state and movement of water in soil is the potential energy, which is due to position or internal condition. Water in soil is subject to several force fields, which cause its potential energy to differ from that of pure, free water. These force fields originate from the attraction of the soil solid phase for water (*matric potential*), the dissolved salts (*osmotic potential*), the action of external *gas pressure potential* (i.e., pneumatic potential), and the gravitational field (*gravitational potential*). These effects may be quantitatively expressed by assigning an individual component potential to each force field. The sum of these potentials is called the *total potential* of soil water. **eau du sol**

S

soil water characteristic The relationship between the soil-water content (by mass or volume) and the soil-water matric potential. Also called soil water characteristic curve and *water retention curve.* **courbe de rétention caractéristique**

soil water diffusivity The *hydraulic conductivity* divided by the differential water capacity, or the flux of water per unit gradient of water content, in the absence of other force fields. **diffusivité de l'eau du sol**

soil water potential The potential energy of a unit quantity of water produced by the interaction of the water with forces such as capillary (i.e., *matric potential*), ion hydration (i.e., *osmotic potential*), and gravity, expressed relative to an arbitrarily selected reference potential. See *soil water.* See *figure.* **potentiel hydrique du sol**

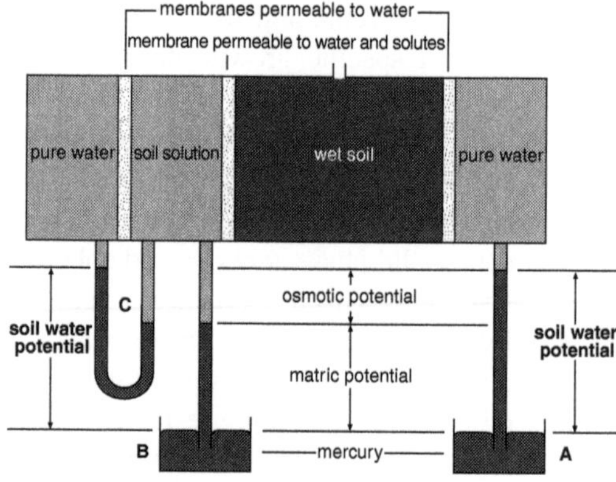

Relationship among osmotic (C), matric (B), and combined (A) **soil water potential** (adapted from Brady and Weil, 1996).

soil wetness The relative water content of a soil. It can be expressed in various ways but usually as *mass wetness* (mass of water relative to mass of dry soil particles) or *volume wetness* (volume of water relative to volume of soil). **teneur en eau du sol**

soil-conserving crops Crops that prevent or retard erosion and maintain or replenish, rather than deplete, soil organic matter. **cultures améliorantes ou de conservation**

soil-depleting crops Crops that, under the usual management, tend to deplete nutrients and organic matter in the soil and permit deterioration of soil structure. **cultures épuisantes pour le sol**

solar angle The angle made by the intersection of the sun's azimuth and a line through true north; varies with time of day, time of year, and geographic position on the Earth's surface. **angle solaire**

solar energy The conversion of direct sunlight into usable forms of energy. **énergie solaire**

solar radiation Energy from the sun; also referred to as short-wave radiation (e.g., ultraviolet radiation, visible radiation, and *infrared radiation).* **rayonnement solaire**

solclime The soil climate. The temperature and moisture conditions of the soil. **pédoclimat, climat du sol**

solid waste Waste material not discarded into surface waters via water treatment systems or directly into the atmosphere (e.g., a component of agricultural, commercial, institutional, municipal, and residential wastes). See *waste.* **déchets solides**

solid waste disposal The ultimate disposal of refuse that cannot be salvaged or recycled. **élimination des déchets solides**

solid waste management The systematic control of the generation, storage, collection, transport, separation, processing, recycling, recovery, and disposal of solid wastes. **gestion des déchets solides**

solid waste management unit The property on which hazardous waste is managed (e.g., a surface impoundment, landfill, incinerator, waste pile, or tank, as well as the adjacent land used for storage, transfer, or preliminary treatment of the waste). **unité de gestion de déchets solides**

solidification (*waste management*) The physical or chemical conversion of liquid or semi-liquid hazardous waste to a solid form before burial to reduce leaching of the waste material and possible migration of waste or its constituents. **solidification**

solifluction A type of *creep* that takes place in regions where the ground freezes to a considerable depth, and as it thaws during the warm seasons the upper thawed position creeps downhill over the frozen material. The soil moves as a viscous liquid down slopes of as little as 2 or 3 degrees, and may carry rocks of considerable size in suspension. **solifluxion**

soligenous Pertaining to *peatlands* affected by water percolating through them and/or carrying minerals into the peatland from outside sources. **soligène**

Solod A great group of soils in the *Solonetzic* order (Canadian system of soil classification) that occur most commonly in the grassland and parkland regions. The soils have a dark-colored surface (Ah) horizon, a prominent eluvial (Ahe or Ae) horizon at least 5 cm thick, a prominent transitional (AB) horizon that breaks readily into blocky aggregates, and a darkly stained *B* (Bnt) horizon over a *C horizon* that is saline and usually calcareous. **solod**

Solodized Solonetz A great group of soils in the *Solonetzic* order (Canadian system of soil classification) that occur most commonly in the grassland and parkland regions and have a variable surface (Ah, Ahe, or Ae) horizon that is underlain by a well developed Ae horizon, a compact prismatic or columnar Bnt horizon, and a *C horizon* that is saline and usually calcareous. **solonetz solodisé**

Solonetz A great group of soils in the *Solonetzic* order (Canadian system of soil classification) that occur most commonly in the grassland and parkland regions and have a variable surface (Ah, Ahe, or Ae) horizon that breaks abruptly into a hard, compact prismatic, or columnar B (Bnt, rarely a Bn) horizon underlain by one or more saline and usually calcareous (Bs, Cs, Csa, Csk, Cca) horizons. They lack a continuous Ae horizon 2.5 cm or more thick. **Solonetz**

Solonetzic An order of soils in the Canadian system of soil classification developed mainly under grass or grass-forest vegetative cover in semiarid to subhumid climates. The soils have a stained brownish solonetzic B (Bnt or Bn) horizon and a saline C horizon. The surface may be one or more of Ap, Ah, or Ae horizons. The order includes the *Solonetz, Solodized Solonetz, Solod,* and *Vertic* great groups. **solonetzique**

solonetzic B A *diagnostic* B horizon (Canadian system of soil classification) characterized by prismatic or columnar primary structure that breaks to blocky secondary structure, by hard to extremely hard consistence when dry, and by ratio of exchangeable Ca to Na of 10 or less. **B solonetzique**

solubility The amount of a solute that dissolves in a given volume of solvent at a given temperature. **solubilité**

solubility product *(Ks)* The product of the molar concentrations of the ions in solution that result from a solid chemical compound partly dissolving in water. Each compound has its own equilibrium dissolved concentration, and the solubility product is a constant for that compound. For example, K_s for an equilibrium between solid ferric hydroxide (FeOH) and dissolved Fe^{-3} and OH ions is 1×10^{-38}, which means that the product of the molar concentrations of Fe^{-3} and OH^- is 1×10^{-38}. Usually expressed as pK_s, which is the negative logarithm of *Ks* The pK_s for ferric hydroxide is 38. **produit de solubilité**

soluble-sodium percentage (*irrigation*) The proportion of sodium ions in solution (e.g., irrigation waters and soil extracts) in relation to the total cation concentration (i.e., [soluble sodium]/[total cation] × 100). **pourcentage de sodium soluble**

solum (plural sola) The upper horizons of a soil in which the parent material has been modified and in which most plant roots are contained. It usually consists of A and B horizons. See *figure.* **solum**

solute A substance dissolved in a solvent to form a solution. **soluté**

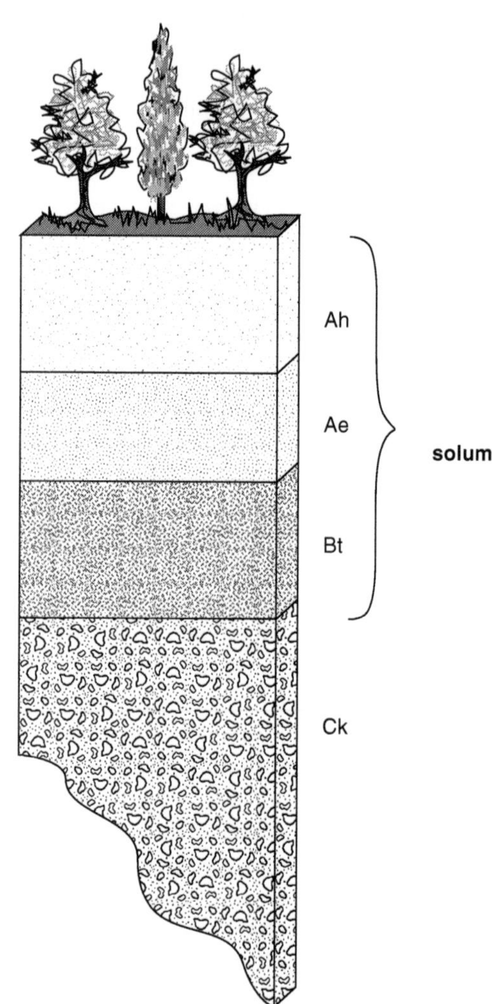

Solum (A.J. VandenBygaart, 2001).

solution A homogeneous mixture of a solute in a solvent. **solution**

solution fertilizer See *fertilizer, solution.* **engrais liquide**

S

solution mining The removal of a mineral deposit by solution in water. Water is injected into the geological strata containing the mineral, and the dissolved material is recovered by wells. **extraction par solution**

solution, soil The aqueous liquid phase of the soil and its solutes, consisting of ions dissociated from the surfaces of the soil particles, and other soluble materials. **solution du sol**

solvent The liquid that dissolves another substance or substances to form a solution. Water is frequently referred to as the universal solvent. The term also applies to *organic solvents* (e.g., benzene, acetone, or gasoline) used to clean (dissolve) oils or grease from machinery, fabrics, or other surfaces, or to extract hydrocarbons from some source material. Many organic solvents are flammable and/or toxic. **solvant**

Sombric Brunisol A great group of soils in the *Brunisolic* order (Canadian system of soil classification). The soils have *moder* Ah horizons more than 5 cm thick and Bm horizons in which the base saturation (NaCl) is usually 65 to 100% and the pH ($CaCl_2$) is usually above or greater than 5.5. **brunisol sombrique**

sombric horizon A subsurface horizon in mineral soils that have formed under free drainage, containing illuvial humus that is neither associated with aluminum, as in a *spodic horizon*, nor dispersed by sodium, as in a *natric horizon*. A subsurface *diagnostic horizon* in the U.S. system of soil taxonomy. **horizon sombrique**

sorption The removal of an ion or molecule from solution by *adsorption* and *absorption*. **sorption**

sorptivity The initial stage of water infiltration. Sorptivity $s = I\,t^{1/2}$ for horizontal infiltration of water, where I is cumulative infiltration and t is time. Sorptivity is dependent on initial and boundary conditions of soil water content, among other factors. **sorptivité**

sorting (*geology*) The separation and segregation of rock fragments according to size and particles, specific gravity, and different shapes by natural processes, mainly the action of running water or wind. **granuloclassement**

sorting coefficient (*geology*) A measure of the degree of *sorting* of sediment particles based on the spread of the distribution of particle sizes. It is defined statistically as the standard deviation of grain size spread. Higher values indicate less homogeneous, poorly sorted mixtures, whereas lower values indicate more homogeneous, better sorted mixtures. **coefficient de classement**

spalling To break up, splinter, or chip; rock *exfoliation*. **délitage**

spatial model A set of rules and procedures for conducting spatial analysis to derive new information that can be analyzed to aid in problem solving and planning. **modèle d'analyse spatiale**

spatial statistics Statistical methodology and theory that accounts for spatial aspects of a spatially distributed data set. Conventional population estimation does not normally account for spatial attributes, except perhaps for spatial identity of subpopulations. **statistiques spatiales, géostatistiques**

spatial variability, soil The variation in soil properties laterally across the landscape, at a given depth, or with a given horizon, or vertically downward through the soil. **variabilité spatiale**

Spearman rank correlation See *correlation, statistical*. **coefficient de corrélation de rang de Spearman**

specialization Restriction of an organism's or a population's activities to a portion of the environment; a trait that enables an organism (or one of its organs) to modify (or differentiate) in order to adapt to a particular function or environment. **spécialisation**

species (*ecology*) A group of organisms that may interbreed and produce fertile offspring having similar structure, habits, and functions. Species ranks next below genus as a fundamental unit in the hierarchy of classification. The name of the species becomes the second word of binomial nomenclature systems. Abbreviated as *sp.* (singular) and *spp.* (plural). (*geology*) A mineral distinguished from others by its unique chemical and physical properties. **espèce**

species composition (*ecology*) The different types, and abundance of each type, of organisms inhabiting a specific area. (*forestry*) The percentage of each recognized tree species comprising forest type, based upon the gross volume (the relative number of stems per hectare or basal area). **composition taxonomique**

species conversion A change from one tree species to another. **conversion d'espèces**

species density The total number of individuals of a species found in a specific area at a certain time, expressed as a specific area. $D = (n/a)/t$, where D is the density, n is the number of individuals, a is the area studied, and t is the time period during which the study was conducted. **densité spécifique**

species diversity The number of species in a community, and their relative abundances. See *Shannon-Weaver index.* **diversité spécifique**

species frequency The proportion of sampling areas in which a specific species is represented. The frequency is independent of the number of individuals of each species located in each sampling area. **fréquence d'espèces**

species richness The number of species found in a sample. **richesse spécifique, diversité des espèces**

species richness index A mathematical expression indicating the number of species in a community relative to the total number of individuals in that community. The number of organisms in each species is not considered in the index. **indice de richesse spécifique**

species, introduced A non-native species; an organism intentionally or accidentally introduced by humans. See *species, native* and *species, naturalized.* **espèce introduite**

species, native A species that occurs naturally within a geographical area. See *species, introduced* and *species, naturalized.* **espèce indigène**

species, naturalized A non-native species that has established as if it were a native species. **espèce naturalisée**

specific activity (*biochemistry*) Number of enzyme activity units per mass of protein; expressed as micromoles of product formed per unit time per milligram of protein. (*radiochemistry*) The radioactivity per mass of material (radioactive + nonradioactive). **activité spécifique**

specific gravity The ratio of the density of a material at a given temperature and pressure to the density of some standard material. The standard for solids and liquids is usually pure water at a temperature of 3.98°C and standard atmospheric pressure; for gases, the standard is usually hydrogen, oxygen, or air at a specified temperature and pressure. **densité, densité relative**

specific heat capacity The amount of heat required to raise a unit mass of a substance by one degree Celsius. Every substance has a characteristic specific heat capacity. **chaleur massique**

specific surface The solid-particle surface area (of a soil or porous medium) divided by the solid-particle mass, expressed in $m^2\ g^{-1}$, or by the solid-particle volume. **surface spécifique**

specific water capacity The change of soil-water mass content with a change in soil-water *matric potential.* **capacité spécifique de rétention en eau**

specific yield (*hydrology*) The volume of water available per unit volume of aquifer, if drawn by gravity. **débit spécifique**

spectral band An interval in the electromagnetic spectrum defined by two wavelengths, frequencies, or wave numbers. **bande spectrale**

spectral colors (*remote sensing*) The continuous band of pure colors in the visible spectrum; divided into seven basic spectral colors (violet, indigo, blue, green, yellow, orange, and red). **couleurs spectrales**

spectral map (*remote sensing*) A classification map in which the classes are based on relative spectra properties rather than the ground cover type represented. **carte spectraloïde**

spectral radiance (*remote sensing*) The radiance of an object or surface described with respect to the distribution of the power across the spectrum. **radiance spectrale**

spectral region (*remote sensing*) A conveniently designated range of wavelengths subdividing the electromagnetic spectrum (e.g., visible region, x-ray region, infrared region, and middle-infrared region). **domaine spectral**

spectral response (*remote sensing*) The response of a material as a function of wavelength to incident electromagnetic energy, particularly in terms of the measurable energy reflected from and emitted by the material. **réponse spectrale**

spectral signature (*remote sensing*) The spectral characterization of an object or class of objects on the Earth's surface; often used in a way that naively oversimplifies the complexity of the spectral representation problem in a natural scene. **signature spectrale**

spectrometer An optical instrument used to measure the apparent electromagnetic radiation emanating from a target in one or more fixed wavelength bands or sequentially through a range of wavelengths. **spectromètre**

spectrophotometer A device used to measure spectral transmittance, spectral reflectance, or relative spectral emittance. **spectrophotomètre**

spectroscopy The use of the absorption, emission, or scattering of *electromagnetic radiation* by atoms or molecules (or atomic or molecular ions) to qualitatively or quantitatively study the atoms or molecules, or to study physical processes. See *atomic absorption spectroscopy, atomic emission spectroscopy, infrared absorption spectroscopy, nuclear magnetic resonance spectroscopy.* **spectroscopie**

spectrum (plural spectra) (1) The entire range of wavelengths of *electromagnetic radiation.* (2) An image or distribution of energy or mass arranged according to frequency or charge. **spectre**

specular reflection The reflectance of electromagnetic energy without scattering or diffusion, as from a surface that is smooth in relation to the wavelengths of incident energy; also called mirror reflection. **réflexion spéculaire**

sphagnic Pertaining to *Organic* soils developed in peat derived mainly from *Sphagnum* spp.; used in describing organic soil families in the Canadian system of soil classification. **sphagnique**

Sphagnum A genus of mosses growing in wet places, some species of which form spongy cushions and which contribute to *bog* formation. ***Sphagnum***

sphagnum peat See *bog peat.* **tourbe de sphaigne**

sphericity The amount by which the shape of a fragment approaches the form of a sphere. **sphéricité**

spheroidal weathering (*geology*) A type of subterranean *chemical weathering* in which jointed rock masses are slowly rounded by the gradual removal of their concentric outer shells, by hydrolysis, to leave an internal spherical boulder. See *exfoliation*. **désagrégation en boules ou sphéroïdale**

spillway A structure over or through which excess or flood flows are discharged. In a controlled spillway, gates control the flow. In an uncontrolled spillway, the elevation of the spillway crest is the only control. **déversoir**

splash erosion A type of *erosion* created by the impact of large raindrops falling at high velocities on the soil surface, causing redistribution of soil particles. It is most intensive under tropical rainstorms in areas that have been recently cleared of vegetation. Soil particles are mechanically detached and transported by surface *runoff*, especially after the surface soil structure has been degraded by fine material filling the pore spaces (i.e., *voids*) in the soil. See *rain splash*. **érosion par éclaboussement**

spodic horizon A mineral soil horizon in the U.S. soil taxonomy that is characterized by the *illuvial* accumulation of amorphous materials composed of aluminum and organic carbon with or without iron. The spodic horizon has a minimum thickness, and a minimum quantity of extractable carbon plus iron plus aluminum in relation to its content of clay. See *Spodosols*. **horizon spodique**

S

Spodosols An order in the U.S. system of soil taxonomy. Mineral soils that have a *spodic horizon* or a *placic* horizon that overlies a *fragipan*. **Spodosols**

spoil (*land reclamation*) (1) The overburden or non-ore material removed while gaining access to the ore or mineral material in surface mining. (2) Debris or waste material from a mine. Mine spoil is often deposited in steep piles and in inverted sequence with that from the lowest layers being placed on top because it was dug out last. Such material may contain toxic heavy metals or sulfides that oxidize and produce extreme acidity in the spoil and in nearby bodies of water. **déblais**

spoil bank (*land reclamation*) Rock waste, banks, and dumps, from the excavation of ditches and strip mines. **remblai détritique, terril**

spore A specialized reproductive cell. Asexual spores germinate without uniting with other cells, whereas sexual spores of opposite mating types unite to form a zygote before germination occurs. **spore**

sporocarp A multi-celled body capable of producing spores. **sporocarpe**

spreader strip A relatively permanent contour strip of variable width planted to a permanent grass or erosion-resistant crop, used to slow down and fan out the runoff from land above the strip. **bande tampon**

spring (1) The season that occurs between winter and summer, usually regarded as the three months of March, April, and May in the northern hemisphere. In astronomical terms, it spans the period from the vernal (spring) equinox to the summer solstice (i.e., 21 March to 21 June). (2) A natural flow of water from the ground at the point where the *water table* intersects the surface. **(1) printemps (2) source**

spring overturn (*limnology*) A physical phenomenon that may take place in a body of water during the early spring. The *overturn* results in a uniformity of the physical and chemical properties of the entire water mass. See *fall overturn*. **renversement printannier**

springtails (*zoology*) Members of the order Collembola, composed of very small insects that live in soil pores, around roots, or on the surface soil. **collemboles**

sprinkler irrigation A method of irrigation in which water is sprayed over the soil surface through nozzles from a pressure system. **irrigation par aspersion**

stability (*ecology*) The ability of the ecosystem to reach or maintain a state of equilibrium with respect to diversity and processes. **stabilité**

stabilization, soil Chemical or mechanical treatment designed to increase or maintain the stability of a mass of soil or otherwise to improve its engineering properties. **stabilisation du sol**

stable age distribution The proportion of individuals in various age classes in a population that has been growing at a constant rate. **distribution par âge stable**

stable equilibrium A condition when a *system* has a tendency to recover its original state after being disturbed by external forces. **équilibre stable**

stable isotope An *isotope* that does not undergo *radioactive decay*. **isotope stable**

stade (*geology*) A recessional phase or stillstand in the overall decline of an icesheet, likely arising from a change in climate. It is marked by various glacial deposits and often by a *recessional moraine*. **stadiaire, stade glaciaire**

stadial moraine See *recessional moraine*. **moraine stadiaire**

stage (1) In *chronostratigraphy*, a time-stratigraphic division of a series that represents the rocks formed during a geologic age. See *biostratigraphy, lithostratigraphy*. (2) A primary subdivision of a glacial epoch in the Pleistocene; the time equivalent of a stratigraphic unit, such as a *formation*, or a grouping of units that may be of glacial or interglacial age. (3) The point to which a landform has evolved during a cycle of erosion (i.e., youth, maturity, senility), marked by distinct developments of landscape features. (4) The level of water in a river channel above a given datum. **(1, 2) étage (3) stade (4) niveau**

stalactite A tapering pendant of concretionary material created by the re-precipitation of carbonate in *calcite* form from percolating groundwater, descending from a cave ceiling, in a karst environment. See *stalagmite*. **stalactite**

stalagmite A columnar concretion ascending from the floor of a cave, formed from the re-precipitation of carbonate in *calcite* form beneath a constant source of groundwater that drips off the lower tip of a *stalactite* or percolates through the roof of a cave in a *karst* environment. It may eventually combine with a stalactite to form a pillar. **stalagmite**

stamen The male reproductive structure of a flower producing the *pollen*; usually composed of anther and filament. **étamine**

staminate A flower that produces *stamens* and, in some cases, a flower that produces stamens but not pistils. **staminé**

stand An area occupied by a collection of plants or trees that are structurally and floristically homogeneous. It is usually applied to forests, where stands of similar trees facilitate logging operations. **peuplement**

stand composition (*forestry*) The proportion of each tree species in a *stand* expressed as a percentage of either the total number, basal area, or volume of all tree species in the stand. **composition du peuplement**

stand conversion (*forestry*) Changing the species composition of a *stand* to more desirable tree species that are less susceptible to damage or mortality from certain insects or diseases. **conversion de peuplement**

stand density (*forestry*) A relative measure of the amount of stocking on a forest area. Often described in terms of stems per hectare. See *stand*. **densité de peuplement**

S

stand development *(forestry)* The part of *stand dynamics* concerned with changes in stand structure over time. **développement du peuplement**

stand dynamics *(forestry)* The study of changes in forest *stand structure* over time, including stand behavior during and after disturbances. **dynamique du peuplement**

stand structure *(forestry)* The distribution of trees in a *stand*, which can be described by species, vertical or horizontal spatial patterns, size of trees or tree parts, age, or a combination of these. **structure de peuplement**

stand tending *(forestry)* A variety of forest management treatments, including spacing, fertilization, pruning, and commercial thinning, carried out at different stages during a stand's development. See *stand*. **éducation du peuplement**

standard A solution of known concentration used to calibrate, or to prove that an analytical procedure is accurate and valid. **étalon**

standard air density The density of dry air at the chosen standard conditions (e.g., at 0°C and 0.101 MPa of pressure, the density of dry air is 1.293 Kg m^{-3}). **masse volumique standard de l'air**

standard cone The cone-shaped tip used at the insertion end of soil *cone penetrometer* probes. For example, the ASAE standard cone is a 30° stainless steel cone having a basal diameter of either 20.27 or 12.83 mm. **cône étalon**

standard curve A curve that graphs concentrations of known *analyte* standard versus the instrument response to the analyte. **courbe d'étalonnage**

standard deviation *(statistics)* The square root of the variance of a data set, which is an indication of the spread of the data set around its mean. In normally distributed sets of moderate size, the interval of the *mean*, plus or minus the standard deviation, includes about two-thirds of the items. **écart type, déviation standard**

standard error *(statistics)* (1) The standard error of the mean is a measure that estimates the average dispersion of the means of statistical samples of a given size about the *mean* of the population. (2) The standard error of estimates is a measure of the average departure of the observed values of the dependent variable from the calculated values. **erreur type**

standard free energy change The change in free energy that will occur for one unit of reaction if the reactants in their standard states are converted to products in their standard states. **changement d'énergie libre standard**

standard free energy of formation The change in free energy that accompanies the formation of one mole of a substance from its constituent elements with all reactants and products in their standard states. **énergie libre standard de formation**

standard method Analytical methods that are recommended by consensus, for specific analyses based on reliability and accuracy. Usually based on interlaboratory comparisons and set by a "standards" body. **méthode standardisée**

standard soil handling procedure (*land reclamation*) A process in the reclamation of pipeline trenches, whereby topsoil is selectively removed in one lift and spoil material is removed in a second lift. Following pipe installation, the topsoil and subsoil materials are replaced in their pre-construction order and depth. Also called two-lift. **méthode standardisée de manipulation du sol**

standard solution A solution whose concentration is accurately known. See *standard*. **solution étalon**

standing crop The quantity of plant biomass in a given area, usually expressed as density (dry mass per unit area) or

S

energy content per unit area. **rendement en biomasse**

starch A complex insoluble carbohydrate; the chief food storage substance of plants; composed of several hundred glucose units ($(C_6H_{10})_5$) and readily broken down enzymatically into these components. **amidon**

starter fertilizer See *fertilizer, starter.* **engrais de démarrage**

state variables The components that make up a system in mathematical models used to describe the environment; they have certain states, or conditions, at a given time (e.g., soil temperature or water content, the distribution of which in a soil profile determine the conduction of heat and water, respectively). **variables d'état**

states of matter The three different forms in which matter can exist (i.e., solid, liquid, or gas). **états de la matière**

Static Cryosol A great group of soils in the Cryosolic order (Canadian system of soil classification). They have developed mainly in coarse textured mineral parent materials, or in recently deposited sediments. They have permafrost within 1 m of the surface but show little or no evidence of cryoturbation. **cryosol statique**

static equilibrium A state achieved by a body or *system* when the balance of forces acting upon it requires no movement in order to maintain or restore the balance; thus, the body or system remains stationary with respect to its surroundings. See *dynamic equilibrium state, equilibrium.* **équilibre statique**

static penetrometer A *penetrometer* that is pushed into the soil at a constant and slow rate. **pénétromètre statique**

stationary growth phase (*microbiology*) The phase in the growth of bacteria in which there is no increase in the number of cells over time. Any cell division that is balanced by cell death. This phase of growth follows the *log growth phase.* See *growth phase.* **phase de croissance stationnaire**

statistical model A model involving variables, constants, parameters, and one or more random components to represent unpredictable fluctuations in the experimental data due to measurement error, equation error, or inherent variability of the objects measured. Also called a random model. **modèle statistique**

statistical quality control A method, based on the quality control of manufactured products, to help control and maintain *soil quality* processes within the range of natural variation. It involves experimental design, process monitoring and control, and ongoing adjustments of the system to maintain a process in an "in-control" state. **contrôle statistique de la qualité**

statistics A scientific approach to collecting, analyzing, and interpreting numerical data related to an aggregate of individuals. Descriptive statistics includes data collection and table and graph construction, together with frequency distributions. Analytical statistics includes correlation, regression, and tests of significance. **statistiques**

steady state See *dynamic equilibrium state, dynamic steady state, steady-state equilibrium.* **régime permanent, équilibre, état stationnaire**

steady state equilibrium The condition of an *open system* in which properties are invariant when considered with reference to a given time scale, but wherein its instantaneous condition may oscillate because of the presence of interacting variables. **état d'équilibre permanent**

steep A type of surface expression of mineral landforms consisting of erosional slopes greater than 70%, occurring on both consolidated and unconsolidated materials. **abrupt, raide**

stem The portion of vascular plants that commonly bears leaves and buds; usually aerial, upright, and elongated, but may be highly modified in structure. Underground stems include *rhizomes* and *corms*. **tige**

stemflow The quantity of precipitation or irrigation water that is intercepted by vegetation and then flows down the stem or trunks to the ground. **ruissellement sur les troncs**

step function A rapid change that occurs between one variable and another, with one of these variables normally being time; uniform conditions exist before and after the rapid change. Step functions occur when a *system* passes across a threshold, thereby initiating a positive feedback mechanism and leading to a state of disequilibrium. At the end of the step, a new equilibrium state will occur only with the establishment of a stabilizing negative-feedback mechanism. **fonction à paliers**

stereo base (*remote sensing*) A line representing the distance and direction between complementary image points on a stereo pair of photos corrected, oriented, and adjusted for comfortable stereoscopic vision under a given stereoscope, or with the unaided eyes. **base stéréoscopique**

stereo pair (*remote sensing*) A pair of photos that overlap in area and are suitable for stereoscopic examination. **couple stéréoscopique**

stereogram (*remote sensing*) A stereo pair of photos or drawings correctly oriented and permanently mounted for stereoscopic examination. **stéréogramme**

stereometer (*remote sensing*) A device for measuring *parallax* difference. **stéréomètre**

stereoscope (*remote sensing*) A binocular optical instrument for viewing two properly oriented photographs, constituting a stereo pair, to obtain a mental impression of a three-dimensional effect. **stéréoscope**

sterile (1) A condition in which a quantity of water, soil, or other substance does not contain viable organisms such as viruses and bacteria. The term, however, is not synonymous with clean. (2) Animals or humans not capable of reproduction because of the absence of gametes. **stérile**

sterilization The process of making *sterile*; the killing of all forms of life. **stérilisation**

steroid One of a class of lipids with a characteristic fused carbon-ring structure that includes cholesterol, hormones, and bile acids. **stéroïde**

Stevenson screen A specially designed housing for meteorological instruments. **abri météorologique, abri de Stevenson**

stewardship Caring for land and associated resources, and passing healthy ecosystems to future generations. **intendance pour les générations futures**

sticky limit The lowest water content at which a soil will stick to a metal blade drawn across the surface of the soil mass. **limite d'adhésivité**

sticky point (1) A condition of consistence at which the soil barely fails to stick to a foreign object. (2) Specifically, the mass water percentage of a well-mixed, kneaded soil that barely fails to adhere to a polished nickel or stainless steel surface when the shearing speed is 5 cm s^{-1}. **point d'adhésivité**

stigma In plants, the female reproductive organs on which *pollen* grains germinate. **stigmate**

stochastic A method of obtaining a solution to a problem in which a random variable is present and because of its presence, various degrees of probability need to be stated. **stochastique**

stochastic process A process in which the dependent variable is random (so that prediction of its value depends on a set of underlying probabilities) and the outcome at any instant is not

S

known with certainty. See *deterministic process.* **processus stochastique**

stochastic-process model See *model.* **modèle de processus stochastique**

stocking rate (*range management*) The number of a specified kind and class of animals grazing a unit of land for a period of time expressed as animals per land area; or the reciprocal, area of land per animal. **taux de chargement**

stoichiometric quantities Quantities of reactants mixed in exactly the correct amounts so that all are used up at the same time. **quantités stoechiométriques**

Stokes' law An equation relating the terminal settling velocity of a smooth, rigid sphere in a viscous fluid of known density and viscosity to the diameter of the sphere when subjected to a known force field. It is used in the *particle-size analysis* of soils by the pipette, hydrometer, or centrifuge methods. The equation is $v = (2\ g\ r^2)(d_1 - d_2)/9\ \eta$, where v, velocity of fall (m s^{-1}); g, acceleration of gravity (m s^{-2}); r, equivalent radius of particles (m); d_1, density of particles (kg m^{-3}); d_2, density of medium (kg m^{-3}); and η, dynamic viscosity of medium (N s m^{-2}). **loi de Stokes**

stolon A horizontal stem that grows along the surface of the soil and produces roots and buds at the nodes (e.g., the runners of the strawberry plant). **stolon**

stoloniferous (*botany*) Having *stolons* or "runners." **stolonifère**

stoma (plural stomata) A small opening bordered by *guard cells* in the epidermis of leaves and stems through which gases and water pass; also used to refer to the entire stomatal apparatus (the guard cells and the pore they form). **stomate**

stone (1) Rock fragment greater than 25 cm in diameter if rounded, and greater than 38 cm along the greater axis if flat. (2) Rock fragments included with boulders, which are considered to be greater than 20 cm in diameter. **pierre**

stoniness The relative proportion of *stones* in or on the soil. *See coarse fragments.* **pierrosité**

stony Containing sufficient stones to interfere with or prevent *tillage.* To be classified as stony, more than 0.1% of the surface of the soil must be covered with stones. The term is used to modify the *soil class* (e.g., stony clay loam or clay loam, stony phase). See *coarse fragments.* **pierreux**

stony land Areas containing sufficient stones to make the use of machinery impractical; usually 15 to 90% of the surface soil is covered with stones. See *stoniness.* **terrain pierreux**

storage pit (wastes) A pit in which solid waste is held prior to processing. **fosse d'entreposage**

stormwater runoff Rain water that runs off land surfaces directly into rivers, lakes, streams, and other waters. **eau d'averse**

stover The dried, cured stems and leaves of tall, coarse grain crops, such as maize and sorghum, after the grain has been removed. **chaume**

straight fertilizer See *fertilizer, straight.* **engrais simple**

straight-chain hydrocarbons Compounds of carbon and hydrogen in which multiple carbon atoms are bonded to each other in a straight line; *aliphatic* compounds. See *aromatic compound.* **hydrocarbures aliphatiques**

strain (*mechanics*) (1) A relative change in the shape and/or volume of a body in response to an applied force, expressed as the ratio of the distortion of a body dimension to some undistorted dimension (not necessarily the same one) (e.g., a change in length in a given direction, per unit original length of the same or some other direction). A deformation that is equal in all directions is called homogeneous strain, whereas that which

S

does not deform equally in all directions, thereby causing linear structures to become curved, is called heterogeneous strain. (*genetics*) (2) A group of organisms of the same species possessing distinctive hereditary characters that distinguish them from other such groups. Strains are maintained through inbreeding or genetic controls to retain or emphasize specific characteristics, as in a domestic animal or agricultural plant. (*microbiology*) (3) A population of microorganisms distinguished biochemically, genetically, or some other way from other organisms of the same taxonomical classification. (*virology*) (4) A virus that has major properties in common with other viruses within its category or type, but which differs in minor characteristics such as vector specificity, or serological or genetic properties. **(1) déformation (2, 3, 4) souche**

stratification (*pedology*) (1) The development of layers within a soil (if soil horizons are being described, the term horizonation is preferred). (*geology*) (2) The arrangement of sediments in layers or *strata* (*stratum*) marked by change in color, texture, dimension of particles, and composition. Stratification usually means layers of sediments that separate readily along bedding planes because of different sizes and kinds of material, or some interruption in deposition that permitted changes to take place before more material was deposited. (*biogeography*) (3) The arrangement of vegetation in layers. (*hydrology*) (4) The formation of distinct temperature layers in a body of water. **(1) horizonation (2, 3, 4) stratification**

stratified Arranged in or composed of *strata* or layers. **stratifié**

stratified drift (or sorted drift) Materials that are distinctly sorted according to size and weight of their component fragments, indicating a medium of transport (water or wind) more fluid than glacier ice; generally refers to *glaciofluvial*, *glaciolacustrine*, or glaciomarine drift. **sédiment stratifié ou trié**

stratified random sample A randomized sample composed of two or more sets of random samples, each drawn from a single homogeneous unit (i.e., stratum) of a heterogeneous population. Stratification is the subdivision of a population into groups or strata, each of which is more homogeneous in respect to the variable measured than the population as a whole. See *random sample*. **échantillon aléatoire stratifié**

stratigraphical correlation The process by which two stratigraphic units or formations may be equated, usually in a time relationship, even though they are spatially separated, through evidence from *lithology, biostratigraphy*, and *paleomagnetism*. **corrélation stratigraphique**

stratigraphy A branch of *geology* that consists of the systemized study, description, and classification of stratified rocks, including their formation, composition, characteristics, arrangement, sequence, age, distribution, and correlation with one another. **stratigraphie**

stratosphere See *atmosphere*. **stratosphère**

stratum (plural strata) (*geology*) A layer or set of successive layers of any deposited surface characterized by certain unifying characteristics, properties, or attributes (e.g., texture, grain size, chemical composition) that distinguish it from adjacent layers. (*meteorology*) An atmospheric layer. (*biology*) A layer of tissue. **strate**

stream A watercourse, with an alluvial sediment bed, formed when water flows on a perennial or intermittent basis between continuous definable banks. **cours d'eau**

stream bank The rising ground bordering a stream channel. **berge**

stream bank erosion control Vegetative or mechanical control of erodible stream banks, including measures to prevent stream banks from caving or sloughing (e.g., jetties, revetments, riprap, and plantings necessary for permanent protection). **lutte contre l'érosion des berges**

stream bank stabilization (*reclamation*) The lining of stream banks with riprap, matting, etc., to control erosion. (*geology*) Natural geological tendency for a stream to mold its banks to conform with the channel of least resistance to flow. **stabilisation des berges**

stream channel The stream bed and banks formed by fluvial processes, including deposited organic debris. **chenal d'un cours d'eau**

stream gaging The quantitative determination of stream flow using measuring instruments at selected locations. See *gaging station*. **jaugeage d'un cours d'eau**

stream gradient The general slope, or rate of vertical drop per unit of length of a flowing stream. **déclivité d'un cours d'eau**

stream load The quantity of solid and dissolved material carried by a stream. **charge totale d'un cours d'eau**

stream recession length The time required for a stream which has experienced a rise to fall to its normal, or *base flow*, stage. **temps de décrue**

streambanks The usual boundaries of a stream channel. **berges, rives**

streambed The bottom of the stream below the usual water surface. **lit du cours d'eau**

streambed erosion The movement of material, causing a lowering or widening of a stream. **érosion du lit**

strength The maximum stress that a material can resist without failing for any given type of loading. **résistance**

stress The directional forces that act on a material and tend to change its dimensions. If two forces are applied towards each other, compressive stress will result; if two forces act away from each other, they will create tensional stress; if the forces act tangentially to each other, tangential stress, or shearing, will occur. When materials are subjected to stress, they will suffer *strain*. **contrainte**

stressed water (*limnology*) A portion of an aquatic environment with poor species diversity due to human actions. **eau polluée**

strike The direction of a horizontal line on the plane of an inclined rock stratum, fault, or cleavage. It is at right angles to the true *dip* of the rock structure. **direction**

strip cropping An erosion control method; the practice of growing crops that require different types of tillage, such as row and permanent grass, in alternate strips along contours or across the slope or the prevailing direction of wind. See *tillage, strip cropping*. **culture en bandes, culture en bandes selon les courbes de niveau**

strip grazing (*range management*) A system of grazing whereby animals are confined to a small area of pasture for a short period of time, usually 1 to 2 days. **pâturage rationné**

strip mining See *surface mining*. **exploitation à ciel ouvert**

strip planting (strip till planting) See *tillage, strip planting (strip till planting)*. **semis en bandes**

strip tillage (partial-width tillage) See *tillage, strip tillage (partial-width tillage)*. **travail du sol en bandes**

stripping (*chemistry*) A method for removing unwanted dissolved gases from water. Stripping techniques involve increasing the surface area of the water to be stripped and maintaining the atmospheric *partial pressure* of the gas(es) to be removed at a low level relative to the partial pressure of the gas dissolved in the water. Oxygen, ammonia, hydrogen sulfide, volatile organic compounds, and carbon dioxide are commonly stripped from water. Also called air stripping. **strippage**

strip-till planting See *tillage, strip-till planting*. **semis sur bandes travaillées**

strobilus A reproductive structure consisting of a number of modified leaves or ovule-bearing scales grouped terminally on a stem; a *cone*. **cône**

stromatolite Laminated carbonate or silicate rocks; organo-sedimentary structures produced by growth, metabolism, trapping, binding, and/or precipitating of sediment by communities of microorganisms, principally *cyanobacteria*. A lithified or fossil form of *microbial mat* communities. **stromatolithe**

strong acid An acid that completely dissociates in an aqueous medium to produce an H^+ ion and the *conjugate base*. **acide fort**

strong base A metal hydroxide salt that completely dissociates into its ions in water. **base forte**

strong electrolyte A material which, when dissolved in water, gives a solution that conducts an electric current very efficiently. **électrolyte fort**

structural charge The charge (usually negative) on a mineral caused by isomorphous substitution within the mineral layer, expressed as moles of charge per kilogram of clay. **charge attribuable aux substitutions**

S

structural geology The scientific study of rock structures, including their form, genesis, and spatial distribution, but not their composition. **géologie structurale**

structure (*general*) An organized body or combination of connected elements. (*pedology*) See *soil structure*. (*geology*) The overall relationships between rock and till masses, together with their large-scale arrangements and dispositions. See *dip, fault, fold, strike, unconformity*. (*petrology*) The more detailed internal relationships between the different parts of rocks or tills. See *bed, cleavage, joint, lamina*. (*biogeography*) The spatial distribution of the vegetation in an association. **structure**

stubble The bottom portion of plants remaining after the top portion has been harvested; also, the portion of the plants, principally grasses, remaining after grazing is completed. **chaume**

stubble crops (1) Crops that develop from the stubble of the previous season. (2) Crops sowed on grain stubble after harvest for turning under the following spring. **(1) regain (2) culture de couverture**

stubble mulch See *tillage, stubble mulch*. **paillis de résidus de culture**

stubble mulch tillage See *tillage, stubble mulch tillage*. **travail du sol préservant les résidus de culture**

study area The total geographical area of land that is studied to obtain information. **aire d'étude**

subaerial Any feature, substance, or process that occurs or operates on the Earth's surface, as distinct from subterranean or submarine phenomena. **subaérien**

subarctic Pertaining to the regions directly adjacent to the Arctic Circle, or to areas with climate, vegetation, and animals similar to those of arctic regions. **subarctique**

subbituminous A coal similar to *bituminous*, but softer and high in moisture. **subbitumineux**

subgrade The soil prepared and compacted to support a structure or a pavement system. **forme, sol de formation**

subgrade modulus (*engineering*) The resistance of soil material to unit area displacement under load. **module**

subgroup, soil A category in the Canadian system of soil classification. These soils are subdivisions of the *great groups,* and therefore each soil is defined specifically. **sous-groupe de sols**

subhedral Minerals with partly developed crystallographic form. **subédrique**

sub-humid Pertaining to a climate in which there is slightly less precipitation than in a humid area, or to a region where this climate prevails and natural vegetation is mostly tall grasses. Annual rainfall varies from 50 cm in cool regions to 150 cm in hot areas. Precipitation is sufficient for the production of many agricultural crops without irrigation or dryland farming. See *arid, semi-arid, humid.* **sub-humide**

sublittoral zone The part of the shore from the lowest water level to the lower boundary of plant growth; transition zone from the littoral to profundal bottom. See *littoral zone, profundal zone.* **région sublittorale**

suborder, soil See *order, soil.* **sous-ordre de sols**

subpopulation Any subset of a population, usually having a specific attribute that distinguishes its members from the rest of the population. **sous-population**

subsere, secondary sere Any point in the seral development of a vegetation climax in which the *succession* toward a *sere* has been halted temporarily by non-climatic factors, both biotic and abiotic (e.g., soil, topography); once these factors are removed, the normal pattern of seral development proceeds. **subsère**

subsidence The gradual settling of the ground resulting from natural processes, as in the sinking of the land relative to sea level in coastal areas, or in the collapse of the ground because of removal of water, coal, or mineral deposits from the underlying strata. Also refers to lowering of the surface in *Organic* soils upon drainage as a consequence of improved aeration, and the loss of organic matter through decomposition. **subsidence, affaissement**

subsoil (1) The soil material found beneath the topsoil (or below plow depth) but above the bedrock. (2) Sometimes used as a general term for the B horizon. **sous-sol**

subsoiling See *tillage, subsoiling.* **sous-solage**

subspecies A taxonomic rank immediately below species, indicating a group of organisms that is geographically isolated from and may display some morphological differences from other populations of a species, but is nevertheless able to interbreed with other such groups within the species where their ranges overlap. See *species.* **sous-espèce**

substituted ring compound A chemical compound consisting of one or more rings of carbon atoms and their attached hydrogen atoms, but with one or more of the hydrogen atoms replaced by another chemical substance (e.g., *polychlorinated biphenyl* is composed of two benzene rings with a varying number of hydrogens replaced by chlorine atoms). **composé cyclique substitué**

substrate (1) The substance, base, or nutrient on which an organism grows. (2) Compounds or substances that are acted upon by enzymes or catalysts and changed to other compounds in the chemical reaction. (3) An underlying layer, such as the subsoil. **substrat**

substratum Any layer lying beneath the soil solum, either conforming or unconforming. **substratum**

subsurface drainage (1) Water flow through permeable soil or rock beneath the surface of the land. (2) Artificial drainage systems, such as tiles, installed below the soil surface to relieve wet soil conditions. **drainage souterrain**

subsurface irrigation A method of irrigation in which water is applied to open ditches or tile lines until the water table is high enough to wet the root zone. Also called subirrigation. **irrigation souterraine**

subsurface tillage See *tillage, subsurface tillage.* **sous-solage**

S

subtropical Pertaining to the latitudinal zones adjacent to but just outside the Tropics of Cancer and Capricorn; more specifically, extending as far polewards as latitudes 40°N and 40°S. In various climatic classifications, descriptive of those areas exhibiting no month with a mean temperature below 6°C. **subtropical**

succession (*ecology*) The sequential development of changes within a plant community as it progresses towards a climax. See *sere*. (*geology*) The vertical sequence of rock strata at a particular locality. **succession**

succession, autogenic Type of *succession* caused by the interactions of organisms in the ecosystem and their modification of intrinsic environmental variables. **succession autogène**

succession, autotrophic Type of succession that occurs if the food web base depends on photosynthetic organisms. **succession autotrophe**

succession, biotic The natural replacement of one or more groups of organisms occupying a specific habitat by new groups; the preceding groups in some ways prepare or favorably modify the habitat for succeeding groups. See *succession*. **succession biotique**

suction See *moisture tension*. **succion**

suction lysimeter A sampling device for the collection of water from the unsaturated zone; a sample is drawn by applying a negative pressure to a porous ceramic cup embedded in the soil layer. **lysimètre à succion**

suitability mapping A habitat interpretation that describes the current potential of a habitat to support a species. Habitat potential is reflected by the present habitat condition or successional stage. **carte de potentiel d'habitat**

sulfate aerosols Particulate matter that consists of sulfur compounds formed by the interaction of sulfur dioxide and sulfur trioxide with other compounds in the atmosphere. Sulfate aerosols are injected into the atmosphere from the combustion of fossil fuels and the eruption of volcanoes. Recent theory suggests that sulfate aerosols may lower the Earth's temperature by reflecting solar radiation (i.e., negative *radiative forcing*). *Global climate models* that incorporate the effects of sulfate aerosols more accurately predict global temperature variations. **aérosols de sulfates ou sulfatés**

sulfur bacteria Anaerobic bacteria that obtain the oxygen needed in metabolism by reducing sulfate ions to hydrogen sulfide or elemental sulfur. Accumulations of sulfur formed in this way are called bacteriogenic ore deposits. See *iron bacteria*. **bactéries sulfureuses, sulfobactéries**

sulfur cycle The sequence of transformations undergone by sulfur wherein it is used by living organisms, transformed upon death and decomposition of the organism, and converted ultimately to its original state of oxidation. Transformation of sulfur resembles in many ways the microbial conversion of nitrogen. **cycle du soufre**

sulfur dioxide (SO_2) (*chemistry*) A molecule made up of one sulfur and two oxygen atoms. (*meteorology*) Sulfur dioxide emitted into the atmosphere through natural and anthropogenic processes is changed, in a complex series of chemical reactions in the atmosphere, to sulfate aerosols. These aerosols result in negative *radiative forcing*, tending to cool the Earth's surface. **dioxyde de soufre**

sulfur hexafluoride (SF_6) (*chemistry*) A molecule made up of one sulfur and six fluorine atoms, used primarily as a gaseous insulator in power breakers in electrical transmission and distribution systems. SF_6 is a very powerful greenhouse gas and has a *global warming potential* of 24,900. **hexafluorure de soufre**

S

summation curve (of particle sizes) A curve showing the cumulative percentage by mass of *particle size* within increasing or decreasing size limits as a function of diameter; the percentage by mass of each size fraction is plotted cumulatively on the ordinate as a function of the total range of diameters represented in the sample plotted on the abscissa. **courbe cumulative (granulométrique)**

summer fallow See *tillage, summer fallow; fallow land.* **jachère d'été**

summer kill (*limnology*) Complete or partial kill of a fish population in ponds or lakes during the warm months, caused by high temperatures, low dissolved oxygen, or toxic substances. **mortalité estivale**

summit The highest point of any landform remnant, hill, or mountain. See *slope morphological classification.* **sommet**

sun synchronous (*remote sensing*) An Earth satellite orbit in which the orbital plane is near polar, and the altitude such that the satellite passes over all places on Earth having the same latitude twice daily at the same local sun time. **orbite héliosynchrone**

supercooling The process of cooling a liquid below its freezing point without it changing to a solid. **surfusion**

superficial deposits Unconsolidated materials of Pleistocene or Holocene age that lie on the land surface. Their origin is mainly unrelated to the underlying *bedrock,* and they have usually been moved to their present position by natural agencies, with the exception of *peat,* which has accumulated *in situ.* Examples are: *alluvium, colluvium, glacial drift,* eolian deposits (see *dune, loess*), *solifluction* deposits. **dépôts superficiels**

supergene A series of genes, often with related functions, so closely placed on a chromosome that virtually no recombination occurs between them. **supergène**

superheating The process of heating a liquid above its boiling point without it boiling. **surchauffe**

supernatant (*waste management*) The clear fluid that is removed from the top of settling tanks or settling ponds after solids have settled out. Also called overflow. **surnageant**

superphosphate A product obtained when phosphate rock is treated with sulfuric or phosphoric acids, or a mixture of the two. **superphosphate**

superphosphate, ammoniated A product obtained when superphosphate is treated with ammonia or with solutions containing ammonia and/or other ammonium-containing compounds. **superphosphate ammonié**

superphosphate, concentrated Superphosphate made with phosphoric acid and usually containing 19 to 21% phosphorus (i.e., containing 44 to 48% P_2O_5); also called triple or treble superphosphate. **superphosphate triple ou concentré**

superphosphate, enriched Superphosphate made with a mixture of sulfuric acid and phosphoric acid; including any grade between 10 and 19% phosphorus (i.e., containing 22 to 44% P_2O_5), commonly 11 to 17% P (i.e., 25 to 36% P_2O_5). **superphosphate enrichi**

superphosphate, normal Superphosphate made by reaction of phosphate rock with sulfuric acid, usually containing 7 to 10% phosphorus (i.e., 16 to 22% P_2O_5); also called ordinary or single superphosphate. **superphosphate normal**

superphosphate, ordinary See *superphosphate, normal.* **superphosphate normal**

superphosphate, single See *superphosphate, normal.* **superphosphate normal**

superphosphoric acid The acid form of polyphosphates, consisting of a mixture of orthophosphoric and polyphosphoric acids. The species distribution varies with concentration, which is typically 31 to 36% P (i.e., 72 to 83% P_2O_5). **acide superphosphorique**

supersaturation A condition in which a solution has more solute dissolved than normally possible under the existing conditions. **sursaturation**

supplemental feeding (*range management*) Supplying concentrates or harvested feed to correct deficiencies of animal diets. **alimentation de complément**

supplemental irrigation Irrigation used as needed to obtain optimum crop growth in areas where rainfall normally supplies most of the crop water requirement. **irrigation de complément**

supralittoral zone That portion of the seashore adjacent to the tidal or spray zone; also called supratidal zone. See *littoral zone*. **région supralittorale**

surface area The area of the solid particles in a given quantity of soil or porous medium. See *specific surface*. **aire superficielle**

surface charge density The excess of negative or positive charge per unit of surface area of soil or soil mineral. **densité de charge surfacique**

surface creep (1) The slow movement of soil and rock debris. (2) The rolling of dislodged soil particles 0.5 to 1.0 mm in diameter by wind along the soil surface. See *saltation*. **saltation**

surface drain A channel to remove surface water from the land. **drain de surface**

surface expression The form (assemblage of slopes) and pattern of forms of parent genetic materials. Classes of surface expression used in soil surveys in Canada for unconsolidated and consolidated mineral materials are: apron, blanket, fan, hummocky, inclined, level, rolling, ridged, step, terraced, and veneer; for organic materials they are: blanket, bowl, domed, floating, horizontal, plateau, ribbed, and sloping. **modelé**

surface flow See *overland flow*. **ruissellement**

surface form See *surface expression*. **forme de terrain**

surface irrigation A method of irrigation in which water is applied directly to the soil surface (e.g., *furrow irrigation, border-strip irrigation),* as opposed to *sprinkler irrigation* and *subsurface irrigation*. **irrigation de surface**

surface mining A method of mining without recourse to shafts and tunnels, in which soil is salvaged and overburden is stripped from the surface in order to expose the economic mineral to be worked by large mechanical excavators and draglines. See *mining*. **exploitation à ciel ouvert**

surface relief Elevations and depressions on the land surface. **topographie**

surface runoff That part of the *runoff* that travels over the soil surface to the nearest water body (stream channel). **ruissellement**

surface sealing The orientation and packing of dispersed soil particles in the immediate surface layer of the soil to render the surface fairly impermeable to water. See *crust*. **colmatage ou obturation de la surface**

surface tension The resistance of a liquid to an increase in its surface area. Defined as either the force acting over the surface of per unit length of surface perpendicular to the force, or the energy required to increase the surface area isothermally by one square meter. It is measured in newtons per meter. **tension superficielle**

surface tillage See *tillage, surface tillage*. **travail du sol superficiel, façon superficielle**

surface water Water exposed to the atmosphere that occupies oceans, rivers, streams, lakes, reservoirs, and wetlands. Also used to describe water that falls to the ground as rain or snow, and does not immediately

S

evaporate or percolate into the ground. **eau de surface**

surface-applied fertilizer Fertilizer applied on top of the ground but not incorporated (usually called *broadcast application*). It is harder for plants to get the nutrients when this method is used, and there is a greater chance of runoff to surface water sources. **engrais non incorporé**

surfactant A substance added to liquid to increase its spreading or wetting properties by reducing its *surface tension* (e.g., detergents). Composed of several phosphate compounds, surfactants are a source of external enrichment thought to speed the eutrophication of lakes. **agent tensio-actif**

surficial erosion See *erosion*. **érosion superficielle**

surrogate property Indirect properties or easily documented observations that can provide an appraisal or estimate of a difficult-to-measure property, or that can provide a substitute when specific data are missing or not available; differs from *pedotransfer functions* in that the property is not obtained by regression analysis. **propriété subrogative**

survey, soil See *soil survey*. **levé des sols, prospection pédologique**

suspended load (*hydrology*) That part of a river's load carried in suspension. **charge en suspension**

suspended sediment Small soil particles that remain in suspension in water for a considerable period of time. **particules en suspension**

suspended solids Small particles of solid pollutants in sewage that contribute to turbidity and resist separation by conventional means. The examination of suspended solids and the *biochemical oxygen demand* test constitute the two main tests for water quality performed at wastewater treatment facilities. **solides ou matière en suspension.**

suspension The containment and support of soil particles or aggregates in air or water that allows their transport in the fluid when it is flowing. In fluids at rest, suspension follows *Stokes' Law*. In wind, this usually refers to particles or aggregates <0.1 mm diameter through the air, usually at a height of >15 cm above the soil surface, for relatively long distances. See *saltation*, *surface creep*. **suspension**

suspension fertilizer See *fertilizer, suspension*. **engrais en suspension**

sustainability (*general*) Lasting character; the ability to continue over time without depletion. (*ecology*) The ability of an ecosystem to maintain ecological processes and functions, biological diversity, and productivity over time. **durabilité**

sustainable agriculture (1) The long-term viability of an agricultural or farming system, based on natural resource protection, protection of adjacent ecosystems, economic viability (including productivity and security), and social acceptability (including natural use and aesthetic quality). (2) A measure of the land's ongoing ability to produce crops adapted to the soil and region without harm to the soil resource or environment. **agriculture durable**

sustainable forest management Management regimes applied to forest land that maintain the productive and renewal capacities as well as the genetic, species, and ecological diversity of forest ecosystems. **gestion durable des forêts**

sustainable yield Maintenance of a continual annual, or periodic, yield of plants or plant products (e.g., grain) from an area; implies management practices that maintain the productive capacity of the land. **rendement durable ou soutenu**

swamp (1) In general, an area saturated with water throughout much of the year, but with the surface of the soil usually not deeply submerged. It is generally characterized by tree or shrub vegetation. See *marsh.* (2) A class in the Canadian wetland classification system (see *wetland classification system, Canadian*) consisting of mineral wetlands or *peatlands* with standing water or water gently flowing through pools or channels. The water table is usually at or near the surface. There is pronounced internal water movement from margin or other mineral sources; hence the waters are rich in nutrients. If peat is present, it is mainly well-decomposed wood, underlain at times by sedge peat. The associated soils are *Mesisols*, *Humisols*, and *Gleysols* (*Hemists, Saprists*, and *aquic* soils). A dense cover of coniferous or deciduous trees or shrubs, herbs, and some mosses characterizes the vegetation. **marécage, milieu marécageux**

swamp peat Peat materials derived mainly from trees such as black spruce, and from *ericaceous* shrubs and feathermosses. Also called *woody peat,* silvic peat, forest peat. (Swamp or forest peat is an organic genetic material in the Canadian system of soil classification). **tourbe marécageuse**

sward A population of herbaceous plants characterized by a relatively short habit of growth and relatively continuous ground cover, including both above- and below-ground parts. **peuplement de graminées**

swath width See *total field of view.* **largeur de couloir couvert, fauchée**

sweep See *tillage, sweep.* **cultivateur ou herse pourvu de socs à ailettes**

swelling hysteresis See *hysteresis.* **hystérèse de gonflement**

swill Semi-liquid waste material consisting of food scraps and free liquids. **déchets alimentaires**

sylvite Crystalline of potassium chloride (KCl). Sylvite is the principal source of potassium for fertilizer use. Usually the sylvite occurs in a mixture with halite (NaCl), which is called sylvinite. **sylvite**

symbiosis An interaction between two different organisms living in close physical association. Often, but not always, mutually beneficial. **symbiose**

symbiotic fixation Conversion of atmospheric nitrogen to protein by heterotrophic bacteria living in association with a host legume. **fixation symbiotique**

sympatric Overlapping spatially, referring to vegetation or animal taxa whose distribution areas overlap at least partly. **sympatrique**

syncline A convex downward *fold* of crustal rocks; the rock strata tip inwards towards a central axis, in contrast to those of an *anticline*. **synclinal**

synecology A subdivision of ecology that deals with the study of the structure, development, and distribution of ecological communities. See *autecology, dynecology*. **synécologie**

synergism The cooperative action of two or more agents in *association* with each other such that the total effect is greater than the sum of their individual effects (e.g., synergisim between organisms, chemicals, or chemical pollutants). **synergie**

synoptic measurements Numerous measurements taken simultaneously over a large area. **mesures synoptiques**

synoptic view (*remote sensing*) The ability to see or otherwise measure widely dispersed areas at the same time and under the same conditions (e.g., the overall view of a large portion of the Earth's surface that can be obtained from satellite altitudes). **vue synoptique**

synthetic aperture radar See *radar, synthetic aperture*. **radar à synthèse d'ouverture, radar à antenne synthétique, radar à ouverture synthétique**

synthetic manure See *compost.* **compost, engrais artificiel**

synthetic variety Advanced generation progenies of a number of clones or

S

lines (or of hybrids among them) obtained by open pollination. **variété synthétique**

system A set of objects together with relationships between the objects and their attributes. Objects are the parameters of systems, including input, process, output, feedback control, and a restriction. Each system parameter may take a variety of values to describe a system state. Attributes are the properties of object parameters; they characterize the parameters of systems and make it possible to assign a value and a dimensional description. Relationships are the bonds that link objects and attributes in the system process. See *general systems theory*. **système**

systematic error An error that always occurs in the same direction. **erreur systématique**

systematic sample A sample consisting of sampling units selected in conformity with some regular pattern. For example, the soil sample from every 50th meter in a row, or transect. **échantillon systématique**

systematics (*general*) The science of classification. (*biology*) The scientific study of the kinds and diversity of organisms, and the relationships between them. Taxonomic classification. **systématique**

systemic Affecting the whole body or portions of the body other than the site of entry of a chemical substance. See *systemic pesticide*. **systémique**

systemic pesticide A *pesticide* that is not localized in any one part of the body but affects the whole or part of a plant or animal after it is injected or taken up from the soil or body surface. **pesticide systémique**

systems analysis The applied study of the functional and structural relationships of phenomena forming a functional unit. **analyse systémique**

systems approach An appraisal of each component of a *system* in terms of the role it plays in the larger system. See *general systems theory*. **approche des systèmes**

systems ecology The study of operations, factors, and processes that influence the association of organisms and their surroundings; using mathematical analyses. **écologie des systèmes**

systems theory See *general systems theory*. **théorie des systèmes**

T

tableland (1) A broad, elevated region with a nearly level or undulating surface of considerable extent. (2) A *plateau* bordered by abrupt cliff-like edges rising sharply from the surrounding lowland; a mesa. **haut plateau**

tactoid The colloidal-sized aggregates of phyllosilicate clay particles that can form under certain conditions of exchangeable cations and ionic strength. **tactoïde**

tagged molecule A molecule containing an atom of a radioactive element used for the purpose of studying the behavior of that molecule. See *radiotracer, tracer*. **molécule marquée**

taiga A belt of coniferous forest circling the land masses of the northern hemisphere between the temperate grasslands and the tundra; often used synonymously with *boreal forests,* although in parts of northern Canada the taiga zone is also occupied by large expanses of *muskeg*. **taïga**

tailings (*agriculture*) Forage material that falls behind the harvesting combine. (*mining*) Second grade or waste material derived when the raw material is screened or processed. (*mining, reclamation*) Mineral refuse from a milling operation, usually deposited from a water medium. See *fine tailings*. **résidus**

talc A trioctahedral magnesium silicate mineral of the formula $Si_4Mg_3O_{10}(OH)_2$. It has a 2:1 type layer structure but without isomorphous substitution, and may occur in soils as an inherited mineral. **talc**

talud A short, steep slope formed gradually at the downslope margin of a field by deposition against a hedge, a stone wall, or other similar barrier. **talud**

talus A sloping heap of loose rock fragments accumulated by gravity at the foot of a cliff or steep slope. See *scree*. **talus**

tannin Any phenolic compound with high molecular weight that contains hydroxyls and other suitable groups to form strong complexes with proteins and other macromolecules (e.g., starch, cellulose). They occur naturally in many forage plants and, upon condensation with protein, form a leatherlike substance that is insoluble and of impaired digestibility. **tanin**

taproot The primary root of a plant formed in direct continuation with the root tip or radicle of the embryo; forms a thick, tapering main root from which arise smaller, lateral branches. **racine pivotante**

taproot system A plant root system dominated by a single large root, normally growing straight downward, from which most of the smaller roots spread out laterally. See *fibrous root system*. **système racinaire pivotant**

tar sand Sandy deposit containing *bitumen*, a viscous, petroleum-like material that has a high sulfur content. Bitumen can be thermally removed after surface mining of sands and upgraded to a synthetic crude oil. Also called *oil sand*. **sables bitumineux**

taxon cycle The cycle of expansion and contraction of the geographic range and population density of a species

T

or higher taxonomic category. **cycle d'un taxon**

TDR See *time-domain reflectometry*. **TDR**

tectonic Pertaining to the internal forces that deform the Earth's crust, whereby landforms are produced by warping, fracturing, and other Earth movements. **tectonique**

tectosilicate A group of silicate minerals, such as quartz or feldspar, having a structure in which the SiO_4 tetrahedra share all four oxygen atoms with neighboring tetrahedra to form a three-dimensional network with a Si:O ratio of 1:2. **tectosilicate**

telluric Pertaining to, or proceeding from, the Earth. **tellurique**

telluric water Water originating from the Earth, carrying dissolved mineral nutrients. See *minerotrophic*. **eau tellurique**

telmatic peat *Minerotrophic* peat developed in very shallow water and consisting mainly of reeds. Also called *reed peat*. **tourbe de roseaux**

temperate deciduous forest A geographic region characterized by distinct seasons, moderate temperatures, and rainfall from 75 to 150 cm per year. These forests are found in eastern North America; eastern Australia; western, central, and eastern Europe; and parts of China and Japan. Typical trees in North American deciduous forests are oak, hickory, maple, ash, and beech. **forêt tempérée à feuilles caduques**

temperature A measure of the average energy of the molecular motion in a body or substance at a certain point. **température**

temperature coefficient The rate of increase in an activity or process over a 10°C increase in temperature. Also referred to as the Q_{10}. **coefficient de température**

temporal change An alteration that occurs over a period of time. **changement temporel**

ten percent rule (*ecology*) The maxim that about 10% of the energy available at one trophic level is passed to and stored in the bodies of organisms at the next-higher trophic level. The large loss of energy between levels is due in part to the incomplete harvesting of food organisms, the respiration requirements of the energy consumers, or the partial excretion of ingested organisms. See *ecological pyramid*. **règle du 10%**

tendril A slender, spiraling outgrowth; a modified leaf, stem, or stipule by which some climbing plants cling for support. **vrille**

tensile strength The load per unit area at which an unconfined cylindrical specimen will fail in a simple tension test. **résistance en traction**

tensiometer A device for measuring the soil-water *matric potential in situ* (from 0 to – 0.08 MPa); a porous, permeable ceramic cup connected through a water-filled tube to a manometer, vacuum gauge, pressure transducer, or other pressure-measuring device. **tensiomètre**

tension Stress caused by opposing forces (i.e., pulling apart). See *compression*. **tension**

tephra Material ejected from volcanoes. **tephra**

terminal moraine (*geology*) A linear ridge of glacial debris marking the maximum limit of an icesheet or glacier. Its inner slope is usually steeper than its outer slope because the former represents the ice-contact slope. See *moraine*. **moraine terminale ou frontale**

terminal settling velocity For a particle falling in a non-turbulent fluid (liquid or gas), the maximum possible velocity reached when the drag, or frictional resistance, on the particle equals the gravitational force on the particle. **vitesse terminale de sédimentation**

ternary diagram A triangular graph used for plotting data simultaneously on three scales in order to illustrate the relative proportions of each. It may be used when the values are percentages, with each of the three sides of the equilateral triangle graduated from 0 to 100. Provided that the three values total 100%, they may be represented by a single point on the diagram. The *textural triangle* is a ternary diagram used in the study of soil texture. See *Appendix B, Figure 2.* **diagramme ternaire**

terrace (1) A flat or gently inclined land surface consisting of a lower or front slope (the riser) and a flat surface (the tread), bounded on its inner margin by another terrace or a steep ascending slope. (2) An *erosion* control structure; raised, level, or nearly level strip of earth constructed on a contour, designed to make sloping land suited to *arable* farming and to convey excess runoff water safely towards a channel. See *erosion control structures.* **terrasse**

terraced Descriptive of a landform (or *surface expression*) consisting of one or more *terraces.* **en terrasse**

terrain The physical features of a tract of land. **terrain**

terrain analysis The scientific interpretation and mapping of the landforms, vegetation, and soils of a given area in relation to the uses to which it may be put (e.g., logging, mining, or road building). **analyse du terrain**

terrestrial Pertaining to land areas as distinct from the marine and atmospheric environments. **terrestre**

terrestrialization Formation of a *mire* system by the filling of a water body with organic remains, usually by gradual extension of peat-forming communities outwards from the shoreline of a lake. **comblement, atternissement**

terric Unconsolidated mineral soil. **terrique**

terric layer An unconsolidated mineral substratum underlying organic soil material. **couche terrique**

terrigenous sediments Sedimentary inorganic deposits that have either been formed and laid down on a land surface (e.g., blown sand) or derived from the land but subsequently mixed in with marine deposits of the *littoral* zone. **sédiments terriques**

tertiary treatment A phase in the treatment of wastewater; the treatment, beyond the *secondary treatment,* or biologic stage, that removes nutrients (e.g., phosphorus and nitrogen) and a high percentage of suspended solids; also called advance waste treatment. **traitement tertiaire**

tetrahedron A three-dimensional figure having four triangular sides. It may be visualized as a pyramid with a triangular base. The tetrahedral arrangement can be represented by three balls placed in a triangle in contact with each other and a fourth ball resting on top above the center of the triangle. The normal spacing of the oxygen ions in tetrahedra and other compact structures is the same as the diameter of an oxygen ion, about 0.28 nanometers. The space inside an oxygen tetrahedron is so small that only a very small cation, such as Si^{4+} or Al^{3+}, can fit inside. Even Al^{3+} is crowded inside oxygen tetrahedra, and is more frequently found in octahedral spaces. See *octahedron.* **tétraèdre**

tetraploid A plant or tissue whose cells have four sets of chromosomes. **tétraploïde**

textural triangle A triangular graph for determining soil-textural classes. See *texture.* See *Appendix B, Figure 2.* **triangle de texture**

texture The relative proportions of the various soil separates in a soil as described by the classes of soil texture. The names of textural soil classes may be modified by adding suitable adjectives when coarse fragments are present in substantial amounts (e.g., stony sandy loam; silt loam, stony phase). For other modifications, see *coarse fragments.* Sand, loamy sand, and sandy loam are further subdivided on the basis of the proportions of the various sand separates present. **texture**

thallus A type of plant body that is not differentiated into root, stem, or leaf. **thalle**

thalweg (1) The line of maximum depth in a stream. The thalweg is the part that has the maximum velocity and causes cutbanks and channel migration. (2) The line following the lowest part of a valley or a stream, whether under water or not. **talweg**

thaw lake A body of water enclosed in a basin that has been formed or enlarged by thawing of frozen ground. **cuvette de dégel**

thawing The action that changes ice or snow into water as a result of temperatures rising above freezing point. **dégel**

thematic map A map designed to demonstrate particular features or concepts, as opposed to a *topographic map*. Thematic maps include soil maps, climatic maps, economic maps, and population maps. **carte thématique ou dérivée**

theodolite An accurate optical instrument for measuring horizontal and vertical angles and used extensively in surveying. It consists of a telescope, mounted so that it can be rotated in the horizontal or vertical plane, a spirit-level, and a compass, all of which are attached to a tripod. **théodolite**

T

theoretical oxygen demand (TOD) The amount of oxygen that is calculated to be required for complete decomposition of a substance. **demande théorique d'oxygène**

theoretical yield *(chemistry)* The maximum amount of a given product that can be formed when the limiting reactant is completely consumed. **rendement théorique**

theory A set of assumptions put forth to explain some aspect of the observed behavior of matter. **théorie**

thermal analysis A method for measuring changes in the chemical or physical properties of materials as a function of temperature. (1) Differential thermal analysis measures the temperature difference between a sample and a standard or reference material. (2) Differential scanning calorimetry measures the difference in heat flow between a sample and reference material. **analyse thermique**

thermal conductivity The rate of transfer of heat to or from a point in the soil under unit thermal gradient. **conductivité thermique**

thermal fracture (*geology*) The action by which rocks crack when subjected to rapid changes of temperature, due to stress fractures resulting from the varying coefficients of expansion of the rock constituents. **thermoclastie**

thermal infrared The preferred term for the middle wavelength ranges of the infrared region, extending roughly from 3 μm at the end of the near infrared to about 15 or 20 μm where the far infrared commences. The thermal band most used in remote sensing extends from 8 to 13 micrometers. **rayonnement infrarouge**

thermal properties Properties of a medium (soil) relative to heat content and heat transfer, such as *thermal conductivity*, specific heat capacity, and thermal diffusivity. **propriétés thermiques**

thermal radiation The electromagnetic radiation emitted by a hot *blackbody*. **rayonnement thermique**

thermal stratification (*limnology*) The formation of layering of water masses or condition of well-defined water temperature zones with depth in a lake or pond. Three layers are usually present, the *epilimnion, thermocline,* and *hypolimnion.* See *fall overturn* and *spring overturn.* **stratification thermique**

thermal treatment of hazardous waste Any treatment of hazardous waste involving exposure of the material to elevated temperatures to change the characteristics of the waste (e.g., incineration, pyrolysis, and microwave discharge). **traitement thermique de déchets dangereux**

thermic A soil temperature regime that has mean annual soil temperatures of 15°C or more, but <22°C and >5°C difference between mean summer and mean winter soil temperatures at 50 cm below the surface. Isothermic is the same except the summer and winter temperatures differ by <5°C. **thermique**

thermistor A small instrument used to measure temperature by monitoring changes in resistance. **thermistance**

thermocline The sharp transition zone of rapid temperature change between the warm epilimnion and cold hypolimnion of stratified bodies of water; temperature change equals or exceeds 1°C for each meter of depth. See *epilimnion, hypolimnion*, and *fall* and *spring overturn*. **thermocline**

thermocouple A union of two conductors. Bars or wires of dissimilar metals joined at their extremities to produce a thermoelectric current. Also called thermo-junction, thermoelectric couple. **thermocouple**

thermodynamics The study of the laws that govern the conversion of energy from one form to another, the direction in which heat will flow, and the availability to do work. See *first law of thermodynamics, second law of thermodynamics*, and *third law of thermodynamics*. **thermodonamique**

thermogenic soils Soils with properties that have been influenced primarily by high soil temperature as the dominant soil-formation factor, developed in subtropical and equatorial regions. **sols thermogéniques**

thermoluminescence The light emitted by many minerals when heated below the temperature of incandescence, resulting from the release of energy stored as electron displacements in the mineral's crystal lattice. The amount of thermoluminescence is proportional to the number of electrons. **thermoluminescence**

thermometer An instrument for measuring temperature. **thermomètre**

thermoperiodism The response of plants to rhythmic fluctuations in temperature. Most plants tend to thrive in an environment of varying temperatures and are kept in check where conditions are constant. **thermopériodisme**

thermophile An organism that grows best at temperatures of 50°C or higher. **thermophile**

thermosequence A sequence of related soils that differ from each other primarily as a result of temperature as a soil-formation factor. **thermoséquence**

therophyte One of the six major floral life-form classes based on the position of the regenerating parts in relation to the exposure of the growing bud to climatic extremes, and without respect to taxonomy. Therophytes are annual plants; after producing seeds, the parent plant dies and the seeds lie dormant until favorable weather conditions return. **thérophyte**

thin section A fragment of rock, mineral, or soil mounted on a glass slide, mechanically ground to a thickness of approximately 0.03 mm, and covered with a glass cover slip to produce a microscopic slide. This reduction renders minerals and soil material transparent or translucent, thus making it possible to study the optical properties and distribution patterns with the aid of a microscope. See *polished section*. **lame mince**

thin-layer chromatography (TLC) A chromatographic technique in which the mobile phase is a solvent, and the stationary phase is a solid adsorbent (e.g., alumina on a flat support such as a glass plate). See *chromatography*. **chromatographie sur couche mince**

thinning (1) (*forestry*) A cutting made in immature tree stands to provide adequate growing space and accelerate diameter growth, but also, by suitable selection, to improve the average form of the remaining trees. (2) (*horticulture*) The removal of buds, flowers, or fruit for superior results. Also, the pruning of certain branches in a tree or shrub, to improve structure, balanced growth, and air circulation. See *pruning*. **(1) éclaircie (2) éclaircissage**

Thiobacillus An aquatic or terrestrial genus of bacteria capable of oxidizing elemental sulfur, sulfide ions, thiosulfates, and other forms of inorganic sulfur to derive the energy needed in metabolism (i.e., they are *chemoautotrophs*). Bacteria belonging to this genus fix carbon dioxide and produce sulfuric acid as an end product. They can increase the acidity of soil or water to levels that result in the destruction of natural environments. ***Thiobacillus***

thiocarbamate Any of a group of soil-applied *herbicides* that restrict plant growth by inhibiting cell division and cell elongation; shoots are more affected than roots. **thiocarbamate**

third law of thermodynamics The entropy of a perfect crystal at *absolute zero* (i.e., 0 K) is zero. **troisième loi de la thermodynamique**

T

thixotropy The property of certain colloidal substances (e.g., bentonitic clay, *quick clay*) to weaken or change from a gel to a sol when disturbed, but to increase in strength upon standing. **thixotropie**

threatened or endangered habitats (1) Ecosystems that are restricted in their distribution over a natural landscape (e.g., freshwater wetlands within certain ecological zones) or are restricted to a specific geographic area or a particular type of local environment. (2) Ecosystems that were previously widespread or common but now occur over a much smaller area due to extensive disturbance or complete destruction by such practices as intensive harvesting or grazing by introduced species, hydro projects, diking, and agricultural conversion. **habitats menacés**

threatened species A species likely to become endangered throughout all or a significant proportion of its distribution if populations decline through mortality (e.g., harvesting, predation) or loss of habitat. See *threatened or endangered habitats*. **espèce menacée**

threshold The maximum or minimum duration or intensity of a stimulus required to produce a response in a system. Also called the critical level. **seuil**

threshold moisture content (*biology*) The minimum moisture condition, measured in terms of either moisture content or moisture stress, at which biological activity just becomes measurable. **humidité critique, point de flétrissement**

threshold salt tolerance (crop) The conductivity of the saturated paste extract at which a plant experiences reduction in growth. **seuil de tolérance au sel**

threshold value (soil quality) The point at which a change in *soil quality* is likely to occur. Also called critical limit. **valeur limite, valeur seuil, seuil**

threshold velocity The minimum velocity at which wind will begin moving particles of sand or other soil material. **vitesse minimale, vitesse limite**

throughfall That portion of precipitation that falls through or drips off a plant canopy. See also *stemflow*. **pluviolessivage**

throw See *tillage, throw*. **projection**

tidal flat An area of nearly flat, barren mud periodically covered by tidal waters. Normally these materials have an excess of soluble salt. A miscellaneous land type. **wadden, bas fonds intertidaux**

tidal marsh A low flat marsh bordering a coast, formed of mud and the resistant mat of roots of salt-tolerant plants, and regularly inundated during high tides. See *salt marsh.* **marais tidal**

tier A depth subdivision used in the classification of *Organic* soils in the Canadian system of soil classification and U.S. system of soil taxonomy. **étage**

Types of tiers include:

surface tier – (*Canadian system*) the upper 40 cm of peat. (*U.S. system*) The upper 30 cm of peat, or the upper 60 cm if fibric or having bulk density less than 0.1 Mg m^{-3}. ***étage supérieur***

middle tier – (*Canadian system*) the tier immediately below the surface tier. It is 80 cm thick or extends to a *lithic* or hydric contact. The great group classification of *Organic* soils is usually based on this tier. (*U.S. system*) Called the subsurface tier, which is 60 cm thick. ***étage intermédiaire***

bottom tier – (*Canadian system*) the tier below the middle tier. It is 40 cm thick or extends to a *lithic* or *hydric* contact. The subgroup classification of Organic soils is based partly on this tier. (*U.S. system*) 40 cm thick unless the control section has its lower boundary at a shallower depth (at a densic, lithic, or *paralithic contact* or a water layer or in permafrost). ***étage inférieur***

tie-ridge See *tillage, tie-ridge.* **billons reliés entre eux de façon à former des bassins**

tight soil A compact, relatively impervious, and tenacious soil or subsoil that may or may not be plastic. See *compaction.* **sol compact ou tenace**

tile drain Pipe placed at suitable depths and spacings in the soil or subsoil to provide water outlets from the soil. The pipe may be concrete, ceramic, fiber, plastic, or any other suitable material. **drain en tuyaux**

till (*geology*) (1) Unstratified glacial drift deposited directly by the ice and consisting of clay, sand, gravel, and boulders intermingled in any proportion. (*soil management*) (2) To plow or cultivate the soil to control weeds or prepare a seedbed. See *tillage.* **(1) till (2) cultiver le sol, labourer**

till fabric The structural and textural composition of a *till*, especially the orientation, dip, and roundness of the included pebbles or boulders. Applied in reconstructing former ice directions and till depositional processes. **trame de till**

till plain An extensive level or undulating land surface covered by glacial till. **plaine de till**

tillability See *tillage, tillability.* **degré de facilité de travail du sol**

tillage The mechanical manipulation of the soil profile for any purpose. In agriculture it is usually restricted to modifying soil conditions and/or managing crop residues and/or weeds and/or incorporating chemicals for crop production. **travail du sol**

Tillage terminology (*Glossary of Soil Science Terms*, 1996, Soil Science Society of America, Madison, WI; reprinted with permission) includes:

anchor – Partially burying foreign materials such as plant residues or paper mulches. ***enfouir, incorporer***

back furrow – The resulting ridge of soil turned up when the first furrow slice is lapped over the previous soil surface when starting the plowing operation. ***ados***

bed planting – A method of planting in which the seed is planted on beds. Often two or more seed rows are planted on each bed. See *tillage, ridge planting.* ***semis sur buttes ou sur lits surélevés***

bed shaper – A soil-handling implement that forms uniform ridges of soil to predetermined shapes. ***billonneuse, buttoir***

bedding – The process of preparing a series of parallel ridges, usually no wider than two crop rows, separated by shallow furrows. The resulting structures are beds. ***billonnage***

block (thinning, checking) – To remove plants from a row with hoes or other cutting devices as a means of reducing and uniformly spacing plants. ***éclaircissage***

broadcast tillage (total surface tillage or full-width tillage) – Manipulation of the entire surface area by tillage implements, as contrasted to partial manipulation in bands or strips. ***travail sur toute la surface de sol***

broadcast planting – A method of planting in which the seeds are uniformly distributed over the entire surface area. ***semis à la volée***

burying – Covering foreign materials or bodies intact, such as drain liners, tile lines, communication wires, or plant residues. ***enfouissement***

chemical fallow (eco-fallow) – a special case of fallowing in which all vegetative growth is killed or prevented by use of chemicals; tillage for other purposes may or may not be used. ***jachère chimique***

chisel – To break up soil using narrow shank-mounted tools. It may be performed at other than the normal plowing depth. Chiseling at depths >40 cm is usually called subsoiling. ***scarifier, sous-soler***

clean tillage (clean culture, clean cultivation) – A that incorporates all residues and prevents growth of all vegetation except the particular crop desired during the growing season. ***labour nettoyant***

combined tillage operations – The simultaneous operation of two or more different types of tillage tools (on the same implement frame) or implements (subsoiler-lister, lister planter, or plow planter) to simplify control or reduce the number of trips over the field. ***opérations combinées de travail du sol***

conservation tillage – Any tillage sequence, the object of which is to minimize or reduce loss of soil and water; operationally, a tillage or tillage and planting combination which leaves a 30% or greater cover of crop residue on the surface. ***pratiques de conservation***

contour tillage – Performing tillage operations and planting on the contour within a given tolerance. ***travail du sol selon les courbes de niveau***

controlled traffic – Tillage in which all operations are performed in fixed paths so that recompaction of soil by traffic (traction or transport) does not occur outside the selected paths. ***circulation limitée***

conventional tillage – Tillage operations normally performed in preparing a seedbed for a given crop grown in a given geographical area. ***travail du sol conventionnel***

crop residue management – Disposition of stubble, stalks, and other crop residues by tillage operations. (1) To remove residues from the soil surface (burying). (2) To anchor residues partially in the surface soil while leaving the residues partially exposed at the surface (mulch tillage). (3) To leave residues entirely at the soil surface intact or cut into smaller pieces. (Residues may be removed by nontillage methods, e.g., harvesting, burning, grazing.) ***gestion des résidus de cultures***

crop residue management system – The operation and management of crop land to maintain stubble, stalks, and other crop residue on the surface to prevent wind and water erosion, to conserve water, and to decrease evaporation. ***système de gestion des résidus de cultures***

cross cultivation – the tillage of a field, orchard, etc., in which the field is cultivated in one direction followed by cultivation at some angle between 10 and 90° from the preceding tillage. ***travail du sol en effectuant des passages croisés***

crushing – Applying forces to the soil surface to destroy the integrity of aggregates or clods. ***écrasement, fragmentation***

cultipack – A broadcast soil crushing and firming operation utilizing wide rollers having corrugated or jagged working surfaces. ***cultitassement, raffermissement du lit de semence***

cultivation – Shallow tillage operations performed to create soil conditions conducive to improved aeration, infiltration, and water conservation, or to control weeds. ***hersage, travail du sol***

cultivation (weeding) – Tillage action which lightly tills the surface 1 to 2 cm of soil for the purpose of destroying weeds. ***sarclage***

cutting – Severing soil by a slicing action that minimizes any other type of failure, such as shear. ***découpage***

dam (pitting, basin listing) – Forming pits, small basins, or waterholding cavities at intervals with appropriate equipment. ***modelage superficiel***

dead furrow – The furrow resulting where land plowed in one direction abuts with land plowed in the opposite direction (i.e., at the completion of each plowed section of a field). ***ados***

dig – To break up, invert, or remove the soil with a spade, plow, or other implement; to bring to the surface (as in harvesting potatoes or disturbing subterranean root and stem structures of weeds) with mechanical tools. ***creuser***

drag – To draw planks or other heavy, rigid implements with wide surfaces across the soil surface in order to crush clods and level or smooth the surface. ***niveler avec une traîne***

eco-fallow – See *tillage, chemical fallow.* ***jachère écologique***

fallow – The practice of leaving land either uncropped and weed free, or with volunteer vegetation during at least one period when a crop would normally be grown; objective may be to control weeds, and to accumulate water and/or available plant nutrients. ***jachère***

firming – A process of achieving a desirable degree of compaction. ***tassement, raffermissement***

flat planting – A planting method in which the seed is planted on harrowed, dragged, or plowed land with a planter that causes minimum disturbance to the soil surface. ***semis superficiel ou à plat***

furrow – An opening left in the soil after a plow or disk has opened a shallow channel at the soil surface. A shallow channel cut in the soil surface, usually between planted rows, for controlling surface water and soil loss, or for conveying irrigation water. ***sillon, raie***

guess row – The rows or inter-row space of adjoining multiple-row equipment passes, where, due to reliance on markers for approximate positioning and guidance of tractor traffic, the inter-row space will vary as the driver deviates from a perfect pattern. ***ligne de guidage***

harrowing – A secondary broadcast tillage operation that pulverizes, smooths, and firms the soil in seedbed preparation, controls weeds, or incorporates material spread on the surface. ***hersage***

hill – To place soil up to and around crops, usually planted in rows. ***buttage***

hoe – To dig, scrape, or the like, with a hoe; also to control weeds or to loosen or rearrange the soil. ***biner, passer la houe***

incorporation – The mixing of materials found on or spread upon the soil surface (e.g., fertilizers, pesticides, or crop residues) into the soil volume via tillage. ***incorporation***

in-row subsoiling – The use of subsoiling in conjunction with traffic control or where the subsoiler tool is an integral part of the planter implement, for the purpose of having zones of maximum soil shattering located directly beneath the planted row in order to maximize root exploration or penetration of a restrictive zone shattered by the subsoiling operation. ***sous-solage***

inversion – Reversal of vertical order of occurrence of layers of soil. ***inversion, bouleversement***

landforming – Tillage operations that move soil to create desired soil configurations. Forming may be done on a large scale such as gully filling or terracing, or on a small scale such as contouring, ridging, or bedding. ***modelage***

land planing – A tillage operation that redistributes small quantities of soil across the soil surface to provide a more nearly level or uniformly sloped surface. ***nivellement, planage***

lift – To separate roots or other crop parts from soil and elevate them to the soil surface or above. ***arracher***

lister planting – A method of planting in which the seed is planted in the bottom of lister furrows, usually simultaneously with the opening of these furrows. ***semis au creux du sillon***

listing (middle breaking) – A tillage and land-forming operation using a tool that turns two furrows laterally in opposite directions, thereby producing beds or ridges. ***forme de buttage ou billonnage***

loosening – Decreasing soil bulk density and increasing porosity due to the application of mechanical forces to the soil. ***ameublissement***

minimum tillage – The minimum use of primary and/or secondary tillage necessary to meet crop production requirements under existing soil and climatic conditions, usually resulting in fewer tillage operations than for conventional tillage. ***travail minimal du sol***

mixing – Blending of soil layers into the soil mass. ***mélange***

moldboard plowing – See *tillage, plowing*. ***labour avec charrue à versoirs***

mulch – (1) Any material such as straw, sawdust, leaves, plastic film, loose soil, etc., that is spread or formed upon the surface of the soil to protect the soil and/or plant roots from the effects of raindrops, soil crusting, freezing, or evaporation. (2) To apply mulch to the soil surface. ***paillis***

mulch farming – A system of tillage and planting operations which maintains a substantial amount of plant residues or other mulch on the soil surface. ***culture sur résidus ou sur paillis***

mulch tillage – Tillage or preparation of the soil in such a way that plant residues or other materials are left to cover the surface; also called mulch farming, trash farming, stubble mulch tillage, plowless farming; operationally, a full-width tillage or tillage and planting combination that leaves >30% of the surface covered with crop residue. ***travail du sol avec conservation des résidus***

narrow row planting – A method of planting in which the seed is planted in uncommonly narrow rows for the given crop (for maize, the narrow row may be 25 to 50 cm). ***plantation en rangs étroits ou serrés***

non-inversive tillage – Tillage that does not mix (or minimizes the mixing of) soil horizons or does not vertically mix soil within a horizon. ***travail sans inversion de sol***

no-tillage (zero tillage) system – A procedure whereby a crop is planted directly into the soil with no primary or secondary tillage since harvest of the previous crop; usually a special planter is necessary to prepare a narrow, shallow seedbed immediately surrounding the seed being planted. No-till is sometimes practiced in combination with subsoiling to facilitate seeding and early root growth, whereby the surface residue is left virtually undisturbed except for a small slot in the path of the subsoil shank. ***zéro-labour, semis-direct, système de culture sans labour***

once-over tillage – A system whereby all tillage preparatory for planting is done in one operation or trip over the field.

préparation du sol par un passage unique d'outils combinés

oriented tillage – Tillage operations that bear specific relations in direction with respect to the sun, prevailing winds, previous tillage operations, or field base lines. ***travail du sol directionnel***

pan – A horizon or layer in soil which is highly compacted, indurated, or very high in clay content relative to the layer immediately above. ***couche indurée***

paraplow – A type of non-inversive subsoiling implement designed to enhance lateral direction of shattering force using broad, angled subsoil lifting surfaces. ***décompacteur de type paraplow***

paratill – A variation on the mounting of paraplow subsoiling implements to allow greater ease of use in row crops, and/or to leave specific non-shattered zones between rows to provide traction and support for vehicle or tractor traffic. ***décompacteur de type paratill***

plowing – A primary broadcast tillage operation performed to shatter soil with partial to complete inversion. ***labour***

plow layer – The greatest depth of soil exhibiting mixing or inversion by surface tillage operations. ***couche de labour, couche de sol arable***

plowless farming – Tilling soil without moldboard plowing so that inversion and/or residue burying is intentionally reduced. ***culture sans labour***

plow pan – A pan created by plowing at the depth of tillage, largely the result of the common practice of dropping the tractor wheels of one side of the tractor into the dead furrow for steering while performing the plowing operation. ***semelle de labour***

plow-planting – The plowing and planting of land in a single trip over the field by drawing both plowing and planting tools with the same power sources. ***labour et semis en un passage (à l'aide d'un train d'outils)***

pre-(post-) emergence tillage – Tillage operations that occur before (after) crop emergence. ***travail du sol en pré- ou post-émergence***

pre-(post-) harvest tillage – Tillage operations that occur before (after) crop harvest. ***travail du sol en pré- ou post-récolte***

pre-(post-) planting tillage – Tillage operations that occur before (after) the crop is planted. ***travail du sol en pré- ou post-semis***

pressure pan (traffic sole, plow pan, tillage pan, traffic pan, plow sole, compacted layer) – An induced subsurface soil horizon or layer having a higher bulk density and lower total porosity than the soil material directly above and below, but similar in particle size analysis and chemical properties. The pan is usually found just below the maximum depth of primary tillage and frequently restricts root development and water movement. ***couche indurée d'origine anthropique, semelle de labour***

primary tillage – Tillage at any time that constitutes the initial, major soil manipulation operation. It is normally a broadcast operation designed to loosen the soil or reduce soil strength, anchor or bury plant materials and fertilizers, and rearrange aggregates. ***préparation primaire du sol***

reduced tillage – A tillage system in which the total number of tillage operations preparatory for seed planting is reduced from that normally used on that particular field or soil. ***travail du sol réduit ou simplifié***

reservoir tillage (damming, pitting, basin listing, furrow diking, dammer diking) – Forming pits, small basins, or water-holding cavities at intervals with a furrow diker or other appropriate equipment. ***modelage superficiel du sol pour former des petits bassins de captage***

residue processing – Operations that cut, crush, shred, or otherwise break (fracture) residues in a step preparatory to tillage, harvesting, or planting operations. ***broyage et déchiquetage des résidus de cultures***

ridge – To form a raised longitudinal mound of soil by a lister or other tillage tool. Ridges wider than required to plant one row of a crop are called beds. ***billon***

ridge planting – A method of planting crops on ridges formed through tillage operations. Usually only one seed row is planted on each ridge. ***semis sur billons***

ridge tillage – A tillage system in which ridges are reformed atop the planted row by cultivation, and the ensuing row crop is planted into ridges formed in the previous growing season. See *tillage, ridge planting*. ***billonnage***

rod weeding – The control or eradication of weeds and soil firming by means of pulling a longitudinally rotating rod below the soil surface. The rod rotates about an axis perpendicular to the line of travel, and pulls or cuts off weeds with minimum disturbance of trash on or near the ground surface. ***désherbage mécanique à l'aide d'un cultivateur à tige rotative***

rolling – A broadcast, secondary tillage operation that crushes clods and compacts or firms and smooths the soil by the action of ground-driven, rotating cylinders. See *tillage, cultipack*. ***roulage***

root bed – The soil profile modified by tillage or amendments for more effective use by plant roots. ***zone racinaire***

rotary hoeing – A tillage operation using ground-driven rotary motion of the tillage tool to shatter and mix soil and control small weed seedlings. ***sarclage ou hersage à la houe rotative***

rotary tilling – A tillage operation employing power-driven rotary motion of the tillage tool to loosen, shatter, and mix soil. ***travail du sol avec bêche ou charrue rotative***

scarifying – To loosen the topsoil aggregates by raking the soil surface with a set of sharp teeth. ***scarifiage léger ou écroûtage***

secondary tillage – Any of a group of separate or distinct tillage operations, following primary tillage, that is designed to provide specific soil conditions for any reason, such as seeding. ***travail secondaire ou superficiel du sol, reprise de labour***

seedbed – The tillage-manipulated soil layer that affects the germination and emergence of crop seeds. ***lit de semence***

shatter – General fragmentation of a rigid or brittle soil mass. ***fragmentation, écrasement***

shearing – Separating parts of a soil mass by applying shearing stresses. ***cisaillement***

slit tillage – Use of narrow straight coulters or knives to open slices of 5 to 10 mm in width in soil that penetrate to beneath a shallow root restrictive layer, allowing precision-planted seeds to develop root systems which penetrate the restrictive layer without requiring large-scale profile disruption or shattering, and the horsepower or energy needed to accomplish such operations. ***travail du sol en bandes étroites***

slit planting (slot planting) – A method of planting crops that involves no seedbed preparation other than opening a fine slit in the soil (usually with a coulter attached to the planter) to place the seed at some intended depth. Herbicides are usually sprayed shortly before, at, or after silt planting is performed in reduced tillage systems. ***semis-direct***

sod planting – A method of planting in sod with little or no tillage. ***semis-direct sur prairie ou sur gazon***

soil management – The combination of all tillage operations, cropping practices, fertilizer, lime, and other treatments conducted on or applied to the soil for the production of plants. ***gestion des sols***

strip cropping (field strip cropping, contour strip cropping) – The practice of growing two or more crops, alternating strips along contours or perpendicular to the prevailing direction of wind or surface water flow. ***culture en bandes, culture en bandes selon les courbes de niveau***

strip planting (strip till planting) – A method of simultaneous tillage and planting in isolated bands of varying width, separated by bands of erect residues essentially undisturbed by tillage. ***semis en bandes***

strip tillage (partial-width tillage) – Tillage operations performed in isolated bands, separated by bands of soil essentially undisturbed by the particular tillage equipment. ***travail du sol en bandes***

strip-till planting – An area 30 to 50 cm wide tilled sufficiently through living mulch or standing residue to form a seedbed for each row. At planting or at first cultivation, the remaining mulch in the row middle is cut loose, killed, or retarded. ***semis sur bandes travaillées***

stubble mulch – The stubble of crops or crop residues left in place on the land as a surface cover before and during the preparation of the seedbed, and at least partly during the growing of a succeeding crop. ***paillis de résidus de cultures***

stubble mulch tillage – See *tillage, mulch tillage,* and *tillage, plowless farming.* ***travail du sol préservant les résidus de culture***

subsoiling – Any treatment to loosen soil with narrow tools below the depth of normal tillage without inversion, and with minimum mixing of the soil. This loosening is usually performed by a lifting action or other displacement of soil dry enough so that shattering occurs. ***sous-solage***

subsurface tillage – Tillage that confines most of its action (usually only fracturing and shattering) to depths below the normal depth of disk cultivation. ***sous-solage***

summer fallow – The prevention of all vegetative growth by shallow tillage in conjunction with or without herbicides during the summer months, in place of growing a crop, in order to store water for use by the next crop or to control weed infestations. ***jachère d'été***

surface tillage – Cultivating or mixing the soil to a shallow depth (i.e., 5 to 10 cm). ***travail du sol superficiel, façon superficielle***

sweep – (1) Tillage with a shallow knife, blade, or sweep cultivating tool which is drawn slightly beneath the soil surface, cutting plant roots and loosening the soil without inverting it, resulting in minimum incorporation of residues into the soil. (2) A type of cultivator shovel that is wing-shaped. ***cultivateur ou herse pourvu de socs à ailettes***

throw – Aerial movement of soil in any direction resulting from momentum imparted to the soil. ***projection***

tie-ridge – Ridges that are tied together at certain intervals by a cross ridge, forming small basins. ***billons reliés entre eux de façon à former des bassins***

tillability – The degree of ease with which a soil may be manipulated for a specific purpose. ***degré de facilité de travail du sol***

tillage action – The specific form or forms of soil manipulation performed by the application of mechanical forces to the soil with a tillage tool, such as cutting, shattering, inversion, or mixing. ***effets découlant du travail du sol***

tillage, deep – A primary tillage operation that manipulates soil to a greater depth than normal plowing. It may be accomplished with a large heavy-duty moldboard or disk plow that inverts the soil, or with a heavy-duty chisel plow that shatters soil. See *tillage, subsoiling.* ***labour profond, labour de défoncementm sous-solage***

T

tillage equipment (tools) – Field tools and machinery designed to lift, invert, stir, and pack soil, reduce the size of clods and uproot weeds (e.g., plows, harrows, disks, cultivators, and rollers). ***outils de travail du sol, instruments aratoires***

tillage operation – The act of applying one or more tillage actions in a distinct mechanical application of force to all or part of the soil mass. ***pratique culturale***

tilth – The physical condition of soil as related to its ease of tillage, fitness as a seedbed, and its impedance to seedling emergence and root penetration. ***état d'ameublissement ou état structural du sol***

trash farming – See *tillage, mulch tillage; tillage, no-tillage system; tillage, zero tillage; tillage, minimum tillage; tillage, plowless farming*. ***culture sur résidus***

turnrow (turn strip, head land) – The land at the margin of a field on which the plow or other equipment may be turned. This land may or may not be planted to a crop. ***tournière, cintre***

vertical mulching – a subsoiling operation in which a vertical band of mulching material is placed into a vertical slit in the soil, for example, immediately behind the soil-opening implement. ***placement vertical d'amendements***

weeding – Tillage action that lightly cultivates the soil for the purpose of destroying weeds. ***sarclage***

wheel track planting – A practice of planting in which the seed is planted in tracks formed by wheels (usually tractor wheels) rolling immediately ahead of the planter. ***semis en ornières***

zero tillage – See *tillage, no-tillage system*. ***zéro-labour, semis-direct***

zone subsoiling – The practice, usually only in row crops, of maximizing subsoil shattering in certain zones along the row while specifically preventing it in trafficked inter-rows, thereby maximizing crop response without impairing traction of vehicles or tractors later entering the field; can be accomplished with in-row subsoilers, but usually seeks a larger shattering zone, such as the type obtained with the paratill. ***sous-solage localisé***

zone tillage – Tillage operations that differentially affect various zones traversed by the machine. ***travail du sol localisé***

tillage equipment (tools) See *tillage, tillage equipment (tools)*. **outils de travail du sol, instruments aratoires**

tillage erosion Displacement of soil downslope by the action of tillage. **érosion ou déplacement de sol causé par le labour**

tillage operation See *tillage, tillage operation*. **pratique culturale**

tillage pan A subsurface soil horizon or layer induced by tillage, having a higher *bulk density* and lower total *porosity* than the soil material directly above and below, but similar in *particle size analysis* and chemical properties. Also called traffic sole, hard pan, plow pan, traffic pan, plow sole, compacted layer. **semelle de labour**

tillage requirement In some soil types, the need for regular tillage to provide an optimum soil structure for crop production. **besoin en labour ou en travail du sol**

tillage, deep See *tillage, tillage, deep*. **labour profond, labour de défoncement**

tiller (*botany*) A subterranean or ground-level lateral shoot, usually erect, contrasted with horizontally spreading *stolons* and *rhizomes*. **talle**

tillite A former *till* that has become compacted and lithified to form a tough sedimentary rock. **tillite**

tilth See *tillage, tilth*. **état d'ameublissement ou état structural du sol**

timber Trees, whether standing, fallen, living, dead, limbed, bucked, or peeled. **bois**

T

time lag Delay in response to a change. **délai**

time series analysis *(statistics)* The separation or decomposition of a time series into its individual components according to some model or set of assumptions. **analyse des séries chronologiques**

time-domain reflectometry (TDR) A method that uses the timing of wave reflections to determine the properties of various materials, such as the dielectric constant of soil as an indication of water content. **réflectométrie à dimension temporelle**

tissue A group of similar cells organized into a structural and functional unit. **tissu**

tissue analysis Laboratory determination of the total content of specified elements (normally plant nutrients) in plant tissue. **analyse de tissus végétaux**

titrant A standard solution of known concentration and composition used in a titration to bring about the endpoint. **solution titrante**

titration A method for analyzing the composition of a solution by adding a known amount of standardized solution (*titrant*) until a reaction (color change, precipitation, conductivity change) occurs. The volume of solution added indicates the concentration of the substance being analyzed. **titration, titrage**

todorokite A black manganese oxide of the formula $(Na, Ca, K, Ba, Mn^{2+})_2 Mn_4O_{12}{\cdot}3H_2O$ that occurs in soils and in the weathered regolith of sediments. It has a tunnel structure. **todorokite**

toe (*engineering*) Terminal edge or edges of a structure. (*mining*) The point of contact between the base of an embankment or spoil and the foundation surface; usually the outer portion of the spoil bank where it contacts the original ground surface. **pied**

toe drain Interceptor drain located near the downstream toe of a structure. **drain de pied**

toeslope (alluvial toeslope) See *slope morphological classification*. **pied ou base du talus**

tolerance (*ecology*) The ability of an organism or biological process to subsist under a given set of environmental conditions. The range of these under which it can subsist, representing its *limits of tolerance*, is called its ecological amplitude. **tolérance**

top dressing See *fertilizer, top-dressed*. **engrais appliqué en post-levée**

topographic map A map showing the topographic features of a land surface, commonly by means of contour lines. It is generally on a sufficiently large scale to show in detail selected constructed and natural features, including relief and such physical and cultural features as vegetation, roads, and drainage. **carte topographique**

topography (1) The configuration of a surface area including its relief, or relative elevations, and the position of its natural and manmade features. (2) The study or description of the physical features of the Earth's surface. (3) The physical features of a district or region, such as represented on maps, taken collectively; especially, the relief and contour of the land. **topographie**

topology The spatial relationships between phenomena rather than those of actual distance or direction, (e.g., relative position and contiguity). **topologie**

toposaic A *photomap* on which topographic or terrain-form lines are shown as on topographic quadrangles. See *planisaic*. **photo-carte topographique**

toposequence A sequence of related soils that differ from each other primarily because of *topography* as a soil-formation factor. **toposéquence**

topsoil (1) The surface soil layer moved or disturbed by cultivation, often enriched in organic matter. (2) Soil material used to topdress roadbanks, lawns, and gardens. **(1) couche de sol labourée ou travaillée (2) sol arable, terre végétale**

Torrards A suborder in the U.S. system of soil taxonomy. *Andisols* that have an aridic soil moisture regime. **Torrands**

Torrerts A suborder in the U.S. system of soil taxonomy. *Vertisols* of arid regions that have wide, deep cracks that remain open throughout the year in most years. **Torrerts**

torric A soil moisture regime defined like *aridic* moisture regime but used in a different category of the U.S. soil taxonomy. **torrique**

Torrox A suborder in the U.S. system of soil taxonomy. *Oxisols* that have a torric soil moisture regime. **Torrox**

tortuosity A dimensionless geometric parameter of porous media, always greater than 1, used commonly in soil porosity and diffusion studies. The average ratio of the actual indirect path to the apparent, or straight flow path. In describing the non-linear nature of soil pores, it is the ratio of the average total length of pore passages to the length of the soil body (e.g., soil core). **tortuosité**

total digestible nutrients (TDN) A standard evaluation of the digestibility of a particular livestock feed, including all the digestible organic nutrients (e.g., protein, fiber, nitrogen-free extract, and fat). The percentage of total digestible nutrients represents the approximate heat or energy value of the feed. **unités nutritives totales, matières digestibles totales**

total dissolved solids (TDS) The total dissolved mineral constituents of water. Measured by filtering sediment from water, evaporating the water, and weighing the suspended solids remaining in residue. **matières totales dissoutes**

total field of view (*remote sensing*) The overall plane angle or linear ground distance covered by a multispectral scanner in the across-track direction, transverse to the direction of travel of the sensor platform. **cône d'analyse total**

total potential (of soil water) The amount of work that must be done per unit quantity of pure water in order to transport it reversibly and isothermally from a pool, at a specified elevation and at atmospheric pressure, to the soil water at the point under consideration. The total potential of soil water consists of the sum of the *matric*, *osmotic*, pressure, and *gravitational* potentials. See *soil water*. **potentiel hydrique total**

total pressure (of soil water) The pressure (positive or negative), in relation to the external gas pressure on the soil water, to which a pool of pure water must be subjected in order to be in equilibrium through a semipermeable membrane with the soil water. Total pressure is therefore equal to the sum of soil *water pressure* and *osmotic pressure*. Total pressure may also be derived from the measurement of the partial pressure of the water vapor in equilibrium with the soil water. It may be identified with the *total potential* when gravitational and external gas pressure potentials can be neglected. **pression totale de l'eau du sol**

total solids A measure of the amount of material that is the sum of the dissolved or suspended material (e.g., particulates) in a water sample. **matières solides totales**

total surface tillage See *broadcast tillage*. **travail sur toute la surface du sol**

total suspended solids A measure of the amount of particulate matter that is suspended in a water sample. **total des solides, matières totales en suspension**

toxic spoil (*land reclamation*) Acid spoil with pH below 4.0; also a spoil having amounts of minerals such as aluminum, manganese, and iron that adversely affect plant growth. See *acid spoil, acid mine drainage, spoil.* **déblais toxiques**

toxicity The capacity of a *toxin* to cause injury, impairment, or death. **toxicité**

toxin A poison produced as the result of the metabolic activities of a living organism and usually capable of inducing antibody formation. **toxine**

trace gas (*meteorology*) Any one of the less common gases found in the Earth's *atmosphere* (e.g., carbon dioxide, water vapor, methane, nitrogen oxides, ozone, and ammonia). Although relatively unimportant in terms of their absolute volume, they have significant effects on the Earth's weather and climate. **gaz à l'état de traces**

trace metals Metals present in low concentrations in air, water, soil, or food chains. **métaux à l'état de traces**

tracer A stable, easily detected substance or a *radioisotope* added to a material in order to follow it in an organism or in the environment, or to detect any physical or chemical changes it undergoes. Tracers are used to study atmospheric dispersion of pollutants, to follow the uptake of fertilizers by plants, and to study the metabolism and excretion of compounds introduced to the human body. See *radiotracer, tagged molecule*. **traceur**

tracheid Thick-walled cell of *xylem* tissue capable of water transport. **trachéide**

traffic pan See *pressure* or *induced pan.* **couche compactée ou indurée**

trafficability The capacity of a soil or a type of terrain to support vehicular traffic. **traficabilité**

training samples (*remote sensing*) The data samples of known identity used to determine decision boundaries in the measurement or feature space prior to classification of the overall set of data. **échantillons d'entraînement**

transduction The transfer of genes through virus vectors (bacteriophages) wherein virus DNA picks up host DNA and transfers it to another host organism. **transduction**

transect A transverse section across a region. It is used to show the spatial relationships existing between vegetation, soils, and relief. **transect, traverse**

transfer RNA (tRNA) A small RNA fragment that finds specific amino acids and attaches them to the protein chain as dictated by the codons in mRNA. **ARN de transfert (ARNt)**

transformation *(statistics)* The conversion of a set of data into a different mathematical form for the purpose of simplifying calculations, enabling the meeting of assumptions that underlie "standard" methods, or to change their distributions so that the data can meet the requirements for a statistical analysis. For example, log, ln, arcsine, square root, and other transformations are applied to data to normalize their distributions for parametric statistical analyses. **transformation, normalisation**

transgenic organism An organism containing genetic material from another organism, usually supplied by molecular biological techniques. **organisme transgénique**

transit time In nutrient cycles, the average time that a substance remains in a particular form; the ratio of biomass to productivity. **période de transition**

transition peatland *Peatland* with vegetation and physical and chemical properties of the *peat* intermediate between *ombrotrophic* and *minerotrophic* peatlands (*bogs* and *fens).* **tourbière de transition**

transitional soil A soil with properties intermediate between those of two different soils and genetically related to them. See *intergrade, soil.* **sol de transition**

translocation The movement of material in solution, suspension, or by organisms from one place to another. (*biology*) In plants, the transport of water,

T

minerals, or food; most often used to refer to food transport. (*genetics*) The interchange of chromosome segments between nonhomologous chromosomes. **translocation**

transmission The amount of radiation of different wavelengths that a filter, lens, or film will transmit. Also called spectral transmission. **transmission**

transmissivity (*hydrology*) The rate at which water can flow through a unit width of an *aquifer* under a unit *hydraulic gradient*. **transmissivité**

transmittance The ratio of the radiant energy transmitted through a body to that incident upon it. **transmittance**

transparency The portion of light that passes through water without distortion or absorption. A measure of the turbidity of water or other liquid. See *turbidity*. **transparence**

transpiration The direct transfer of water as a gas from plant leaves to the atmosphere, mainly through the stomata. Transpiration combined with evaporation from the soil is called *evapotranspiration*. **transpiration**

transpiration efficiency Grams of *net primary production* per 1000 grams of water transpired by plants. **efficacité de transpiration**

transplant A *seedling* that has been transplanted one or more times in the nursery. **plant repiqué**

transport The movement, shifting, or carrying away of sediment or of any loose, broken, or weathered material, either as solid particles or in solution, by a natural agent between a point of *erosion* and a site of *deposition*. The agents of transport may be running water (*saltation, solution, suspension*), wind, glacier ice, marine waves, tides, and currents. Transport is sometimes distinguished from movement induced simply by gravity (*landslide, mass movement*), although *creep, solifluction,* earth-flows, mud-flows, and slumping have been referred to as gravity transport. **transport**

trash farming See *tillage, trash farming*. **culture sur résidus**

treatment tank (*wastewater management*) A watertight tank designed to retain sewage long enough for satisfactory bacterial decomposition of the solids to take place; septic tanks and aerobic sewage treatment tanks. **cuve de traitement**

tree A woody perennial that reaches a mature height of at least 2.4 m (except genetic dwarfs) and has a well-defined stem and a definite crown shape. Sometimes there is no clear-cut distinction between a small tree and a large shrub. See *shrub*. **arbre**

tree farm A privately owned area dedicated by the owner to the production of wood crops. **propriété forestière de production, ferme forestière**

tree line The upper limit of tree growth, either of altitude or latitude; its position can be affected by local factors of soil, exposure, and aspect. **ligne des arbres**

tree ring analysis See *dendrochronology*. **analyse des anneaux de croissance**

trenching A method of drilling waste disposal whereby a back-hoe is used to construct deep, narrow trenches, confined to the drill lease area. Liquids or solids are squeezed out of the sump as stockpiled soils are slowly introduced into the sump to displace contained liquids or solids that flood into the trenches. Soil excavated from the next trench is cast on top of the material in the active trench, resulting in dilution and stabilization. **excavation en tranchées**

trend surface A mathematical expression fitted to spatially distributed data to produce a three-dimensional surface by means of a least-sum-of-squares regression. Trend-surface analysis is a method by which height or depth data at points on a surface can be plotted by constructing surfaces of

T

increasing complexity. **surface de tendance, surface polynomiale**

trend surface analysis See *trend surface.* **établissement d'une surface de tendance**

triangulation (*surveying*) A method of measuring a land area by extending a survey from a base line of known length and measuring the angles in a network of triangles which include the base line as one side. **triangulation**

triaxial shear test A method for measuring soil shear strength. A test in which a cylindrical specimen of soil, encased in an impervious membrane to provide a confining pressure, is subjected to an increasing axial load. **test de cisaillement triaxial**

triazine Any of a group of herbicides that inhibit *photosynthesis* and cause yellowing of leaves and eventual death of leaf tissue; uptake can occur through leaves or by the roots. **triazine**

tributary Secondary or branch of a stream, drain, or other channel that contributes flow to the primary or main channel. **affluent**

trickling filter (*wastewater* management) A filter used in the biologic or *secondary treatment* of wastewater, consisting of a bed of rocks or stones that encourages and supports bacterial growth. Sewage is trickled over the bed, enabling the bacteria to break down organic wastes. **lit bactérien**

trihalomethanes (THMs) A group of low-molecular-weight, halogenated hydrocarbons that include *chloroform*, bromoform, bromodichloromethane, and dibromochloromethane. Small amounts of THMs have been detected in raw water collected from surface sources used as a public water supply, and concentrations have been shown to be increased during the chlorination phase of the water purification process. Larger increases during chlorination have been recorded in water containing suspended particles and/or humic substances. See *halogen.* **trihalométhanes**

trioctahedral An octahedral sheet or a mineral containing such a sheet that has all of the sites filled, usually by divalent ions such as magnesium or ferrous iron. See *phyllosilicate mineral terminology* and *dioctahedral.* **trioctaédrique**

triple bond A bond in which three pairs of electrons are shared by two atoms. **liaison triple**

triploid Referring to a cell or organism containing three sets of chromosomes. See *ploidy*. **triploïde**

tripton The dead suspended particulate matter in aquatic habitats; the nonliving portion of the *seston.* **tripton**

Tropepts A suborder in the U.S. system of soil taxonomy. *Inceptisols* that have a mean annual soil temperature of 8°C or more, and less than 5°C difference between mean summer and mean winter temperatures at a depth of 50 cm below the surface. Tropepts may have an *ochric epipedon* and a *cambic horizon*, or an *umbric epipedon*, or a *mollic epipedon* under certain conditions, but no *plaggen epipedon*, and are not saturated with water for periods long enough to limit their use for most crops. **Tropepts**

trophic Pertaining to food or nutrition; relating to the processes of energy and nutrient transfer from one or more organisms to others in an *ecosystem.* **trophique**

trophic level A feeding level within a *food web.* The primary producers form the first trophic level; primary consumers, the second; secondary consumers, the third, etc. **niveau trophique**

trophic status Nutrient status; availability of nutrients to plants. See *oligotrophic*, *mesotrophic*, and *eutrophic.* **condition trophique**

trophic structure Organization of the community based on feeding relationships among its species. **structure trophique**

trophogenic region The superficial layer of a lake in which organic production from mineral substances takes place on the basis of light energy and photosynthetic activity. **zone trophogène**

tropical rainforest Forest that resembles the equatorial rainforest in its vegetation structure and soil characteristics, but because of its location (between 7 and 23.5° latitude on windward coasts) has marked seasonality; a long wet season alternates with a season of lower rainfall (but not drought) in which temperatures also are lower, resulting in a season of much slower tree growth. Tropical rainforest is typical of the coastlands of Burma, Western India, Vietnam, the Philippines, Eastern Brazil, Central America, and the West Indies. **forêt tropicale humide**

tropism A response to an external stimulus in which the direction of the movement is usually determined by the direction from which the most intense stimulus comes. See *geotropism, heliotropism*. **tropisme**

tropophyte A plant that adopts alternating growth patterns with seasonally changing weather; it acts as a *xerophyte* during the dry season but becomes a hygrophyte during the rainy season. **tropophyte**

troposphere See *atmosphere*. **troposphère**

tropospheric ozone Ozone (O_3) located in the troposphere; it plays a significant role in the greenhouse gas effect and urban smog. See *ozone*. **ozone troposphérique**

tropospheric ozone precursor Gases that influence the rate at which ozone is created and destroyed in the atmosphere (e.g., carbon monoxide, nitrogen oxides, and nonmethane volatile organic compounds). **précurseur d'ozone troposphérique**

true north The direction of geographic north from the observer (i.e., the direction along a meridian towards the North Pole) in contrast to the grid north and the magnetic north. **nord géographique**

truncated soil Having lost all or part of the upper soil horizon or horizons. **tronqué**

trunk valley The main valley occupied by the dominant stream of a river system. **vallée principale**

t-test A parametric statistical test used to establish the significance of the difference in the means of two samples of data measured on an interval scale. It can be used to compare differences in both independent samples and paired samples, and to decide whether to reject or retain the *null hypothesis*. **test t**

tuber An underground, food-storing stem (e.g., potato). Roots and shoots grow from growth buds, or eyes, scattered over the surface. **tubercule**

tuberous root A swollen root that serves as a nutrient storage organ (e.g., sweet potato); produces fibrous roots to take in moisture and nutrients. New growth buds appear on the base of the old stem where the stem joins the tuberous root. **racine tuberculeuse, racine tubéreuse**

tufa A sedimentary deposit consisting mainly of calcium carbonate ($CaCO_3$) formed around a spring of calcareous groundwater. It is found mainly in limestone regions where it fills in cavities, builds *stalactites* and *stalagmites,* and cements superficial gravel to produce calcrete. In the vicinity of hot springs, a type of tufa is known as travertine. Tufa is deposited when water saturated with $CaCO_3$ and CO_2 is subjected to an increase in temperature or a decrease in pressure. Loss of water by evaporation will also cause it to be deposited. **calcaire tuffacé**

tuff Volcanic ash, usually more or less stratified and in various states of consolidation. **tuf**

tundra Southern or altitudinal extent of treeless areas in northern North America and northern Eurasia, lying principally along the Arctic Circle and on the northern side of the coniferous forests; the mean monthly temperature is below the freezing point for most of the year; winters are long and severe, summers are short and warm, and the subsoil may be permanently frozen. **toundra**

tundra soil Any soil in the *tundra* region. **sol de toundra**

Turbels A suborder in the U.S. system of soil taxonomy. Turbels are soils that show marked influence of *cryoturbation* (more than one-third of the active layer portion of the pedon) such as irregular, broken, or distorted horizon boundaries and involutions, and areas with patterned ground. They commonly contain tongues of mineral and organic horizons, organic and mineral intrusions, and oriented rock fragments. Organic matter is accumulated on top of the permafrost, and ice wedges are a common feature in Turbels. These soils occur primarily in the zone of continuous permafrost. **Turbels**

Turbic Cryosol A great group of soils in the Cryosolic order (Canadian system of soil classification). They are mineral soils that have permafrost within 2 m of the surface and show marked evidence of cryoturbation laterally within the active layer, as indicated by disrupted or mixed or broken horizons, or displaced material, or a combination of both. **cryosol turbique**

turbidity (*limnology*) Cloudiness caused by suspended matter (e.g., sediment) in water or some other fluid as determined by the relative light transmission of the suspension. Measured by light transmittance (0 to 100 %) or absorbance using a turbidimeter. (*atmosphere*) Hazy condition of the atmosphere that reduces its transparency to radiation due to the presence of particulates or pollutants. **turbidité**

turbulence, turbulent flow (*meteorology*) The net forward movement of air in an irregular, eddying flow. It is an important atmospheric mechanism for mixing and dispersal, near the soil surface. **turbulence, écoulement turbulent**

turbulent velocity That velocity above which *turbulent flow* will always exist and below which the flow may either be turbulent or *laminar*. **vitesse limite de régime turbulent**

turf (1) Layer of surface soil matted by grass and plant roots; *sod*. (2) A block of peat used as a fuel. **gazon**

turgid Fully expanded. **turgescent**

turnover (*ecology*) The replacement of old biomass by new biomass in an ecosystem; the rate of productivity, divided by standing crop or biomass, expressed as T=P/B, where P is productivity (in units of mass per area-time) and B is biomass (in units of mass per area). Turnover, T, has units of 1/time. **renouvellement**

turnover time (*ecology*) The average time (t) required for the biomass (B) in an ecosystem to replace itself; the inverse of *turnover*, expressed as, t=B/P, where P is the productivity (in units of mass per area-time) and B has units of mass per area. (*microbiology*) The time required to metabolize a specific substance in soil or a body of water. **temps de renouvellement**

turnrow (turn strip, head land) See *tillage, turnrow*. **tournière, cintre**

tussock A dense clump or large tuft of sedge or grass. **tusco, butte**

two way analysis of variance (*statistics*) An *analysis of variance* in which the total sum of squares is expressed as the sum of the treatment sum of squares, the block sum of squares, the error sum of squares, and no other. **analyse de variance à deux critères de classification**

type The general form, structure, or character that distinguishes a group or class of objects or organisms. Among the terms commonly employed are: holotype, a single specimen chosen in order to illustrate the main character of a species; paratype, additional specimens chosen along with the holotype to show other species' characteristics not exhibited by the holotype; syntypes, a series of type specimens of equal status, chosen to illustrate the range of variation within a species. **type**

Type I error *(statistics)* When judging the results of a scientific study, the rejection of the null hypothesis when in fact the null hypothesis is true. See *Type II error*. **erreur de type I**

Type II error *(statistics)* When judging the results of a scientific study, the acceptance of the null hypothesis when in fact the null hypothesis is false. See *Type I error*. **erreur de type II**

type, soil (1) A unit in the natural system of *soil classification*; a subdivision of a soil series consisting of, or describing, soils that are alike in all characteristics including the texture of the *A horizon*. (2) In Europe, the term is roughly equivalent to a soil great group. **type de sol**

U

ubiquitous organisms Organisms that can tolerate a wide range of environmental conditions or variation, or are so active or numerous as to exist in all types of environments. **organisme ubiquiste**

Udalfs A suborder in the U.S. system of soil taxonomy. *Alfisols* that have a *udic* soil moisture regime and mesic or warmer soil temperature regimes. Udalfs generally have brownish colors throughout, and are not saturated with water for periods long enough to limit their use for most crops. **Udalfs**

Udands A suborder in the U.S. system of soil taxonomy. *Andisols* that have a *udic* soil moisture regime. **Udands**

Uderts A suborder in the U.S. system of soil taxonomy. *Vertisols* of relatively humid regions having wide, deep cracks that usually remain open continuously for <60 days or intermittently for periods that total <90 days. **Uderts**

udic A soil moisture regime (used in U.S. system of soil taxonomy) that is neither dry for as long as 90 cumulative days nor for as long as 60 consecutive days in the 90 days following the summer solstice at periods when the soil temperature at 50 cm below the surface is above 5°C. **udique**

Udolls A suborder in the U.S. system of soil taxonomy. *Mollisols* that have a *udic* soil moisture regime with mean annual soil temperatures of 8°C or more. Udolls have no calcic or gypsic horizon, and are not saturated with water for periods long enough to limit their use for most crops. **Udolls**

Udults A suborder in the U.S. system of soil taxonomy. *Ultisols* that have low or moderate amounts of organic carbon, reddish or yellowish argillic horizons, and a udic soil moisture regime. Udults are not saturated with water for periods long enough to limit their use for most crops. **Udults**

ultimate factors Aspects of the environment that are directly important to the well-being of an organism (e.g., food). **facteurs essentiels ou ultimes**

Ultisols An order in the U. S. system of soil taxonomy. Mineral soils that have an *argillic* horizon with a base saturation of <35% when measured at pH 8.2. Ultisols have a mean annual soil temperature of 8°C or higher. **Ultisols**

ultrabasic rock Igneous rock that consists primarily of ferromagnesian minerals and has a low percentage of silica and feldspar; mainly *plutonic rock* with relatively few fine-grained volcanic varieties. Also called ultramafic rock. **roche ultrabasique**

ultraviolet radiation (UV radiation) Electromagnetic radiation having wavelengths between that of violet light and long x-rays (i.e., 400 to 4 nm). Ultraviolet radiation is classified into three ranges according to its effect on human skin: UV-A (320 to 400 nm) is not harmful in normal doses and is used to treat skin problems; UV-B (280 to 320 nm) is responsible for a wide range of potentially damaging human and animal health effects, primarily related to the skin, eyes, and immune system; UV-C (230 to 280 nm) is

dangerous to plants and animals, but this part of the UV spectrum is completely absorbed by stratospheric ozone and does not reach the Earth's surface. **rayonnement ultraviolet (rayonnement UV)**

Umbrepts A suborder in the U.S. system of soil taxonomy. *Inceptisols* formed in cold or temperate climates that commonly have an *umbric epipedon*, but they may have a *mollic* or an *anthropic* epipedon 25 cm or more thick under certain conditions. These soils are not dominated by amorphous materials and are not saturated with water for periods long enough to limit their use for most crops. **Umbrepts**

umbric epipedon A surface layer of mineral soil (U.S. system of soil taxonomy) that has the same requirements as the *mollic epipedon* with respect to color, thickness, organic carbon content, consistence, structure, and phosphorus content, but that has a base saturation <50% when measured at pH 7. **épipédon umbrique**

unavailable nutrient A plant nutrient that is present in the soil but cannot be taken up by the roots because it has not been released from the rock by weathering or from organic matter by decomposition. **élément nutritif non disponible**

unavailable water Water that is present in the soil but cannot be taken up by plant roots because it is strongly adsorbed (see *adsorption*) onto the surface of soil particles. **eau non disponible**

unconformity (*geology*) A hiatus or break representing an unspecified period of time in a geological sequence, marked by a major break in sedimentation or by a structural planar surface separating younger rocks above from older rocks below. The surface of an unconformity may be a surface of denudation or of nondeposition. Different types of unconformity are angular unconformity, in which upper and lower beds exhibit different *dips* and *strikes* from each other; disconformity (also called parallel unconformity), where the upper and lower beds dip in the same direction and by the same amount; a non-depositional unconformity, in which a break in a sedimentary sequence is indistinguishable except for changes in fossil characteristics; and a nonconformity (also called heterolithic unconformity), in which sediments overlie a denuded surface of igneous and/or metamorphic rocks. **discordance**

unconsolidated sediment Sediment that is loosely arranged or unstratified, or in which particles are not cemented together, occurring either at the surface or at depth. **non consolidé**

undercutting Removal of material at the base of a steep slope, overfall, or cliff by falling water, a stream, wind erosion, or wave action, resulting in a steepened slope or an overhanging cliff. **sapement**

underflow (*hydrology*) The downstream movement of groundwater through permeable rock beneath a riverbed. **inféroflux, sous-écoulement**

undergrazing (*range management*) An intensity of grazing so light that the forage is not utilized to best advantage. **sous-pâturage**

underground development waste Waste rock mixtures of coal, shale, claystone, siltstone, sandstone, limestone, or related materials that are excavated, moved, and disposed of during development and preparation of areas incident to underground mining activities. **déchets miniers souterrains**

underground mining *See mining*. **exploitation souterraine**

underground runoff Water flowing toward stream channels after infiltration into the ground. See *seepage*. **écoulement souterrain**

undergrowth Seedlings, shoots, and small saplings under an existing stand of trees. **sous-bois**

underseeding Sowing a secondary crop with the main crop to provide soil cover after the primary crop is harvested. **semis sous couvert végétal, sous-ensemensement**

understory Any plants growing under the canopy formed by other plants, particularly herbaceous and shrub vegetation under a tree canopy. Also, used in reference only to trees and other woody species growing under a more or less continuous cover of branches and foliage formed collectively by the upper portions of adjacent trees and other woody growth. **sous-étage**

undifferentiated group (*U.S. soil survey*) A kind of map unit used in soil surveys comprising two or more taxa components that are not consistently associated geographically. Similar to *soil complex*. **groupe de sols non différenciés, unité indifférenciée**

undifferentiated organic material Organic (peat) material of unknown origin. (Undifferentiated organic material is a genetic material in the Canadian system of soil classification). **matériau organique non différencié**

undifferentiated soil map unit (*Canadian soil survey*) A soil mapping unit in which two or more soil units occur, but not in a regular geographic association. For example, the steep phases of two or more soils might be shown as a unit on a map because slope is the prime characteristic. See *soil association* and *complex, soil*. **unité cartographique de sols non differenciés**

undisturbed ecosystem An ecosystem that has not been influenced by human activity. **écosystème non perturbé ou naturel**

undulating (*geomorphology*) A landform (or *surface form*) with a regular sequence of gentle slopes that merge from rounded concavities into broad, rounded convexities producing a wavelike pattern of low local relief. Slopes are generally less than 0.8 km long and have gradients of 2 to 5%. **ondulé**

uneven-aged stand A stand of trees containing three or more age classes. In a balanced uneven-aged stand, each age class is represented by approximately equal areas, providing a balanced distribution of diameter classes. **peuplement inéquienne**

unhulled seed Any seed normally covered by a hull (e.g., bracts or other coating) and from which the hull has not been removed. **semence non-mondée, semence non-décortiquée**

unicellular One-celled; refers to an organism, the entire body of which consists of a single cell. **unicellulaire**

unidimensional shrinkage Shrinkage that occurs exclusively in the vertical direction or one dimension. **retrait unidirectionel**

Unified Soil Classification System (*engineering*) A classification system based on the identification of soils according to their particle size, gradation, plasticity index, and liquid limit. **système de classification unifié des sols, classification U.S.C.S.**

uniform flow (*hydrology*) Steady flow when the mean velocity and cross-sectional area are equal at all sections of a *reach*. **écoulement uniforme**

unimodal distribution A statistical expression denoting that the frequency curve representing the data *distribution* has a single maximum. See *bimodal distribution*. **distribution unimodale**

unit cell The smallest repeating unit of a *lattice*. **unité de base**

unit hydrograph The *hydrograph* of 2.5 cm of direct (i.e., surface plus *interflow*) *runoff* resulting from rainfall uniformly distributed over the *drainage basin* and of uniform intensity

throughout a specified period of time. **hydrogramme unitaire**

unitary shrinkage Shrinkage equivalent to the change in water volume. **retrait unitaire**

Universal Polar Stereographic coordinates (UPS coordinates) A metric system used for map projections polewards of the 80°N and 80°S lines of latitude (i.e., polewards of *UTM* grid). The term refers to the Grid System, which is superimposed upon a polar stereographic map projection within the circle formed by the 80° lines of latitude. Its horizontal grid lines are parallel with meridians of 90° E and 90° W longitude, while its vertical gridlines are parallel 500 km squares. The pole is the origin of the grid, situated at the center of the projection, but the false origin is transferred 2000 km W and 2000 km S in the northern hemisphere and 2000 km W and 2000 km S in the southern hemisphere. **coordonnées du système de quadrillage stéréographique polaire universel, coordonnées UPS**

universal soil loss equation (USLE) An equation used to predict A, the average annual soil loss per unit area per year (e.g., Mg ha^{-1} $year^{-1}$), and defined as A = RKLSPC, where R is the rainfall factor, K is the soil erodibility factor, L is the length of slope, S is the percent slope, P is the conservation practice factor, and C is the cropping and management factor. See *erosion*. **équation universelle de perte de sol**

Universal Transverse Mercator coordinates (UTM coordinates) A map coordinate system covering the world from 80° north to 80° south with 60 north-south zones, each covering 6° of longitude and divided into 8° latitude sections. The zones overlap 0.5° on each side. Each zone has an individual origin, and the coordinates are read in meters east and meters north of the origin. A square of land enclosed by incremental UTM coordinates (between 5355 and 5356 North and between 243 and 244 East) is 1 km on a side and contains an area of 100 hectares. **coordonnées de la projection transversale universelle de Mercator (coordonnées UTM)**

unrubbed fiber See *fiber*. **fibre non frottée**

unsaturated flow The movement of water in a soil that is not filled to capacity with water. The moving force is a negative pressure potential gradient. See *soil water*. **écoulement non-saturé ou en milieu non-saturé**

unsaturated hydraulic conductivity The *hydraulic conductivity* associated with *unsaturated flow* that occurs under a negative pressure potential. **conductivité hydraulique non-saturée ou en milieu non-saturé**

unsaturated zone The zone above the water table in an aquifer; the vadose zone. **zone non-saturée**

unstable (*chemistry*) Descriptive elements or compounds that react easily or spontaneously to form other elements or compounds (e.g., ozone [O_3] is an extremely unstable gas because it reacts readily with many materials). **instable**

unstable equilibrium A condition in a *system* when a small displacement leads to an even greater displacement; usually terminated by the achievement of a new stage of *stable equilibrium*. **équilibre instable**

unstratified drift The unsorted *moraine* and *till* deposits laid down by ice. **diamicton**

upland (1) Extensive region of high ground lying above a coastal plain or floodplain. (2) The high ground of a region, as distinguished from valley and plains. Also, used to distinguish locally elevated, well drained lands from adjacent, low lying, often poorly drained lands. See *lowland*. **haute-terre**

upper plastic limit See *liquid limit.* **limite supérieure de plasticité**

upper subsoil The soil material found immediately below the topsoil. **sous-sol supérieur**

up-scaling The use of information gathered at a small scale to derive information at larger scales. A challenging procedure, because new interactions and processes can emerge in the latter that were not present at the smaller scale. **mise à l'échelle**

urban forestry The cultivation and management of trees and forests for their present and potential contributions to the physiological, sociological, and economic well-being of urban society. **foresterie urbaine**

urban land Areas so altered or obstructed by urban works or structures that identification of soils is not feasible. A *miscellaneous land type.* **terrain urbain, zone urbaine**

urban runoff Storm water from city streets and gutters. **écoulement urbain**

urban waste The entire waste stream from the urban area; sometimes used in contrast to rural waste. See *waste.* **déchets urbains**

Ustalfs A suborder in the U.S. system of soil taxonomy. *Alfisols* that have an *ustic* soil moisture regime and *mesic* or warmer soil temperature regimes. Ustalfs are brownish or reddish throughout, and are not saturated with water for periods long enough to limit their use for most crops. **Ustalfs**

Ustands A suborder in the U.S. system of soil taxonomy. *Andisols* that have an *ustic* soil moisture regime. **Ustands**

Usterts A suborder in the U.S. system of soil taxonomy. *Vertisols* of temperate or tropical regions that have wide, deep cracks that usually remain open for periods that total > 90 days but do not remain open continuously throughout the year, and have either a mean annual soil temperature of 22°C or more, or a mean summer and mean winter soil temperature at 50 cm below the surface that differ by < 5°C, or have cracks that open and close more than once during the year. **Usterts**

ustic A soil moisture regime (used in U.S. system of soil taxonomy) that is intermediate between *aridic* and *udic* and is common in temperate subhumid or semiarid regions, or in tropical and subtropical regions with a monsoon climate. A limited amount of water is available for plants, however it occurs at times when the soil temperature is optimum for plant growth. **ustique**

Ustolls A suborder in the U.S. system of soil taxonomy. *Mollisols* that have an *ustic* soil moisture regime and *mesic* or warmer soil temperature regimes. Ustolls may have a *calcic, petrocalcic,* or *gypsic* horizon, and are not saturated with water for periods long enough to limit their use for most crops. **Ustolls**

Ustox A suborder in the U.S. system of soil taxonomy. *Oxisols* that have an *ustic* moisture regime and either hyperthermic or isohyperthermic soil temperature regimes or have <20 kg organic carbon in the surface cubic meter. **Ustox**

Ustults A suborder in the U.S. system of soil taxonomy. *Ultisols* that have low or moderate amounts of organic carbon, are brownish or reddish throughout, and have an *ustic* soil moisture regime. **Ustults**

V

vacuole A space or cavity in the cell, usually filled with some substance. **vacuole**

vadose water Water of the *vadose zone.* **eau vadose**

vadose zone The aerated region of soil above the permanent water table. **zone vadose**

valence That property of an element that is measured in terms of the number of gram atoms of hydrogen that one gram atom of that element will combine with or displace (e.g., the valence of oxygen in water, H_2O, is 2). **valence**

valley Any linear, depressional, or low-lying land area bounded by higher ground, commonly traversed by a stream or river that receives the drainage of the surrounding heights. **vallée**

valley fill (*geomorphology*) The unconsolidated rock waste derived from the denudation of surrounding uplands and deposited in a valley. See *colluvium.* (*mining*) The placement of overburden material from adjacent contour or mountaintop mines in compacted layers in narrow, steep-sided valleys so that surface drainage is possible. **remblayage de vallée**

value, color One of the three variables of color. It expresses the relative lightness of color, which is approximately a function of the square root of the total amount of light. See *Munsell color system.* **luminosité, couleur**

van der Waals' forces The weak attraction exerted by all atoms on one another, resulting from the mutual interaction of the electrons and nuclei of the atoms. See *intermolecular forces.* **forces de van der Waals**

van't Hoff factor The ratio of moles of particles in solution to moles of solute dissolved. **facteur de van't Hoff**

vapor The gaseous phase of substances that are normally either liquids or solids at atmospheric temperature and pressure (e.g., steam and phenolic compounds). **vapeur**

vapor density The ratio of the density of a pure gas or vapor to the density of hydrogen or the density of air. **densité de vapeur**

vapor flow The gaseous flow of water vapor in soils from a moist or warm zone of higher potential to a drier or colder zone of lower potential. **écoulement de vapeur**

vapor pressure The pressure exerted by a gas or vapor. These pressures are experimentally determined by establishing an equilibrium between the gas and liquid phases of the substance in a closed vessel at a specific temperature. The higher the vapor pressure, the greater the tendency of the liquid to evaporate. **pression de vapeur**

vaporize To change into a vapor; to evaporate. **vaporiser**

variable A measurable item or number capable of taking different values. The types of variables include a *dependent variable,* whose value is determined or constrained by the values assumed by other variables; an *independent variable,* whose value can be freely chosen within a permitted range, and whose variation constrains or determines the value assumed by the dependent variable; an endogenous variable, whose value is to be determined by forces operating within the model under consideration; and an exogenous variable, whose value is determined by forces outside the model and is unexplained by the model. **variable**

variance A number that measures the *dispersion* within a data population or sample, expressed as the arithmetic mean of the squares of the deviation from the distribution mean. The greater the variance, the bigger the dispersion. **variance**

variant, soil A soil whose properties are believed to be sufficiently different from other known soils to justify a new series name, but comprising such a limited geographic area that creation of a new series is not justified. **variante de sol**

variate One measure or estimate among a set of values for a single *variable*. **variable**

varied flow (*hydrology*) Non-uniform flow in which depth of flow changes along the length of a *reach*. **écoulement varié**

variegated foliage Leaves that are edged, splotched, spotted, or patterned with color other than the background color. **feuillage panaché**

variogram (*statistics*) A graphical representation of the correlation between spatial observations. The spatial distribution of variables such as soil properties can be determined by measuring those properties at different locations and then determining the spatial patterns through extrapolation by variogram analysis. In investigating spatial data, measurements at points close together are usually more similar than those further apart. A variogram summarizes the variance of the difference in pairs of measurements and the distance of the corresponding points from each other. It indicates how different, on average, the measurements are at various distances apart. Quantifying this spatial variability is useful in understanding its form or in planning a sample survey of an area. **variogramme**

varve A distinct band representing the annual deposit of sedimentary materials, regardless of origin. It usually consists of two layers—a thick, light-colored layer of silt and fine sand laid down in the spring and summer, and a thin, dark-colored layer of clay laid down in the fall and winter. The material may be of any origin, but the term is most often used in connection with glacial lake sediments, because low temperatures are important in delaying the settling of the clay particles. The salts of seawater prevent the formation of varves of this kind. The electrolytes in seawater cause flocculation, resulting in a homogeneous mass. **varve**

vascular Pertaining to any plant tissue or region consisting of or giving rise to conducting tissue (e.g., *xylem*, *phloem*, vascular cambium). **vasculaire**

vascular bundle Tissue in a plant responsible for transport of materials from the roots up and from the leaves down. **faisceau vasculaire ou libéro-ligneux**

vector A physical quantity with both magnitude and direction. A vector is often depicted as an arrow line drawn stage by stage from an initial point to a final point (e.g., to mark a tidal current's velocity and flow direction over time). Such a technique of monitoring change of magnitude and direction is called vector analysis. **vecteur**

vegetation Plants in general; the plant life or plant cover in an area. **végétation**

vegetation type A plant community with distinguishable characteristics. **type de végétation**

vegetative (*botany*) (1) Non-reproductive plant parts (e.g., leaf and stem), in contrast to reproductive plant parts (e.g., flower and seed) in the developmental stages of plant growth. (2) The non-reproductive stage in plant development. (3) Of, relating to, or involving propagation by asexual processes (e.g., budding, cutting, division, or grafting). Also referring to non-reproductive plant parts. **(1, 2) végétatif (3) multiplication végétative**

vegetative cell The growing form of a microbial cell, as opposed to a resistant resting form. **cellule végétative**

vegetative material Plant parts or tissues used to produce vegetative propagules through asexual means. **matériel végétatif**

vegetative state (*botany*) A stage in plant development prior to the appearance of fruiting structures. **état ou stade végétatif**

vein (1) (*geology*) A mineral-bearing lode. Also, a mineralized zone having a more or less regular development in length, width, and depth, giving it a tabular form. (2) (*botany*) A vascular bundle forming part of the framework of the conducting and supporting tissue of a leaf or other expanded organ. **(1) filon (2) nervure**

velocity head (*hydrology*) *Head* due to the velocity of a moving fluid. **hauteur-vitesse**

venation The arrangement of veins in leaves, petals, and some fruit. **nervation**

veneer (*geology, geomorphology*) A mantle of unconsolidated materials too thin to mask the minor irregularities of the underlying unit surface. A veneer is generally less than 1 m in thickness. **placage**

ventifact (*geology*) Any loose stone or pebble that has been shaped, worn, faceted, cut, or polished by the abrasive or sandblast action of wind-blown sand, generally under desert or polar conditions. **ventifact, caillou éolisé, pierre éolisée**

vermiculite A highly charged (averaging about 159 $cmol_c\,kg^{-1}$ for soil vermiculites but very widely ranging) layer silicate of the 2:1 type that is formed from mica. It is characterized by adsorption preference for potassium, ammonium, and cesium over smaller exchange cations. It may be di- or tri-octahedral. **vermiculite**

vernalization (*botany*) Treatment of germinating seeds or seedlings with low (or high) temperatures to induce flowering at maturity. **vernalisation**

vertebrate (*zoology*) An animal that has an internal skeletal system. See *invertebrate*. **vertébré**

vertic horizon A *diagnostic horizon* (Canadian system of soil classification) affected by *argillipedoturbation*, as manifested by disruption and mixing caused by shrinking and swelling of the soil mass. It is characterized by irregularly shaped, randomly oriented intrusions of displaced materials within the solum, and by vertical cracks, often containing sloughed-in surface materials. **horizon vertique**

Vertic Solonetz A great group of soils in the Solonetzic order (Canadian system of soil classification). They are mineral soils that have horizons that are characteristic of any of the other Solonetzic great groups, but additionally have properties that indicate an intergrading to the Vertisolic order. They have a slickenside horizon, the upper boundary of which occurs within 1 m of the surface. They may also have a weak vertic horizon. **solonetz vertique**

vertical mulching See *tillage, vertical mulching*. **placement vertical d'amendements**

vertical photograph See *aerial photograph*. **photographie verticale, cliché vertical**

vertical shrinkage The shrinkage-induced change in length of a soil in the vertical direction. Also called surface subsidence if it occurs exclusively at the soil surface. **retrait vertical**

vertical zonation Zonation of soils arising from climate and vegetation that changes with altitude of the land. See *zonation, latitude zonation*. **zonalité verticale**

Vertisol A great group of soils in the *Vertisolic* order (Canadian system of soil classification). The soils lack a mineral-organic (Ah) horizon more than

10 cm in thickness, or if cultivated, have an Ap *color* value of ≥3.5 (dry) and a chroma usually >1.5 (dry). The *A horizon* is not clearly distinguishable from the rest of the solum. **vertisol**

Vertisols An order in the U.S. system of soil taxonomy. Mineral soils that have 30% or more clay, deep wide cracks when dry, and either *gilgai* microrelief, intersecting *slickensides*, or wedge-shaped structural aggregates tilted at an angle from the horizon. **vertisols**

Vertisolic An order of soils in the Canadian system of soil classification that occur in heavy-textured materials (>60% clay, of which at least half is smectite) and have a shrink-swell character. They lack the degree of horizon development diagnostic of soils of the other soil orders, and the surface (Ah) horizon, when dry, has massive structure and is hard. The presence of *slickensides*, *gilgai*, and severe disruption within the control section reflects the dominant soil forming processes of cracking, mass movement, and pedoturbation. It consists of the *Vertisol* and *Humic Vertisol* great groups. **vertisolique**

very coarse sand See *soil separates*, *texture*. **sable très grossier**

very fine sand See *soil separates*, *texture*. **sable très fin**

very fine sandy loam See *soil separates*, *texture*. **loam sableux très fin**

vesicle A globular structure, formed intracellularly, by *vesicular arbuscular* mycorrhizal fungi. **vésicule**

vesicular arbuscular A common *endomycorrhizal* association produced by phycomycetous fungi of the family *Endogonaceae*. The host range includes most agricultural and horticultural crops. **vésiculo-arbusculaire**

virgin forest A forest essentially unaffected by human activity (usually referring to mature or over-mature stages). **forêt vierge**

virgin soil Soil that has never been cultivated. **sol vierge**

virus A minute organism, resembling certain molecules of protein, found in the cells of other organisms and frequently causing disease. **virus**

viscous Pertaining to a substance that is sticky and adhesive, and flows slowly (e.g., asphalt, wax, and certain types of lava). **visqueux**

visible wavelengths The radiation range in which the human eye is sensitive, approximately 0.4 to 0.7 micrometers. **longueurs d'ondes visibles**

Vitrands A suborder in the U.S. system of soil taxonomy. *Andisols* that have 1500 kPa water retention of <15% on air dry, <30% on undried samples throughout 60% of the thickness either: a) within 60 cm of the soil surface or top of an organic layer with andic properties, whichever is shallower if there is no *lithic*, *paralithic contact*, *duripan*, or *petrocalcic horizon* within that depth; or b) between the mineral soil surface or top of an organic layer with *andic* properties, whichever is shallower and a *lithic layer,* paralithic contact, duripan, or petrocalcic horizon. **Vitrands**

void Space in soil mass that is not occupied by solid mineral matter. This space may be occupied by gaseous or liquid material. **vide**

void ratio The ratio of the volume of void space to the volume of solid particles in a given soil mass. **indice des vides**

volatile hydrocarbons Organic compounds composed of carbon and hydrogen that evaporate rapidly at room temperatures (e.g., gasoline, methanol, and benzene). **hydrocarbures volatils**

volatile organic compounds (VOCs) A category of volatile organic compounds with relatively high vapor pressures, a major category of air contaminants. Most VOCs are carbon-hydrogen compounds (hydrocarbons), but they also may be aldehydes, ketones, chlorinated hydrocarbons, and others. Thousands of individual compounds exist, including the unburnt hydrocarbon compounds emitted from automobiles or industrial processes and the organic solvents lost to evaporation from household, commercial, or industrial cleaning and painting operations, and other activities. Some VOCs participate in the atmospheric reactions that lead to photochemical air pollution, and excessive exposure to certain individual compounds is associated with skin irritation, central nervous system depression, and/or an increased risk of cancer. **composés organiques volatils**

volcanic ash Fine pyroclastic matter (<2 mm diameter) originating from a volcanic eruption. The ash particles are sometimes ejected high into the atmosphere and carried long distances by the wind. Ash deposits are unconsolidated, with particles that are small, angular, and easily weathered. Soils formed from volcanic ash generally have high clay contents. **cendre volcanique**

volume wetness A measure of *soil wetness*. The volume fraction of soil water (or *volumetric water content*). The soil water content computed as a percentage of the total volume of the soil. **humidité volumétrique**

volumetric water content The soil-water content expressed as the volume of water per unit bulk volume of soil. **teneur en eau volumétrique**

von Post humification scale A scale describing peat in varying stages of decomposition, ranging from H_1, which is unconverted, to H_{10}, which is completely decomposed.

The stages are:

H_1 – completely unhumified peat; upon pressing in the hand, gives off only colorless clear water.

H_2 – almost completely unhumified peat; upon pressing, gives off almost clear but yellow-brown water.

H_3 – very little humified peat; upon pressing, gives off distinctly turbid water but the residue is not mushy.

H_4 – poorly humified peat; upon pressing, gives off strongly turbid water. The residue is somewhat mushy.

H_5 – partially humified peat; the plant remains are recognizable but not distinct. Upon pressing, some of the substance passes between the fingers, together with mucky water. The residue in the hand is strongly mushy.

H_6 – partially humified peat; the plant remains are not distinct. Upon pressing, one-third (at the most) of the peat passes between the fingers. The residue is strongly mushy, but the plant remains stand out better than in the unpressed peat.

H_7 – well humified peat; upon pressing, about half of the peat passes between the fingers. If water separates, it is soupy and very dark in color.

H_8 – well humified peat; the plant remains are not recognizable. Upon pressing, about two-thirds of the peat passes between the fingers. If water separates at all, it is soupy. The remains consist mainly of more resistant root fibers, etc.

H_9 – very well humified peat; hardly any plant remains are apparent. Upon pressing, nearly all of the peat passes between the fingers like a homogeneous mush.

H_{10} – completely humified peat; no plant remains are apparent. Upon pressing, all of the peat passes between the fingers. **échelle d'humification de von Post**

vugh (*soil micromorphology*) A relatively large *void*, usually irregular and not normally interconnected with other voids of comparable size. Vughs appear as discrete entities at the magnifications at which they are recognized. **cavité**

W

W layer Zone of saturation in a waterlogged soil (Canadian system of soil classification). **couche W**

warmwater fish Fish that normally survive, grow, and reproduce in warm water, 25 to 32°C. **poisson d'eaux chaudes**

wash The process of downslope movement of fine material by water; coarse alluvial material. **ruissellement**

washboard moraine (*geomorphology*) A type of *moraine* in which the relatively low ridges are closely spaced and parallel to each other (i.e., resembling a washboard). **moraine bosselée**

washoff Materials transported from a land or soil surface by overland flow; often used to describe soil materials transported off runoff test plots. **ruissellement**

washout Destruction of land or structures by a torrential stream or river. It can refer to bank erosion, levee breaching, dam bursting, and road or bridge destruction. **ravinement**

waste Material that has no original value or no value for human use. **déchets, résidus**

waste processing An operation in which the physical or chemical properties of wastes are changed (e.g., shredding, compaction, composting, and incineration). **traitement des déchets**

waste sources Agricultural, residential, commercial, and industrial activities that generate wastes. **sources de déchets ou de résidus**

waste treatment Any physical or chemical process that makes waste more compatible or acceptable to humans and their environment. **traitement des déchets**

wasteland Land not suitable for, or capable of, producing materials or services of value. A *miscellaneous land type*. **terrain inutilisable**

wastewater Water carrying wastes from homes, businesses, and industries; a mixture of water and dissolved or suspended inorganic or organic solids. See *sewage, domestic sewage.* **eaux usées ou résiduaires**

water application efficiency (*irrigation*) Ratio of the volume of water stored in the root zone of a soil during irrigation to the volume of water applied. **efficacité d'irrigation**

water balance A measure or account of the amount of water entering and the amount leaving a system (e.g., a given volume of soil), including the amount stored within the system. **bilan hydrique**

water content, soil The amount of water lost from the soil when it is dried to constant mass at 105°C; expressed either as the mass of water per unit mass of dry soil or as the volume of water per unit bulk volume of soil. The relationship between water content and soil water pressure is called the soil *water retention curve*. See *soil wetness, water retention.* **teneur en eau du sol**

water cycle See *hydrologic cycle.* **cycle hydrologique**

water demand Water requirements for a particular purpose (e.g., irrigation, power, municipal supply, plant transpiration, or storage). **besoin en eau d'irrigation**

water disposal system The complete system for removing excess water from land, causing minimum erosion. For sloping land, it may include a terrace system, terrace outlet channels, dams, and grassed waterways. For level land, it may include surface drains or both surface and subsurface drains. **système de drainage, système d'évacuation d'eau**

water enrichment See *eutrophication*. **enrichissement de l'eau**

water flow velocity The volume of water transported per unit of time and per unit of cross-sectional area normal to the direction of water flow. **vitesse d'écoulement de l'eau**

water holding capacity The ability of soil to hold water. The water-holding capacity of sandy soils is usually considered to be low, while that of clayey soils is high. **capacité de rétention en eau**

water management The planned development, distribution, and use of water resources. **gestion de l'eau**

water pressure The pressure (positive or negative), in relation to the external gas pressure on soil water, to which a solution identical in composition with the soil water must be subjected in order to be in equilibrium through a porous permeable wall with the soil water. It may be identified with the *matric potential*. See *soil water*. **pression de l'eau**

water quality The chemical, physical, and biological condition of water related to a specific use (e.g., human consumption). Often evaluated on the basis of critical concentrations of chemicals or organisms. **qualité de l'eau**

water requirement The number of grams of water a plant uses to produce a gram of dry matter; also called the *transpiration* ratio. These values are expressed as ratios (typically several hundred to one). They are increased by wind, high temperatures, and low humidity, and are decreased by factors that increase productivity. **coefficient de transpiration, besoin en eau**

water resources The supply of water in a given area or basin interpreted in terms of availability of surface and underground water. **ressources en eau**

water retention The relationship between soil-water matric potential and soil water content (by mass or by volume) is represented graphically as the *soil water characteristic* curve or the soil *water retention curve*. **rétention en eau**

water retention curve A graph showing the soil water content (by mass or by volume) versus applied tension or pressure. Points on the graph are usually obtained by increasing or decreasing the applied tension or pressure over a specified range. **courbe de rétention en eau**

water softener An apparatus designed to remove divalent metal ions (e.g., calcium, magnesium, and iron) from water, often replacing the divalent or trivalent ions with the monovalent sodium ion. See *ion exchange*. **adoucisseur d'eau**

water sprout A side shoot of a plant, originating from an adventitious bud on the trunk or main branches of a tree. **gourmand**

water suction An obsolete term now replaced by *matric potential*. See *soil water*, *soil water potential*. **succion d'eau**

water table The upper surface of groundwater or that level below which the soil is saturated with water. **nappe phréatique, surface de saturation**

water track Vegetation types marking the path of mineral-influenced waters through a *peatland*. On air photos, water tracks contrast sharply with the adjacent *swamps* or forested peatlands. **chenal d'écoulement végétalisé**

water treatment The processing of source water (well water or surface water) for distribution in a public drinking water system. **traitement des eaux**

water use efficiency Crop production per unit of water used, irrespective of water source, expressed in units of weight per unit of water depth per unit area. The concept of utilization applies to both dryland and irrigated agriculture. **efficacité d'utilisation de l'eau**

water vapor (*chemistry*) The gaseous form of water. (*meteorology*) The most abundant greenhouse gas, playing an important part in the natural *greenhouse effect*, because the warming influence of greenhouse gases leads to a positive water vapor feedback. Water vapor also helps to regulate the Earth's temperature, because clouds form when excess water vapor in the atmosphere condenses to form ice, water droplets, and precipitation. **vapeur d'eau**

water, interstitial Water that exists in the interstices or voids in soil, rock, or other porous media. **eau intersticielle**

water, juvenile Water from the interior of the Earth that is new or that has never been part of the general groundwater system. **eau juvénile**

watercourse A natural stream or source of supply of water, which may or may not contain water (e.g., river, creek, gulch, or artificial channel). See *waterway*. **cours d'eau**

water-filled pore space See *degree of saturation*. **taux de saturation en eau**

waterlogged Saturated or nearly saturated with water. **gorgé ou saturé d'eau**

watershed (1) The area contained within a drainage divide above a specified point on a stream. (2) The area drained by a given stream. **bassin versant ou hydrographique**

watershed assessment Evaluation of the present state of *watersheds* and the cumulative impact of proposed development on peak flows, suspended sediment, bedload, and stream channel stability within the watershed. **évaluation de bassin versant**

watershed integrity The stable, overall physical condition of the *watershed* (i.e., soils, bedrock, landforms, drainage ways) within which transfers of energy, matter, and especially water, occur. **intégrité du bassin versant**

watershed lag (*hydrology*) The time from the center of the mass of effective rainfall to peak of the *hydrograph*. **temps de réponse**

watershed management The planned use of drainage basins in accordance with predetermined objectives. **gestion de bassin versant**

water-stable aggregate A soil aggregate that is stable to the action of water, such as falling drops or agitation as in *wet-sieving* analysis. **agrégat stable à l'eau ou hydrostable**

waterwall incinerator An energy recovery device used to process municipal waste. The combustion chamber of the incinerator is lined with steel tubes containing circulating water; the combustion heat boils the water, and the steam can be sold or used to turn turbines in an electric generator. **incinérateur générateur de vapeur**

waterway A natural course or constructed channel for the flow of water. **voie d'eau, cours d'eau**

wavelength The mean distance between maximums (or minimums) of a roughly periodic pattern; the least distance between particles moving in the same phase of oscillation in a wave disturbance; wavelength = velocity/frequency. **longueur d'onde**

wax A solid or semi-solid substance. There are two types: mineral waxes, which are mixtures of hydrocarbons with a high molecular weight (e.g., *paraffins*), and waxes secreted by plants and animals, which contain esters of *fatty acid* and function as a protective coating. **cire**

weak acid An acid that dissociates only slightly in aqueous solution. **acide faible**

weak base A base that reacts with water to produce hydroxide ions to only a slight extent in aqueous solution. **base faible**

weak electrolyte A material that, when dissolved in water, gives a solution that conducts only a small electric current. **électrolyte faible**

weather The specific condition of the atmosphere at a particular place and time, measured in terms of wind, temperature, humidity, atmospheric pressure, cloudiness, precipitation, etc. In most places, weather can change hourly, daily, and seasonally. See *climate*. **temps**

weathering (*geology*) The complex combination of physical, chemical, and organic processes that decompose, disintegrate, and alter rocks and minerals at or near the Earth's surface. Weathering can be subdivided into *chemical weathering*, *mechanical weathering,* and *organic weathering*. **météorisation, altération**

weathering, basal surface of (*geology*) A line of variable depth beneath the Earth's surface that marks the lower limit of active weathering (i.e., weathering front); the transition which marks the change from sound rock (beneath) to weathered rock (above). **front d'altération**

weeding See *tillage, weeding*. **sarclage**

weep-holes Openings left in retaining walls, aprons, linings, or foundations to permit drainage and reduce pressure. **barbacanes, chantepleures**

weighted average For a series of recorded observations, the sum of the products of the frequency of certain values and the value of the observation, divided by the total number of observations. **moyenne pondérée**

weir A device for measuring or regulating the flow of water. **déversoir**

well-graded soil (*engineering*) A soil in which the particles are well distributed over a wide range in size or diameter. Such a soil's density and bearing properties can normally be easily increased by compaction. See *poorly graded soil*. **sol bien gradué**

Wentworth scale (*geology*) A scale used to classify the particle size of sediments. The grades are: >256 mm diameter, boulders; 64 to 256 mm, cobbles; 2 to 64 mm, pebbles; 0.0625 to 2 mm, sand; 0.004 to 0.0625 mm, silt; <0.004 mm, clay. See *soil texture*. **échelle granulométrique de Wentworth**

wet deposition The introduction of acidic material to the soil or surface waters by sulfuric and nitric acids dissolved in rain or snow. See *dry deposition*. **déposition humide**

wet digestion (*waste management*) A *stabilization* process in which mixed solid organic wastes are placed in an open digestion pond to decompose anaerobically. **digestion par voie humide**

wet milling (*waste management*) The mechanical reduction of the volume of solid wastes, that has been wetted to soften its paper and cardboard constituents. **broyage en conditions humides**

wet-bulb temperature The temperature reading from a thermometer with a wetted wick surrounding its bulb. The evaporative loss of latent heat from the wick lowers the temperature reading. Used with the *dry bulb temperature* and a table to compute *relative humidity*. **température du thermomètre mouillé**

wetland Land having the water table at, near, or above the land surface, or that is saturated for long enough periods to promote wetland or aquatic processes as indicated by hydric soils, hydrophytic vegetation, and various kinds of biological activity that are adapted to the wet environment. Wetlands include peatlands and areas that are influenced by excess water but which, for climatic, edaphic, or biotic reasons, produce little or no peat. Shallow open water, generally less than 2 m deep, is also included in wetlands. **milieu ou terre humide**

wetland class See *wetland classification system, Canadian.* **classe de milieu humide**

wetland classification system, Canadian A system for categorizing wetlands in Canada consisting of three hierarchical levels: class, form, and type. Five wetland classes are recognized on the basis of overall genetic origin of wetland ecosystems. Seventy wetland forms are differentiated on the basis of surface morphology, surface pattern, water type, and morphology of underlying mineral soil. Wetland types are classified according vegetation and physiognomy. **système canadien de classification des milieux humides**

wetland form See *wetland classification system, Canadian.* **aspect des milieux humides**

wetland type See *wetland classification system, Canadian.* **type de milieu humide**

wet-sieving A technique used to assess *aggregate distribution* and stability to *slaking* forces, and to provide a measure of *aggregate stability* to water. **tamisage à l'eau, tamisage humide**

wettability A measure of the ability of a soil to receive and infiltrate water. In contrast, the infiltration of water into soil pores can be impeded where the contact angle is greater than zero, resulting in poor wettability. **mouillabilité**

wetting agent A chemical that reduces the *surface tension* of water and enables it to soak into porous material more readily. **agent mouillant**

wheel track planting See *tillage, wheel track planting.* **semis en ornières**

whirling psychrometer An instrument for measuring the *relative humidity* of the air. **psychromètre à fronde**

white box system (*modeling*) A system in which most of the internal structures (e.g., storage, flows) are known, making it possible to know in detail the linkages between a given input and a given output. See *black box system, gray box system.* **système de type boîte blanche**

white rot (1) A common disease of onions, leeks, and shallots, caused by the fungi *Sclerotium cepivorum.* (2) A type of timber decay in which the cellulose, hemicellulose, and lignin in the wood are decomposed, leaving the wood soft, white, and fibrous. See *brown rot.* **(1) pourriture blanche (2) carie blanche**

white rot fungus Fungus that attacks wood and causes white rot. See *white rot.* **pourriture blanche**

wilderness An area of land generally greater than 1000 ha that predominantly retains its natural character and on which the impact of humans is transitory and, in the long run, substantially unnoticeable. **région sauvage**

wildfire An unplanned or unwanted natural or human-caused fire, or a prescribed fire that threatens to escape its bounds. **feu de friches**

wildlife Any animal species living unrestrained or free-roaming in the wild. Some definitions include plants, fungi, algae, and bacteria. **espèces sauvages**

wildlife management The application of scientific and technical principles to wildlife populations and habitats to maintain such populations (particularly mammals, birds, and fish) essentially for recreational and/or scientific purposes. **gestion ou aménagement de la faune**

wilting The loss of turgidity in plant tissue; the intake of water is insufficient to replace that lost by transpiration or other means, thus causing a deflation of plant cells. **flétrissement**

wilting point See *permanent wilting point.* **point de flétrissement**

wind A horizontal movement of air in relation to the surface of the Earth. **vent**

wind abrasion A process of erosion in which windblown particles of rock material scour and wear away exposed surfaces of any kind. **abrasion éolienne**

wind erosion equation An equation for predicting *E*, the average annual soil loss due to wind in mass per unit area per year, defined as $E=IKCLV$, where *I* is the soil erodibility factor, *K* is the soil ridge roughness factor, *C* is the local climatic factor, *L* is the field width, and *V* is the vegetative factor. **équation de l'érosion éolienne**

wind-blown deposits See *loess*. **dépôts éoliens**

windbreak A planting of trees, shrubs, or other vegetation, or a human-made porous barrier, usually perpendicular to the prevailing wind direction, erected to protect soil, crops, homesteads, and roads against wind and the drifting of soil and snow. Often called a shelterbelt. See *erosion control structures*. **brise-vent, haie brise-vent**

windrow (*waste management*) A long, narrow pile of organic waste (e.g., compost, manure). Windrows are formed as part of the composting process to facilitate aeration. In large-scale operations, the design allows convenient access by machines, which turn and mix the material periodically. (*agronomy*) A long row of cut hay that allows for more uniform and thorough drying. (*forestry*) An accumulation of slash, branchwood, and debris on a harvested cutblock created to clear the ground for regeneration. **andain**

windthrow See *blowdown*. **chablis**

wing wall Sidewalls of a structure used to prevent eroding and sloughing of banks or channels, and to direct and confine overfall. **mur en aile**

Winogradsky column (*microbiology*) A glass column with an anaerobic lower zone and an aerobic upper zone, which allows growth of microorganisms under conditions similar to those found in nutrient-rich water and sediment. **colonne de Winogradsky**

winter kill (*limnology*) The death of fish in a body of water during a prolonged period of ice and snow cover; caused by oxygen exhaustion due to respiration and lack of photosynthesis. (*botany*) The death of plants and trees due to cold or freezing conditions. **mortalité hivernale**

wood Secondary *xylem* of dicotyledons and conifers, which provides mechanical support to the plant. **bois**

wood pulp The basic primary materials from which most papers are made, consisting of small, loose wood fibers mixed with water. **pâte de bois**

woodland Any land used primarily for growing trees and shrubs (e.g., forests, trees, plantations, shelterbelts, windbreaks, wide hedgerows containing woodland species for wildlife food or cover, stream banks and other banks with woodland cover). **boisé**

woodland management The management of established woodlands and plantations, including all measures designed to improve the quality and quantity of woodland growing stock and to maintain litter and herbaceous ground cover for soil, water, and other resource conservation (e.g., planting, improvement cutting, thinning, pruning, slash disposal, and protection from fire and grazing). **aménagement ou gestion de boisé**

woodlot A small area of land occupied by trees. **terrain boisé**

Wood-Werkman phenomenon The uptake and incorporation into organic substances of CO_2 by heterotrophs; tracer data show that heterotrophs, although obtaining most of their cell carbon from organic matter, do obtain some from CO_2. **phénomène de Wood-Werkman**

woody peat *Peat* containing many woody remains of trees and shrubs, mainly (in North America). *Picea mariana* and *Abies amabilis*; generally occurs as a transition between *minerotrophic* peat or *dystrophic* mineral soils and raised bogs. Also called forest peat, silvic peat. See *carr*. **tourbe ligneuse**

X

xenobiotic Compounds alien to biological systems. Usually refers to human-made compounds that are resistant or recalcitrant to biodegradation and/or decomposition. **xénobiotique**

Xeralfs A suborder in the U.S. system of soil taxonomy. *Alfisols* that have a *xeric* soil moisture regime. Xeralfs are brownish or reddish throughout. **Xéralfs**

Xerands A suborder in the U.S. system of soil taxonomy. *Andisols* that have a xeric soil moisture regime. **Xérands**

Xererts A suborder in the U.S. system of soil taxonomy. Vertisols that have a thermic, mesic, or frigid temperature regime, and if not irrigated, cracks that remain both 5 cm or more wide through a thickness of 25 cm or more within 50 cm of the mineral soil surface for 60 or more consecutive days during 90 days following the summer solstice, and closed 60 or more consecutive days during the 90 days following the winter solstice. **Xérerts**

xeric A soil moisture regime (used in U.S. system of soil taxonomy) common to Mediterranean cliamtes that have moist cool winters and warm dry summers. A limited amount of moisture is present but does not occur at optimum periods for plant growth. Irrigation or summerfallow is commonly necessary for crop production. **xérique**

Xerolls A suborder in the U.S. system of soil taxonomy. *Mollisols* that have a *xeric* soil moisture regime. Xerolls may have a *calcic*, *petrocalcic*, or *gypsic* horizon, or a duripan. **Xérolls**

xerophyte A plant adapted to prolonged moisture deficiency. **xérophyte**

xerosere The sequential development of a plant community (see *sere*) in a very dry habitat. See *lithosere, psammosere*. **xérosère**

Xerults A suborder in the U.S. system of soil taxonomy. *Ultisols* that have low or moderate amounts of organic carbon, are brownish or reddish throughout, and have a xeric soil moisture regime. **Xérults**

x-ray A form of electromagnetic radiation having a wavelength shorter than that of ultraviolet light and usually longer than that of gamma rays. This form of radiation is called ionizing since it has excellent penetrating ability and produces ions within material as it interacts with the atoms of the material through which it passes. **rayon-x**

x-ray diffraction An analytical technique for establishing the structures of crystalline solids by directing x-rays of a single wavelength at a crystal and obtaining a diffraction pattern from which inter-atomic spaces can be determined. **diffraction des rayons-x**

xylem Woody tissue that conducts water and inorganic salts throughout the plant and provides it with mechanical support. In leaves, flowers, and young stems, xylem is present in conjunction with *phloem* in the form of conducting strands called vascular bundles. Roots have a central core of **xylem.** See *figure*. **xylème**

xylophagous Feeding on wood. **xylophage**

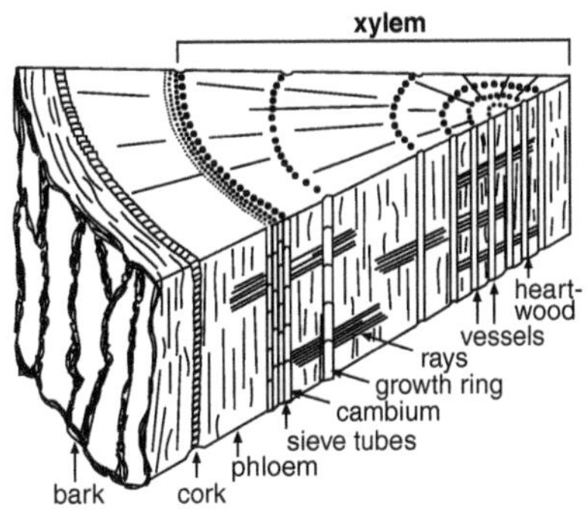

Xylem (adapted from Dunster and Dunster, 1996).

Y

yard waste Grass clippings, prunings, and other discarded material from yards and gardens. See *waste*. **résidus de jardin**

yeasts A group of fungi that produce characteristic budding cells rather than *hyphae*. **levures**

yield The amount of a specified substance (e.g., grain, total dry matter) produced per unit area. (*hydrology*) (1) The amount of water that can be taken continuously from a lake. (2) The amount of organic matter (plant and animal) produced by a lake. **rendement**

Z

zeolite A group, or any member of this group, of hydrous aluminosilicate minerals containing sodium and calcium as major cations or, less commonly, barium, beryllium, lithium, potassium, magnesium, and strontium; characterized by the ratio (Al+Si):O=1:2. Zeolites have an open tetrahedral framework structure with large cations capable of ion exchange and loosely held water molecules that allow reversible dehydration. They fuse and swell when strongly heated. Zeolites occur as well-formed crystals in cavities in basalt and authigenic minerals, especially in beds of tuff. Natural and artificial zeolites are used extensively as water softeners. **zéolite**

zero population growth (ZPG) A condition in which a population in a given location neither increases nor decreases over time (i.e., increases due to births and immigration are balanced with decreases caused by deaths and emigration). **croissance nulle**

zero tillage See *tillage, zero tillage.* **zéro-labour, semis-direct**

zero-order reaction A chemical reaction in which the rate of reaction is independent of a reactant's concentration or the concentration of any other chemicals present. See *rate constant.* **réaction d'ordre zéro**

zeta potential See *electrokinetic zeta potential.* **potentiel zêta**

Zingg bench terrace A type of bench *terrace* designed for dryland moisture conservation consisting of an earthen embankment and a bench-leveled part of the terrace interval immediately above the ridge. Runoff water from the sloping area is retained on the leveled area and absorbed by the soil. **terrasse en gradins**

zonal soil (1) Any one of the *great groups* of soils having well-developed soil characteristics that reflect the zonal influence of climate and living organisms, mainly vegetation, as active factors of soil genesis (in the Canadian system of soil classification). (2) A soil characteristic of a large area or zone. **sol zonal**

zonation Differentiation of areas of soils having common characteristics as a consequence of having developed under the similar *soil formation factors* of the climate and vegetation in that zone. **zonalité**

zone In general, a region defined by specific limits. (*ecology*) An area characterized by similar flora or fauna; a belt or area to which certain species are limited. (*engineering*) In earth dams, a segment of the earthfill containing similar materials. (*climate*) A distinction based on the global temperature belts. (*biostratigraphy*) A fundamental division demarcated by fossils used to correlate rock successions. (*geology*) A spatial division of an area of metamorphic rocks, based on index minerals, used to define areas of different metamorphic grade. See *zone, soil.* **zone**

zone subsoiling See *tillage, zone subsoiling.* **sous-solage localisé**

zone tillage See *tillage, zone tillage.* **travail du sol localisé**

zone, soil An area in which the dominant or *zonal soils* reflect the influence of climate and vegetation, and form a natural land pattern with other soils that exhibit the zonal influence only weakly or not at all. The soil zone is not a taxonomic unit (in the Canadian system of soil classification) but may be used as a cartographic unit. **zone de sols**

zooecology The branch of ecology concerned with the relationships between animals and their environment. **zooécologie**

zoogeography The study of the geographical distribution of animals. **zoogéographie**

zooplankton Animal forms of plankton. Unattached microscopic animals of plankton having minimal capability for locomotion. See *plankton, phytoplankton.* **zooplancton**

zoospore A motile reproductive spore common to aquatic fungi. **zoospore**

zygospore A thick-walled resting spore formed by conjugation of gametes or, in the Zygomycetes, by fusion of similar *gametangia.* See *figure.* **zygospore**

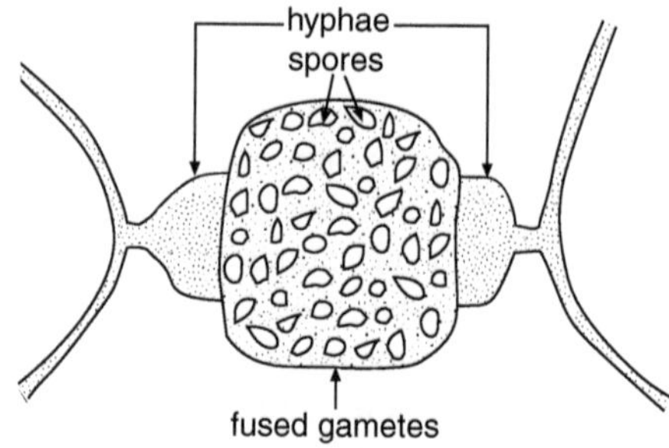

Zygospore

zymogenous flora Microorganims that respond quickly by enzyme production and growth when simple (i.e., easily decomposable) organic materials become available. **flore zymogène**

ENGLISH/FRENCH INDEX OF TERMS BY SUBJECT AREA/DISCIPLINE

AGRONOMY, AGRICULTURE, FERTILITY, PLANT NUTRITION

English	French
absorption, active	**absorption active**
absorption, passive	**absorption passive**
acid-forming fertilizer	**engrais acidifiant**
acidulation	**acidulation**
agrology	**agrologie**
agronomic practices	**pratiques agronomiques ou agricoles**
agronomy	**agronomie**
amendment, soil	**amendement du sol**
ammoniation	**ammoniation**
ammonium nitrate	**nitrate d'ammonium**
ammonium phosphate	**phosphate d'ammonium**
ammonium sulfate	**sulfate d'ammonium**
antagonism	**antagonisme**
available nutrient	**élément assimilable**
banding	**application en bande**
basic fertilizer	**engrais de base**
basic slag	**scories**
biuret	**biuret**
broadcast	**application à la volée**
bulk blending	**mélange en vrac**
bulk fertilizer	**engrais en vrac**
bulk-blend fertilizer	**engrais de mélange**
calcium sulfate extractant	**sulfate de calcium**
calibration	**réglage**
chemical fertilizer	**engrais minéral**
chemigation	**fertigation**
citrate-insoluble phosphorus	**phosphore insoluble au citrate**
citrate-soluble phosphorus	**phosphore soluble au citrate**
closed handling system	**système de manutention fermé**
coated fertilizer	**engrais enrobé**
conditioned fertilizer	**engrais conditionné**
critical nutrient concentration	**seuil critique**
crop nutrient requirement	**besoin en éléments nutritifs**
crust	**croûte**
deficiency symptom	**symptôme de carence**
EDTA	**EDTA**
essential element	**élément essentiel ou constitutif**
farming system	**système de production**
fertigation	**irrigation fertilisante**
fertility, soil	**fertilité du sol**
fertilization	**fertilisation**
fertilization, foliar	**fertilisation foliaire**
fertilizer	**engrais**
fertilizer, acid-forming	**engrais acidifiant**
fertilizer, bulk-blend	**engrais de mélange**
fertilizer, chemical	**engrais minéral**
fertilizer, coated	**engrais enrobé**
fertilizer, conditioned	**engrais conditionné**
fertilizer, granular	**engrais granulaire**
fertilizer, liquid	**engrais liquide, ammoniaque anhydre**
fertilizer, nongranular	**engrais perlé**
fertilizer, solution	**engrais liquide**
fertilizer, straight	**engrais simple**
fertilizer, suspension	**engrais en suspension**
fertilizer analysis	**composition élémentaire d'un engrais**
fertilizer fixation	**rétrogradation d'engrais**
fertilizer formula	**formule d'engrais**
fertilizer grade	**formule d'engrais, teneur d'un engrais**
fertilizer ratio	**rapport d'éléments fertilisants**
fertilizer requirement	**besoin en engrais**
fertilizer value	**valeur fertilisante**
fertilizer, bulk-blended	**engrais de mélange en vrac**
fertilizer, complete	**engrais ternaire**

fertilizer, compound **engrais composé**
fertilizer, controlled-release **engrais à libération lente**
fertilizer, fluid **engrais liquide**
fertilizer, injected **engrais injecté**
fertilizer, inorganic **engrais minéral**
fertilizer, manufactured **engrais de synthèse**
fertilizer, mixed **engrais composé**
fertilizer, organic **engrais organique**
fertilizer, pop-up **engrais de démarrage**
fertilizer, salt index **indice de salinité de l'engrais**
fertilizer, sidedressed **engrais appliqué en bande en post-levée**
fertilizer, slow-release **engrais à libération lente**
fertilizer, starter **engrais de démarrage**
fertilizer, suspension **engrais en suspension**
fertilizer, top-dressed **engrais en post-levée**
fixed phosphorus **phosphore fixé**
foliar diagnosis **diagnostic foliaire**
formulation **forme**
frits **forme frittée**
fritted trace element **oligo-éléments frittés**
granular fertilizer **engrais granulaire**
granulation **granulation**
green manure **engrais vert**
guano **guano**
impoverishment, soil **appauvrissement du sol**
incorporation (into soil) **incorporation**
Liebig's law **loi de Liebig, loi du minimum**
liquid fertilizer **engrais liquide**
luxury uptake **consommation de luxe**
macroelement **élément majeur**
macronutrient **élément majeur**
mass flow (nutrient) **écoulement massique**
mass flow rate **débit massique**
microelement **élément mineur**
micronutrient **oligo-élément**
minor element **élément mineur**
nitric phosphates **nitrophosphates**
nitrogen requirement **besoin en azote**
nongranular fertilizer **engrais non granulaire**
nonpressure solution **solution azotée**
nutrient **élément nutritif**
nutrient antagonism **antagonisme entre éléments nutritifs**
nutrient balance **équilibre nutritif**
nutrient budget **bilan nutritif**
nutrient concentration **concentration en éléments nutritifs**
nutrient content **teneur en éléments nutritifs**
nutrient cycle **cycle des éléments nutritifs**
nutrient diffusion **diffusion des éléments nutritifs**
nutrient interaction **interaction entre éléments nutritifs**
nutrient retention **rétention d'un élément nutritif**
nutrient stress **stress nutritif**
nutrient uptake **prélèvement d'un élément nutritif**
nutrient-supplying power of soils **capacité de libération des éléments nutritifs du sol**
optimum level **niveau optimal**
organic fertilizer **engrais organique**
orthophosphate **orthophosphate**
particulate phosphate **phosphate particulaire**
phosphate **phosphate**
phosphate rock **roche phosphatée**
phosphoric acid **acide phosphorique**
phosphorite **phosphorite**
phosphorus **phosphore**
phosphorus, total **phosphore total**
plant food **élément nutritif minéral**
plant nutrient **élément nutritif**
plant nutrition **nutrition des plantes**
potash **potasse**
potassium-supplying power of soils **capacité de libération du potassium du sol**
prilled fertilizer **engrais perlé**
primary nutrient **élément majeur, élément nutritif primaire**
quench **surfertilisation**
residual fertility **fertilité résiduelle**
residual nitrate **nitrate résiduel**
reversion **rétrogradation**
ridge **billon**
rock phosphate **phosphate naturel**
rolling **roulage**
root interception **absorption racinaire directe**
rotation **rotation**
secondary nutrient **élément nutritif secondaire**
seed **semence, semer**
slow release fertilizer **engrais à libération lente**
soil fertility **fertilité du sol**
solution fertilizer **engrais liquide**
starter fertilizer **engrais de démarrage**
straight fertilizer **engrais simple**
superphosphate **superphosphate**

superphosphate, ammoniated **superphosphate ammonié**
superphosphate, concentrated **superphosphate triple ou concentré**
superphosphate, enriched **superphosphate enrichi**
superphosphate, normal **superphosphate normal**
superphosphate, ordinary **superphosphate normal**
superphosphate, single **superphosphate normal**
superphosphoric acid **acide superphosphorique**
surface-applied fertilizer **engrais non incorporé**
suspension fertilizer **engrais en suspension**
till **cultiver le sol, labourer**
top dressing **engrais appliqué en post-levée**
unavailable nutrient **élément nutritif non disponible**
water use efficiency **efficacité de l'utilisation de l'eau**

BIOCHEMISTRY

α-amino acid **acide α-aminé**
acetylene-reduction assay **essai de réduction de l'acétylène**
adaptive enzyme (enzyme induction) **enzyme induite**
adenosine triphosphate (ATP) **adénosine triphosphate (ATP)**
adenosine diphosphate (ADP) **adénosine diphosphate (ADP)**
amino acid **acide aminé**
ammonium fixation **fixation d'ammonium**
ammonia volatilization **volatilisation de l'ammoniaque**
ammonification **ammonification**
anabolism **anabolisme**
anaerobic decomposition **décomposition anaérobie**
anaerobic respiration **respiration anaérobie**
analog **substance analogue**
antibody **anticorps**
antigen **antigène**
antitoxin **antitoxine**
assimilation **assimilation**
assimilation efficiency **efficacité d'assimilation**
autolysis **autolyse**
auxin **auxine**
biochemistry **biochimie**
biological interchange **transformation biologique**
biophile **élément biophile**
biosynthesis **biosynthèse**
biotransformation **détoxication**
C3 plant **plante ou végétation de type C3**
C4 plant **plante ou végétation de type C4**
CAM plant **plante ou végétation de type CAM**
carbon cycle **cycle du carbone**
chemodenitrification **chimio-dénitrification**
coenzyme **coenzyme**
conjugated metabolites **métabolites conjugués**
constitutive enzyme **enzyme de constitution**
cytochrome **cytochrome**
cytokinin **cytokinine**
cytoplasm **cytoplasme**
deamination **désamination**
decomposition **décomposition**
dehydrogenase **déshydrogénase**
denaturation **dénaturation**
denature **dénaturer**
denitrification **dénitrification**
deoxyribonucleic acid (DNA) **acide désoxyribonucléique (ADN)**
desulfuration **désulfuration**
dinitrogen fixation **fixation de l'azote atmosphérique**
dissimilation **désassimilation**
dissimilatory reduction of nitrate **réduction non-assimilatoire des nitrates**
endoenzyme, or intracellular enzyme **enzyme endogène**
enzyme **enzyme**
fat (glyceride) **graisse (glycéride)**
feedback inhibition **inhibition rétroactive**
fixation **fixation**
genetics **génétique**
genotype **génotype**
gibberellins **gibbérellines**
glycogen **glycogène**
gross productivity **productivité brute**

growth regulator **régulateur de croissance**
haploid **haploïde**
hemoglobin **hémoglobine**
heterotroph **hétérotrophe**
heterotrophic **hétérotrophe**
heterotrophic nitrification **nitrification hétérotrophe**
heterotrophy **hétérotrophie**
heterozygous **hétérozygote**
histones **histones**
hormone **hormone**
humic substance **substance humique**
humin **humine**
immobilization **immobilisation, organisation**
labile **labile**
leghemoglobin **leghémoglobine**
lignin **lignine**
lipid **lipide**
lysosome **lysosome**
metabolism **métabolisme**
metabolite **métabolite**
Michaelis-Menton equation **équation de Michaelis-Menton**
mineralization **minéralisation**
nitrate reduction **réduction du nitrate**
nitrification **nitrification**
nitrogen assimilation **assimilation de l'azote**
nitrogen cycle **cycle de l'azote**
nitrogen fixation **fixation biologique de l'azote**
nitrogenase **nitrogénase**
nucleic acid **acide nucléique**
organelle **organite cellulaire**
osmoregulation **régulation osmotique**
osmosis **osmose**
osmotic gradient **gradient osmotique**
osmotic lysis **lyse osmotique**
oxidative phosphorylation **phosphorylation oxydative**
oxygen debt **déficit en oxygène**
peptidoglycan **peptidoglycane**
permease **perméase**
phospholipids **phospholipides**
phosphorus cycle **cycle du phosphore**
photosynthate **photosynthat**
photosynthesis **photosynthèse**
plasmid (episome) **plasmide (épisome)**
ploidy **ploïdie**
polyploidy **polyploïdie**
protease **protéase**
protein **protéine**
proteolytic enzyme **enzyme protéolytique**
protoplasm **protoplasme**
putrefaction **putréfaction**
putrescible **putrescible**
Q_{10} **Q_{10}**
respiration **respiration**
respiration/biomass ratio **respiration spécifique**
respiratory quotient (RQ) **quotient respiratoire**
respiratory reduction of nitrate **réduction respiratoire des nitrates**
ribosome **ribosome**
ribonucleic acid (RNA) **acide ribonucléique (ARN)**
ridge **dorsale**
secondary metabolite **métabolite secondaire**
siderophore **sidérophore**
starch **amidon**
sulfur cycle **cycle du soufre**
supergene **supergène**
temperature coefficient **coefficient de température**
transfer RNA (tRNA) **ARN de transfert (ARNt)**
translocation **translocation**
transpiration **transpiration**
triploid **triploïde**
water requirement **coefficient de transpiration, besoin en eau**

BOTANY, HORTICULTURE, RANGE SCIENCE

abaxial **abaxial**
acid detergent fiber (ADF) **fibre au détergent acide (ADF)**
acid detergent fiber digestibility **digestibilité de la fibre au détergent acide**
acidophilic **acidiphile, acidophile**
actual use (range-pasture) **utilisation courante (parcours-pâturage)**
adaxial **adaxial**
adventitious roots **racines adventives**
aerobic **aérobie**
aflatoxin **aflatoxine**
after-ripening **postmaturation**
agroforestry **agroforesterie, agrosylviculture**
agro-silvo-pastoral **agro-sylvo-pastoral**
agrostology **agrostologie**

alkaloid **alcaloïde**
anaerobic **anaérobie**
angiosperm **angiosperme**
animal unit (A.U.) **unité animale (U.A.), unité-gros-bétail (U.G.B.)**
animal unit month (A.U.M.) **unité animale-mois (U.A.M.)**
annual plant **plante annuelle**
anther **anthère**
anthesis **anthèse**
anthocyanins **anthocyanines**
anti-oxidant **antioxydant**
apical meristem **méristème apical**
arboretum **arboretum**
ash **cendre**
auxin **auxine**
biennial plant **plante bisannuelle**
binomial nomenclature **nomenclature binominale**
bioassay **test biologique, épreuve biologique**
biogeochemical cycle **cycle biogéochimique**
biomass **biomasse**
bloat **ballonnement, météorisation**
bloom, early **début floraison**
bloom, full **pleine floraison**
blooming **floraison**
bone meal **poudre d'os**
boot stage **gonflement**
brown rot **pourriture brune, carie brune**
browse **brout**
browse line **ligne d'abroutissement**
bryophyte **bryophyte**
bud **bourgeon**
buffalo wallow **dépression à bisons**
bulb **bulbe**
bunch grass **graminée cespiteuse**
callus **cal**
calorie **calorie**
cambium **cambium**
canker **chancre**
canopy **couvert**
carbohydrate **glucide, hydrate de carbone**
carbohydrates, nonstructural **glucides non-structuraux**
carbohydrates, structural **glucides structuraux**
carotenoids **caroténoïdes**
caryopsis **caryopse**
catkin **chaton**
cecum **caecum**
cell **cellule**
cell division **division cellulaire**
cell wall **paroi cellulaire**
cell wall constituents **constituants de la paroi cellulaire**
cellulase **cellulase**
cellulose **cellulose**
chlorophyll **chlorophylle**
chloroplast **chloroplaste**
chromosome **chromosome**
class **classe**
clone **clone**
coated seed **semence pelliculée**
companion crop **culture associée, culture compagne, plante-abri**
cone **cône**
conifer **conifère**
coniferous **coniférien**
continuous stocking **pâturage continu, pâturage libre**
cool-season plant **plante de climat frais**
corm **cormus**
cortex **cortex**
cotyledons **cotylédons**
cross-pollination **pollinisation croisée**
cryptogam **cryptogame**
culm **tige**
cultivar **cultivar**
cuticle **cuticule**
cutin **cutine**
deciduous plant **plante à feuilles caduques**
decomposer **décomposeur**
decreaser **espèce en évolution régressive**
dendritic **dendriforme, dendritique**
dendrochronology **dendrochronologie**
dendrology **dendrologie**
detritus **détritus**
dibble **plantoir**
dicotyledon **dicotylédone**
diffusion pressure gradient **gradient de pression de diffusion, tension sur la colonne d'eau**
dioecious **dioïque**
diploid **diploïde**
diurnal **diurne, journalier, quotidien**
diurnal temperature variations **variations journalières de température**
dormancy **dormance**
dough stage **stade pâteux**
dry matter (DM) **matière sèche (MS)**
Dutch elm disease **maladie hollandaise de l'orme**

endolithic **endolithique**
endophyte **endophyte**
ensilage **ensilage**
epidermal cells **cellules épidermiques**
epidermis **épiderme**
ergot **ergot**
ericaceous **éricacé**
evergreen plant **plante à feuillage persistant**
exclosure **exclos**
exudate **exsudat**
family **famille**
feed **aliment des animaux**
feedlot **parc d'engraissement**
fibrous root system **système racinaire fasciculé**
fire blight **brûlure bactérienne**
flag leaf **feuille étendard, dernière feuille, feuille de l'épi**
flower **fleur**
flushing **alimentation intensive, flushing**
fodder **fourrage**
foliage **feuillage**
forage **fourrage, fourrager**
forage quality **qualité des fourrages**
forb **plante herbacée dicotylédone**
fresh mulch **paillis frais**
fruiting body **organe de fructification**
genus **genre**
geotropism **géotropisme**
germination **germination**
gibberellin **gibbérelline**
girdling **annélation, incision annulaire**
graft **greffe**
graminoid **graminiforme**
grass **graminée**
grass tetany (hypomagnaesemia) **tétanie d'herbage (hypomagnésiémie)**
grasslike plant **plante graminiforme**
graze **pâturer**
grazing capacity **charge limite d'un pâturage**
grazing distribution **distribution du pâturage**
grazing management **conduite du pâturage**
grazing period **temps de pâturage**
grazing pressure **taux de charge des pâturages**
grazing system **système de pâturage**
green-chopped **affourragement vert**
grove **bocage**
growth ring **anneau de croissance**
guard cells **cellules de garde, cellules stomatiques**
guttation **guttation**
gymnosperm **gymnosperme**
halophyte **halophyte**
hardiness **rusticité**
hay **foin**
haylage **ensilage mi-fané, ensilage préfané**
heartwood **bois de coeur**
heath **plante éricacée**
heliotropism **héliotropisme**
hemicellulose **hémicellulose**
herbaceous **herbacé**
herbage **herbage**
heredity **hérédité**
horticulture **horticulture**
hybrid **hybride**
hydrophyte **hydrophyte**
imperfect flower **fleur incomplète**
improved pasture **pâturage amélioré**
in situ ***in situ***
in vitro ***in vitro***
in vivo ***in vivo***
increaser **espèce en évolution progressive**
inflorescence **inflorescence, exertion**
ingesta **ingesta**
inoculant **inoculant**
intensive grazing management **conduite de pâturage intensive**
intermittent grazing **pâturage intermittent**
kernel **grain**
kingdom **règne**
lamina (plural laminae) **lamina**
landscape **paysage**
lateral **racine latérale ou secondaire**
leaf curl **enroulement de la feuille**
legume **légumineuse**
Lemna gibba ***Lemna gibba***
ley **prairie temporaire**
lichen **lichen**
lignin **lignine**
lipid **lipide**
litter **litière**
livestock **bétail**
macroscopic **macroscopique**
mast **paisson**
maturity stage **stade de maturité**
meadow **pré**
meiosis **méiose**
meristem **méristème**
mesophyll **mésophylle**
mesophyte **mésophyte**
milk stage **stade laiteux**

mitochondrion **mitochondrie**
mitosis **mitose**
monocotyledon **monocotylédone**
monoecious **monoïque**
moss **mousse**
mucigel **mucigel**
mulch **paillis, mulch**
mycotoxin **mycotoxine**
near infrared reflectance spectroscopy (NIRS) **spectroscopie de réflexion dans le proche infrarouge**
neurotoxin **neurotoxine**
neutral detergent fiber (NDF) **fibre au détergent neutre, fibre détergente neutre**
nitrogen-fixing plant **plante fixatrice d'azote**
noxious species **espèce nuisible**
nutrition **nutrition**
open range **terrain de parcours libre**
orchard **verger**
order **ordre**
ornamental **plante ornementale**
overgrazed range **terrain de parcours surpâturé**
overgrazing **surpâturage**
overstocking **surcharge pastorale**
paddock **enclos**
palaeobotany **paléobotanique**
palatability **palatabilité, appétibilité**
palisade parenchyma **parenchyme palissadique**
palynology **palynologie**
panicle **panicule**
parenchyma **parenchyme**
pasture **pâturage**
pasture improvement **amélioration du pâturage**
pasture, tame **pâturage cultivé ou amélioré**
pasture, temporary **pâturage temporaire**
perennial plant **plante pérenne ou vivace**
perfect flower **fleur complète**
perlite **perlite**
permanent pasture **pâturage permanent**
phenology **phénologie**
phenotype **phénotype**
phloem **phloème**
photoperiod **photopériode**
photoperiodism **photopériodisme**
photorespiration **photorespiration**
photosensitization **photosensibilisation**
photosynthetic quotient **quotient chlorophyllien**
phototropism **phototropisme**
phyllosphere **phylloplan**
phylogeny **phylogénie**
phylum **phylum**
physiological drought **sécheresse physiologique**
phytogeography **phytogéographie**
phytolith **phytolithe**
phytometer **phytomètre, plante indicatrice**
phytotoxic **phytotoxique**
pistil **pistil**
pistillate **pistillé**
pith **coeur**
plant **plante**
plastid **plaste**
pollen **pollen**
pollen analysis **analyse pollinique**
pollination **pollinisation**
pomology **pomologie**
pot-bound **feutrage radiculaire**
primary growth **croissance primaire**
procumbent **procombant**
propagation **multiplication**
propagule **propagule**
protein, crude **protéine brute, matière azotée totale**
proximate analysis **analyse immédiate**
prussic acid **acide prussique ou cyanhydrique**
psammophyte **psammophyte**
Pteridophyta ***Pteridophyta***
pulse **graines de légumineuses**
pulse labeling **marquage bref, marquage de courte durée**
radicle **radicule**
range **parcours, parcours naturel, aire de distribution géographique**
range improvement **amélioration du parcours, amélioration du parcours naturel**
range management **conduite des parcours**
range science **science des parcours**
range seeding **semis des parcours**
rangeland **terrain de parcours**
rangeland health **santé du terrain de parcours**
ration **ration fourragère, ration**
ration, balanced **ration équilibrée**
rhizomatous **rhizomateux**
rhizome **rhizome**
rhizosphere **rhizosphère**
ridge planting **semis sur billon**

root **racine**
root cap **coiffe**
root hairs **poils absorbants**
root hardy **plante vivace**
root pruning **élagage ou taille des racines**
root zone **zone ou profil racinaire**
rootstock **porte-greffe**
rotation grazing **pâturage en rotation**
rotation pasture **pâturage en rotation**
ruderal vegetation **végétation rudérale**
rumen **rumen**
ruminant **ruminant**
rumination **rumination**
salt lick **salignon**
sap **sève**
sap test **essai ou test de sève**
scarification **scarification**
sclerenchyma **sclérenchyme**
sclerophyllous **sclérophylle**
section **section**
sedge **carex**
seed **semence**
seed coat **tégument**
seedling **semis, plantule**
seep **suintement**
self-pollinated **autopollinisé**
self-pruning **élagage naturel**
senescence **vieillissement, sénescence**
sheath **gaine**
shifting agriculture (slash-and-burn agriculture) **agriculture itinérante (agriculture sur brûlis)**
shrub **arbuste, arbrisseau**
silage **ensilage**
slash-and-burn agriculture **agriculture sur brûlis**
sod **gazon**
sod grasses **graminées à gazon**
species **espèces**
spore **spore**
sporocarp **sporocarpe**
stamen **étamine**
staminate **staminé**
stem **tige**
stigma **stigmate**
stocking rate **taux de chargement**
stolon **stolon**
stoloniferous **stolonifère**
stoma (plural stomata) **stomate**
strip grazing **pâturage rationné**
strobilus **cône**
stubble **chaume**
supplemental feeding **alimentation de complément**
sward **peuplement de graminées**
synthetic variety **variété synthétique**
tannin **tanin**
taproot **racine pivotante**
taproot system **système racinaire pivotant**
tendril **vrille**
tetraploid **tétraploïde**
thallus **thalle**
thermoperiodism **thermopériodisme**
therophyte **thérophyte**
thinning **éclaircissage**
tiller **talle**
tissue **tissu**
total digestible nutrients (TDN) **unités nutritives totales, matières digestibles totales**
toxicity **toxicité**
toxin **toxine**
tracheid **trachéide**
transgenic organism **organisme transgénique**
tree **arbre**
tree ring analysis **analyse des anneaux de croissance**
tropism **tropisme**
tuber **tubercule**
tuberous root **racine tuberculeuse, racine tubéreuse**
turf **gazon**
turgid **turgescent**
type **type**
undergrazing **sous-pâturage**
unhulled seed **semence non-mondée, semence non-décortiquée**
vacuole **vacuole**
variegated foliage **feuillage panaché**
vascular **vasculaire**
vascular bundle **faisceau vasculaire ou libéro-ligneux**
vegetative **végétatif, multiplication végétative**
vegetative state **état ou stade végétatif**
vein **nervure**
venation **nervation**
vernalization **vernalisation**
water sprout **gourmand**
white rot **pourriture blanche, carie blanche**
wilting **flétrissement**
wood **bois**
xylem **xylème**

CHEMISTRY

absolute alcohol **alcool absolu**
absolute temperature **température absolue**
absolute zero **zéro absolu**
absorbance **absorbance**
absorptance **absorptance, facteur d'absorption**
absorption **absorption**
absorption band **bande d'absorption**
absorptivity **absorptivité**
acid **acide**
acid dissociation constant (K_a) **constante de dissociation d'un acide**
acid gas **gaz acide**
acid soil **sol acide**
acid-base indicator **indicateur acido-basique**
acidic **acide**
acidic cation **cation acidique**
acidic solution **solution acide**
acidimetry **acidimétrie**
acidity constant **constante d'acidité**
acidity, exchangeable **acidité d'échange**
acidity, residual **acidité résiduelle**
acidity, salt-replaceable **acidité échangeable par un sel**
acidity, total **acidité totale**
activation energy **énergie d'activation**
activity **activité**
adiabatic process **transformation adiabatique**
adsorption **adsorption**
adsorption complex **complexe d'adsorption**
adsorption isotherm **isotherme d'adsorption**
alcohol **alcool**
aldehyde **aldéhyde**
aliphatic compound **composé aliphatique**
aliquot **aliquote**
alkali **alcali**
alkali metals **métaux alcalins**
alkaline **alcalin**
alkaline soil **sol alcalin**
alkaline solution **solution alcaline**
alkaline-earth metals **métaux alcalino-terreux**
alkalinity **alcalinité**
alkalinity, soil **alcalinité du sol**
alkalinity, total **alcalinité totale**
alkane **alcane**
alkene **alcène**
alkyl group **groupement alcoyle ou alkyle**
alkyne **alkine**
alloy **alliage**
alpha decay **désintégration alpha**
alpha particle **particule alpha**
aluminum **aluminium**
amines **amines**
ammonia **ammoniaque**
amphoteric substance **substance amphotère**
analysis **analyse**
analyte **analyte**
anhydrite **anhydrite**
anion **anion**
anion exchange capacity **capacité d'échange anionique**
anion exclusion **exclusion anionique**
anionic resin **résine anionique**
aqua regia **eau régale**
aqueous **aqueux**
aqueous solution **solution aqueuse**
aromatic compound **composé aromatique**
Arrhenius equation **équation d'Arrhénius**
ash **cendre**
atmophile elements **éléments atmosphiles ou atmophiles**
atom **atome**
atom percent **pourcentage atomique**
atomic absorption (AA) spectroscopy **spectroscopie d'absorption atomique**
atomic bond **liaison atomique**
atomic mass unit **unité de masse atomique**
atomic number **nombre atomique**
atomic radius **rayon atomique**
atomic weight **masse atomique**
atomic-emission spectroscopy (AES) **spectroscopie d'émission atomique**
atomizer **atomiseur**
autocatalysis **autocatalyse**
Avogadro's law **loi d'Avogadro**
Avogadro's constant **constante d'Avogadro**
background samples **échantillons de référence, échantillons témoins**
base **base**
base dissociation constant (k_b) **constante de dissociation d'une base**
base metal **métal**
base saturation percentage **pourcentage ou taux de saturation en bases**
basic **basique**
becquerel **becquerel**
Beer-Lambert law **loi de Beer-Lambert**
benzene **benzène**
benzine **benzine**

beta counter **compteur bêta**
beta particle **particule bêta**
beta radiation **radiation bêta**
bicarbonate **bicarbonate**
bicarbonate alkalinity **alcalinité bicarbonatée**
bidentate ligand **ligand bidenté**
blank **blanc, essai à blanc**
boiling point **point d'ébullition**
bomb calorimeter **bombe calorimétrique**
bond energy **énergie de liaison**
bond length **longueur de liaison**
Boyle's law **loi de Boyle**
breakthrough curve **courbe de fuite**
Brownian motion **mouvement brownien**
buffer **tampon**
buffer capacity **capacité tampon**
buffer compounds, soil **complexe tampon du sol**
buffer power **pouvoir tampon**
buffered (soil) **sol tamponné**
buffered solution **solution tamponnée**
calcine **calcine**
calcium carbonate equivalent **équivalent carbonate de calcium**
calibration **étalonnage**
calibration curve **courbe d'étalonnage**
calorific value **pouvoir calorifique**
carbohydrate **glucide, hydrate de carbone**
carbon dating **datation au carbone**
carbon-14 (^{14}C) **carbone 14 (^{14}C)**
carbonate **carbonate**
carbonate buffer system **système de tampon carbonate**
carbonic acid **acide carbonique**
carboxyl group **carboxyle**
carboxylic acid **acide carboxylique**
carrier gas **gaz vecteur**
catalyst **catalyseur**
cathode **cathode**
cathodic protection **protecteur cathodique**
cation **cation**
cation exchange capacity (CEC) **capacité d'échange cationique**
caustic soda **soude caustique**
cell potential (electromotive force) **force électromotrice**
centrifuge **centrifuge**
change of state **changement d'état**
Charles' law **loi de Charles**
chelate **chélat**
cheluviation **chéluviation**
chemical bond **liaison chimique**
chemical equilibrium **équilibre chimique**
chemical potential **potentiel chimique**
chemical stoichiometry **stoechiométrie chimique**
chemisorption **adsorption chimique, chimisorption**
chemistry, soil **chimie du sol**
chirality **chiralité**
chlorides **chlorures**
chlorinated **chloré**
chloroform **chloroforme**
chromatography **chromatographie**
closed system **système fermé**
coagulation **coagulation**
colligative properties **propriétés colligatives**
colorimetry **colorimétrie**
combustible liquid **liquide combustible**
combustion **combustion**
common ion effect **effet d'ion commun**
complex **complexe**
complex ion **ion complexe**
composite sample **échantillon composite**
compound **composé**
concentration **concentration**
concentration gradient **gradient de concentration**
condensation reaction **réaction de condensation**
conjugate acid **acide conjugué**
conjugate acid-base pair **paire acide-base conjuguée**
conjugate base **base conjuguée**
constant-potential surface **surface à potentiel constant**
coordinate bond **liaison de coordination**
covalent bond **liaison covalente**
cracking **cracking, craquage**
crosslinking **liaison transversale**
crucible furnace **fournaise à creusets**
cyclotron **cyclotron**
Dalton's law **loi de Dalton**
decay product **produit de désintégration**
degradation **dégradation**
dehydrogenation **déshydrogénation**
deionization **désionisation**
deionized water **eau déionisée**
deliquescence **déliquescence**
desiccant **desséchant, dessicant**
desorption **désorption**
detection limit **limite de détection**

detergent **détergent**
deuterium **deutérium**
dialysis **dialyse**
diamagnetism **diamagnétisme**
differential rate law **loi de vitesse différentielle**
differential thermal analysis **analyse thermale différentielle**
diffuse double layer **double couche diffuse**
diffusion **diffusion**
diffusion (nutrient) **diffusion d'un élément nutritif**
diffusion coefficient **coefficient de diffusion**
diffusivity **diffusivité**
diluent **diluant**
dilution **dilution**
dilution factor **facteur de dilution**
dimer **dimère**
dipole **dipôle**
dipole moment **moment dipolaire**
dipole-dipole attraction **attraction dipolaire**
direct photolysis **photolyse directe**
disaccharide **disaccharide**
disintegration **désintégration**
dissociation **dissociation**
dissociation constant **constante de dissociation**
dissolved gases **gaz dissous**
distillation **distillation**
distilled water **eau distillée**
disulfide linkage **liaison disulfide**
double bond **liaison double**
dry weight basis **à base de poids sec**
Dumas method **méthode de Dumas**
effective calcium carbonate equivalent **équivalent effectif de carbonate de calcium**
effective cation exchange capacity **capacité d'échange cationique effective**
E_H **E_H (potentiel redox)**
electrical conductivity (EC) **conductivité électrique**
electricity **électricité**
electrochemistry **électrochimie**
electrodialysis **électrodialyse**
electrokinetic (zeta) potential **potentiel électrocinétique ou potentiel zêta**
electrolysis **électrolyse**
electrolyte **électrolyte**
electrolytic **électrolytique**
electrolytic cell **cellule électrolytique**
electromagnetic radiation **radiation électromagnétique**
electron **électron**
electron capture detector (ECD) **détecteur à conduction**
electrophoresis **électrophorèse**
element **élément**
elutriation **élutriation**
enantiomers **énantiomères**
endothermic **endothermique**
energy **énergie**
enrichment **enrichissement**
enthalpy **enthalpie**
entropy **entropie**
equation of state **équation caractéristique**
equilibrium constant **constante d'équilibre**
equivalent or equivalent weight **équivalent-gramme**
ester **ester**
exchange complex **complexe d'échange**
exchangeable anion **anion échangeable**
exchangeable base **base échangeable**
exchangeable cation **cation échangeable**
exchangeable cation percentage **pourcentage de cations échangeables**
exchangeable nutrient **élément nutritif échangeable**
exothermic **exothermique**
exponential decay **décroissance exponentielle**
extract, soil **extrait de sol**
extractant **solution d'extraction**
fats **gras**
fatty acid **acide gras**
field blanks **blancs d'essai**
filtration **filtration**
fire point **point d'inflammation spontanée**
first law of thermodynamics **première loi de la thermodynamique**
first-order reaction **réaction de premier ordre**
fission **fission**
fixed ammonium **ammonium fixé**
flash point **point éclair**
fluorescence **fluorescence**
flux **flux**
formula weight **masse moléculaire**
Fourier transform **transformation de Fourier**
Frasch process **procédé de Frasch**
free energy **énergie libre**
free radical **radical libre**
freezing point **point de congélation**
frequency **fréquence**

Freundlich isotherm **formule de Freundlich**
fuel cell **pile à combustible**
functional group **groupement fonctionnel**
furans **furannes**
fusion **fusion**
galvanic cell **pile galvanique**
galvanizing **galvanisation**
gamma ray **rayon gamma**
gas **gaz**
gas chromatography (GC) **chromatographie en phase gazeuse**
gas-phase absorption spectrum **spectre d'absorption de la phase gazeuse**
Geiger-Muller counter (Geiger counter) **compteur à pointe de Geiger**
geometrical *(cis-trans)* isomerism **isomérie géométrique cis-trans**
Gibbs free energy **énergie libre de Gibbs**
glass electrode **électrode de verre**
glycosidic linkage **liaison glycosidique**
Graham's law **loi de Graham**
gravimetric analysis **analyse gravimétrique**
ground state **état fondamental**
group (of the periodic table) **groupement**
Haber process **procédé de Haber**
half-life **période, demi-vie**
half-reactions **demi-réactions**
halogen **halogène**
halogenation **halogénation**
heat of combustion **chaleur de combustion**
heat of condensation **chaleur latente de condensation**
heat of hydration **chaleur d'hydratation**
heat of solution **chaleur de dissolution**
heat of vaporization **chaleur latente de vaporisation**
heavy atom **atome lourd**
heavy hydrogen **hydrogène lourd (deutérium)**
heavy metal **métal lourd**
heavy water **eau lourde**
Henry's law **loi de Henry**
Henry's law constant **constante de la loi de Henry**
heterogeneous equilibrium **équilibre hétérogène**
high performance liquid chromatography (HPLC) **chromatographie liquide à haute performance**
homogeneous equilibrium **équilibre homogène**
homologous series **série homologue**
hydrate **hydrate**
hydrated lime **chaux hydratée**
hydration **hydratation**
hydrocarbon **hydrocarbure**
hydrocarbon derivative **dérivé d'hydrocarbure**
hydrodynamic dispersion **dispersion hydrodynamique**
hydrogen bond **liaison hydrogène**
hydrogen bonding **liaison hydrogène**
hydrogen sulfide (H_2S) **sulfure d'hydrogène**
hydrogenation reaction **réaction d'hydrogénation**
hydrogen-ion concentration **concentration en ions hydrogène**
hydrolysis **hydrolyse**
hydrometallurgy **hydrométallurgie**
hydronium ion **ion hydronium**
hydrophilic **hydrophile**
hydrophobic **hydrophobe**
hydroxide **hydroxyde**
hyperosmotic **hyperosmotique**
hypersaline **hypersalin**
hypoosmotic **hypoosmotique**
hypotonic **hypotonique**
ideal gas **gaz parfait**
ideal gas law (universal gas equation) **loi des gaz parfaits**
ideal solution **solution idéale**
immiscible **immiscible**
in situ ***in situ***
in vitro ***in vitro***
in vivo ***in vivo***
inactivation **inactivation**
incongruent solution **dissolution non congruente**
indicator **indicateur**
inert gas **gaz inerte**
infrared absorption spectroscopy (IR) **spectroscopie d'absorption dans l'infrarouge**
inorganic **inorganique**
instrument blank **blanc d'instrument**
intermolecular forces **forces de van der Waals**
ion **ion**
ion activity **activité ionique**
ion exchange **échange d'ions**
ion pairing **formation d'une paire ionique**

ion selective electrode **électrode à ion spécifique**
ion selectivity **sélectivité ionique**
ionic bond **liason ionique**
ionic bonding **liaison ionique**
ionic compound (binary) **combinaison ionique**
ionic diffusion **diffusion ionique**
ionic solid **solide ionique**
ionic strength **force ionique**
ionic substitution **substitution ionique**
ionization **ionisation**
ionization energy **énergie d'ionisation**
ionizing radiation **radiation ionisante**
ion-selective electrode **électrode à ions spécifique**
irradiated **irradié**
isoelectric point **point isoélectrique**
isoelectronic ions **ions isoélectriques**
isomer **isomère**
isotactic chain **chaîne isotactique**
isotonic solutions **solutions isotoniques**
isotope dilution **dilution isotopique**
isotope **isotope**
isotopic fractionation **fractionnement isotopique**
isotopically exchangeable ion **ion isotopiquement échangeable**
ketone **cétone**
Kjeldahl method **méthode Kjeldahl**
Kjeldahl nitrogen **azote Kjeldahl**
laboratory blank **blanc de laboratoire**
Langmuir isotherm **isotherme de Langmuir**
law of conservation of energy **loi de la conservation de l'énergie**
Le Chatelier's principle **principe de Le Chatelier**
Lewis structure **structure de Lewis**
ligand **ligand (coordinat)**
lime **chaux**
lime potential **potentiel de chaux**
lime requirement **besoin en chaux**
lime-soda process **procédé chaux-soude**
linear accelerator **accélérateur linéaire**
linear alkyl sulfonate (LAS) **alkylsulfonate linéaire**
liquid chromatography **chromatographie en phase liquide**
lithophile element **élément lithophile**
lone pair **paire ionique libre**
mass number **nombre de masse**
mass spectrometer **spectromètre de masse**
material blank **blanc de matériau**
melting point **point de fusion**
membrane filter **filtre à membrane**
mercaptans **mercaptans**
metal **métal**
metalloids (semimetals) **métalloïdes**
metallurgy **métallurgie**
methyl mercury **méthyle de mercure**
microwave **micro-onde, hyperfréquence**
miscible **miscible**
mixture **mélange**
molality **molalité**
molar concentration **concentration molaire**
molar heat capacity **chaleur molaire**
molar mass **masse molaire**
molar volume **volume molaire**
molarity **molarité**
mole fraction **fraction molaire**
mole ratio (stoichiometry) **rapport molaire**
molecular equation **équation moléculaire**
molecular formula **formule moléculaire**
molecular structure **structure moléculaire**
molecular weight **poids moléculaire**
molecule **molécule**
monodentate (unidentate) ligand **coordinat monodenté**
monomer **monomère**
monoprotic acid **monoacide**
monosaccharide (simple sugar) **monosaccharide**
muffle furnace **fournaise à mouffles**
natural abundance **abondance naturelle**
natural radioactivity **radioactivité naturelle**
nephelometer **néphélomètre**
Nernst equation **équation de Nernst**
neutral soil **sol neutre**
neutralization **neutralisation**
neutron **neutron**
nitrate (NO_3) **nitrate**
nitric acid **acide nitrique**
nitric oxide (NO) **oxyde nitrique**
nitrite **nitrite**
nitrogen **azote**
nitrogen dioxide (NO_2) **dioxyde d'azote**
nitrosamines **nitrosamines**
nonpolar solvent **solvant non polaire**
normal solution **solution normale**
nuclear energy **énergie nucléaire**
nuclear fission **fission nucléaire**

nuclear fusion **fusion nucléaire**
nuclear magnetic resonance (NMR) spectroscopy **spectroscopie de résonance magnétique nucléaire**
nuclear reactor **réacteur nucléaire**
nucleon **nucléon**
nucleus **noyau**
nuclide **nucléide**
optical activity **activité optique**
organic **organique**
organic acid **acide organique**
organic carbon **carbone organique**
organic chemistry **chimie organique**
organic phosphorus **phosphore organique**
organic solvent **solvant organique**
organochlorine **organochloré**
Ostwald process **procédé d'Ostwald**
outgassing **dégazage**
oven-dry soil **sol séché à l'étuve**
oxidant **oxydant**
oxidation **oxydation**
oxidation state **état d'oxydation**
oxidation-reduction (redox) reaction **réaction d'oxydo-réduction**
oxides of nitrogen, NOx **oxydes d'azote, NOx**
oxidizing agent (electron acceptor) **agent oxydant**
paraffins **paraffines**
partial pressure **pression partielle**
particle accelerator **accélérateur de particules**
partition coefficient **coefficient de partage ou de répartition**
PCBs **BPC**
pentachlorophenol (PCP) **pentachlorophénol (PCP)**
peptide **peptide**
peptide bond **liaison peptidique**
period **période**
periodic table **tableau périodique**
petrochemicals **dérivés du pétrole**
pH curve (titration curve) **courbe de titration**
pH scale **échelle pH**
pH, soil **pH du sol**
pH_C **pH_C**
pH-dependent cation exchange capacity **capacité d'échange cationique dépendante du pH**
pH-dependent charge **charge dépendante du pH**
phenols **phénols**
photochemical **photochimique**
photoelectric cell **cellule photoélectrique**
photoelectric effect **effet photoélectrique**
photoelectron **photoélectron**
photon **photon**
phthalates **phtalates**
physisorption **physisorption**
pK_a **pK_a**
pK_b **pK_b**
plant analysis **analyse végétale**
point of zero net charge **point de charge nette nulle**
polar covalent bond **liaison covalente polaire**
polar molecule **molécule polaire**
polar solvent **solvant polaire**
polarity **polarité**
polychlorinated biphenyls (PCBs) **biphényles polychlorés (BPC)**
polychlorinated dibenzofurans (PCDFs, furans) **dibenzofurannes polychlorés (DFCP, furannes)**
polycyclic **polycyclique**
polycyclic aromatic hydrocarbons (PAH) **hydrocarbures aromatiques polycycliques (HAP)**
polymers **polymères**
polymerization **polymérisation**
polynuclear aromatic hydrocarbons (PAH) **hydrocarbures aromatiques polynucléaires (HAP)**
polypeptide **polypeptide**
polyprotic acid **acide polyprotique**
polysaccharide **polysaccharide**
polyvinyl chloride (PVC) **chlorure de polyvinyle**
porphyrin **porphyrine**
precipitant **précipitant**
precipitate **précipité**
pressure filter **filtre à pression**
pressure membrane **membrane de tensiomètre**
proton **proton**
purine **purine**
pyrimidine **pyrimidine**
pyrolysis **pyrolise**
pyrophoric **pyrophorique**
qualitative analysis **analyse qualitative**
quality assurance **assurance de qualité**
quality control **contrôle de qualité**
quantitative analysis **analyse quantitative**
quantum **quantum**
radical **radical**

radioactive decay **dégradation radioactive**
radioactive isotope **isotope radioactif**
radioactive series **série radioactive**
radioactivity **radioactivité**
radiocarbon **radiocarbone**
radiocarbon dating **datation au radiocarbone**
radiochemistry **radiochimie**
radioisotope **radio-isotope**
radionuclide **radionucléide**
radiotracer **radiotraceur**
radon (Rn) **radon**
Raoult's law **loi de Raoult**
rate constant **constante de vitesse**
rate-determining step **étape déterminante de la vitesse**
reactant **réactif**
reaction rate **vitesse de réaction**
reaction, soil **réaction du sol**
reagent **réactif**
reagent blank **blanc de réactif**
redox potential **potentiel d'oxydoréduction**
reducing agent (electron donor) **agent réducteur (donneur d'électron)**
reducing environment **environnement réducteur**
reduction **réduction**
reference electrode **électrode de référence**
reflux **reflux**
reverse osmosis **osmose inversée**
ring compound **composé cyclique**
salt **sel**
salt bridge **pont de sel**
saturated hydrocarbon **hydrocarbure saturé**
saturated soil paste **pâte de sol à saturation**
saturation percentage **pourcentage de saturation**
saturation-paste extract **extrait de pâte à saturation**
scintillation counter **compteur à scintillation**
second law of thermodynamics **deuxième loi de la thermodynamique**
selective precipitation **précipitation sélective**
semipermeable membrane **membrane semi-perméable**
serial dilution **suspension-dilution, dilution en série**
silica gel **gel de silice**
single bond **liaison simple**
size-exclusion chromatography (SEC) **chromatographie sur gel ou par exclusion**
smelting **fusion**
soil acidity **acidité du sol**
soil sample **échantillon de sol**
soil test **analyse de sol**
solubility **solubilité**
solubility product (K_s) **produit de solubilité**
solute **soluté**
solution **solution**
solution, soil **solution du sol**
solvent **solvant**
sorption **sorption**
specific activity **activité spécifique**
specific heat capacity **chaleur massique**
spectrometer **spectromètre**
spectro-photometer **spectrophotomètre**
spectroscopy **spectroscopie**
spectrum (plural spectra) **spectre**
specular reflection **réflexion spéculaire**
stable isotope **isotope stable**
standard **étalon**
standard curve **courbe d'étalonnage**
standard free energy change **changement d'énergie libre standard**
standard free energy of formation **énergie libre standard de formation**
standard method **méthode standardisée**
standard solution **solution étalon**
states of matter **états de la matière**
steroid **stéroïde**
stoichiometric quantities **quantités stoechiométriques**
straight-chain hydrocarbons **hydrocarbures aliphatiques**
stripping **stripping, strippage**
strong acid **acide fort**
strong base **base forte**
strong electrolyte **électrolyte fort**
substituted ring compound **composé cyclique substitué**
supercooling **surfusion**
superheating **surchauffe**
supersaturation **sursaturation**
surface charge density **densité de charge surfacique**
surface tension **tension superficielle**
surfactant **agent tensio-actif**
tagged molecule **molécule marquée**
theoretical yield **rendement théorique**
thermal analysis **analyse thermique**
thermodynamics **thermodynamique**
thin-layer chromatography (TLC) **chromatographie sur couche mince**

third law of thermodynamics **troisième loi de la thermodynamique**
tissue analysis **analyse de tissus végétaux**
titrant **solution titrante**
titration **titration, titrage**
total solids **matières solides totales**
total suspended solids **total des solides, matières totales en suspension**
trace metals **métaux à l'état de traces**
tracer **traceur**
trihalomethanes (THMs) **trihalométhanes**
triple bond **liaison triple**
ultraviolet radiation (UV radiation) **rayonnement ultraviolet (rayonnement UV)**
unstable **instable**
valence **valence**
van der Waals' forces **forces de van der Waals**
van't Hoff factor **facteur de van't Hoff**
vapor pressure **pression de vapeur**
vaporize **vaporiser**
volatile hydrocarbons **hydrocarbures volatils**
volatile organic compounds (VOCs) **composés organiques volatils**
volumetric water content **teneur en eau volumétrique**
water softener **adoucisseur d'eau**
wax **cire**
weak acid **acide faible**
weak base **base faible**
weak electrolyte **électrolyte faible**
wetting agent **agent mouillant**
zero-order reaction **réaction d'ordre zéro**
zeta potential **potentiel zêta**

CONSERVATION, SOIL MANAGEMENT, TILLAGE

agri-environmental indicator **indicateur agro-environnemental**
agroecosystem **agroécosystème, écosystème agricole**
agroecosystem complexity **diversité ou complexité d'un agroécosystème**
agronomically sustainable yield **rendement agronomique durable**
anchor **enfouir, incorporer**
arable **arable**
attribute of soil quality **attribut de la qualité des sols**
back furrow **ados**
bed planting **semis sur buttes ou sur lits surélevés**
bed shaper **billonneuse, buttoir**
bedding **billonnage, formation de plate-bandes**
benchmark site **site de référence**
bisect **coupe, profil racinaire**
block **bloc**
blown-out land **terrain de déflation**
blowout **cuvette de déflation**
boundary screen **écran de protection**
broadcast planting **semis à la volée**
broadcast tillage **labour complet**
broadcast tillage (total surface tillage, full-width tillage) **travail sur toute la surface de sol**
buffer strip **bande ou zone tampon**
burying **enfouissement**
catch crop **culture dérobée**
channel stabilization **stabilisation des berges**
chemical fallow (eco-fallow) **jachère chimique**
chisel **scarifier, sous-soler**
chisel cultivator **cultivateur, chisel, charrue scarificatrice**
chisel planting **travail à l'aide d'un cultivateur ou d'un chisel**
chiseling **préparation de sol avec cultivateur ou chisel, scarification, sous-solage**
clean tillage **labour nettoyant**
combined tillage operations **opérations combinées de travail du sol**
companion crop **culture associée, culture compagne**
conservation **conservation**
conservation plan **plan ou programme de conservation**
conservation practice **pratique de conservation**
conservation tillage **pratique de conservation**
continuous cropping **culture continue, monoculture**
contour cultivation **culture en courbes de niveau**
contour furrows **canaux de dérivation**

contour strip cropping **culture en bandes suivant les courbes de niveau**
contour tillage **travail du sol suivant les courbes de niveau**
controlled traffic **circulation limitée**
controlling variable **facteur limitatif**
conventional tillage **travail du sol conventionnel**
cover crop **culture de couverture**
crop land **terre arable, terre en culture**
crop residue **résidus de cultures**
crop residue management **gestion des résidus de cultures**
crop residue management system **système de gestion des résidus de cultures**
crop rotation **rotation culturale ou des cultures**
cropping intensity **assolement**
cropping pattern **système de culture**
cropping system **système de culture**
cross cultivation **travail du sol en effectuant des passages croisés**
cross-slope farming **culture en contre-pente**
crushing **écrasement, fragmentation**
cultipack **cultitassement, raffermissement du lit de semence**
cultivated land **terre cultivée**
cultivation **hersage, travail du sol**
cultivation (weeding) **sarclage**
cultural vegetation **plante cultivée**
cutting **découpage**
cybernetic systems **systèmes cybernétiques**
dam (pitting, basin listing) **modelage superficiel**
dead furrow **ados, sillon terminal, raie de curage**
demonstration area **site de démonstration, site d'essais**
depleted soil **sol épuisé, sol appauvri**
depth, effective soil **profondeur de la couche arable**
detachment **détachement**
differential erosion **érosion différentielle**
dig **creuser**
down-scaling **réduction ou diminution d'échelle**
drag **niveler avec une traîne**
drill seeding **semis en ligne à espacement rapproché**
dryland farming **aridoculture, culture en région sèche**
duckfoot **en patte d'oie**
dust mulch **sol pulvérisé**
dynamometer **dynamomètre**
eco-fallow **jachère écologique**
ecosystem **écosystème**
ecosystem health **santé des écosystèmes**
ecosystem management **gestion des écosystèmes**
ecosystem perspective **vision écosystémique**
effective soil depth **profondeur effective de sol**
erode **éroder**
erosion **érosion**
erosion classes **classes d'érosion**
erosion control **méthodes de lutte contre l'érosion**
erosion control structures **structures de lutte contre l'érosion**
erosion pavement **dallage d'érosion**
erosion potential **érosion potentielle**
erosion risk, actual **risque ou potentiel d'érosion**
erosion risk, inherent **risque inhérent d'érosion**
erosion surface **surface érodée**
erosion, accelerated **érosion accélérée, érosion anthropique**
erosion, differential **érosion différentielle**
erosion, geological **érosion géologique, érosion naturelle**
erosivity **érosivité**
extended rotation **rotation longue ou allongée**
extrinsic factor **facteur extrinsèque**
exudate **exsudat**
fallow **jachère**
fallow land **terre en jachère**
farming system **système de production**
filter strip **bande filtrante**
firming **tassement, raffermissement**
flat planting **semis superficiel ou à plat**
flexible cropping **assollement facultatif ou souple**
fragile land **terre vulnérable ou fragile**
furrow **sillon, raie**
furrow slice **couche de labour, couche travaillée**
grass barrier **bande enherbée**
grassed waterway **voie d'eau engazonnée**
green belt **ceinture verte, ceinture de verdure**
green manure **engrais vert**
groin **épi**

guess row **ligne de guidage**
gullied land **terrain raviné**
gully **ravin**
gully control plantings **stabilisation de ravins, végétalisation de ravins**
gully erosion **érosion par ravinement**
gully reclamation **remise en état de ravins**
gypsum requirement **besoin en gypse**
harrowing **hersage**
hill **buttage**
hoe **biner, passer la houe**
incorporation **incorporation**
indicator **indicateur**
in-row subsoiling **sous-solage**
intensive cropping **culture intensive**
intercropping **cultures intercalaires**
interplanting **cultures intercalaires, plantation intercalaire**
inter-rill erosion **érosion en nappe ou en couche**
interseeding **semis en intercalaire ou en plante-abri**
inversion **inversion, bouleversement**
K factor **facteur K**
L factor **facteur L**
land planing **nivellement, planage**
land quality **qualité des terres**
land resting **jachère, friche temporaire**
landforming **modelage**
lift **arracher**
liming **chaulage**
limiting nutrient **élément déficient ou limitant**
lister planting **semis au creux du sillon**
listing (middle breaking) **forme de buttage ou billonnage**
loosening **ameublissement**
marginal land **terre marginale, terre à faible potentiel agricole**
maximum sustainable yield **rendement optimal, rendement maximal durable**
minimum data set **ensemble de données minimal**
minimum tillage **travail minimal du sol**
mixed farm **ferme ou entreprise agricole diversifiée**
mixing **mélange**
moldboard plowing **labour avec charrue à versoirs**
monoculture **monoculture**
mulch **mulch, paillis**
mulch farming **culture sur résidus ou sur paillis**
mulch tillage **travail du sol avec conservation des résidus, culture sur paillis**
narrow row planting **plantation en rangs étroits ou serrés**
natural erosion **érosion naturelle**
non-inversive tillage **travail sans inversion de sol**
normal erosion **érosion normale**
no-tillage (zero tillage) **système zéro-labour, semis-direct, système de culture sans labour**
nurse crop **plante-abri**
off-site damage costs **coût des dommages hors-ferme**
once-over tillage **préparation du sol par un passage unique d'outils combinés**
organic farming **agriculture biologique, agriculture organique, agriculture écologique**
oriented tillage **travail du sol directionnel**
overseeding **sursemis**
P factor **facteur P**
pan **couche indurée**
paraplow **décompacteur de type paraplow**
paratill **décompacteur de type paratill**
pedotransfer function **fonction de pédotransfert**
permanent crop cover **couverture végétale permanente**
physical rooting conditions **conditions physiques d'enracinement**
plant residue **résidus végétaux**
plow layer **couche de labour, couche de sol arable**
plow pan **semelle de labour**
plow sole **semelle de labour**
plowing **labour**
plowless farming **culture sans labour**
plow-plant **semis direct**
plow-planting **labour et semis en un passage (à l'aide d'un train d'outils)**
pre-(post-) emergence tillage **travail du sol en pré- ou post-émergence**
pre-(post-) harvest tillage **travail du sol en pré- ou post-récolte**

pre-(post-) planting tillage **travail du sol en pré- ou post-semis**
pressure pan (traffic sole, plow pan, tillage pan, traffic pan, plow sole, compacted layer) **couche indurée d'origine anthropique, semelle de labour**
primary tillage **préparation primaire du sol**
prime agricultural land **terre agricole de première qualité**
productivity **productivité**
proxy property **propriété subrogative**
puddled soil **sol sans stucture ou à structure massive**
R factor **facteur R**
rain splash **éclaboussement**
rain wash **délavage par les pluies**
raindrop erosion **érosion par éclaboussement**
reduced tillage **travail du sol réduit ou simplifié**
reference site **site de référence, site témoin, site repère**
renewable energy **énergie renouvelable**
renewable resource **ressource renouvelable**
reservoir tillage (damming, pitting, basin listing, furrow diking, dammer diking) **modelage superficiel du sol pour former des petits bassins de captage**
residue **résidu**
residue management **gestion des résidus de cultures**
residue processing **broyage et déchiquetage des résidus de cultures**
resource (natural) **ressource naturelle**
resource management system **système de gestion des ressources**
resource monitoring **suivi de l'évolution des ressources**
ridge **billon**
ridge planting **semis sur billons**
ridge tillage **billonnage**
rill **rigole**
ripping **sous-solage**
rod weeding **désherbage mécanique à l'aide d'un cultivateur à tige rotative**
rolling **roulage**
root bed **profil racinaire**
rotary hoeing **sarclage ou hersage à la houe rotative**
rotary tilling **travail du sol avec bêche ou charrue rotative**
rotation **rotation**
row crop **culture en rangées ou en lignes, grande culture**
runoff **ruissellement**
runoff plot **parcelle de mesure d'érosion**
S factor **facteur S**
saltation **saltation**
scarifying **scarifiage léger, écroûtage**
secondary tillage **travail secondaire ou superficiel du sol, reprise de labour**
seedbed **lit de semence**
seed-drilling **semis à l'aide d'un planteur**
sequential cropping **cultures successives**
shatter **fragmentation, écrasement**
shatter belt **zone de faille**
shearing **cisaillement**
sheet erosion **érosion en nappe ou inter-rigoles**
sheet piling **structure de confinement**
shelterbelt **haie brise-vent**
slit planting (slot planting) **semis-direct**
slit tillage **travail du sol en bandes étroites**
snow hedge **trappe à neige (haie)**
snow management **gestion de la neige**
sod planting **semis-direct sur prairie ou sur gazon**
soil amendment **amendement du sol**
soil biological quality **qualité biologique des sols**
soil chemical quality **qualité chimique des sols**
soil conservation **conservation du sol**
soil degradation **dégradation des sols**
soil function **rôle ou fonction du sol**
soil health **santé des sols**
soil improvement **amélioration des sols**
soil loss **perte de sol**
soil loss equation **équation de perte de sol**
soil loss tolerance **perte de sol tolérable**
soil management **gestion des sols**
soil management group **groupe de gestion des sols**
soil organic matter quality **qualité de la matière organique du sol**
soil pedological quality **qualité pédologique des sols**
soil physical quality **qualité physique des sols**
soil productivity **productivité des sols**
soil quality **qualité des sols**
soil quality attribute **attribut de la qualité des sols**

soil quality control **gestion de la qualité des sols**
soil quality framework **cadre d'évaluation de la qualité des sols**
soil quality indicator **indicateur de la qualité des sols**
soil quality monitoring program **programme de surveillance de la qualité des sols**
soil quality susceptibility, index of **indice de vulnérabilité des sols**
soil quality, assessment of **évaluation de la qualité des sols**
soil quality, dynamic **propriété dynamique du sol**
soil quality, inherent **qualité inhérente ou intrinsèque du sol**
soil redistribution **redéposition de sol**
soil resiliency **résilience du sol**
soil transportation **transport de particules de sol**
soil variability **variabilité (spatio-temporelle) des sols**
soil-conserving crops **cultures améliorantes ou de conservation**
soil-depleting crops **cultures épuisantes pour le sol**
splash erosion **érosion par éclaboussement**
spreader strip **bande tampon**
statistical quality control **contrôle statistique de la qualité**
strip cropping **culture en bandes, culture en bandes selon les courbes de niveau**
strip planting (strip till planting) **semis en bandes**
strip tillage (partial-width tillage) **travail du sol en bandes**
strip-till planting **semis sur bandes travaillées**
stubble **chaume**
stubble crops **regain, culture de couverture**
stubble mulch **paillis de résidus de culture**
stubble mulch tillage **travail du sol préservant les résidus de culture**
subsoiling **sous-solage**
subsurface tillage **sous-solage**
summer fallow **jachère d'été**
surface creep **saltation**
surface tillage **travail du sol superficiel, façon superficielle**
surficial erosion **érosion superficielle**
surrogate property **propriété subrogative**
suspension **suspension**
sustainable agriculture **agriculture durable**
sustainable yield **rendement durable ou soutenu**
sweep **cultivateur ou herse pourvu de socs à ailettes**
terrace **terrasse**
threshold value (soil quality) **valeur limite, valeur seuil, seuil**
threshold velocity **vitesse minimale, vitesse limite**
throw **projection**
tie-ridge **billons reliés entre eux de façon à former des bassins**
till **cultiver le sol, labourer**
tillability **degré de facilité de travail du sol**
tillage **travail du sol**
tillage action **effets découlant du travail du sol**
tillage equipment (tools) **outil de travail du sol, instrument aratoire**
tillage erosion **érosion ou déplacement de sol causé par le labour**
tillage operation **pratique culturale**
tillage requirement **besoin en labour ou en travail du sol**
tillage, deep **labour profond, labour de défoncement, sous-solage**
tilth **état d'ameublissement ou état structural du sol**
topsoil **couche de sol labourée ou travaillée, sol arable, terre végétale**
total surface tillage **travail sur toute la surface du sol**
trash farming **culture sur résidus**
turnrow (turn strip, head land) **tournière, cintre**
underseeding **semis sous couvert végétal, sous-ensemensement**
universal soil loss equation **équation universelle de perte de sol**
up-scaling **mise à l'échelle**
vertical mulching **placement vertical d'amendements**
weeding **sarclage**
wheel track planting **semis en ornières**
windbreak **brise-vent, haie brise-vent**
zero tillage **zéro-labour, semis-direct**
zone subsoiling **sous-solage localisé**
zone tillage **travail du sol localisé**

ECOLOGY

abiotic **abiotique**
acclimation **acclimatement**
acclimatization **acclimatation**
adaptation **adaptation**
adaptive zone **zone adaptative**
age structure **structure d'âge**
agronomy **agronomie**
allele **allèle**
allelopathy **allélopathie**
allochthonous **allochtone**
allogenic succession **succession allogène**
allopatric speciation **spéciation allopatrique**
alpha diversity **diversité alpha**
alpine **alpin**
alpine biome **biome alpin**
alpine tundra **toundra alpine**
amensalism **allélopathie**
amplitude **amplitude**
antagonism **antagonisme**
arctic **arctique**
arctic tundra **toundra arctique**
association **association**
associative dinitrogen fixation **fixation associative d'azote**
autecology **auto-écologie**
autochthonous **autochtone**
autopoiesis **autopoièse**
autotoxic **autotoxique**
autotroph **autotrophe**
autotrophy **autotrophie**
barren **dénudé ou stérile**
barren lands **terrains improductifs**
belt **zone**
beta diversity **diversité bêta**
biochore **biochore**
biocoenosis **biocénose**
biodiversity **biodiversité**
biofacies **biofaciès**
biogenic **biogénique**
biogeochemical cycle **cycle biogéochimique**
biogeographic province **province biogéographique**
biogeography **biogéographie**
bioindicator **bioindicateur**
biological availability **biodisponibilité**
biological productivity **productivité biologique**
biomass **biomasse**
biomass energy **énergie de la biomasse**
biome **biome**
biosphere **biosphère**
biota **biote**
biota influence **influence biotique**
biotic **biotique**
biotic factors **facteurs biotiques**
biotic potential **potentiel biotique**
biotope **biotope**
biotype **biotype**
biozone **biozone**
border **bande**
boreal **boréal**
boreal forest **forêt boréale**
bovine **bovin**
broad-leaved deciduous forest **forêt feuillue**
broad-leaved evergreen forest **forêt feuillue à feuillage persistant**
brucellosis **brucellose**
calciphobe, calcifuge **calcifuge**
calciphyte **calciphyte**
casual species **espèce sporadique**
chapparral **chaparral**
character plant **plante spécimen**
chemical rooting conditions **conditions chimiques d'enracinement**
climax **climax**
cline **gradient géographique**
co-dominant trees **arbres codominants**
coevolution **coévolution**
coexistence **coexistence**
cohort **cohorte**
cold desert **désert froid**
community **communauté**
community-unit hypothesis **hypothèse des communautés**
competition **compétition**
competitive exclusion **exclusion compétitive**
consilience **connexité**
consociation **consociation**
continuum **continuum**
counteradaptation **contradaptation**
counterevolution **contrévolution**
cover, percent **pourcentage de couverture ou de recouvrement**
critical area **site fragile**
cryophilous **cryophyle**
cryophyte **cryophyte**
cryovegetation **cryovégétation**
cycle **cycle**

deciduous forest, temperate **forêt tempérée à feuilles caduques**
deep ecology **écologisme radical**
desert **désert**
desert biome **biome de désert**
desertification **désertification**
dieback **dépérissement**
direct competition **compétition directe**
dispersal **dispersion**
dispersion **dispersion**
distribution **distribution**
disturbance **perturbation**
diversity **diversité**
diversity index **indice de diversité**
dominance **dominance**
dominance-diversity curve **relation abondance-diversité**
driving force **force motrice**
dynamic equilibrium state **état d'équilibre dynamique**
dynecology **dynécologie**
ecoclimate **écoclimat**
ecodistrict **écodistrict**
ecologic **écologique**
ecologic counterparts **contreparties écologiques**
ecologic efficiency **efficacité écologique**
ecological classification **classification écologique**
ecological density **densité écologique**
ecological equivalents **équivalents écologiques**
ecological factor **facteur écologique**
ecological niche **niche écologique**
ecological process **processus écologique**
ecological pyramid **pyramide écologique**
ecological release **expansion écologique**
ecological risk assessment **évaluation du risque écologique**
ecological succession **succession écologique**
ecology **écologie**
ecology, soil **écologie du sol**
ecoregion **écorégion**
ecospecies **écoespèce**
ecosphere **écosphère**
ecosystem **écosystème**
ecosystem composition **composition écosystémique**
ecosystem function **fonction écosystémique**
ecosystem structure **structure écosystémique**
ecosystem sustainability **durabilité écosystémique**
ecotone **écotone**
ecotope **écotope**
ecozone **écozone**
edaphic **édaphique**
edaphic climax **climax édaphique**
edaphology **édaphologie**
edge effect **effet de bordure**
element **élément**
endemic **endémique**
endemic species **espèce endémique**
endogenous **endogène**
energy budget **bilan énergétique**
energy flow **flux d'énergie**
environment **environnement**
environment management system **système de gestion de l'environnement**
environmental resistance **résistance environnementale**
environmental sustainability **durabilité de l'environnement**
environmentally sustainable agriculture **agriculture durable du point de vue de l'environnement, environnement durable en agriculture**
epiphyte **épiphyte**
estivation **estivation**
eutroph **eutrophe**
eutrophic **eutrophe**
evenness **régularité**
evolution **évolution**
evolutionary opportunism **opportunisme évolutif**
evolutionary time **période d'évolution**
exotic **exotique**
exploit **exploiter**
exploitation competition **competition d'exploitation**
exploitation equilibrium **équilibre d'exploitation**
exponential growth **croissance exponentielle**
exposure **exposition**
exposure characterization **condition d'exposition**
extinction **extinction**
extrinsic factor **facteur extrinsèque**
fauna **faune**
flora **flore**
food chain **chaîne alimentaire**
food cycle **cycle alimentaire**
food web **chaîne alimentaire ou de prédation**
forcing functions **fonctions motrices**

forcing variable **variable de forçage**
forest **forêt**
formation-class **formation végétale**
founder principle **principe fondateur**
free-living **libre**
fresh water biome **biome d'eau douce**
Gaia hypothesis **hypothèse de Gaia**
gamma diversity **diversité gamma**
globally stable **système résilient**
glycophyte **glycophyte**
gradient analysis **analyse des gradients**
grassland **prairie**
grassland biome **biome de prairie**
gross primary productivity (GPP) **productivité primaire brute**
guild **guilde**
habitat **habitat**
habitat breadth **amplitude de l'habitat**
habitat function **fonction de l'habitat**
habitat needs **besoins en habitat**
habitat niche **niche d'habitat**
habitat selection **sélection d'habitat**
habitat type **type d'habitat**
habitat, critical **habitat critique**
habitat, preferred **habitat préféré**
harvest index **indice de récolte**
heath **lande**
heliophyte **héliophyte**
heterogeneous **hétérogène**
heterotroph **hétérotrophe**
heterotrophic ecosystem **écosystème hétérotrophe**
Holarctic **holarctique**
holistic **holistique**
holistic ecology **écologie holistique**
homogeneous **homogène**
host **hôte**
immigrant **immigrant**
indicator plants **plantes indicatrices**
indigenous **indigène**
indirect competition **compétition indirecte**
interaction **interaction**
intraspecific **intraspécifique**
introduced species **espèce introduite**
keystone species **espèce clé ou pivot**
k-selected **individu à strategie k**
land ecosystem **écosystème terrestre**
landscape ecology **écologie du paysage**
life cycle **cycle de vie**
limiting factor **facteur limitant**
limits of tolerance **limites de tolérance**
limnology **limnologie**
line plot survey **inventaire par échantillonnage en ligne**
line-intercept method **méthode d'intersection**
lithophyte **lithophyte**
lithosere **lithosère**
lithosphere **lithosphère**
macroflora **macroflore**
macroorganism **macroorganisme**
macrophyte **macrophyte**
mangrove **mangrove**
marine biome **biome océanique**
marsh **marécage**
marsh, tidal **marais tidal**
meadow **pré**
Mediterranean scrub biome **biome de maquis méditerranéen**
microcosm **microcosme**
microsere **microsère**
morph **morphe**
mutualism **mutualisme**
natural disturbance **perturbation naturelle**
natural ecosystem **écosystème naturel**
natural increase **croissance naturelle**
natural law **loi naturelle**
natural resources **ressources naturelles**
natural selection **sélection naturelle**
natural sink **puits naturel**
natural vegetation **végétation naturelle**
net aboveground productivity (NAP) **productivité épigée nette**
net community productivity (NCP) **productivité communautaire nette**
net primary productivity (NPP) **productivité primaire nette**
net production efficiency **efficacité de production nette**
niche **niche**
occurrence **occurrence, présence**
old age **vieillesse**
ontogeny **ontogénie**
opportunistic species **espèce oportuniste**
organism **organisme**
orophyte **orophyte**
overpopulation **surpopulation**
oxyphyte **oxyphyte**
paleoecology **paléoécologie**
photosynthesis **photosynthèse**
phylogeny **phylogénie, phylogenèse**
physical–chemical environment **environnement physico-chimique**

phytoecology **phytoécologie**
pioneer **végétation ou plante pionnière**
plagiosere (plagioclimax) **plagiosère**
plant competition **compétition végétale**
plant succession **succession végétale**
population **population**
population dynamics **dynamique de population**
potential natural plant community **végétation naturelle potentielle**
prairie **prairie**
primary productivity **productivité primaire**
primary succession **succession primaire**
principle of allocation **principe d'allocation**
proclimax **proclimax**
producer **producteur**
productivity **productivité**
progeny **progénie**
provenance **provenance**
provenance test **test de provenance**
proximate factors **facteurs proximaux**
psammon **psammon**
psammosere **psammosère**
pyramid of biomass **pyramide de biomasse**
pyramid of energy **pyramide d'énergie**
quadrat **quadrat**
radioecology **radioécologie**
range **aire de distribution géographique, parcours, parcours naturel**
range of variability **variabilité**
rare **rare**
recruitment **recrutement**
refuge **refuge**
regulated population **population régulée**
relict **relique**
resident species **espèce résidente**
resilience **résilience**
resistance **résistance**
richness **richesse**
r-selected **individu à stratégie r**
salt marsh **marais salant**
saturation response **réaction de saturation**
savanna (savannah, savana) **savanne**
scale **échelle**
sciophilous **sciaphile**
sciophyte **sciophyte**
scrub **broussaille**
secondary consumer **consommateur secondaire**
secondary productivity **productivité secondaire**
secondary succession **succession secondaire**
section **section**
segetal vegetation **végétation ségétale**
selective enrichment **enrichissement sélectif**
semi-desert **région semi-désertique**
senescence **vieillissement**
seral community **groupement végétal préclimatique**
seral stage **stade séral**
serclimax **serclimax**
sere **sère**
Shannon-Weaver index **indice de Shannon-Weaver**
Shelford's law **loi de Shelford**
shrub **arbuste**
sigmoid growth **croissance sigmoïde**
site **site, station**
site factors **facteurs de station**
site productivity **productivité ou potentiel du site**
site quality **qualité de station**
specialization **spécialisation**
species **espèce**
species composition **composition taxonomique**
species density **densité spécifique**
species diversity **diversité spécifique**
species frequency **fréquence d'espèces**
species richness **richesse spécifique, diversité des espèces**
species richness index **indice de richesse spécifique**
species, introduced **espèce introduite**
species, native **espèce indigène**
species, naturalized **espèce naturalisée**
stability **stabilité**
stable age distribution **distribution par âge stable**
standing crop **rendement en biomasse**
strain **souche**
stratification **stratification**
study area **aire d'étude**
subarctic **subarctique**
subpopulation **sous-population**
subsere, secondary sere **subsère**
subtropical **subtropical**
succession **succession**
succession, autogenic **succession autogène**
succession, autotrophic **succession autotrophe**
succession, biotic **succession biotique**

sustainability **durabilité**
symbiosis **symbiose**
sympatric **sympatrique**
synecology **synécologie**
synergism **synergie**
synoptic measurements **mesures synoptiques**
system **système**
systematics **systématique**
systems ecology **écologie des systèmes**
taiga **taïga**
taxon cycle **cycle d'un taxon**
ten percent rule **règle du 10%**
threatened species **espèce menacée**
threshold moisture content **humidité critique, point de flétrissement**
tidal marsh **marais tidal**
time lag **délais**
tolerance **tolérance**
transit time **période de transition**
transpiration efficiency **efficacité de transpiration**
tree line **ligne des arbres**
trophic **trophique**
trophic level **niveau trophique**
trophic structure **structure trophique**
tropical rainforest **forêt tropicale humide**
tropophyte **tropophyte**
tundra **toundra**
turnover **renouvellement**
turnover time **temps de renouvellement**
tussock **tusco, butte**
ubiquitous organisms **organismes ubiquistes**
ultimate factors **facteurs essentiels ou ultimes**
undisturbed ecosystem **écosystème non perturbé ou naturel**
uneven-aged stand **peuplement inéquien**
vegetation **végétation**
vegetation type **type de végétation**
virgin forest **forêt vierge**
xerophyte **xérophyte**
xerosere **xérosère**
xylophagous **xylophage**
zero population growth (ZPG) **croissance nulle**
zone **zone**

ENVIRONMENTAL SOIL SCIENCE, AGRICHEMICALS, PESTICIDES, POLLUTANTS

acid deposition **déposition acide**
acid rain **pluie acide**
active ingredient **matière active**
additive effects **effets additifs**
agrichemical **produit agrochimique**
algicide **algicide**
amide **amide**
ammate **ammate**
atomic waste **déchet atomique**
band application **traitement en bandes**
basal application **traitement à la base des tiges, traitement basal**
bed **lit**
benzene hexachloride (BHC) **hexachlorure de benzène**
benzoic **herbicide benzoïque**
biocide **biocide**
biological control **lutte biologique**
botanical pesticide **pesticide botanique ou d'origine végétale**
broadcast application **application à la volée**
calibration **réglage**
chemosterilant **stérilisant chimique**
chlorinated hydrocarbon **hydrocarbure chloré**
contact herbicide **herbicide de contact**
contact pesticide **pesticide de contact**
critical load **charge critique**
DDT (dichloro-diphenyl-trichloro-ethane) **DDT (dichloro-diphényl-trichloréthane)**
dinitroaniline **dinitroaniline**
dinitrophenol **dinitrophénol**
diphenyl ethers **éthers diphényle**
dissipation capacity **capacité de dissipation**
fogging **nébulisation, atomisation, brumisage**
fumigant **fumigant**
fumigation **fumigation**
fungicide **fongicide**
fungistat **fongistatique**
herbicide **herbicide**
infection **infection**
infestation **infestation**
inheritance **héritage**
insecticide **insecticide**
integrated pest management (IPM) **lutte intégrée**
lindane **lindane**
manure **fumier, fumure, engrais**

mitigation **atténuation**
monitoring **suivi, monitoring, surveillance**
negligible residue **résidu négligeable**
organomercurial **composé organomercuriel**
organophosphate **organophosphate**
persistence **persistance, stabilité**
persistent pesticide **pesticide rémanent**
pest **organisme nuisible, ennemi des cultures, ravageur**
pest resistant varieties **variétés ou cultivars résistants aux organismes nuisibles**
pesticide **pesticide**
pesticide residue **résidu de pesticide**
pesticide tolerance **tolérance aux pesticides**
phenoxy **herbicide de type hormonal, phénoxy**
pollution **pollution**
pyrethroid **pyréthrinoïde**
residual insecticide **insecticide à effet rémanent**
rodenticide **rodenticide**
rotenone **roténone**
soft pesticide **pesticide non-persistant**
soil sterilization **stérilisation du sol**
stemflow **ruissellement sur les troncs**
systemic **systémique**
systemic pesticide **pesticide systémique**
thiocarbamate **thiocarbamate**
throughfall **pluviolessivage**
triazine **triazine**
water quality **qualité de l'eau**

FORESTRY

adaptive management **gestion adaptative**
additive effects **effets additifs**
afforestation **boisement**
artificial regeneration **régénération artificielle**
bareroot seedling **plant à racines nues**
biogeoclimatic classification system **système de classification biogéoclimatique**
biogeoclimatic unit **unité biogéoclimatique**
biogeoclimatic zone **zone biogéoclimatique**
biological control **lutte biologique**
biological herbicide **herbicide biologique**
biota **biote**
blowdown **chablis**
bole **tronc**
botanical forest products **produits forestiers botaniques**
broad-leaved deciduous forest **forêt feuillue**
broad-leaved evergreen forest **forêt feuillue à feuillage persistant**
brushing **dégagement**
buffer strip **bande de protection**
buffer zone **zone tampon**
Canadian forest fire weather index system **système canadien de l'indice forêt-météo**
canopy cover **surface de la cime**
chlorosis **chlorose**
clearcut **coupe totale, rase ou à blanc**
clearcutting silvicultural system **système sylvicole par coupe totale**
clearcutting with reserves **coupe totale avec réserves**
climax forest **forêt climacique**
clinometer **clinomètre**
coarse woody debris **débris ligneux grossiers**
conifer **conifère**
coniferous forest **forêt coniférienne**
conservation **conservation**
conservation biology **biologie de conservation**
container seedling **semis en récipient**
contour map **carte topographique**
coppice (coppicing) **taillis**
corridor **corridor**
critical wildlife habitat **habitat faunique critique**
crown **cime**
crown closure **fermeture du couvert**
cumulative effects **effets cumulatifs**
cut-over forest **friche**
declination (magnetic) **déclinaison magnétique**
deforestation **déboisement**
degradation **dégradation**
disc trencher **trancheuse à disque**
drag scarification **scarifiage par traînage**
drainage basin **bassin versant, bassin hydrographique**
duff **matière organique, humus**
ecological balance **équilibre écologique**
ecological classification **classification écologique**
ecological health **santé écologique**
ecological integrity **intégrité écologique**
ecological reserve **réserve écologique**

ecosystem management **gestion écosystémique**
ecosystem productivity **productivité de l'écosystème**
ecotone **écotone**
edatope **édatope**
edge **bordure**
edge effect **effet de bordure**
environmentally sensitive areas **zones écologiquement vulnérables**
evergreen **végétation à feuillage persistant**
extension services **vulgarisation**
extirpation **extirpation**
farm forestry **foresterie paysane ou rurale**
fire management **gestion des incendies ou des feux**
firebreak **pare-feu**
floodplain **périmètre ou plaine d'inondation**
fluvial processes **processus fluviatiles**
foliar analysis **analyse foliaire**
forest cover **couvert forestier**
forest cover map **carte forestière**
forest cover type **strate forestière**
forest ecology **écologie forestière**
forest fire **incendie forestier**
forest floor **couche holorganique, couverture morte, litière**
forest health **santé des forêts**
forest health agents **facteurs de santé des forêts**
forest health treatments **traitements pour améliorer la santé des forêts**
forest influences **influences forestières**
forest interior conditions **conditions intérieures des forêts**
forest inventory **inventaire forestier**
forest land **terrain forestier**
forest management **aménagement forestier**
forest practice **pratique forestière**
forest profile **profil des forêts**
forest renewal **régénération forestière**
forest regeneration **régénération forestière**
forest resources **ressources forestières**
forest service road **chemin forestier**
forestry **foresterie**
fragmentation **fragmentation**
genetic diversity **diversité génétique**
genotype **génotype**
ground fire **feu de surface**
ground truthing **vérification au sol, vérification de terrain**
groundwater **eau souterraine**
habitat **habitat**
habitat enhancement **amélioration de l'habitat, aménagement de l'habitat**
habitat management **gestion de l'habitat**
hardwoods **feuillus**
healthy ecosystem **écosystème sain**
hydroseeding **ensemencement hydraulique**
hypsometer **hypsomètre**
impact assessment **étude d'impacts**
improvement cutting **coupe d'amélioration**
indicator species **espèce indicatrice, bio-indicateur**
integrated resource management **gestion intégrée des ressources**
interpretive forest site **site d'interprétation forestier**
keystone species **espèce clé**
land use planning **planification de l'utilisation des terres**
landscape unit **unité de paysage**
litter layer **litière**
management plan **plan directeur**
microclimate **microclimat**
mineral soil **sol minéral**
mixed forest **forêt mixte**
monoculture **monoculture**
mortality **mortalité**
natural disturbance regime **régime de perturbations naturelles**
natural range barrier **barrière à la répartition naturelle**
natural regeneration **régénération naturelle**
new forestry **nouvelle foresterie**
old growth forest **vieille forêt**
old-growth forest attributes **attributs des vieilles forêts**
overstory **étage dominant**
overstory species **espèce de l'étage dominant**
overtopped **dominé**
overtopping **domination**
partial cutting **coupe partielle**
patch **polygone**
pest **organisme nuisible, ennemi des cultures, ravageur**
pest incidence **incidence des ravageurs**
pesticide **pesticide**
phenotype **phénotype**
pioneer **végétation ou plante pionnière**
plantation, forest **plantation forestière**
planting **plantation**

plot **parcelle**
plug seedling **semis en contenant**
preservation **préservation**
production forest **forêt de production**
productive forest land **forêt productive**
protected areas **territoires protégés**
protection forest **forêt de protection**
provenance **provenance**
pruning **élagage, taille**
pulp **pâte**
pulpwood **bois à pâte**
pure forest **peuplement pur**
reforestation **reboisement**
remote automatic weather station **station météorologique automatique**
reserves **arbres en réserve**
resilience **résilience**
restoration **restauration**
right-of-way **droit de passage**
riparian **riparien, riverain**
risk rating, forestry (assessment) **évaluation du risque**
roguing **rebut de sélection**
rotation **révolution**
rotation age **âge d'exploitabilité**
salvage harvesting **coupe de récupération**
sapling **gaule**
sapwood **aubier**
sawtimber **bois de sciage**
scalping **scalpage**
scarification **scarification**
screefing **scalpage**
second growth **forêt de seconde venue**
sedimentation **sédimentation**
seed orchard **verger à graines**
seed source **origine des graines**
seed tree **semencier**
seed tree silvicultural system **mode de régénération par coupe avec réserve de semenciers**
seedbed **lit de semis**
seedling **semis**
seedlot **lot de semences**
selective cutting **coupe d'écrémage ou sélective**
sensitive slope **pente sensible ou fragile**
sensitive watershed **bassin versant sensible ou fragile**
settlement pond, settling pond **étang de décantation**
silvics **écologie forestière appliquée**
silvicultural system **pratique ou régime sylvicole**
silviculture **sylviculture**
site **site, station**
site index **indice de qualité de station**
skid trail **sentier de débardage**
skidder **débusqueuse**
skidding **débardage**
slide **glissement**
slope failure **glissement de talus**
slope stability **stabilité de la pente**
softwood **résineux**
soil productivity **productivité du sol**
stand **peuplement**
stand composition **composition du peuplement**
stand conversion **conversion de peuplement**
stand density **densité de peuplement**
stand development **développement du peuplement**
stand dynamics **dynamique du peuplement**
stand structure **structure de peuplement**
stand tending **éducation du peuplement**
stewardship **intendance pour les générations futures**
stream **cours d'eau**
stream bank **berge**
stream channel **chenal d'un cours d'eau**
stream gradient **déclivité d'un cours d'eau**
streambed **lit du cours d'eau**
subspecies **sous-espèce**
subsurface drainage **drainage souterrain**
suitability mapping **carte de potentiel d'habitat**
sustainability **durabilité**
sustainable forest management **gestion durable des forêts**
temperate deciduous forest **forêt tempérée à feuilles caduques**
terrain **terrain**
thinning **éclaircie**
threatened or endangered habitats **habitats menacés**
timber **bois**
tolerance **tolérance**
topography **topographie**
transplant **plant repiqué**
tree farm **propriété forestière de production, ferme forestière**
undergrowth **sous-bois**

understory **sous-étage**
urban forestry **foresterie urbaine**
vegetative material **matériel végétatif**
water management **gestion de l'eau**
water quality **qualité de l'eau**
water resources **ressources en eau**
watercourse **cours d'eau**
watershed **bassin versant ou hydrographique**
watershed assessment **évaluation de bassin versant**
watershed integrity **intégrité du bassin versant**
watershed management **gestion de bassin versant**
wilderness **région sauvage**
wildfire **feu de friches**
wildlife management **gestion ou aménagement de la faune**
windrow **andain**
windthrow **chablis**
wood pulp **pâte de bois**
woodland **boisé**
woodland management **aménagement ou gestion de boisé**
woodlot **terrain boisé**

GEOMORPHOLOGY, GEOLOGY, CIVIL ENGINEERING

acidic rock **roche acide**
adhesion **adhésion**
alluvial **alluvial**
alluvial terrace **terrasse alluviale**
alluvium **alluvion**
anastomosing **anastomosé**
angle of dip **angle de chute, pendage**
angle of repose (rest) **angle de repos**
angular **anguleux**
angularity **forme anguleuse**
anisotropy **anisotropie**
anomaly **anomalie**
anthracite **anthracite**
anticline **anticlinal**
applanation **aplanissement**
apron **plaine d'épandage**
areal map **carte de surface**
aspect **orientation, exposition**
asphalt **asphalte**
attitude **disposition**
attribute **variable qualitative, attribut**
authigenic **authigène**
backslope **revers**
back-swamp **dépression latérale humide**
badland **bad land**
banding **litage métamorphique, litage**
bar **barre, levée**
barchan **barkhane**
barrier beach **cordon littoral**
barrier lake **lac de barrage**
basal till **till de fond**
basalt **basalte**
base course **base, assise**
base level **niveau de base d'érosion**
basement **socle**
basic rock **roche basique**
beach **plage**
beach deposit **dépôt de plage**
beach ridge **crête de plage**
esker, beaded **esker en chapelet**
bearing capacity **capacité portante, capacité de charge**
bed **lit**
bedded **lité**
bedding **enterrage, litage**
bedrock **assise rocheuse, roc sous-jacent, roc**
bench **replat, plate-forme**
bench terrace **terrasse en gradins**
berm **berme**
biaxial compression **compression biaxe**
biochronology **stratigraphie paléontologique**
biogenic sediment **sédiment organogène**
biostratigraphy **biostratigraphie**
bitumen **bitume**
bituminous **bitumineux**
blanket deposit **gisement stratoïde**
blind drain **drain en pierres, pierrée**
blind inlet **orifice de percolation**
blinding material **pierrée**
block **bloc**
block-cut method **foudroyage**
borehole **puits de forage**
boulder pavement **pavage de pierres, dallage de pierres**
BP **BP, avant le présent**
break of slope **rupture de pente**
breccia **brèche**
brittleness **fragilité**
broad-base terrace **terrasse à large base**
brow **front de colline, sommet**

brown coal **lignite, houille brune**
buried topography **paléorelief enfoui**
butte **butte**
calcareous rock **roche calcaire**
calcification **calcification**
calcify **calcifier**
calcitic dolomite **dolomie calcaire**
calcitic limestone **calcaire calcitique**
caldera **caldeira, caldera**
California bearing ratio **indice portant californien**
canyon **canyon**
cap-rock **roche couverture**
carbon ratio **rapport isotopique du carbone**
carbonaceous **carboné**
catena **caténa**
cement **ciment**
cementation **cimentation**
cemented **cimenté**
chalk **craie**
channel bar **levée de chenal**
channel terrace **terrasse en canaux**
channel-fill deposit **remplissage de chenaux**
chronological **chronologique**
chronostratigraphy **chronostratigraphie**
clast **claste**
clastation **fragmentation**
clastic **clastique**
clastic rock **roche clastique**
clastic sediment **sédiment clastique**
claystone **argilite**
cleavage **clivage**
cleavage plane **plan de clivage**
coal **charbon, houille**
coal ash **cendre de houille**
coal gas **gaz de houille**
coal gasification **gazéification du charbon**
coal seam **filon de charbon**
coalification **houillification, carbonification**
coarse-grained **à grain grossier**
coast **côte**
coastal erosion **érosion côtière**
coastal plain **plaine côtière**
coastal zone **zone côtière**
coefficient of friction **coefficient de frottement**
colluvial **colluvial**
colluvial slope **pente colluviale**
colluvium **colluvion**
compactibility **compactibilité**
compression **compression**
compressional wave **onde de compression**
core **noyau, carotte de sondage**
core drilling **carottage**
concavity **concavité**
concordant **concordant**
cone **cône**
conformable **concordant**
conformity **conforme**
congeliflution **gélifluxion**
conglomerate **conglomérat**
consolidate **consolider**
consolidation **consolidation**
consolidation test **essai de compressibilité, essai oedométrique**
continent **continent**
continental crust **croûte continentale**
continental drift **dérive des continents**
continental glacier **inlandsis, glacier continental**
contorted drift **moraine de poussée**
contraction **contraction**
convexity **convexité**
coprolite **coprolithe, coprolite**
cordillera **cordillère**
core **noyau, carotte de sondage**
Coriolis force **force de Coriolis**
stratigraphical correlation **corrélation stratigraphique**
cote **côte**
coulee **coulée**
country-rock **roche encaissante**
karst, covered **karst couvert**
crater **cratère**
creep **reptation**
crest **crête**
crevasse **crevasse**
crevasse filling **remplissage de crevasse**
critical angle **angle critique**
critical density **densité critique**
critical-path analysis **méthode du chemin critique**
cross section **section transversale, coupe transversale**
cross-bedding **lits entrecroisés**
cross-stratification **stratification entrecroisée**
crushing strength **résistance à l'écrasement**
crust **croûte, écorce terrestre**
cryology **cryologie**
cryomorphology **cryomorphologie**
cryoplanation **cryoplanation, géliplanation**
cryoplankton **cryoplancton**

cryosphere **cryosphère**
cupriferous **cuprifère**
cut and fill **cut and fill**
cycle of erosion **cycle d'érosion**
cyclic salt **embrun salé**
dam **barrage**
datum **donnée**
datum elevation **plan horizontal de référence**
datum level **niveau de référence**
dead-ice features **modelé de glace morte**
debris **débris**
debris cone **cône de déjection**
debris fall **éboulis**
debris fan **cône de déjection, cône de débris**
debris flow **coulée de débris**
deflation **déflation**
deflation basin or hollow **creux de déflation**
deformation **déformation**
degradation **aplanissement**
degree-day **degré-jour**
delta **delta**
deltaic **deltaïque**
denudation **dégradation**
deposit **dépôt**
deposition **sédimentation, accumulation**
detrital **détritique**
diagenesis **diagénèse**
diatomaceous earth **terre de diatomées**
diatomite **diatomite**
dike **digue, dyke, filon intrusif**
dip **pendage**
disconformity **discordance parallèle**
discontinuity **discontinuité**
discordance **discordance**
dispersion **dispersion**
dissected **disséqué**
dissection **dissection**
divide **ligne de partage des eaux, interfluve**
doline **doline**
dolomite **dolomite, dolomie**
dolomitic lime **chaux dolomitique**
draw **vallon, ravine**
drawdown **rabattement**
drift **sédiment, dépôt**
drumlin **drumlin**
dune **dune**
dune complex **champ de dunes**
dune, mobile **dune active**
dune, stabilized **dune stabilisée**
dyke **dyke, filon intrusif**
earth **terre**
earth flow **coulée de boue**
earth pillar **cheminée de fée, demoiselle coiffée**
earth science **science de la terre, géosciences**
earth slide **glissement de terrain**
earth tremor **faible secousse sismique**
earthquake **tremblement de terre, séisme**
effective stress **contrainte effective**
elastic **élastique**
elastic deformation **déformation élastique**
elastic limit **limite d'élasticité**
elasticity **élasticité**
elevation **élévation, altitude**
embankment **endiguement**
emergent shoreline **ancienne ligne de rivage, ligne de rivage perchée**
en echelon **en échelons**
end moraine **moraine terminale**
englacial **intraglaciaire**
eolian **éolien**
epicenter **épicentre**
epigenetic (epigenic) process **processus épigénétique**
epoch **époque**
equifinality **équifinalité**
equiplanation **équiplanation**
era **ère**
erratic **erratique**
escarpment **escarpement**
esker **esker**
estuarine **estuarien**
estuary **estuaire**
eustasy, eustatism **eustatisme, eustasie**
eustatic **eustatique**
evaporite **évaporite**
exogenetic **exogène**
fabric, soil **organisation du sol, état structural du sol**
till fabric **trame de till**
facet **facette**
facies **faciès**
failure **rupture**
fallout **dépôt atmosphérique, retombée radioactive**
fault **faille**
faulting **formation de faille**
ferrallitic soil **sol ferrallitique**
ferrallitization **ferrallitisation**
ferriferous **ferrifère**
field **champ**
fill **décharge, remblai**

fine grained **à grain fin**
fissile **fissile**
fissile bedding **litage fissile, en feuillet**
fissure **fissure**
flood-plain meander scar **échancrure de méandre de plaine alluviale**
fluted moraine **moraine cannelée**
fluvial **fluvial, fluviatile**
fluvial cycle of erosion **cycle d'érosion fluviale**
fluvial deposits **dépôts fluviatiles**
fluvioglacial **fluvio-glaciaire**
fly ash **cendres volantes**
fold **pli**
foot **pied**
foothills **piedmont, contrefort**
footing **semelle**
footslope **base de pente**
formation **formation**
fossil **fossile**
fossil fuel **combustible fossile**
fracture **fracture**
frozen ground **gélisol**
gangue **gangue**
gas, natural **gaz naturel**
gasification **gazéification**
gelifluction **gélifluxion**
gelisol **gélisol**
geo- **géo**
geobotanical prospecting **prospection géobotanique**
geocryology **géocryologie**
geodesy **géodésie**
geodetic surveying **relevé géodésique**
geographic province **province géographique**
geography **géographie**
geography, physical **géographie physique**
geologic erosion **érosion géologique**
geologic hazard **risque géologique**
geological map **carte géologique**
geological time **temps géologique**
geology **géologie**
geomagnetic field **champ géomagnétique**
geomagnetic poles **pôles géomagnétiques**
geomagnetism **géomagnétisme**
geomorphic **géomorphologique**
geomorphic cycle **cycle géomorphologique, cycle d'érosion**
geomorphic surface **unité géomorphologique**
geomorphological map **carte géomorphologique**
geomorphology **géomorphologie**
geophysics **géophysique**
geosphere **géosphère**
geosyncline **géosynclinal**
geothermal **géothermique**
geothermal energy **énergie géothermique**
glacial **glaciaire**
glacial spillway **déversoir de lac glaciaire**
glacial boulder **bloc glaciaire, erratique**
glacial drainage channel **chenal d'eau de fonte**
glacial drift **sédiment glaciaire, matériau de transport glaciaire, dépôt glaciaire**
glacial epoch **époque glaciaire**
glacial erosion **érosion glaciaire**
glacial erratic **blocerratique**
glacial geology **glaciogéologie**
glacial lake **lac glaciaire**
glacial maximum **maximum glaciaire, pléniglaciaire**
glacial minimum **minimum glaciaire**
glacial recession **retrait, recul glaciaire**
glacial retreat **retrait glaciaire**
glacial scour **affouillement glaciaire**
glacial till **till glaciaire**
glaciated **glacié**
glaciation **glaciation**
glacier **glacier**
glacier milk **lait de glacier**
glacier retreat **retrait de glacier**
glacio-eustasy **glacio-eustasie**
glaciofluvial **fluvio-glaciaire**
glaciolacustrine **glacio-lacustre**
glaciology **glaciologie**
gradation **aplanissement, granulométrie**
grade **pente, niveau du sol**
graded **à l'équilibre, classé, gradué**
graded terrace **terrasse régularisée**
grain **grain**
gravel **gravier**
gravimeter **gravimètre**
gravity **gravité**
ground moraine **moraine de fond**
Gutenberg discontinuity **discontinuité de Gutenberg**
hamada **hamada**
hard rock **roche dure**
hardness scale **échelle de dureté**
head **promontoire**
heave **soulèvement, rejet horizontal**
hiatus **hiatus, lacune**

hill **colline**
hill creep **reptation**
hogback, hog's-back ridge **hogback, dos d'âne**
hollow **creux**
hoodoo **cheminée de fée, demoiselle coiffée**
hot spring **source thermale**
hummock **button, thufur**
hummocky **en bosses et creux**
hummocky moraine **moraine bosselée**
hypogene **hypogénique**
hypsithermal **hypsithermal, altithermal**
hypsography **hypsométrie**
ice age **âge glaciaire**
icecap **calotte glaciaire**
ice-contact deposit **sédiment de contact glaciaire**
ice-dammed lake **lac de barrage glaciaire**
ice rafting **transport glaciel**
ice-sheet **inlandsis**
inclination **pendage**
inclinometer **clinomètre**
indurated **induré**
indurated layer **couche indurée**
inselberg **inselberg**
interbed **couche interstratifiée**
intercalated **intercalé**
interfluve **interfluve**
interglacial **interglaciaire**
interstade **interstade**
interstratified **interstratifié**
intertidal zone **zone intertidale**
intrusive **intrusive**
isobath **isobathe**
isostasy **isostasie**
isotropy **isotropie**
isthmus **isthme**
joint **joint, diaclase**
joint planes **plans de séparation ou de diaclase**
juvenile water **eau juvénile**
kame **kame**
kame complex, kame field **complexe juxtaglaciaire**
kame moraine **moraine de kame**
kame terrace **terrasse juxtaglaciaire, terrasse de kame**
kame-and-kettle terrain **topographie en bosses et creux**
kaolinite, kaolin **kaolinite**
karst **karst**
karst plain **plaine karstique**
karst topography **topographie superficielle de karst**
karst valley **ouvala**
kerogen **kérogène, kérobitume**
kettle, kettle hole **kettle**
kettle lake **lac de kettle**
kettle moraine **moraine à kettles**
knickpoint **rupture de pente, brisure**
knob **mamelon**
knob-and-kettle topography **paysage en bosses et creux**
knoll **colline**
lacustrine **lacustre**
lacustrine deposit **dépôt lacustre**
lacustrine plain **plaine lacustre**
lag **dépôt résiduel, résidu de déflation**
lake plain **plaine lacustre**
lake rampart **bourrelet lacustre, bourrelet glaciel**
lake terrace **terrasse lacustre**
lamina (plural laminae) **dépôt laminique**
laminar **laminaire**
laminar flow **écoulement laminaire**
land leveling **profilage**
landform **forme de terrain**
landform classification **classification des formes de terrain, unité de relief**
landscape **paysage**
landslide, landslip **glissement de terrain**
lateral moraine **moraine latérale**
lateral planation **aplanissement des interfluves**
lava flow **coulée de lave**
lee **côté sous le vent**
lens, lensing **stratification lenticulaire**
levee **levée, digue**
level rod **mire**
leveling **nivellement**
lignite **lignite**
lineament **linéament**
lineation **linéation**
liquefaction **liquéfaction spontanée**
lithification, lithifaction **lithification**
litho- **litho**
lithology **lithologie**
lithosphere **lithosphère**
lithostratigraphic unit **division lithostratigraphique**
lithostratigraphy **lithostratigraphie**
Little Ice Age **petit âge glaciaire**

loading **chargement**
lodgment till **till de fond**
loess **loess**
log **diagraphie**
longitudinal dune **dune longitudinale**
longitudinal fault **faille longitudinale**
lowland **basse terre**
macrofossil **macrofossile**
mafic **mafique**
magma **magma**
magnetic anomaly **anomalie magnétique**
declination (magnetic) **déclinaison magnétique**
magnetic poles **pôles magnétiques**
magnetometer **magnétomètre**
mantle **manteau**
marine **marin**
marine terrace **terrasse marine**
mass transport **transport en masse**
mass wasting **mouvement en masse**
massive **structure massive, homophane**
matrix **matrice**
M-discontinuity **discontinuité M ou de Mohorovicic**
meander **méandre**
meander scar **bras mort**
medial moraine, median moraine **moraine médiane**
meltwater **eau de fonte**
meltwater channel **chenal d'eau de fonte**
meridian **méridien**
mesa **mesa**
metamorphic rock **roche métamorphique**
metamorphism **métamorphisme**
meteorite **météorite**
microfossil **microfossile**
microrelief **microrelief**
Mohorovicic discontinuity **discontinuité de Mohorovicic**
Mohs scale **échelle de Mohs**
monadnock **monadnock**
morainal **morainique**
moraine kame **moraine de kame**
morphochronology **morphochronologie**
morphographic map **carte physiographique**
morphologic unit **unité morphologique**
morphology **morphologie**
morphometry **morphométrie**
mosaic **mosaïque**
mountain **montagne**
mountain range **chaîne de montagnes**
mountain-building **orogénèse, formation des montagnes**
mudflow **coulée boueuse**
mudslide **coulée de boue**
natural gas **gaz naturel**
nonconformity **discordance majeure**
nondestructive testing (NDT) **essai non destructif**
Ohm's law **loi d'Ohm**
oil pool **gisement de pétrole**
oil sand **sables bitumineux**
oil shale **pyroschiste**
oolith **oolithe, oolite**
organic reef **récif organique**
orogenic **orogénique**
orogeny **orogenèse**
outcrop **affleurement**
outwash **épandage fluvio-glaciaire**
outwash plain **plaine d'épandage fluvio-glaciaire**
over-consolidated soil deposit **dépôt de sol surcompacté**
oxbow lake **lac en croissant**
paleomagnetism **paléomagnétisme**
parabolic dune **dune parabolique**
paralithic contact **contact paralithique**
pedestal **pilier d'érosion**
pediment **pédiment**
pediplain **pédiplaine**
peneplain **pénéplaine**
peneplanation **pénéplanation**
penetration resistance **résistance à la pénétration**
penetrometer **pénétromètre**
perfect elasticity **élasticité parfaite**
perfect plasticity **plasticité parfaite**
periglacial **périglaciaire**
period **période**
petrofabric analysis **analyse de la fabrique**
petrogenesis **pétrogenèse**
petroleum **pétrole**
petrology **pétrologie**
phase **phase**
phi scale **échelle granulométrique logarithmique des unités phi**
physical geology **géologie physique**
physiography **physiographie**
piedmont **piémont**
pisolith **pisolite, pisolithe**
pit and mound topography **topographie en bosses et creux**

pitch **bitumen, inclinaison**
pitted outwash **plaine d'épandage piquée**
placer deposit **dépôt placérien**
plain **plaine**
planimetric map **carte planimétrique**
plastic **plastique**
plastic soil **sol plastique**
plate **plaque**
plate tectonics **tectonique des plaques**
plateau **plateau**
playa **playa**
plunge **plongement, inclinaison**
plutonic rock **roche plutonique**
point bar **banc arqué**
polycyclic landscape **paysage polycyclique**
polygenetic landscape **paysage polygénique**
poorly graded soil **matériau de granulométrie médiocre**
preconsolidation pressure (or prestress) **pression de contrainte**
pressure melting point **pression de fusion *in situ***
pressure wave **onde de pression**
primary (p) wave **onde primaire**
primary energy **énergie primaire**
prime meridian **méridien origine**
principal meridian **méridien principal**
principal stresses **contraintes principales**
proglacial **proglaciaire**
province **province**
push moraine **moraine de poussée**
pyroclastics **pyroclastique**
Q-wave **onde de Love, onde Q**
rafting **transport glaciel, transport par corps flottant**
raised beach **plage soulevée**
range **chaîne, répartition**
recessional moraine **moraine de retrait**
rectilinear slope **pente rectiligne**
regional metamorphism **métamorphisme régional**
regolith **régolithe**
regrading **remodelage de terrain**
relic geomorph **lambeau ou résidu d'érosion**
relic sediments **sédiments reliques**
relict landform **forme relique ou héritée**
relief **relief**
repose **repos**
reservoir rock **roche-magasin, roche-réservoir**
residual material **matériau résiduel**
residual shrinkage **retrait résiduel**
resistance **résistance**
resistivity **résistivité**
reworked **remanié**
rheology **rhéologie**
Richter scale **échelle de Richter**
ridge **dorsale**
riprap **perré, enrochement de protection**
river terrace **terrasse fluviatile**
river valley **vallée fluviale**
river wash **épandage fluviatile**
rock **roche, rocher**
rock creep **reptation**
rock drumlin **drumlin à noyau rocheux**
rotational landslide **glissement rotationnel**
roundness **arrondi, émoussé**
rubble **blocaille**
sand bank **banc de sable**
sand lens **lentille sableuse**
sand-bearing method **essai de chargement à la boîte de sable**
sand-dune **dune de sable**
sandstone **grès**
sandur **sandr, plaine d'épandage fluvio-glaciaire**
scarp **escarpement**
schist **schiste**
schistosity **schistosité**
scoria **scorie**
scoria land **champs de scories**
scour **affouillement, curage, décapage**
scree **talus d'éboulis**
second bottom **terrasse inférieure, basse terrasse**
secondary rocks or minerals **roches ou minéraux secondaires**
section **coupe, lame mince**
sediment **sédiment**
sediment basin **bassin de sédimentation**
sediment delivery ratio **rapport de production de sédiments**
sediment discharge **débit solide**
sediment grade sizes **granulométrie des matériaux**
sediment load **charge solide**
sedimentary **sédimentaire**
sedimentary rock **roche sédimentaire**
sedimentation **sédimentation**
sedimentology **sédimentologie**
segregation **ségrégation**
seismic focus, seismic origin **foyer sismique**

seismic survey **relevé sismique**
seismic waves **ondes sismiques**
seismic zone **zone sismique**
seismograph **sismographe**
seismology **sismologie, séismologie**
shear force **force de cisaillement**
shear strength **résistance au cisaillement**
shear stress **contrainte de cisaillement**
shingle **galet de plage**
shoulder **épaulement**
shrinkage index **indice de retrait**
shrinkage limit **limite de retrait**
sial **sial**
side slope **pente latérale, pente de talus**
siliceous **siliceux**
silicification **silicification**
sill **filon-couche, seuil**
siltstone **grès fin, siltstone**
sima **sima**
sinkhole **aven**
slide **glissement**
slip **glissement, rejet net, glissement basal**
slope **pente**
slope catchment area **bassin hydrographique, bassin versant**
slope gradient (angle) **inclinaison de la pente**
slope morphological classification **classification morphologique des versants**
slope morphology **morphologie des versants**
slope plan curvature **courbure de niveau**
slope profile curvature **courbure dans le sens de la pente**
slope stability **stabilité de la pente**
slope wash **érosion de pente**
slough **éboulis**
slump **glissement**
soil mechanics **mécanique des sols**
solution mining **extraction par solution**
sorting **granuloclassement**
sorting coefficient **coefficient de classement**
spalling **délitage**
specific gravity **densité, densité relative**
specific surface **surface spécifique**
stabilization, soil **stabilisation du sol**
stade **stadiaire, stade glaciaire**
stadial moraine **moraine stadiaire**
stage **étage, stade, niveau**
sticky limit **limite d'adhésivité**
stone **pierre**
strain **déformation**
stratification **stratification**
stratified **stratifié**
stratified drift (or sorted drift) **sédiment stratifié ou trié**
stratigraphy **stratigraphie**
stratum (plural strata) **strate**
strength **résistance**
stress **contrainte**
strike **direction**
structural geology **géologie structurale**
structure **structure**
subaerial **subaérien**
subbituminous **subbitumineux**
subgrade **forme, sol de formation**
subgrade modulus **module**
subsidence **subsidence, affaissement**
summit **sommet**
superficial deposits **dépôts superficiels**
surface relief **topographie**
suspension **suspension**
syncline **synclinal**
tableland **haut plateau**
talus **talus**
tar sand **sables bitumineux**
tectonic **tectonique**
telluric **tellurique**
tensile strength **résistance en traction**
tension **tension**
tephra **tephra**
terminal moraine **moraine terminale ou frontale**
terrace **terrasse**
terrain **terrain**
terrain analysis **analyse du terrain**
terrestrial **terrestre**
tidal flat **wadden, bas fonds intertidaux**
till **till**
till plain **plaine de till**
tillite **tillite**
toe slope position **pied ou base du talus**
topographic map **carte topographique**
topography **topographie**
toposequence **toposéquence**
trafficability **traficabilité**
transport **transport**
triangulation **triangulation**
ultrabasic rock **roche ultrabasique**
unconformity **discordance**
unconsolidated sediment **non consolidé**
undercutting **sapement**

Unified Soil Classification System **système de classification unifié des sols, classification U.S.C.S.**
unstratified drift **diamicton**
upland **haute terre**
valley **vallée**
valley fill **remblayage de vallée**
varve **varve**
vein **filon**
veneer **placage**
volcanic ash **cendre volcanique**
wash **ruissellement**
washboard moraine **moraine bosselée**
washoff **ruissellement**
wash-out **ravinement**
water, juvenile **eau juvénile**
well-graded soil **sol bien gradué**
Wentworth scale **échelle granulométrique de Wentworth**
wind-blown deposits **dépôts éoliens**
Zingg bench terrace **terrasse en gradins**

HYDROLOGY, LIMNOLOGY, WATER QUALITY

amictic lake **lac amictique**
aquatic **aquatique**
aquatic plant **plante aquatique**
aquiclude **aquiclude**
aquifer **formation aquifère**
aquifer, confined **nappe artésienne**
aquifer, unconfined **aquifère non captive**
aquifuge **aquifuge**
aquitard **couche semi-perméable capacitive**
artesian spring **source artésienne**
artesian water **eau artésienne**
artesian well **puits artésien**
aufwuchs **algue épiphytique**
base flow **débit de base**
basin **bassin**
beaded drainage **lacs orientés**
bed load **charriage**
bed material **matériau du lit**
benthic **benthique**
benthic region **région benthique**
benthos **benthos**
biochemical oxygen demand (BOD) **demande biochimique en oxygène (DBO)**
biological methylation **méthylation biologique**
biomanipulation **biomanipulation**
bloom **fleur d'eau**
BOD **DBO**
BOD_5 **DBO_5**
bottom load **charriage**
brackish **saumâtre**
carbonate hardness **dureté carbonatée**
catchment **bassin hydrographique, bassin versant, bassin d'alimentation**
centripetal drainage pattern **tracé centripète de drainage**
channel **canal, chenal, lit**
channel density **densité de drainage**
channel flow **écoulement en chenal**
channel storage **emmagasinement dans un cours d'eau**
channelization **canalisation**
chemical oxygen demand (COD) **demande chimique en oxygène (DCO)**
chloramine **chloramine**
chute **rapide**
closed basin **bassin fermé**
COD **DCO**
coldwater fish **poisson d'eaux froides**
concentrated flow **écoulement concentré**
cone of depression **cône de dépression**
confined ground water **nappe captive**
confining layer **couche encaissante**
critical depth **profondeur critique**
critical velocity **vitesse critique**
cultural eutrophication **eutrophisation**
current meter **moulinet**
dam **barrage**
delayed runoff **écoulement retardé**
dendritic drainage **drainage dendritique**
density stratification **stratification**
depression storage **emmagasinement dans les dépressions**
deranged drainage pattern **tracé de drainage dérangé**
desiccation, soil **dessèchement**
de-silting area **zone de dessablement**
detention basin **bassin de retenue**
diatom **diatomée**
dimictic lake **lac dimictique**
discharge **débit**
dissolved load **charge en solution**
dissolved organic carbon (DOC) **carbone organique dissous (COD)**

dissolved organic matter (DOM) **matière organique dissoute (MOD)**
dissolved oxygen (DO) **oxygène dissous (OD)**
dissolved solid **matière solide en solution**
divergence **divergence**
drainage **drainage**
drainage basin **bassin versant, bassin hydrographique**
drainage density **densité de drainage**
drainage divide **ligne de faîte, ligne de partage des eaux**
drainage pattern **tracé de drainage**
drainage structures **ouvrages de drainage**
drawdown **rabattement**
dystrophic **dystrophe**
dystrophic lake **lac dystrophe**
elevation head **charge d'eau d'élévation**
endoreic drainage **écoulement endoréique**
enrichment **enrichissement**
ephemeral stream **cours d'eau intermittent**
epilimnion **épilimnion**
equi-potential line **ligne équipotentielle**
euphotic zone **zone euphotique**
eutrophic **eutrophe**
eutrophic lake **lac eutrophe**
eutrophication **eutrophisation**
external drainage **drainage superficiel**
fall overturn **renversement automnal**
fishway **passe à poissons**
flash flood **crue éclair**
flood **crue**
flood control **lutte contre l'inondation**
flood frequency **fréquence de crue**
flood peak **débit de pointe de crue**
flood plain **plaine alluviale**
floodway **canal de crue ou de dérivation**
flume **canal**
forest hydrology **hydrologie forestière**
fresh water **eau douce**
gage or gauge **jauge**
gaging station **station de jaugeage**
gradient **gradient**
ground water **eau souterraine**
ground water flow **écoulement souterrain**
ground water level **niveau de l'eau souterraine**
ground water reservoir **réservoir aquifère**
ground water runoff **ruissellement souterrain**
ground water-table **nappe phréatique**
groundwater velocity **vitesse de l'eau souterraine**
hard water **eau dure**
hardness **dureté**
head **pression hydrostatique, hauteur, tête de réseau**
headwater **cours supérieur d'une rivière**
holomictic lake **lac holomictique**
hydraulic gradient **gradient hydralique**
hydraulic pressure **pression hydraulique**
hydraulic radius **rayon hydraulique**
hydraulics **hydraulique**
hydric **hydrique**
hydrodynamics **hydrodynamique**
hydrogeology **hydrogéologie**
hydrograph **hydrogramme**
hydrologic **hydrologique**
hydrologic budget **bilan hydrologique**
hydrologic cycle **cycle hydrologique**
hydrologic equation **équation hydrologique**
hydrologic model **modèle hydrologique**
hydrology **hydrologie**
hydrosphere **hydrosphère**
hydrostatic pressure **pression hydrostatique**
hydrostatics **hydrostatique**
hypolimnion **hypolimnion**
hypoxia **hypoxie**
initial storage **rétention initiale**
intake area **région d'alimentation**
interception **interception**
interflow **écoulement divergent**
intermittent stream **cours d'eau temporaire ou intermittent**
internal soil drainage **drainage interne du sol**
irrigation flume **canal sur appuis, canal surélevé, canal de jaugeage**
Jackson turbidity unit (JTU) **unité de turbidité Jackson (UTJ)**
lake **lac**
laminar flow **écoulement laminaire**
lateral **canalisation secondaire**
lentic **lentique**
limnetic **limnétique**
limnetic zone **région limnétique**
limnetic, limnic **limnétique**
limnology **limnologie**
limnophyte **limnophyte**
littoral zone **région ou zone littorale**
load **charge**
lotic **lotique**
macroorganic matter **matière organique particulaire ou grossière**
mean velocity **vitesse moyenne**

measuring weir **déversoir**
meromictic lake **lac méromictique**
meroplankton **méroplancton**
mesotrophic **mésotrophe**
microplankton **microplancton**
monomictic lake **lac monomictique**
nannoplankton **nannoplancton**
nauplius **nauplie**
nekton **necton**
neritic zone **région néritique**
neuston **neuston**
nonuniform flow **écoulement varié**
oligotrophic lake **lac oligotrophe**
organic load **charge organique**
outlet **sortie d'eau**
overdraft **surexploitation**
overland flow **ruissellement**
overturn **renversement**
oxygen demand **demande en oxygène**
oxygen depletion **épuisement d'oxygène**
paludal **paludien**
paralic **paralique**
particulate matter **matière particulaire**
particulate organic matter (POM) **matière organique particulaire (MOP)**
peak discharge **débit maximal**
pelagic zone **région ou milieu pélagique**
perched water table **nappe d'eau suspendue ou perchée**
perennial stream **cours d'eau pérenne**
periodic drift **dérive périodique**
periphyton **périphyton**
permanent hardness of water **dureté permanente de l'eau**
photic zone **zone photique**
phreatic water **eau souterraine, nappe phréatique**
phycosphere **phycosphère**
phytoplankton **phytoplancton**
piezometer **piézomètre**
piezometric surface **surface piézométrique**
plankton **plancton**
plankton bloom **fleur d'eau planctonique**
polymictic lake **lac polymictique**
pond **étang**
potable water **eau potable**
potamology **potamologie**
potamon zone **potamon**
pot-hole **marmite torrentielle**
pressure head **hauteur piézométrique**
profundal zone **zone profonde**
pycnocline **pycnocline**
quick sand **sable mouvant**
radial drainage **drainage radial**
rainfall excess **pluie nette**
raw water **eau brute**
reach **tronçon**
receiving waters **eaux réceptrices**
recharge **alimentation**
red tide **marée rouge**
reservoir **réservoir**
retention **rétention**
return flow **écoulement restitué**
riparian **riparien, riverain**
rithron zone **rithron**
river basin **bassin versant**
river bed **lit**
river system **réseau hydrographique**
rock flour **farine de roche**
runoff **écoulement**
safe yield **débit de sécurité**
salt-water intrusion **invasion d'eau salée**
saturated zone **zone de saturation**
sediment oxygen demand (SOD) **demande d'oxygène des sédiments**
seepage **suintement**
seiche **seiche**
self-purification **auto-épuration**
sessile **sessile**
seston **seston**
settling basin **bassin de décantation**
sheet flow **écoulement en nappe**
silting **envasement**
sink **puits, cuvette**
slough **marécage, mare vaseuse, bourbier**
specific yield **débit spécifique**
spillway **déversoir**
spring **printemps, source**
spring overturn **renversement printannier**
stage **stade**
stormwater runoff **eau d'averse**
stratification **stratification**
stream gaging **jaugeage d'un cours d'eau**
stream load **charge totale d'un cours d'eau**
stream recession length **temps de décrue**
streambanks **berges, rives**
streambed erosion **érosion du lit**
stressed water **eau polluée**
sublittoral zone **région sublittorale**
summer kill **mortalité estivale**
supralittoral zone **région supralittorale**
surface flow **ruissellement**

surface runoff **ruissellement**
surface water **eau de surface**
suspended load **charge en suspension**
suspension **suspension**
thalweg **talweg**
thaw lake **cuvette de dégel**
theoretical oxygen demand (TOD) **demande théorique d'oxygène**
thermal stratification **stratification thermique**
thermocline **thermocline**
total dissolved solids (TDS) **matières totales dissoutes**
transmissivity **transmissivité**
transparency **transparence**
tributary **affluent**
tripton **tripton**
trophogenic region **zone trophogène**
trunk valley **vallée principale**
turbidity **turbidité**
turbulent flow **écoulement turbulent**
turbulent velocity **vitesse limite de régime turbulent**
underflow **inféroflux, sous-écoulement**
underground runoff **écoulement souterrain**
uniform flow **écoulement uniforme**
unit hydrograph **hydrogramme unitaire**
urban runoff **écoulement urbain**
vadose water **eau vadose**
vadose zone **zone vadose**
varied flow **écoulement varié**
velocity head **hauteur-vitesse**
warmwater fish **poisson d'eaux chaudes**
water cycle **cycle hydrologique**
water enrichment **enrichissement de l'eau**
water resources **ressources en eau**
water table **surface de saturation**
watercourse **cours d'eau**
watershed **bassin versant**
watershed lag **temps de réponse**
waterway **voie d'eau, cours d'eau**
weir **déversoir**
winter kill **mortalité hivernale**
yield **rendement**
zooplankton **zooplancton**

IRRIGATION

border dike **levée de planche**
border ditch **fossé de planche**
centrifugal pump **pompe centrifuge**
check dam **barrage submersible**
closed drain **tuyau fermé**
conjunctive water use **utilisation combinée des eaux souterraines et de surface**
consumptive use **évapotranspiration**
contour ditch **rigole de niveau**
contour flooding **inondation en contour**
controlled drainage **drainage contrôlé**
crib dam **barrage en encoffrement**
debris dam **barrage de retenue des débris**
detention dam **barrage de retenue du sol**
ditch **fossé**
diversion dam **digue, barrage de dérivation**
drag **entrave, résistance à l'écoulement**
drain **drain, canal de drainage**
drain tile **tuyau de drainage**
drop spillway **déversoir vertical**
drop structure **chute**
dugout pond **fosse-réservoir**
earth dam **barrage en terre**
emergency spillway **évacuateur de secours**
farm pond **étang fermier**
fertigation **irrigation fertilisante, fertigation**
furrow dams **barrages de rigoles**
head ditch **canal de distribution**
irrigable land **terre irrigable**
irrigation **irrigation**
irrigation efficiency **efficacité d'irrigation**
irrigation requirement, consumptive **besoin des cultures en eau d'irrigation**
irrigation, basin **irrigation par bassin**
irrigation, border-strip **irrigation par calants ou à la planche**
irrigation, center-pivot **irrigation par pivot central**
irrigation, check-basin **irrigation par bassin de retenue**
irrigation, contour-furrow **irrigation par rigoles d'infiltration suivant les courbes de niveau**
irrigation, corrugation **irrigation par infiltration ou par billons**
irrigation, drip **irrigation au goutte-à-goutte**
irrigation, flood **irrigation par submersion**
irrigation, furrow **irrigation par rigoles**
irrigation, wild flooding **irrigation par submersion non-contrôlée**
irrigation, winter **irrigation d'hiver**

leaching fraction **facteur de perte par lessivage**
leaching requirement **besoin en lessivage**
mole drain **galerie-taupe**
open drain **tranchée de drainage**
pumping station **poste de pompage**
rock-fill dam **barrage en enrochement**
salt balance **bilan de sels**
sausage dam **barrage en gahions**
sluice **canal à écoulement rapide**
sodium estimation **estimation de la teneur en sodium**
soluble-sodium percentage **pourcentage de sodium soluble**
spillway **déversoir**
sprinkler irrigation **irrigation par aspersion**
subsurface irrigation **irrigation souterraine**
supplemental irrigation **irrigation de complément**
surface irrigation **irrigation de surface**
tile drain **drain en tuyaux**
water application efficiency **efficacité d'irrigation**
water demand **besoin en eau d'irrigation**
weep-holes **barbacane, chantepleure**

METEOROLOGY, CLIMATOLOGY, AIR QUALITY

absorption of radiation **absorption du rayonnement**
acid rain **pluie acide**
actinometer **pyranomètre, actinomètre**
advection **advection**
aerosols **aérosols**
air frost **gel atmosphérique**
air pollution **pollution de l'air**
albedo **albédo**
anemometer **anémomètre**
anthropogenic **anthropogénique**
area source **source diffuse ou étendue**
arid **aride**
atmosphere **atmosphère**
bar **bar, megabarye**
barometer **baromètre**
barometric pressure **pression barométrique**
borehole **puits de forage**
carbon equivalent (CE) **équivalent en carbone**
carbon dioxide (CO_2) **dioxyde de carbone**
carbon dioxide fertilization **effet fertilisant du dioxyde de carbone**
carbon dioxide equivalent (CDE) **équivalent en dioxyde de carbone**
carbon sequestration **immobilisation ou stockage du carbone**
carbon sinks **puits de carbone**
catalytic converter **convertisseur catalytique**
chlorofluorocarbons and related compounds **chlorofluorocarbures**
climate **climat**
climate change **changements climatiques**
climate feedback **rétroaction climatique**
climate lag **retard climatique**
climate modeling **modélisation du climat**
climate model **modèle climatique**
climate system **système climatique**
climatic region **région climatique**
climatology **climatologie**
condensation **condensation**
convection **convection**
cryosphere **cryosphère**
deforestation **déboisement**
deposition **dépôt**
deposition velocity **vitesse de dépôt**
desertification **désertification**
dew **rosée**
dew point **point de rosée**
diffuse radiation **rayonnement diffus**
drought **sécheresse**
dry-bulb temperature **température du thermomètre sec**
dry deposition **dépôt sec**
dry scrubber **épurateur à sec**
dust **poussière**
effective precipitation **précipitation effective**
El Niño **El Niño**
electrostatic precipitator **précipitateur électrostatique**
emission **émission**
emission inventory **inventaire des émissions**
emission factor **coefficient d'émission**
emission rate **taux d'émission**
enhanced greenhouse effect **effet de serre amplifié**
evaporation **évaporation**
evaporimeter **évaporimètre**

evapotranspiration **évapotranspiration**
fallout **dépôt atmosphérique, retombée radioactive**
feedback mechanism **mécanisme de rétroaction**
flue dust **poussière de combustion**
flue gas **gaz de combustion**
flue gas scrubber **épurateur de gaz de combustion**
fluorocarbons **fluorocarbures**
forcing mechanism **mécanisme de forçage**
fossil fuel **combustible fossile**
frost **gel**
frost-free period **saison sans gel**
general circulation model (GCM) **modèle de circulation générale**
geosphere **géosphère**
global warming **réchauffement planétaire**
greenhouse effect **effet de serre**
greenhouse gases **gaz à effet de serre**
ground frost **gel au sol**
growing season **saison de croissance**
halocarbon **halocarbone**
heat **chaleur**
humid **humide**
humidity, absolute **humidité absolue**
humidity, relative **humidité relative**
hydrofluorocarbons (HFCs) **hydrofluorocarbures**
hydrometeorology **hydrométéorologie**
hydrosphere **hydrosphère**
hygrometer **hygromètre**
indoor air pollution **pollution intérieure des locaux**
infrared radiation **rayonnement infrarouge**
insolation **rayonnement solaire direct incident**
ionosphere **ionosphère**
Koppen's climatic classification **classification climatique de Koppen**
lag **retard de réponse**
lysimeter **lysimètre**
mesosphere **mésosphère**
meteorology **météorologie**
methane (CH_4) **méthane**
microclimate **microclimat**
micrometeorology **micrométéorologie**
mixing ratio **rapport de mélange**
net radiation **rayonnement net, bilan radiatif**
nitrogen oxides (NOx) **oxydes d'azote (NOx)**
nitrous oxide (N_2O) **protoxyde d'azote (N_2O)**
normal atmospheric pressure **pression atmosphérique normale**
ozone **ozone**
photolysis **photolyse**
pollution **pollution**
potential evapotranspiration **évapotranspiration potentielle**
precipitation interception **interception des précipitations**
precipitation **précipitations**
precipitation efficiency (effectiveness) **efficacité des précipitations**
psychrometer **psychromètre**
radiation **rayonnement**
radiative forcing **forçage radiatif**
rain **pluie**
rain-gauge **pluviomètre**
rainfall **hauteur d'eau, hauteur pluviométrique**
rainfall intensity **intensité pluviale**
rainfall duration **durée de la pluie**
rainfall frequency **fréquence de la pluie**
rainfall interception **interception de la pluie**
reflected radiation **rayonnement réfléchi**
residence time **temps de résidence**
ridge **dorsale**
saturation mixing ratio **rapport de mélange de saturation**
semi-arid **semi-aride**
shade temperature **température sous abri**
sink **puits, cuvette**
snow **neige**
snow density **densité de la neige**
snowpack **couverture de neige**
solar angle **angle solaire**
solar energy **énergie solaire**
solar radiation **rayonnement solaire**
standard air density **masse volumique standard de l'air**
Stevenson screen **abri météorologique, abri de Stevenson**
stratosphere **stratosphère**
sub-humid **sub-humide**
sulfate aerosols **aérosols de sulfates ou sulfatés**
sulfur dioxide (SO_2) **dioxyde de soufre**
sulfur hexafluoride (SF_6) **hexafluorure de soufre**
temperature **température**
thawing **dégel**
thermometer **thermomètre**

trace gas **gaz à l'état de traces**
troposphere **troposphère**
tropospheric ozone precursor **précurseur d'ozone troposphérique**
tropospheric ozone (O_3) **ozone troposphérique**
turbulence, turbulent flow **turbulence, écoulement turbulent**
water vapor **vapeur d'eau**
weather **temps**
wet deposition **déposition humide**
wet-bulb temperature **température du thermomètre mouillé**
whirling psychrometer **psychromètre à fronde**
wind **vent**
global warming potential (GWP) **potentiel de réchauffement planétaire**

MICROBIOLOGY

abiontic enzymes **enzymes abiontiques**
acetylene-block assay **test de blocage à l'acétylène**
actinomycetes **actinomycète**
adenylate energy charge ratio (EC) **charge énergétique**
aerobe **organisme aérobie**
aerobic **aérobie**
aerobic decomposition **décomposition aérobie**
agar **agar**
akinete **akinète**
alga (plural algae) **algue**
algal bloom **prolifération algale**
algology **algologie**
ammonification **ammonification**
amoeba (plural amoebae) **amibe**
anaerobe **organisme anaérobie**
anaerobic **anaérobie**
antibiosis **antibiose**
antibiotic **antibiotique**
arbuscule **arbuscule**
ascospores **ascospores**
aseptic **aseptique**
assimilatory nitrate reduction **réduction assimilatoire des nitrates**
autoclave **autoclave**
autotrophic nitrification **nitrification autotrophe**
axenic **axénique**
bacillus **bacille**
bacterial plate count **numération bactérienne sur plaque**
bactericidal **bactéricide**
bacteriophage **bactériophage**
bacteriostatic **bactériostatique**
bacterium (plural bacteria) **bactérie**
bacteroid **bactéroïde**
barophile **barophile**
basidiospore **basidiospore**
basidium **baside**
batch culture **culture en batch, culture discontinue**
binary fission **fission binaire**
biochemical oxygen demand (BOD) **demande chimique en oxygène**
biochemistry, soil **biochimie du sol**
biological denitrification **dénitrification biologique**
bioluminescence **bioluminescence**
bioremediation **biorestauration**
biosequence **bioséquence**
botulism **botulisme**
calcareous algae **algue calcaire**
capsule **capsule**
carrier **porteur**
catabolism **catabolisme**
chemoautotroph **chimiotrophe**
chemodenitrification **dénitrification chimique**
chemoheterotroph **chimio-hétérotrophe**
chemolithotroph **chimiolithotrophe**
chemoorganotroph **chimio-organotrophe**
chemostat **chémostat**
chemotaxis **chimiotaxie, chimiotactisme**
chemotrophy **chimiotrophie**
chlamydospore **chlamydospore**
chromatid **chromatide**
chromatin **chromatine**
ciliate **cilié**
cilium (plural cilia) **cil**
coccus **coque**
coenocyte **coenocyte**
coenocytic organism **organisme coenocytique**
coliform **coliforme, colibacille**
coliphage **coliphage**
colonization **colonisation**
colony **colonie**

colony count **numération de colonies bactériennes**
cometabolism **cométabolisme**
conidium (plural conidia) **conidie**
conjugation **conjugaison**
contaminate **contaminer**
counting chamber **cellule de comptage**
culture **culture**
culture dish **boîte de Pétri**
culture media **milieu de culture**
death phase **phase de déclin**
diatom **diatomée**
diatomaceous earth **terre de diatomées**
diazotroph **diazotrophe**
direct count **numération directe, décompte direct**
direct plating **ensemensement direct**
ectomycorrhiza **ectomycorhize**
ectotrophic mycorrhiza **mycorhize ectotrophe**
electron-transport chain **chaîne de transport d'électrons**
endogenous **endogène**
endomycorrhiza **endomycorhize**
endotherm **endotherme**
endotoxin **endotoxine**
endotrophic **endotrophe**
endotrophic mycorrhiza **mycorhize endotrophe**
enrichment culture **façon culturale d'enrichissement**
enteric bacteria **entébactérie, entérobactérie**
epifluorescence **épifluorescence**
Escherichia ***Escherichia***
eukaryote **eucaryote**
eukaryotic **eucaryote**
exoenzyme **exoenzyme**
exogenous **exogène**
exotoxin **exotoxine**
exponential growth phase **phase de croissance exponentielle**
extracellular enzyme **enzyme extracellulaire**
exudate **exsudat**
facultative **facultatif**
facultative anaerobe **anaérobie facultatif**
facultative organism **organisme facultatif**
fecal bacteria **bactéries fécales**
fecal coliform **coliforme, colibacille fécal**
fecal streptococcus **streptocoque fécal**
fermentation **fermentation**
filamentous algae **algue filamenteuse**
flagellate **flagellé**
flagellum (plural flagella) **flagelle**
fluorescein **fluorescéine**
fluorescent antibody **anticorps fluorescent**
fumigation, soil **fumigation du sol**
fungus (plural fungi) **champignon**
gametangium **gamétange**
gamete **gamète**
genophore **génophore**
geometric growth **croissance géométrique**
geometric rate of increase **taux géométrique d'augmentation**
germicide **germicide**
Golgi complex **appareil de Golgi**
Gram negative/Gram positive **Gram positive/Gram négative**
Gram stain **coloration de Gram**
growth phase **phase de croissance**
halophile **halophile**
Hartig net **filet de Hartig**
heterocyst **hétérocyste**
holophytic **autotrophe**
holoplankton **holoplancton**
homologous chromosomes **chromosomes homologues**
homozygous **homozygote**
hormogonium **hormogonie**
hybridization **hybridation**
hyperbolic reaction **réaction hyperbolique**
hypha (plural hyphae) **hyphe**
incubation **incubation**
incubation period **période d'incubation**
indicator bacteria **bactéries indicatrices**
indicator microorganisms **micro-organismes indicateurs**
infection **infection**
infectious **infectieux**
inhibition **inhibition**
innate capacity for increase (r_o) **capacité de croissance innée**
inoculate **inoculer**
inoculation **inoculation**
inoculum **inoculum**
intracellular **intracellulaire**
intrinsic rate of increase (r_m) **taux de croissance intrinsèque**
iron bacteria **bactéries ferrugineuses, ferrobactéries**
k_m **k_m, constante de Michaelis-Menten**
lag growth phase **phase de latence**
lamina (plural laminae) **lamina**
leaf mold **terreau de feuilles**

leaf spot **helminthosporiose**
lithotroph **lithotrophe**
lithotrophy **lithotrophie**
log growth phase **phase logarithmique de croissance**
lysis **lyse**
macrofauna, soil **macrofaune du sol**
medium **milieu**
mesofauna, soil **mésofaune du sol**
mesophile **mésophile**
methanotroph **méthanotrophe**
microaerophile **micro-aérophile**
microbe **microbe**
microbial biomass **biomasse microbienne**
microbial load **charge microbienne**
microbial mat **feutre ou mat microbien, natte microbienne**
microbial oxidation **oxydation biologique**
microbial population **population microbienne**
microbiology **microbiologie**
microbiology, soil **microbiologie du sol**
microbiota **microbiote**
microfauna, soil **microfaune du sol**
microflora **microflore**
microhabitat, soil **microhabitat du sol**
microorganism **micro-organisme**
microscopic **microscopique**
microsite, soil **microsite du sol**
mixotrophy **mixotrophe**
Monod equation **équation de Monod**
most probable number (MPN) **nombre le plus probable**
motile **motile**
mycelium (plural mycelia) **mycélium**
myco **myco**
mycology **mycologie**
mycorrhiza **mycorhize**
mycotoxin **mycotoxine**
myxamoeba **myxamibe**
nitrogen-fixing bacteria **bactéries fixatrices d'azote**
nodule **nodule, nodosité**
nodule bacteria **bactéroïde**
nodule, soil **nodule de sol**
non-symbiotic fixation **fixation non symbiotique**
obligate **stricte**
organotroph **organotrophe**
organotrophic **organotrophe**
organotrophy **organotrophie**
osmotroph **osmotrophe**
partial sterilization **stérilisation partielle**
pasteurization **pasteurisation**
pathogen **agent pathogène**
pathogenic **pathogène**
Petri dish **boîte de Pétri**
phosphobacteria **phosphobactéries**
phosphorescence **phosphorescence**
photoautotroph **photo-autotrophe**
photolithotroph **photolithotrophe**
phototrophic **phototrophe**
phototropism **phototropisme**
plankton **plancton**
plant growth promoting rhizobacteria (PGPR) **rhizobactéries favorisant la croissance des plantes**
plasmodium **plasmodium**
plate count **numération sur plaque**
polymerase chain reaction (PCR) **réaction en chaîne de la polymérase**
prokaryote **procaryote**
prokaryotic **procaryotique**
propagule **propagule**
protist **protiste**
protoctist **protoctiste**
protoplast **protoplaste**
protozoan (plural protozoa) **protozoaire**
pseudomonad **pseudomonade**
pseudopodium (plural pseudopodia) **pseudopode**
psychrophile **psychrophile**
pure culture **culture pure**
radiation sterilization **radappertisation**
reducers **décomposeurs**
restriction enzyme **enzyme de restriction**
rhizobium (plural rhizobia) **rhizobium**
rhizocylinder **rhizocylindre**
rhizoplane **rhizoplan**
rhizosphere **rhizosphère**
root nodule **nodule racinaire**
root rot **pourriture des racines**
Salmonella ***Salmonella***
saprobe **saprobionte**
saprobic **saprobe**
saprophyte **saprophyte**
saprophytic competence **capacité saprophyte**
septate **septé**
siderophore **sidérophore**
spore **spore**
stationary growth phase **phase de croissance stationnaire**
sterile **stérile**

sterilization **stérilisation**
strain **souche**
stromatolite **stromatolithe**
substrate **substrat**
sulfur bacteria **bactéries sulfureuses, sulfobactéries**
symbiosis **symbiose**
symbiotic fixation **fixation symbiotique**
synergism **synergie**
thermophile **thermophile**
Thiobacillus ***Thiobacillus***
transduction **transduction**
unicellular **unicellulaire**
vegetative cell **cellule végétative**
vesicle **vésicule**
vesicular arbuscular **vésiculo-arbusculaire**
virus **virus**
white rot fungus **pourriture blanche**
Winogradsky column **colonne de Winogradsky**
Wood-Werkman phenomenon **phénomène de Wood-Werkman**
xenobiotic **xénobiotique**
yeasts **levures**
zoospore **zoospore**
zygospore **zygospore**
zymogenous **zymogène**

MINERALOGY, GEOCHEMISTRY, WEATHERING

a axis **axe a**
Abbe refractometer **réfractomètre d'Abbe**
abrasion **abrasion**
Adobe soil **terre à briques**
allophane **allophane**
aluminosilicate **silicate d'aluminium**
ammonium entrapment **fixation ou intégration de l'ammonium en position interfeuillet**
ammonium fixation **fixation d'ammonium**
amorphous mineral **minéral amorphe**
amphibole **amphibole**
andesite **andésite**
anhedral **allotriomorphe**
apatite **apatite**
argillaceous **argileux**
asbestos **amiante**
asymmetrical **asymétrique**
attapulgite **attapulgite**
b axis **axe b**
bauxite **bauxite**
beidellite **beidellite**
bentonite **bentonite**
biogeochemical cycle **cycle biogéochimique**
biogeochemistry **biogéochimie**
biotite **biotite**
birefringence **biréfringence**
birefringent **biréfringent**
birnessite **birnessite**
boehmite **boehmite**
bisiallitization **bisiallitisation**
bog iron **fer des marais**
Bragg's equation **équation de Bragg**
broken-edge bond **liaison attribuable à un bris de lien en périphérie de minéral**
c axis **axe c**
calcite **calcite**
calcium bentonite **bentonite calcaire**
capillary **capillaire**
carbonate **carbonate**
carbonation **carbonatation**
cardhouse structure **structure en château de cartes**
ceramic **céramique**
chemical weathering **altération chimique météorique, météorisation chimique**
chert **chert**
china clay **kaolinite**
chlorite **chlorite**
clay **argile**
clay domain **domaine argileux, champ des argiles**
clay mineral **minéral argileux**
clay mineralogy **minéralogie des argiles**
Clerici solution **solution de Clérici**
comminution **fragmentation**
congelifraction **gélifraction, gélivation, cryoclastie**
constant-charge surface **surface à charge constante**
constant-potential surface **surface à potentiel variable**
continuous reaction series **suite réactionnelle continue**
corrasion **corrasion**
corrosion **corrosion**

crossed nicols **nicols croisés, lumière polarisée analysée**
cryptocrystalline **cryptocristallin**
crystal **cristal**
crystal chemistry **chimie des cristaux**
crystal lattice **réseau cristallin**
crystal morphology **morphologie des cristaux**
crystal structure **structure des cristaux**
crystalline **cristallin**
crystalline rock **roche cristalline**
crystalline solid **solide cristallin**
crystallization **cristallisation**
crystallography **cristallographie**
decomposition **décomposition**
deep weathering **altération profonde**
desilication **désilicification**
differential weathering **altération différentielle ou sélective**
diffraction **diffraction**
diffraction pattern **patron de diffraction**
diffraction spacing **distance interfeuillet**
diffuse double layer **double couche diffuse**
dimorphism **bimorphisme**
dioctahedral **dioctaédrique**
disintegration **désagrégation**
double refraction **biréfringence**
d-spacing **espace interfeuillet**
electron diffraction pattern **patron de diffraction des rayons-X**
electron microprobe **microsonde électronique**
electron microscope **microscope électronique**
euhedral **euèdre**
exfoliation **desquamation, exfoliation**
extinction **extinction**
extinction angle **angle d'extinction**
fat clay **argile plastique**
feldspar **feldspath**
feldspathic **feldspathique**
felsic **felsique**
ferrihydrite **ferrihydrite**
ferrolysis **ferrolyse**
ferromagnesian **ferromagnésien**
ferruginous **ferrugineux**
fine clay **argile fine**
fireclay **argile réfractaire**
fluorapatite **fluoroapatite**
free iron oxides **oxydes de fer libres**
free oxides **oxydes libres**
freeze-thaw action **action gel-dégel**
frost weathering **gélivation, gélifraction, cryoclastie**
fuller's earth **argile de fuller**
geochemical cycle **cycle géochimique**
geochemistry **géochimie**
gibbsite **gibbsite**
glauconite **glauconie, glauconite**
gneiss **gneiss**
goethite **goethite**
goniometer **goniomètre**
granite **granite**
granular **granulaire**
granular disintegration **désagrégation granulaire**
graywacke **grauwacke**
gypsum **gypse**
halite **chlorure de sodium, halite**
halloysite **halloysite**
heavy liquid **liqueur lourde**
heavy media separation **flottation ou séparation gravimétrique**
heavy mineral **minéral lourd**
hematite **hématite**
holocrystalline **holocristallin**
horneblende **hornblende**
hyaline **hyalin**
hydromica **hydromica**
hydromuscovite **muscovite hydratée ou illite**
hydrous **hydraté**
hydrous mica **mica hydraté**
hydroxy-aluminum interlayers **polymère d'hydoxyaluminium en position inter-feuillet**
hydroxy-interlayered vermiculite **vermiculite chloritisée**
hydroxylapatite **hydroxyapatite**
igneous **ignée**
igneous rock **roche ignée**
illite **illite**
imogolite **imogolite**
index of refraction **indice de réfraction**
inheritance **héritage**
intergrade minerals **intergrades**
interlayer **position inter-feuillet**
intermediate rock **roche mafique**
interstratified clay mineral **minéral argileux interstratifié**
iridescence **iridescence, irisation**
iron **fer**
iron oxides **oxydes de fer**
iron pyrites **pyrites**
ironstone **terre de fer, roche ferrugineuse**
isomorphism **isomorphisme**

isomorphous substitution **substitution isomorphe**
jarosite **jarosite**
kaolin **kaolin**
kaolin group **groupe du kaolin**
kaolinite **kaolinite**
kaolinitic **kaolinitique**
K-bentonite **bentonite potassique**
K-feldspar **feldspath potassique**
lag **dépôt résiduel, résidu de déflation**
lamella **lamelle**
lamellar **lamellaire**
latolization **latolisation**
lattice **réseau**
lattice energy **énergie de réseau ou réticulaire**
lattice structure **structure maillée, réticulée ou en treillis**
layer **couche**
layer charge **charge électrique de la couche**
layer silicate mineral **phyllosilicate**
lepidocrocite **lépidocrocite**
light mineral **minéral léger**
limestone **calcaire**
limonite **limonite**
lithiophorite **lithiophorite**
lyotropic series **série lyotropique**
maghemite **maghémite**
magnesian limestone **calcaire magnésien, dolomite**
magnesite **magnésite**
magnetite **magnétite**
manganese oxides **oxydes de manganèse**
marble **marbre**
mechanical weathering **désagrégation mécanique**
mica **mica**
mica schist **schiste micacé, micaschiste**
micaceous **micacé**
micelle **micelle**
microcline **microcline**
microcrystalline **microcristallin**
microgranite **microgranite**
mineral **minéral**
mineral deposit **dépôt, gisement**
mineralogical analysis **analyse minéralogique**
mineralogy **minéralogie**
mineralogy, soil **minéralogie des sols**
mixed-layer mineral **minéral interstatifié**
montmorillonite **montmorillonite**
montmorillonite-saponite group **groupe des smectites**
mud rock **argilite**
mudstone **argilite**
muscovite **muscovite**
neoformation **néogenèse, néoformation**
nesosilicate **nésosilicate**
non-silicate **non-silicaté**
nontronite **nontronite**
octahedron **octaèdre**
oligoclase **oligoclase**
olivine **olivine**
onyx **onyx**
oolite **oolite**
oolith **oolithe, oolite**
oolitic limestone **calcaire oolithique**
opacity **opacité**
opaque **opaque**
organic weathering **altération biologique**
orthoclase **orthoclase**
orthogneiss **orthogneiss**
oxide mineral **oxyde**
pallid zone **lithomarge, zone pâle**
palygorskite **palygorskite**
paragneiss **paragneiss**
patina **patine**
pegmatite **pegmatite**
permanent charge **charge permanente**
phenocryst **phénocristal**
phyllite **phyllite**
phyllosilicate **phyllosilicate**
physical weathering **désagrégation physique**
phytoliths **phytolithe**
pitchblende **pitchblende**
plagioclase **plagioclase**
plane of atoms **plan atomique**
polarized light microscopy **microscopie à lumière polarisée**
porphyritic **porphyrique**
porphyroblastic **porphyroblastique**
porphyry **porphyre**
potassium fixation **fixation du potassium**
primary mineral **minéral primaire**
pseudomorph **pseudomorphe**
pumice **pierre ponce**
pyrite **pyrite**
pyrophosphate **pyrophosphate**
pyrophyllite **pyrophyllite**
pyroxene **pyroxène**
quartz **quartz**
quartzitic schist **schiste quartzique**

quicksilver **mercure**
refractory **réfractaire**
residuum **matériau résiduel**
resistant rock **roche résistante**
rhyolite **rhyolite**
rock **roche, rocher**
rock crystal **cristal de roche**
rock nodules **nodules rocheux**
rock weathering **altération des roches**
saprolite **saprolithe, altérite, saprolite**
scanning electron microscope (SEM) **microscope électronique à balayage**
schist **schiste**
secondary mineral **minéral secondaire**
sepiolite **sépiolite**
serpentine **serpentine**
sesquioxide **sesquioxyde**
shale **shale, phyllade**
sheet of polyhedra **couche polyédrique (tétraédrique ou octaédrique)**
silica **silice**
silica, crystalline **silice cristalline**
silica-alumina ratio **rapport silice-alumnae**
silica-sesquioxide ratio **rapport silice-sesquioxyde**
silicate **silicate, minéral silicaté**
siliceous **siliceux**
slate **ardoise**
smectite **smectite**
soapstone **stéatite, pierre à savon**
soil mineral **minéral du sol**
soil mineralogy **minéralogie des sols**
spalling **délitage**
spheroidal weathering **désagrégation en boules ou sphéroïdale**
stalactite **stalactite**
stalagmite **stalagmite**
structural charge **charge attribuable aux substitutions**
subhedral **subédrique**
sylvite **sylvite**
tactoid **tactoïde**
talc **talc**
tectosilicate **tectosilicate**
tetrahedron **tétraèdre**
thermal fracture **thermoclastie**
thermoluminescence **thermoluminescence**
todorokite **todorokite**
trioctahedral **trioctaédrique**
tufa **calcaire tuffacé**
tuff **tuf**
unit cell **unité de base**
ventifact **ventifact, caillou éolisé, pierre éolisée**
vermiculite **vermiculite**
weathering **météorisation, altération**
weathering, basal surface of **front d'altération**
wind abrasion **abrasion éolienne**
x-ray diffraction **diffraction des rayons-x**
zeolite **zéolite**

MODELING, STATISTICS

abscissa **abscisse**
accuracy **exactitude**
algorithm **algorithme**
alpha error **erreur de première espèce, erreur de type I**
analysis of variance **analyse de variance**
analytical model **modèle analytique**
ANOVA **analyse de variance**
arithmetic mean **moyenne arithmétique**
attribute **variable qualitative, attribut**
autocorrelation **autocorrélation**
average **moyenne**
beta error **erreur bêta, erreur de type II**
bias **biais, erreur systématique**
bimodal distribution **distribution bimodale**
biometrics **biométrie**
biostatistics **biostatistique**
black box **boîte noire**
black box system **système de type boîte noire**
box model **modèle de la boîte**
cascading system **système en cascade**
categorical variable **variable qualitative**
causality **causalité**
central tendency **tendance centrale**
chi-square test **test du khi-carré**
classification **classification, classement**
cluster analysis **analyse en classification automatique, analyse de groupement**
coefficient of correlation (r) **coefficient de corrélation**
coefficient of determination (r^2) **coefficient de détermination**
coefficient of multiple correlation (R) **coefficient de corrélation multiple**

coefficient of multiple determination (R^2) **coefficient de détermination multiple**
coefficient of variation **coefficient de variation**
complete block design **plan d-expérience à blocs complets**
completely randomized design **plan d'expérience complètement aléatoire**
computer simulation model **modèle de simulation par ordinateur**
conceptual model **modèle conceptuel**
confidence coefficient **coefficient de confiance**
confidence interval **intervalle de confiance**
confidence limits **limites de confiance**
constant **constante**
continuous variable **variable continue**
continuous-flow system **système à écoulement continu**
control group **groupe témoin, groupe contrôle**
control system **système de contrôle**
correlation **corrélation**
correlation, statistical **corrélation statistique**
covariance **covariance**
cumulative distribution **distribution cumulée ou cumulative**
curve fitting **ajustement d'une courbe**
cybernetics **cybernétique**
damping **amortissement**
decision analysis model **modèle d'analyse de décision**
decision region **région décisionnelle**
decision rule **critère de décision**
decision tree **arbre de décision**
deduction **déduction**
degrees of freedom **degrés de liberté**
dependent variable **variable dépendante**
deterministic model **modèle déterministe**
deterministic process **processus déterministe**
deviation, standard **écart type, déviation standard**
discrete variable **variable discrète ou discontinue**
discriminant function **fonction discriminante**
dispersion diagram **diagramme de dispersion**
distribution **distribution**
double sample **échantillon double**
effect **effet**
eigenvalues **valeurs propres**
empirical **empirique**
endogenous variable **variable endogène**
equilibrium **équilibre**
error **erreur**
exogenous variable **variable exogène**
experimental design **dispositif ou plan expérimental**
experimental unit **unité expérimentale**
extrapolation **extrapolation**
factor **facteur**
factor analysis **analyse factorielle**
factorial experiment **plan d'expérience factoriel**
false negative **faux négatif**
false positive **faux positif**
feedback **feedback, rétroaction**
feedback control system **système à rétrocontrôle, système asservi**
field trial **essai en champ**
Fourier analysis **analyse de Fourier**
frame **base de sondage**
frequency **fréquence**
frequency band **bande de fréquences**
frequency curve **courbe de fréquence**
frequency distribution **distribution de fréquence, distribution statistique**
F-test **test *F***
Gaussian curve **courbe de Gauss, courbe gaussienne**
Gaussian distribution **distribution de Gauss, distribution gaussienne**
general systems theory **théorie du système général**
geometric mean **moyenne géométrique**
geometric series **progression géométrique**
geometric standard deviation **écart type géométrique**
grab sample **échantillon instantané**
graph **graphique**
gray box system **système de type boîte grise**
heuristic approach **approche heuristique**
hierarchy **hiérarchie**
histogram **histogramme**
hypothesis **hypothèse**
iconic model **modèle iconique**
incomplete block design **plan d'expérience à blocs incomplets**
independent variable **variable indépendante**
induction **induction**
information theory **théorie de l'information**
inputs and outputs **entrées et sorties**

interaction **interaction**
interpolation **interpolation**
interval **intervalle**
inverse square law **loi de l'inverse des carrés**
irreversible process **processus irréversible**
iteration **itération**
judgment sample **échantillon choisi à dessein, échantillon au jugé**
kriging **krigeage**
kurtosis **aplatissement, kurtosis**
law **loi**
least squares method **méthode des moindres carrés**
linear programming **programmation linéaire**
loading **saturation**
log **log**
logistic curve **courbe logistique**
log-normal distribution **distribution log-normale**
Markov chain **chaîne de Markov**
Markov process **processus de Markov**
mass diagram **courbe des valeurs cumulées**
mean **moyenne**
mean deviation **écart moyen**
measure of central tendency **mesure de la tendance centrale**
mechanistic **mécaniste**
median **médiane**
mode **mode, valeur dominante**
model **modèle**
model building **modélisation**
modeling **modélisation**
moment measure **mesure des moments**
Monte Carlo method **méthode de Monte-Carlo**
morphostasis **morphostase**
multiple correlation **corrélation multiple**
multivariate analysis **analyse multivariable, analyse à plusieurs variables**
natural event system **système à événement naturel**
negative correlation **corrélation négative**
negative feedback **rétroaction négative, feedback négatif**
nested experiment **expérience en tiroirs**
nested sampling **échantillonnage hiérarchique**
network **réseau**
noise **bruit**
nominal variable **variable qualitative**
nomograph **nomogramme**
non-linear function **fonction non linéaire**
non-linear regression **régression non linéaire, régression curvilinéaire**
non-parametric statistical test **test statistique non paramétrique**
normal **normale**
normal distribution **distribution normale**
null hypothesis **hypothèse nulle**
one-way analysis of variance **analyse de variance à un critère de classification**
open system **système ouvert**
ordinal variable **variable ordinale ou ordonnée**
ordinate **ordonnée**
ordination **ordination**
origin **origine**
parabola **parabole**
paradigm **paradigme**
parameter **paramètre**
parametric statistical tests **tests paramétriques**
partial correlation **corrélation partielle**
percent error **pourcentage d'erreur**
percentile **percentile, centile**
physical system **système physique**
pilot project **projet pilote**
plane surface system **système à surface plane**
point distribution **distribution ponctuelle**
poised equilibrium **équilibre indifférent**
Poisson distribution **distribution de Poisson**
positional probability **probabilité de position**
positive association **association positive**
positive feedback **rétroaction positive**
potential model **modèle de potentiel**
power of test **puissance d'un test**
precision **précision**
predictive **prédictif**
prescriptive **prescriptif**
probability **probabilité**
probability distribution **distribution de probabilité**
probability sample **échantillon probabiliste**
process-response system **système de régulation des procédés**
process-response model **modèle de régulation des procédés**
product-moment correlation **corrélation du moment des produits**
qualitative variable **variable qualitative**
quantitative variable **variable quantitative**
quartile **quartile**

quasi-equilibrium **quasi-équilibre**
random error **erreur aléatoire**
random sample **échantillon aléatoire**
randomization **randomisation**
randomized blocks **blocs aléatoires**
random-walk model **modèle de cheminement aléatoire**
range **étendue, plage**
ranking **classement**
rate variables **variables de vitesse**
regression **régression**
regulator **régulateur**
regulatory response **réponse de régulation**
replicate **répétition**
replication **répétition**
response **réponse**
response time **temps de réponse**
reversible process **processus réversible**
sample **échantillon**
sample plot **placette d'échantillonnage, parcelle d'échantillon**
sampling **échantillonnage**
sampling frame **base de sondage**
sampling unit **unité d'échantillonnage**
scattergram **diagramme de dispersion**
scientific method **méthode scientifique**
secondary response **réaction ou réponse secondaire**
self regulation **autorégulation**
semi-variance **semi-variance**
semivariogram **semivariogramme**
serial samples **prélèvements sériés**
servomechanism **servomécanisme**
significance, statistical **signification statistique**
sill **seuil**
simulation **simulation**
simulation model **modèle de simulation**
skewness **asymétrie**
spatial model **modèle d'analyse spatiale**
spatial statistics **statistiques spatiales, géostatistiques**
spatial variability, soil **variabilité spatiale**
Spearman rank correlation **coefficient de corrélation de rang de Spearman**
stable equilibrium **équilibre stable**
standard deviation **écart type, déviation standard**
standard error **erreur type**
state variables **variables d'état**
static equilibrium **équilibre statique**
statistical model **modèle statistique**
statistics **statistiques**
steady state **régime permanent**
steady-state equilibrium **état d'équilibre permanent**
step function **fonction à paliers**
stochastic **stochastique**
stochastic process **processus stochastique**
stochastic-process model **modèle de processus stochastique**
stratified random sample **échantillon aléatoire stratifié**
system **système**
systematic error **erreur systématique**
systematic sample **échantillon systématique**
systems analysis **analyse de systèmique**
systems approach **approche systémique**
systems theory **théorie des systèmes**
temporal change **changement temporel**
ternary diagram **diagramme ternaire**
theory **théorie**
threshold **seuil**
time series analysis **analyse des séries chronologiques**
transformation **transformation, normalisation**
trend surface **surface de tendance, surface polynomiale**
trend surface analysis **établissement d'une surface de tendance**
t-test **test t**
two way analysis of variance **analyse de variance à deux critères de classification**
Type I error **erreur de type I**
Type II error **erreur de type II**
unimodal distribution **distribution unimodale**
unstable equilibrium **équilibre instable**
variable **variable**
variance **variance**
variate **variable**
variogram **variogramme**
vector **vecteur**
weighted average **moyenne pondérée**
white box system **système de type boîte blanche**

PEAT, PEATLANDS

allochthonous peat **tourbe allochtone**
amorphous peat **tourbe amorphe**
autochtonous peat **tourbe autochtone**
black mud **boue noire**
bog **tourbière ombrotrophe**
bog peat **tourbe de sphaigne**
brown moss peat **tourbe de mousse brune**
Canadian wetland classification system **système canadien de classification des milieux humides**
carex peat **tourbe de *Carex***
carpet **tapis, radeau flottant**
carr **fourré marécageux, carr**
collapse scar bog, collapse scar fen **tourbière effondrée, fen effondré**
coprogenic **coprogène**
coprogenous earth **terre coprogène**
cumulose deposits **dépôts cumuliques**
diatomaceous earth **terre de diatomées**
diversity index **indice de diversité**
dopplerite **dopplerite**
drainage basin **bassin versant, bassin hydrographique**
dy **bourbe dystrophe**
dystrophic **dystrophe**
fen **tourbière minérotrophe**
fen peat **tourbe minérotrophe**
fen soils **sols de fen**
fiber **fibre**
rubbed fiber **fibre frottée**
unrubbed fiber **fibre non frottée**
fibric (peat) **fibrique**
fibric layer **couche fibrique**
fibric materials **matériaux fibriques**
Fibrisol **fibrisol**
Fibrist **Fibrist**
fibrous **fibreux**
filling in **comblement**
forest peat **tourbe forestière**
forested fen or treed fen **fen arboré**
groundwater **eau souterraine**
gyttja **bourbe eutrophe**
herb **herbe**
humic peat **tourbe humique**
hummock **butte, massif**
hydromorphic **hydromorphe**
hydrophyte **hydrophyte**
limnic **limnique**
limno layer **couche limnique**
lithic layer **couche lithique**
marsh **marais**
marsh gas **gaz des marais**
mesic **mésique**
mesophyte **mésophyte**
mesotrophic **mésotrophe**
minerotrophic **minérotrophe**
mire **tourbière, milieu tourbeux, bourbier**
moss peat **mousse de tourbe**
mounded **à buttes**
muck **terre tourbeuse**
muskeg **muskeg, plée, savane**
oligotrophic **oligotrophe**
ombrogenous **ombrogène**
ombrotrophic **ombrotrophe**
organic soil **sol organique**
organic soil materials **matériaux de sol organique**
palsa **palse**
paludification **paludification**
peat **tourbe**
peat bog **tourbière**
peat moss **mousse de tourbe**
peat mound **butte tourbeuse**
peat plateau **plateau tourbeux**
peat stratigraphy **stratigraphie de tourbière**
peatland **milieu tourbeux, tourbière**
permafrost **pergélisol**
permafrost table **limite du pergélisol**
reed peat **tourbe de roseaux**
rheophilous **rhéophile**
rheotrophic **rhéotrophe**
sedge peat **tourbe de *Carex***
sedge-reed peat **tourbe d'herbacées**
sedimentary peat **terre coprogène**
shrub **arbuste**
silvic **silvique**
slough **marécage, mare vaseuse, bourbier**
soligenous **soligène**
sphagnic **sphagnique**
Sphagnum ***Sphagnum***
sphagnum peat **tourbe de sphaigne**
swamp **marécage, milieu marécageux**
swamp peat **tourbe marécageuse**
telluric water **eau tellurique**
telmatic peat **tourbe de roseaux**
terrestrialization **comblement, atternissement**
terric **terrique**
terric layer **couche terrique**

terrigenous sediments **sédiments terriques**
transition peatland **tourbière de transition**
trophic status **condition trophique**
undifferentiated organic material **matériau organique non différencié**
unsaturated zone **zone non saturée, zone vadose**
von Post humification scale **échelle d'humification de von Post**
water table **nappe phréatique**
water track **chenal d'écoulement végétalisé**
waterlogged **saturé d'eau**
watershed **bassin versant ou hydrographique**
wetland **milieu ou terre humide**
wetland class **classe de milieu humide**
wetland form **aspect des milieux humides**
wetland type **type de milieu humide**
woody peat **tourbe ligneuse**

PHYSICS, SOIL WATER

absorbed water **eau absorbée**
access tube **tube d'accès**
adsorbed water **eau adsorbée**
aerate **aérer**
agrohydrology **hydrologie agricole**
air dry **séché à l'air**
air entry value **point d'entrée d'air**
air porosity **porosité d'air, teneur en air**
air-dry mass **masse sèche à l'air**
air-filled porosity **porosité d'air, teneur en air**
albedo **albédo**
anaerobic **anaérobie**
anisotropic soils **sols anisotropes**
antecedent moisture **humidité antérieure**
anthric saturation **saturation anthrique**
apparent cohesion **cohésion apparente**
apparent density **masse volumique apparente, densité apparente**
apparent specific gravity **densité spécifique apparente**
atmospheric pressure **pression atmosphérique**
Atterberg limits **limites d'Atterberg**
available moisture **eau disponible**
available water (capacity) **eau disponible**
basic intake rate **taux d'infiltration limite**
bound water **eau liée**
bulk area **aire brute**
bulk density, soil **masse volumique apparente, densité apparente**
bulk specific gravity **poids spécifique apparent**
bulk volume **volume brut**
bypass flow **écoulement préférentiel**
capillarity **capillarité**
capillary action **action capillaire**
capillary fringe **frange capillaire**
capillary interstice **interstice capillaire**
capillary rise **remontée capillaire**
capillary water **eau capillaire**
capillary zone **zone capillaire**
cavitation **cavitation**
coefficient of linear extensibility (COLE) **coefficient d'extensibilité linéaire (indice COLE)**
colloid **colloïde**
compactibility **compactibilité**
compaction **compaction, compactage, tassement**
compressibility **compressibilité**
compressibility index **indice de compressibilité**
compression **compression**
compressive strength **résistance à la compression**
concentrated flow **écoulement concentré**
condensation **condensation**
conductance **conductance**
conduction, thermal or heat **conduction thermique**
cone index **indice de résistance à la pénétration de la pointe**
cone penetrometer **pénétromètre à cône**
consistence **consistance**
convection **convection**
critical angle **angle critique**
critical density **masse volumique critique**
critical pressure **pression critique**
critical void ratio **indice des vides critique**
crushing strength **résistance à l'écrasement**
crust **croûte**
cumulative infiltration **infiltration cumulée ou cumulative**
Darcy's law **loi de Darcy**
deep percolation **percolation profonde**
deflation **déflation**
deflocculate **défloculer**
degree of saturation **taux de saturation**

density **masse volumique**
desication crack **fissure ou fente de retrait**
detachment **détachement**
differential water capacity **capacité différentielle de rétention de l'eau**
diffusivity **diffusivité**
direct shear test **essai de cisaillement direct**
discharge curve **courbe des débits, courbe d'étalonnage**
dispersion **dispersion**
dispersivity **dispersibilité**
drainage **drainage**
dry bulk density **masse volumique apparente sèche, densité apparente sèche**
dry mass **masse sèche**
dynamic viscosity **viscosité dynamique**
dynamometer **dynamomètre**
effective precipitation **précipitation effective**
effective stress **contrainte effective**
electrokinetic (zeta) potential **potentiel électrocinétique, potentiel zêta**
emulsion **émulsion**
equivalent depth **profondeur équivalente**
equivalent diameter **diamètre équivalent**
equivalent radius **rayon équivalent**
erodibility **érodabilité ou érodibilité**
erodible **érodable**
erosion enrichment ratio (ER) **rapport d'enrichissement érosif**
erosive velocity **vitesse érosive**
extract electrical conductivity, ECe **conductivité électrique d'un extrait saturé**
fall cone penetrometer **pénétromètre à cône tombant**
field capacity **capacité au champ**
film water **eau pelliculaire**
fine earth **terre fine**
fine sand **sable fin**
fine texture **texture fine**
fines **fraction fine**
firm **ferme**
flocculation **floculation**
flow velocity **vitesse d'écoulement, débit**
flux **flux, fondant**
flux density **densité de flux**
free water **eau libre**
friable **friable**
friction cone penetrometer **pénétromètre à cône à friction latérale**
frost, concrete **masse cryoconsolidée**
frost, honeycomb **gel alvéolaire**
furrow erosion **érosion des rigoles d'irrigation**
furrow mulching **paillage des rigoles d'irrigation**
gamma probe **sonde à rayons gamma**
gamma-ray attenuation **atténuation du rayonnement gamma**
gas pressure potential **potentiel pneumatique**
grading curve **courbe granulométrique**
grain density **densité particulaire**
grain-size analysis **analyse granulométrique, analyse mécanique**
grain-size distribution **granulométrie**
gravitational potential **potentiel d'eau libre, potentiel gravitaire**
gravitational water **eau libre, eau gravitaire**
gravity flow **écoulement par gravité**
groundwater hydrology **hydrogéologie, hydrologie des eaux souterraines**
head **pression hydrostatique, hauteur**
heat capacity **capacité calorifique**
heat flux density **densité du flux thermique**
heat of immersion **chaleur d'immersion**
heat sink **puits thermique**
heavy clay **argile lourde**
hydraulic conductivity **conductivité hydraulique**
hydraulic gradient **gradient hydraulique**
hydraulic head **charge hydraulique**
hydraulic potential **potentiel hydraulique**
hydric soil **sol hydromorphe, sol hydrique**
hydrodynamic dispersion **dispersion hydrodynamique**
hydrodynamic dispersion coefficient **coefficient de dispersion hydrodynamique**
hydrologic cycle **cycle hydrologique**
hydrology **hydrologie**
hydrometer **hydromètre**
hydrometry **hydrométrie**
hydrophilic **hydrophile**
hydrophobic **hydrophobe**
hydrophobicity **hydrophobicité**
hygroscopic **hygroscopique**
hygroscopic water **eau hygroscopique**
hysteresis **hystérèse**
impedance **impédance**
impeded drainage **drainage entravé**
impermeability **imperméabilité**
impervious **imperméable**
infiltrability **taux maximal d'infiltration**
infiltration **infiltration**

infiltration flux (or rate) **taux d'infiltration**
infiltrometer **infiltromètre**
infrared (IR) **infrarouge**
intake rate **taux d'infiltration**
intermittent stream **cours d'eau temporaire ou intermittent**
internal friction **friction interne**
interstitial water **eau intersticielle**
intrinsic permeability **perméabilité intrinsèque**
irrigation flume **canal sur appuis, canal surélevé, canal de jaugeage**
isodyne **isodyne**
isotropic shrinkage **retrait isotrope**
kinematic viscosity **viscosité cinématique**
kinetic energy **energie cinétique**
latent heat **chaleur latente**
liquefaction **liquéfaction spontanée**
liquid limit **limite de liquidité**
lower plastic limit **limite de plasticité inférieure**
lysimeter **lysimètre**
macropore **macropore**
macropore flow **écoulement préférentiel**
macroporosity **macroporosité**
manometer **manomètre**
mass movement **mouvement de masse**
mass wetness **teneur en eau gravimétrique ou massique**
matric potential **potentiel matriciel**
mechanical analysis **analyse mécanique**
medium texture **texture moyenne**
miscible displacement **déplacement de fluides miscibles**
moderately coarse texture **texture modérément grossière**
moderately fine texture **texture modérément fine**
Mohr circle **cercle de Mohr**
Mohr envelope **enveloppe de Mohr**
moisture release curve **courbe de désorption**
moisture retention curve **courbe de rétention en eau**
moisture tension (or suction) **tension de l'eau du sol, potentiel hydrique**
moisture-retention curve **courbe de rétention**
neutron probe **sonde à neutrons**
non-limiting water range **étendue des teneurs en eau non-limitantes**
osmotic potential **potentiel osmotique**
osmotic pressure **pression osmotique**
oxygen diffusion rate (ODR) **taux de diffusion de l'oxygène**
packing voids **vides d'entassement**
particle density **densité particulaire**
particle size **calibre**
particle-size analysis **analyse mécanique ou granulométrique**
particle-size distribution **distribution granulométrie**
ped **ped**
penetrability **pénétrabilité**
penetration resistance **résistance à la pénétration**
perched water table **surface de saturation suspendue**
percolation (of soil water) **percolation**
permanent wilting point **point de flétrissement permanent**
permeability **perméabilité**
permeameter **perméamètre**
physical properties of soils **propriétés physiques des sols**
physical weathering **désagrégation ou altération physique**
plastic limit **limite de plasticité**
plasticity index **indice de plasticité**
plasticity range **étendue du comportement plastique**
pore **pore**
pore ice **glace de remplissage des pores, glace interstitielle**
pore space **espace poral**
pore water pressure **pression de l'eau porale**
pore water velocity **vitesse d'écoulement de l'eau porale**
pore-size distribution **distribution volumétrique des pores**
porosity **porosité**
precipitation interception **interception des précipitations**
preferential flow **écoulement préférentiel**
pressure membrane **membrane de tensiomètre**
primary particles **particules primaires, particules élémentaires**
quick clay **argile sensible**
radiant heat **chaleur rayonnante**
rainfall erosivity index **indice d'érosivité par la pluie**
redistribution **redistribution**
reflectance **réflectance**

relative compaction **compaction relative**
residual shrinkage **retrait résiduel**
retardation factor **facteur de ralentissement**
rill erosion **érosion en rigoles, en ruisselets ou en filets**
root mean square roughness (RMS) **rugosité efficace, rugosité moyenne quadratique**
saltation flux **flux de saltation**
sand **sable**
sandy **sableux**
sandy clay **argile sableuse**
sandy clay loam **loam sablo-argileux**
sandy loam **loam sableux**
saturate **saturer**
saturated flow **écoulement saturé ou à saturation**
saturated hydraulic conductivity **conductivité hydraulique saturée ou à saturation**
saturation **saturation**
saturation content **teneur en eau de l'extrait de saturation**
screen **crible**
sensitivity **sensibilité**
settling velocity **vitesse de sédimentation ou de décantation**
shear **cisaillement**
shear plane **surface de cisaillement**
shear strength **résistance au cisaillement**
shear stress **contrainte de cisaillement**
shearing **cisaillement**
sheet erosion **érosion en nappe ou inter-rigoles**
shrinkage coefficient, soil **coefficient de retrait**
shrinkage index **indice de retrait**
shrinkage limit **limite de retrait**
shrinkage, soil structural **retrait structural**
shrink-swell potential **potentiel de retrait et de gonflement**
sieve analysis **analyse granulométrique par tamisage**
silt **limon**
silt loam **loam limoneux**
silty clay **argile limoneuse**
silty clay loam **loam limono-argileux**
skeleton grains **grains du squelette**
skew planes **fentes de retrait irrégulières**
soil aeration **aération du sol**
soil air **air du sol**
soil core **carotte de sol**
soil loss tolerance **perte de sol tolérable**
soil mechanics **mécanique des sols**
soil moisture **humidité du sol**
soil physics **physique du sol**
soil pores **pores du sol**
soil probe **sonde pédologique**
soil separates **fractions du sol**
soil strength **résistance du sol**
soil suction **succion du sol**
soil temperature **température du sol**
soil water **eau du sol**
soil water characteristic **courbe de rétention caractéristique**
soil water diffusivity **diffusivité de l'eau du sol**
soil water potential **potentiel hydrique du sol**
soil wetness **teneur en eau du sol**
solclime **pédoclimat, climat du sol**
sorptivity **sorptivité**
specific gravity **densité, densité relative**
specific surface **surface spécifique**
specific water capacity **capacité spécifique de rétention en eau**
sphericity **sphéricité**
standard cone **cône étalon**
static penetrometer **pénétromètre statique**
stemflow **ruissellement sur les troncs**
sticky point **point d'adhésivité**
Stokes' law **loi de Stokes**
stone **pierre**
strain **déformation**
subsurface drainage **drainage souterrain**
suction **succion**
suction lysimeter **lysimètre à succion**
summation curve (of particle sizes) **courbe cumulative (granulométrique)**
surface area **aire superficielle**
surface drain **drain de surface**
surface sealing **colmatage ou obturation de la surface**
suspension **suspension**
swelling hysteresis **hystérèse de gonflement**
tactoid **tactoïde**
TDR **TDR**
tensile strength **résistance en traction**
tensiometer **tensiomètre**
terminal settling velocity **vitesse terminale de sédimentation**
textural triangle **triangle de texture**
texture **texture**
thermal analysis **analyse thermique**
thermal conductivity **conductivité thermique**

thermal properties **propriétés thermiques**
thermistor **thermistance**
thermocouple **thermocouple**
thixotropy **thixotropie**
throughfall **pluviolessivage**
tight soil **sol compact ou tenace**
tillability **degré de facilité de travail du sol**
tillage pan **semelle de labour**
time-domain reflectometry **réflectométrie à dimension temporelle**
tortuosity **tortuosité**
total potential (of soil water) **potentiel hydrique total, pression totale de l'eau du sol**
triaxial shear test **test de cisaillement triaxial**
unavailable water **eau non disponible**
unidimensional shrinkage **retrait unidirectionel**
unitary shrinkage **retrait unitaire**
universal soil loss equation (USLE) **équation universelle de perte de sol**
unsaturated flow **écoulement non-saturé ou en milieu non-saturé**
unsaturated hydraulic conductivity **conductivité hydraulique non-saturée ou en milieu non-saturé**
unsaturated zone **zone non-saturée**
upper plastic limit **limite supérieure de plasticité**
vadose zone **zone vadose**
vapor **vapeur**
vapor density **densité de vapeur**
vapor flow **écoulement de vapeur**
vertical shrinkage **retrait vertical**
very coarse sand **sable très grossier**
very fine sand **sable très fin**
very fine sandy loam **loam sableux très fin**
viscous **visqueux**
void **vide**
void ratio **indice des vides**
volume wetness **humidité volumétrique**
volumetric water content **teneur en eau volumétrique**
water balance **bilan hydrique**
water content, soil **teneur en eau du sol**
water flow velocity **vitesse d'écoulement de l'eau**
water holding capacity **capacité de rétention en eau**
water pressure **pression de l'eau**
water retention **rétention en eau**
water retention curve **courbe de rétention en eau**
water suction **succion d'eau**
water table **nappe phréatique, surface de saturation**
water use efficiency **efficacité d'utilisation de l'eau**
water, interstitial **eau intersticielle**
water-filled pore space **taux de saturation en eau**
waterlogged **gorgé ou saturé d'eau**
water release curve **courbe de désorption en eau**
water retention curve **courbe de rétention en eau**
wettability **mouillabilité**
wilting point **point de flétrissement**
wind erosion equation **équation de l'érosion éolienne**
zeta potential **potentiel zêta**

RECLAMATION, REMEDIATION

acid mine drainage **drainage minier acide**
acid spoil **déblais acides**
approximate original contour **configuration approximative d'origine**
area mining **exploitation minière de surface**
artificial brine **saumure artificielle**
ash **cendre**
backfill **remblayer, remblai**
bedrock spoil **déblais rocheux**
bench **gradin**
berm **berme**
borrow pit **carrière d'emprunt**
brush matting **paillassonnage en branches**
capping **recouvrement**
contour mining **exploitation suivant les courbes de niveau**
contour stripping **décapage suivant les courbes de niveau**
contour surface mining **mine à ciel ouvert exploitée suivant les courbes de niveau**
core trench **tranchée pour mur écran**
cover soil **terre de recouvrement**
cut and fill **déblai-remblai**
disturbed area **zone perturbée**
disturbed land **terrain perturbé**
drill cuttings **déblais de forage**

drilling fluid **fluide de forage**
drilling mud **boue de forage**
drilling waste fluid **résidus liquides de forage**
drilling waste solid **résidus solides de forage**
encapsulation **encapsulation**
equivalent land capability **possibilité d'utilisation équivalente des terres**
fill **remblai**
filling **remblaiement**
filter cloth **toile filtrante**
fine tailings (fine tails, sludge) **résidus fins**
fines **éléments fins, particules fines**
gabion **gabion**
gob **remblai**
grade stabilization structure **ouvrage de stabilisation de pente**
gravel envelope **enveloppe de gravier**
gravel filter **filtre de gravier**
ground cover **couvert végétal**
haul road **chemin d'exploitation**
holding pond **étang de retenue**
hydraulic dredging **dragage hydraulique**
hydraulic mining **abattage hydraulique**
hydroseeding **ensemencement hydraulique**
interception channel **fossé de crête, fossé d'interception**
interceptor drain **drain d'interception**
land reclamation **mise en valeur ou restauration des terres**
mine **mine**
mine drainage **drainage de mine**
mine dump **terril**
mine wash **boue de mines**
mined land **terrain exploité**
minesoil **sol minier**
mining **extraction ou exploitation minière**
mining by-products **résidus miniers**
mix, bury and cover (drilling wastes) **technique d'élimination des résidus de forage**
mud **boue**
native prairie **prairie naturelle**
native species **espèce indigène**
natural revegetation **remise en végétation naturelle**
natural seeding **ensemencement naturel**
naturalized plant **plante naturalisée**
non-native species **espèce introduite**
oil wasteland **terrain d'épandage de pétrole**
open-pit mine **mine ou exploitation à ciel ouvert**
open-cast mining **exploitation à ciel ouvert**
ore **minerai**
orphan lands **terrains orphelins**
overburden **morts-terrains, stérile**
overstripping **surexploitation du sol**
pit **fosse, puits**
process water **eau de traitement**
prospecting **prospection**
quarry **carrière**
reclamation **bonification ou restauration des terres**
reconstructed profile **profil reconstitué**
rehabilitation **remise en valeur**
relief drain **drain de décharge**
restoration **restauration**
revegetation **revégétation, remise en végétation, végétalisation**
revetment **revêtement**
slickens **boue de mines fine**
solidification **solidification**
spoil **déblais**
spoil bank **remblai détritique, terril**
standard soil handling procedure **méthode standardisée de manipulation du sol**
stream bank erosion control **lutte contre l'érosion des berges**
stream bank stabilization **stabilisation des berges**
strip mining **exploitation à ciel ouvert**
surface mining **exploitation à ciel ouvert**
tailings **résidus**
toe **pied**
toe drain **drain de pied**
toxic spoil **déblais toxiques**
trenching **excavation en tranchées**
underground development waste **déchets miniers souterrains**
underground mining **exploitation souterraine**
upper subsoil **sous-sol supérieur**
wing wall **mur en aile**

SOIL GENESIS, PEDOLOGY, AND CLASSIFICATION

A horizon **horizon A**
AB horizon **horizon AB**
ABC soil **sol ABC**
AC horizon **horizon AC**

AC soil **sol AC**
active layer **couche active**
agric horizon **horizon agrique**
agroclimate **agroclimat**
agroecological resource area **aire de ressource agroécologique**
agroecological resource region **région de ressource agroécologique**
agroecological resource zone **zone de ressource agroécologique**
albic horizon **horizon albique**
Albolls **Albolls**
Alfisols **Alfisols**
alkalization **alcalinisation**
allitization **allitisation**
alpine soil **sol alpin**
anaerobic (soil) **anaérobie (sol)**
Andepts **Andepts**
Andisols **Andisols**
Andosol **Andosol**
anthropic epipedon **épipédon anthropique**
anthropic soil **sol anthropique**
apedal **apédal**
Aqualfs **Aqualfs**
Aquands **Aquands**
Aquents **Aquents**
Aquerts **Aquerts**
aquic **aquique**
Aquods **Aquods**
Aquolls **Aquolls**
Aquox **Aquox**
Aquults **Aquults**
Arents **Arents**
Argids **Argids**
argillic **argilique**
argillic horizon **horizon argilique**
argillipedoturbation **argilipédoturbation**
aridic **aridique**
Aridisols **Aridisols**
artificial soil body **sol anthropique**
associate, soil **sol associé**
auger, soil **tarière**
azonal soil **sol azonal**
B Horizon **horizon B**
BE horizon **horizon BE**
biosequence **bioséquence**
biostasis **biostasie**
bisequa **biséquums**
Black Chernozem **chernozem noir**
blanket **couverture**
blocky **polyédrique**
Boralfs **Boralfs**
Borolls **Borolls**
bottom tier **étage inférieur**
boulder **blocs rocheux, bloc**
Brown Chernozem **chernozem brun**
Brunisolic **brunisolique**
brunification **brunification**
buried soil **sol fossile, sol enfoui**
C horizon **horizon C**
calcan **calcane**
calcareous **calcaire**
calcareous classes **classes calcaires**
calcareous soil **sol calcaire**
calcic horizon **horizon calcique**
calcids **calcids**
calcification **calcification**
calcium carbonate equivalent **équivalent de carbonate de calcium**
calcrete **calcrète**
caliche **caliche**
cambic horizon **horizon cambique**
cambids **cambids**
capability **possibilité**
capability class, soil **classe de possibilités des sols**
category **catégorie**
catena **caténa, chaîne de sols**
chambers **vacuoles reliées, chambres**
channeled (eroded) **cannelé, raviné**
channery **en plaquettes**
Chernozemic **chernozémique**
chernozemic A **A chernozémique**
cherty **cherteux**
chroma **chroma, saturation**
chronosequence **chronoséquence**
class, soil **classe de sols**
clay **argile**
clay (1:1) **argile (1:1)**
clay (2:1) **argile (2:1)**
clay film (skin) **pellicule argileuse, enrobement, film argileux**
clay loam **loam argileux**
clayey **argileux**
claypan **claypan**
climosequence **climoséquence**
coarse fragments **fragments grossiers**
coarse sand **sable grossier**
coarse texture **texture grossière**
coating **revêtement, enduit, enrobement**
cobble **galet, caillou**
cobblestone **caillou roulé, galet**

cobbly **caillouteux**
co-dominant **composantes significatives (d'une unité cartographique)**
collapsible soil **sol sensible au tassement ou à l'affaissement**
colluvial **colluvial**
colluvium **colluvion**
color **couleur**
columnar structure **structure colomnaire**
complex, soil **complexe de sols**
compound unit **unité composée**
concrete frost **masse cryoconsolidée**
conglomerate **conglomérat**
continuous permafrost **pergélisol continu**
control section, soil **coupe témoin d'un sol, profil témoin**
coppice mound **tertre avec taillis**
correlation **corrélation**
coulee **chenal de fusion glaciaire**
cradle knoll **butte de chablis**
crotovina **crotovina**
Cryands **Cryands**
Cryerts **Cryerts**
Cryids **Cryids**
Cryods **Cryods**
cryogenic soil **sol cryogénique**
cryopedology **cryopédologie**
Cryosolic **cryosolique**
cryoturbation **cryoturbation**
cumulization **accumulation**
Dark Brown Chernozem **chernozem brun foncé**
Dark Gray Chernozem **chernozem gris foncé**
dealkalization **déalcalinisation**
decalcification **décalcification**
deposition **déposition**
desalinization **désalinisation**
desert crust **croûte désertique**
desert pavement **pavé désertique, reg**
desert polish **poli désertique**
desiccation **dessication**
desiccation crack **fissure ou fente de dessication**
desilication **désilicification**
diagnostic horizon **horizon diagnostique**
dispersive clay **argile dispersée, peptisée, collosol**
dry permafrost **pergélisol sec**
duff **matte, humus peu décomposé**
duff mull **humus intermédiaire**
duric **durique**
duricrust **croûte pédologique concrétionné**
Durids **Durids**
durinodes **durinodes**
duripan **duripan**
Dystric Brunisol **brunisol dystrique**
E horizon **horizon E**
earth hummock **butte gazonnée, thufur**
ecodistrict **écodistrict**
ecoelement **écoélément**
ecological approach **approche écologique**
ecological land classification **classification écologique du territoire**
ecological land classification system, Canadian **système canadien de classification écologique du territoire**
ecoprovince **écoprovince**
ecoregion **écorégion**
ecosection **écosection**
ecosite **écosite**
ecozone **ecozone**
efflorescence **efflorescence**
eluvial horizon **horizon éluvial**
eluviation **éluviation**
Entisols **Entisols**
epipedon **épipédon**
erosion, surficial **érosion de surface**
euic **euique**
Eutric Brunisol **brunisol eutrique**
exhumed paleosol **paléosol exhumé**
F horizon **horizon F**
factors of soil formation **facteurs de formation du sol**
family, soil **famille de sols**
fan **éventail, glacis**
ferricrete **conglomérat à ciment ferrugineux**
Ferrods **Ferrods**
fersiallic soil **sol fersiallique**
fersiallitization **fersiallitisation**
Ferro-Humic Podzol **podzol ferro-fumique**
fine texture **texture fine**
flaggy **en dalles**
flagstone **dalle**
fluted moraine **moraine cannelée**
Fluvents **Fluvents**
fluvial **fluvial, fluviatile**
fluvioeolian **fluvio-éolien**
fluviolacustrine **fluvio-lacustre**
folic **folique**
Folisol **folisol**
folistic epipedon **épipédon folistique**
Folists **Folists**

forest floor **couche holorganique, couverture morte, litière**
forest soils **sols forestiers**
fossil soil **sol fossile**
fragic **fragique**
fragipan **fragipan**
frost boil **ventre de boeuf**
frost crack **fissure de gel, fente de gel**
frost creep **reptation des sols causée par le gel**
frost free period **période sans gel**
frost heave **soulèvement par le gel**
frost line **profondeur maximum du gel, seuil du gel**
gelic material **matériau gélique**
Gelisols **Gélisols**
genesis, soil **genèse ou formation des sols**
geography, soil **géographie des sols**
gilgai **gilgai**
glacial **glaciaire**
glaciofluvial **fluvio-glaciaire**
glaciolacustrine **glacio-lacustre**
gleyed soil **sol gleyifié**
Gleysol **gleysol**
Gleysolic **gleysolique**
gleyzation, gleysation **gleyification**
glossic horizon **horizon glossique**
gradient **gradient, inclinaison**
gravelly **graveleux**
Gray Brown Luvisol **luvisol brun gris**
Gray Luvisol **luvisol gris**
great group **grand groupe**
growing season **saison de croissance**
gullying **ravinement**
gypsic horizon **horizon gypsique**
Gypsids **Gypsids**
H horizon **horizon H**
H layer **couche H**
halomorphic soil **sol halomorphe**
hardening **durcissement**
hardpan **carapace, horizon induré**
heavy soil **sol lourd**
hemic material **matériau hémique**
Hemists **Hémists**
hierarchical **hiérarchique**
Histels **Histels**
histic epipedon **épipédon histique**
Histosols **Histosols**
honeycomb frost **gel à structure alvéolaire**
horizon, soil **horizon du sol, horizon pédologique**
hue **teinte, tonalité, gamme**
Humic Gleysol **gleysol humique**
humic layer **couche humique**
Humic Podzol **podzol humique**
Humic Regosol **règosol humique**
Humic Vertisol **vertisol humique**
Humisol **humisol**
hummock **button, thufur**
hummocky **moutonné, en creux et bosses**
Humods **Humods**
Humo-Ferric Podzol **podzol humo-ferrique**
Humox **Humox**
Humults **Humults**
hydric layer **couche hydrique**
hydric soil **sol hydromorphe, sol hydrique**
hydrogenic soil **sol hydrogénique**
hydromorphic soil **sol hydromorphe**
hydrophobic soil **sol hydrophobe**
ice wedge **fente de remplissage de glace**
ice-thrust **de chevauchement glaciaire**
ice-thrust moraine **moraine de chevauchement**
ice-wedge polygon **polygone de fente de gel**
illuvial horizon **horizon illuvial**
illuviation **illuviation**
impeded drainage **drainage restreint**
impeding horizon **couche ou horizon d'impédance**
Inceptisols **Inceptisols**
inclined **incliné**
indurated layer **couche indurée**
inheritance **héritage**
inorganic soil **sol minéral**
intergrade, soil **sol intergrade, sol de transition**
intrazonal soil **sol intrazonal**
iron pan **alios**
kandic horizon **horizon kandique**
krotovina **krotovina**
L horizon **horizon L**
lacustrine **lacustre**
lag **dépôt résiduel, résidu de déflation**
lake **lac**
land system **système paysager**
land system inventory **inventaire ou cartographie des systèmes paysagers**
landform **forme de terrain**
landscape **paysage**
landscape model **modèle de paysage**
laterite **latérite**
laterization **latérisation**

latitude zonation **zone de latitude, zonalité horizontale**
latosol **latosol**
latosolization **latosolisation**
leaching **lessivage, lixiviation**
legend **légende**
lessivage **lessivage, éluviation**
leucinization **leucinisation**
level **plat, horizontal, subhorizontal**
LFH horizon **horizon LFH**
lime concretion **concrétion calcaire**
lithic contact **contact lithique**
lithic layer **couche lithique**
lithosequence **lithoséquence**
litter **litière**
littering **litiérage**
loam **loam**
loamy **loameux**
loamy coarse sand **sable grossier loameux**
loamy fine sand **sable fin loameux**
loamy sand **sable loameux**
loamy very fine sand **sable très fin loameux**
loosening **ameublissement**
luvic Gleysol **gleysol luvique**
luvisolic **luvisolique**
map, electronic **carte numérique**
marine **marin**
marl **marne**
massive **structure massive**
medium texture **texture moyenne**
Melanic Brunisol **brunisol mélanique**
melanic epipedon **épipédon mélanique**
melanization **mélanisation**
mesic layer **couche mésique**
Mesisol **mésisol**
midden **tertre, butte-témoin**
middle tier **étage intermédiaire**
mineral soil **sol minéral**
modal profile **profil modal, typique ou représentatif**
mollic epipedon **épipédon mollique**
Mollisols **Mollisols**
monolith, soi **monolithe**
moraine **moraine**
morphology, soil **morphologie du sol**
mottled zone **zone marbrée, tachetée, ou marmorisée**
mottles **marbrures, mouchetures, taches**
mottling **marmorisation, marbrures**
muck soil **terre tourbeuse, terre organique, terre noire**
mudstone **mudstone**
Munsell color system **code de couleurs Munsell**
natric horizon **horizon natrique**
neoformation **néoformation**
nonsoil **non-sol**
O horizon **horizon O**
Ochrepts **Ochrepts**
ochric epipedon **épipédon ochrique**
order, soil **ordre de sols**
Organic **organique**
Organic cryosol **cryosol organique**
organic deposit **dépôt organique**
Orthels **Orthels**
Orthents **Orthents**
Orthic **orthique**
Orthids **Orthids**
Orthods **Orthods**
Orthox **Orthox**
ortstein **ortstein**
oxic horizon **horizon oxique**
Oxisols **Oxisols**
paleosol **paléosol**
paleosol, buried **paléosol enfoui**
paleosol, exhumed **paléosol exhumé**
paludization **paludication**
pan, genetic **pan d'origine naturelle**
pan **couche indurée**
paralithic **paralithique**
parent material, genetic material **matériau parental ou originel**
parent rock **roche-mère**
patterned (ribbed) **côtelé, strié**
peat **tourbe**
peat soil **sol tourbeux**
ped **ped**
pedality **pédalité, pédicité**
pediment **pédiment**
pedogenesis **pédogenèse**
pedogenic **pédogénétique**
pedology **pédologie**
pedon **pédon**
pedoturbation **pédoturbation**
pergelic **pergélique**
permafrost **pergélisol**
permafrost table **limite du pergélisol**
Perox **Perox**
perudic **perudique**
petrocalcic horizon **horizon pétrocalcique**
petrogypsic horizon **horizon pétrogypsique**
phytomorphic soils **sols phytomorphes**

pingo **pingo**
placic **placique**
plaggen epipedon **épipédon de plaggen**
plain **plaine**
plaggepts **plaggepts**
plateau **plateau**
plinthite **plinthite**
plow pan **semelle de labour**
Podzolic **podzolique**
podzolic B **B podzolique**
podzolization **podzolisation**
polygenic (polygenetic) soil **sol polygénique**
polygonal ground **sol polygonal**
polypedon **polypédon**
profile, soil **profil de sol**
Psamments **Psamments**
R layer **couche R**
recharge area **zone d'alimentation en eau**
Regosol **régosol**
Regosolic **régosolique**
Rendolls **Rendolls**
rendzina **rendzine**
residual soil **sol résiduel**
reticulate mottling **marbrures réticulées**
ridged **dorsal**
rill **ravineau, rigole**
rilled **en rigoles, raviné**
ripening **maturation, vieillissement**
rolling **vallonné**
rough broken **escarpement, ravin**
rubifaction **rubéfaction**
salic horizon **horizon salique**
Salids **Salids**
sand **sable**
sand wedge **fente de remplissage de sable**
sapric material **matériau saprique**
Saprists **Saprists**
saprolite **saprolithe**
section **coupe, lame mince**
self-mulching soil **sol à autogranulation ou à autofoisonnement**
separates, soil **fractions granulométriques**
sequum **séquum**
series, soil **série de sols**
shaly **schisteux, en feuillet**
silcrete **silcrète, croûte siliceuse**
slaty **ardoisier**
slickensides **surfaces de glissement, miroirs de faille**
slopewash **érosion de pente, dépôt de pente**
sloping **en pente**
slough **marécage, mare vaseuse, bourbier**
softrock **roche tendre**
soil classification **classification des sols**
Soil Classification, Canadian System **système canadien de classification des sols**
Soil Classification, FAO/UNESCO soil units **système de classification des sols FAO-UNESCO, Carte mondiale des sols**
Soil Classification, U.S. Soil Taxonomy **système américain de classification des sols, Soil Taxonomy**
soil complex **complexe de sols**
soil correlation area **aire de corrélation des sols**
soil creep **reptation de sol**
soil drainage classes **classes de drainage du sol**
soil family **famille de sols**
soil formation factors **facteurs de formation des sols**
soil genesis **genèse des sols, pédogenèse**
soil geography **géographie des sols**
soil individual **individu-sol**
soil inventory meta data **métadonnées d'inventaire pédologique**
soil landscape model **modèle de pédopaysage**
soil map delineation **délimitation cartographique des sols**
soil map unit **unité cartographique des sols**
soil model **modèle pédologique**
soil order **ordre de sols**
soil series **série de sols**
solifluction **solifluxion**
Solod **Solod**
Solodized Solonetz **solonetz solodisé**
Solonetz **solonetz**
Solonetzic **solonetzique**
solonetzic B **B solonetzique**
solum (plural sola) **solum**
Sombric Brunisol **brunisol sombrique**
sombric horizon **horizon sombrique**
spodic horizon **horizon spodique**
Spodosols **Spodosols**
static cryosol **cryosol statique**
steep **abrupt, raide**
stones **pierres**
stoniness **pierrosité**
stony **pierreux**
stratification **horizonation**
subgroup, soil **sous-groupe de sols**
suborder, soil **sous-ordre de sols**

subsoil **sous-sol**
substratum **substratum**
surface expression **modelé**
surface form **forme de terrain**
surface tier **étage supérieur**
talud **talud**
terraced **en terrace**
thermogenic soils **sols thermogéniques**
thermosequence **thermoséquence**
tidal flat **wadden, bas fonds intertidaux**
tier **étage**
toposequence **toposéquence**
Torrands **Torrands**
Torrerts **Torrerts**
torric **torrique**
Torrox **Torrox**
traffic pan **couche compactée ou indurée**
transitional soil **sol de transition**
translocation **translocation**
Tropepts **Tropepts**
truncated **tronqué**
Turbic cryosol **cryosol turbique**
tundra soil **sol de toundra**
Turbels **Turbels**
type, soil **type de sol**
Udalfs **Udalfs**
Udands **Udands**
Uderts **Uderts**
udic **udique**
Udolls **Udolls**
Udults **Udults**
Ultisols **Ultisols**
Umbrepts **Umbrepts**
umbric epipedon **épipédon umbrique**
undifferentiated **groupe de sols non différenciés, unité indifférenciée**
undulating **ondulé**
upland **haute-terre**
Ustalfs **Ustalfs**
Ustands **Ustands**
Usterts **Usterts**
ustic **ustique**
Ustolls **Ustolls**
Ustox **Ustox**
Ustults **Ustults**
valley **vallée**
value, color **luminosité, couleur**
variant, soil **variante de sol**
vertic horizon **horizon vertique**
Vertic Solonetz **solonetz vertique**
vertical zonation **zonalité verticale**
Vertisol **vertisol**
Vertisolic **vertisolique**
Vertisols **Vertisols**
virgin soil **sol vierge**
Vitrands **Vitrands**
W layer **couche W**
wetland **milieu humide, terre humide**
Xeralfs **Xéralfs**
Xerands **Xérands**
Xererts **Xérerts**
xeric **xérique**
Xerolls **Xérolls**
Xerults **Xérults**
zonal soil **sol zonal**
zonation **zonalité**
zone, soil **zone de sols**

SOIL INORGANIC/ORGANIC CHEMISTRY, SALINITY

acid **acide**
acid soil **sol acide**
acidic **acide**
acidic cation **cation acidique**
acidity constant **constante d'acidité**
acidity, exchangeable **acidité d'échange**
acidity, residual **acidité résiduelle**
acidity, salt-replaceable **acidité échangeable par un sel**
acidity, total **acidité totale**
active organic matter **matière organique active**
adsorption **adsorption**
adsorption complex **complexe d'adsorption**
alkali **alcali**
alkali metals **métaux alcalins**
alkali soil **sol à alcalis, sol alcalin**
alkaline **alcalin**
alkaline soil **sol alcalin**
alkaline solution **solution alcaline**
alkaline-earth metals **métaux alcalino-terreux**
alkalinity **alcalinité**
alkalinity, soil **alcalinité du sol**
alkalinity, total **alcalinité totale**
alkalization **alcalinisation**
aluminum **aluminium**
amines **amines**
ammonia **ammoniaque**

anhydrite **anhydrite**
anion **anion**
aromatic compound **composé aromatique**
ash **cendre**
base **base**
base metal **métal**
base saturation percentage **pourcentage ou taux de saturation en bases**
buffer capacity **capacité tampon**
buffer compounds, soil **complexe tampon du sol**
buffer power **pouvoir tampon**
buffered (soil) **sol tamponné**
calcium carbonate equivalent **équivalent carbonate de calcium**
carbon (C) **carbone**
carbon dating **datation au carbone**
carbon pool **bassin ou pool de carbone**
carbon:nitrogen ratio (C:N ratio) **rapport carbone:azote**
carbon-14 (^{14}C) **carbone 14 (^{14}C)**
cation **cation**
cation exchange capacity (CEC) **capacité d'échange cationique**
cheluviation **chéluviation**
chitin **chitine**
composite sample **échantillon composite**
desalinization **désalinisation**
detritus **détritus**
dry weight basis **à base de poids sec**
effective calcium carbonate equivalent **équivalent effectif de carbonate de calcium**
effective cation exchange capacity **capacité d'échange cationique effective**
exchange complex **complexe d'échange**
exchangeable anion **anion échangeable**
exchangeable base **base échangeable**
exchangeable cation **cation échangeable**
exchangeable cation percentage **pourcentage de cations échangeables**
exchangeable nutrient **élément nutritif échangeable**
exchangeable sodium fraction **fraction de sodium échangeable**
exchangeable sodium percentage (ESP) **pourcentage de sodium échangeable, sodivité**
exchangeable sodium ratio (ESR) **rapport de sodium échangeable**
extract, soil **extrait de sol**
fatty acid **acide gras**
fulvic acid **acide fulvique**
gravimetric analysis **analyse gravimétrique**
heavy fraction (organic matter) **fraction lourde ou dense de la matière organique**
heavy metal **métal lourd**
humic **humique**
humic acid **acide humique**
humification **humification**
humus **humus**
hydrocarbon **hydrocarbure**
hydrophilic **hydrophile**
hydrophobic **hydrophobe**
inert organic matter **matière organique inerte**
ion **ion**
ion activity **activité ionique**
ion exchange **échange d'ion**
ionic bond **liaison ionique**
ionic strength **force ionique**
ionic substitution **substitution ionique**
ion-selective electrode **électrode à ion spécifique**
isotopically exchangeable ion **ion isotopiquement échangeable**
Kjeldahl method **méthode Kjeldahl**
Kjeldahl nitrogen **azote Kjeldahl**
light fraction (organic matter) **fraction légère de la matière organique**
lime **chaux**
lime potential **potentiel de chaux, potential calcaire**
lime requirement **besoin en chaux**
monosaccharide (simple sugar) **monosaccharide, sucre simple, ose**
nitrate (NO_3) **nitrate**
nitrite **nitrite**
nitrogen **azote**
nonsaline-alkali soil **sol alcalin non-salin**
organic **organique**
organic acid **acide organique**
organic carbon **carbone organique**
organic matter, soil **matière organique du sol**
organic nitrogen **azote organique**
organic phosphorus **phosphore organique**
oven-dry soil **sol séché à l'étuve**
pH curve (titration curve) **courbe de titration**
pH scale **échelle pH**
pH, soil **pH du sol**
pH_C **pH_C**
pH-dependent cation exchange capacity **capacité d'échange cationique dépendante du pH**

pH-dependent charge **charge dépendante du pH**
plant analysis **analyse végétale**
point of zero net charge **point de charge nette nulle**
polyvinyl chloride (PVC) **chlorure de polyvinyle**
radiocarbon dating **datation au radiocarbone**
raw humus **humus brut**
reaction, soil **réaction du sol**
reducing environment **environnement réducteur**
resistivity **résistivité**
resistivity survey **levé de résistivité**
salination **salinisation**
saline **salin**
saline seep **suintement salin**
saline soil **sol salin**
saline-alkali soil **sol salin à alcalis**
saline-sodic soil **sol salin sodique**
salinity **salinité**
salinity control **lutte contre la salinisation**
salinity risk index **indice du risque de salinité**
salinity, soil **salinité du sol**
salinization **salinisation**
salt **sel**
salt balance **bilan de sels**
salt bridge **pont de sel**
salt flat **fond salin**
salt pan **étang salé**
salt-affected soil **sol altéré par le sel, sol salsodique**
saturated soil paste **pâte de sol à saturation**
saturation percentage **pourcentage de saturation**
saturation-paste extract **extrait de pâte à saturation**
slick spots **taches lisses**
sodic soil **sol sodique**
sodication **sodisation**
sodicity **sodicité**
sodium adsorption ratio, adjusted **rapport corrigé d'adsorption du sodium**
sodium-adsorption ratio (SAR) **rapport d'adsorption du sodium**
soil acidity **acidité du sol**
soil sample **échantillon de sol**
soil test **analyse de sol**
solution, soil **solution du sol**
stover **chaume**
surface charge density **densité de charge de surface**
threshold salt tolerance (crop) **seuil de tolérance au sel**
trace metals **métaux à l'etat de traces**

SOIL STRUCTURE, MICROMORPHOLOGY

aggregate **agrégat**
aggregate distribution **distribution des agrégats**
aggregate stability **stabilité des agrégats**
aggregated **structure agrégée ou fragmentaire**
aggregation **agrégation**
alban **albane**
apedal **apédal**
argillan **argilane**
bioturbation **bioturbation**
clay-organo complex **complexe argilo-organique**
clod **motte**
coating **revêtement, enduit, enrobement**
cohesion **cohésion**
concretion **concrétion**
crumb structure **structure grumeleuse ou granuleuse**
crust **croûte**
cutan **cutane, revêtement, patine**
disperse **disperser**
dry aggregate **agrégat sec**
fabric, soil **organisation du sol, état structural du sol**
fecal pellets **turricules**
ferran **ferrane**
ferri-argillan **ferri-argilane**
granular **granulaire**
granule **granule**
gypsan **gypsane**
humus form **forme d'humus**
Kubiena box or tin **échantillonneur de Kubiena, boîte d'échantillonnage de Kubiena**
lamina (plural laminae) **dépôt laminique**
macro-aggregate **macroagrégat**
mangan **mangane**
massive **structure massive**
matran **matrane**

micro-aggregate **microagrégat**
mor (or raw humus) **mor, humus brut**
mull **mull**
mull-like moder **moder à tendance mull**
organan **organane**
organo-mineral complex **complexe organo-minéral**
ped **ped**
pedological feature **trait ou caractéristique pédologique**
photomicrograph **photographie microscopique**
plasma **plasma**
platy soil structure **structure lamellaire**
polished section **lame polie**
prismatic soil structure **structure prismatique**
sesquan **sesquane**
single-grain structure **structure particulaire**
skeletan **squelettane**
skeleton grains **grains du squelette**
skew planes **fentes de retrait irregulières**
slaking **éclatement**
slickenside **surface de glissement, miroir de faille**
s-matrix (of a soil material) **fond matriciel**
soil conditioner **conditionneur de sol**
soil micromorphology **micromorphologie du sol**
soil pores **pores du sol**
soil stabilization **stabilisation du sol**
soil structural form **état structural du sol**
soil structural resiliency **résilience structurale du sol**
soil structural stability **stabilité structurale du sol**
soil structure **structure du sol**
soil structure grade **grade ou netteté de la structure du sol**
soil structure index **mesure ou indicateur de la structure du sol**
soil structure type **type de structure du sol**
surface sealing **colmatage ou obturation de la surface**
thin section **lame mince**
void **vide**
vugh **cavité**
water-stable aggregate **agrégat stable à l'eau ou hydrostable**
wet-sieving **tamisage à l'eau, tamisage humide**

SOIL SURVEY, REMOTE SENSING, LAND USE INTERPRETATION

aerial photograph **photographie aérienne**
atmospheric attenuation **atténuation atmosphérique**
atmospheric windows **fenêtres atmosphériques**
attitude **orientation**
attribute **variable qualitative, attribut**
base map **carte de base, géobase, fond de carte**
benchmark **repère, point géodésique**
benchmark site **site de référence, site repère**
broad reconnaissance soil map **carte pédologique de reconnaissance générale**
clinometer **clinomètre**
compiled map **carte dérivée, carte synthèse**
composite map **carte composée, plurifactorielle ou multicouche**
contour **courbe de niveau, élévation**
contour interval **intervalle d'élévation, équidistance**
contour line **courbe de niveau, isohypse**
contour map **carte topographique**
control **témoin, réseau de points de contrôle**
core sample **échantillon non dérangé, carotte de sondage**
delineation **délimitation**
detailed soil map **carte pédologique détaillée**
elevation **élévation, altitude**
ephemeral data **données accessoires ou auxiliaires**
exploratory soil map **carte pédologique exploratoire**
fiducial mark **repère de cadre**
free lime test **test de carbonates libres**
generalized soil map **carte pédologique généralisée ou de synthèse**
geographic information system (GIS) **système d'information géographique**
graphic scale **échelle graphique**
grid **grille, maille**
ground data **réalité de terrain, données de terrain**
index map **carte-index**
intensity, map **densité de délimitations cartographiques**

intermediate scale map **carte à moyenne échelle ou à échelle intermédiaire**
interpretation, soil survey **interprétation des données pédologiques**
interpretive map **carte interprétative**
isopach **isopaque**
isopleth **isoplète**
land **terre, terrain**
land capability **possibilités d'utilisation des terres, capabilité des terres**
land classification **classification des terres**
land use **utilisation des terres ou des sols**
land use capability **possibilités d'utilisation des terres**
land use planning **aménagement du territoire, planification de l'utilisation du territoire**
landscape **paysage**
legend **légende**
made land **terrain anthropique, terre rapportée**
map projection **projection cartographique**
map resolution **résolution cartographique**
map scale **échelle cartographique**
map series **série de cartes**
map unit, soil **unité cartographique**
map, large-scale **carte à grande échelle**
map, medium-scale **carte à moyenne échelle**
map, small-scale **carte à petite échelle**
map, soil **carte pédologique**
mapping unit **unité cartographique**
matching **raccordement**
miscellaneous land type **type de terrain divers**
phototriangulation **phototriangulation**
planimetry **planimétrie**
planisaic **photo-carte planimétrique**
plottable error **erreur du tracé cartographique**
point gage **pointe limnimétrique droite**
reconnaissance **reconnaissance**
reconnaissance soil map **carte pédologique de reconnaissance, semi-détaillée**
relief map **carte de relief, carte topographique**
riverwash **batture**
rockland **terrain rocheux**
schematic soil map **carte pédologique schématique**
soil association **association de sols**
soil auger **tarière**
soil boundary **limite de sol**
soil capability map **carte de possibilités des sols**
soil complex **complexe de sols**
soil correlation **corrélation des sols**
soil interpretations **interprétation des sols**
soil map **carte des sols, carte pédologique**
soil map delineation **délimitation cartographique des sols**
soil map unit **unité cartographique de sols**
soil survey **prospection pédologique, levé des sols**
spectral band **bande spectral**
spectral colors **couleurs spectrales**
spectral map **carte spectraloïde**
spectral radiance **radiance spectrale**
spectral region **domaine spectrale**
spectral response **réponse spectrale**
spectral signature **signature spectrale**
stony land **terrain pierreux**
survey, soil **levé des sols, prospection pédologique**
thematic map **carte thématique ou dérivée**
toposaic **photo-carte topographique**
transect **transect, traverse**
undifferentiated soil map unit **unité cartographique de sols non differenciés**
Universal Polar Stereographic coordinates (UPS coordinates) **coordonnées du système de quadrillage stéréographique polaire universel (coordonnées UPS)**
urban land **terrain urbain, zone urbaine**
vertical photograph **photographie verticale, cliché vertical**
wasteland **terrain inutilisable**

WASTE MANAGEMENT, COMPOSTING

activated carbon **charbon activé ou actif**
artificial manure **fumier artificiel, compost**
back end system **système de récupération**
backwashing **lavage à contre-courant, lavage par retour d'eau, nettoyage inversé**
bagasse **bagasse**
ballistic separator **séparateur par projection**
batch method **méthode en lots**
Beccari process **procédé de Beccari**
bedding, animal **litière**

bioconversion **transformation biologique**
biological additives **activateurs ou additifs biologiques**
biological wastewater treatment **traitement biologique secondaire**
biostabilizer **bioréacteur**
byproduct **sous-produit**
cap **couche imperméabilisante**
carbon filtration **filtration au carbone**
cell **cellule d'entreposage, unité de stockage**
cesspool **fosse septique, puisard, puits absorbant**
charcoal filter **filtre au charbon**
chlorination **chloration, javellisation**
chlorine residual **chlore résiduel**
clarification **clarification, décantation**
clarifier **clarificateur, décanteur**
clinker **scories, mâchefer**
coal processing waste **résidus de conditionnement de charbon**
coliform **coliforme**
coliform index **niveau de coliformes**
comminution **dilacération**
compaction **compaction, compactage, tassement**
completed test **épreuve de complétion**
compost **compost**
confirmed test **épreuve de confirmation**
continuous-feed reactor (composting) **digesteur à alimentation en continu**
continuous-flow microbiological system **système microbiologique à régime continu**
cover material **matériel de recouvrement**
debris basin **bassin de décantation ou de déjection**
dewatering **déshydratation, assèchement**
diffused air **aération par diffusion d'air**
digester **digesteur**
digestion **digestion**
disinfection **désinfection**
dispersant **agent dispersant**
disposal field **champ d'épuration**
disposal pond **lagune, étang d'épuration**
domestic sewage **eaux usées domestiques**
domestic waste **eaux usées domestiques**
domestic water **eau de consommation**
effluent **effluent**
evaporation pond **étang d'évaporation, lagune**
Fairfield-Hardy digester **digesteur ou bioréacteur Fairfield-Hardy**
fecal material **matières fécales, excréments**
fill **décharge**
filter sand **sable filtrant**
filtration **filtration**
final clarifier **clarificateur terminal**
finished water **eau traitée**
fixation **fixation**
fixed carbon **charbon calciné**
floc **floc**
flocculation **floculation**
free liquids **lixiviat**
hazardous waste **déchets dangereux**
lagoon **lagune, étang**
land farming **épandage sur le sol**
landfill **site d'enfouissement, décharge**
landfill trench method **enfouissement par la méthode de la tranchée**
landfill, secure **site d'enfouissement à accès contrôlé**
Lantz process **procédé de distillation de Lantz**
leaching field **champ d'épuration ou de percolation**
limited water-soluble substance **substance peu soluble à l'eau**
liner **revêtement**
liquid manure **lisier**
liquid waste **effluent**
loading **charge**
loading capacity **capacité de charge**
loading schedule **calendrier d'épandage**
primary treatment **traitement primaire**
rapid sand filtration **filtration rapide sur sable**
raw sewage **eaux usées**
reactive waste **déchets réactifs**
recycling **valorisation, recyclage**
refuse reclamation **récupération d'ordures**
reinforced manure **fumier enrichi**
rendering **équarrissage**
resource recovery **tri, tri à la source**
retention time **temps de rétention**
reuse **valorisation, recyclage**
sand filter **filtre à sable**
sanitary landfill **site d'enfouissement sanitaire**
secondary clarifier **décanteur secondaire**
secondary treatment **traitement secondaire**
sedimentation **sédimentation, décantation**
sedimentation tank **décanteur statique**
seepage pit **fosse de rétention**

septage **boues activées**
septic tank **fosse septique**
septic tank absorption field **champ d'épuration**
settleable solids **matières décantables**
settling chamber **chambre de sédimentation**
settling pond **bassin de décantation**
settling tank **bac de décantation**
sewage **eaux usées**
sewage lagoon **bassin de stabilisation des eaux usées**
sewage sludge **boues de station d'épuration**
sewage treatment plant **usine d'épuration des eaux usées**
sewer **égout**
sewerage **assainissement**
sludge **boues**
sludge digestion **digestion de boues**
sludge disposal **élimination de boues, mise en décharge**
sludge loading **charge de boues autorisée**
soil absorption system **site d'épandage sur sol**
soil flushing **curage ou nettoyage du sol**
solid waste **déchets solides**
solid waste disposal **élimination des déchets solides**
solid waste management **gestion des déchets solides**
solid waste management unit **unité de gestion de déchets solides**
solidification **solidification**
storage pit (wastes) **fosse d'entreposage**
supernatant **surnageant**
suspended sediment **particules en suspension**
suspended solids **solides ou matière en suspension**
swill **déchets alimentaires**
synthetic manure **compost, engrais artificiel**
tertiary treatment **traitement tertiaire**
thermal treatment of hazardous waste **traitement thermique de déchets dangereux**
treatment tank **cuve de traitement**
trickling filter **lit bactérien**
urban waste **déchets urbains**
waste **déchets, résidus**
waste processing **traitement des déchets**
waste sources **sources de déchets ou de résidus**
waste treatment **traitement des déchets**
wastewater **eaux usées ou résiduaires**
water disposal system **système de drainage, système d'évacuation d'eau**
water treatment **traitement des eaux**
waterwall incinerator **incinérateur générateur de vapeur**
wet digestion **digestion par voie humide**
wet milling **broyage en conditions humides**
windrow **andain**
yard waste **résidus de jardin**

ZOOLOGY

annelid **annélide**
apatetic **apatétique**
arthropod **arthropode**
basal metabolism **métabolisme de base**
bloodworm **ver de vase**
commensalism **commensalisme**
consumer **consommateur**
coprophagous **coprophage**
detritivore **détritivore**
detritus food web **réseau trophique détritivore**
ecological efficiency **efficacité écologique**
euryhaline **euryhalin**
exudate **exsudat**
family **famille**
gastropod **gastéropode, gastropode**
holozoic **holozoïque**
homeostasis **homéostasie**
instar **stade larvaire**
invertebrate **invertébré**
larva **larve**
macroconsumer **macroconsommateur**
megafauna **mégafaune**
mesobiota **mésobiote**
methemoglobinemia **méthémoglobinémie**
mite **mite**
molt **mue**
myriapod **myriapode**
herbivore **herbivore**
nematode **nématode**
nymph **nymphe**
omnivore **omnivore**
palaeozoology **paléozoologie**
parasite **parasite**
phagocyte **phagocyte**
phagotroph **phagotrophe**

primary consumer **consommateur primaire**
pupa **pupe**
quiescent **quiescent**
springtails **collemboles**
strain **souche**
vein **nervure**
vertebrate **vertébré**
zooecology **zooécologie**
zoogeography **zoogéographie**

FRENCH/ENGLISH INDEX OF TERMS BY SUBJECT AREA/DISCIPLINE

AGRONOMIE, AGRICULTURE, FERTILITÉ DES SOLS, NUTRITION DES PLANTES (AGRONOMY, AGRICULTURE, FERTILITY, PLANT NUTRITION)

absorption active absorption, active
absorption passive absorption, passive
absorption racinaire directe root interception
acide phosphorique phosphoric acid
acide superphosphorique superphosphoric acid
acidulation acidulation
agrologie agrology
agronomie agronomy
amendement du sol amendment, soil
ammoniaque anhydre fertilizer, liquid
ammoniation ammoniation
antagonisme antagonism
antagonisme entre éléments nutritifs nutrient antagonism
appauvrissement du sol impoverishment, soil
application à la volée broadcast
application en bande banding
besoin en azote nitrogen requirement
besoin en éléments nutritifs crop nutrient requirement
besoin en engrais fertilizer requirement
bilan nutritif nutrient budget
billon ridge
biuret biuret
capacité de libération des éléments nutritifs du sol nutrient-supplying power of soils
capacité de libération du potassium du sol potassium-supplying power of soils
composition élémentaire d'un engrais fertilizer analysis
concentration en éléments nutritifs nutrient concentration
consommation de luxe luxury uptake
croûte crust
cultiver le sol till
cycle des éléments nutritifs nutrient cycle
débit massique mass flow rate
diagnostic foliaire foliar diagnosis
diffusion des éléments nutritifs nutrient diffusion
écoulement massique mass flow (nutrient)
EDTA EDTA
efficacité de l'utilisation de l'eau water use efficiency
élément assimilable available nutrient
élément essentiel ou constitutif essential element
élément majeur macroelement, macronutrient, major element
élément nutritif nutrient, plant nutrient
élément nutritif minéral plant food
élément nutritif primaire primary nutrient
élément nutritif secondaire secondary nutrient
élément nutritif non disponible unavailable nutrient
élément majeur primary nutrient
élément mineur microelement
engrais fertilizer
engrais à libération lente fertilizer, controlled-release; fertilizer, slow-release
engrais acidifiant acid-forming fertilizer, fertilizer, acid-forming

engrais appliqué en bande en post-levée fertilizer, sidedressed
engrais appliqué en post-levée top dressing
engrais composé fertilizer, compound
engrais composé fertilizer, mixed
engrais conditionné conditioned fertilizer
engrais conditionné fertilizer, conditioned
engrais de base basic fertilizer
engrais de démarrage fertilizer, pop-up
engrais de démarrage fertilizer, starter
engrais de démarrage starter fertilizer
engrais de mélange bulk-blend fertilizer
engrais de mélange fertilizer, bulk-blend
engrais de mélange en vrac fertilizer, bulk-blended
engrais de synthèse fertilizer, manufactured
engrais en post-levée fertilizer, top-dressed
engrais en suspension fertilizer, suspension; suspension fertilizer
engrais en vrac bulk fertilizer
engrais enrobé coated fertilizer; fertilizer, coated
engrais granulaire fertilizer, granular; granular fertilizer
engrais injecté fertilizer, injected;
engrais liquide fertilizer, fluid; fertilizer, solution; liquid fertilizer; solution fertilizer
engrais minéral chemical fertilizer; fertilizer, chemical; fertilizer, inorganic
engrais non granulaire nongranular fertilizer
engrais non incorporé surface-applied fertilizer
engrais organique fertilizer, organic; organic fertilizer
engrais perlé fertilizer, nongranular; prilled fertilizer
engrais simple fertilizer, straight; straight fertilizer
engrais ternaire fertilizer, complete
engrais vert green manure
équilibre nutritif nutrient balance
fertigation chemigation
fertilisation fertilization
fertilisation foliaire fertilization, foliar
fertilité du sol fertility, soil; soil fertility
fertilité résiduelle residual fertility
forme formulation
forme frittée frits
formule d'engrais fertilizer formula; fertilizer grade
granulation granulation
guano guano
incorporation incorporation (into soil)
indice de salinité de l'engrais fertilizer, salt index
interaction entre éléments nutritifs nutrient interaction
irrigation fertilisante fertigation
labourer till
loi de Liebig Liebig's law
loi du minimum Liebig's law
mélange en vrac bulk blending
nitrate d'ammonium ammonium nitrate
nitrate résiduel residual nitrate
nitrophosphates nitric phosphates
niveau optimal optimum level
nutrition des plantes plant nutrition
oligo-élément micronutrient
oligo-éléments frittés fritted trace element
orthophosphate orthophosphate
phosphate phosphate
phosphate d'ammonium ammonium phosphate
phosphate naturel rock phosphate
phosphate particulaire particulate phosphate
phosphore phosphorus
phosphore fixé fixed phosphorus
phosphore insoluble au citrate citrate-insoluble phosphorus
phosphore soluble au citrate citrate-soluble phosphorus
phosphore total phosphorus, total
phosphorite phosphorite
potasse potash
pratiques agronomiques ou agricoles agronomic practices
prélèvement d'un élément nutritif nutrient uptake
rapport d'éléments fertilisants fertilizer ratio
réglage calibration
rétention d'un élément nutritif nutrient retention
rétrogradation reversion
rétrogradation d'engrais fertilizer fixation
roche phosphatée phosphate rock
rotation rotation
roulage rolling
scories basic slag
semer seed
seuil critique critical nutrient concentration
solution azotée nonpressure solution

stress nutritif nutrient stress
sulfate d'ammonium ammonium sulfate
sulfate de calcium calcium sulfate extractant
superphosphate superphosphate
superphosphate ammonié superphosphate, ammoniated
superphosphate enrichi superphosphate, enriched
superphosphate normal superphosphate, normal, superphosphate, ordinary; superphosphate, single
superphosphate triple ou concentré superphosphate, concentrated
surfertilisation quench
symptôme de carence deficiency symptom
système de manutention fermé closed handling system
système de production farming system
teneur d'un engrais fertilizer grade
teneur en éléments nutritifs nutrient content
valeur fertilisante fertilizer value

BIOCHIMIE (BIOCHEMISTRY)

acide α-aminé α-amino acid
acide aminé amino acid
acide désoxyribonucléique (ADN) deoxyribonucleic acid (DNA)
acide nucléique nucleic acid
acide ribonucléique (ARN) ribonucleic acid (RNA)
adénosine diphosphate (ADP) adenosine diphosphate (ADP)
adénosine triphosphate (ATP) adenosine triphosphate (ATP)
amidon starch
ammonification ammonification
anabolisme anabolism
anticorps antibody
antigène antigen
antitoxine antitoxin
ARN de transfert (ARNt) transfer RNA (tRNA)
assimilation assimilation
assimilation de l'azote nitrogen assimilation
autolyse autolysis
auxine auxin
besoin en eau water requirement
biochimie biochemistry
biosynthèse biosynthesis
chimio-dénitrification chemodenitrification
coefficient de température temperature coefficient
coefficient de transpiration water requirement
coenzyme coenzyme
cycle de l'azote nitrogen cycle
cycle du carbone carbon cycle
cycle du phosphore phosphorus cycle
cycle du soufre sulfur cycle
cytochrome cytochrome
cytokinine cytokinin
cytoplasme cytoplasm
décomposition decomposition
décomposition anaérobie anaerobic decomposition
déficit en oxygène oxygen debt
dénaturation denaturation
dénaturer denature
dénitrification denitrification
désamination deamination
désassimilation dissimilation
déshydrogénase dehydrogenase
désulfuration desulfuration
détoxication biotransformation
dorsale ridge
efficacité d'assimilation assimilation efficiency
élément biophile biophile
enzyme enzyme
enzyme de constitution constitutive enzyme
enzyme endogène endoenzyme, or intracellular enzyme
enzyme induite adaptive enzyme (enzyme induction)
enzyme protéolytique proteolytic enzyme
équation de Michaelis-Menton Michaelis-Menton equation
essai de réduction de l'acétylène acetylene-reduction assay
fixation fixation
fixation biologique de l'azote nitrogen fixation
fixation d'ammonium ammonium fixation
fixation de l'azote atmosphérique dinitrogen fixation
génétique genetics
génotype genotype

gibbérellines gibberellins
glycogène glycogen
gradient osmotique osmotic gradient
graisse (glycéride) fat (glyceride)
haploïde haploid
hémoglobine hemoglobin
hétérotrophe heterotrophic
hétérotrophe heterotroph
hétérotrophie heterotrophy
hétérozygote heterozygous
histones histones
hormone hormone
humine humin
immobilisation immobilization
inhibition rétroactive feedback inhibition
labile labile
leghémoglobine leghemoglobin
lignine lignin
lipide lipid
lyse osmotique osmotic lysis
lysosome lysosome
métabolisme metabolism
métabolite metabolite
métabolite secondaire secondary metabolite
métabolites conjugués conjugated metabolites
minéralisation mineralization
nitrification nitrification
nitrification hétérotrophe heterotrophic nitrification
nitrogénase nitrogenase
organisation immobilization
organite cellulaire organelle
osmose osmosis
peptidoglycan peptidoglycane
perméase permease
phospholipides phospholipids
phosphorylation oxydative oxidative phosphorylation
photosynthat photosynthate
photosynthèse photosynthesis
plante ou végétation de type C3 C3 plant
plante ou végétation de type C4 C4 plant
plante ou végétation de type CAM CAM plant
plasmide (épisome) plasmid (episome)
ploïdie ploidy
polyploïdie polyploidy
productivité brute gross productivity
protéase protease
protéine protein
protoplasme protoplasm
putréfaction putrefaction
putrescible putrescible
Q_{10} Q_{10}
quotient respiratoire respiratory quotient (RQ)
réduction du nitrate nitrate reduction
réduction non-assimilatoire des nitrates dissimilatory reduction of nitrate
réduction respiratoire des nitrates respiratory reduction of nitrate
régulateur de croissance growth regulator
régulation osmotique osmoregulation
respiration respiration
respiration anaérobie anaerobic respiration
respiration spécifique respiration/biomass ratio
ribosome ribosome
sidérophore siderophore
substance analogue analog
substance humique humic substance
supergène supergene
transformation biologique biological interchange
translocation translocation
transpiration transpiration
triploïde triploid
volatilisation de l'ammoniaque ammonia volatilization

BOTANIQUE, HORTICULTURE, SCIENCE DES PARCOURS (BOTANY, HORTICULTURE, RANGE SCIENCE)

abaxial abaxial
acide cyanhydrique prussic acid
acide prussique prussic acid
acidiphile acidophilic
adaxial adaxial
aérobie aerobic
affourragement vert green-chopped
aflatoxine aflatoxin
agriculture itinérante shifting agriculture (slash-and-burn agriculture)
agriculture sur brûlis slash-and-burn agriculture

agroforesterie, agrosylviculture agroforestry
agrostologie agrostology
agro-sylvo-pastoral agro-silvo-pastoral
aire de distribution géographique range
alcaloïde alkaloid
aliment des animaux feed
alimentation de complément supplemental feeding
alimentation intensive, flushing flushing
amélioration du parcours range improvement
amélioration du parcours naturel range improvement
amélioration du pâturage pasture improvement
anaérobie anaerobic
analyse des anneaux de croissance tree ring analysis
analyse immédiate proximate analysis
analyse pollinique pollen analysis
angiosperme angiosperm
anneau de croissance growth ring
annélation girdling
anthère anther
anthèse anthesis
anthocyanines anthocyanins
antioxydant anti-oxidant
appétibilité palatability
arboretum arboretum
arbre tree
arbrisseau shrub
arbuste shrub
autopollinisé self-pollinated
auxine auxin
ballonnement bloat
bétail livestock
biomasse biomass
bocage grove
bois wood
bois de coeur heartwood
bourgeon bud
brout browse
brûlure bactérienne fire blight
bryophyte bryophyte
bulbe bulb
caecum cecum
cal callus
calorie calorie
cambium cambium
carex sedge
carie blanche white rot
carie brune brown rot
caroténoïdes carotenoids
caryopse caryopsis
cellulase cellulase
cellule cell
cellules de garde guard cells
cellules épidermiques epidermal cells
cellules stomatiques guard cells
cellulose cellulose
cendre ash
chancre canker
charge limite d'un pâturage grazing capacity
chaton catkin
chaume stubble
chlorophylle chlorophyll
chloroplaste chloroplast
chromosome chromosome
classe class
clone clone
coeur pith
coiffe root cap
conduite de pâturage intensive intensive grazing management
conduite des parcours range management
conduite du pâturage grazing management
cône cone; strobilus
conifère conifer
coniférien coniferous
constituants de la paroi cellulaire cell wall constituents
cormus corm
cortex cortex
cotylédons cotyledons
couvert canopy
croissance primaire primary growth
cryptogame cryptogam
cultivar cultivar
culture associée companion crop
culture compagne companion crop
cuticule cuticle
cutine cutin
cycle biogéochimique biogeochemical cycle
début floraison bloom, early
décomposeur decomposer
dendriforme dendritic
dendritique dendritic
dendrochronologie dendrochronology
dendrologie dendrology
dépression à bisons buffalo wallow
dernière feuille flag leaf

détritus detritus
dicotylédone dicotyledon
digestibilité de la fibre au détergent acide acid detergent fiber digestibility
dioïque dioecious
diploïde diploid
distribution du pâturage grazing distribution
diurne diurnal
division cellulaire cell division
dormance dormancy
éclaircissage thinning
élagage naturel self-pruning
élagage ou taille des racines root pruning
enclos paddock
endolithique endolithic
endophyte endophyte
enroulement de la feuille leaf curl
ensilage ensilage; silage
ensilage mi-fané haylage
ensilage préfané haylage
épiderme epidermis
épreuve biologique bioassay
ergot ergot
éricacé ericaceous
espèce en évolution progressive increaser
espèce en évolution régressive decreaser
espèce nuisible noxious species
espèce species
essai ou test de sève sap test
étamine stamen
état ou stade végétatif vegetative state
exclos exclosure
exertion inflorescence
exsudat exudate
faisceau libéro-ligneux vascular bundle
faisceau vasculaire vascular bundle
famille family
feuillage foliage
feuillage panaché variegated foliage
feuille de l'épi flag leaf
feuille étendard flag leaf
feutrage radiculaire pot-bound
fibre au détergent acide (ADF) acid detergent fiber (ADF)
fibre au détergent neutre (fibre détergente neutre) neutral detergent fiber (NDF)
flétrissement wilting
fleur flower
fleur complète perfect flower
fleur incomplète imperfect flower
floraison blooming
flushing flushing
foin hay
fourrage fodder; forage
fourrager forage
gaine sheath
gazon sod; turf
genre genus
géotropisme geotropism
germination germination
gibbérelline gibberellin
glucide carbohydrate
glucides non-structuraux carbohydrates, nonstructural
glucides structuraux carbohydrates, structural
gonflement boot stage
gourmand water sprout
gradient de pression de diffusion diffusion pressure gradient
grain kernel
graines de légumineuses pulse
graminée grass
graminée cespiteuse bunch grass
graminées à gazon sod grasses
graminiforme graminoid
greffe graft
guttation guttation
gymnosperme gymnosperm
halophyte halophyte
héliotropisme heliotropism
hémicellulose hemicellulose
herbacé herbaceous
herbage herbage
hérédité heredity
horticulture horticulture
hybride hybrid
hydrate de carbone carbohydrate
hydrophyte hydrophyte
in situ *in situ*
in vitro *in vitro*
in vivo *in vivo*
incision annulaire girdling
inflorescence inflorescence
ingesta ingesta
inoculant inoculant
journalier diurnal
lamina lamina (plural laminae)
légumineuse legume
Lemna gibba *Lemna gibba*
lichen lichen
ligne d'abroutissement browse line

lignine lignin
lipide lipid
litière litter
macroscopique macroscopic
maladie hollandaise de l'orme dutch elm disease
marquage bref pulse labeling
marquage de coute durée pulse labeling
matière azotée totale protein, crude
matière sèche (MS) dry matter (DM)
matières digestibles totales total digestible nutrients (TDN)
méiose meiosis
méristème meristem
méristème apical apical meristem
mésophylle mesophyll
mésophyte mesophyte
météorisation bloat
mitochondrie mitochondrion
mitose mitosis
monocotylédone monocotyledon
monoïque monoecious
mousse moss
mucigel mucigel
mulch mulch
multiplication propagation
multiplication végétative vegetative
mycotoxine mycotoxin
nervation venation
nervure vein
neurotoxine neurotoxin
nomenclature binominale binomial nomenclature
nutrition nutrition
ordre order
organe de fructification fruiting body
organisme transgénique transgenic organism
paillis mulch
paillis frais fresh mulch
paisson mast
palatabilité palatability
paléobotanique palaeobotany
palynologie palynology
panicule panicle
parc d'engraissement feedlot
parcours range
parcours naturel range
parenchyme parenchyma
parenchyme palissadique palisade parenchyma
paroi cellulaire cell wall
pâturage pasture
pâturage amélioré improved pasture
pâturage continu continuous stocking
pâturage cultivé ou amélioré pasture, tame
pâturage en rotation rotation grazing; rotation pasture
pâturage intermittent intermittent grazing
pâturage libre continuous stocking
pâturage permanent permanent pasture
pâturage rationné strip grazing
pâturage temporaire pasture, temporary
pâturer graze
paysage landscape
perlite perlite
peuplement de graminées sward
phénologie phenology
phénotype phenotype
phloème phloem
photopériode photoperiod
photopériodisme photoperiodism
photorespiration photorespiration
photosensibilisation photosensitization
phototropisme phototropism
phylloplan phyllosphere
phylogénie phylogeny
phylum phylum
phytogéographie phytogeography
phytolithe phytolith
phytomètre phytometer
phytotoxique phytotoxic
pistil pistil
pistillé pistillate
plante plant
plante à feuillage persistant evergreen plant
plante à feuilles caduques deciduous plant
plante annuelle annual plant
plante bisannuelle biennial plant
plante de climat frais cool-season plant
plante éricacée heath
plante fixatrice d'azote nitrogen-fixing plant
plante graminiforme grasslike plant
plante herbacée dicotylédone forb
plante indicatrice phytometer
plante ornementale ornamental
plante pérenne ou vivace perennial plant
plante vivace root hardy
plante-abri companion crop
plantoir dibble
plantule seedling
plaste plastid
pleine floraison bloom, full

poils absorbants root hairs
pollen pollen
pollinisation pollination
pollinisation croisée cross-pollination
pomologie pomology
porte-greffe rootstock
postmaturation after-ripening
poudre d'os bone meal
pourriture blanche white rot
pourriture brune brown rot
prairie temporaire ley
pré meadow
procombant procumbent
propagule propagule
protéine brute protein, crude
psammophyte psammophyte
Ptéridophyta *Pteridophyta*
qualité des fourrages forage quality
quotidien diurnal
quotient chlorophyllien photosynthetic quotient
racine root
racine latérale ou secondaire lateral root
racine pivotante taproot
racine tuberculeuse tuberous root
racine tubéreuse tuberous root
racines adventives adventitious roots
radicule radicle
ration ration
ration équilibrée ration, balanced
ration fourragère ration
règne kingdom
rhizomateux rhizomatous
rhizome rhizome
rhizosphère rhizosphere
rumen rumen
ruminant ruminant
rumination rumination
rusticité hardiness
salignon salt lick
santé du terrain de parcours rangeland health
scarification scarification
science des parcours range science
sclérenchyme sclerenchyma
sclérophylle sclerophyllous
sécheresse physiologique physiological drought
section section
semence seed
semence non-décortiquée unhulled seed
semence non-mondée unhulled seed
semence pelliculée coated seed
semis seedling
semis des parcours range seeding
semis sur billon ridge planting
sénescence senescence
sève sap
sous-pâturage undergrazing
spectroscopie de réflexion dans le proche infrarouge near infrared reflectance spectroscopy (NIRS)
spore spore
sporocarpe sporocarp
stade de maturité maturity stage
stade laiteux milk stage
stade pâteux dough stage
staminé staminate
stigmate stigma
stolon stolon
stolonifère stoloniferous
stomate stoma (plural stomata)
suintement seep
surcharge pastorale overstocking
surpâturage overgrazing
système de pâturage grazing system
système racinaire fasciculé fibrous root system
système racinaire pivotant taproot system
talle tiller
tanin tannin
taux de charge des pâturages grazing pressure
taux de chargement stocking rate
tégument seed coat
temps de pâturage grazing period
tension sur la colonne d'eau diffusion pressure gradient
terrain de parcours rangeland
terrain de parcours libre open range
terrain de parcours surpâturé overgrazed range
test biologique bioassay
tétanie d'herbage (hypomagnésiémie) grass tetany (hypomagnaesemia)
tétraploïde tetraploid
thalle thallus
thermopériodisme thermoperiodism
thérophyte therophyte
tige culm; stem
tissu tissue
toxicité toxicity

toxine toxin
trachéide tracheid
tropisme tropism
tubercule tuber
turgescent turgid
type type
unité animale (U.A.) animal unit (A.U.)
unité animale-mois (U.A.M.) animal unit month (A.U.M.)
unité-gros-bétail (U.G.B.) animal unit (A.U.)
unités nutritives totales total digestible nutrients (TDN)
utilisation courante (parcours-pâturage) actual use (range-pasture)
vacuole vacuole
variations journalières de température diurnal temperature variations
variété synthétique synthetic variety
vasculaire vascular
végétatif vegetative
végétation rudérale ruderal vegetation
verger orchard
vernalisation vernalization
vieillissement senescence
vrille tendril
xylème xylem
zone ou profil racinaire root zone

CHIMIE (CHEMISTRY)

à base de poids sec dry weight basis
abondance naturelle natural abundance
absorbance absorbance, absorptance
absorption absorption
absorptivité absorptivity
accélérateur de particules particle accelerator
accélérateur linéaire linear accelerator
acide acid, acidic
acide carbonique carbonic acid
acide carboxylique carboxylic acid
acide conjugué conjugate acid
acide faible weak acid
acide fort strong acid
acide gras fatty acid
acide nitrique nitric acid
acide organique organic acid
acide polyprotique polyprotic acid
acidimétrie acidimetry
acidité d'échange acidity, exchangeable
acidité du sol soil acidity
acidité échangeable par un sel acidity, salt-replaceable
acidité résiduelle acidity, residual
acidité totale acidity, total
activité activity
activité ionique ion activity
activité optique optical activity
activité spécifique specific activity
adoucisseur d'eau water softener
adsorption adsorption
adsorption chimique chemisorption
agent mouillant wetting agent
agent oxydant oxidizing agent (electron acceptor)
agent réducteur (donneur d'électron) reducing agent (electron donor)
agent tensio-actif surfactant
alcali alkali
alcalin alkaline
alcalinité alkalinity
alcalinité bicarbonatée bicarbonate alkalinity
alcalinité du sol alkalinity, soil
alcalinité totale alkalinity, total
alcane alkane
alcène alkene
alcool alcohol
alcool absolu absolute alcohol
aldéhyde aldehyde
aliquote aliquot
alkine alkyne
alkylsulfonate linéaire linear alkyl sulfonate (LAS)
alliage alloy
aluminium aluminum
amines amines
ammoniaque ammonia
ammonium fixé fixed ammonium
analyse analysis
analyse de sol soil test
analyse de tissus végétaux tissue analysis
analyse gravimétrique gravimetric analysis
analyse qualitative qualitative analysis
analyse quantitative quantitative analysis
analyse thermale différentielle differential thermal analysis
analyse thermique thermal analysis

analyse végétale plant analysis
analyte analyte
anhydrite anhydrite
anion anion
anion échangeable exchangeable anion
aqueux aqueous
assurance de qualité quality assurance
atome atom
atome lourd heavy atom
atomiseur atomizer
attraction dipolaire dipole-dipole attraction
autocatalyse autocatalysis
azote nitrogen
azote Kjeldahl Kjeldahl nitrogen
bande d'absorption absorption band
base base
base conjuguée conjugate base
base échangeable exchangeable base
base faible weak base
base forte strong base
basique basic
becquerel becquerel
benzène benzene
benzine benzine
besoin en chaux lime requirement
bicarbonate bicarbonate
biphényles polychlorés (BPC) polychlorinated biphenyls (PCBs)
blanc blank
blanc d'instrument instrument blank
blanc de laboratoire laboratory blank
blanc de matériau material blank
blanc de réactif reagent blank
blancs d'essai field blanks
bombe calorimétrique bomb calorimeter
BPC PCBs
calcine calcine
capacité d'échange anionique anion exchange capacity
capacité d'échange cationique effective effective cation exchange capacity
capacité d'échange cationique cation exchange capacity (CEC)
capacité d'échange cationique dépendante du pH pH-dependent cation exchange capacity
capacité tampon buffer capacity
carbonate carbonate
carbone 14 (^{14}C) carbon-14 (^{14}C)
carbone organique organic carbon
carboxyle carboxyl group
catalyseur catalyst
cathode cathode
cation cation
cation acidique acidic cation
cation échangeable exchangeable cation
cellule électrolytique electrolytic cell
cellule photoélectrique photoelectric cell
cendre ash
centrifuge centrifuge
cétone ketone
chaîne isotactique isotactic chain
chaleur d'hydratation heat of hydration
chaleur de combustion heat of combustion
chaleur de dissolution heat of solution
chaleur latente de condensation heat of condensation
chaleur latente de vaporisation heat of vaporization
chaleur massique specific heat capacity
chaleur molaire molar heat capacity
changement d'énergie libre standard standard free energy change
changement d'état change of state
charge dépendante du pH pH-dependent charge
chaux lime
chaux hydratée hydrated lime
chélat chelate
chéluviation cheluviation
chimie du sol chemistry, soil
chimie organique organic chemistry
chimisorption chemisorption
chiralité chirality
chloré chlorinated
chloroforme chloroform
chlorure de polyvinyle polyvinyl chloride (PVC)
chlorures chlorides
chromatographie chromatography
chromatographie en phase gazeuse gas chromatography (GC)
chromatographie en phase liquide liquid chromatography
chromatographie liquide à haute performance high performance liquid chromatography (HPLC)
chromatographie sur couche mince thin-layer chromatography (TLC)
chromatographie sur gel ou par exclusion size-exclusion chromatography (SEC)
cire wax

coagulation coagulation
coefficient de diffusion diffusion coefficient
coefficient de partage ou de répartition partition coefficient
colorimétrie colorimetry
combinaison ionique ionic compound (binary)
combustion combustion
complexe complex
complexe d'adsorption adsorption complex
complexe d'échange exchange complex
complexe tampon du sol buffer compounds, soil
composé compound
composé aliphatique aliphatic compound
composé aromatique aromatic compound
composé cyclique ring compound
composé cyclique substitué substituted ring compound
composés organiques volatils volatile organic compounds (VOCs)
compteur à pointe de Geiger Geiger-Muller counter (Geiger counter)
compteur à scintillation scintillation counter
compteur bêta beta counters
concentration concentration
concentration en ions hydrogène hydrogen-ion concentration
concentration molaire molar concentration
conductivité électrique electrical conductivity (EC)
constante d'acidité acidity constant
constante d'Avogadro Avogadro's constant
constante d'équilibre equilibrium constant
constante de dissociation dissociation constant
constante de dissociation d'un acide acid dissociation constant (K_a)
constante de dissociation d'une base base dissociation constant (k_b)
constante de la loi de Henry Henry's law constant
constante de vitesse rate constant
contrôle de qualité quality control
coordinat monodenté monodentate (unidentate) ligand
courbe d'étalonnage calibration curve, standard curve
courbe de fuite breakthrough curve
courbe de titration pH curve (titration curve)
cracking cracking
craquage cracking
cyclotron cyclotron
datation au carbone carbon dating
datation au radiocarbone radiocarbon dating
décroissance exponentielle exponential decay
dégazage outgassing
dégradation degradation
dégradation radioactive radioactive decay
déliquescence deliquescence
demi-réactions half-reactions
demi-vie half-life
densité de charge surfacique surface charge density
dérivé d'hydrocarbure hydrocarbon derivative
dérivés du pétrole petrochemicals
déshydrogénation dehydrogenation
désintégration disintegration
désintégration alpha alpha decay
désionisation deionization
désorption desorption
desséchant desiccant
dessicant desiccant
détecteur à conduction electron capture detector (ECD)
détergent detergent
deutérium deuterium
deuxième loi de la thermodynamique second law of thermodynamics
dialyse dialysis
diamagnétisme diamagnetism
dibenzofurannes polychlorés (DFCP, furannes) polychlorinated dibenzofurans (PCDFs, furans)
diffusion diffusion
diffusion d'un élément nutritif diffusion (nutrient)
diffusion ionique ionic diffusion
diffusivité diffusivity
diluant diluent
dilution dilution
dilution en série serial dilution
dilution isotopique isotope dilution
dimère dimer
dioxyde d'azote nitrogen dioxide (NO_2)
dipôle dipole
disaccharide disaccharide
dispersion hydrodynamique hydrodynamic dispersion
dissociation dissociation

dissolution non congruente incongruent solution
distillation distillation
double couche diffuse diffuse double layer
eau déionisée deionized water
eau distillée distilled water
eau lourde heavy water
eau régale aqua regia
échange d'ions ion exchange
échantillon composite composite sample
échantillon de sol soil sample
échantillons de référence background samples
échantillons témoins background samples
échelle pH pH scale
effet d'ion commun common ion effect
effet photoélectrique photoelectric effect
E_H (potentiel redox) E_H
électricité electricity
électrochimie electrochemistry
électrode à ion spécifique ion-selective electrode
électrode de référence reference electrode
électrode de verre glass electrode
électrodialyse electrodialysis
électrolyse electrolysis
électrolyte electrolyte
électrolyte faible weak electrolyte
électrolyte fort strong electrolyte
électrolytique electrolytic
électron electron
électrophorèse electrophoresis
élément element
élément lithophile lithophile element
élément nutritif échangeable exchangeable nutrient
éléments atmosphiles ou atmophiles atmophile elements
élutriation elutriation
énantiomères enantiomers
endothermique endothermic
énergie energy
énergie d'activation activation energy
énergie d'ionisation ionization energy
énergie de liaison bond energy
énergie libre free energy
énergie libre de Gibbs Gibbs free energy
énergie libre standard de formation standard free energy of formation
énergie nucléaire nuclear energy
enrichissement enrichment
enthalpie enthalpy
entropie entropy
environnement réducteur reducing environment
équation caractéristique equation of state
équation d'Arrhénius Arrhenius equation
équation de Nernst Nernst equation
équation moléculaire molecular equation
équilibre chimique chemical equilibrium
équilibre hétérogène heterogeneous equilibrium
équilibre homogène homogeneous equilibrium
équivalent carbonate de calcium calcium carbonate equivalent
équivalent effectif de carbonate de calcium effective calcium carbonate equivalent
équivalent-gramme equivalent or equivalent weight
essai à blanc blank
ester ester
étalon standard
étalonnage calibration
étape déterminante de la vitesse rate-determining step
état d'oxydation oxidation state
état fondamental ground state
états de la matière states of matter
exclusion anionique anion exclusion
exothermique exothermic
extrait de pâte à saturation saturation-paste extract
extrait de sol extract, soil
facteur d'absorption absorptance
facteur de dilution dilution factor
facteur de van't Hoff van't Hoff factor
filtration filtration
filtre à membrane membrane filter
filtre à pression pressure filter
fission fission
fission nucléaire nuclear fission
fluorescence fluorescence
flux flux
force électromotrice cell potential (electromotive force)
force ionique ionic strength
forces de van der Waals van der Waals' forces; intermolecular forces
formation d'une paire ionique ion pairing
formule de Freundlich Freundlich isotherm
formule moléculaire molecular formula

fournaise à creusets crucible furnace
fournaise à mouffles muffle furnace
fraction molaire mole fraction
fractionnement isotopique isotopic fractionation
fréquence frequency
furannes furans
fusion fusion, smelting
fusion nucléaire nuclear fusion
galvanisation galvanizing
gaz gas
gaz acide acid gas
gaz dissous dissolved gases
gaz inerte inert gas
gaz parfait ideal gas
gaz vecteur carrier gas
gel de silice silica gel
glucide carbohydrate
gradient de concentration concentration gradient
gras fats
groupement group (of the periodic table)
groupement alcoyle ou alkyle alkyl group
groupement fonctionnel functional group
halogénation halogenation
halogène halogen
hydratation hydration
hydrate hydrate
hydrate de carbone carbohydrate
hydrocarbure hydrocarbon
hydrocarbure saturé saturated hydrocarbon
hydrocarbures aliphatiques straight-chain hydrocarbons
hydrocarbures aromatiques polycycliques (HAP) polycyclic aromatic hydrocarbons (PAH)
hydrocarbures aromatiques polynucléaires (HAP) polynuclear aromatic hydrocarbons (PAH)
hydrocarbures volatils volatile hydrocarbons
hydrogène lourd (deutérium) heavy hydrogen
hydrolyse hydrolysis
hydrométallurgie hydrometallurgy
hydrophile hydrophilic
hydrophobe hydrophobic
hydroxyde hydroxide
hyperfréquence microwave
hyperosmotique hyperosmotic
hypersalin hypersaline
hypoosmotique hypoosmotic
hypotonique hypotonic
immiscible immiscible
in situ *in situ*
in vitro *in vitro*
in vivo *in vivo*
inactivation inactivation
indicateur indicator
indicateur acido-basique acid-base indicator
inorganique inorganic
instable unstable
ion ion
ion complexe complex ion
ion hydronium hydronium ion
ion isotopiquement échangeable isotopically exchangeable ion
ionisation ionization
ions isoélectriques isoelectronic ions
irradié irradiated
isomère isomer
isomérie géométrique cis-trans geometrical *(cis-trans)* isomerism
isotherme d'adsorption adsorption isotherm
isotherme de Langmuir Langmuir isotherm
isotope isotope
isotope radioactif radioactive isotope
isotope stable stable isotope
liaison atomique atomic bond
liaison chimique chemical bond
liaison covalente covalent bond
liaison covalente polaire polar covalent bond
liaison de coordination coordinate bond
liaison disulfide disulfide linkage
liaison double double bond
liaison glycosidique glycosidic linkage
liaison hydrogène hydrogen bond; hydrogen bonding
liaison ionique ionic bonding
liaison peptidique peptide bond
liaison simple single bond
liaison transversale crosslinking
liaison triple triple bond
ligand (coordinat) ligand
ligand bidenté bidentate ligand
limite de détection detection limit
liquide combustible combustible liquid
loi d'Avogadro Avogadro's law
loi de Beer-Lambert Beer-Lambert law
loi de Boyle Boyle's law
loi de Charles Charles' law
loi de Dalton Dalton's law
loi de Graham Graham's law

loi de Henry Henry's law
loi de la conservation de l'énergie law of conservation of energy
loi de Raoult Raoult's law
loi de vitesse différentielle differential rate law
loi des gaz parfaits ideal gas law (universal gas equation)
longueur de liaison bond length
masse atomique atomic weight
masse molaire molar mass
masse moléculaire formula weight
matières solides totales total solids
matières totales en suspension total suspended solids
mélange mixture
membrane de tensiomètre pressure membrane
membrane semi-perméable semipermeable membrane
mercaptans mercaptans
métal base metal, metal
métal lourd heavy metal
métalloïdes metalloids (semimetals)
métallurgie metallurgy
métaux à l'état de traces trace metals
métaux alcalino-terreux alkaline-earth metals
métaux alcalins alkali metals
méthode de Dumas Dumas method
méthode Kjeldahl Kjeldahl method
méthode standardisée standard method
méthyle de mercure methyl mercury
micro-onde microwave
miscible miscible
molalité molality, molarity
molécule molecule
molécule marquée tagged molecule
molécule polaire polar molecule
moment dipolaire dipole moment
monoacide monoprotic acid
monomère monomer
monosaccharide monosaccharide (simple sugar)
mouvement brownien Brownian motion
néphélomètre nephelometer
neutralisation neutralization
neutron neutron
nitrate nitrate (NO_3)
nitrite nitrite
nitrosamines nitrosamines
nombre atomique atomic number
nombre de masse mass number
noyau nucleus
nucléide nuclide
nucléon nucleon
organique organic
organochloré organochlorine
osmose inversée reverse osmosis
oxydant oxidant
oxydation oxidation
oxyde nitrique nitric oxide (NO)
oxydes d'azote, NOx oxides of nitrogen, NOx
paire acide-base conjuguée conjugate acid-base pair
paire ionique libre lone pair
paraffines paraffins
particule alpha alpha particle
particule bêta beta particle
pâte de sol à saturation saturated soil paste
pentachlorophénol (PCP) pentachlorophenol (PCP)
peptide peptide
période half-life, period
pH du sol pH, soil
pH_C pH_C
phénols phenols
phosphore organique organic phosphorus
photochimique photochemical
photoélectron photoelectron
photolyse directe direct photolysis
photon photon
phtalates phthalates
physisorption physisorption
pile à combustible fuel cell
pile galvanique galvanic cell
pK_a pK_a
pK_b pK_b
poids moléculaire molecular weight
point d'ébullition boiling point
point d'inflammation spontanée fire point
point de charge nette nulle point of zero net charge
point de congélation freezing point
point de fusion melting point
point éclair flash point
point isoélectrique isoelectric point
polarité polarity
polycyclique polycyclic
polymères polymers
polymérisation polymerization

polypeptide polypeptide
polysaccharide polysaccharide
pont de sel salt bridge
porphyrine porphyrin
potentiel chimique chemical potential
potentiel d'oxydoréduction redox potential
potentiel de chaux lime potential
potentiel électrocinétique ou potentiel zêta electrokinetic (zeta) potential
potentiel zêta zeta potential
pourcentage atomique atom percent
pourcentage de cations échangeables exchangeable cation percentage
pourcentage de saturation saturation percentage
pourcentage de saturation en bases base saturation percentage
pouvoir calorifique calorific value
pouvoir tampon buffer power
précipitant precipitant
précipitation sélective selective precipitation
précipité precipitate
première loi de la thermodynamique first law of thermodynamics
pression de vapeur vapor pressure
pression partielle partial pressure
principe de Le Chatelier Le Chatelier's principle
procédé chaux-soude lime-soda process
procédé d'Ostwald Ostwald process
procédé de Frasch Frasch process
procédé de Haber Haber process
produit de désintégration decay product
produit de solubilité solubility product (K_s)
propriétés colligatives colligative properties
protecteur cathodique cathodic protection
proton proton
purine purine
pyrimidine pyrimidine
pyrolise pyrolysis
pyrophorique pyrophoric
quantités stoechiométriques stoichiometric quantities
quantum quantum
radiation bêta beta radiation
radiation électromagnétique electromagnetic radiation
radiation ionisante ionizing radiation
radical radical
radical libre free radical
radioactivité radioactivity
radioactivité naturelle natural radioactivity
radiocarbone radiocarbon
radiochimie radiochemistry
radio-isotope radioisotope
radionucléide radionuclide
radiotraceur radiotracer
radon radon (Rn)
rapport molaire mole ratio (stoichiometry)
rayon atomique atomic radius
rayon gamma gamma ray
rayonnement ultraviolet (rayonnement UV) ultraviolet radiation (UV radiation)
réacteur nucléaire nuclear reactor
réactif reactant
réactif reagent
réaction d'hydrogénation hydrogenation reaction
réaction d'ordre zéro zero-order reaction
réaction d'oxydo-réduction oxidation-reduction (redox) reaction
réaction de condensation condensation reaction
réaction de premier ordre first-order reaction
réaction du sol reaction, soil
réduction reduction
réflexion spéculaire specular reflection
reflux reflux
rendement théorique theoretical yield
résine anionique anionic resin
sel salt
sélectivité ionique ion selectivity
série homologue homologous series
série radioactive radioactive series
sol acide acid soil
sol alcalin alkaline soil
sol neutre neutral soil
sol séché à l'étuve oven-dry soil
sol tamponné buffered (soil)
solide ionique ionic solid
solubilité solubility
soluté solute
solution solution
solution acide acidic solution
solution alcaline alkaline solution
solution aqueuse aqueous solution
solution d'extraction extractant
solution du sol solution, soil
solution étalon standard solution
solution idéale ideal solution

solution normale normal solution
solution tamponnée buffered solution
solution titrante titrant
solutions isotoniques isotonic solutions
solvant solvent
solvant non polaire nonpolar solvent
solvant organique organic solvent
solvant polaire polar solvent
sorption sorption
soude caustique caustic soda
spectre spectrum (plural spectra)
spectre d'absorption de la phase gazeuse gas-phase absorption spectrum
spectromètre spectrometer
spectromètre de masse mass spectrometer
spectrophotomètre spectro-photometer
spectroscopie spectroscopy
spectroscopie d'absorption atomique atomic absorption (AA) spectroscopy
spectroscopie d'absorption dans l'infra rouge infrared absorption spectroscopy (IR)
spectroscopie d'émission atomique atomic-emission spectroscopy (AES)
spectroscopie de résonance magnétique nucléaire nuclear magnetic resonance (NMR) spectroscopy
stéroïde steroid
stoechiométrie chimique chemical stoichiometry
strippage stripping
stripping stripping
structure de Lewis Lewis structure
structure moléculaire molecular structure
substance amphotère amphoteric substance
substitution ionique ionic substitution
sulfure d'hydrogène hydrogen sulfide (H_2S)
surchauffe superheating
surface à potentiel constant constant-potential surface
surfusion supercooling
sursaturation supersaturation
suspension-dilution serial dilution
système de tampon carbonate carbonate buffer system
système fermé closed system
tableau périodique periodic table
tampon buffer
taux de saturation en bases base saturation percentage
température absolue absolute temperature
teneur en eau volumétrique volumetric water content
tension superficielle surface tension
thermodynamique thermodynamics
titrage titration
titration titration
total des solides total suspended solids
traceur tracer
transformation adiabatique adiabatic process
transformation de Fourier Fourier transform
trihalométhanes trihalomethanes (THMs)
troisième loi de la thermodynamique third law of thermodynamics
unité de masse atomique atomic mass unit
valence valence
vaporiser vaporize
vitesse de réaction reaction rate
volume molaire molar volume
zéro absolu absolute zero

CONSERVATION, GESTION DES SOLS, TRAVAIL DU SOL (CONSERVATION, SOIL MANAGEMENT, TILLAGE)

ados back furrow, dead furrow
agriculture biologique organic farming
agriculture durable sustainable agriculture
agriculture écologique organic farming
agriculture organique organic farming
agroécosystème agroecosystem
amélioration des sols soil improvement
amendement du sol soil amendment
ameublissement loosening
arable arable
aridoculture dryland farming
arracher lift
assolement cropping intensity
assollement facultatif ou souple flexible cropping
attribut de la qualité des sols attribute of soil quality, soil quality attribute
bande enherbée grass barrier
bande filtrante filter strip
bande ou zone tampon buffer strip

bande tampon spreader strip
besoin en gypse gypsum requirement
besoin en labour ou en travail du sol tillage requirement
billon ridge
billonnage bedding, ridge tillage
billonneuse bed shaper
billons reliés entre eux de façon à former des bassins tie-ridge
biner hoe
bloc block
bouleversement inversion
brise-vent windbreak
broyage et déchiquetage des résidus de cultures residue processing
buttage hill
buttoir bed shaper
cadre d'évaluation de la qualité des sols soil quality framework
canaux de dérivation contour furrows
ceinture de verdure green belt
ceinture verte green belt
charrue scarificatrice chisel cultivator
chaulage liming
chaume stubble
chisel chisel cultivator
cintre turnrow (turn strip, head land)
circulation limitée controlled traffic
cisaillement shearing
classes d'érosion erosion classes
conditions physiques d'enracinement physical rooting conditions
conservation conservation
conservation du sol soil conservation
contrôle statistique de la qualité statistical quality control
couche de labour furrow slice, plow layer
couche de sol arable plow layer
couche de sol labourée ou travaillée topsoil
couche indurée pan
couche indurée d'origine anthropique pressure pan (traffic sole, plow pan, tillage pan, traffic pan, plow sole, compacted layer)
couche travaillée furrow slice
coupe, profil racinaire bisect
coût des dommages hors-ferme off-site damage costs
couverture végétale permanente permanent crop cover
creuser dig
cultitassement cultipack
cultivateur chisel cultivator
cultivateur ou herse pourvu de socs à ailettes sweep
cultiver le sol till
culture associée companion crop
culture compagne companion crop
culture continue continuous cropping
culture de couverture cover crop, stubble crops
culture dérobée catch crop
culture en bandes strip cropping
culture en bandes selon les courbes de niveau strip cropping
culture en bandes suivant les courbes de niveau contour strip cropping
culture en contre-pente cross-slope farming
culture en courbes de niveau contour cultivation
culture en rangées ou en lignes row crop
culture en région sèche dryland farming
culture intensive intensive cropping
culture sans labour plowless farming
culture sur paillis mulch tillage
culture sur résidus trash farming
culture sur résidus ou sur paillis mulch farming
cultures améliorantes ou de conservation soil-conserving crops
cultures épuisantes pour le sol soil-depleting crops
cultures intercalaires intercropping, interplanting
cultures successives sequential cropping
cuvette de déflation blowout
dallage d'érosion erosion pavement
décompacteur de type paraplow paraplow
décompacteur de type paratill paratill
découpage cutting
dégradation des sols soil degradation
degré de facilité de travail du sol tillability
délavage par les pluies rain wash
désherbage mécanique à l'aide d'un cultivateur à tige rotative rod weeding
détachement detachment
diversité ou complexité d'un agroécosystème agroecosystem complexity
dynamomètre dynamometer
éclaboussement rain splash
écosystème ecosystem

écosystème agricole agroecosystem
écran de protection boundary screen
écrasement crushing, shatter
écroûtage scarifying
effets découlant du travail du sol tillage action
élément déficient ou limitant limiting nutrient
en patte d'oie duckfoot
énergie renouvelable renewable energy
enfouir anchor
enfouissement burying
engrais vert green manure
ensemble de données minimal minimum data set
épi groin
équation de perte de sol soil loss equation
équation universelle de perte de sol universal soil loss equation
éroder erode
érosion erosion
érosion accélérée erosion, accelerated
érosion anthropique erosion, accelerated
érosion différentielle differential erosion; erosion, differential
érosion en nappe ou en couche inter-rill erosion
érosion en nappe ou inter-rigoles sheet erosion
érosion géologique erosion, geological
érosion naturelle erosion, geological; natural erosion
érosion normale normal erosion
érosion ou déplacement de sol causé par le labour tillage erosion
érosion par éclaboussement splash erosion raindrop erosion
érosion par ravinement gully erosion
érosion potentielle erosion potential
érosion superficielle surficial erosion
érosivité erosivity
état d'ameublissement ou état structural du sol tilth
évaluation de la qualité des sols soil quality, assessment of
exsudat exudate
façon superficielle surface tillage
facteur extrinsèque extrinsic factor
facteur K K factor
facteur L L factor
facteur limitatif controlling variable
facteur P P factor
facteur R R factor
facteur S S factor
ferme ou entreprise agricole diversifiée mixed farm
fonction de pédotransfert pedotransfer function
formation de plate-bandes bedding
forme de buttage ou billonnage listing (middle breaking)
fragmentation crushing, shatter
friche temporaire land resting
gestion de la neige snow management
gestion de la qualité des sols soil quality control
gestion des écosystèmes ecosystem management
gestion des résidus de cultures crop residue management, residue management
gestion des sols soil management
grande culture row crop
groupe de gestion des sols soil management group
haie brise-vent shelterbelt; windbreak
hersage cultivation, harrowing
incorporation incorporation
incorporer anchor
indicateur indicator
indicateur agro-environnemental agri-environmental indicator
indicateur de la qualité des sols soil quality indicator
indice de vulnérabilité des sols soil quality susceptibility, index of
instruments aratoires tillage equipment (tools)
inversion inversion
jachère fallow, land resting
jachère chimique chemical fallow (eco-fallow)
jachère d'été summerfallow
jachère écologique eco-fallow
labour plowing
labour avec charrue à versoirs moldboard plowing
labour complet broadcast tillage
labour de défoncement tillage, deep
labour et semis en un passage (à l'aide d'un train d'outils) plow-planting
labour nettoyant clean tillage
labour profond tillage, deep

labourer till
ligne de guidage guess row
lit de semence seedbed
mélange mixing
méthodes de lutte contre l'érosion erosion control
mise à l'échelle up-scaling
modelage landforming
modelage superficiel dam (pitting, basin listing)
modelage superficiel du sol pour former des petits bassins de captage reservoir tillage (damming, pitting, basin listing, furrow diking, dammer diking)
monoculture continuous cropping, monoculture
mulch mulch
niveler avec une traîne drag
nivellement land planing
opérations combinées de travail du sol combined tillage operations
outils de travail du sol tillage equipment (tools)
paillis mulch
paillis de résidus de culture stubble mulch
parcelle de mesure d'érosion runoff plot
passer la houe hoe
perte de sol soil loss
perte de sol tolérable soil loss tolerance
placement vertical d'amendements vertical mulching
plan ou programme de conservation conservation plan
planage land planing
plantation en rangs étroits ou serrés narrow row planting
plantation intercalaire interplanting
plante cultivée cultural vegetation
plante-abri nurse crop
pratique culturale tillage operation
pratique de conservation conservation practice, conservation tillage
préparation de sol avec cultivateur ou chisel chiseling
préparation du sol par un passage unique d'outils combinés once-over tillage
préparation primaire du sol primary tillage
productivité des sols soil productivity
productivité productivity
profil racinaire root bed
profondeur de la couche arable depth, effective soil
profondeur effective de sol effective soil depth
programme de surveillance de la qualité des sols soil quality monitoring program
projection throw
propriété dynamique du sol soil quality, dynamic
propriété subrogative surrogate property, proxy property
qualité biologique des sols soil biological quality
qualité chimique des sols soil chemical quality
qualité de la matière organique du sol soil organic matter quality
qualité des sols soil quality
qualité des terres land quality
qualité inhérente ou intrinsèque du sol soil quality, inherent
qualité pédologique des sols soil pedological quality
qualité physique des sols soil physical quality
raffermissement firming
raffermissement du lit de semence cultipack
raie furrow
raie de curage dead furrow
ravin gully
redéposition de sol soil redistribution
réduction ou diminution d'échelle down-scaling
regain stubble crops
remise en état de ravins gully reclamation
rendement agronomique durable agronomically sustainable yield
rendement durable ou soutenu sustainable yield
rendement maximal durable maximum sustainable yield
rendement optimal maximum sustainable yield
reprise de labour secondary tillage
résidu residue
résidus de cultures crop residue
résidus végétaux plant residue
résilience du sol soil resiliency
ressource naturelle resource (natural)
ressource renouvelable renewable resource

rigole rill
risque inhérent d'érosion erosion risk, inherent
risque ou potentiel d'érosion erosion risk, actual
rôle ou fonction du sol soil function
rotation rotation
rotation culturale ou des cultures crop rotation
rotation longue ou allongée extended rotation
roulage rolling
ruissellement runoff
saltation saltation, surface creep
santé des écosystèmes ecosystem health
santé des sols soil health
sarclage cultivation (weeding); weeding
sarclage ou hersage à la houe rotative rotary hoeing
scarifiage léger scarifying
scarification chiseling
scarifier chisel
semelle de labour plow pan; plow sole; pressure pan (traffic sole, plow pan, tillage pan, traffic pan, plow sole, compacted layer)
semis à l'aide d'un planteur seed-drilling
semis à la volée broadcast planting
semis au creux du sillon lister planting
semis direct plow-plant
semis en bandes strip planting (strip till planting)
semis en intercalaire ou en plante-abri interseeding
semis en ligne à espacement rapproché drill seeding
semis en ornières wheel track planting
semis sous couvert végétal underseeding
semis superficiel ou à plat flat planting
semis sur bandes travaillées strip-till planting
semis sur billons ridge planting
semis sur buttes ou sur lits surélevés bed planting
semis-direct no-tillage (zero tillage), slit planting (slot planting), zero tillage
semis-direct sur prairie ou sur gazon sod planting
seuil threshold value (soil quality)
sillon furrow
sillon terminal dead furrow
site d'essais demonstration area
site de démonstration demonstration area
site de référence benchmark site, reference site
site repère reference site
site témoin reference site
sol appauvri depleted soil
sol arable topsoil
sol épuisé depleted soil
sol pulvérisé dust mulch
sol sans stucture ou à structure massive puddled soil
sous-ensemensement underseeding
sous-solage chiseling, in-row subsoiling, ripping, subsoiling, subsurface tillage
sous-solage localisé zone subsoiling
sous-soler chisel
stabilisatiation de ravins gully control plantings
stabilisation des berges channel stabilization
structure de confinement sheet piling
structures de lutte contre l'érosion erosion control structures
suivi de l'évolution des ressources resource monitoring
surface érodée erosion surface
sursemis overseeding
suspension suspension
système zéro-labour no-tillage (zero tillage)
système de culture cropping pattern, cropping system
système de culture sans labour no-tillage (zero tillage)
système de gestion des résidus de cultures crop residue management system
système de gestion des ressources resource management system
système de production farming system
systèmes cybernétiques cybernetic systems
tassement firming
terrain de déflation blown-out land
terrain raviné gullied land
terrasse terrace
terre à faible potentiel agricole marginal land
terre agricole de première qualité prime agricultural land
terre arable crop land
terre cultivée cultivated land
terre en culture crop land
terre en jachère fallow land

terre marginale marginal land
terre végétale topsoil
terre vulnérable ou fragile fragile land
tournière turnrow (turn strip, head land)
transport de particules de sol soil transportation
trappe à neige (haie) snow hedge
travail à l'aide d'un cultivateur ou d'un chisel chisel planting
travail du sol cultivation, tillage
travail du sol avec bêche ou charrue rotative rotary tilling
travail du sol avec conservation des résidus mulch tillage
travail du sol conventionnel conventional tillage
travail du sol directionnel oriented tillage
travail du sol en bandes strip tillage (partial-width tillage)
travail du sol en bandes étroites slit tillage
travail du sol en effectuant des passages croisés cross cultivation
travail du sol en pré- ou post-émergence pre-(post) emergence tillage
travail du sol en pré- ou post-récolte pre-(post) harvest tillage
travail du sol en pré- ou post-semis pre-(post) planting tillage
travail du sol localisé zone tillage
travail du sol préservant les résidus de culture stubble mulch tillage
travail du sol réduit ou simplifié reduced tillage
travail du sol suivant les courbes de niveau contour tillage
travail du sol superficiel surface tillage
travail minimal du sol minimum tillage
travail sans inversion de sol non-inversive tillage
travail secondaire ou superficiel du sol secondary tillage
travail sur toute la surface de sol broadcast tillage (total surface tillage, full-width tillage); total surface tillage
valeur limite threshold value (soil quality)
valeur seuil threshold value (soil quality)
variabilité (spatio-temporelle) des sols soil variability
végétalisation de ravins gully control plantings
vision écosystémique ecosystem perspective
vitesse limite threshold velocity
vitesse minimale threshold velocity
voie d'eau engazonnée grassed waterway
zéro-labour zero tillage
zone de faille shatter belt

ÉCOLOGIE (ECOLOGY)

abiotique abiotic
acclimatation acclimatization
acclimatement acclimation
adaptation adaptation
agriculture durable du point de vue de l'environnement environmentally sustainable agriculture
agronomie agronomy
aire d'étude study area
aire de distribution géographique range
allèle allele
allélopathie allelopathy, amensalism
allochtone allochthonous
alpin alpine
amplitude amplitude
amplitude de l'habitat habitat breadth
analyse des gradients gradient analysis
antagonisme antagonism
arbres codominants co-dominant trees
arbuste shrub
arctique arctic
association association
autochtone autochthonous
auto-écologie autecology
autopoïèse autopoiesis
autotoxique autotoxic
autotrophe autotroph
autotrophie autotrophy
bande border
besoins en habitat habitat needs
bilan énergétique energy budget
biocénose biocoenosis
biochore biochore
biodisponibilité biological availability
biodiversité biodiversity
biofaciès biofacies
biogénique biogenic
biogéographie biogeography

bioindicateur bioindicator
biomasse biomass
biome biome
biome alpin alpine biome
biome d'eau douce fresh water biome
biome de désert desert biome
biome de maquis méditerranéen Mediterranean scrub biome
biome de prairie grassland biome
biome océanique marine biome
biosphère biosphere
biote biota
biotique biotic
biotope biotope
biozone biozone
boréal boreal
bovin bovine
broussaille scrub
brucellose brucellosis
butte tussock
calcifuge calciphobe, calcifuge
calciphyte calciphyte
chaîne alimentaire food chain
chaîne alimentaire ou de prédation food web
chaparral chapparral
classification écologique ecological classification
climax climax
climax édaphique edaphic climax
coévolution coevolution
coexistence coexistence
cohorte cohort
communauté community
compétition competition
competition d'exploitation exploitation competition
compétition directe direct competition
compétition indirecte indirect competition
compétition végétale plant competition
composition écosystémique ecosystem composition
composition taxonomique species composition
condition d'exposition exposure characterization
conditions chimiques d'enracinement chemical rooting conditions
connexité consilience
consociation consociation
consommateur secondaire secondary consumer
continuum continuum
contradaptation counteradaptation
contreparties écologiques ecologic counterparts
contrévolution counterevolution
croissance exponentielle exponential growth
croissance naturelle natural increase
croissance nulle zero population growth (ZPG)
croissance sigmoïde sigmoid growth
cryophyle cryophilous
cryophyte cryophyte
cryovégétation cryovegetation
cycle cycle
cycle alimentaire food cycle
cycle biogéochimique biogeochemical cycle
cycle d'un taxon taxon cycle
cycle de vie life cycle
délais time lag
densité écologique ecological density
densité spécifique species density
dénudé ou stérile barren
dépérissement dieback
désert desert
désert froid cold desert
désertification desertification
dispersion dispersal, dispersion
distribution distribution
distribution par âge stable stable age distribution
diversité diversity
diversité alpha alpha diversity
diversité bêta beta diversity
diversité des espèces species richness
diversité gamma gamma diversity
diversité spécifique species diversity
dominance dominance
durabilité sustainability
durabilité de l'environnement environmental sustainability
durabilité écosystémique ecosystem sustainability
dynamique de population population dynamics
dynécologie dynecology
échelle scale
écoclimat ecoclimate
écodistrict ecodistrict
écoespèce ecospecies

écologie ecology
écologie des systèmes systems ecology
écologie du paysage landscape ecology
écologie du sol ecology, soil
écologie holistique holistic ecology
écologique ecologic
écologisme radical deep ecology
écorégion ecoregion
écosphère ecosphere
écosystème ecosystem
écosystème hétérotrophe heterotrophic ecosystem
écosystème naturel natural ecosystem
écosystème non perturbé ou naturel undisturbed ecosystem
écosystème terrestre land ecosystem
écotone ecotone
écotope ecotope
écotype ecotype
écozone ecozone
édaphique edaphic
édaphologie edaphology
effet de bordure edge effect
efficacité de production nette net production efficiency
efficacité de transpiration transpiration efficiency
efficacité écologique ecologic efficiency
élément element
endémique endemic
endogène endogenous
énergie de la biomasse biomass energy
enrichissement sélectif selective enrichment
environnement environment
environnement durable en agriculture environmentally sustainable agriculture
environnement physico-chimique physical–chemical environment
épiphyte epiphyte
équilibre d'exploitation exploitation equilibrium
équivalents écologiques ecological equivalents
espèce species
espèce clé ou pivot keystone species
espèce endémique endemic species
espèce indigène species, native
espèce introduite introduced species, species, introduced
espèce menacée threatened species
espèce naturalisée species, naturalized
espèce oportuniste opportunistic species
espèce résidente resident species
espèce sporadique casual species
estivation estivation
état d'équilibre dynamique dynamic equilibrium state
eutrophe eutroph, eutrophic
évaluation du risque écologique ecological risk assessment
évolution evolution
exclusion compétitive competitive exclusion
exotique exotic
expansion écologique ecological release
exploiter exploit
exposition exposure
extinction extinction
facteur écologique ecological factor
facteur extrinsèque extrinsic factor
facteur limitant limiting factor
facteurs biotiques biotic factors
facteurs de station site factors
facteurs essentiels ou ultimes ultimate factors
facteurs proximaux proximate factors
faune fauna
fixation associative d'azote associative dinitrogen fixation
flore flora
flux d'énergie energy flow
fonction de l'habitat habitat function
fonction écosystémique ecosystem function
fonctions motrices forcing functions
force motrice driving force
forêt forest
forêt boréale boreal forest
forêt feuillue broad-leaved deciduous forest
forêt feuillue à feuillage persistant broad-leaved evergreen forest
forêt tempérée à feuilles caduques deciduous forest, temperate
forêt tropicale humide tropical rainforest
forêt vierge virgin forest
formation végétale formation-class
fréquence d'espèces species frequency
glycophyte glycophyte
gradient géographique cline
groupement végétal préclimatique seral community
guilde guild
habitat habitat
habitat critique habitat, critical

habitat préféré habitat, preferred
héliophyte heliophyte
hétérogène heterogeneous
hétérotrophe heterotroph
holarctique Holarctic
holistique holistic
homogène homogeneous
hôte host
humidité critique threshold moisture content
hypothèse de Gaia Gaia hypothesis
hypothèse des communautés community-unit hypothesis
immigrant immigrant
indice de diversité diversity index
indice de récolte harvest index
indice de richesse spécifique species richness index
indice de Shannon-Weaver Shannon-Weaver index
indigène indigenous
individu à strategie k k-selected
individu à stratégie r r-selected
influence biotique biota influence
interaction interaction
intraspécifique intraspecific
inventaire par échantillonnage en ligne line plot survey
lande heath
libre free-living
ligne des arbres tree line
limites de tolérance limits of tolerance
limnologie limnology
lithophyte lithophyte
lithosère lithosere
lithosphère lithosphere
loi de Shelford Shelford's law
loi naturelle natural law
macroflore macroflora
macroorganisme macroorganism
macrophyte macrophyte
mangrove mangrove
marais salant salt marsh
marais tidal marsh, tidal
marais tidal tidal marsh
marécage marsh
mesures synoptiques synoptic measurements
méthode d'intersection line-intercept method
microcosme microcosm
microsère microsere
morphe morph
mutualisme mutualism
niche niche
niche d'habitat habitat niche
niche écologique ecological niche
niveau trophique trophic level
occurrence occurrence
ontogénie ontogeny
opportunisme évolutif evolutionary opportunism
organisme organism
organismes ubiquistes ubiquitous organisms
orophyte orophyte
oxyphyte oxyphyte
paléoécologie paleoecology
parcours range
parcours naturel range
période d'évolution evolutionary time
période de transition transit time
perturbation disturbance
perturbation naturelle natural disturbance
peuplement inéquien uneven-aged stand
photosynthèse photosynthesis
phylogénie, phylogenèse phylogeny
phytoécologie phytoecology
plagiosère plagiosere (plagioclimax)
plante spécimen character plant
plantes indicatrices indicator plants
point de flétrissement threshold moisture content
population population
population régulée regulated population
potentiel biotique biotic potential
pourcentage de couverture ou de recouvrement cover, percent
prairie grassland, prairie
pré meadow
présence occurrence
principe d'allocation principle of allocation
principe fondateur founder principle
processus écologique ecological process
proclimax proclimax
producteur producer
productivité productivity
productivité biologique biological productivity
productivité communautaire nette net community productivity (NCP)
productivité épigée nette net aboveground productivity (NAP)
productivité ou potentiel du site site productivity
productivité primaire primary productivity

productivité primaire brute gross primary productivity (GPP)
productivité primaire nette net primary productivity (NPP)
productivité secondaire secondary productivity
progénie progeny
provenance provenance
province biogéographique biogeographic province
psammon psammon
psammosère psammosere
puits naturel natural sink
pyramide d'énergie pyramid of energy
pyramide de biomasse pyramid of biomass
pyramide écologique ecological pyramid
quadrat quadrat
qualité de station site quality
radioécologie radioecology
rare rare
réaction de saturation saturation response
recrutement recruitment
refuge refuge
région semi-désertique semi-desert
règle du 10% ten percent rule
régularité evenness
relation abondance-diversité dominance-diversity curve
relique relict
rendement en biomasse standing crop
renouvellement turnover
résilience resilience
résistance resistance
résistance environnementale environmental resistance
ressources naturelles natural resources
richesse richness
richesse spécifique species richness
savanne savanna (savannah, savana)
sciaphile sciophilous
sciophyte sciophyte
section section
sélection d'habitat habitat selection
sélection naturelle natural selection
serclimax serclimax
sère sere
site site
site fragile critical area
souche strain
sous-population subpopulation
spécialisation specialization
spéciation allopatrique allopatric speciation
stabilité stability
stade séral seral stage
station site
stratification stratification
structure d'âge age structure
structure écosystémique ecosystem structure
structure trophique trophic structure
subarctique subarctic
subsère subsere, secondary sere
subtropical subtropical
succession succession
succession allogène allogenic succession
succession autogène succession, autogenic
succession autotrophe succession, autotrophic
succession biotique succession, biotic
succession écologique ecological succession
succession primaire primary succession
succession secondaire secondary succession
succession végétale plant succession
surpopulation overpopulation
symbiose symbiosis
sympatrique sympatric
synécologie synecology
synergie synergism
systématique systematics
système system
système de gestion de l'environnement environment management system
système résilient globally stable
taïga taiga
temps de renouvellement turnover time
terrains improductifs barren lands
test de provenance provenance test
tolérance tolerance
toundra tundra
toundra alpine alpine tundra
toundra arctique arctic tundra
trophique trophic
tropophyte tropophyte
tusco tussock
type d'habitat habitat type
type de végétation vegetation type
variabilité range of variability
variable de forçage forcing variable
végétation vegetation
végétation naturelle natural vegetation
végétation naturelle potentielle potential natural plant community
végétation ou plante pionnière pioneer
végétation ségétale segetal vegetation
vieillesse old age
vieillissement senescence
xérophyte xerophyte
xérosère xerosere

xylophage xylophagous
zone belt, zone
zone adaptative adaptive zone

SCIENCES DU SOL ET DE L'ENVIRONNEMENT, PRODUITS AGROCHIMIQUES, PESTICIDES, POLLUANTS (ENVIRONMENTAL SOIL SCIENCE, AGRICHEMICALS, PESTICIDES, POLLUTANTS)

algicide algicide
amide amide
ammate ammate
application à la volée broadcast application
atomisation fogging
atténuation mitigation
biocide biocide
brumisage fogging
capacité de dissipation dissipation capacity
charge critique critical load
composé organomercuriel organomercurial
DDT (dichloro-diphényl-trichloréthane) DDT (dichloro-diphenyl-trichloro-ethane)
déchet atomique atomic waste
déposition acide acid deposition
dinitroaniline dinitroaniline
dinitrophénol dinitrophenol
effets additifs additive effects
engrais manure
ennemi des cultures pest
éthers diphényle diphenyl ethers
fongicide fungicide
fongistatique fungistat
fumier manure
fumigant fumigant
fumigation fumigation
fumure manure
herbicide herbicide
herbicide benzoïque benzoic
herbicide de contact contact herbicide
herbicide de type hormonal phenoxy
héritage inheritance
hexachlorure de benzène benzene hexachloride (BHC)
hydrocarbure chloré chlorinated hydrocarbon
infection infection
infestation infestation
insecticide insecticide
insecticide à effet rémanent residual insecticide
lindane lindane
lit bed
lutte biologique biological control
lutte intégrée integrated pest management (IPM)
matière active active ingredient
monitoring monitoring
nébulisation fogging
organisme nuisible pest
organophosphate organophosphate
persistance persistence
pesticide pesticide
pesticide botanique ou d'origine végétale botanical pesticide
pesticide de contact contact pesticide
pesticide non-persistant soft pesticide
pesticide rémanent persistent pesticide
pesticide systémique systemic pesticide
phénoxy phenoxy
pluie acide acid rain
pluviolessivage throughfall
pollution pollution
produit agrochimique agrichemical
pyréthrinoïde pyrethroid
qualité de l'eau water quality
ravageur pest
réglage calibration
résidu négligeable negligible residue
résidu de pesticide pesticide residue
rodenticide rodenticide
roténone rotenone
ruissellement sur les troncs stemflow
stabilité persistence
stérilisant chimique chemosterilant
stérilisation du sol soil sterilization
suivi monitoring
surveillance monitoring
systémique systemic
thiocarbamate thiocarbamate
tolérance aux pesticides pesticide tolerance
traitement à la base des tiges basal application
traitement basal basal application
traitement en bandes band application
triazine triazine
variétés ou cultivars résistants aux organismes nuisibles pest resistant varieties

FORESTERIE (FORESTRY)

âge d'exploitabilité rotation age
amélioration de l'habitat habitat enhancement
aménagement de l'habitat habitat enhancement
aménagement forestier forest management
aménagement ou gestion de boisé woodland management
analyse foliaire foliar analysis
andain windrow
arbres en réserve reserves
attributs des vieilles forêts old-growth forest attributes
aubier sapwood
bande de protection buffer strip
barrière à la répartition naturelle natural range barrier
bassin hydrographique drainage basin
bassin versant drainage basin
bassin versant ou hydrographique watershed
bassin versant sensible ou fragile sensitive watershed
berge stream bank
bio-indicateur indicator species
biologie de conservation conservation biology
biote biota
bois timber
bois à pâte pulpwood
bois de sciage sawtimber
boisé woodland
boisement afforestation
bordure edge
carte de potentiel d'habitat suitability mapping
carte forestière forest cover map
carte topographique contour map
chablis blowdown, windthrow
chemin forestier forest service road
chenal d'un cours d'eau stream channel
chlorose chlorosis
cime crown
classification écologique ecological classification
clinomètre clinometer
composition du peuplement stand composition
conditions intérieures des forêts forest interior conditions
conifère conifer
conservation conservation
conversion de peuplement stand conversion
corridor corridor
couche holorganique forest floor
coupe d'amélioration improvement cutting
coupe d'écrémage ou sélective selective cutting
coupe de récupération salvage harvesting
coupe partielle partial cutting
coupe totale clearcut
coupe totale avec réserves clearcutting with reserves
cours d'eau stream, watercourse
couvert forestier forest cover
couverture morte forest floor
débardage skidding
déboisement deforestation
débris ligneux grossiers coarse woody debris
débusqueuse skidder
déclinaison magnétique declination (magnetic)
déclivité d'un cours d'eau stream gradient
dégagement brushing
dégradation degradation
densité de peuplement stand density
développement du peuplement stand development
diversité génétique genetic diversity
domination overtopping
dominé overtopped
drainage souterrain subsurface drainage
droit de passage right-of-way
durabilité sustainability
dynamique du peuplement stand dynamics
eau souterraine groundwater
éclaircie thinning
écologie forestière forest ecology
écologie forestière appliquée silvics
écosystème sain healthy ecosystem
écotone ecotone
édatope edatope
éducation du peuplement stand tending
effet de bordure edge effect
effets additifs additive effects
effets cumulatifs cumulative effects
élagage pruning

ennemi des cultures pest
ensemencement hydraulique hydroseeding
équilibre écologique ecological balance
espèce clé keystone species
espèce de l'étage dominant overstory species
espèce indicatrice indicator species
espèces sauvages wildlife
étage dominant overstory
étang de décantation settlement pond, settling pond
étude d'impacts impact assessment
évaluation de bassin versant watershed assessment
évaluation du risque risk rating, forestry (assessment)
extirpation extirpation
facteurs de santé des forêts forest health agents
ferme forestière tree farm
fermeture du couvert crown closure
feu de friches wildfire
feu de surface ground fire
feuillus hardwoods
foresterie forestry
foresterie paysane ou rurale farm forestry
foresterie urbaine urban forestry
forêt climacique climax forest
forêt coniférienne coniferous forest
forêt de production production forest
forêt de protection protection forest
forêt de seconde venue second growth
forêt feuillue broad-leaved deciduous forest
forêt feuillue à feuillage persistant broad-leaved evergreen forest
forêt mixte mixed forest
forêt productive productive forest land
forêt tempérée à feuilles caduques temperate deciduous forest
fragmentation fragmentation
friche cut-over forest
gaule sapling
génotype genotype
gestion adaptative adaptive management
gestion de bassin versant watershed management
gestion de l'eau water management
gestion de l'habitat habitat management
gestion des incendies ou des feux fire management
gestion durable des forêts sustainable forest management
gestion écosystémique ecosystem management
gestion intégrée des ressources integrated resource management
gestion ou aménagement de la faune wildlife management
glissement slide
glissement de talus slope failure
habitat habitat
habitat faunique critique critical wildlife habitat
habitats menacés threatened or endangered habitats
herbicide biologique biological herbicide
humus duff
hypsomètre hypsometer
incendie forestier forest fire
incidence des ravageurs pest incidence
indice de qualité de station site index
influences forestières forest influences
intégrité du bassin versant watershed integrity
intégrité écologique ecological integrity
intendance pour les générations futures stewardship
inventaire forestier forest inventory
lit de semis seedbed
lit du cours d'eau streambed
litière forest floor, litter layer
lot de semences seedlot
lutte biologique biological control
matériel végétatif vegetative material
matière organique duff
microclimat microclimate
mode de régénération par coupe avec réserve de semenciers seed tree silvicultural system
monoculture monoculture
mortalité mortality
nouvelle foresterie new forestry
organisme nuisible pest
origine des graines seed source
parcelle plot
pare-feu firebreak
pâte pulp
pâte de bois wood pulp
pente sensible ou fragile sensitive slope
périmètre ou plaine d'inondation floodplain
pesticide pesticide
peuplement stand
peuplement pur pure forest

phénotype phenotype
plan directeur management plan
planification de l'utilisation des terres land use planning
plant à racines nues bareroot seedling
plant repiqué transplant
plantation planting
plantation forestière plantation, forest
polygone patch
pratique forestière forest practice
pratique ou régime sylvicole silvicultural system
préservation preservation
processus fluviatiles fluvial processes
productivité de l'écosystème ecosystem productivity
productivité du sol soil productivity
produits forestiers botaniques botanical forest products
profil des forêts forest profile
propriété forestière de production tree farm
provenance provenance
qualité de l'eau water quality
rase ou à blanc clearcut
ravageur pest
reboisement reforestation
rebut de sélection roguing
régénération artificielle artificial regeneration
régénération forestière forest regeneration, forest renewal
régénération naturelle natural regeneration
régime de perturbations naturelles natural disturbance regime
région sauvage wilderness
réserve écologique ecological reserve
résilience resilience
résineux softwood
ressources en eau water resources
ressources forestières forest resources
restauration restoration
révolution rotation
riparien riparian
riverain riparian
santé des forêts forest health
santé écologique ecological health
scalpage scalping, screefing
scarifiage par traînage drag scarification
scarification scarification
sédimentation sedimentation
semencier seed tree
semis seedling
semis en contenant plug seedling
semis en récipient container seedling
sentier de débardage skid trail
site site
site d'interprétation forestier interpretive forest site
sol minéral mineral soil
sous-bois undergrowth
sous-espèce subspecies
sous-étage understory
station site
stabilité de la pente slope stability
station météorologique automatique remote automatic weather station
strate forestière forest cover type
structure de peuplement stand structure
surface de la cime canopy cover
sylviculture silviculture
système canadien de l'indice forêt-météo Canadian forest fire weather index system
système de classification biogéoclimatique biogeoclimatic classification system
système sylvicole par coupe totale clearcutting silvicultural system
taille pruning
taillis coppice (coppicing)
terrain terrain
terrain boisé woodlot
terrain forestier forest land
territoires protégés protected areas
tolérance tolerance
topographie topography
traitements pour améliorer la santé des forêts forest health treatments
trancheuse à disque disc trencher
tronc bole
unité biogéoclimatique biogeoclimatic unit
unité de paysage landscape unit
végétation à feuillage persistant evergreen
végétation ou plante pionnière pioneer
verger à graines seed orchard
vérification au sol ground truthing
vérification de terrain ground truthing
vieille forêt old growth forest
vulgarisation extension services
zone biogéoclimatique biogeoclimatic zone
zone tampon buffer zone
zones écologiquement vulnérables environmentally sensitive areas

GÉOMORPHOLOGIE, GÉOLOGIE, GÉNIE CIVIL (GEOMORPHOLOGY, GEOLOGY, CIVIL ENGINEERING)

à grain fin fine grained
à grain grossier coarse-grained
à l'équilibre graded
accumulation deposition
adhésion adhesion
affaissement subsidence
affleurement outcrop
affouillement scour
affouillement glaciaire glacial scour
âge glaciaire ice age
alluvial alluvial
alluvion alluvium
altithermal hypsithermal
altitude elevation
analyse de la fabrique petrofabric analysis
analyse du terrain terrain analysis
anastomosé anastomosing
ancienne ligne de rivage emergent shoreline
angle critique critical angle
angle de chute angle of dip
angle de repos angle of repose (rest)
anguleux angular
anisotropie anisotropy
anomalie anomaly
anomalie magnétique magnetic anomaly
anthracite anthracite
anticlinal anticline
aplanissement applanation, degradation, gradation
aplanissement des interfluves lateral planation
argilite claystone
arrondi roundness
asphalte asphalt
assise base course
assise rocheuse bedrock
attribut attribute
authigène authigenic
aven sinkhole
bad land badland
banc arqué point bar
banc de sable sand bank
barkhane barchan
barrage dam
barre bar
bas fonds intertidaux tidal flat
basalte basalt
base base course
base de pente footslope
basse terrasse second bottom
basse terre lowland
bassin de sédimentation sediment basin
bassin hydrographique slope catchment area
bassin versant slope catchment area
berme berm
billonage bedding
biostratigraphie biostratigraphy
bitume bitumen
bitumen pitch
bitumineux bituminous
bloc block
bloc glaciaire glacial boulder
blocaille rubble
blocerratique glacial erratic
bourrelet glaciel lake rampart
bourrelet lacustre lake rampart
BP, avant le présent BP
bras mort meander scar
brèche breccia
brisure knickpoint
buttage bedding
butte butte
button hummock
calcaire calcitique calcitic limestone
calcification calcification
calcifier calcify
caldeira caldera
caldera caldera
calotte glaciaire icecap
canyon canyon
capacité de charge bearing capacity
capacité portante bearing capacity
carboné carbonaceous
carbonification coalification
carottage core drilling
carotte de sondage core
carte de surface areal map
carte géologique geological map
carte géomorphologique geomorphological map
carte physiographique morphographic map
carte planimétrique planimetric map
carte topographique topographic map
caténa catena
cendre de houille coal ash

cendre volcanique volcanic ash
cendres volantes fly ash
chaîne de montagnes mountain range
chaîne range
champ field
champ de dunes dune complex
champ géomagnétique geomagnetic field
champs de scories scoria land
charbon coal
charge solide sediment load
chargement loading
chaux dolomitique dolomitic lime
cheminée de fée earth pillar, hoodoo
chenal d'eau de fonte glacial drainage channel, meltwater channel
chronologique chronological
chronostratigraphie chronostratigraphy
ciment cement
cimentation cementation
cimenté cemented
classé graded
classification des formes de terrain landform classification
classification morphologique des versants slope morphological classification
claste clast
clastique clastic
clinomètre inclinometer
clivage cleavage
coefficient de classement sorting coefficient
coefficient de frottement coefficient of friction
colline hill, knoll
colluvial colluvial
colluvion colluvium
combustible fossile fossil fuel
compactibilité compactibility
complexe juxtaglaciaire kame complex, kame field
compression compression
compression biaxe biaxial compression
concavité concavity
concordant concordant, conformable
cône cone
cône de débris debris fan
cône de déjection debris cone, debris fan
conforme conformity
conglomérat conglomerate
consolidation consolidation
consolider consolidate
contact paralithique paralithic contact
continent continent
contraction contraction
contrainte stress
contrainte de cisaillement shear stress
contrainte effective effective stress
contraintes principales principal stresses
contrefort foothills
convexité convexity
coprolite coprolite
coprolithe coprolite
cordillère cordillera
cordon littoral barrier beach
corrélation stratigraphique stratigraphical correlation
côte coast, cote
côté sous le vent lee
couche indurée indurated layer
couche interstratifiée interbed
coulée coulee
coulée de boue earthflow, mudslide
coulée de débris debris flow
coulée de lave lava flow
coupe section
coupe transversale cross section
courbure dans le sens de la pente slope profile curvature
courbure de niveau slope plan curvature
craie chalk
cratère crater
crête crest
crête de plage beach ridge
creux hollow
creux de déflation deflation basin or hollow
crevasse crevasse
croûte crust
croûte continentale continental crust
cryologie cryology
cryomorphologie cryomorphology
cryoplanation cryoplanation
cryoplancton cryoplankton
cryosphère cryosphere
cuprifère cupriferous
curage scour
cut and fill cut and fill
cycle d'érosion cycle of erosion, geomorphic cycle
cycle d'érosion fluviale fluvial cycle of erosion
cycle géomorphologique geomorphic cycle
dallage de pierres boulder pavement
débit solide sediment discharge

débris debris
décapage scour
décharge fill
déclinaison magnétique declination (magnetic)
déflation deflation
déformation deformation, strain
déformation élastique elastic deformation
dégradation denudation
degré-jour degree-day
délitage spalling
delta delta
deltaïque deltaic
demoiselle coiffée earth pillar, hoodoo
densité specific gravity
densité critique critical density
densité relative specific gravity
dépôt deposit, drift
dépôt atmosphérique fallout
dépôt de plage beach deposit
dépôt de sol surcompacté over-consolidated soil deposit
dépôt glaciaire glacial drift
dépôt lacustre lacustrine deposit
dépôt laminique lamina (plural laminae)
dépôt placérien placer deposit
dépôt résiduel lag
dépôts éoliens wind-blown deposits
dépôts fluviatiles fluvial deposits
dépôts superficiels superficial deposits
dépression latérale humide back-swamp
dérive des continents continental drift
détritique detrital
déversoir de lac glaciaire glacial spillway
diaclase joint
diagénèse diagenesis
diagraphie log
diamicton unstratified drift
diatomite diatomite
digue dike, levee
direction strike
discontinuité discontinuity
discontinuité de Gutenberg Gutenberg discontinuity
discontinuité de Mohorovicic Mohorovicic discontinuity
discontinuité M ou de Mohorovicic M-discontinuity
discordance discordance, unconformity
discordance majeure nonconformity
discordance parallèle disconformity
dispersion dispersion
disposition attitude
dissection dissection
disséqué dissected
division lithostratigraphique lithostratigraphic unit
doline doline
dolomie dolomite
dolomie calcaire calcitic dolomite
dolomite dolomite
donnée datum
dorsale ridge
dos d'âne hogback, hog's-back ridge
drain en pierres blind drain
drumlin drumlin
drumlin à noyau rocheux rock drumlin
dune dune
dune active dune, mobile
dune de sable sand-dune
dune longitudinale longitudinal dune
dune parabolique parabolic dune
dune stabilisée dune, stabilized
dyke dike, dyke
eau de fonte meltwater
eau juvénile juvenile water; water, juvenile
éboulis debris fall, slough
échancrure de méandre de plaine alluviale flood-plain meander scar
échelle de dureté hardness scale
échelle de Mohs Mohs scale
échelle de Richter Richter scale
échelle granulométrique de Wentworth Wentworth scale
échelle granulométrique logarithmique des unités phi phi scale
écorce terrestre crust
écoulement laminaire laminar flow
élasticité elasticity
élasticité parfaite perfect elasticity
élastique elastic
élévation elevation
embrun salé cyclic salt
émoussé roundness
en bosses et creux hummocky
en échelons en echelon
en feuillet fissile bedding
endiguement embankment
énergie géothermique geothermal energy
énergie primaire primary energy
enrochement de protection riprap
enterrage bedding

éolien	eolian
épandage fluviatile	river wash
épandage fluvio-glaciaire	outwash
épaulement	shoulder
épicentre	epicenter
époque	epoch
époque glaciaire	glacial epoch
équifinalité	equifinality
équiplanation	equiplanation
ère	era
érosion côtière	coastal erosion
érosion de pente	slope wash
érosion géologique	geologic erosion
érosion glaciaire	glacial erosion
erratique	erratic, glacial boulder
escarpement	escarpment, scarp
esker	esker
esker en chapelet	esker, beaded
essai de chargement à la boîte de sable	sand-bearing method
essai de compressibilité	consolidation test
essai non destructif	nondestructive testing (NDT)
essai oedométrique	consolidation test
estuaire	estuary
estuarien	estuarine
étage	stage
état structural du sol	fabric, soil
eustasie	eustasy, eustatism
eustatique	eustatic
eustatisme	eustasy, eustatism
évaporite	evaporite
exogène	exogenetic
exposition	aspect
extraction par solution	solution mining
facette	facet
faciès	facies
faible secousse sismique	earth tremor
faille	fault
faille longitudinale	longitudinal fault
ferrallitisation	ferrallitization
ferrifère	ferriferous
filon	vein
filon de charbon	coal seam
filon intrusif	dike, dyke
filon-couche	sill
fissile	fissile
fissure	fissure
fluvial	fluvial
fluviatile	fluvial
fluvio-glaciaire	fluvioglacial, glaciofluvial
force de cisaillement	shear force
force de Coriolis	Coriolis force
formation	formation
formation de faille	faulting
formation des montagnes	mountain-building
forme	subgrade
forme anguleuse	angularity
forme de terrain	landform
forme relique ou héritée	relict landform
fossile	fossil
foudroyage	block-cut method
foyer sismique	seismic focus, seismic origin
fracture	fracture
fragilité	brittleness
fragmentation	clastation
front de colline	brow
galet de plage	shingle
gangue	gangue
gaz de houille	coal gas
gaz naturel	gas, natural; natural gas
gazéification	gasification
gazéification du charbon	coal gasification
gélifluxion	congelifluction, gelifluction
géliplanation	cryoplanation
gélisol	frozen ground, gelisol
géo	geo-
géocryologie	geocryology
géodésie	geodesy
géographie	geography
géographie physique	geography, physical
géologie	geology
géologie physique	physical geology
géologie structurale	structural geology
géomagnétisme	geomagnetism
géomorphologie	geomorphology
géomorphologique	geomorphic
géophysique	geophysics
géosciences	earth science
géosphère	geosphere
géosynclinal	geosyncline
géothermique	geothermal
gisement de pétrole	oil pool
gisement stratoïde	blanket deposit
glaciaire	glacial
glaciation	glaciation
glacié	glaciated
glacier	glacier
glacier continental	continental glacier
glacio-eustasie	glacio-eustasy
glaciogéologie	glacial geology
glacio-lacustre	glaciolacustrine

glaciologie glaciology
glissement slide, slip, slump
glissement basal slip
glissement de terrain earth slide; landslide, landslip
glissement rotationnel rotational landslide
gradué graded
grain grain
granuloclassement sorting
granulométrie gradation
granulométrie des matériaux sediment grade sizes
gravier gravel
gravimètre gravimeter
gravité gravity
grès sandstone
grès fin siltstone
hamada hamada
haut plateau tableland
haute terre upland
hiatus hiatus
hogback hogback, hog's-back ridge
homophane massive
houille coal
houille brune brown coal
houillification coalification
hypogénique hypogene
hypsithermal hypsithermal
hypsométrie hypsography
inclinaison pitch, plunge
inclinaison de la pente slope gradient (angle)
indice de retrait shrinkage index
indice portant californien California bearing ratio
induré indurated
inlandsis continental glacier, icesheet
inselberg inselberg
intercalé intercalated
interfluve divide, interfluve
interglaciaire interglacial
interstade interstade
interstratifié interstratified
intraglaciaire englacial
intrusive intrusive
isobathe isobath
isostasie isostasy
isotropie isotropy
isthme isthmus
joint joint
kame kame
kaolinite kaolinite, kaolin
karst karst
karst couvert karst, covered
kérobitume kerogen
kérogène kerogen
kettle kettle, kettle hole
lac de barrage barrier lake
lac de barrage glaciaire ice-dammed lake
lac de kettle kettle lake
lac en croissant oxbow lake
lac glaciaire glacial lake
lacune hiatus
lacustre lacustrine
lait de glacier glacier milk
lambeau ou résidu d'érosion relic geomorph
lame mince section
laminaire laminar
lentille sableuse sand lens
levée bar, levee
levée de chenal channel bar
ligne de partage des eaux divide
ligne de rivage perchée emergent shoreline
lignite brown coal, lignite
limite d'adhésivité sticky limit
limite d'élasticité elastic limit
limite de retrait shrinkage limit
linéament lineament
linéation lineation
liquéfaction spontanée liquefaction
lit bed
litage banding, bedding
litage fissile fissile bedding
litage métamorphique banding
lité bedded
lithification lithification, lithifaction
litho litho-
lithologie lithology
lithosphère lithosphere
lithostratigraphie lithostratigraphy
lits entrecroisés cross-bedding
loess loess
loi d'Ohm Ohm's law
macrofossile macrofossil
mafique mafic
magma magma
magnétomètre magnetometer
mamelon knob
manteau mantle
marin marine
matériau de granulométrie médiocre poorly graded soil
matériau de transport glaciaire glacial drift

matériau résiduel residual material
matrice matrix
maximum glaciaire glacial maximum
méandre meander
mécanique des sols soil mechanics
méridien meridian
méridien origine prime meridian
méridien principal principal meridian
mesa mesa
métamorphisme metamorphism
métamorphisme régional regional metamorphism
météorite meteorite
méthode du chemin critique critical-path analysis
microfossile microfossil
microrelief microrelief
minimum glaciaire glacial minimum
mire level rod
modelé de glace morte dead-ice features
module subgrade modulus
monadnock monadnock
montagne mountain
moraine à kettles kettle moraine
moraine bosselée hummocky moraine, washboard moraine
moraine cannelée fluted moraine
moraine de fond ground moraine
moraine de kame kame moraine, moraine kame
moraine de poussée contorted drift, push moraine
moraine de retrait recessional moraine
moraine latérale lateral moraine
moraine médiane medial moraine, median moraine
moraine stadiaire stadial moraine
moraine terminale end moraine
moraine terminale ou frontale terminal moraine
morainique morainal
morphochronologie morphochronology
morphologie morphology
morphologie des versants slope morphology
morphométrie morphometry
mosaïque mosaic
mouvement en masse mass wasting
niveau stage
niveau de base d'érosion base level
niveau de référence datum level
niveau du sol grade
nivellement leveling
non consolidé unconsolidated sediment
noyau core
onde de compression compressional wave
onde de Love Q-wave
onde de pression pressure wave
onde primaire primary (p) wave
onde Q Q-wave
ondes sismiques seismic waves
oolite oolith
oolithe oolith
organisation du sol fabric, soil
orientation aspect
orifice de percolation blind inlet
orogenèse mountain-building, orogeny
orogénique orogenic
ouvala karst valley
paléomagnétisme paleomagnetism
paléorelief enfoui buried topography
pavage de pierres boulder pavement
paysage landscape
paysage en bosses et creux knob-and-kettle topography
paysage polycyclique polycyclic landscape
paysage polygénique polygenetic landscape
pédiment pediment
pédiplaine pediplain
pendage angle of dip, dip, inclination
pénéplaine peneplain
pénéplanation peneplanation
pénétromètre penetrometer
pente grade, slope
pente colluviale colluvial slope
pente de talus side slope
pente latérale side slope
pente rectiligne rectilinear slope
périglaciaire periglacial
période period
perré riprap
petit âge glaciaire Little Ice Age
pétrogenèse petrogenesis
pétrole petroleum
pétrologie petrology
phase phase
physiographie physiography
pied foot
pied ou base du talus toe slope position
piedmont foothills
piémont piedmont
pierre stone
pierrée blind drain, blinding material

pilier d'érosion pedestal
pisolite pisolith
pisolithe pisolith
placage veneer
plage beach
plage soulevée raised beach
plaine plain
plaine côtière coastal plain
plaine d'épandage apron
plaine d'épandage fluvio-glaciaire outwash plain, sandur
plaine d'épandage piquée pitted outwash
plaine de till till plain
plaine karstique karst plain
plaine lacustre lacustrine plain, lake plain
plan de clivage cleavage plane
plan horizontal de référence datum elevation
plans de séparation ou de diaclase joint planes
plaque plate
plasticité parfaite perfect plasticity
plastique plastic
plateau plateau
plate-forme bench
playa playa
pléniglaciaire glacial maximum
pli fold
plongement plunge
pôles géomagnétiques geomagnetic poles
pôles magnétiques magnetic poles
pression de contrainte preconsolidation pressure (or prestress)
pression de fusion in situ pressure melting point
processus épigénétique epigenetic (epigenic) process
profilage land leveling
proglaciaire proglacial
promontoire head
prospection géobotanique geobotanical prospecting
province province
province géographique geographic province
puits de forage borehole
pyroclastique pyroclastics
pyroschiste oil shale
rabattement drawdown
rapport de production de sédiments sediment delivery ratio
rapport isotopique du carbone carbon ratio
ravine draw
ravinement wash-out
récif organique organic reef
recul glaciaire glacial recession
régolithe regolith
rejet horizontal heave
rejet net slip
relevé géodésique geodetic surveying
relevé sismique seismic survey
relief relief
remanié reworked
remblai fill
remblayage de vallée valley fill
remodelage de terrain regrading
remplissage de chenaux channel-fill deposit
remplissage de crevasse crevasse filling
répartition range
replat bench
repos repose
reptation creep, hill creep, rock creep
résidu de déflation lag
résistance resistance, strength
résistance à l'écrasement crushing strength
résistance à la pénétration penetration resistance
résistance au cisaillement shear strength
résistance en traction tensile strength
résistivité resistivity
retombée radioactive fallout
retrait glacial recession
retrait de glacier glacier retreat
retrait glaciaire glacial retreat
retrait résiduel residual shrinkage
revers backslope
rhéologie rheology
risque géologique geologic hazard
roc bedrock
roc sous-jacent bedrock
roche rock
roche acide acidic rock
roche basique basic rock
roche calcaire calcareous rock
roche clastique clastic rock
roche couverture cap-rock
roche dure hard rock
roche encaissante country-rock
roche métamorphique metamorphic rock
roche plutonique plutonic rock
roche sédimentaire sedimentary rock
roche-magasin reservoir rock
rocher rock
roche-réservoir reservoir rock

roches ou minéraux secondaires secondary rocks or minerals
roche ultrabasique ultrabasic rocks
ruissellement wash, washoff
rupture failure
rupture de pente break of slope, knickpoint
sables bitumineux oil sand, tar sand
sandr sandur
sapement undercutting
schiste schist
schistosité schistosity
science de la terre earth science
scorie scoria
section transversale cross section
sédiment drift, sediment
sédiment clastique clastic sediment
sédiment de contact glaciaire ice-contact deposit
sédiment glaciaire glacial drift
sédiment organogène biogenic sediment
sédiment stratifié ou trié stratified drift (or sorted drift)
sédimentaire sedimentary
sédimentation deposition, sedimentation
sédimentologie sedimentology
sédiments reliques relic sediments
ségrégation segregation
séisme earthquake
séismologie seismology
semelle footing
seuil sill
sial sial
siliceux siliceous
silicification silicification
siltstone siltstone
sima sima
sismographe seismograph
sismologie seismology
socle basement
sol bien gradué well-graded soil
sol de formation subgrade
sol ferrallitique ferrallitic soil
sol plastique plastic soil
sommet brow, summit
soulèvement heave
source thermale hot spring
stabilisation du sol stabilization, soil
stabilité de la pente slope stability
stade stage
stade glaciaire stade
stadiaire stade
strate stratum (plural strata)
stratification stratification
stratification entrecroisée cross-stratification
stratification lenticulaire lens, lensing
stratifié stratified
stratigraphie stratigraphy
stratigraphie paléontologique biochronology
structure structure
structure massive massive
subaérien subaerial
subbitumineux subbituminous
subsidence subsidence
surface spécifique specific surface
suspension suspension
synclinal syncline
système de classification unifié des sols, classification U.S.C.S. Unified Soil Classification System
talus talus
talus d'éboulis scree
tectonique tectonic
tectonique des plaques plate tectonics
tellurique telluric
temps géologique geological time
tension tension
tephra tephra
terrain terrain
terrasse terrace
terrasse à large base broad-base terrace
terrasse alluviale alluvial terrace
terrasse de kame kame terrace
terrasse en canaux channel terrace
terrasse en gradins bench terrace, Zingg bench terrace
terrasse fluviatile river terrace
terrasse inférieure second bottom
terrasse juxtaglaciaire kame terrace
terrasse lacustre lake terrace
terrasse marine marine terrace
terrasse régularisée graded terrace
terre earth
terre de diatomées diatomaceous earth
terrestre terrestrial
thufur hummock
till till
till de fond basal till, lodgment till, lodgement till
till glaciaire glacial till
tillite tillite
topographie surface relief, topography

topographie en bosses et creux kame-and-kettle terrain, pit and mound topography
topographie superficielle de karst karst topography
toposéquence toposequence
traficabilité trafficability
trame de till till fabric
transport transport
transport en masse mass transport
transport glaciel ice rafting, rafting
transport par corps flottant rafting
tremblement de terre earthquake
triangulation triangulation
unité de relief landform classification
unité géomorphologique geomorphic surface
unité morphologique morphologic unit
vallée valley
vallée fluviale river valley
vallon draw
variable qualitative attribute
varve varve
wadden tidal flat
zone côtière coastal zone
zone intertidale intertidal zone
zone sismique seismic zone

HYDROLOGIE, LIMNOLOGIE, QUALITÉ DE L'EAU (HYDROLOGY, LIMNOLOGY, WATER QUALITY)

affluent tributary
algue épiphytique aufwuchs
alimentation recharge
aquatique aquatic
aquiclude aquiclude
aquifère non captive aquifer, unconfined
aquifuge aquifuge
auto-épuration self-purification
barrage dam
bassin basin
bassin d'alimentation catchment
bassin de décantation settling basin
bassin de retenue detention basin
bassin fermé closed basin
bassin hydrographique catchment, drainage basin
bassin versant catchment, drainage basin, river basin, watershed
benthique benthic
benthos benthos
berges streambanks
bilan hydrologique hydrologic budget
biomanipulation biomanipulation
bourbier slough
canal channel, flume
canal de crue ou de dérivation floodway
canal de jaugeage irrigation flume
canal sur appuis irrigation flume
canal surélevé irrigation flume
canalisation channelization
canalisation secondaire lateral
carbone organique dissous (COD) dissolved organic carbon (DOC)
charge load
charge d'eau d'élévation elevation head
charge en solution dissolved load
charge en suspension suspended load
charge organique organic load
charge totale d'un cours d'eau stream load
charriage bed load, bottom load
chenal channel
chloramine chloramine
cône de dépression cone of depression
couche encaissante confining layer
couche semi-perméable capacitive aquitard
cours d'eau watercourse, waterway
cours d'eau intermittent ephemeral stream
cours d'eau pérenne perennial stream
cours d'eau temporaire ou intermittent intermittent stream
cours supérieur d'une rivière headwater
crue flood
crue éclair flash flood
cuvette sink
cuvette de dégel thaw lake
cycle hydrologique hydrologic cycle, water cycle
DBO BOD
DBO_5 BOD_5
DCO COD
débit discharge
débit de base base flow
débit de pointe de crue flood peak
débit de sécurité safe yield
débit maximal peak discharge
débit spécifique specific yield

demande biochimique en oxygène (DBO) biochemical oxygen demand (BOD)
demande chimique en oxygène (DCO) chemical oxygen demand (COD)
demande d'oxygène des sédiments sediment oxygen demand (SOD)
demande en oxygène oxygen demand
demande théorique d'oxygène theoretical oxygen demand (TOD)
densité de drainage channel density, drainage density
dérive périodique periodic drift
dessèchement desiccation, soil
déversoir measuring weir, spillway, weir
diatomée diatom
divergence divergence
drainage drainage
drainage dendritique dendritic drainage
drainage interne du sol internal soil drainage
drainage radial radial drainage
drainage superficiel external drainage
dureté hardness
dureté carbonatée carbonate hardness
dureté permanente de l'eau permanent hardness of water
dystrophe dystrophic
eau artésienne artesian water
eau brute raw water
eau d'averse stormwater runoff
eau de surface surface water
eau douce fresh water
eau dure hard water
eau polluée stressed water
eau potable potable water
eau souterraine ground water
eau souterraine phreatic water
eau vadose vadose water
eaux réceptrices receiving waters
écoulement runoff
écoulement concentré concentrated flow
écoulement divergent interflow
écoulement en chenal channel flow
écoulement en nappe sheet flow
écoulement endoréique endoreic drainage
écoulement laminaire laminar flow
écoulement restitué return flow
écoulement retardé delayed runoff
écoulement souterrain ground water flow, underground runoff
écoulement turbulent turbulent flow
écoulement uniforme uniform flow
écoulement urbain urban runoff
écoulement varié nonuniform flow, varied flow
emmagasinement dans les dépressions depression storage
emmagasinement dans un cours d'eau channel storage
enrichissement enrichment
enrichissement de l'eau water enrichment
envasement silting
épilimnion epilimnion
épuisement d'oxygène oxygen depletion
équation hydrologique hydrologic equation
érosion du lit streambed erosion
étang pond
eutrophe eutrophic
eutrophisation cultural eutrophication, eutrophication
farine de roche rock flour
fleur d'eau bloom
fleur d'eau planctonique plankton bloom
formation aquifère aquifer
fréquence de crue flood frequency
gradient gradient
gradient hydralique hydraulic gradient
hauteur head
hauteur piézométrique pressure head
hauteur-vitesse velocity head
hydraulique hydraulics
hydrique hydric
hydrodynamique hydrodynamics
hydrogéologie hydrogeology
hydrogramme hydrograph
hydrogramme unitaire unit hydrograph
hydrologie hydrology
hydrologie forestière forest hydrology
hydrologique hydrologic
hydrosphère hydrosphere
hydrostatique hydrostatics
hypolimnion hypolimnion
hypoxie hypoxia
inféroflux underflow
interception interception
invasion d'eau salée salt-water intrusion
jauge gage or gauge
jaugeage d'un cours d'eau stream gaging
lac lake
lac amictique amictic lake
lac dimictique dimictic lake
lac dystrophe dystrophic lake
lac eutrophe eutrophic lake

lac holomictique holomictic lake
lac méromictique meromictic lake
lac monomictique monomictic lake
lac oligotrophe oligotrophic lake
lac polymictique polymictic lake
lacs orientés beaded drainage
lentique lentic
ligne de faîte drainage divide
ligne de partage des eaux drainage divide
ligne équipotentielle equi-potential line
limnétique limnetic, limnetic, limnic
limnologie limnology
limnophyte limnophyte
lit channel, river bed
lotique lotic
lutte contre l'inondation flood control
mare vaseuse slough
marécage slough
marée rouge red tide
marmite torrentielle pot-hole
matériau du lit bed material
matière organique dissoute (MOD) dissolved organic matter (DOM)
matière organique particulaire (MOP) particulate organic matter (POM)
matière organique particulaire ou grossière macroorganic matter
matière particulaire particulate matter
matière solide en solution dissolved solid
matières totales dissoutes total dissolved solids (TDS)
méroplancton meroplankton
mésotrophe mesotrophic
méthylation biologique biological methylation
microplancton microplankton
modèle hydrologique hydrologic model
mortalité estivale summer kill
mortalité hivernale winter kill
moulinet current meter
nannoplancton nannoplankton
nappe artésienne aquifer, confined
nappe captive confined ground water
nappe d'eau suspendue ou perchée perched water table
nappe phréatique ground water-table, phreatic water
nauplie nauplius
necton nekton
neuston neuston
niveau de l'eau souterraine ground water level
ouvrages de drainage drainage structures
oxygène dissous (OD) dissolved oxygen (DO)
paludien paludal
paralique paralic
passe à poissons fishway
périphyton periphyton
phycosphère phycosphere
phytoplancton phytoplankton
piézomètre piezometer
plaine alluviale flood plain
plancton plankton
plante aquatique aquatic plant
pluie nette rainfall excess
poisson d'eaux chaudes warmwater fish
poisson d'eaux froides coldwater fish
potamologie potamology
potamon potamon zone
pression hydraulique hydraulic pressure
pression hydrostatique head, hydrostatic pressure
printemps spring
profondeur critique critical depth
puits sink
puits artésien artesian well
pycnocline pycnocline
rabattement drawdown
rapide chute
rayon hydraulique hydraulic radius
région benthique benthic region
région d'alimentation intake area
région limnétique limnetic zone
région néritique neritic zone
région ou milieu pélagique pelagic zone
région ou zone littorale littoral zone
région sublittorale sublittoral zone
région supralittorale supralittoral zone
rendement yield
renversement overturn
renversement automnal fall overturn
renversement printannier spring overturn
réseau hydrographique river system
réservoir reservoir
réservoir aquifère ground water reservoir
ressources en eau water resources
rétention retention
rétention initiale initial storage
riparien riparian
rithron rithron zone

riverain riparian
rives streambanks
ruissellement overland flow, surface flow, surface runoff
ruissellement souterrain ground water runoff
sable mouvant quick sand
saumâtre brackish
seiche seiche
sessile sessile
seston seston
sortie d'eau outlet
source spring
source artésienne artesian spring
sous-écoulement underflow
stade stage
station de jaugeage gaging station
stratification density stratification, stratification
stratification thermique thermal stratification
suintement seepage
surexploitation overdraft
surface de saturation water table
surface piézométrique piezometric surface
suspension suspension
talweg thalweg
temps de décrue stream recession length
temps de réponse watershed lag
tête de réseau head
thermocline thermocline
tracé centripète de drainage centripetal drainage pattern
tracé de drainage drainage pattern
tracé de drainage dérangé deranged drainage pattern
transmissivité transmissivity
transparence transparency
tripton tripton
tronçon reach
turbidité turbidity
unité de turbidité Jackson (UTJ) Jackson turbidity unit (JTU)
vallée principale trunk valley
vitesse critique critical velocity
vitesse de l'eau souterraine groundwater velocity
vitesse limite de régime turbulent turbulent velocity
vitesse moyenne mean velocity
voie d'eau waterway
zone de dessablement de-silting area
zone de saturation saturated zone
zone euphotique euphotic zone
zone photique photic zone
zone profonde profundal zone
zone trophogène trophogenic region
zone vadose vadose zone
zooplancton zooplankton

IRRIGATION (IRRIGATION)

barbacane weep-holes
barrage de dérivation diversion dam
barrage de retenue des débris debris dam
barrage de retenue du sol detention dam
barrage en encoffrement crib dam
barrage en enrochement rock-fill dam
barrage en gahions sausage dam
barrage en terre earth dam
barrage submersible check dam
barrages de rigoles furrow dams
besoin des cultures en eau d'irrigation irrigation requirement, consumptive
besoin en eau d'irrigation water demand
besoin en lessivage leaching requirement
bilan de sels salt balance
canal à écoulement rapide sluice
canal de distribution head ditch
canal de drainage drain
chantepleure weep-holes
chute drop structure
déversoir spillway
déversoir vertical drop spillway
digue diversion dam
drain drain
drain en tuyaux tile drain
drainage contrôlé controlled drainage
efficacité d'irrigation irrigation efficiency, water application efficiency
entrave drag
estimation de la teneur en sodium sodium estimation
étang fermier farm pond
évacuateur de secours emergency spillway
évapotranspiration consumptive use
facteur de perte par lessivage leaching fraction
fertigation fertigation
fossé ditch

fossé de planche border ditch
fosse-réservoir dugout pond
galerie-taupe mole drain
inondation en contour contour flooding
irrigation irrigation
irrigation par bassin de retenue irrigation, check-basin
irrigation au goutte-à-goutte irrigation, drip
irrigation de complément supplemental irrigation
irrigation de surface surface irrigation
irrigation d'hiver irrigation, winter
irrigation fertilisante fertigation
irrigation par aspersion sprinkler irrigation
irrigation par bassin irrigation, basin
irrigation par calants ou à la planche irrigation, border-strip
irrigation par infiltration ou par billons irrigation, corrugation
irrigation par pivot central irrigation, center-pivot
irrigation par rigoles irrigation, furrow
irrigation par rigoles d'infiltration suivant les courbes de niveau irrigation, contour-furrow
irrigation par submersion irrigation, flood
irrigation par submersion non-contrôlée irrigation, wild flooding
irrigation souterraine subsurface irrigation
levée de planche border dike
pompe centrifuge centrifugal pump
poste de pompage pumping station
pourcentage de sodium soluble soluble-sodium percentage
résistance à l'écoulement drag
rigole de niveau contour ditch
terre irrigable irrigable land
tranchée de drainage open drain
tuyau de drainage drain tile
tuyau fermé closed drain
utilisation combinée des eaux souterraines et de surface conjunctive water use

MÉTÉROLOGIE, CLIMATOLOGIE, QUALITÉ DE L'AIR (METEOROLOGY, CLIMATOLOGY, AIR QUALITY)

abri de Stevenson Stevenson screen
abri météorologique Stevenson screen
absorption du rayonnement absorption of radiation
advection advection
aérosols aerosols
aérosols de sulfates ou sulfatés sulfate aerosols
albédo albedo
anémomètre anemometer
angle solaire solar angle
anthropogénique anthropogenic
aride arid
atmosphère atmosphere
bar bar
baromètre barometer
bilan radiatif net radiation
chaleur heat
changements climatiques climate change
chlorofluorocarbures chlorofluorocarbons and related compounds
classification climatique de Koppen Koppen's climatic classification
climat climate
climatologie climatology
coefficient d'émission emission factor
combustible fossile fossil fuel
condensation condensation
convection convection
convertisseur catalytique catalytic converter
couverture de neige snowpack
cryosphère cryosphere
cuvette sink
déboisement deforestation
dégel thawing
densité de la neige snow density
déposition humide wet deposition
dépôt deposition
dépôt atmosphérique fallout
dépôt sec dry deposition
désertification desertification
dioxyde de carbone carbon dioxide (CO_2)
dioxyde de soufre sulfur dioxide (SO_2)
dorsale ridge
durée de la pluie rainfall duration
écoulement turbulent turbulence, turbulent flow
effet de serre greenhouse effect
effet de serre amplifié enhanced greenhouse effect

effet fertilisant du dioxyde de carbone carbon dioxide fertilization
efficacité des précipitations precipitation efficiency (effectiveness)
El Niño El Niño
émission emission
énergie solaire solar energy
épurateur à sec dry scrubber
épurateur de gaz de combustion flue gas scrubber
équivalent en carbone carbon equivalent (CE)
équivalent en dioxyde de carbone carbon dioxide equivalent (CDE)
évaporation evaporation
évaporimètre evaporimeter
évapotranspiration evapotranspiration
évapotranspiration potentielle potential evapotranspiration
fluorocarbures fluorocarbons
forçage radiatif radiative forcing
fréquence de la pluie rainfall frequency
gaz à effet de serre greenhouse gases
gaz à l'état de traces trace gas
gaz de combustion flue gas
gel frost
gel atmosphérique air frost
gel au sol ground frost
géosphère geosphere
halocarbone halocarbon
hauteur d'eau rainfall
hauteur pluviométrique rainfall
hexafluorure de soufre sulfur hexafluoride (SF_6)
humide humid
humidité absolue humidity, absolute
humidité relative humidity, relative
hydrofluorocarbures hydrofluorocarbons (HFCs)
hydrométéorologie hydrometeorology
hydrosphère hydrosphere
hygromètre hygrometer
immobilisation ou stockage du carbone carbon sequestration
intensité pluviale rainfall intensity
interception de la pluie rainfall interception
interception des précipitations precipitation interception
inventaire des émissions emission inventory
ionosphère ionosphere
lysimètre lysimeter
masse volumique standard de l'air standard air density
mécanisme de forçage forcing mechanism
mécanisme de rétroaction feedback mechanism
megabarye bar
mésosphère mesosphere
météorologie meteorology
méthane methane (CH_4)
microclimat microclimate
microméteorologie micrometeorology
modèle climatique climate model
modèle de circulation générale general circulation model (GCM)
modélisation du climat climate modeling
neige snow
oxydes d'azote (NOx) nitrogen oxides (NOx)
ozone ozone
ozone troposphérique tropospheric ozone (O_3)
photolyse photolysis
pluie rain
pluie acide acid rain
pluviomètre rain-gauge
point de rosée dew point
pollution pollution
pollution de l'air air pollution
pollution intérieure des locaux indoor air pollution
potentiel de réchauffement planétaire global warming potential (GWP)
poussière dust
poussière de combustion flue dust
précipitateur électrostatique electrostatic precipitator
précipitation effective effective precipitation
précipitations precipitation
précurseur d'ozone troposphérique tropospheric ozone precursor
pression atmosphérique normale normal atmospheric pressure
pression barométrique barometric pressure
protoxyde d'azote (N_2O) nitrous oxide (N_2O)
psychromètre psychrometer
psychromètre à fronde whirling psychrometer
puits sink
puits de carbone carbon sinks
puits de forage borehole
pyranomètre, actinomètre actinometer
rapport de mélange mixing ratio
rapport de mélange de saturation saturation mixing ratio
rayonnement radiation

rayonnement diffus diffuse radiation
rayonnement infrarouge infrared radiation
rayonnement net net radiation
rayonnement réfléchi reflected radiation
rayonnement solaire solar radiation
rayonnement solaire direct incident insolation
réchauffement planétaire global warming
région climatique climatic region
retard climatique climate lag
retard de réponse lag
retombée radioactive fallout
rétroaction climatique climate feedback
rosée dew
saison de croissance growing season
saison sans gel frost-free period
sécheresse drought
semi-aride semi-arid
source diffuse ou étendue area source
stratosphère stratosphere
sub-humide sub-humid
système climatique climate system
taux d'émission emission rate
température temperature
température du thermomètre mouillé wet-bulb temperature
température du thermomètre sec dry-bulb temperature
température sous abri shade temperature
temps weather
temps de résidence residence time
thermomètre thermometer
troposphère troposphere
turbulence turbulence, turbulent flow
vapeur d'eau water vapor
vent wind
vitesse de dépôt deposition velocity

MICROBIOLOGIE (MICROBIOLOGY)

actinomycète actinomycetes
aérobie aerobic
agar agar
agent pathogène pathogen
akinète akinete
algologie algology
algue alga (plural algae)
algue calcaire calcareous algae
algue filamenteuse filamentous algae
amibe amoeba (plural amoebae)
ammonification ammonification
anaérobie anaerobic
anaérobie facultatif facultative anaerobe
antibiose antibiosis
antibiotique antibiotic
anticorps fluorescent fluorescent antibody
appareil de Golgi Golgi complex
arbuscule arbuscule
ascospores ascospores
aseptique aseptic
autoclave autoclave
autotrophe holophytic
axénique axenic
bacille bacillus
bactéricide bactericidal
bactérie bacterium (plural bacteria)
bactériostatique bacteriostatic
bactéries fécales fecal bacteria
bactéries ferrugineuses iron bacteria
bactéries fixatrices d'azote nitrogen-fixing bacteria
bactéries indicatrices indicator bacteria
bactéries sulfureuses sulfur bacteria
bactériophage bacteriophage
bactéroïde bacteroid, nodule bacteria
barophile barophile
baside basidium
basidiospore basidiospore
biochimie du sol biochemistry, soil
bioluminescence bioluminescence
biomasse microbienne microbial biomass
biorestauration bioremediation
bioséquence biosequence
boîte de Pétri culture dish, Petri dish
botulisme botulism
capacité de croissance innée innate capacity for increase (r_o)
capacité saprophyte saprophytic competence
capsule capsule
catabolisme catabolism
cellule de comptage counting chamber
cellule végétative vegetative cell
chaîne de transport d'électrons electron-transport chain
champignon fungus (plural fungi)
charge énergétique adenylate energy charge ratio (EC)
charge microbienne microbial load
chémostat chemostat

chimio-hétérotrophe chemoheterotroph
chimiolithotrophe chemolithotroph
chimio-organotrophe chemoorganotroph
chimiotactisme chemotaxis
chimiotaxie chcmotaxis
chimiotrophe chemoautotroph
chimiotrophie chemotrophy
chlamydospore chlamydospore
chromatide chromatid
chromatine chromatin
chromosomes homologues homologous chromosomes
cil cilium (plural cilia)
cilié ciliate
coenocyte coenocyte
colibacille coliform
colibacille fécal fecal coliform
coliforme coliform, fecal coliform
coliphage coliphage
colonie colony
colonisation colonization
colonne de Winogradsky Winogradsky column
coloration de Gram Gram stain
cométabolisme cometabolism
conidie conidium (plural conidia)
conjugaison conjugation
contaminer contaminate
coque coccus
croissance géométrique geometric growth
culture culture
culture discontinue batch culture
culture en batch batch culture
culture pure pure culture
décomposeurs reducers
décomposition aérobie aerobic decomposition
décompte direct direct count
demande biochimique en oxygène biochemical oxygen demand (BOD)
dénitrification biologique biological denitrification
dénitrification chimique chemodenitrification
diatomée diatom
diazotrophe diazotroph
ectomycorhize ectomycorrhiza
endogène endogenous
endomycorhize endomycorrhiza
endotherme cndotherm
endotoxine endotoxin
endotrophe endotrophic
ensemensement direct direct plating
entébactérie enteric bacteria
entérobactérie enteric bacteria
enzyme de restriction restriction enzyme
enzyme extracellulaire extracellular enzyme
enzymes abiontiques abiontic enzymes
épifluorescence epifluorescence
équation de Monod Monod equation
Escherichia *Escherichia*
eucaryote eukaryote, eukaryotic
exoenzyme exoenzyme
exogène exogenous
exotoxine exotoxin
exsudat exudate
façon culturale d'enrichissement enrichment culture
facultatif facultative
fermentation fermentation
ferrobactéries iron bacteria
feutre ou mat microbien microbial mat
filet de Hartig Hartig net
fission binaire binary fission
fixation non symbiotique non-symbiotic fixation
fixation symbiotique symbiotic fixation
flagelle flagellum (plural flagella)
flagellé flagellate
fluorescéine fluorescein
fumigation du sol fumigation, soil
gamétange gametangium
gamète gamete
génophore genophore
germicide germicide
Gram positive/Gram négative Gram negative/Gram positive
halophile halophile
helminthosporiose leaf spot
hétérocyste heterocyst
holoplancton holoplankton
homozygote homozygous
hormogonie hormogonium
hybridation hybridization
hyphe hypha (plural hyphae)
incubation incubation
infectieux infectious
infection infection
inhibition inhibition
inoculation inoculation
inoculer inoculate
inoculum inoculum

intracellulaire intracellular
k_m, constante de Michaelis-Menten k_m
lamina lamina (plural laminae)
levures yeasts
lithotrophe lithotroph
lithotrophie lithotrophy
lyse lysis
macrofaune du sol macrofauna, soil
mésofaune du sol mesofauna, soil
mésophile mesophile
méthanotrophe methanotroph
micro-aérophile microaerophile
microbe microbe
microbiologie microbiology
microbiologie du sol microbiology, soil
microbiote microbiota
microfaune du sol microfauna, soil
microflore microflora
microhabitat du sol microhabitat, soil
micro-organisme microorganism
micro-organismes indicateurs indicator microorganisms
microscopique microscopic
microsite du sol microsite, soil
milieu medium
milieu de culture culture media
mixotrophe mixotrophy
motile motile
mycélium mycelium (plural mycelia)
myco myco
mycologie mycology
mycorhize mycorrhiza
mycorhize ectotrophe ectotrophic mycorrhiza
mycorhize endotrophe endotrophic mycorrhiza
mycotoxine mycotoxin
myxamibe myxamoeba
natte microbienne microbial mat
nitrification autotrophe autotrophic nitrification
nodosité nodule
nodule nodule
nodule de sol nodule, soil
nodule racinaire root nodule
nombre le plus probable most probable number (MPN)
numération bactérienne sur plaque bacterial plate count
numération de colonies bactériennes colony count
numération directe direct count
numération sur plaque plate count
organisme aérobie aerobe
organisme anaérobie anaerobe
organisme coenocytique coenocytic organism
organisme facultatif facultative organism
organotrophe organotroph, organotrophic
organotrophie organotrophy
osmotrophe osmotroph
oxydation biologique microbial oxidation
pasteurisation pasteurization
pathogène pathogenic
période d'incubation incubation period
phase de croissance growth phase
phase de croissance exponentielle exponential growth phase
phase de croissance stationnaire stationary growth phase
phase de déclin death phase
phase de latence lag growth phase
phase logarithmique de croissance log growth phase
phénomène de Wood-Werkman Wood-Werkman phenomenon
phosphobactéries phosphobacteria
phosphorescence phosphorescence
photo-autotrophe photoautotroph
photolithotrophe photolithotroph
phototrophe phototrophic
phototropisme phototropism
plancton plankton
plasmodium plasmodium
population microbienne microbial population
porteur carrier
pourriture blanche white rot fungus
pourriture des racines root rot
procaryote prokaryote
procaryotique prokaryotic
prolifération algale algal bloom
propagule propagule
protiste protist
protoctiste protoctist
protoplaste protoplast
protozoaire protozoan (plural protozoa)
pseudomonade pseudomonad
pseudopode pseudopodium (plural pseudopodia)
psychrophile psychrophile
radappertisation radiation sterilization

réaction en chaîne de la polymérase polymerase chain reaction (PCR)
réaction hyperbolique hyperbolic reaction
réduction assimilatoire des nitrates assimilatory nitrate reduction
rhizobactéries favorisant la croissance des plantes plant growth promoting rhizobacteria (PGPR)
rhizobium rhizobium (plural rhizobia)
rhizocylindre rhizocylinder
rhizoplan rhizoplane
rhizosphère rhizosphere
Salmonella *Salmonella*
saprobe saprobic
saprobionte saprobe
saprophyte saprophyte
septé septate
sidérophore siderophore
souche strain
spore spore
stérile sterile
stérilisation sterilization
stérilisation partielle partial sterilization
streptocoque fécal fecal streptococcus
stricte obligate
stromatolithe stromatolite
substrat substrate
sulfobactéries sulfur bacteria
symbiose symbiosis
synergie synergism
taux de croissance intrinsèque intrinsic rate of increase (r_m)
taux géométrique d'augmentation geometric rate of increase
terre de diatomées diatomaceous earth
terreau de feuilles leaf mold
test de blocage à l'acétylène acetylene-block assay
thermophile thermophile
Thiobacillus *Thiobacillus*
transduction transduction
unicellulaire unicellular
vésicule vesicle
vésiculo-arbusculaire vesicular arbuscular
virus virus
xénobiotique xenobiotic
zoospore zoospore
zygospore zygospore
zymogène zymogenous

MINÉRALOGIE, GÉOCHIMIE, ALTÉRATION (MINERALOGY, GEOCHEMISTRY, WEATHERING)

abrasion abrasion
abrasion éolienne wind abrasion
action gel-dégel freeze-thaw action
allophane allophane
allotriomorphe anhedral
altération weathering
altération biologique organic weathering
altération chimique météorique chemical weathering
altération des roches rock weathering
altération différentielle ou sélective differential weathering
altération profonde deep weathering
altérite saprolite
amiante asbestos
amphibole amphibole
analyse minéralogique mineralogical analysis
andésite andesite
angle d'extinction extinction angle
apatite apatite
ardoise slate
argile clay
argile de fuller fuller's earth
argile fine fine clay
argile plastique fat clay
argile réfractaire fireclay
argileux argillaceous
argilite mud rock
argilite mudstone
asymétrique asymmetrical
attapulgite attapulgite
axe a a axis
axe b b axis
axe c c axis
bauxite bauxite
beidellite beidellite
bentonite bentonite
bentonite calcaire calcium bentonite
bentonite potassique K-bentonite
bimorphisme dimorphism
biogéochimie biogeochemistry

biotite biotite
biréfringence birefringence, double refraction
biréfringent birefringent
birnessite birnessite
bisiallitisation bisiallitization
boehmite boehmite
caillou éolisé ventifact
calcaire limestone
calcaire magnésien magnesian limestone
calcaire oolithique oolitic limestone
calcaire tuffacé tufa
calcite calcite
capillaire capillary
carbonatation carbonation
carbonate carbonate
céramique ceramic
champ des argiles clay domain
charge attribuable aux substitutions structural charge
charge électrique de la couche layer charge
charge permanente permanent charge
chert chert
chimie des cristaux crystal chemistry
chlorite chlorite
chlorure de sodium halite
corrasion corrasion
corrosion corrosion
couche layer
couche polyédrique (tétraédrique ou octaédrique) sheet of polyhedra
cristal crystal
cristal de roche rock crystal
cristallin crystalline
cristallisation crystallization
cristallographie crystallography
cryoclastie congelifraction, frost weathering
cryptocristallin cryptocrystalline
cycle biogéochimique biogeochemical cycle
cycle géochimique geochemical cycle
décomposition decomposition
délitage spalling
dépôt mineral deposit
dépôt résiduel lag
désagrégation disintegration
désagrégation en boules ou sphéroïdale spheroidal weathering
désagrégation granulaire granular disintegration
désagrégation mécanique mechanical weathering
désagrégation physique physical weathering
désilicification desilication
desquamation exfoliation
diffraction diffraction
diffraction des rayons-x x-ray diffraction
dioctaédrique dioctahedral
distance interfeuillet diffraction spacing
dolomite magnesian limestone
domaine argileux clay domain
double couche diffuse diffuse double layer
énergie de réseau ou réticulaire lattice energy
équation de Bragg Bragg's equation
espace interfeuillet *d*-spacing
euèdre euhedral
exfoliation exfoliation
extinction extinction
feldspath feldspar
feldspath potassique K-feldspar
feldspathique feldspathic
felsique felsic
fer iron
fer des marais bog iron
ferrihydrite ferrihydrite
ferrolyse ferrolysis
ferromagnésien ferromagnesian
ferrugineux ferruginous
fixation d'ammonium ammonium fixation
fixation du potassium potassium fixation
fixation ou intégration de l'ammonium en position interfeuillet ammonium entrapment
flottation ou séparation gravimétrique heavy media separation
fluoroapatite fluorapatite
fragmentation comminution
front d'altération weathering, basal surface of
gélifraction congelifraction, frost weathering
gélivation congelifraction, frost weathering
géochimie geochemistry
gibbsite gibbsite
gisement mineral deposit
glauconie glauconite
glauconite glauconite
gneiss gneiss
goethite goethite
goniomètre goniometer
granite granite
granulaire granular
grauwacke graywacke

groupe des smectites montmorillonite-saponite group
groupe du kaolin kaolin group
gypse gypsum
halite halitc
halloysite halloysite
hématite hematite
héritage inheritance
holocristallin holocrystalline
hornblende horneblende
hyalin hyaline
hydraté hydrous
hydromica hydromica
hydroxyapatite hydroxylapatite
ignée igneous
illite illite
imogolite imogolite
indice de réfraction index of refraction
intergrade intergrade minerals
iridescence iridescence
irisation iridescence
isomorphisme isomorphism
jarosite jarosite
kaolin kaolin
kaolinite china clay, kaolinite
kaolinitique kaolinitic
lamellaire lamellar
lamelle lamella
latolisation latolization
lépidocrocite lepidocrocite
liaison attribuable à un bris de lien en périphérie de minéral broken-edge bond
limonite limonite
liqueur lourde heavy liquid
lithiophorite lithiophorite
lithomarge pallid zone
lumière polarisée analysée crossed nicols
maghémite maghemite
magnésite magnesite
magnétite magnetite
marbre marble
matériau résiduel residuum
mercure quicksilver
météorisation weathering
météorisation chimique chemical weathering
mica mica
mica hydraté hydrous mica
micacé micaceous
micaschiste mica schist
micelle micelle
microcline microcline
microcristallin microcrystalline
microgranite microgranite
microscope électronique electron microscope
microscope électronique à balayage scanning electron microscope (SEM)
microscopie à lumière polarisée polarized light microscopy
microsonde électronique electron microprobe
minéral mineral
minéral amorphe amorphous mineral
minéral argileux clay mineral
minéral argileux interstratifié interstratified clay mineral
minéral du sol soil mineral
minéral interstatifié mixed-layer mineral
minéral léger light mineral
minéral lourd heavy mineral
minéral primaire primary mineral
minéral secondaire secondary mineral
minéral silicaté silicate
minéralogie mineralogy
minéralogie des argiles clay mineralogy
minéralogie des sols soil mineralogy
montmorillonite montmorillonite
morphologie des cristaux crystal morphology
muscovite muscovite
muscovite hydratée ou illite hydromuscovite
néoformation neoformation
néogenèse neoformation
nésosilicate nesosilicate
nicols croisés crossed nicols
nodules rocheux rock nodules
non-silicaté non-silicate
nontronite nontronite
octaèdre octahedron
oligoclase oligoclase
olivine olivine
onyx onyx
oolite oolite
oolite oolith
oolithe oolith
opacité opacity
opaque opaque
orthoclase orthoclase
orthogneiss orthogneiss
oxyde oxide mineral
oxyde de fer et d'aluminium sesquioxide
oxydes de fer iron oxides

oxydes de manganèse manganese oxides
oxydes de fer libres free iron oxides
oxydes libres free oxides
palygorskite palygorskite
paragneiss paragneiss
patine patina
patron de diffraction diffraction pattern
patron de diffraction des rayons-x x-ray diffraction pattern
pegmatite pegmatite
phénocristal phenocryst
phyllade shale
phyllite phyllite
phyllosilicate phyllosilicate; layer silicate mineral
phytolithe phytoliths
pierre à savon soapstone
pierre éolisée ventifact
pierre ponce pumice
pitchblende pitchblende
plagioclase plagioclase
plan atomique plane of atoms
polymère d'hydoxyaluminium en position inter-feuillet hydroxy-aluminum interlayers
porphyre porphyry
porphyrique porphyritic
porphyroblastique porphyroblastic
position inter-feuillet interlayer
pseudomorphe pseudomorph
pyrite iron pyrite, pyrite
pyrophosphate pyrophosphate
pyrophyllite pyrophyllite
pyroxène pyroxene
quartz quartz
rapport silice-alumine silica-alumina ratio
rapport silice-sesquioxyde silica-sesquioxide ratio
réfractaire refractory
réfractomètre d'Abbe Abbe refractometer
réseau lattice
réseau cristallin crystal lattice
résidu de déflation lag
rhyolite rhyolite
roche rock
roche cristalline crystalline rock
roche ferrugineuse ironstone
roche ignée igneous rock
roche mafique intermediate rock
roche résistante resistant rock
rocher rock
saprolite saprolite
saprolithe saprolite
schiste schist
schiste micacé mica schist
schiste quartzique quartzitic schist
sépiolite sepiolite
série lyotropique lyotropic series
serpentine serpentine
sesquioxyde sesquioxide
shale shale
silicate silicate
silicate d'aluminium aluminosilicate
silice silica
silice cristalline silica, crystalline
siliceux siliceous
smectite smectite
solide cristallin crystalline solid
solution de Clérici Clerici solution
stalactite stalactite
stalagmite stalagmite
stéatite soapstone
structure des cristaux crystal structure
structure en château de cartes cardhouse structure
structure maillée, réticulée ou en treillis lattice structure
subédrique subhedral
substitution isomorphe isomorphous substitution
suite réactionnelle continue continuous reaction series
surface à charge constante constant-charge surface
surface à potentiel variable constant-potential surface
sylvite sylvite
tactoïde tactoid
talc talc
tectosilicate tectosilicate
terre à briques Adobe soil
terre de fer ironstone
tétraèdre tetrahedron
thermoclastie thermal fracture
thermoluminescence thermoluminescence
todorokite todorokite
trioctaédrique trioctahedral
tuf tuff
unité de base unit cell
ventifact ventifact
vermiculite vermiculite

vermiculite chloritisée hydroxy-interlayered vermiculite
zéolite zeolite
zone pâle pallid zone

MODÉLISATION, STATISTIQUES (MODELING, STATISTICS)

abscisse abscissa
ajustement d'une courbe curve fitting
algorithme algorithm
amortissement damping
analyse à plusieurs variables multivariate analysis
analyse de Fourier Fourier analysis
analyse de groupement cluster analysis
analyse de variance analysis of variance, ANOVA
analyse de variance à deux critères de classification two way analysis of variance
analyse de variance à un critère de classification one-way analysis of variance
analyse des séries chronologiques time series analysis
analyse en classification automatique cluster analysis
analyse factorielle factor analysis
analyse multivariable multivariate analysis
analyse systèmique systems analysis
aplatissement, kurtosis kurtosis
approche heuristique heuristic approach
approche systémique systems approach
arbre de décision decision tree
association positive positive association
asymétrie skewness
attribut attribute
autocorrélation autocorrelation
autorégulation self regulation
bande de fréquences frequency band
base de sondage frame, sampling frame
biais bias
biométrie biometrics
biostatistique biostatistics
blocs aléatoires randomized blocks
boîte noire black box
bruit noise
causalité causality
centile percentile
chaîne de Markov Markov chain
changement temporel temporal change
classement classification, ranking
classification classification
coefficient de confiance confidence coefficient
coefficient de corrélation coefficient of correlation (r)
coefficient de corrélation de rang de Spearman Spearman rank correlation
coefficient de corrélation multiple coefficient of multiple correlation (R)
coefficient de détermination coefficient of determination (r^2)
coefficient de détermination multiple coefficient of multiple determination (R^2)
coefficient de variation coefficient of variation
constante constant
corrélation correlation
corrélation du moment des produits product-moment correlation
corrélation multiple multiple correlation
corrélation négative negative correlation
corrélation partielle partial correlation
corrélation statistique correlation, statistical
courbe de fréquence frequency curve
courbe de Gauss Gaussian curve
courbe des valeurs cumulées mass diagram
courbe gaussienne Gaussian curve
courbe logistique logistic curve
covariance covariance
critère de décision decision rule
cybernétique cybernetics
déduction deduction
degrés de liberté degrees of freedom
déviation standard deviation, standard; standard deviation
diagramme de dispersion dispersion diagram, scattergram
diagramme ternaire ternary diagram
dispositif ou plan expérimental experimental design
distribution distribution
distribution bimodale bimodal distribution

distribution cumulée ou cumulative cumulative distribution
distribution de fréquence frequency distribution
distribution de Poisson Poisson distribution
distribution de probabilité probability distribution
distribution log-normale log-normal distribution
distribution normale normal distribution
distribution ponctuelle point distribution
distribution statistique frequency distribution
distribution unimodale unimodal distribution
écart moyen mean deviation
écart type deviation, standard, standard deviation
écart type géométrique geometric standard deviation
échantillon sample
échantillon aléatoire random sample
échantillon aléatoire stratifié stratified random sample
échantillon au jugé judgment sample
échantillon choisi à dessein judgment sample
échantillon double double sample
échantillon instantané grab sample
échantillon probabiliste probability sample
échantillon systématique systematic sample
échantillonnage sampling
échantillonnage hiérarchique nested sampling
effet effect
empirique empirical
entrées et sorties inputs and outputs
équilibre equilibrium
équilibre indifférent poised equilibrium
équilibre instable unstable equilibrium
équilibre stable stable equilibrium
équilibre statique static equilibrium
erreur error
erreur aléatoire random error
erreur bêta beta error
erreur de première espèce alpha error
erreur de type I Type I error; alpha error
erreur de type II Type II error; beta error
erreur systématique bias, systematic error
erreur type standard error
essai en champ field trial
établissement d'une surface de tendance trend surface analysis
état d'équilibre permanent steady state equilibrium
étendue range
exactitude accuracy
expérience en tiroirs nested experiment
extrapolation extrapolation
facteur factor
faux négatif false negative
faux positif false positive
feedback feedback
feedback négatif negative feedback
fonction à paliers step function
fonction discriminante discriminant function
fonction non linéaire non-linear function
fréquence frequency
géostatistiques spatial statistics
graphique graph
groupe contrôle control group
groupe témoin control group
hiérarchie hierarchy
histogramme histogram
hypothèse hypothesis
hypothèse nulle null hypothesis
induction induction
interaction interaction
interpolation interpolation
intervalle interval
intervalle de confiance confidence interval
itération iteration
krigeage kriging
limites de confiance confidence limits
log log
loi law
loi de l'inverse des carrés inverse square law
mécaniste mechanistic
médiane median
mesure de la tendance centrale measure of central tendency
mesure des moments moment measure
méthode de Monte-Carlo Monte Carlo method
méthode des moindres carrés least squares method
méthode scientifique scientific method
mode mode
modèle model
modèle analytique analytical model
modèle conceptuel conceptual model
modèle d'analyse de décision decision analysis model
modèle d'analyse spatiale spatial model

modèle de cheminement aléatoire random-walk model
modèle de la boîte box model
modèle de potentiel potential model
modèle de processus stochastique stochastic-process model
modèle de régulation des procédés process-response model
modèle de simulation simulation model
modèle de simulation par ordinateur computer simulation model
modèle déterministe deterministic model
modèle iconique iconic model
modèle statistique statistical model
modélisation model building, modeling
morphostase morphostasis
moyenne average, mean
moyenne arithmétique arithmetic mean
moyenne géométrique geometric mean
moyenne pondérée weighted average
nomogramme nomograph
normale normal
normalisation transformation
ordination ordination
ordonnée ordinate
origine origin
parabole parabola
paradigme paradigm
paramètre parameter
parcelle d'échantillon sample plot
percentile percentile
placette d'échantillonnage sample plot
plage range
plan d'expérience à blocs complets complete block design
plan d'expérience à blocs incomplets incomplete block design
plan d'experience complètement aléatoire completely randomized design
plan d'expérience factoriel factorial experiment
pourcentage d'erreur percent error
précision precision
prédictif predictive
prélèvements sériés serial samples
prescriptif prescriptive
probabilité probability
probabilité de position positional probability
processus de Markov Markov process
processus déterministe deterministic process
processus irréversible irreversible process
processus réversible reversible process
processus stochastique stochastic process
programmation linéaire linear programming
progression géométrique geometric series
projet pilote pilot project
puissance d'un test power of test
quartile quartile
quasi-équilibre quasi-equilibrium
randomisation randomization
réaction ou réponse secondaire secondary response
régime permanent steady state
région décisionnelle decision region
régression regression
régression curvilinéaire non-linear regression
régression non linéaire non-linear regression
régulateur regulator
répétition replicate, replication
réponse response
réponse de régulation regulatory response
réseau network
rétroaction feedback
rétroaction négative negative feedback
rétroaction positive positive feedback
saturation loading
semi-variance semi-variance
semivariogramme semivariogram
servomécanisme servomechanism
seuil sill, threshold
signification statistique significance, statistical
simulation simulation
statistiques statistics
statistiques spatiales spatial statistics
stochastique stochastic
surface de tendance trend surface
surface polynomiale trend surface
système system
système à écoulement continu continuous-flow system
système à événement naturel natural event system
système à rétrocontrôle feedback control system
système à surface plane plane surface system
système asservi feedback control system
système de contrôle control system
système de régulation des procédés process-response system

système de type boîte blanche white box system
système de type boîte grise gray box system
système de type boîte noire black box system
système en cascade cascading system
système ouvert open system
système physique physical system
temps de réponse response time
tendance centrale central tendency
test du khi-carré chi-square test
test *F* *F*-test
test statistique non paramétrique non-parametric statistical test
test t t-test
tests paramétriques parametric statistical tests
théorie theory
théorie de l'information information theory
théorie des systèmes systems theory
théorie du système général general systems theory
transformation transformation
unité d'échantillonnage sampling unit
unité expérimentale experimental unit
valeur dominante mode
valeurs propres eigenvalues
variabilité spatiale spatial variability, soil
variable variable, variate
variable continue continuous variable
variable dépendante dependent variable
variable discrète ou discontinue discrete variable
variable endogène endogenous variable
variable exogène exogenous variable
variable indépendante independent variable
variable ordinale ou ordonnée ordinal variable
variable qualitative attribute, categorical variable, nominal variable, qualitative variable
variable quantitative quantitative variable
variables d'état state variables
variables de vitesse rate variables
variance variance
variogramme variogram
vecteur vector

TOURBE, MILIEUX TOURBEUX (PEAT, PEATLANDS)

à buttes mounded
arbuste shrub
aspect des milieux humides wetland form
atternissement terrestrialization
bassin hydrographique drainage basin
bassin versant drainage basin
bassin versant ou hydrographique watershed
boue noire black mud
bourbe dystrophe dy
bourbe eutrophe gyttja
bourbier mire, slough
butte tourbeuse peat mound
butte hummock
carr carr
chenal d'écoulement végétalisé water track
classe de milieu humide wetland class
comblement filling in, terrestrialization
condition trophique trophic status
coprogène coprogenic
couche fibrique fibric layer
couche limnique limno layer
couche lithique lithic layer
couche terrique terric layer
dépôts cumuliques cumulose deposits
dopplerite dopplerite
dystrophe dystrophic
eau souterraine groundwater
eau tellurique telluric water
échelle d'humification de von Post von Post humification scale
fen arboré forested fen or treed fen
fen effondré collapse scar bog, collapse scar fen
fibre fiber
fibre frottée rubbed fiber
fibre non frottée unrubbed fiber
fibreux fibrous
fibrique fibric (peat)
fibrisol Fibrisol
Fibrist Fibrist
fourré marécageux carr
gaz des marais marsh gas
herbe herb
hydromorphe hydromorphic
hydrophyte hydrophyte

indice de diversité diversity index
limite du pergélisol permafrost table
limnique limnic
marais marsh
mare vaseuse slough
marécage slough
marécage swamp
massif hummock
matériau organique non différencié undifferentiated organic material
matériaux de sol organique organic soil materials
matériaux fibriques fibric materials
mésique mesic
mésophyte mesophyte
mésotrophe mesotrophic
milieu marécageux swamp
milieu ou terre humide wetland
milieu tourbeux mire, peatland
minérotrophe minerotrophic
mousse de tourbe moss peat, peat moss
muskeg muskeg
nappe phréatique water table
oligotrophe oligotrophic
ombrogène ombrogenous
ombrotrophe ombrotrophic
palse palsa
paludification paludification
pergélisol permafrost
plateau tourbeux peat plateau
plée muskeg
radeau flottant carpet
rhéophile rheophilous
rhéotrophe rheotrophic
saturé d'eau waterlogged
savane muskeg
sédiments terriques terrigenous sediments
silvique silvic
sol organique organic soil
soligène soligenous
sols de fen fen soils
sphagnique sphagnic
Sphagnum *Sphagnum*
stratigraphie de tourbière peat stratigraphy
système canadien de classification des milieux humides Canadian wetland classification system
tapis carpet
terre coprogène coprogenous earth, sedimentary peat
terre de diatomées diatomaceous earth
terre tourbeuse muck
terrique terric
tourbe peat
tourbe allochtone allochthonous peat
tourbe amorphe amorphous peat
tourbe autochtone autochtonous peat
tourbe d'herbacées sedge-reed peat
tourbe de Carex carex peat, sedge peat
tourbe de mousse brune brown moss peat
tourbe de roseaux reed peat, telmatic peat
tourbe de sphaigne bog peat, sphagnum peat
tourbe forestière forest peat
tourbe humique humic peat
tourbe ligneuse woody peat
tourbe marécageuse swamp peat
tourbe minérotrophe fen peat
tourbière mire, peat bog, peatland
tourbière de transition transition peatland
tourbière effondrée collapse scar bog, collapse scar fen
tourbière minérotrophe fen
tourbière ombrotrophe bog
type de milieu humide wetland type
zone non saturée unsaturated zone
zone vadose unsaturated zone

PHYSIQUE DU SOL, EAU DU SOL (PHYSICS, SOIL WATER)

action capillaire capillary action
aération du sol soil aeration
aérer aerate
air du sol soil air
aire brute bulk area
aire superficielle surface area
albédo albedo
anaérobie anaerobic
analyse granulométrique grain-size analysis
analyse granulométrique par tamisage sieve analysis
analyse mécanique grain-size analysis, mechanical analysis
analyse mécanique ou granulométrique particle-size analysis
analyse thermique thermal analysis

angle critique critical angle
argile limoneuse silty clay
argile lourde heavy clay
argile sableuse sandy clay
argile sensible quick clay
atténuation du rayonnement gamma gamma-ray attenuation
bilan hydrique water balance
calibre particle size
canal de jaugeage irrigation flume
canal sur appuis irrigation flume
canal surélevé irrigation flume
capacité au champ field capacity
capacité calorifique heat capacity
capacité de rétention en eau water holding capacity
capacité différentielle de rétention de l'eau differential water capacity
capacité spécifique de rétention en eau specific water capacity
capillarité capillarity
carotte de sol soil core
cavitation cavitation
cercle de Mohr Mohr circle
chaleur d'immersion heat of immersion
chaleur latente latent heat
chaleur rayonnante radiant heat
charge hydraulique hydraulic head
cisaillement shear, shearing
climat du sol solclime
coefficient d'extensibilité linéaire (indice COLE) coefficient of linear extensibility (COLE)
coefficient de dispersion hydrodynamique hydrodynamic dispersion coefficient
coefficient de retrait shrinkage coefficient, soil
cohésion apparente apparent cohesion
colloïde colloid
colmatage ou obturation de la surface surface sealing
compactage compaction
compactibilité compactibility
compaction compaction
compaction relative relative compaction
compressibilité compressibility
compression compression
condensation condensation
conductance conductance
conduction thermique conduction, thermal or heat
conductivité électrique d'un extrait saturé extract electrical conductivity, ECe
conductivité hydraulique hydraulic conductivity
conductivité hydraulique non-saturée ou en milieu non-saturé unsaturated hydraulic conductivity
conductivité hydraulique saturée ou à saturation saturated hydraulic conductivity
conductivité thermique thermal conductivity
cône étalon standard cone
consistance consistence
contrainte de cisaillement shear stress
contrainte effective effective stress
convection convection
courbe cumulative (granulométrique) summation curve (of particle sizes)
courbe d'étalonnage discharge curve
courbe de désorption moisture release curve
courbe de désorption en eau water release curve
courbe de rétention moisture-retention curve
courbe de rétention caractéristique soil water characteristic
courbe de rétention en eau moisture retention curve, water retention curve
courbe des débits discharge curve
courbe granulométrique grading curve
cours d'eau temporaire ou intermittent intermittent stream
crible screen
croûte crust
cycle hydrologique hydrologic cycle
débit flow velocity
déflation deflation
défloculer deflocculate
déformation strain
degré de facilité de travail du sol tillability
densité specific gravity
densité apparente apparent density, bulk density, soil
densité apparente sèche dry bulk density
densité de flux flux density
densité de vapeur vapor density
densité du flux thermique heat flux density
densité particulaire grain density, particle density
densité relative specific gravity
densité spécifique apparente apparent specific gravity

déplacement de fluides miscibles miscible displacement
désagrégation ou altération physique physical weathering
détachement detachment
diamètre équivalent equivalent diameter
diffusivité diffusivity
diffusivité de l'eau du sol soil water diffusivity
dispersibilité dispersivity
dispersion dispersion
dispersion hydrodynamique hydrodynamic dispersion
distribution granulométrie particle-size distribution
distribution volumétrique des pores pore-size distribution
drain de surface surface drain
drainage drainage
drainage entravé impeded drainage
drainage souterrain subsurface drainage
dynamomètre dynamometer
eau absorbée absorbed water
eau capillaire capillary water
eau disponible available moisture, available water (capacity)
eau du sol soil water
eau gravitaire gravitational water
eau hygroscopique hygroscopic water
eau intersticielle interstitial water, water, interstitial
eau libre free water, gravitational water
eau liée bound water
eau non disponible unavailable water
eau pelliculaire film water
écoulement concentré concentrated flow
écoulement de vapeur vapor flow
écoulement non-saturé ou en milieu non-saturé unsaturated flow
écoulement par gravité gravity flow
écoulement préférentiel bypass flow, macropore flow, preferential flow
écoulement saturé ou à saturation saturated flow
efficacité d'utilisation de l'eau water use efficiency
émulsion emulsion
energie cinétique kinetic energy
enveloppe de Mohr Mohr envelope
équation de l'érosion éolienne wind erosion equation
équation universelle de perte de sol universal soil loss equation (USLE)
érodabilité ou érodibilité erodibility
érodable erodible
érosion des rigoles d'irrigation furrow erosion
érosion en nappe, inter-rigoles sheet erosion
érosion en rigoles, en ruisselets ou en filets rill erosion
espace poral pore space
essai de cisaillement direct direct shear test
étendue des teneurs en eau non-limitantes non-limiting water range
étendue du comportement plastique plasticity range
facteur de ralentissement retardation factor
fentes de retrait irrégulières skew planes
ferme firm
fissure ou fente de retrait desication crack
floculation flocculation
flux flux
flux de saltation saltation flux
fondant flux
fraction fine fines
fractions du sol soil separates
frange capillaire capillary fringe
friable friable
friction interne internal friction
gel alvéolaire frost, honeycomb
glace de remplissage des pores pore ice
glace interstitielle pore ice
gorgé ou saturé d'eau waterlogged
gradient hydraulique hydraulic gradient
grains du squelette skeleton grains
granulométrie grain-size distribution
hauteur head
humidité antérieure antecedent moisture
humidité du sol soil moisture
humidité volumétrique volume wetness
hydrogéologie groundwater hydrology
hydrologie hydrology
hydrologie agricole agrohydrology
hydrologie des eaux souterraines groundwater hydrology
hydromètre hydrometer
hydrométrie hydrometry
hydrophile hydrophilic
hydrophobe hydrophobic
hydrophobicité hydrophobicity
hygroscopique hygroscopic
hystérèse hysteresis

hystérèse de gonflement swelling hysteresis
impédance impedance
imperméabilité impermeability
imperméable impervious
indice d'érosivité par la pluie rainfall erosivity index
indice de compressibilité compressibility index
indice de plasticité plasticity index
indice de résistance à la pénétration de la pointe cone index
indice de retrait shrinkage index
indice des vides void ratio
indice des vides critique critical void ratio
infiltration infiltration
infiltration cumulée ou cumulative cumulative infiltration
infiltromètre infiltrometer
infrarouge infrared (IR)
interception des précipitations precipitation interception
interstice capillaire capillary interstice
isodyne isodyne
limite de liquidité liquid limit
limite de plasticité plastic limit
limite de plasticité inférieure lower plastic limit
limite de retrait shrinkage limit
limite supérieure de plasticité upper plastic limit
limites d'Atterberg Atterberg limits
limon silt
liquéfaction spontanée liquefaction
loam limoneux silt loam
loam limono-argileux silty clay loam
loam sableux sandy loam
loam sableux très fin very fine sandy loam
loam sablo-argileux sandy clay loam
loi de Darcy Darcy's law
loi de Stokes Stokes' law
lysimètre lysimeter
lysimètre à succion suction lysimeter
macropore macropore
macroporosité macroporosity
manomètre manometer
masse cryoconsolidée frost, concrete
masse sèche dry mass
masse sèche à l'air air-dry mass
masse volumique density
masse volumique apparente apparent density, bulk density, soil
masse volumique apparente sèche dry bulk density
masse volumique critique critical density
mécanique des sols soil mechanics
membrane de tensiomètre pressure membrane
mouillabilité wettability
mouvement de masse mass movement
nappe phréatique water table
paillage des rigoles d'irrigation furrow mulching
particules élémentaires primary particles
particules primaires primary particles
ped ped
pédoclimat solclime
pénétrabilité penetrability
pénétromètre à cône cone penetrometer
pénétromètre à cône à friction latérale friction cone penetrometer
pénétromètre à cône tombant fall cone penetrometer
pénétromètre statique static penetrometer
percolation percolation (of soil water)
percolation profonde deep percolation
perméabilité permeability
perméabilité intrinsèque intrinsic permeability
perméamètre permeameter
perte de sol tolérable soil loss tolerance
physique du sol soil physics
pierre stone
pluviolessivage throughfall
poids spécifique apparent bulk specific gravity
point d'entrée d'air air entry value
point d'adhésivité sticky point
point de flétrissement wilting point
point de flétrissement permanent permanent wilting point
pore pore
pores du sol soil pores
porosité porosity
porosité d'air air porosity, air-filled porosity
potentiel de retrait et de gonflement shrink-swell potential
potentiel d'eau libre gravitational potential
potentiel électrocinétique electrokinetic (zeta) potential
potentiel gravitaire gravitational potential
potentiel hydraulique hydraulic potential
potentiel hydrique du sol soil water potential

potentiel hydrique total total potential (of soil water)
potentiel matriciel matric potential
potentiel osmotique osmotic potential
potentiel pneumatique gas pressure potential
potentiel zêta electrokinetic (zeta) potential, zeta potential
précipitation effective effective precipitation
pression atmosphérique atmospheric pressure
pression critique critical pressure
pression de leau water pressure
pression de l'eau porale pore water pressure
pression hydrostatique head
pression osmotique osmotic pressure
pression totale de l'eau du sol total pressure (of soil water)
profondeur équivalente equivalent depth
propriétés physiques des sols physical properties of soils
propriétés thermiques thermal properties
puits thermique heat sink
rapport d'enrichissement érosif erosion enrichment ratio (ER)
rayon équivalent equivalent radius
redistribution redistribution
réflectance reflectance
réflectométrie à dimension temporelle time-domain reflectometry
remontée capillaire capillary rise
résistance à l'écrasement crushing strength
résistance à la compression compressive strength
résistance à la pénétration penetration resistance
résistance au cisaillement shear strength
résistance du sol soil strength
résistance en traction tensile strength
rétention en eau water retention
retrait isotrope isotropic shrinkage
retrait résiduel residual shrinkage
retrait structural shrinkage, soil structural
retrait unidirectionel unidimensional shrinkage
retrait unitaire unitary shrinkage
retrait vertical vertical shrinkage
rugosité efficace root mean square roughness (RMS)
rugosité moyenne quadratique root mean square roughness (RMS)
ruissellement sur les troncs stemflow
sable sand
sable fin fine sand
sable très fin very fine sand
sable très grossier very coarse sand
sableux sandy
saturation saturation
saturation anthrique anthric saturation
saturer saturate
séché à l'air air dry
semelle de labour tillage pan
sensibilité sensitivity
sol compact ou tenace tight soil
sol hydrique hydric soil
sol hydromorphe hydric soil
sols anisotropes anisotropic soils
sonde à neutrons neutron probe
sonde à rayons gamma gamma probe
sonde pédologique soil probe
sorptivité sorptivity
sphéricité sphericity
succion suction
succion d'eau water suction
succion du sol soil suction
surface de cisaillement shear plane
surface de saturation water table
surface de saturation suspendue perched water table
surface spécifique specific surface
suspension suspension
tactoïde tactoid
tassement compaction
taux d'infiltration infiltration flux (or rate), intake rate
taux d'infiltration limite basic intake rate
taux de diffusion de l'oxygène oxygen diffusion rate (ODR)
taux de saturation degree of saturation
taux de saturation en eau water-filled pore space
taux maximal d'infiltration infiltrability
TDR TDR
température du sol soil temperature
teneur en air air porosity, air-filled porosity
teneur en eau de l'extrait de saturation saturation content
teneur en eau du sol soil wetness, water content, soil
teneur en eau gravimétrique ou massique mass wetness
teneur en eau volumétrique volumetric water content

tensiomètre tensiometer
tension de l'eau du sol, potentiel hydrique moisture tension (or suction)
terre fine fine earth
test de cisaillement triaxial triaxial shear test
texture texture
texture fine fine texture
texture modérément fine moderately fine texture
texture modérément grossière moderately coarse texture
texture moyenne medium texture
thermistance thermistor
thermocouple thermocouple
thixotropie thixotropy
tortuosité tortuosity
triangle de texture textural triangle
tube d'accès access tube
vapeur vapor
vide void
vides d'entassement packing voids
viscosité cinématique kinematic viscosity
viscosité dynamique dynamic viscosity
visqueux viscous
vitesse d'écoulement de l'eau water flow velocity
vitesse d'écoulement de l'eau porale pore water velocity
vitesse de sédimentation ou de décantation settling velocity
vitesse d'écoulement flow velocity
vitesse érosive erosive velocity
vitesse terminale de sédimentation terminal settling velocity
volume brut bulk volume
zone capillaire capillary zone
zone non-saturée unsaturated zone
zone vadose vadose zone

RESTAURATION ET REMISE EN VALEUR DES TERRES, MESURES CORRECTIVES (RECLAMATION, REMEDIATION)

abattage hydraulique hydraulic mining
berme berm
bonification ou restauration des terres reclamation
boue mud
boue de forage drilling mud
boue de mines mine wash
boue de mines fine slickens
carrière quarry
carrière d'emprunt borrow pit
cendre ash
chemin d'exploitation haul road
configuration approximative d'origine approximate original contour
couvert végétal ground cover
déblai de forage drill cuttings
déblai-remblai cut and fill
déblais spoil
déblais acides acid spoil
déblais rocheux bedrock spoil
déblais toxiques toxic spoil
décapage suivant les courbes de niveau contour stripping
déchets miniers souterrains underground development waste
dragage hydraulique hydraulic dredging
drain de décharge relief drain
drain de pied toe drain
drain d'interception interceptor drain
drainage de mine mine drainage
drainage minier acide acid mine drainage
eau de traitement process water
éléments fins fines
encapsulation encapsulation
ensemencement hydraulique hydroseeding
ensemencement naturel natural seeding
enveloppe de gravier gravel envelope
espèce indigène native species
espèce introduite non-native species
étang de retenue holding pond
excavation en tranchées trenching
exploitation à ciel ouvert open-cast mining, strip mining, surface mining
exploitation minière de surface area mining
exploitation souterraine underground mining
exploitation suivant les courbes de niveau contour mining
extraction ou exploitation minière mining
filtre de gravier gravel filter
fluide de forage drilling fluid
fosse pit
fossé de crête interception channel
fossé d'interception interception channel
gabion gabion

gradin bench
lutte contre l'érosion des berges stream bank erosion control
méthode standardisée de manipulation du sol standard soil handling procedure
mine mine
mine à ciel ouvert exploitée suivant les courbes de niveau contour surface mining
mine ou exploitation à ciel ouvert open-pit mine
minerai ore
mise en valeur ou restauration des terres land reclamation
morts-terrains overburden
mur en aile wing wall
ouvrage de stabilisation de pente grade stabilization structure
paillassonnage en branches brush matting
particules fines fines
pied toe
plante naturalisée naturalized plant
possibilité d'utilisation équivalente des terres equivalent land capability
prairie naturelle native prairie
profil reconstitué reconstructed profile
prospection prospecting
puits pit
recouvrement capping
remblai fill, backfill, gob
remblai détritique spoil bank
remblaiement filling
remblayer backfill
remise en valeur rehabilitation
remise en végétation revegetation
remise en végétation naturelle natural revegetation
résidus tailings
résidus fins fine tailings (fine tails, sludge)
résidus liquides de forage drilling waste fluid
résidus miniers mining by-products
résidus solides de forage drilling waste solid
restauration restoration
revégétation revegetation
revêtement revetment
saumure artificielle artificial brine
sol minier minesoil
solidification solidification
sous-sol supérieur upper subsoil
stabilisation des berges stream bank stabilization
stérile overburden
surexploitation du sol overstripping
technique d'élimination des résidus de forage mix, bury and cover (drilling wastes)
terrain exploité mined land
terrain perturbé disturbed land
terrain d'épandage de pétrole oil wasteland
terrains orphelins orphan lands
terre de recouvrement cover soil
terril mine dump
terril spoil bank
toile filtrante filter cloth
tranchée pour mur écran core trench
végétalisation revegetation
zone perturbée disturbed area

PÉDOGENESE, PÉDOLOGIE, CLASSIFICATION DES SOLS (SOIL GENESIS, PEDOLOGY, AND CLASSIFICATION)

A chernozémique chernozemic A
abrupt steep
accumulation cumulization
agroclimat agroclimate
aire de corrélation des sols soil correlation area
aire de ressource agroécologique agroecological resource area
Albolls Albolls
alcalinisation alkalization
Alfisols Alfisols
alios iron pan
allitisation allitization
ameublissement loosening
anaérobie (sol) anaerobic (soil)
Andepts Andepts
Andisols Andisols
Andosol Andosol
apédal apedal
approche écologique ecological approach
Aqualfs Aqualfs
Aquands Aquands
Aquents Aquents
Aquerts Aquerts

aquique aquic
Aquods Aquods
Aquolls Aquolls
Aquox Aquox
Aquults Aquults
ardoisier slaty
Arents Arents
Argids Argids
argile clay
argile (1:1) clay (1:1)
argile (2:1) clay (2:1)
argile dispersée, peptisée ou collosol dispersive clay
argileux clayey
argilipédoturbation argillipedoturbation
argilique argillic
aridique aridic
Aridisols Aridisols
B podzolique podzolic B
B solonetzique solonetzic B
bas fonds intertidaux tidal flat
bioséquence biosequence
biostasie biostasis
biséquums bisequa
bloc boulder
bloc rocheux boulder
Boralfs Boralfs
Borolls Borolls
bourbier slough
brunification brunification
brunisol dystrique Dystric Brunisol
brunisol eutrique Eutric Brunisol
brunisol mélanique Melanic Brunisol
brunisol sombrique Sombric Brunisol
brunisolique Brunisolic
butte de chablis cradle knoll
butte gazonnée earth hummock
butte-témoin midden
button hummock
caillou cobble
caillou roulé cobblestone
caillouteux cobbly
calcaire calcareous
calcane calcan
calcids calcids
calcification calcification
calcrète calcrete
caliche caliche
cambids cambids
cannelé channeled (eroded)
carapace hardpan
carte numérique map, electronic
catégorie category
caténa catena
chaîne de sols catena
chambres chambers
chenal de fusion glaciaire coulee
chernozem brun Brown Chernozem
chernozem brun foncé Dark Brown Chernozem
chernozem gris foncé Dark Gray Chernozem
chernozem noir Black Chernozem
chernozémique Chernozemic
cherteux cherty
chroma chroma
chronoséquence chronosequence
classe de possibilités des sols capability class, soil
classe de sols class, soil
classes calcaires calcareous classes
classes de drainage du sol soil drainage classes
classification des sols soil classification
classification écologique du territoire ecological land classification
claypan claypan
climoséquence climosequence
code de couleurs Munsell Munsell color system
colluvial colluvial
colluvion colluvium
complexe de sols complex, soil, soil complex
composantes significatives (d'une unité cartographique) co-dominant
concrétion calcaire lime concretion
conglomérat conglomerate
conglomérat à ciment ferrugineux ferricrete
contact lithique lithic contact
corrélation correlation
côtelé patterned (ribbed)
couche active active layer
couche compactée ou indurée traffic pan
couche H H layer
couche holorganique forest floor
couche humique humic layer
couche hydrique hydric layer
couche indurée indurated layer, pan
couche lithique lithic layer
couche mésique mesic layer
couche ou horizon d'impédance impeding horizon
couche R R layer

couche W W layer
couleur color
coupe section
coupe témoin d'un sol control section, soil
couverture blanket
couverture morte forest floor
cryosol organique Organic Crysol
cryosol statique Static Cryosol
cryosol turbique Turbic Cryosol
crotovina crotovina
croûte désertique desert crust
croûte pédologique concrétionné duricrust
croûte siliceuse silcrete
Cryands Cryands
Cryerts Cryerts
Cryids Cryids
Cryods Cryods
cryopédologie cryopedology
cryosolique Cryosolic
cryoturbation cryoturbation
dalle flagstone
de chevauchement glaciaire ice-thrust
déalcalinisation dealkalization
décalcification decalcification
délimitation cartographique des sols soil map delineation
déposition deposition
dépôt de pente slopewash
dépôt organique organic deposit
dépôt résiduel lag
désalinisation desalinization
désilicification desilication
dessication desiccation
dorsal ridged
drainage restreint impeded drainage
durcissement hardening
Durids Durids
durinodes durinodes
duripan duripan
durique duric
écodistrict ecodistrict
écoélément ecoelement
écoprovince ecoprovince
écorégion ecoregion
écosection ecosection
écosite ecosite
ecozone ecozone
efflorescence efflorescence
éluviation eluviation, lessivage
en creux et bosses hummocky
en dalles flaggy
en feuillet shaly
en pente sloping
en plaquettes channery
en rigoles rilled
en terrace terraced
enduit coating
enrobement clay film (skin), coating
Entisols Entisols
épipédon epipedon
épipédon anthropique anthropic epipedon
épipédon de plaggen plaggen epipedon
épipédon folistique folistic epipedon
épipédon histique histic epipedon
épipédon mélanique melanic epipedon
épipédon mollique mollic epipedon
épipédon ochrique ochric epipedon
épipédon umbrique umbric epipedon
équivalent de carbonate de calcium calcium carbonate equivalent
érosion de pente slopewash
érosion de surface erosion, surficial
escarpement rough broken
étage tier
étage inférieur bottom tier
étage intermédiaire middle tier
étage supérieur surface tier
euique euic
éventail fan
facteurs de formation des sols soil formation factors
famille de sols family, soil, soil family
fente de gel frost crack
fente de remplissage de glace ice wedge
fente de remplissage de sable sand wedge
Ferrods Ferrods
fersiallitisation fersiallitization
film argileux clay film (skin)
fissure de gel frost crack
fissure ou fente de dessication desiccation crack
Fluvents Fluvents
fluvial fluvial
fluviatile fluvial
fluvio-éolien fluvioeolian
fluvio-glaciaire glaciofluvial
fluvio-lacustre fluviolacustrine
folique folic
folisol Folisol
Folists Folists
forme de terrain landform, surface form
fractions granulométriques separates, soil

fragipan fragipan
fragique fragic
fragments grossiers coarse fragments
galet cobble, cobblestone
gamme hue
gel à structure alvéolaire honeycomb frost
Gélisols Gelisols
genèse des sols soil genesis
genèse ou formation des sols genesis, soil
géographie des sols geography, soil, soil geography
gilgai gilgai
glaciaire glacial
glacio-lacustre glaciolacustrine
glacis fan
gleyification gleyzation, gleysation
gleysol Gleysol
gleysol humique Humic Gleysol
gleysol luvique Luvic Gleysol
gleysolique Gleysolic
gradient gradient
grand groupe great group
graveleux gravelly
groupe de sols non différenciés undifferentiated
Gypsids Gypsids
haute-terre upland
Hémists Hemists
héritage inheritance
hiérarchique hierarchical
Histels Histels
Histosols Histosols
horizon A A horizon
horizon AB AB horizon
horizon AC AC horizon
horizon agrique agric horizon
horizon albique albic horizon
horizon argilique argillic horizon
horizon B B horizon
horizon BE BE horizon
horizon C C horizon
horizon calcique calcic horizon
horizon cambique cambic horizon
horizon diagnostique diagnostic horizon
horizon du sol horizon, soil
horizon E E horizon
horizon éluvial eluvial horizon
horizon F F horizon
horizon glossique glossic horizon
horizon gypsique gypsic horizon
horizon H H horizon
horizon illuvial illuvial horizon
horizon induré hardpan
horizon kandique kandic horizon
horizon L L horizon
horizon LFH LFH horizon
horizon natrique natric horizon
horizon O O horizon
horizon oxique oxic horizon
horizon pédologique horizon, soil
horizon pétrocalcique petrocalcic horizon
horizon pétrogypsique petrogypsic horizon
horizon salique salic horizon
horizon sombrique sombric horizon
horizon spodique spodic horizon
horizon vertique vertic horizon
horizonation stratification
horizontal level
humisol Humisol
Humods Humods
Humox Humox
Humults Humults
humus intermédiaire duff mull
humus peu décomposé duff
illuviation illuviation
Inceptisols Inceptisols
inclinaison gradient
incliné inclined
individu-sol soil individual
inventaire ou cartographie des systèmes paysager land system inventory
krotovina krotovina
lac lake
lacustre lacustrine
lame mince section
latérisation laterization
latérite laterite
latosol latosol
latosolisation latosolization
légende legend
lessivage leaching, lessivage
leucinisation leucinization
limite du pergélisol permafrost table
lithoséquence lithosequence
litiérage littering
litière forest floor, litter
lixiviation leaching
loam loam
loam argileux clay loam
loameux loamy
luminosité value, color
luvisol brun gris Gray Brown Luvisol

luvisol gris Gray Luvisol
luvisolique Luvisolic
marbrures mottles, mottling
marbrures réticulées reticulate mottling
mare vaseuse slough
marécage slough
marin marine
marmorisation mottling
marne marl
masse cryoconsolidée concrete frost
matériau gélique gelic material
matériau hémique hemic material
matériau parental ou originel parent material, genetic material
matériau saprique sapric material
matte duff
maturation ripening
mélanisation melanization
mésisol Mesisol
métadonnées d'inventaire pédologique soil inventory meta data
milieu humide wetland
miroirs de faille slickensides
modelé surface expression
modèle de paysage landscape model
modèle de pédopaysage soil landscape model
modèle pédologique soil model
Mollisols Mollisols
monolithe monolith, soi
moraine moraine
moraine cannelée fluted moraine
moraine de chevauchement ice-thrust moraine
morphologie du sol morphology, soil
mouchetures mottles
moutonné hummocky
mudstone mudstone
néoformation neoformation
non-sol nonsoil
Ochrepts Ochrepts
ondulé undulating
ordre de sols order, soil, soil order
organique Organic
Orthels Orthels
Orthents Orthents
Orthids Orthids
orthique Orthic
Orthods Orthods
Orthox Orthox
ortstein ortstein
Oxisols Oxisols
paléosol paleosol
paléosol enfoui paleosol, buried
paléosol exhumé exhumed paleosol; paleosol exhumed
paludication paludization
pan d'origine naturelle pan, genetic
paralithique paralithic
pavé désertique desert pavement
paysage landscape
ped ped
pédalité pedality
pédicité pedality
pédiment pediment
pédogenèse soil genesis, pedogenesis
pédogénétique pedogenic
pédologie pedology
pédon pedon
pédoturbation pedoturbation
pellicule argileuse clay film (skin)
pergélique pergelic
pergélisol permafrost
pergélisol continu continuous permafrost
pergélisol sec dry permafrost
période sans gel frost free period
Perox Perox
perudique perudic
pierres stones
pierreux stony
pierrosité stoniness
pingo pingo
placique placic
Plaggepts Plaggepts
plaine plain
plat level
plateau plateau
plinthite plinthite
podzol ferro-fumique Ferro-Humic Podzol
podzol humique Humic Podzol
podzol humo-ferrique Humo-Ferric Podzol
podzolique Podzolic
podzolisation podzolization
poli désertique desert polish
polyédrique blocky
polygone de fente de gel ice-wedge polygon
polypédon polypedon
possibilité capability
profil de sol profile, soil
profil modal, typique ou représentatif modal profile
profil témoin control section, soil
profondeur maximum du gel frost line

Psamments Psamments
raide steep
ravin rough broken
raviné channeled (eroded), rilled
ravineau rill
ravinement gullying
reg desert pavement
région de ressource agroécologique agroecological resource region
régosol Regosol
regosol humique Humic Regosol
régosolique Regosolic
Rendolls Rendolls
rendzine rendzina
reptation de sol soil creep
reptation des sols causée par le gel frost creep
résidu de déflation lag
revêtement coating
rigole rill
roche tendre softrock
roche-mère parent rock
rubéfaction rubifaction
sable sand
sable fin loameux loamy fine sand
sable grossier coarse sand
sable grossier loameux loamy coarse sand
sable loameux loamy sand
sable très fin loameux loamy very fine sand
saison de croissance growing season
Salids Salids
Saprists Saprists
saprolithe saprolite
saturation chroma
schisteux shaly
semelle de labour plow pan
séquum sequum
série de sols series, soil; soil series
seuil du gel frost line
silcrète silcrete
sol à autogranulation ou à autofoisonnement self-mulching soil
sol ABC ABC soil
sol AC AC soil
sol alpin alpine soil
sol anthropique anthropic soil, artificial soil body
sol associé associate, soil
sol azonal azonal soil
sol calcaire calcareous soil
sol cryogénique cryogenic soil
sol de toundra tundra soil
sol de transition intergrade, soil; transitional soil
sol enfoui buried soil
sol fersiallique fersiallic soil
sol fossile buried soil, fossil soil
sol gleyifié gleyed soil
sol halomorphe halomorphic soil
sol hydrique hydric soil
sol hydrogénique hydrogenic soil
sol hydromorphe hydric soil, hydromorphic soil
sol hydrophobe hydrophobic soil
sol intergrade intergrade, soil
sol intrazonal intrazonal soil
sol lourd heavy soil
sol minéral inorganic soil, mineral soil
sol polygénique polygenic (polygenetic) soil
sol polygonal polygonal ground
sol résiduel residual soil
sol sensible au tassement ou à l'affaissement collapsible soil
sol tourbeux peat soil
sol vierge virgin soil
sol zonal zonal soil
solifluxion solifluction
solod Solod
solonetz Solonetz
solonetz solodisé Solodized Solonetz
solonetz vertique Vertic Solonetz
solonetzique Solonetzic
sols forestiers forest soils
sols phytomorphes phytomorphic soils
sols thermogéniques thermogenic soils
solum solum (plural sola)
soulèvement par le gel frost heave
sous-groupe de sols subgroup, soil
sous-ordre de sols suborder, soil
sous-sol subsoil
Spodosols Spodosols
strié patterned (ribbed)
structure colomnaire columnar structure
structure massive massive
subhorizontal level
substratum substratum
surfaces de glissement slickensides
système américain de classification des sols, Soil Taxonomy Soil Classification, U.S. Soil Taxonomy
système canadien de classification des sols Soil Classification, Canadian System

système canadien de classification écologique du territoire ecological land classification system, Canadian
système de classification des sols FAO-UNESCO, Carte mondiale des sols Soil Classification, FAO/UNESCO soil units
système paysager land system
taches mottles
talud talud
tarière auger, soil
teinte hue
terre humide wetland
terre noire muck soil
terre organique muck soil
terre tourbeuse muck soil
tertre midden
tertre avec taillis coppice mound
texture fine fine texture
texture grossière coarse texture
texture moyenne medium texture
thermoséquence thermosequence
thufur earth hummock, hummock
tonalité hue
toposéquence toposequence
Torrands Torrands
Torrerts Torrerts
torrique torric
Torrox Torrox
tourbe peat
translocation translocation
tronqué truncated
Tropepts Tropepts
Turbels Turbels
type de sol type, soil
Udalfs Udalfs
Udands Udands
Uderts Uderts
udique udic
Udolls Udolls
Udults Udults
Ultisols Ultisols
Umbrepts Umbrepts
unité cartographique des sols soil map unit
unité composée compound unit
unité indifférenciée undifferentiated
Ustalfs Ustalfs
Ustands Ustands
Usterts Usterts
ustique ustic
Ustolls Ustolls
Ustox Ustox
Ustults Ustults
vacuoles reliées chambers
vallée valley
vallonné rolling
variante de sol variant, soil
ventre de boeuf frost boil
vertisol Vertisol
vertisol humique Humic Vertisol
vertisolique Vertisolic
Vertisols Vertisols
vieillissement ripening
Vitrands Vitrands
wadden tidal flat
Xéralfs Xeralfs
Xérands Xerands
Xérerts Xererts
xérique xeric
Xérolls Xerolls
Xérults Xerults
zonalité zonation
zonalité horizontale latitude zonation
zonalité verticale vertical zonation
zone d'alimentation en eau recharge area
zone de latitude latitude zonation
zone de ressource agroécologique agroecological resource zone
zone de sols zone, soil
zone marbrée, tachetée, ou marmorisée mottled zone

CHIMIE MINÉRALE/ORGANIQUE DU SOL, SALINITÉ (SOIL INORGANIC/ORGANIC CHEMISTRY, SALINITY)

à base de poids sec dry weight basis
acide acid
acide acidic
acide fulvique fulvic acid
acide gras fatty acid
acide humique humic acid
acide organique organic acid
acidité d'échange acidity, exchangeable
acidité du sol soil acidity
acidité échangeable par un sel acidity, salt-replaceable
acidité résiduelle acidity, residual

acidité totale acidity, total
activité ionique ion activity
adsorption adsorption
alcali alkali
alcalin alkaline
alcalinisation alkalization
alcalinité alkalinity
alcalinité du sol alkalinity, soil
alcalinité totale alkalinity, total
aluminium aluminum
amines amines
ammoniaque ammonia
analyse de sol soil test
analyse gravimétrique gravimetric analysis
analyse végétale plant analysis
anhydrite anhydrite
anion anion
anion échangeable exchangeable anion
azote nitrogen
azote Kjeldahl Kjeldahl nitrogen
azote organique organic nitrogen
base base
base échangeable exchangeable base
bassin ou pool de carbone carbon pool
besoin en chaux lime requirement
bilan de sels salt balance
capacité d'échange cationique effective effective cation exchange capacity
capacité d'échange cationique cation exchange capacity (CEC)
capacité d'échange cationique dépendante du pH pH-dependent cation exchange capacity
capacité tampon buffer capacity
carbone carbon (C)
carbone 14 (^{14}C) carbon-14 (^{14}C)
carbone organique organic carbon
cation cation
cation acidique acidic cation
cation échangeable exchangeable cation
cendre ash
charge dépendante du pH pH-dependent charge
chaume stover
chaux lime
chéluviation cheluviation
chitine chitin
chlorure de polyvinyle polyvinyl chloride (PVC)
complexe d'adsorption adsorption complex
complexe d'échange exchange complex
complexe tampon du sol buffer compounds, soil
composé aromatique aromatic compound
constante d'acidité acidity constant
courbe de titration pH curve (titration curve)
datation au carbone carbon dating
datation au radiocarbone radiocarbon dating
densité de charge de surface surface charge density
désalinisation desalinization
détritus detritus
échange d'ions ion exchange
échantillon composite composite sample
échantillon de sol soil sample
échelle pH pH scale
électrode à ion spécifique ion-selective electrode
élément nutritif échangeable exchangeable nutrient
environnement réducteur reducing environment
équivalent carbonate de calcium calcium carbonate equivalent
équivalent effectif de carbonate de calcium effective calcium carbonate equivalent
étang salé salt pan
extrait de pâte à saturation saturation-paste extract
extrait de sol extract, soil
fond salin salt flat
force ionique ionic strength
fraction de sodium échangeable exchangeable sodium fraction
fraction légère de la matière organique light fraction (organic matter)
fraction lourde ou dense de la matière organique heavy fraction (organic matter)
humification humification
humique humic
humus humus
humus brut raw humus
hydrocarbure hydrocarbon
hydrophile hydrophilic
hydrophobe hydrophobic
indice du risque de salinité salinity risk index
ion ion
ion isotopiquement échangeable isotopically exchangeable ion
levé de résistivité resistivity survey
liaison ionique ionic bond

lutte contre la salinisation salinity control
matière organique active active organic matter
matière organique du sol organic matter, soil
matière organique inerte inert organic matter
métal base metal
métal lourd heavy metal
métaux alcalino-terreux alkaline-earth metals
métaux alcalins alkali metals
métaux à l'etat de traces trace metals
méthode Kjeldahl Kjeldahl method
monosaccharide monosaccharide (simple sugar)
nitrate nitrate (NO_3)
nitrite nitrite
organique organic
ose monosaccharide (simple sugar)
pâte de sol à saturation saturated soil paste
pH du sol pH, soil
pH_C pH_C
phosphore organique organic phosphorus
point de charge nette nulle point of zero net charge
pont de sel salt bridge
potential calcaire lime potential
potentiel de chaux lime potential
pourcentage cationique échangeable exchangeable cation percentage
pourcentage de saturation saturation percentage
pourcentage de sodium échangeable exchangeable sodium percentage (ESP)
pourcentage ou taux de saturation en bases base saturation percentage
pouvoir tampon buffer power
rapport carbone:azote carbon:nitrogen ratio (C:N ratio)
rapport corrigé d'adsorption du sodium sodium adsorption ratio, adjusted
rapport d'adsorption du sodium sodium-adsorption ratio (SAR)
rapport de sodium échangeable exchangeable sodium ratio (ESR)
réaction du sol reaction, soil
résistivité resistivity
salin saline
salinisation salination, salinization
salinité salinity
salinité du sol salinity, soil
sel salt
seuil de tolérance au sel threshold salt tolerance (crop)
sodicité sodicity
sodisation sodication
sodivité exchangeable sodium percentage (ESP)
sol à alcalis alkali soil
sol acide acid soil
sol alcalin alkali soil, alkaline soil
sol alcalin non-salin nonsaline-alkali soil
sol altéré par le sel salt-affected soil
sol salin saline soil
sol salin à alcalis saline-alkali soil
sol salin sodique saline-sodic soil
sol salsodique salt-affected soil
sol séché à l'étuve oven-dry soil
sol sodique sodic soil
sol tamponné buffered (soil)
solution alcaline alkaline solution
solution du sol solution, soil
substitution ionique ionic substitution
sucre simple monosaccharide (simple sugar)
suintement salin saline seep
taches lisses slick spots

STRUCTURE DU SOL, MICROMORPHOLOGIE (SOIL STRUCTURE, MICROMORPHOLOGY)

agrégat aggregate
agrégat sec dry aggregate
agrégat stable à l'eau ou hydrostable water-stable aggregate
agrégation aggregation
albane alban
apédal apedal
argilane argillan
bioturbation bioturbation
boîte d'échantillonnage de Kubiena Kubiena box or tin
cavité vugh
cohésion cohesion
colmatage ou obturation de la surface surface sealing

complexe argilo-organique clay-organo complex
complexe organo-minéral organo-mineral complex
concrétion concretion
conditionneur de sol soil conditioner
croûte crust
cutane cutan
dépôt laminique lamina (plural laminae)
disperser disperse
distribution des agrégats aggregate distribution
échantillonneur de Kubiena Kubiena box or tin
éclatement slaking
enduit coating
enrobement coating
état structural du sol fabric, soil; soil structural form
fentes de retrait irregulières skew planes
ferrane ferran
ferri-argilane ferri-argillan
fond matriciel s-matrix (of a soil material)
forme d'humus humus form
grade ou netteté de la structure du sol soil structure grade
grains du squelette skeleton grains
granulaire granular
granule granule
gypsane gypsan
humus brut mor (or raw humus)
lame mince thin section
lame polie polished section
macroagrégat macro-aggregate
mangane mangan
matrane matran
mesure ou indicateur de la structure du sol soil structure index
microagrégat micro-aggregate
micromorphologie du sol soil micromorphology
miroir de faille slickenside
moder à tendance mull mull-like moder
mor mor (or raw humus)
motte clod
mull mull
organane organan
organisation du sol fabric, soil
patine cutan
ped ped
photographie microscopique photomicrograph
plasma plasma
pores du sol soil pores
résilience structurale du sol soil structural resiliency
revêtement coating, cutan
sesquane sesquan
squelettane skeletan
stabilisation du sol soil stabilization
stabilité des agrégats aggregate stability
stabilité structurale du sol soil structural stability
structure agrégée ou fragmentaire aggregated
structure du sol soil structure
structure grumeleuse ou granuleuse crumb structure
structure lamellaire platy soil structure
structure massive massive
structure particulaire single-grain structure
structure prismatique prismatic soil structure
surface de glissement slickenside
tamisage à l'eau wet-sieving
tamisage humide wet-sieving
trait ou caractéristique pédologique pedological feature
turricules fecal pellets
type de structure du sol soil structure type
vide void

PROSPECTION DES SOLS, TÉLÉDÉTECTION, INTERPRÉTATION DES DONNÉES PÉDOLOGIQUES (SOIL SURVEY, REMOTE SENSING, LAND USE INTERPRETATION)

aménagement du territoire land use planning
association de sols soil association
altitude altitude, elevation
atténuation atmosphérique atmospheric attenuation
attribut attribute
bande spectrale spectral band

batture riverwash
capabilité des terres land capability
carotte de sondage core sample
carte à grande échelle map, large-scale
carte à moyenne échelle map, medium-scale
carte à moyenne échelle ou à échelle intermédiaire intermediate scale map
carte à petite échelle map, small-scale
carte composée, plurifactorielle ou multicouche composite map
carte de base base map
carte de possibilités des sols soil capability map
carte de relief relief map
carte dérivée compiled map
carte des sols soil map
carte interprétative interpretive map
carte pédologique map, soil; soil map
carte pédologique de reconnaissance générale broad reconnaissance soil map
carte pédologique de reconnaissance, semi-détaillée reconnaissance soil map
carte pédologique détaillée detailed soil map
carte pédologique exploratoire exploratory soil map
carte pédologique généralisée ou de synthèse generalized soil map
carte pédologique schématique schematic soil map
carte spectraloïde spectral map
carte synthèse compiled map
carte thématique ou dérivée thematic map
carte topographique contour map, relief map
carte-index index map
classification des terres land classification
cliché vertical vertical photograph
clinomètre clinometer
complexe de sols soil complex
coordonnées du système de quadrillage stéréographique polaire universel (coordonnées UPS) Universal Polar Stereographic coordinates (UPS coordinates)
corrélation des sols soil correlation
couleurs spectrales spectral colors
courbe de niveau contour, contour line
délimitation delineation
délimitation cartographique des sols soil map delineation
densité de délimitations cartographiques intensity, map
domaine spectral spectral region
données accessoires ou auxiliaires ephemeral data
données de terrain ground data
carotte de sondage core sample
échelle cartographique map scale
échelle graphique graphic scale
élévation contour, elevation
équidistance contour interval
erreur du tracé cartographique plottable error
fenêtres atmosphériques atmospheric windows
fond de carte base map
géobase base map
grille grid
interprétation des données pédologiques interpretation, soil survey
interprétation des sols soil interpretations
intervalle d'élévation contour interval
isohypse contour line
isopaque isopach
isoplète isopleth
légende legend
levé des sols soil survey
limite de sol soil boundary
maille grid
orientation attitude
paysage landscape
photo-carte planimétrique planisaic
photo-carte topographique toposaic
photographie aérienne aerial photograph
photographie verticale vertical photograph
phototriangulation phototriangulation
planification de l'utilisation du territoire land use planning
planimétrie planimetry
point géodésique benchmark
pointe limnimétrique droite point gage
possibilités d'utilisation des terres land capability, land use capability
projection cartographique map projection
prospection pédologique soil survey; survey, soil
raccordement matching
radiance spectrale spectral radiance
réalité de terrain ground data
reconnaissance reconnaissance
repère benchmark
repère de cadre fiducial mark
réponse spectrale spectral response
réseau de points de contrôle control

résolution cartographique map resolution
série de cartes map series
signature spectrale spectral signature
site de référence benchmark site
site repère benchmark site
système d'information géographique geographic information system (GIS)
tarière soil auger
témoin control
terrain land
terrain anthropique made land
terrain inutilisable wasteland
terrain pierreux stony land
terrain rocheux rockland
terrain urbain urban land
terre rapportée made land
terre land
test de carbonates libres free lime test
transect transect
traverse transect
type de terrain divers miscellaneous land type
unité cartographique map unit, soil; mapping unit
unité cartographique de sols soil map unit
unité cartographique de sols non differenciés undifferentiated soil map unit
utilisation des terres ou des sols land use
variable qualitative attribute
zone urbaine urban land

GESTION DES DÉCHETS, COMPOSTAGE (WASTE MANAGEMENT, COMPOSTING)

activateurs ou additifs biologiques biological additives
aération par diffusion d'air diffused air
agent dispersant dispersant
andain windrow
assainissement sewerage
assèchement dewatering
bac de décantation settling tank
bagasse bagasse
bassin de décantation settling pond
bassin de décantation ou de déjection debris basin
bassin de stabilisation des eaux usées sewage lagoon
bioréacteur biostabilizer
boues sludge
boues activées septage
boues de station d'épuration sewage sludge
broyage en conditions humides wet milling
calendrier d'épandage loading schedule
capacité de charge loading capacity
cellule d'entreposage cell
chambre de sédimentation settling chamber
champ d'épuration disposal field, septic tank absorption field
champ d'épuration ou de percolation leaching field
charbon activé ou actif activated carbon
charbon calciné fixed carbon
charge loading
charge de boues autorisée sludge loading
chloration chlorination
chlore résiduel chlorine residual
clarificateur clarifier
clarificateur terminal final clarifier
clarification clarification
coliforme coliform
compactage compaction
compaction compaction
compost artificial manure, compost, synthetic manure
couche imperméabilisante cap
curage ou nettoyage du sol soil flushing
cuve de traitement treatment tank
décantation clarification, sedimentation
décanteur clarifier
décanteur secondaire secondary clarifier
décanteur statique sedimentation tank
décharge fill, landfill
déchets waste
déchets alimentaires swill
déchets dangereux hazardous waste
déchets réactifs reactive waste
déchets solides solid waste
déchets urbains urban waste
déshydratation dewatering
désinfection disinfection
digesteur digester
digesteur à alimentation en continu continuous-feed reactor (composting)
digesteur ou bioréacteur Fairfield-Hardy Fairfield-Hardy digester
digestion digestion
digestion de boues sludge digestion
digestion par voie humide wet digestion
dilacération comminution
eau de consommation domestic water
eau traitée finished water

eaux usées raw sewage, sewage
eaux usées domestiques domestic sewage, domestic waste
eaux usées ou résiduaires wastewater
effluent effluent, liquid waste
égout sewer
élimination de boues sludge disposal
élimination des déchets solides solid waste disposal
enfouissement par la méthode de la tranchée landfill trench method
engrais artificiel synthetic manure
épandage sur le sol land farming
épreuve de complétion completed test
épreuve de confirmation confirmed test
équarrissage rendering
étang lagoon
étang d'épuration disposal pond
étangs d'évaporation evaporation ponds
excréments fecal material
filtration filtration
filtration au carbone carbon filtration
filtration rapide sur sable rapid sand filtration
filtre à sable sand filter
filtre au charbon charcoal filter
fixation fixation
floc floc
floculation flocculation
fosse d'entreposage storage pit (wastes)
fosse de rétention seepage pit
fosse septique cesspool, septic tank
fumier artificiel artificial manure
fumier enrichi reinforced manure
gestion des déchets solides solid waste management
incinérateur générateur de vapeur waterwall incinerator
javellisation chlorination
lagune disposal pond, evaporation pond, lagoon
lavage à contre courant backwashing
lavage par retour d'eau backwashing
lisier liquid manure
lit bactérien trickling filter
litière bedding, animal
lixiviat free liquids
mâchefer clinker
matériel de recouvrement cover material
matières décantables settleable solids
matières fécales fecal material
méthode en lots batch method
mise en décharge sludge disposal
nettoyage inversé backwashing
niveau de coliformes coliform index
particules en suspension suspended sediment
procédé de Beccari Beccari process
procédé de distillation de Lantz Lantz process
puisard cesspool
puits absorbant cesspool
récupération d'ordures refuse reclamation
recyclage recycling, reuse
résidus waste
résidus de conditionnement de charbon coal processing waste
résidus de jardin yard waste
revêtement liner
sable filtrant filter sand
scories clinker
sédimentation sedimentation
séparateur par projection ballistic separator
site d'enfouissement landfill
site d'enfouissement à accès contrôlé landfill, secure
site d'enfouissement sanitaire sanitary landfill
site d'épandage sur sol soil absorption system
solides ou matière en suspension suspended solids
solidification solidification
sources de déchets ou de résidus waste sources
sous-produit by-product
substance peu soluble à l'eau limited water-soluble substance
surnageant supernatant
système d'évacuation d'eau water disposal system
système de drainage water disposal system
système de récupération back end system
système microbiologique à régime continu continuous-flow microbiological system
tassement compaction
temps de rétention retention time
traitement biologique secondaire biological wastewater treatment
traitement des déchets waste processing, waste treatment
traitement des eaux water treatment
traitement primaire primary treatment
traitement secondaire secondary treatment
traitement tertiaire tertiary treatment
traitement thermique de déchets dangereux thermal treatment of hazardous waste
transformation biologique bioconversion
tri, tri à la source resource recovery

unité de gestion de déchets solides solid waste management unit
unité de stockage cell
usine d'épuration des eaux usées sewage treatment plant
valorisation recycling, reuse

ZOOLOGIE (ZOOLOGY)

annélide annelid
apatétique apatetic
arthropode arthropod
collemboles springtails
commensalisme commensalism
consommateur consumer
consommateur primaire primary consumer
coprophage coprophagous
détritivore detritivore
efficacité écologique ecological efficiency
euryhalin euryhaline
exsudat exudate
famille family
gastéropode gastropod
gastropode gastropod
herbivore herbivore
holozoïque holozoic
homéostasie homeostasis
invertébré invertebrate
larve larva
macroconsommateur macroconsumer
mégafaune megafauna
mésobiote mesobiota
métabolisme de base basal metabolism
méthémoglobinémie methemoglobinemia
mite mite
mue molt
myriapode myriapod
nématode nematode
nervure vein
nymphe nymph
omnivore omnivore
paléozoologie palaeozoology
parasite parasite
phagocyte phagocyte
phagotrophe phagotroph
pupe pupa
quiescent quiescent
réseau trophique détritivore detritus food web
souche strain
stade larvaire instar
ver de vase bloodworm
vertébré vertebrate
zooécologie zooecology
zoogéographie zoogeography

References: Sources of Terms

Acton, D.F. and Gregorich, L.J. (Eds.). 1995. *The Health of Our Soils – Toward Sustainable Agriculture in Canada.* Centre for Land and Biological Resources Research, Research Branch, Agriculture and Agri-Food Canada, Ottawa.

Allaby, M. (Ed.). 1998. *Oxford Dictionary of Plant Sciences.* Oxford University Press, Oxford, U.K.

American Association for Testing and Materials. 1999. *Annual Book of ASTM Standards. Section 4. Construction. Volume 04.08. Soil and rock (I): D420 – D4914.* American Society for Testing and Materials, West Conshohocken, PA.

Bailey, A.W. 1987. Effect of grazing on forage and beef production from rough fescue rangeland in central Alberta. *Farming for the Future,* Alberta Agriculture. Edmonton, Alberta. (Glossary pp. 59-60.)

Baldock, J.A. and Nelson, P.N. 2000. Soil organic matter. Pages B25-84. In M.E. Sumner (Ed.) *Handbook of Soil Science.* CRC Press, Boca Raton, FL.

Basher, L. R. 1997. Is pedology dead and buried? *Australian Journal of Soil Research* 35: 979-994.

Bates, R.L. and Jackson, J.A. (Eds.) 1984. *Dictionary of Geological Terms.* American Geological Institute. Anchor Books, Doubleday, New York.

Blackmore, S. and E. Tootill (Eds.). 1984. *The Facts on File Dictionary of Botany.* Facts on File, New York.

Brady, N.C. and Weil, R.R. 1996. *The Nature and Properties of Soils.* Prentice-Hall Inc. Upper Saddle River, New Jersey.

Brock, T.D. 1994. *Biology of Microorganisms.* 7th ed. Prentice Hall, Englewood Cliffs, New Jersey.

British Columbia Ministry of Forests. *Glossary of Forestry.* Available at *http://www.for.gov.bc.ca/PAB/PUBLCTNS/GLOSSARY/GLOSSARY.HTM* (verified April, 2001).

Canadian Society of Soil Science. 1976. *Glossary of Terms in Soil Science.* Publication 1459, Research Branch, Canada Department of Agriculture, Ottawa.

Cole, G.A. 1994. *Textbook of Limnology.* C. V. Mosby, St. Louis, MO.

Commission Canadienne de Pédologie. 1978. Le système Canadien de classification des sols. Publication 1646. Direction générale de la recherche. Ministère de l'Agriculture du Canada, Ottawa.

Commission I (Soil Physics). 1976. Soil Physics Terminology. In *Bulletin of the International Society of Soil Science,* 49,1.

Cormier, C. 1992. *Canadian Quaternary Vocabulary. Terminology Bulletin 209.* Department of the Secretary of State of Canada. Ottawa.

Daintith, J. (Ed.). 1988. *The Facts On File Dictionary of Chemistry.* Facts On File, New York.

Daintith, J. (Ed.). 1996. *Oxford Dictionary of Chemistry.* Oxford University Press, Oxford, U.K.

De Santo, R.S. 1978. *Concepts of Applied Ecology.* Heidelberg Science Library, Springer-Verlag, New York. (Glossary pp. 216-304.)

Donahue, R.L., Miller, R.W., and Shickluna, J.C. 1983. *Soils: An Introduction to Soils and Plant Growth,* 5th ed. Prentice-Hall, Englewood Cliffs, New Jersey. (Glossary pp. 630-656.)

Dunster, J. A. and Dunster, K. J. 1996 *Dictionary of Natural Resource Management.* UBC Press, Vancouver, B. C.

EMAP (Environmental Monitoring and Assessment Program). 1993. Environmental Protection Agency. Environmental Monitoring and Assessment Program Master Glossary. EPA/620/R-93/013, Research Triangle Park, N.C.: U.S. Environmental Protection Agency, Office of Research and Development, Environmental Monitoring and Assessment Program.

Fitzpatrick, E.A. 1983. *Soils, Their Formation, Classification and Distribution.* Longman, London, U.K.

Forage Information System. Glossary of Forage Terms. Department of Crop and Soil Science, Oregon State University, Corvallis. Available at *http://forages.orst.edu/main.cfm?PageID=60* (verified April, 2001).

Foth, H.D. 1978. *Fundamentals of Soil Science,* 6th ed. John Wiley & Sons, New York. (Glossary pp. 409-427.)

Freund, J.E. and Williams, F.J. *Dictionary/Outline of Basic Statistics.* McGraw-Hill, New York.

Groupe de Travail National sur les Terres Humides. 1988. Terres humides du Canada. Série de la classification écologique du territoire #24. Direction du développement durable. Service Canadien de la Faune, Environnement Canada, Ottawa, Ontario et Polyscience Publications, Montréal.

Hach Company. 1992. *Soil and Irrigation Water Interpretation Manual*, 36-40. (Some terms taken or adapted from Western Fertilizer Handbook, 7th Ed. Soil Improvement Committee, California Fertilizer Assoc.

Hausenbuiller, R.L. 1985. *Soil Science, Principles and Practices.* 3rd ed. Wm. C. Brown, Dubuque. (Glossary pp. 571-591.)

Hole, F.D. and Campbell, J.B. 1985. *Soil Landscape Analysis.* Rowman and Allanheld, Totowa, New Jersey. (Glossary pp. 161-179.)

Ironside, G.R. 1982. Glossary of wildlife-related terms used in ecological land surveys - preliminary draft. Land/Wildlife Integration. Ecological Land Classification Series, No. 11. Lands Directorate, Environment Canada, Ottawa, 47-153

Johannsen, C.J. and Sanders, J.L. (Eds.). 1982. *Remote Sensing for Resource Management.* Soil Conservation Society of America, Ankeny, Iowa. (Glossary pp. 637-655.)

Kaufman, M.R., Graham, R.T., Boyd, D.A. Jr., Moir, W.H., Perry, L., Reynolds, R.T., Bassett, R.L., Mehlhop, P., Edminster, C.B., Black, W.M., and Corn, P.S. 1994. An Ecological Basis for Ecosystem Management. Gen. Tech. Rep. RM-246. USDA Forest Service, Rocky Mountain Forest and Range Experiment Station, Fort Collins, Colorado. (Glossary pp. 16-17; some terms taken or adapted from Jensen, M.E. and Bougeron, P.S. 1993. Eastside forest ecosystem health assessment, Vol. II, Ecosystem management: principles and applications. USDA Forest Service, Northern Region, Missoula, Montana, 7-15)

Keith, L.H. (Ed.). 1991. *Compilation of E.P.A.'s Sampling and Analysis Methods.* Lewis Publishers, Chelsea, Michigan. (Glossary pp. 799-803.)

Kendall, M.G. and Buckland, W.R. 1971. *A Dictionary of Statistical Terms.* Published for the International Statistical Institute, Longman, London.

Lozet, J. and Mathieu, C. 1997. *Dictionnaire de Science du Sol.* 3rd ed. Lavoisier, Paris.

McNaughton, S.J. and Wolf, L.W. 1979. *General Ecology*, 2nd ed. Holt, Rinehart and Winston, New York. (Glossary pp. 685-694.)

Michel, J.P. and Fairbridge, R.W. 1992. *Dictionary of Earth Sciences.* Masson, Paris and John Wiley and Sons, New York.

National Wetlands Working Group. 1997. *The Canadian Wetland Classification System.* 2nd ed. B.G. Warner and Rubec, C.D.A., Eds. Wetlands Research Centre, University of Waterloo, Ontario.

Paul, E.A. and Clark, F.E. 1989. *Soil Microbiology and Biochemistry.* Academic Press, San Diego.

Powter, C.B. (compiler). 2000. Glossary of Reclamation Terms Used in Alberta. 6th ed. Alberta Environment, Environmental Sciences Division, Edmonton. Report No. ESD/LM/00-3.

Prairie Pools Inc. 1994. The Farm Environmental Assessment Guide. Alberta Wheat Pool. Calgary. (Glossary pp. 35-36.)

Rambler, M.B., Margulis, L., and Fester, R., (Eds.) 1989. *Global Ecology, Toward a Science of the Biosphere.* Academic Press, Boston. (Glossary pp. 149-169.)

Raven, P.H., Evert, R.F., and Eichhorn, S.E. 1999. *Biology of Plants.* 6th ed. W.H. Freeman: Worth Publishers, New York.

Schwarz, C.F., Thor, E.C., and Elsner, G.A. 1976. Wildland Planning Glossary. USDA Forest Serv. Gen. Tech. Rep. PSW-13. Pacific Southwest Forest and Range Exp. Stn., Berkeley.

Soil Classification Working Group. 1998. *The Canadian System of Soil Classification.* 3rd ed. NRC Research Press, Ottawa.

Soil Conservation Society of America. 1982. Resource Conservation Glossary. Soil Conservation Society of America. Ankeny, Iowa.

Soil Science Society of America. 1997. Glossary of Soil Science Terms: 1996. Soil Science Society of America. Madison, Wisconsin.

Stanek, W. and Worley, I.A. 1983. A Terminology of Virgin Peat and Peatlands. C.H. Fuchsman and Spigarelli, S.A., (Eds.), International Symposium on Peat Utilization. Bemidji State University Center for Environmental Studies, Bemidji, MN, 75–102.

Stevenson, L.H. and B. Wyman. 1991. *The Facts On File Dictionary of Environmental Science.* Facts On File, New York.

Stokes, B.J., Ashmore, C., Rawlins, C.L., and Sirois, D.L. 1989. Glossary of Terms Used in Timber Harvesting and Forest Engineering. USDA Forest Serv. Gen. Tech. Rep. SO-73. Southern Forest Exp. Stn., Auburn, Alabama. (Glossary pp. 334-342.)

Sylvia, D. (compiler). 2000. Soil Microbiology Terms. Available at *http://dmsylvia.ifas.ufl.edu/glossary.htm.*

Termium Plus. Translation service. Public Works and Government Services Canada, Translation bureau. Available at *http://termiumplus.translationbureau.gc.ca/site/english/welcome.html.*

Troeh, F.R. and Thompson, L.M. 1993. *Soils and Soil Fertility,* 5th ed. Oxford University Press, New York. (Glossary pp. 419-442.)

Usher, G. 1970. *A Dictionary of Botany, Including Terms used in Biochemistry, Soil Science and Statistics.* Constable. London.

Whittow, J. 1984. *The Penguin Dictionary of Physical Geography.* Penguin Books, London.

Wiken, E. (compiler). 1986. Terrestrial Ecozones of Canada. Ecological Land Classification Series, No. 19. Lands Directorate, Environment Canada. Ottawa. (Glossary p. 25.)

Zumdahl, S.S. 1989. *Chemistry,* 2nd ed. D.C. Heath and Company, Lexington, MA. (Glossary pp. A29-A41.)

References: Sources of Illustrations

Figures for: Relationship among osmotic, matric, and combined soil water potential; Volume of solids, water, and air in a loam soil at saturation, field capacity, and wilting point. Adapted from Brady, N.C. and Weil, R.R. 1996. *The Nature and Properties of Soils.* Prentice-Hall, Upper Saddle River, New Jersey.

Figure for: Stable carbon isotope ratios. Adapted from Boutton, T.W. 1991. Stable Carbon Isotope Ratios of Natural Materials: II. Atmospheric, Terrestrial, Marine, and Freshwater Environments. In D.C. Coleman and B. Fry (Eds.), *Carbon Isotope Techniques.* Academic Press, New York.

Figure for: The rhizosphere. Adapted from Bolton, H. and Frederickson, J.K. 1993. In Metting, F.B. (Ed.) *Soil Microbial Ecology: Applications in Agricultural and Environmental Management.* Marcel Dekker, New York.

Figure for: Typical growth phases of a bacterial population. Adapted from Brock, T.D., Smith, D.W., and Madigan, M.T. 1984. *Biology of Microorganisms.* Prentice-Hall, Englewood Cliffs, NJ.

Figures for: Cambium; Illuviation; Mass movement of rock, soil and organic matter initiated by creep, fall, flow, slide, and slump; Parts of a root; Perched water table; Types of folds. Adapted from Dunster, J. and Dunster, K. 1996. *Dictionary of Natural Resource Management.* CAB International, Wallingford, U.K.

Figure for: Terrestrial Ecozones of Canada (Appendix E). Adapted from Ecological Stratification Working Group. 1996. *A National Ecological Framework for Canada.* Agriculture and Agri-Food Canada, Research Branch, Centre for Land and Biological Resources Research and Environment Canada, State of Environment Directorate, Ottawa.

Figure for: The greenhouse effect. Adapted from Janzen, H.H., Desjardins, R.L., Asselin, J.M.R. and Grace, B. (Eds.) 1998. *The Health of Our Air: Toward Sustainable Agriculture in Canada.* Research Branch, Agriculture and Agri-Food Canada, Ottawa.

Figures for: A soil bacterium; Conidia; Hypha; Mycelium; Representatives of the soil protozoa. Adapted from Killham, K. 1994. *Soil Ecology.* Cambridge University Press. Cambridge, U.K.

Figures for: Nutrient diffusion; Root interception. Adapted from Ontario Ministry of Agriculture, Food and Rural Affairs. 1998. *Soil Fertility Handbook. Publication 611.* Queen's Printer for Ontario. Toronto.

Figure for: Association of arbuscular mycorrhizal fungi and soil aggregate of a plant root. Adapted from Paul, E.A. and Clark, F.E. 1996. *Soil Microbiology and Biochemistry.* 2nd ed. Academic Press. London.

Figures for: Different degrees of soil development found in a catena; Shoulder, backslope, and footslope landform elements; Flowlines in areas of profile plan curvature; Basic slope morphological units. Pennock, D.J. 2001. Personal communication. University of Saskatchewan. Saskatoon, Saskatchewan.

Figure for: The global carbon cycle. Adapted from Schmiel, D. 1995. Terrestrial ecosystems and the carbon cycle. *Global Change Biol.* 1:77-91.

Figures for: Soil texture classes; Types, kinds, and classes of soil structure (both in Appendix B). Reproduced with permission from Soil Classification Working Group. 1998. *The Canadian System of Soil Classification.* 3rd ed. NRC Research Press, Ottawa, Canada.

Figure for: The nitrogen cycle in soil. Adapted from Stevenson, F.J. (Ed.) 1982. *Nitrogen in Agricultural Soils. Number 22.* American Society of Agronomy. Madison, Wisconsin.

Figures for: Buried soil; Solum. VandenBygaart, A.J. 2001. Personal communication. Department of Land Resource Science, University of Guelph, Guelph, Ontario.

Appendix A

UNITS

TABLE A.1
SI Prefixes

Factor	Name	Symbol
10^{24}	yotta	Y
10^{21}	zetta	Z
10^{18}	exa	E
10^{15}	peta	P
10^{12}	tera	T
10^{9}	giga	G
10^{6}	mega	M
10^{3}	kilo	k
10^{2}	hecto	h
10^{1}	deka	da
10^{-1}	deci	d
10^{-2}	centi	c
10^{-3}	milli	m
10^{-6}	micro	μ
10^{-9}	nano	n
10^{-12}	pico	p
10^{-15}	femto	f
10^{-18}	atto	a
10^{-21}	zepto	z
10^{-24}	yocto	y

TABLE A.2
SI Base Units

Base Quantity	SI Base Unit	
	Name	Symbol
length	meter	m
mass	kilogram	kg
time	second	s
electric current	ampere	A
thermodynamic temperature	kelvin	K
amount of substance	mole	mol
luminous intensity	candela	cd

TABLE A.3
Conversion Factors for SI and non-SI Units (Soil Science Society of America, 1997)

To Convert Column 1 into Column 2, Multiply by	Column 1 SI Unit	Column 2 non-SI Units	To Convert Column 2 into Column 1, Multiply by
	Length		
0.621	kilometer, km (10^3 m)	mile, mi	1.609
1.094	meter, m	yard, yd	0.914
3.28	meter, m	foot, ft	0.304
3.28	meter, m	foot, ft	0.304
1.0	micrometer, µm (10^{-6}m)	micron, µ	1.0
3.28	meter, m	foot, ft	0.304
3.94×10^{-2}	millimeter, mm (10^{-3} m)	inch, in	25.4
10	nanometer, nn (10^{-9} m)	Angstrom, Å	0.1
	Area		
2.47	hectare, ha	acre	0.405
247	square kilometer, km^2 (10^3 m)^2	acre	4.05×10^{-3}
0.386	square kilometer, km^2 (10^3 m)^2	square mile, mi^2	2.590
2.47×10^4	square meter, m^2	acre	4.05×10^3
10.76	square meter, m^2	square foot, ft^2	9.29×10^{-2}
1.55×10^{-3}	square millimeter, mm^2 (10^{-3} m)^2	square inch, in^2	645
	Volume		
9.73×10^{-3}	cubic meter, m^3	acre-inch	102.8
35.3	cubic meter, m^3	cubic foot, ft^3	2.83×10^{-2}
6.10×10^4	cubic meter, m^3	cubic inch, in^3	1.64×10^{-5}
2.84×10^{-2}	liter, L (10^{-3} m^3)	bushel, bu	35.24
1.057	liter, L (10^{-3} m^3)	quart (liquid), qt	0.946
3.53×10^{-2}	liter, L (10^{-3} m^3)	cubic foot, ft^3	28.3
0.265	liter, L (10^{-3} m^3)	gallon	3.78
33.78	liter, L (10^{-3} m^3)	ounce (fluid), oz	2.96×10^{-2}
2.11	liter, L (10^{-3} m^3)	pint (fluid), pt	0.473
	Mass		
2.20×10^{-3}	gram, g (10^{-3} kg)	pound, lb	454
3.52×10^{-2}	gram, g (10^{-3} kg)	ounce (avdp), oz	28.4
2.205	kilogram, kg	pound, lb	0.454
0.01	kilogram, kg	quintal (metric), q	100
1.10×10^{-3}	kilogram, kg	ton (2000 lb), ton	907
1.102	megagram, Mg (tonne)	ton (U.S.), ton	0.907
1.102	tonne, t	ton (U.S.), ton	0.907
	Yield and Rate		
0.893	kilogram per hectare, kg ha^{-1}	pound per acre, lb $acre^{-1}$	1.12
7.77×10^{-2}	kilogram per cubic meter, kg m^3	pound per bushel, lb bu^{-1}	12.87
1.49×10^{-2}	kilogram per hectare, kg ha^{-1}	bushel per acre, 60 lb	67.19
1.59×10^{-2}	kilogram per hectare, kg ha^{-1}	bushel per acre, 56 lb	62.71
1.86×10^{-2}	kilogram per hectare, kg ha^{-1}	bushel per acre, 48 lb	53.75
0.107	liter per hectare, L ha^{-1}	gallon per acre	9.35
893	tonne per hectare, t ha^{-1}	pound per acre, lb $acre^{-1}$	1.12×10^{-3}

continued

TABLE A.3 (continued)
Conversion Factors for SI and non-SI Units (Soil Science Society of America, 1997)

To Convert Column 1 into Column 2, Multiply by	Column 1 SI Unit	Column 2 non-SI Units	To Convert Column 2 into Column 1, Multiply by
893	megagram per hectare, Mg ha^{-1}	pound per acre, lb $acre^{-1}$	1.12×10^{-3}
0.446	megagram per hectare, Mg ha^{-1}	ton (2000 lb) per acre, ton $acre^{-1}$	2.24
2.24	meter per second, m s^{-1}	mile per hour	0.447
	Specific Surface		
10	square meter per kilogram, m^2 kg^{-1}	square centimeter per gram, cm^2 g^{-1}	0.1
1000	square meter per kilogram, m^2 kg^{-1}	square millimeter per gram, mm^2 g^{-1}	0.001
	Pressure		
9.90	megapascal, MPa (10^6 Pa)	atmosphere	0.101
10	megapascal, MPa (10^6 Pa)	bar	0.1
1.00	megagram, per cubic meter, Mg M^{-3}	gram per cubic centimeter, g cm^{-3}	1.00
2.09×10^{-2}	Pascal, Pa	pound per square foot, lb ft^{-2}	47.9
1.45×10^{-4}	Pascal, Pa	pound per square inch lb in^{-2}	6.90×10^3
1.45×10^{-4}	Pascal, Pa	pound per square inch lb in^{-2}	6.90×10^3
	Temperature		
1.00 (K - 273)	kelvin, K	Celsius, °C	1.00 (°C + 273)
(9/5 °C) + 32	Celsius, °C	Fahrenheit, °F	5/9 (°F – 32)
	Energy, Work, Quantity of Heat		
9.52×10^{-4}	joule, J	British thermal unit, Btu	1.05×10^3
0.239	joule, J	calorie, cal	4.19
10^1	joule, J	Erg	10^{-7}
0.735	joule, J	Foot-pound	1.36
2.387×10^{-5}	joule per square meter, J m^{-2}	calorie per square centimeter (langley)	4.19×10^4
10^5	newton, N	Dyne	10^{-5}
1.43×10^{-3}	watt per square meter, W m^{-2}	calorie per square centimeter minute (irradiance), cal cm^{-2} min^{-1} minute (irradiance), cal cm^{-2} min^{-1}	698
	Transpiration and Photosynthesis		
3.60×10^{-2}	milligram per square meter second, mg m^{-2} s^{-1}	gram per square decimeter hour, g dm^{-2} h^{-1}	27.8
5.56×10^{-3}	milligram (H_20) per square meter second, mg m^2 s^{-1}	micromole (H_20) per square centimeter second, µmol cm^{-2} s^{-1}	180
10^{-4}	milligram per square meter second, mg m^{-2} s^{-1}	milligram per square centimeter second, mg cm^{-2} s^{-1}	104

continued

TABLE A.3 (continued)
Conversion Factors for SI and non-SI Units (Soil Science Society of America, 1997)

To Convert Column 1 into Column 2, Multiply by	Column 1 SI Unit	Column 2 non-SI Units	To Convert Column 2 into Column 1, Multiply by
35.97	milligram per square meter second, mg m^{-2} s^{-1}	milligram per square decimeter hour, mg dm^{-2} h^{-1}	2.78×10^{-2}
	Plane Angle		
57.3	radian, rad	degrees (angle), °	1.75×10^{-2}
	Electrical Conductivity, Electricity, and Magnetism		
10	siemen per meter, S m^{-1}	millimho per centimeter, mmho cm^{-1}	0.1
10^4	tesla, T	gauss, G	10^{-4}
	Water Measurement		
9.73×10^{-3}	cubic meter, m^3	acre-inch, acre-in	102.8
9.81×10^{-3}	cubic meter per hour, m^3 h^{-1}	cubic foot per second, ft^3 s^{-1}	101.9
4.40	cubic meter per hour, m^3 h^{-1}	U.S. gallon per minute, gal min^{-1}	0.227
8.11	hectare meter, ha m	acre-foot, acre-ft	0.123
97.28	hectare meter, ha m	acre-inch, acre-in	1.03×10^{-2}
8.1×10^{-2}	hectare centimeter, ha cm	acre-foot, acre-ft	12.33
	Concentrations		
1	centimole per kilogram, cmol kg^{-1}	milliequivalent per 100 grams, meq 100 g^{-1}	1
0.1	gram per kilogram, g kg^{-1}	percent, %	10
1	milligram per kilogram, mg kg^{-1}	parts per million, ppm	1
	Radioactivity		
2.7×10^{-11}	becquerel, Bq	curie, Ci	3.7×10^{10}
2.7×10^{-2}	becquerel per kilogram, Bq kg^{-1}	picocurie per gram, pCi g^{-1}	37
100	gray, Gy (absorbed dose)	rad, rd	0.01
100	sievert, Sv (equivalent dose)	rem (roentgen equivalent man)	0.01
	Plant Nutrient Conversion		
	Elemental	Oxide	
2.29	P	P_2O_5	0.437
1.20	K	K_2O	0.830
1.39	Ca	CaO	0.715
1.66	Mg	MgO	0.602

Appendix B

SOIL PROPERTIES

TABLE B.1
Types and Classes of Soil Structure

Type	Kind	Class	Size (mm)
Structureless: no observable aggregation or no definite orderly arrangement around natural lines of weakness	**Single grain structure:** loose, incoherent mass of individual particles as in sands		
	Amorphous (massive) structure: a coherent mass showing no evidence of any distinct arrangement of soil particles		
Blocklike: soil particles are arranged around a point and bounded by flat or rounded surfaces	**Blocky (angular blocky):** faces rectangular and flattened, vertices sharply angular	Fine blocky	<10
		Medium blocky	10-20
		Coarse blocky	20-50
		Very coarse blocky	>50
	Subangular blocky: faces subrectangular, vertices mostly oblique, or subrounded	Fine subangular blocky	<10
		Medium subangular blocky	10-20
		Coarse subangular blocky	20-50
		Very coarse subangular blocky	>50
	Granular: spheroidal and characterized by rounded vertices	Fine granular	<2
		Medium granular	2-5
		Coarse granular	5-10
Platelike: soil particles are arranged around a horizontal plane and generally bounded by relatively flat horizontal surfaces	**Platy structure:** horizontal planes more or less developed	Fine platy	<2
		Medium platy	2-5
		Coarse platy	>5
Prismlike: soil particles are arranged around a vertical axis and bounded by relatively flat vertical surfaces	**Prismatic structure:** vertical faces well defined, and edges sharp	Fine prismatic	<20
		Medium prismatic	20-50
		Coarse prismatic	50-100
		Very coarse prismatic	>100
	Columnar structure: vertical edges near top of columns not sharp; columns flat-topped, round-topped or irregular	Fine columnar	<20
		Medium columnar	20-50
		Coarse columnar	50-100
		Very coarse columnar	>100

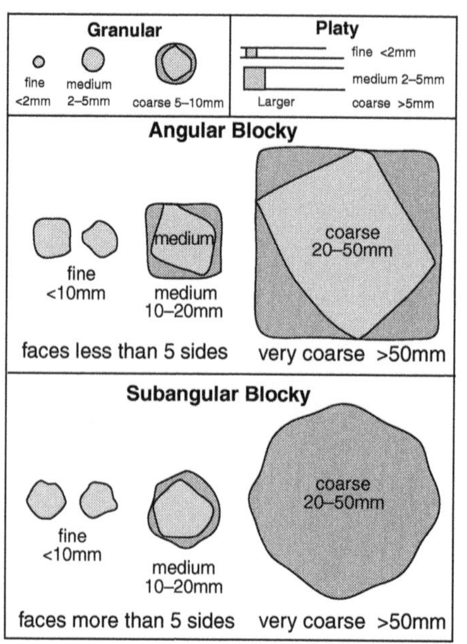

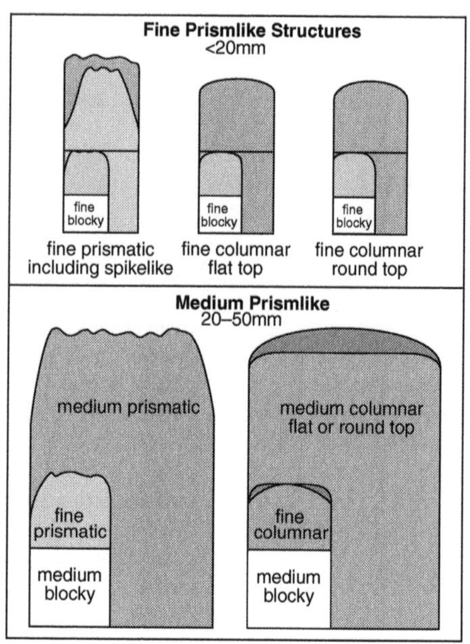

FIGURE B.1 Types, kinds, and classes of **soil structure.** Reproduced from *Soil Classification Working Group* (1998), with permission of NRC Research Press, Ottawa.

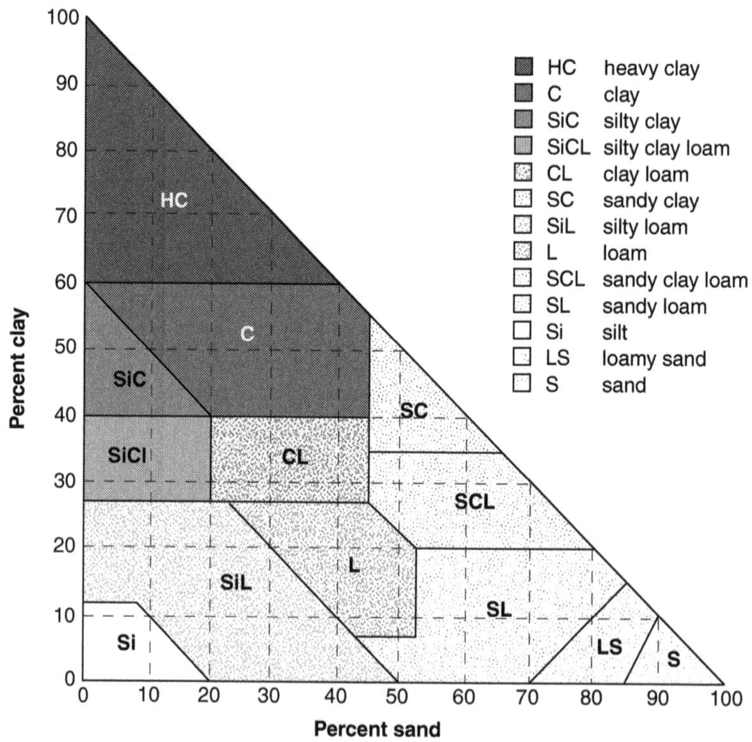

FIGURE B.2 Soil texture classes. Percentages of clay and sand in the main textural classes of soil; the remainder of each class is silt. Reproduced from *Soil Classification Working Group* (1998), with permission of NRC Research Press, Ottawa.

Appendix C

GEOLOGICAL TIMESCALE

TABLE C.1
The Geological Time Scale and Major Geological and Biological Events

Era	Period		Years BP (Millions)	Epoch	Major Events
Cenozoic	Quaternary		0.01	Recent	The modern epoch; age of human beings
			2	Pleistocene	Worldwide glaciation; appearance of Homo sapiens
	Tertiary		7	Pliocene	Himalayan and late Rocky Mountain orogenies; early homonid evolution
			26	Miocene	Alpine orogenies; grazing animals and apes evolve
			37	Oligocene	Alpine orogenies, esp. European Alps and Himalayas; mammalian evolution
			53	Eocene	Alpine orogenies; Australia and Antarctica separate; widespread mammals
			65	Paleocene	Alpine orogenies; mammalian evolution; early primates
Mesozoic	Cretaceous		136		Initial Rocky Mountain orogeny; dinosaurs become extinct; angiosperms and insects emerge
	Jurassic		190		Early mammals and birds; gymnosperms dominant
	Triassic		225		First dinosaurs, birds and mammals appear; gymnosperms and ferns dominant
Paleozoic	Permian		280		Assembly of supercontinent Pangea complete; glaciation; reptiles evolve
	Carboniferous	Pennsylvanian	280		Extensive seas, coal swamps and glaciation; age of amphibians; insects and reptiles evolve; extensive forests
		Mississippian	320		
	Devonian		346		Appalachian orogenies; age of fishes; land plants evolve
	Silurian		395		Land plants evolve; extensive glaciation
	Ordovician		430		First fish; molluscs, land plants and first fungi evolve
	Cambrian		500		First shelled animals appear; vertebrates evolve

continued

TABLE C.1 (continued)
The Geological Time Scale and Major Geological and Biological Events

Era	Period	Years BP (Millions)	Epoch	Major Events
Precambrian	Proterozoic	2500		Earth's crust forms; eukaryotes appear about 1.5 billion years ago, and multi-celled animals about 0.7 billion years ago.
	Archean	4600		Earliest bacteria and algae; earliest fossil record aged about 3.8 billion years

Appendix D

SOIL CLASSIFICATION

TABLE D.1
Soil Classification at Levels of Order, Great Group, and Subgroup

Order	Great Group	Subgroup
Brunisolic	Melanic Brunisol	Orthic Melanic Brunisol
		Eluviated Melanic Brunisol
		Gleyed Melanic Brunisol
		Gleyed Eluviated Melanic Brunisol
	Eutric Brunisol	Orthic Eutric Brunisol
		Eluviated Eutric Brunisol
		Gleyed Eutric Brunisol
		Gleyed Eluviated Eutric Brunisol
	Sombric Brunisol	Orthic Sombric Brunisol
		Eluviated Sombric Brunisol
		Duric Sombric Brunisol
		Gleyed Sombric Brunisol
		Gleyed Eluviated Sombric Brunisol
	Dystric Brunisol	Orthic Dystric Brunisol
		Eluviated Dystric Brunisol
		Duric Dystric Brunisol
		Gleyed Dystric Brunisol
		Gleyed Eluviated Dystric Brunisol
Chernozemic	Brown Chernozem	Orthic Brown Chernozem
		Rego Brown Chernozem
		Calcareous Brown Chernozem
		Eluviated Brown Chernozem
		Solonetzic Brown Chernozem
		Vertic Brown Chernozem
		Gleyed Brown Chernozem
		Gleyed Rego Brown Chernozem
		Gleyed Calcareous Brown Chernozem
		Gleyed Eluviated Brown Chernozem
		Gleyed Solonetzic Brown Chernozem
		Gleyed Vertic Brown Chernozem
	Dark Brown Chernozem	Orthic Dark Brown Chernozem
		Rego Dark Brown Chernozem
		Calcareous Dark Brown Chernozem
		Eluviated Dark Brown Chernozem
		Solonetzic Dark Brown Chernozem
		Vertic Dark Brown Chernozem
		Gleyed Dark Brown Chernozem
		Gleyed Rego Dark Brown Chernozem
		Gleyed Calcareous Dark Brown Chernozem

continued

TABLE D.1 (CONTINUED)
Soil Classification at Levels of Order, Great Group, and Subgroup

Order	Great Group	Subgroup
		Gleyed Eluviated Dark Brown Chernozem
		Gleyed Solonetzic Dark Brown Chernozem
		Gleyed Vertic Dark Brown Chernozem
Chernozemic	Black Chernozem	Orthic Black Chernozem
		Rego Black Chernozem
		Calcareous Black Chernozem
		Eluviated Black Chernozem
		Solonetzic Black Chernozem
		Vertic Black Chernozem
		Gleyed Black Chernozem
		Gleyed Rego Black Chernozem
		Gleyed Calcareous Black Chernozem
		Gleyed Eluviated Black Chernozem
		Gleyed Solonetzic Black Chernozem
		Gleyed Vertic Black Chernozem
	Dark Gray Chernozem	Orthic Dark Gray Chernozem
		Rego Dark Gray Chernozem
		Calcareous Dark Gray Chernozem
		Solonetzic Dark Gray Chernozem
		Vertic Dark Gray Chernozem
		Gleyed Dark Gray Chernozem
		Gleyed Rego Dark Gray Chernozem
		Gleyed Calcareous Dark Gray Chernozem
		Gleyed Solonetzic Dark Gray Chernozem
		Gleyed Vertic Dark Gray Chernozem
Cryosolic	Turbic Cryosol	Orthic Eutric Turbic Cryosol
		Orthic Dystric Turbic Cryosol
		Brunisolic Eutric Turbic Cryosol
		Brunisolic Dystric Turbic Cryosol
		Gleysolic Turbic Cryosol
		Regosolic Turbic Cryosol
		Histic Eutric Turbic Cryosol
		Histic Dystric Turbic Cryosol
		Histic Regosolic Turbic Cryosol
	Static Cryosol	Orthic Eutric Static Cryosol
		Orthic Dystric Static Cryosol
		Brunisolic Eutric Static Cryosol
		Brunisolic Dystric Static Cryosol
		Luvisolic Static Cryosol
		Gleysolic Static Cryosol
		Regosolic Static Cryosol
		Histic Eutric Static Cryosol
		Histic Dystric Static Cryosol
		Histic Regosolic Static Cryosol
	Organic Cryosol	Fibric Organic Cryosol
		Mesic Organic Cryosol
		Humic Organic Cryosol
		Terric Fibric Organic Cryosol

continued

TABLE D.1 (CONTINUED)
Soil Classification at Levels of Order, Great Group, and Subgroup

Order	Great Group	Subgroup
		Terric Mesic Organic Cryosol
		Terric Humic Organic Cryosol
		Glacic Organic Cryosol
Gleysolic	Luvic Gleysol	Solonetzic Luvic Gleysol
		Fragic Luvic Gleysol
		Humic Luvic Gleysol
		Fera Luvic Gleysol
		Orthic Luvic Gleysol
		Vertic Luvic Gleysol
Gleysolic	Humic Gleysol	Solonetzic Humic Gleysol
		Fera Humic Gleysol
		Orthic Humic Gleysol
		Rego Humic Gleysol
		Vertic Humic Gleysol
	Gleysol	Solonetzic Gleysol
		Fera Gleysol
		Orthic Gleysol
		Rego Gleysol
		Vertic Gleysol
Luvisolic	Gray Brown Luvisol	Orthic Gray Brown Luvisol
		Brunisolic Gray Brown Luvisol
		Podzolic Gray Brown Luvisol
		Vertic Gray Brown Luvisol
		Gleyed Gray Brown Luvisol
		Gleyed Brunisolic Gray Brown Luvisol
		Gleyed Podzolic Gray Brown Luvisol
		Gleyed Vertic Gray Brown Luvisol
	Gray Luvisol	Orthic Gray Luvisol
		Dark Gray Luvisol
		Brunisolic Gray Luvisol
		Podzolic Gray Luvisol
		Solonetzic Gray Luvisol
		Fragic Gray Luvisol
		Vertic Gray Luvisol
		Gleyed Gray Luvisol
		Gleyed Dark Gray Luvisol
		Gleyed Brunisolic Gray Luvisol
		Gleyed Podzolic Gray Luvisol
		Gleyed Solonetzic Gray Luvisol
		Gleyed Fragic Gray Luvisol
		Gleyed Vertic Gray Luvisol
Organic	Fibrisol	Typic Fibrisol
		Mesic Fibrisol
		Humic Fibrisol
		Limnic Fibrisol
		Cumulic Fibrisol
		Terric Fibrisol
		Terric Mesic Fibrisol

continued

TABLE D.1 (CONTINUED)
Soil Classification at Levels of Order, Great Group, and Subgroup

Order	Great Group	Subgroup
		Terric Humic Fibrisol
		Hydric Fibrisol
	Mesisol	Typic Mesisol
		Fibric Mesisol
		Humic Mesisol
		Limnic Mesisol
		Cumulic Mesisol
		Terric Mesisol
		Terric Fibric Mesisol
		Terric Humic Mesisol
		Hydric Mesisol
Organic	Humisol	Typic Humisol
		Fibric Humisol
		Mesic Humisol
		Limnic Humisol
		Cumulic Humisol
		Terric Humisol
		Terric Fibric Humisol
		Terric Mesic Humisol
		Hydric Humisol
	Folisol	Hemic Folisol
		Humic Folisol
		Lignic Folisol
		Histic Folisol
Podzolic	Humic Podzol	Orthic Humic Podzol
		Ortstein Humic Podzol
		Placic Humic Podzol
		Duric Humic Podzol
		Fragic Humic Podzol
	Ferro-Humic Podzol	Orthic Ferro-Humic Podzol
		Ortstein Ferro-Humic Podzol
		Placic Ferro-Humic Podzol
		Duric Ferro-Humic Podzol
		Fragic Ferro-Humic Podzol
		Luvisolic Ferro-Humic Podzol
		Sombric Ferro-Humic Podzol
		Gleyed Ferro-Humic Podzol
		Gleyed Ortstein Ferro-Humic Podzol
		Gleyed Sombric Ferro-Humic Podzol
	Humo-Ferric Podzol	Orthic Humo-Ferric Podzol
		Ortstein Humo-Ferric Podzol
		Placic Humo-Ferric Podzol
		Duric Humo-Ferric Podzol
		Fragic Humo-Ferric Podzol
		Luvisolic Humo-Ferric Podzol
		Sombric Humo-Ferric Podzol
		Gleyed Humo-Ferric Podzol
		Gleyed Ortstein Humo-Ferric Podzol

continued

TABLE D.1 (CONTINUED)
Soil Classification at Levels of Order, Great Group, and Subgroup

Order	Great Group	Subgroup
		Gleyed Sombric Humo-Ferric Podzol
Regosolic	Regosol	Orthic Regosol
		Cumulic Regosol
		Gleyed Regosol
		Gleyed Cumulic Regosol
	Humic Regosol	Orthic Humic Regosol
		Cumulic Humic Regosol
		Gleyed Humic Regosol
		Gleyed Cumulic Humic Regosol
Solonetzic	Solonetz	Brown Solonetz
		Dark Brown Solonetz
		Black Solonetz
		Alkaline Solonetz
		Gleyed Brown Solonetz
		Gleyed Dark Brown Solonetz
		Gleyed Black Solonetz
	Solodized Solonetz	Brown Solodized Solonetz
		Dark Brown Solodized Solonetz
		Black Solodized Solonetz
		Dark Gray Solodized Solonetz
		Gray Solodized Solonetz
		Gleyed Brown Solodized Solonetz
		Gleyed Dark Brown Solodized Solonetz
		Gleyed Black Solodized Solonetz
		Gleyed Dark Gray Solodized Solonetz
		Gleyed Gray Solodized Solonetz
	Solod	Brown Solod
		Dark Brown Solod
		Black Solod
		Dark Gray Solod
		Gray Solod
		Gleyed Brown Solod
		Gleyed Dark Brown Solod
		Gleyed Black Solod
		Gleyed Dark Gray Solod
		Gleyed Gray Solod
	Vertic Solonetz	Brown Vertic Solonetz
		Dark Brown Vertic Solonetz
		BlackVertic Solonetz
		Gleyed Brown Vertic Solonetz
		Gleyed Dark Brown Vertic Solonetz
		Gleyed BlackVertic Solonetz
Vertisolic	Vertisol	Orthic Vertisol
		Gleyed Vertisol
		Gleysolic Vertisol
		Humic Vertisol
		Orthic Humic Vertisol
		Gleyed Humic Vertisol
		Gleysolic Humic Vertisol

Source: *Soil Classification Working Group*, 1998.

TABLE D.2
Correlation of Horizon Definitions and Designations

Canadian	U.S.	FAO	Comments
O	O	H	(Can. limit): organic horizon (O) >17% organic C*
Of	Oi	H	(U.S. and FAO limits): lower limit of organic horizons ranges
Om	Oe	H	proportionately from 20% organic matter (OM) with 0% clay to 30%
Oh	Oa	H	OM with >50% clay
Oco	Oa	H	Coprogenous limnic material
L-F	Oi-Oe	O	Generally not saturated with water for prolonged periods
L-H	Oi-Oa	O	
F-H	Oe-Oa	O	
A	A	A	(Can. limit): ≤17% organic C; (U.S. and FAO limits): upper limit of
Ah	A	Ah	OM ranges proportionately from 20% OM with 0% clay to 30% OM
Ahe	AE	(Ah-E)	with clay >50%
Ae	E	E	
Ap	Ap	Ap	
AB	AB or EB	AB or EB	Transitional horizons
BA	BA or BE	BA or BE	
A and B	A and B	A/B	Interfingered horizons
AC	AC	A/C	
B	B	B	
Bt	Bt	Bt	
Bf	Bs	Bs	(Can. limit): specific limits; (U.S. and FAO limits): no specific limit
Bhf	Bhs	Bhs	(Can. limit): >5% organic C
Bgf	Bgs	Bgs	
Bh	Bh	Bh	(Can. limit): specific C to Fep ratio; (U.S. and FAO limits): no specific C to Fep ratio
Bn	Bn	Bn	
Bm	Bw	Bw	
C	C	C	
HC	2C	IIC	
R	R	R	
W	-	-	Water
Other suffixes			
			May be used with A, B, or C horizons
b	b	b	(Can. and U.S. limit): buried; (FAO limit): buried or bisequa
c	m	m	Slightly altered by hydrolysis, oxidation, and/or solution
ca	k	k	Accumulation of carbonates
-	y	y	Accumulation of gypsum
cc	m	c	cemented, ireversible
g	g	g or r	(FAO): g - mottling, - strong reduction
j	-	-	juvenile, weak expression of development
k	-	-	Indicates presence of carbonate
-	v	-	Plinthite
-	q	q	Silica accumulation
s	z	z	Visible salts

continued

TABLE D.2 (CONTINUED)
Correlation of Horizon Definitions and Designations

Canadian	U.S.	FAO	Comments
ss	ss	-	Indicates presence of slickensides
sa	y or z	y or z	(Can. limit): includes gypsum
			(U.S. and FAO limits): y - gypsum, z - other more soluble salts
-	o	-	Residual sequioxide concentration
u	-	-	Turbic
-	-	u	Unspecified
v	-	-	Vertic horizon
x	x	x	Fragipan
y	-	-	Cryoturbation
z	f	i	Permafrost layer

* 17% organic C is equivalent to about 30% organic matter.
Source: *Soil Classification Working Group*, 1998.

TABLE D.3
Correlation of United States and FAO Diagnostic Horizons with Nearest Canadian Equivalents

U.S.	FAO	Canadian	Comments
Mollic Epipedon	Mollic A	Chernozemic A	With high base status
Anthropic Epipedon	Mollic A	Cultivated Chernozemic A	
Umbric Epipedon	Umbric A	Ah	With low base status
Histic Epipedon	Histic H	Of, Om, Oh	
Ochric Epipedon	Ochric A	light-colored A	
Plaggen Epipedon	-	Ap	
Albic horizon	Albic E	Ae	
Argillic horizon	Argillic B	Bt	
Agric horizon	Argillic B	Illuvial B	Formed under cultivation
Natric horizon	Natric B	Bn or Brit	
Spodic horizon	Spodic B	podzolic B	
Cambic horizon	Cambic B	Bm, Bg, Btj	
Oxic horizon	Oxic B	-	
Duripan	m	c	
Durinodes	-	cc	
Fragipan	x	Fragipan	
Calcic horizon	Calcic horizon	Bca or Cca	
Petrocalcic	Bkm	Bcac or Ccac	
Gypsic	Gypsic	Asa, Bsa, Csa	(Can. limit): only if sa horizon is dominantly $CaS0_4$
Salic	-	Asa, Bsa, Csa	
Placic	Thin iron pan	Placic	
Plinthite	Plinthite	-	
Lithic contact	-	Lithic contact	
Paralithic contact	-	IICc	
g	Gleyic horizon	g	
-	sulfuric horizon	-	Low pH, jarosite mottles

Source: *Soil Classification Working Group*, 1998.

TABLE D.4
Taxonomic Correlation at the Canadian Order and Great Group Levels[1]

Canadian System	U.S. SOIL TAXONOMY	FAO System
Chernozemic	Borolls	Kastanozem, Chernozem, Greyzem, Phaeozem
Brown Chernozem	Aridic Boroll subgroups	Kastanozem (aridic)
Dark Brown Chernozem	Typic Boroll subgroups	Kastanozem (Haplic)
Black Chernozem	Udic Boroll subgroups	Chernozem
Dark Gray Chernozem	Boralfic Boroll subgroups, Albolls	Greyzem
Solonetzic	Natric great groups, Mollisols and Alfisols	Solonetz
Solonetz	Natric great groups, Mollisols and Alfisols	Mollic, Haplic, or Gleyic Solonetz
Solodized Solonetz	Natric great groups, Mollisols and Alfisols	Mollic, Haplic, or Gleyic Solonetz
Solod	Glossic Natriborolls, Natralbolls	Solodic Planosol
Vertic Solonetz	Haplocryerts	Sodic Vertisol
Luvisolic	Boralfs and Udalfs	Luvisol
Gray Brown Luvisol	Hapludalfs or Glossudalfs	Albic Luvisol, Haplic Luvisol
Gray Luvisol	Boralfs	Albic Luvisol, Gleyic Luvisol
Podzolic	Spodosols, some Inceptisols	Podzol
Humic Podzol	Cryaquods, Humods	Humic Podzol
Ferro-Humic Podzol	Humic Cryorthods, Humic Haplorthods	Orthic Podzol
Humo-Ferric Podzol	Cryorthods, Haplorthods	Orthic Podzol
Brunisolic	Inceptisols, some Psamments	Cambisol
Melanic Brunisol	Cryochrepts, Eutrochrepts, Hapludolls	Cambisol, Eutric Cambisol
Eutric Brunisol	Cryochrepts, Eutrochrepts	Eutric Cambisol, Calcic Cambisol
Sombric Brunisol	Humbric Dystrochrepts	Dystric Cambisol, Umbric Cambisol
Dystric Brunisol	Dystrochrepts, Cryochrepts	Dystric Cambisol
Regosolic	Entisols	Fluvisol, Regosol
Regosol	Entisols	Regosol
Humic Regosol	Entisols	Fluvisol, Regosol
Gleysolic	Aqu-suborders	Gleysol, Planosol
Humic Gleysol	Aquolls, Humaquepts	Mollic, Umbric, Calcic Gleysol
Gleysol	Aquents, Fluvents, Aquepts	Eutric, Dystric Gleysol
Luvic Gleysol	Argialbolls, Argiaquolls, Aqualfs	Planosol
Organic	Histosols	Histosol
Fibrisol	Fibrists	Histosol
Mesisol	Hemists	Histosol
Humisol	Saprists	Histosol
Folisol	Folists	Histosol
Cryosolic	Gelisols	Cryosol
Turbic Cryosol	Turbels	Cryosol
Static Cryosol	Orthels	Cryosol
Organic Cryosol	Histels	Cryic Histosol
Vertisolic	Cryerts	Vertisol
Vertisol	Haplocryerts	Calcic Vertisol, Eutric Vertisol
Humic Vertisol	Humicryerts	Dystric Vertisol

[1] Only the nearest equivalents are indicated.

Source: *Soil Classification Working Group*, 1998.

Appendix E

ECOZONES OF CANADA

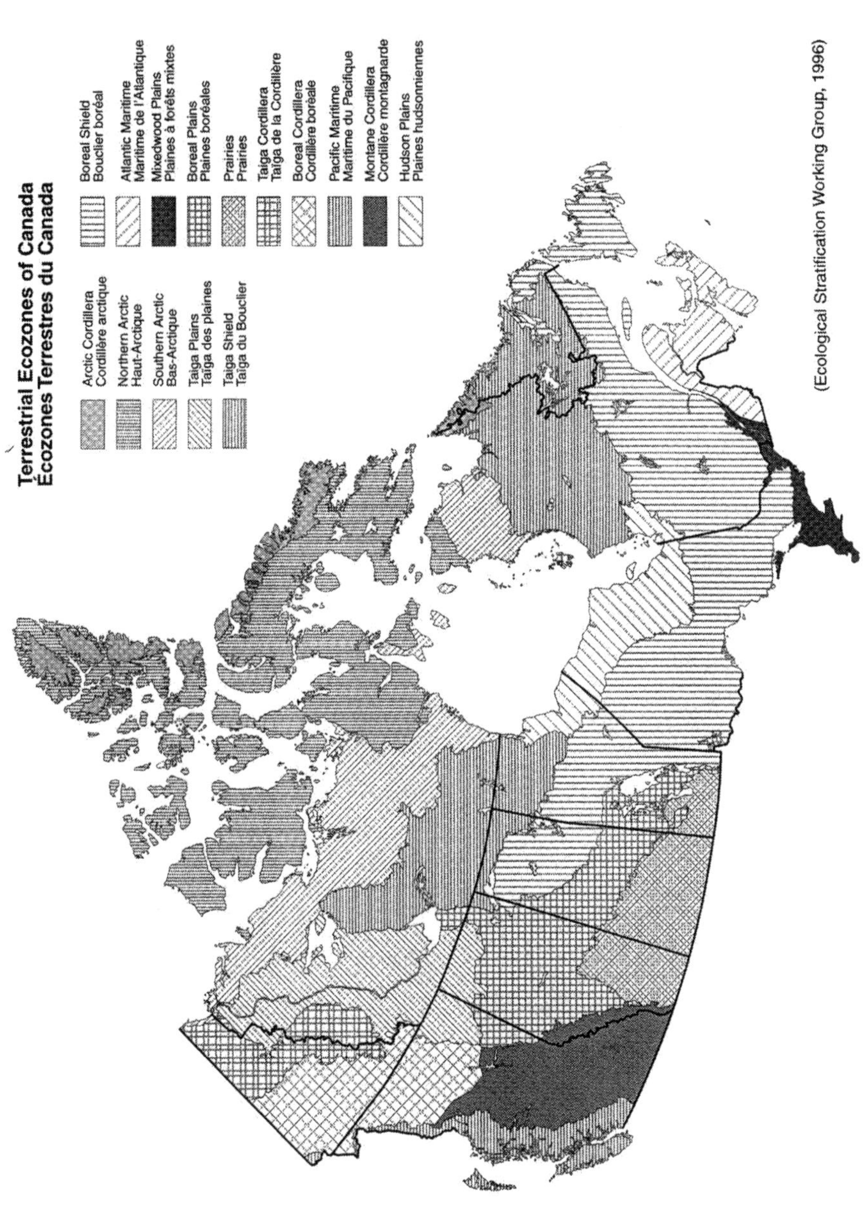

FIGURE E.1 Terrestrial Ecozones of Canada

Milton Keynes UK
Ingram Content Group UK Ltd.
UKHW052028071024
449327UK00027B/2474